# Lecture Notes in Mechanical Engineering

**Lecture Notes in Mechanical Engineering (LNME)** publishes the latest developments in Mechanical Engineering—quickly, informally and with high quality. Original research reported in proceedings and post-proceedings represents the core of LNME. Volumes published in LNME embrace all aspects, subfields and new challenges of mechanical engineering. Topics in the series include:

- Engineering Design
- Machinery and Machine Elements
- Mechanical Structures and Stress Analysis
- Automotive Engineering
- Engine Technology
- Aerospace Technology and Astronautics
- Nanotechnology and Microengineering
- Control, Robotics, Mechatronics
- MEMS
- Theoretical and Applied Mechanics
- Dynamical Systems, Control
- Fluid Mechanics
- Engineering Thermodynamics, Heat and Mass Transfer
- Manufacturing
- Precision Engineering, Instrumentation, Measurement
- Materials Engineering
- Tribology and Surface Technology

To submit a proposal or request further information, please contact the Springer Editor of your location:

**China:** Ms. Ella Zhang at ella.zhang@springer.com
**India:** Priya Vyas at priya.vyas@springer.com
**Rest of Asia, Australia, New Zealand:** Swati Meherishi at swati.meherishi@springer.com
**All other countries:** Dr. Leontina Di Cecco at Leontina.dicecco@springer.com

To submit a proposal for a monograph, please check our Springer Tracts in Mechanical Engineering at https://link.springer.com/bookseries/11693 or contact Leontina.dicecco@springer.com

**Indexed by SCOPUS. All books published in the series are submitted for consideration in Web of Science.**

N. Lakshmi Narasimhan · Mahmoud Bourouis ·
Vasudevan Raghavan
Editors

# Recent Advances in Energy Technologies

## Select Proceedings of ICEMT 2021

Springer

*Editors*
N. Lakshmi Narasimhan
Department of Mechanical Engineering
Sri Sivasubramaniya Nadar College
of Engineering
Chennai, Tamil Nadu, India

Mahmoud Bourouis
Department of Mechanical Engineering
Universitat Rovira i Virgili
Tarragona, Spain

Vasudevan Raghavan
Department of Mechanical Engineering
Indian Institute of Technology Madras
Chennai, Tamil Nadu, India

ISSN 2195-4356 ISSN 2195-4364 (electronic)
Lecture Notes in Mechanical Engineering
ISBN 978-981-19-3466-7 ISBN 978-981-19-3467-4 (eBook)
https://doi.org/10.1007/978-981-19-3467-4



This Springer imprint is published by the registered company Springer Nature Singapore Pte Ltd.
The registered company address is: 152 Beach Road, #21-01/04 Gateway East, Singapore 189721, Singapore

# Organizing Committee

## External Members

Dr. G. Venkatarathnam, Institute Chair Professor, Department of Mechanical Engineering, IIT Madras, Chennai, Tamil Nadu, India
Dr. Mahmoud Bourouis, Professor, Department of Mechanical Engineering, Universitat Rovira i Virgili, Tarragona, Spain
Dr. Vasudevan Raghavan, Professor, Department of Mechanical Engineering, IIT Madras, Chennai, TN, India

## Internal Members

Dr. N. Nallusamy, (formerly) Professor, Department of Mechanical Engineering, SSNCE, India, presently Registrar, Shiv Nadar University Chennai, India
Dr. M. Suresh, Associate Professor, Department of Mechanical Engineering, SSNCE, India
Dr. S. Rajkumar, Associate Professor, Department of Mechanical Engineering, SSNCE, India
Dr. R. Prakash, Associate Professor, Department of Mechanical Engineering, SSNCE, India
Dr. K. L. Harikrishna, Associate Professor, Department of Mechanical Engineering, SSNCE, India
Dr. S. Suresh Kumar, Associate Professor, Department of Mechanical Engineering, SSNCE, India
Dr. K. S. Vijay Sekar, Associate Professor, Department of Mechanical Engineering, SSNCE, India
Dr. M. S. Alphin, Associate Professor, Department of Mechanical Engineering, SSNCE, India

Dr. S. Vijayan, Associate Professor, Department of Mechanical Engineering, SSNCE, India

Dr. A. K. Lakshminarayanan, Associate Professor, Department of Mechanical Engineering, SSNCE, India

Dr. L. Poovazhagan, Associate Professor, Department of Mechanical Engineering, SSNCE, India

Dr. R. Damodaram, Associate Professor, Department of Mechanical Engineering, SSNCE, India

Dr. K. Babu, Associate Professor, Department of Mechanical Engineering, SSNCE, India

Dr. M. Nalla Mohamed, Associate Professor, Department of Mechanical Engineering, SSNCE, India

Dr. M. Selvaraj, Associate Professor, Department of Mechanical Engineering, SSNCE, India

Dr. K. Jayakumar, Associate Professor, Department of Mechanical Engineering, SSNCE, India

Dr. K. S. Jayakumar, Associate Professor, Department of Mechanical Engineering, SSNCE, India

Dr. B. Anand Ronald, Associate Professor, Department of Mechanical Engineering, SSNCE, India

Dr. D. Ananthapadmanaban, Associate Professor, Department of Mechanical Engineering, SSNCE, India

Dr. G. Satheesh Kumar, Associate Professor, Department of Mechanical Engineering, SSNCE, India

Dr. C. Arun Prakash, Assistant Professor, Department of Mechanical Engineering, SSNCE, India

Mr. B. Jayakishan, Assistant Professor, Department of Mechanical Engineering, SSNCE, India

Dr. G. Selvakumar, Associate Professor, Department of Mechanical Engineering, SSNCE, India

Dr. A. S. Ramana, Associate Professor, Department of Mechanical Engineering, SSNCE, India

Dr. M. Dhananchezian, Associate Professor, Department of Mechanical Engineering, SSNCE, India

Dr. R. Rajeswari, Associate Professor, Department of Mechanical Engineering, SSNCE, India

Dr. R. Vimal Samsingh, Associate Professor, Department of Mechanical Engineering, SSNCE, India.

Mr. D. Ebenezer, Assistant Professor, Department of Mechanical Engineering, SSNCE, India

# Foreword

It is a delight to have a book of this kind published by Springer as a collection of select peer-reviewed papers that were presented during our prestigious International Conference on Energy and Materials Technologies (ICEMT 2021) organized by our department of Mechanical Engineering at SSN college of Engineering during August 20–21, 2021. This book specifically focuses on the thematic area of energy and allied topics. It is heartening to note that all the chapters of this book reflect the dedicated research works of authors across different laboratories and institutions. Indeed, the contents are very interesting covering the relevant research topics on domains like alternate fuels, solar photovoltaics, solar thermal systems, energy storage, HVAC, sustainable energy and so on. I appreciate all the editors for their sincere efforts and great work of compilation. I also appreciate the reviewers for the reviews and good selection of papers. I appreciate all the authors for their technical contributions. I hope the book reaches a wide spectrum of audience across institutions to industries and serve as a best source of knowledge. My best wishes to the publishers, Springer Nature, for the earnest and sincere efforts on having the book published on time.

Dr. K. S. Vijay Sekar
Professor and Head
Department of Mechanical Engineering
Sri Sivasubramaniya Nadar College
of Engineering
Chennai, Tamil Nadu, India

# Preface

Energy and materials are undoubtedly the two big pillars of science. The future of scientific and technological progress of the world depends strongly on the advancements in the above areas. The key contributions of several researchers in the past have laid a strong base for the scientific community to seek sustainable solutions to the problems and challenges faced by the mankind at present. In this context, we are happy to contribute our part to the progress of science and technology through a publication of this kind. This book is a collection of select research papers on energy and allied topics that were presented during the virtual International Conference on Energy and Materials Technologies (ICEMT 2021) organized by the Department of Mechanical Engineering, SSN College of Engineering, Tamil Nadu, India, during August 20–21, 2021. We received about 160 research papers on energy and related topics out of which about 110 papers were invited for oral presentation. Post the presentation, we selected and recommended about 44 papers for possible publication as a book in the prestigious Springer's Lecture Notes in Mechanical Engineering following a thorough review process. The articles in this book present many interesting research findings of the contributing authors. The papers have been organized under twelve different chapters covering the topics such as solar thermal, solar photovoltaics, alternate fuels, energy storage, energy management, waste heat recovery, sustainable energy systems, solar HVAC systems, batteries and electric vehicles, heat exchangers and heat pipes, combustion and emissions, and miscellaneous applications. The articles throw light on the current trends and state-of-the-art research covering simulations/analysis/experiments/models, etc., carried out either at system or at process level or both. We hope that this book would serve as a knowledge addition and motivate researchers across institutes, laboratories, and industries for further research and development. We do hope that the book will also serve as a handy tool/guide to all the professionals and students as well.

Chennai, India — N. Lakshmi Narasimhan
Tarragona, Spain — Mahmoud Bourouis
Chennai, India — Vasudevan Raghavan

# Acknowledgments

Finally, the much-awaited book has come to its visible form for the earnest and enthusiastic audience, thanks to the selfless support of several dedicated hearts. As editors, we express our deep sense of gratitude to all the expert reviewers for their meticulous and careful reviews of the research articles presented here. Special thanks are always due from our side to all the respective authors who had contributed their research papers. We also thank the SSN institution and its management for all the support and encouragement. The help of our colleagues and student volunteers is duly recognized by the editors with heartfelt thanks. We wish to place on record our humble note of thanks to Prof. G. Venkatarathnam, IIT Madras, for the timely suggestions and guidance. We duly acknowledge with gratitude the excellent support of the conference organizing committee members, session chairs, and session coordinators. We express our sincere thanks to Dr. K. S. Vijay Sekar, HoD/Mech., SSNCE, and Dr. V. E. Annamalai, Principal, SSNCE, for the constant support and guidance. Also, we extend our deep sense of gratitude to Dr. Ramkumar P., Associate Professor, IIT Madras, and Dr. Subbu Sethuvenkataraman, Research Group Leader, CSIRO Energy Business Division, Australia, for their keynote address during the conference. We wish to file our special note of thanks to Springer Nature and Springer publishing team for the excellent support and guidance at various stages during the preparation of this book.

N. Lakshmi Narasimhan
Mahmoud Bourouis
Vasudevan Raghavan

# About the Institution

SSN Institutions, founded by Padma Bhushan Dr. Shiv Nadar, Chairman, HCL Technologies, stands out as a premier center of higher learning with a mission of pursuing excellence in education and research. The institutions, with their diverse and dynamic community of students, offer a distinctive combination of some of the finest graduate, undergraduate, and research programs, accomplished faculty, world-class facilities, and a residential campus set on a sprawling 250 acres of sylvan surroundings. SSN Institutions provides a variety of stimulating environments for intellectual development, free thinking, and personal growth, challenging its students with dynamic learning opportunities and equipping them with the skills, insights, attitudes, and practical experiences that are necessary to take up responsibilities in the society.

While students at SSN immerse themselves in academics, the college has a lot in store for them outside the classroom. Student life includes participation in sports, recreational and co-curricular activities, and culturals. In short, at SSN, students will find an academic and social environment where everyone— from faculty members to peers—help shape their future. SSN is a home to aesthetically designed buildings with state-of-the-art computer and Internet facilities, modern workshops, seminar halls, auditoriums, and well-stocked libraries, sports, and games fields in addition to an indoor stadium with gymnasium.

SSNCE is accredited by NAAC with A+ grade. Also, NBA, AICTE, New Delhi, has accredited seven UG and four PG programs. Within a span of 22 years, SSN has grown into one of the top educational institutions in the country and become an autonomous institute in the year 2018. Some of the salient features of SSN are: ranked 45th in engineering category in India rankings 2021 by NIRF, MHRD; the state-of-the-art infrastructure with over 1.8 million square feet of buildings; spacious hostels, sports facilities, and faculty quarters within the campus; a student body of 4300+ students, 500 PhD scholars, and 280 faculty members, 75% of whom are PhDs.

As an exemplary institution of learning, SSN follows an admission policy that strongly favors merit, even as it enables access to education for students from all strata of society through appropriate scholarships. The institution boasts of a strong alumni network with alumni events held every year serving as a platform for past students to give back to SSN and share their experiences with its present fellow students.

With so much to offer, it is only natural that students of SSN get a unique opportunity to carve a niche for themselves in their chosen field of study that enables them to become well-rounded and discerning citizens, fully qualified for their chosen professions in the workplace.

# About the Department

The department of mechanical engineering was established in the year 2007. The department offers B.E. (Mechanical Engineering) from the academic year 2007 to 2008, M.E. (Manufacturing Engineering) from the academic year 2012 to 2013, and M.E. (Energy Engineering) from the academic year 2013 to 2014. The department became an approved research center of Anna University in the year 2012. The UG program of the department has been accredited by the prestigious NBA. Our vision is to be an eminent Centre of Excellence in the field of mechanical engineering, where teaching, learning, and research synergize to deliver technical education and scientific research for the public good. Our mission is to create efficient mechanical engineers who can compete at the global level. We intend to do this by creating a research hub engaged in contemporary research in association with industries. The projects and hands-on activities are designed to foster a spirit of entrepreneurship among students and to enable a few to launch their own ventures.

The department has qualified and experienced faculty members; many of them are with Ph.D. degree. The faculty members actively engage in research and constantly publish papers in international and national journals. The department regularly organizes technical workshops for the faculty members to expose them to emerging areas. The department has state-of-the art facilities for various laboratories, department library, and classrooms to support e-learning. The department has a well-equipped centralized workshop facility that caters to the needs of various departments. Guest lectures and industrial visits are periodically arranged for the students to supplement their curriculum. We strive for all-round excellence in students, encouraging them in all extracurricular activities.

The department has research projects funded by the Department of Science and Technology (DST), All India Council for Technical Education (AICTE), Science and Engineering Research Board, and Naval Research Board (NRB) to the tune of around Rs. 3.15 crores for conducting research in friction stir welding, magnetic molding of composite materials, vibration studies, robotics, abrasive reclamation, and finite element analysis of composites. AICTE sponsored Industry Institute Partnership Cell (IIPC), and Design/Fabrication Labs support additional industry-relevant learning.

The major research areas are friction stir welding, grinding, composite materials, metal cutting, energy, heat transfer, thermal energy storage, evaporative cooling, alternate fuels, battery thermal management, vapor absorption refrigeration systems, waste management, etc.

# About the Conference

Energy and materials are both inseparable and form the essential core of science and engineering. Being mutually complimentary, they are the basis for everything in this vast Universe. Exciting research is happening elsewhere focusing the development of state-of-the-art technologies, development of ultra-new and smart systems, and so on in the two areas. It is high time that researchers working in the areas of energy and materials get linked and share their R&D with each other and the outside world for the (re)emergence of a safe and pollution free environment. Considering the above, this conference was organized by us with an objective of setting a platform for researchers/scientists/industries/academia / scholars / students across the globe to exchange ideas and share their current research and findings for the benefit of the community and environment at large. The conference was organized online and received a great response from researchers worldwide. We received about 350+ abstracts, and about 230 were selected for oral presentation. There were about 110 papers on energy and the rest on materials. As organizers, we are extremely happy to share that the conference was well received. There were about 22 parallel sessions that covered all the oral presentations in the two days with two keynote address by two eminent speakers. We selected around 100 papers in each of the themes, viz. energy and materials. Following a thorough review process, we shortlisted about 50 papers in each of the themes for possible publication as two separate book volumes under the prestigious book series, "Lecture Notes in Mechanical Engineering" published by Springer Nature. All the sessions were chaired by experts across the field, and we awarded one best paper for each session. The conference commenced with an official inaugural and ended with a formal valedictory presided by the Head of the Institution, SSNCE and graced by the presence of the entire Organizing Committee and all the Authors and general participants. Overall ICEMT 2021 was a great knowledge sharing experience and wish to thank everyone for making the Conference memorable. We look forward to organize the second version of ICEMT in the near future and look forward to a great support from prospective researchers.

# ICEMT 2021 Keynote Speakers and Topics

| Keynote Speaker | Organization | Day | Topic of the Keynote |
|---|---|---|---|
| Dr. P. Ramkumar | Associate Professor, Department of Mechanical Engineering, IIT Madras, India. | August 20, 2021 | **Micro Texturing Enabled Engineering Surface** |
| Dr. Subbu Sethuvenkataraman | Research Group Leader, CSIRO Energy Business Division, Australia. | August 21, 2021 | **Energy Sector in Transition—Some Challenges and Opportunities** |

The conference organizers are pleased to express their sincere gratitude to both the speakers for the excellent presentation and wisdom shared.

# Contents

## Alternate Fuels



## Automotive Applications



## Combustion and Emissions

# About the Editors

**Dr. N. Lakshmi Narasimhan** is currently serving as Associate Professor in the Department of Mechanical Engineering at Sri Sivasubramaniya Nadar College of Engineering, Tamil Nadu, India. He has about 20+ years of academic experience and play an active role in strengthening the industry–institute interactions at SSNCE. He obtained his Ph.D. from IIT Madras, India, specializing in cryogenic mixed refrigerant coolers (cryocoolers). He did his Masters, M.S. (by research) from Anna University, Chennai, specializing in thermal energy storage. He has published about 18 research papers in international journals of high repute and 15 international conference papers. He is a recipient of awards from both industries and academia for his joint industrial R&D and academic research. He has carried out consultancy projects for reputed industries. He is a life member of the IEI (India) and National Society of Fluid Mechanics and Fluid Power. His areas of research include cryogenic refrigerators, heat transfer, thermal energy storage, electronic cooling, and Li-Ion battery thermal management.

**Dr. Mahmoud Bourouis** is currently a permanent professor at the Department of Mechanical Engineering, Rovira i Virgili University, Tarragona, Spain. He obtained his Ph.D. in Chemical Engineering from the National Polytechnic School of Toulouse in France. His major areas of research interests include absorption heat pumps, heat and mass exchangers, energy conversion systems, and renewable energies. He has published over 70 indexed papers. Currently, he is an editorial board member of the *International Journal MDPI-Energies* (Section "Energy and Environment").

**Prof. Vasudevan Raghavan** received his Ph.D. from Indian Institute of Technology (IIT) Madras in 2004 and worked as a postdoctoral research fellow and research assistant professor at the University of Nebraska-Lincoln for around three years before joining IIT Madras as Assistant Professor in the Department of Mechanical Engineering. He is currently working as Professor at IIT Madras. He has been working on transient two-phase numerical modeling of isolated droplet vaporization and combustion, renewable fuels, flame spread over liquid fuel pools, detailed chemical kinetics studies on homogeneous and heterogeneous flames, simulations of non-reactive and reactive particle laden flows, and coal and biomass gasification and combustion. He has authored more than 150 peer-reviewed journal and conference articles. He has guided 11 Ph.Ds. and 18 M.S. scholars till date. He is in the editorial board of *Fire Safety Journal.*

# Solar Thermal Systems

# Performance Improvement of Compound Parabolic Collector Using Dual Receivers

Diptanshu K. Pise, Pranjali S. Deole, and Sandeep S. Joshi

## Abbreviations

| | |
|---|---|
| $2\theta_a$ | Solar swing angle or acceptance angle |
| $C$ | Concentration ratio |
| CPC | Compound parabolic collector |
| $D$ | Diameter of receiver (m) |
| $D_i$ | Inner diameter of receiver (m) |
| $D_o$ | Outer diameter of receiver (m) |
| $h$ | Height of CPC (m) |
| $H$ | Height of focus (m) |
| $H_p$ | Distance between centres of receiver to edge of parabola (m) |
| $\dot{m}$ | Mass flow rate of water (Kg/s) |
| $W$ | Aperture width (m) |
| $\beta$ | Optimum tilt angle |
| $\delta$ | Declination angle |
| $\theta$ | Angle of incidence |
| $\theta$ | Half acceptance angle |
| $\phi$ | Latitude angle |
| $\omega$ | Hour angle |
| $\omega_s$ | Hour angle corresponding to sunrise and sunset on a horizontal surface |

D. K. Pise (✉) · P. S. Deole · S. S. Joshi
Department of Mechanical Engineering, Shri Ramdeobaba College of Engineering and Management, Nagpur, Maharashtra 440013, India
e-mail: dkpise@gmail.com


N. L. Narasimhan et al. (eds.), *Recent Advances in Energy Technologies*, Lecture Notes in Mechanical Engineering, https://doi.org/10.1007/978-981-19-3467-4_1

## 1 Introduction

Solar energy is the ultimate source of renewable energy. Solar energy is being used for several applications. For medium and high temperature, concentrated systems are preferred. In the concentrated systems, for medium-temperature applications, line focusing collectors such as compound parabolic collector, parabolic trough collectors and linear Fresnel reflectors are widely used. The compound parabolic collector is the combination of two parabolas [1]. The solar radiations falling on the reflectors of CPC reflected towards the receiver. The CPC can be used without any tracking mechanism; hence, tracking errors can be avoided [1–4]. The concentration ratio of CPC is low, but it has ability to receive both beam and diffuse radiations [2, 5–7]. The CPC is used for water heating [1, 8, 9], industrial process heating [3, 8, 10], power generation [1, 11], solar distillation [1, 12] and solar cocking [8]. CPC with evacuated tube has optical efficiency of 65% and thermal efficiency of 50%. The efficiency can be enhanced by 15% using proper sun tracking [2]. Various techniques to improve the performance of CPC are as shown in Fig. 1.

The performance of compound parabolic collector can be improved by either effective collection of solar energy [13] or by reducing the heat losses from the receiver [1, 14, 15]. There are several ways for effective collection of solar energy. The accurate design of collector [5, 8, 16–18] is one of the basic ways to collect the radiations effectively [19, 20]. The material of reflector [21] has more importance, while designing the collector, mirrors [1, 3, 8] having the reflectivity of 85% [8] are widely used in compound parabolic collector and anodized aluminium sheets are also been used as a reflector in CPC [18, 22]. The orientation of collector [3, 8, 17] plays an important role in non-tracking collectors. In general, two-dimensional compound parabolic collector is slopping towards south for better performance, and reflectors are facing towards east and west direction [17]. The useful heat gain depends on specific heat of fluid; hence, different heat transfer fluids are being used [2, 11]. The water is commonly used as a working fluid in CPC because of its availability and ease for direct applications. The concentration ratio decides the performance of CPC, useful heat gain and flow rate of inlet fluid. The fluid can be flown by natural circulation [8], forced circulation [3] and depends on the heat requirement. The optimum tilt angle [17] is different for different location; it can be calculated mathematically [23]. The effective collection of solar energy also depends on the position of receiver [16, 24, 25] as well as types of receivers. The four types of receivers, viz. evacuated tube receiver [2, 5, 14, 17, 26], metal tube receiver [3, 8, 27, 28], metal tube with secondary reflector [27] and finned tube receiver [5] are widely used. Depending on the application, one of the receivers can be used. Evacuated tube receivers are most popular receivers among all.

It has been observed that, for an accurately designed stationary compound parabolic collector with single linear receiver, reflected radiations are getting lost to the atmosphere without absorption by the receiver. The peripheral area near the receiver tube can be used to collect the reflected untapped radiations. With this motivation, use of dual receivers for a CPC is suggested in the present study.

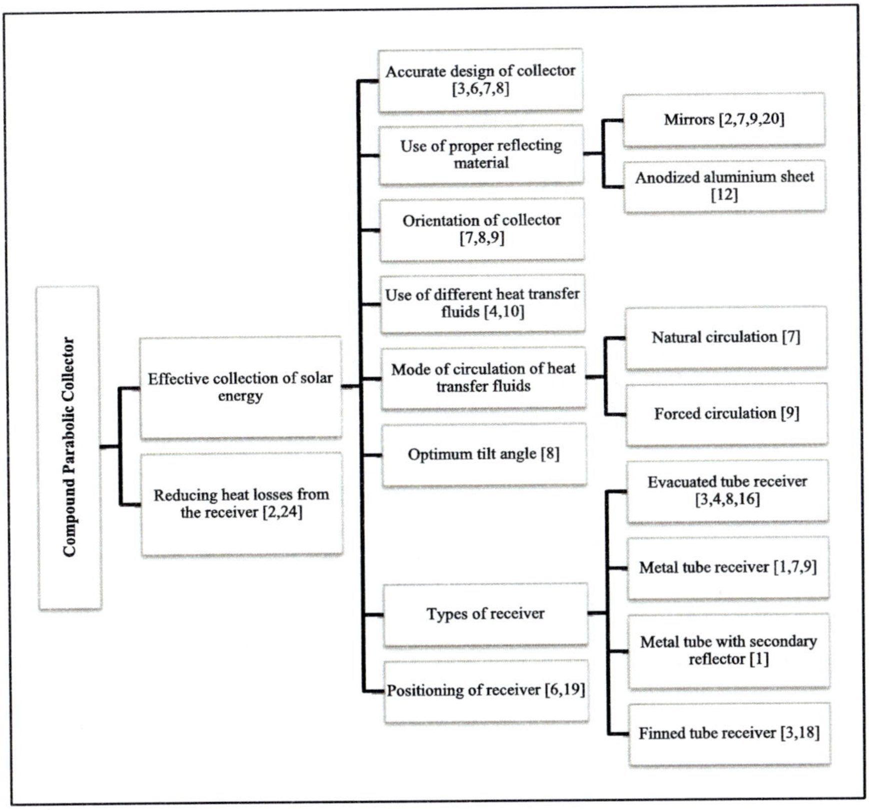


**Fig. 1** Different ways to improve the performance of compound parabolic collector

## 2 Design of CPC

A front view of the compound parabolic collector used in this work is shown in Fig. 2, and positioning of mirror strips is shown in Fig. 3. As suggested in the literature, an acceptance angle of 17.5° would give eight or more hours of collections [17]. The relations required for calculating the concentration ratio, declination angle, solar swing angle, height of focus, aperture width, height of CPC are given in Eqs. (1–6) [16, 17, 29, 30].

Concentration ratio ($C$)

$$C = \frac{1}{\sin \theta_a} \tag{1}$$

Declination angle ($\delta$)

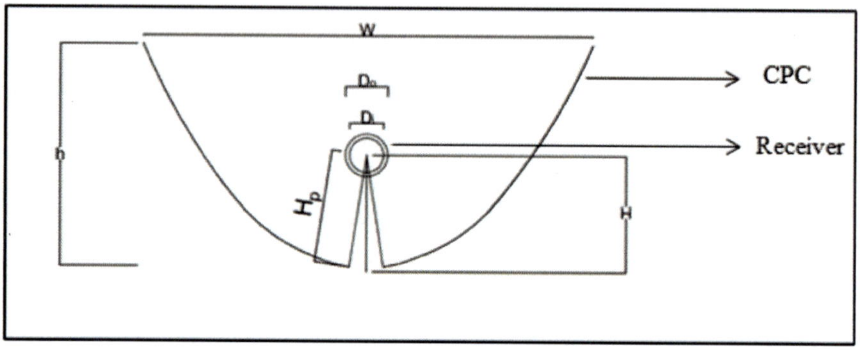


**Fig. 2** Schematic of CPC ($D_i$ = inner diameter of receiver, $D_o$ = outer diameter of receiver, $H$ = height of focus, $W$ = aperture width, $h$ = height of CPC)

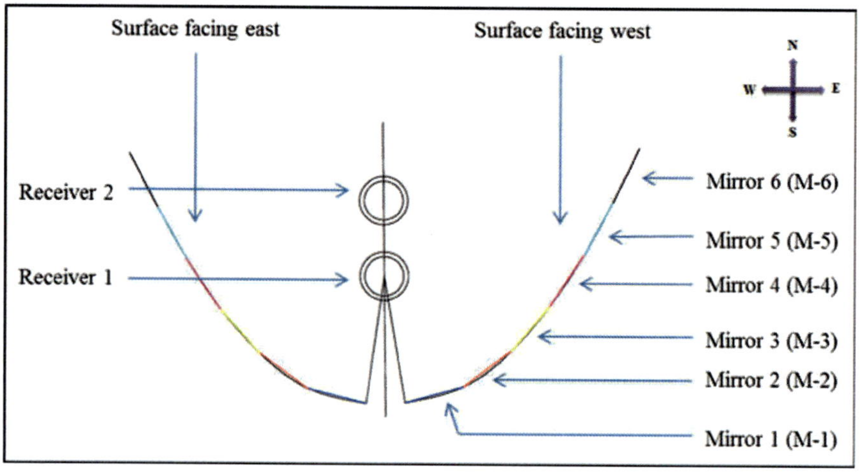


**Fig. 3** CPC showing mirror strips, surface facing east, surface facing west and dual receivers

$$\delta = 23.45 \sin\left[\frac{360}{365}(284 + n)\right] \tag{2}$$

Solar swing angle or acceptance angle ($2\theta_a$)

$$2\theta_a = \left|\delta - \tan^{-1}\left(\frac{\tan \delta}{\cos \omega t}\right)\right| \tag{3}$$

Height of focus ($H$)

$$H = H_p \cos \theta \tag{4}$$

**Table 1** Design parameters of CPC

| Parameter | Value |
|---|---|
| Mounting | N–S |
| Aperture area | 1.099 $m^2$ |
| Aperture width ($W$) | 0.6108 m |
| Height of focus ($H$) | 0.1527 m |
| Height of CPC ($h$) | 0.30536 m |

Aperture width ($W$)

$$W = 4H \tag{5}$$

Height of CPC ($h$)

$$h = \frac{w}{2} \tag{6}$$

Table 1 shows the geometrical parameters of CPC used in this study.

## 3 Instrument Used to Measured Solar Radiation

A pyranometer as shown in (Fig. 4), is used to measure the incident solar radiation falling on collector. The pyranometer has the following specifications.

Specifications of the pyranometer:

Make: Kipp and Zones
Measuring range: 0–2000 W/$m^2$
Operating temperature: $-10$ to 60 °C.

## 4 Theoretical Analyses to Identify the Effective Peripheral Area of Receiver

In general, the solar radiations fall on the compound parabolic collectors and reflected to the focus where receiver is fixed. It has been observed that, for an accurately designed stationary compound parabolic collector with single linear receiver, reflected radiations are getting lost to the atmosphere without absorption by the receiver. The peripheral area near the receiver tube can be used to collect the reflected untapped radiations. The receiver position is fixed by using law of reflection [31]. At first, for the location Nagpur (21° N, 79° E), the incident angle for beam radiations is estimated for the inclined mirror strips of the CPC. The mirrors are considered as a flat surface with fixed slopes. Figure. 3 shows the basics of law of reflection, which

Fig. 4 Pyranometer

is used to determine the receiver position of second receiver. There are six mirror strips on each side of the CPC as shown in Figs. 5a, b.

The angle of incident of beam radiations which is equal to angle of reflection of radiations by the mirrors can be calculated by using following relation [17].

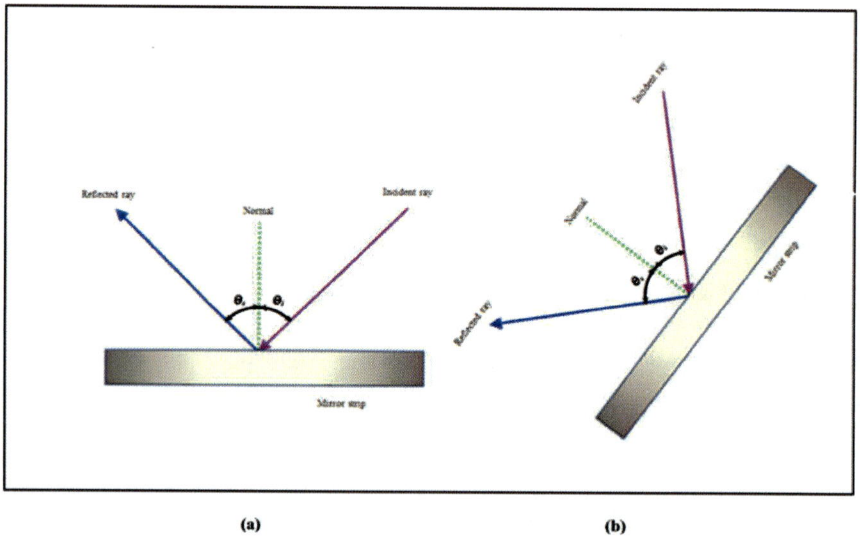


**Fig. 5** Law of reflection **a** for horizontal mirror **b** for inclined mirror

$$\cos\theta = \sin\phi(\sin\delta\cos\beta + \cos\delta\cos\gamma\cos\omega\sin\beta) + \cos\phi(\cos\delta\cos\omega\cos\beta - \sin\delta\cos\gamma\sin\beta) + \cos\delta\sin\gamma\sin\omega\sin\beta \tag{7}$$

where

$\theta$ = angle of incident or reflection, $\delta$ = declination angle, $\beta$ = tilt angle, $\gamma$ = azimuth angle, $\omega$ = hour angle.

According to law of reflection for flat plate, the angle of incident is equal to angle of reflection [31]. The angle of incident of beam radiations is calculated for entire year for 10 a.m. to 5 p.m. Table 2 shows the monthly average of incident and reflected angles for 1, 2, 3 p.m. There are 12 mirror strips on each parabola. For the simplicity of calculation, two strips are considered as one. The mirror strips are fixed at an angle of 15°, 37°, 50°, 56°, 61°, 63°, respectively, from bottom to top of the parabola. Hour angle is taken positive before 12 p.m. and negative after 12 p.m. The azimuth angle is taken as 90° for surface facing west and −90° for surface facing east [17]. The radiation intensity is higher at 1 p.m. to 3 p.m. The angle of incident and reflection is sketched on front view of compound parabolic collector. Figure 3 shows the front view of angles formed by incident rays and reflected rays on each mirror strips placed in CPC.

Based on the calculated angles of reflection of beam radiations by all the mirrors, the useful peripheral area of the receiver has been identified. The second receiver can be placed in this identified useful peripheral area along with the first receiver. Figures 6a–c show the identified position for second receiver.

From the above theoretical analysis, it is found that the radiations fall on the receiver, fixed according to the geometric relation [16], i.e. 15.27 cm from the bottom of the CPC as well as the maximum time radiations are seen above the fixed receiver too.

## 5 Experimentations

To validate the theoretical claims, a pilot experiment has been carried out. The experiment is performed at Nagpur (India) location (21° N, 79° E) in the month of March. The CPC is inclined at an angle of 21° and is installed facing towards south. The mirror strips are facing east and west as shown in figure. Four temperature indicators are placed on the axis of CPC to measure the temperature of reflected radiations. Figure 7 shows the position of temperature indicator, and Fig. 8 shows the temperature range at each point.

Figure. 8 shows that reflected radiations are getting lost to the atmosphere with single receiver. The peripheral area near the receiver tube can be used to collect the reflected untapped radiations for experiments; two receivers have been fixed to the CPC. The system consists of CPC, two evacuated tube receivers, fresh water tank,

**Table 2** Monthly incident and reflected angle for (i) 1 p.m., (ii) 2 p.m., (iii) 3 p.m.

| | | Surface facing west mirrors[a] Angle of refection of beam radiations (degrees) | | | | | | Surface facing east mirrors[a] Angle of refection of beam radiations (degrees) | | | | | |
|---|---|---|---|---|---|---|---|---|---|---|---|---|---|
| (i) 1 p.m. | Month | M-1 | M-2 | M-3 | M-4 | M-5 | M-6 | M-1 | M-2 | M-3 | M-4 | M-5 | M-6 |
| | Jan | 41 | 44 | 50 | 53 | 56 | 57 | 51 | 65 | 74 | 79 | 82 | 76 |
| | Feb | 34 | 38 | 45 | 49 | 53 | 54 | 45 | 61 | 72 | 76 | 81 | 74 |
| | Mar | 23 | 31 | 40 | 45 | 49 | 51 | 38 | 57 | 68 | 74 | 78 | 72 |
| | Apr | 11 | 25 | 37 | 42 | 47 | 49 | 32 | 53 | 66 | 71 | 76 | 70 |
| | May | 2 | 23 | 36 | 42 | 47 | 49 | 29 | 51 | 64 | 70 | 75 | 69 |
| | Jun | 3 | 23 | 36 | 42 | 47 | 49 | 29 | 51 | 64 | 70 | 75 | 69 |
| | Jul | 2 | 23 | 36 | 42 | 47 | 49 | 29 | 51 | 64 | 70 | 75 | 69 |
| | Aug | 7 | 24 | 36 | 42 | 47 | 49 | 31 | 52 | 65 | 71 | 76 | 70 |
| | Sep | 18 | 28 | 38 | 44 | 48 | 50 | 36 | 55 | 67 | 73 | 78 | 71 |
| | Oct | 30 | 36 | 43 | 48 | 51 | 53 | 43 | 60 | 70 | 76 | 80 | 73 |
| | Nov | 39 | 43 | 49 | 52 | 55 | 57 | 50 | 64 | 73 | 78 | 82 | 76 |
| | Dec | 43 | 46 | 51 | 54 | 57 | 58 | 53 | 66 | 75 | 79 | 83 | 77 |
| (ii) 2 p.m. | Month | M-1 | M-2 | M-3 | M-4 | M-5 | M-6 | M-1 | M-2 | M-3 | M-4 | M-5 | M-6 |
| | Jan | 43 | 39 | 40 | 42 | 45 | 46 | 61 | 77 | 87 | 88 | 84 | 89 |
| | Feb | 36 | 31 | 34 | 37 | 40 | 41 | 56 | 74 | 85 | 89 | 85 | 90 |
| | Mar | 27 | 21 | 27 | 31 | 35 | 37 | 51 | 71 | 83 | 88 | 87 | 88 |
| | Apr | 17 | 12 | 22 | 28 | 32 | 34 | 46 | 67 | 80 | 86 | 89 | 86 |
| | May | 13 | 9 | 22 | 28 | 33 | 35 | 43 | 65 | 78 | 84 | 89 | 84 |
| | Jun | 13 | 11 | 23 | 29 | 34 | 36 | 43 | 65 | 78 | 83 | 88 | 84 |
| | Jul | 13 | 10 | 22 | 28 | 33 | 35 | 43 | 65 | 78 | 84 | 89 | 84 |
| | Aug | 15 | 10 | 22 | 27 | 32 | 34 | 45 | 66 | 79 | 85 | 90 | 85 |
| | Sep | 23 | 17 | 25 | 29 | 34 | 35 | 49 | 69 | 82 | 87 | 88 | 87 |
| | Oct | 33 | 28 | 32 | 35 | 38 | 40 | 54 | 73 | 85 | 90 | 86 | 89 |
| | Nov | 42 | 37 | 39 | 41 | 43 | 44 | 60 | 77 | 87 | 88 | 84 | 90 |
| | Dec | 46 | 41 | 42 | 44 | 46 | 47 | 62 | 78 | 88 | 87 | 84 | 89 |
| (iii) 3 p.m. | Month | M-1 | M-2 | M-3 | M-4 | M-5 | M-6 | M-1 | M-2 | M-3 | M-4 | M-5 | M-6 |
| | Jan | 50 | 38 | 35 | 35 | 35 | 36 | 73 | 90 | 79 | 74 | 70 | 74 |
| | Feb | 44 | 31 | 28 | 28 | 29 | 30 | 69 | 88 | 80 | 75 | 71 | 75 |
| | Mar | 36 | 20 | 17 | 19 | 22 | 23 | 64 | 85 | 83 | 77 | 72 | 76 |
| | Apr | 30 | 10 | 8 | 13 | 18 | 20 | 60 | 82 | 86 | 80 | 75 | 79 |
| | May | 27 | 6 | 9 | 14 | 19 | 21 | 57 | 79 | 88 | 82 | 77 | 81 |
| | Jun | 27 | 9 | 12 | 17 | 21 | 23 | 56 | 78 | 89 | 83 | 78 | 82 |
| | Jul | 27 | 7 | 10 | 16 | 20 | 22 | 57 | 79 | 88 | 83 | 78 | 81 |

(continued)

**Table 2** (continued)

| | | | | | | | | | | | | |
|---|---|---|---|---|---|---|---|---|---|---|---|---|
| Aug | 29 | 7 | 7 | 13 | 18 | 20 | 59 | 81 | 86 | 81 | 76 | 79 |
| Sep | 34 | 16 | 13 | 16 | 20 | 21 | 62 | 84 | 84 | 78 | 73 | 77 |
| Oct | 42 | 27 | 24 | 25 | 27 | 28 | 67 | 87 | 81 | 76 | 71 | 75 |
| Nov | 49 | 36 | 33 | 33 | 34 | 34 | 72 | 90 | 79 | 74 | 70 | 74 |
| Dec | 52 | 40 | 37 | 37 | 37 | 38 | 74 | 89 | 79 | 74 | 70 | 74 |

[a]Refer Figs. 4a, b for positions of mirrors M-1 to M-6

insulated tank, rubber cork, etc. The fresh water is supplied from fresh water tank placed above the insulated storage tank. Figure. 9 shows the actual setup of fresh water inlet condition with two receivers.

The continuous fresh water is supplied from upper tank having the mass flow rate of 5 kg/hr to the inlet of lower evacuated tube, and outlet water is supply to inlet of upper tube receiver. The temperature of outlet water is recorded. The rubber cork which is used as a packing material as well as the insulator has two holes, one is inlet, and another is outlet. The fresh water is mixed with heated water in the evacuated tube. Due to inlet pressure and thermos-syphon effect, the heated water rises up to the upper tube. The piping arrangement of water circulation in the setup is as shown in Fig. 10.

Figure. 11 shows the comparison of temperature in single receiver and dual receivers. The performance that is shown for average radiation intensity is 906 W/m$^2$.

## 6 Discussions

It has been observed that, for an accurately designed stationary compound parabolic collector with single linear receiver, reflected radiations are getting lost to the atmosphere without absorption by the receiver. In the current study, a concept of use of two receivers for a compound parabolic collector is proposed. The useful peripheral area near the single receiver is identified using theoretical analysis. The analysis is based on estimation of angle of incidence of beam radiations on the collector and subsequent reflections. To validate the theoretical claims, a pilot experimental study has also been conducted.

## 7 Conclusion

During experiments, for average irradiance of 906 W/m$^2$, the maximum outlet temperature of water is found to be 76.60 °C with single receiver and it was 95.18 °C for dual receivers.

**a**

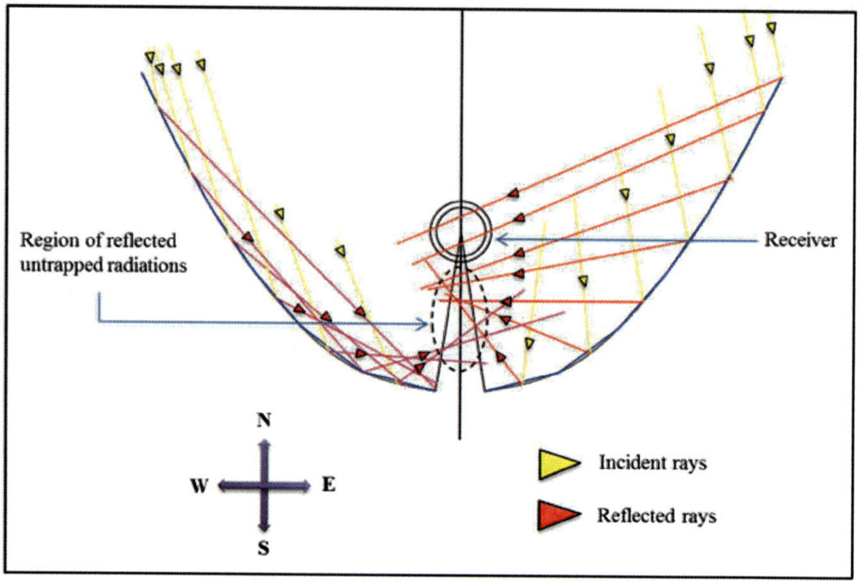


**b**

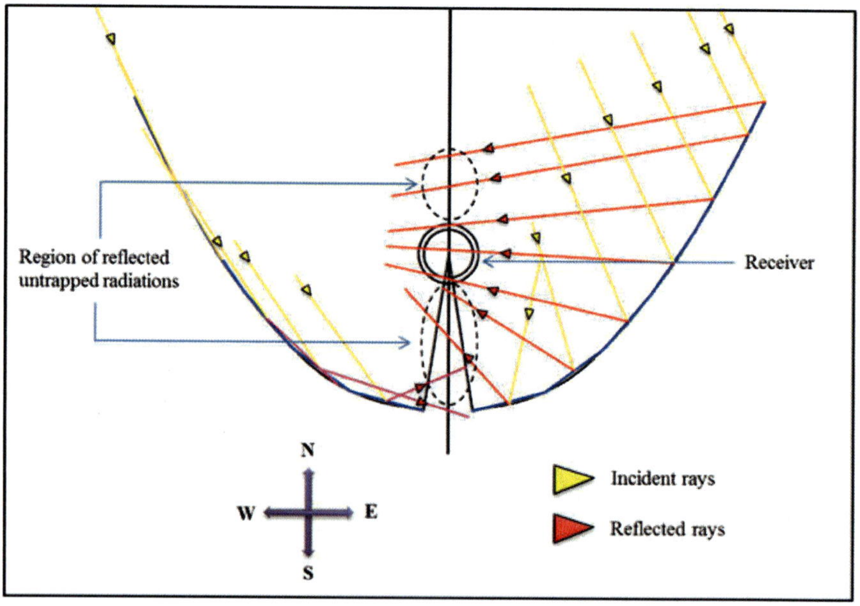


**Fig. 6** **a** Region of reflected untapped radiations at 1 p.m. **b** region of reflected untapped radiations at 2 p.m. **c** region of reflected untapped radiations at 3 p.m.

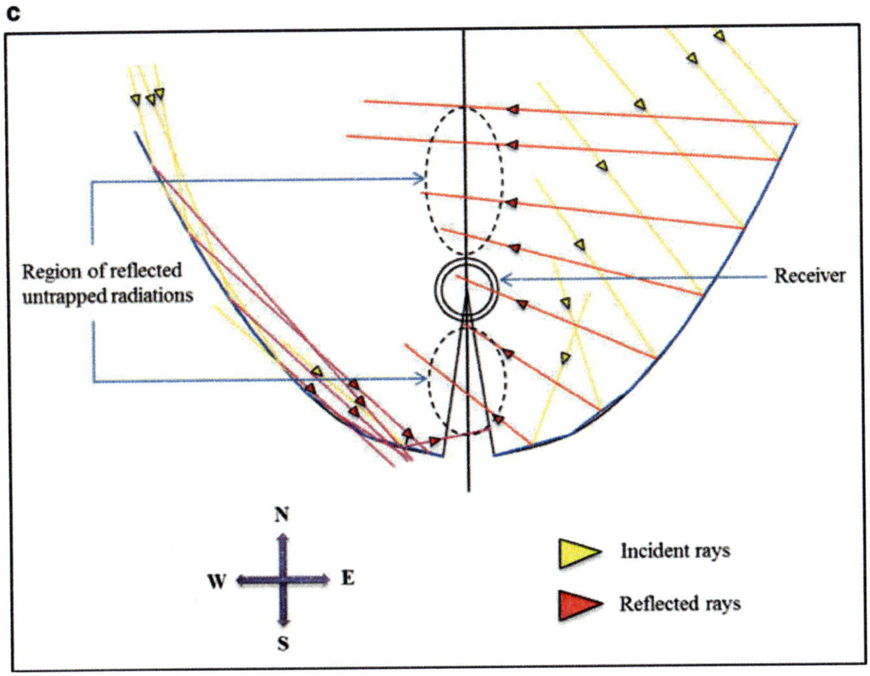


**Fig. 6** (continued)

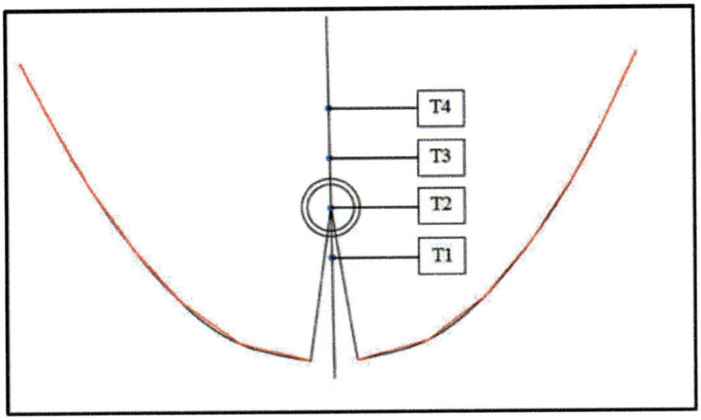


**Fig. 7** Position of temperature indicator in CPC

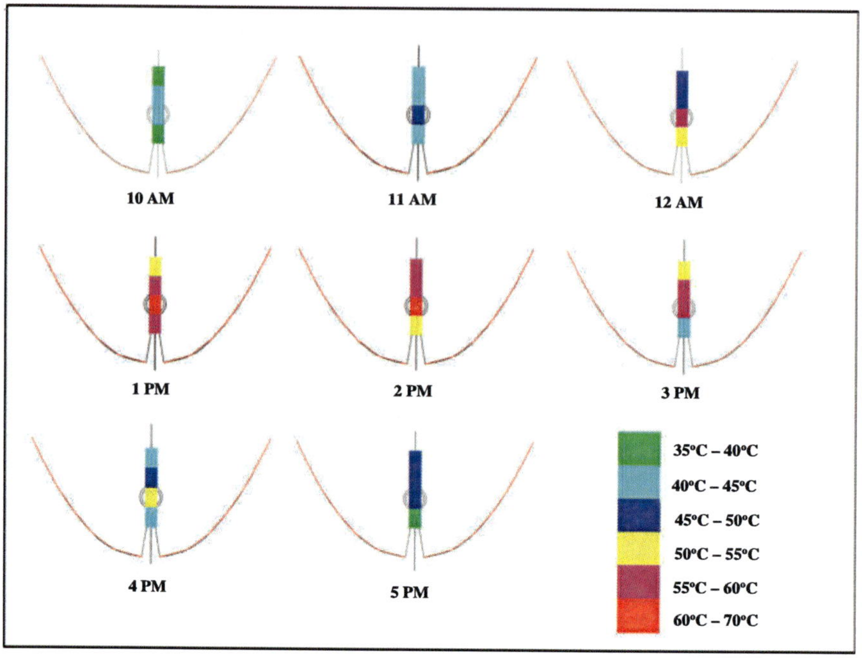


**Fig. 8** Temperature at different position on the axis of CPC

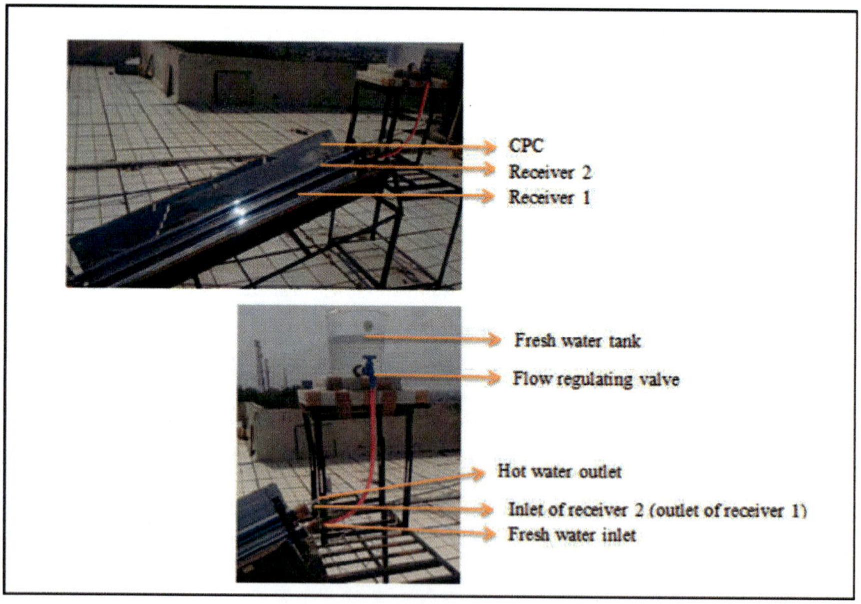


**Fig. 9** Experimental setup of fresh water inlet condition using two receivers

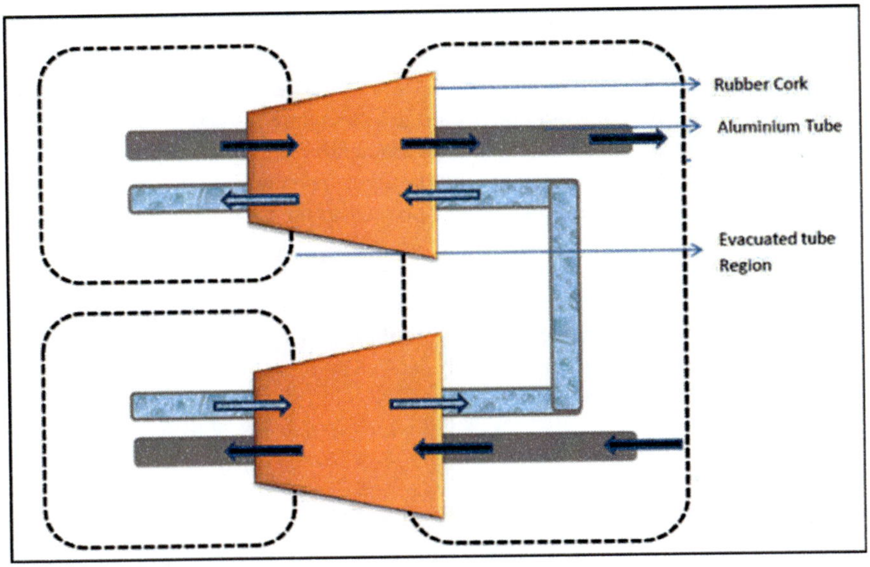


**Fig. 10** Line diagram of water circulation

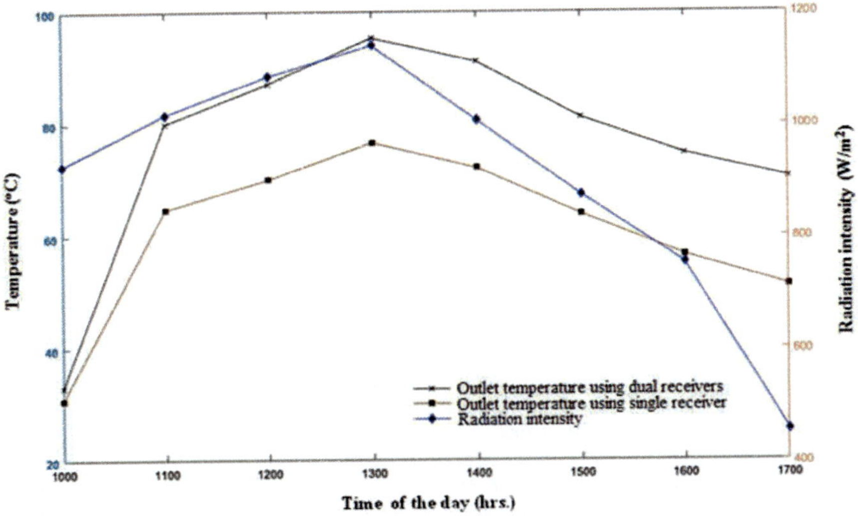


**Fig. 11** Comparison of temperature in single receiver and dual receivers

The current study prevails that the dual receiver system performs better as compared to the CPC with single receiver. In the suggested modification (use of two receivers), one additional receiver is required to be attached in the same CPC system. Thus, the modified system would be economical and more effective as compared to the conventional one. A probable drawback associated with this modification might be the shading of the additional receiver on the mirrors. However, as the receiver diameter is small, the shading effect will be negligible. Given the additional heat gain, the shading effect can be neglected.

The primary objective of this study is to put forward the concept of dual receivers for compound parabolic collector. The present study is based on the analytical estimations and the pilot experiments. It is strongly recommended to carry out further experiments to ascertain the gain by dual receiver system.

## 8 Future Scope

The effective utilization of concentrator can be achieved by using different receiver arrangements or using more than one receiver.

## References

1. Xu D et al (2013) Compound parabolic concentrators in solar thermal applications: a review. In: Proceedings of the ASME 2013 7th international conference on energy sustainability ES2013 July 14–19, 2013, Minneapolis, MN, USA
2. Bellos E et al (2016) Design, simulation and optimization of a compound parabolic collector. Sustainable Energy Technol Assess 16:53–63
3. Jadhav AS et al (2012) Performance analysis of a novel and cost effective CPC system. In: Energy conversion and management, November 2012, pp 56–65
4. McLoughlin OA et al (2004) Photocatalytic disinfection of water using low cost compound parabolic collectors. Sol Energy 77:625–633
5. Gao HY et al (2011) Thermal performance analysis of a novel compound parabolic concentrator solar collector. In: 2011 International conference on transportation, mechanical, and electrical engineering (TMEE), December 16–18, Changchun
6. Zhongyuan S et al (2017) Modeling and simulation of ray tracing for compound parabolic thermal solar collector. Int Commun Heat Mass Transfer 87:169–174
7. Tanveer M et al (2013) Solar assisted photo degradation of wastewater by compound parabolic collectors: review of design and operational parameters. Renew Sustain Energy Rev 24:534–543
8. Gudekar AS et al (2013) Cost effective design of compound parabolic collector for steam generation. Sol Energy 90:43–50
9. Essabbani T et al (2015) Numerical computation of thermal performance of a simulation of a solar domestic hot water system. Appl Sol Energy 51:22
10. Gu X et al (2014) Analysis of a new compound parabolic concentrator-based solar collector designed for methanol reforming, November 2014, vol 136/041012-1-9. ASME
11. Antonelli M et al (2014) Analysis of a low concentration solar plant with compound parabolic collectors and a rotary expander for electricity generation. Energy Procedia 45:170–179

12. Sampathkumar K et al (2013) The experimental investigation of a solar still coupled with an evacuated tube collector. Energy Sources, Part A: Recov Util Environ Effects 35(3):261–270
13. Joshi SS et al (2018) Performance analysis of photovoltaic thermal system using silicone oil spectrum filter. Appl Sol Energy 54:4
14. Cherian R et al (2016) Performance enhancement of solar water heater using compound parabolic reflector and numerical simulation of thermal losses. 978-1-4673–9925–8/16. IEEE
15. Prapas DE et al (1987) Thermal design of compound parabolic concentrating solar-energy collectors, May 1987, vol 109/161–168. ASME
16. Waghmare SA et al (2016) Design and ray tracing of a compound parabolic collector with tubular receiver. Sol Energy 137:165–172
17. Sukhatme SP, Nayak JK. Solar energy principle of thermal collection and storage, 3rd edn. Tata McGraw-Hill Education, New Delhi
18. Shanmugam S et al (2014) Modeling and analysis of a solar parabolic dish thermoelectric generator. Energy Sources, Part A: Recov Util Environ Effects 36(14):1531–1539
19. Oshchepkov MY et al (2015) Thermal stratification in storage tanks of integrated collector storage solar water heaters. Appl. Sol. Energy 51:74
20. Klychev SI et al (2013) Thermal characteristics of tubular receivers of solar radiation line concentrators. Appl Sol Energy 49:235
21. Jamali H et al (2019) Investigation and review of mirrors reflectance in parabolic trough solar collectors (PTSCs). Energy Rep 5:145–158
22. Joshi SS (2018) A project report "Development of experimental facilities of indoor solar cooking system for further research" submitted at RCOEM, Nagpur, Jan 2018
23. Kothdiwala AF et al. The effect of variation of angle of inclination on the performance of low-concentration-ratio compound parabolic concentrating solar collectors. Solar Energy 55(4):301–309
24. Waghmare SA et al (2018) Optimization of receiver height in compound parabolic collector by optical analysis and experimental method. Optik 157:1331–1341
25. Waghmare SA et al. Experimentation of surface areal irradiance method for solar flux measurement in compound parabolic collector. Optik—Int J Light Electron Optics. https://doi.org/10.1016/j.ijleo.2017.08.111
26. Geete A et al (2018) Exergy analyses of fabricated compound parabolic solar collector with evacuated tubes at different operating conditions: Indore (India). The Institution of Engineers (India)
27. Hu Z et al (2016) Analysis of compound parabolic collector with different cavity structures. 978-1-4673-9194-8/16/31.00. IEEE
28. Riazi MR et al (2009) An experimental evaluation of copper, steel and polypropylene tubes in solar water heaters with thermosyphonic flow. Appl Sol Energy 45:65
29. Kuo C-W et al (2014) The design and optical analysis of compound parabolic collector. Procedia Eng 79:258–262
30. Zheng W et al (2016) Numerical and experimental investigation on a new type of compound parabolic concentrator solar collector. Energy Convers Manage 129:11–22
31. Bhattacharjee PR (2005) The generalized vectorial laws of reflection and refraction. Eur J Phys 26(2005):901–911

# Experimental and Theoretical Analysis of $CaCl_2$ Liquid Desiccant System with Solar Regeneration

**D. Sarukasan, S. Sudharson, K. Thirumavalavan, M. Prahadeeswaran, and S. Ajeeth Austin**

## 1 Introduction

In recent decades, halogenated chlorofluorocarbons have been used in air conditioning and to make airborne splashes, blowing experts in place of froths and pressing materials, utilizing solvents, and as refrigerants. But the presence of harmful gases that deplete the ozone layer results in global warming. So researchers turned their research into alternative refrigerants. On this, vapor compression systems are commonly used in refrigerant preparation systems. The limitation of this system operates on closed cycles which require high-power sources [1, 2]. To overcome this problem, a combination of conventional refrigeration systems known as desiccant systems has been employed. In this system, it contains both solid and liquid desiccants, but due to operational feasibility and regeneration temperature, liquid desiccants are commonly used. To deduce load, temperature, and humidity, this system is used. The use of liquid desiccant stored as chemical energy instead of thermal energy [3–5].

The two main systems used in this liquid desiccant are the dehumidifier and regenerator [6]. The investigations on fluid desiccant systems mainly focused on counter-stream dehumidifiers for their high-dehumidification productivity. Despite the mass exchange execution of the cross-stream type dehumidifier being lower contrasted with that of the counter-stream type, the cross-stream setup offers various benefits [7–9]. According to Zhang et al. [10], combined desiccant dehumidification in heat pumps occurs when one heat pump handles the air, while another handles pre-cooling and reheating. The joined system was able to serve in both summers and winters, with COPs, which were approximately 97% and 20% larger, respectively,

D. Sarukasan (✉) · S. Sudharson · M. Prahadeeswaran · S. A. Austin
Department of Mechanical Engineering, Anna University, Chennai, India
e-mail: saru31kasan@gmail.com

K. Thirumavalavan
Anna University, Chennai, India



N. L. Narasimhan et al. (eds.), *Recent Advances in Energy Technologies*, Lecture Notes in Mechanical Engineering, https://doi.org/10.1007/978-981-19-3467-4_2

than conventional systems. According to Zhang et al. [11], contributing a mixture of air and water-cooled condensers to a liquid desiccant-based heat pump system allows for the use of absorbed condenser heat for desiccant regeneration. The results indicated that there is an 18 and 35% increase in COP were observed. According to Mansuriya et al. [12] and Yinglin et al. [13], experimented with a liquid desiccant vapor compression hybrid (LDCH) system with theoretical modification in LDCH by adding an auxiliary regenerator to reduce the loss that occurred in the evaporator. The result shows that the evaporator cooling capacity loss from 8.2 to 1.5% with the modification.

Sanaye and Taheri [14] designed a modified LDHP system of integrated heat exchangers to negate its need for electrical heaters while using calcium chloride as a liquid desiccant. Liu et al. [15] found that solution flow rate reduces the retrieval rate and pump efficiency. The execution of a counter-flow HPLD system can lead to 30% improvements in performance co-efficient. Yang et al. [16] experimented that implementing internal heating in the system enhanced efficiency by 34.6% with an improved amplitude of 45.2%. Shen et al. [17] concluded that with heat pipe heat exchanger (HPHE), the recovery heat ratio is optimized to 25% and 26.5% which are 26.5 and 27% contrasted and regenerators without HPHE. Chen et al. [18] formulated that 23.5% of energy has been saved with the help of hybrid liquid desiccant dehumidification and evaporative cooling system. Ren et al. [19] stated that the inclusion of additives in liquid desiccants enhanced the heat and mass transfer performance increasing the heating capability with reducing vapor pressure.

Lin et al. [20] conducted an analysis on a cross-flow flat membrane with the liquid desiccant dehumidifier. The results show a developed CFD model with continuity, energy made for dehumidified air temperature too, humidity correspondingly. Cho et al. [4] concluded that increasing inlet air humidity from 10.1 to 22.7 g/kg simultaneously increased the efficiency of dehumidification and enthalpy in counter-flow from 51.2 and 40.2% to 82.4 and 72.5%. Likewise, in the cross-flow dehumidifier (CFD) improved as of 62.3% and 46.3% to 63.4% and 55.0% with increased temperature from 15.2 to 31.1 °C. Vivekh et al. [21] reduced the absorption of power sources by up to 50% by using complete fresh air systems. When recirculation 70% of air depresses, the total cooling load from 15 to 50% when treated with the coated heat exchanger of 20–30% from the chiller. Chen et al. stated that growth in the number of heat transfer and mass transfer units had a significant increase in dehumidification. When heat transfer increased from 0.4 to 7, improved the effectiveness by 0.35–0.95.

Due to the obvious vapor pressure difference between air and desiccant, desiccant material can absorb moisture from the air. Because of its low viscosity and ease of handling, the liquid desiccant is desired over a solid desiccant. Vertical film plate dehumidifier is preferred over packed bed type (normally used), because pressure drop on airside will be less, and carryover problem is reduced. So dehumidification performance is obtained for different conditions (concentration and temperature) while regeneration performance for various temperatures in vertical film type dehumidifier. A liquid desiccant system is combined with the solar thermal collector to improve the performance.

# 2 Materials and Method

## 2.1 *Materials*

Calcium chloride is a salt solution which has very low-vapor pressure thus capable of absorbing more quantity of water vapor from the air. Higher concentration and low temperature of desiccant will reduce the dew point. The concentration of calcium is usually varied from (30–45)%; if the concentration is more than 50%, crystallization will take place. Calcium chloride in mixing with water is a highly exothermic reaction; thus a huge amount of heat is released as temperature during reaction increases more than 50 °C [22, 23]. Due to developing the performance of the regeneration system, calcium chloride is purchased from Jain Industrial Chemicals, Park Town, Chennai.

## 2.2 *Description of System*

Desiccant cooling systems can be used as self-contained units or in conjunction with a hybrid vapor compression system (VCS) or a vapor absorption system (VAS). Figure 1 shows a desiccant cooling system (DCS) consists of four key constituents: an absorber, an indirect/direct evaporative cooler, a regenerator, and an air-air heat exchanger. In a hybrid system, a desiccant dehumidifier handles latent cooling load, and VCS/VAS handles sensible cooling load [24]. The proposed hybrid desiccant cooling mainly consists following components: dehumidifier, regenerator, and solar thermal collector. Due to this hybrid system, there is an efficiency improvement [25, 26]. Figure 2 shows the vertical film design-based cross-flow dehumidification system that delivers the desiccant fluid and airflow on the vapor absorption cooling system. In this analysis, the desiccant fluid flows in a perpendicular direction, and the ambient air passes through the system using a blower. The specifications of the dehumidifier are also noted in Table 1.

In a dehumidifier, the liquid desiccant is made to flow in a vertical plate from a storage tank at a lower temperature (18–24 °C) and higher concentration (45%); hence mass transfer takes place from air to desiccant due to low-vapor pressure as a result, the dehumidifier's specific humidity ratio falls near the outlet. Desiccants coming out is pumped to a regenerator to remove the moisture absorbed by the solution [4]. In regenerator, the desiccant solution is made to flow at higher temperature (40–60 °C) in a vertical plate by spending the low-grade thermal energy, the vapor pressure of the desiccant solution increases, the mass transfer takes place from solution to air, moisture gets evaporated, and flow along with air thus humidity ratio of air increases [27]. Solar flat-plate thermal collector provides the heat energy for regeneration of desiccant solar incident energy is converted into heat energy due to it; the carrier fluid (water) is heated. The hot carrier fluid is pumped to the regenerator, wherever the desiccant solution is heated to the required temperature using a heat exchanger. After

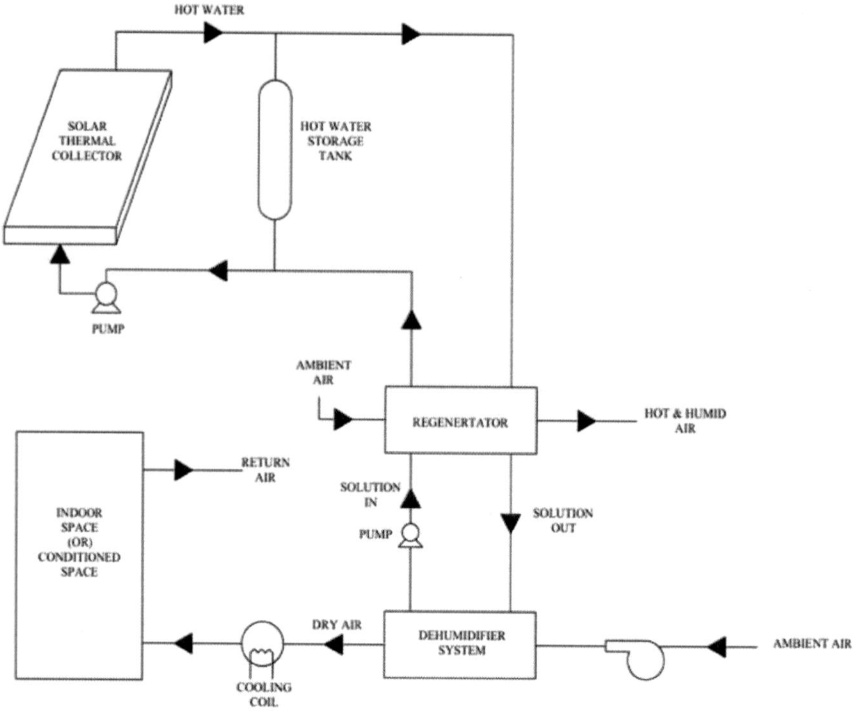


**Fig. 1** Schematic diagram of liquid desiccant system

**Fig. 2** Vertical film design

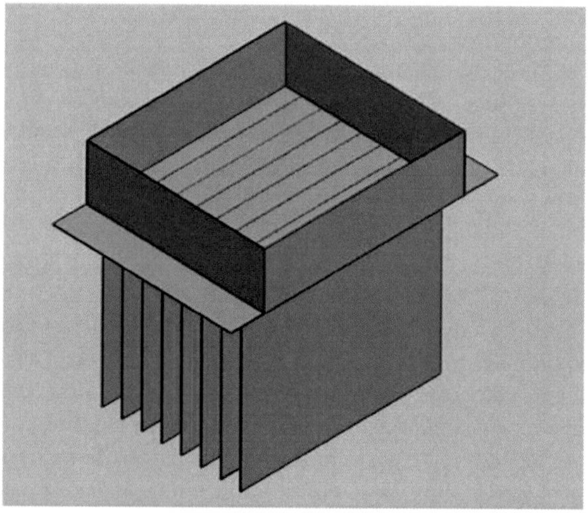

**Table 1** Specification of dehumidifier

| S. No. | Specification | Value |
|---|---|---|
| 1 | Length | 300 mm |
| 2 | Width | 270 mm |
| 3 | Height | 250 mm |
| 4 | Plate thickness | 1 mm |
| 5 | No. of vertical plates | 10 |
| 6 | Air flow rate | 300 $m^3$/hr |
| 7 | Ambient conditions | 28 °C and 75% RH |

regeneration of the desiccant solution, feedback to the dehumidifier is to complete cyclically [28].

# 3 Result and Discussion

## *3.1 Analysis of Desiccant Liquid Solution*

While calcium chloride is owned as a desiccant solution in the proposed dehumidification/regenerator system, different concentration of the solution is prepared and fed into the system. Require concentration is obtained by mixing the correct proportion of desiccant salt in water. The total mass of desiccant available is 4 kg, so the quantity of water is varied to obtain the required concentration represented in Table 2. The desiccant solution is mixed with water, due to exothermic reaction, the temperature of the solution increases drastically more than 50 °C, and solution is kept undistributed for a certain time for cooling of desiccant to ambient condition [29]. Then, solution has tested the variation relative humidity with the temperature of range 10–70 °C using a heater in a closed envelope. Relative increases with temperature because the moisture in air increases [30].

**Table 2** Quantity for different concentration

| S. No. | Concentration (%) | Quantity of desiccant salt (kg) | Quantity of water (kg) |
|---|---|---|---|
| 1 | 35 | 4 | 7 |
| 2 | 40 | 4 | 6 |
| 3 | 45 | 4 | 5 |

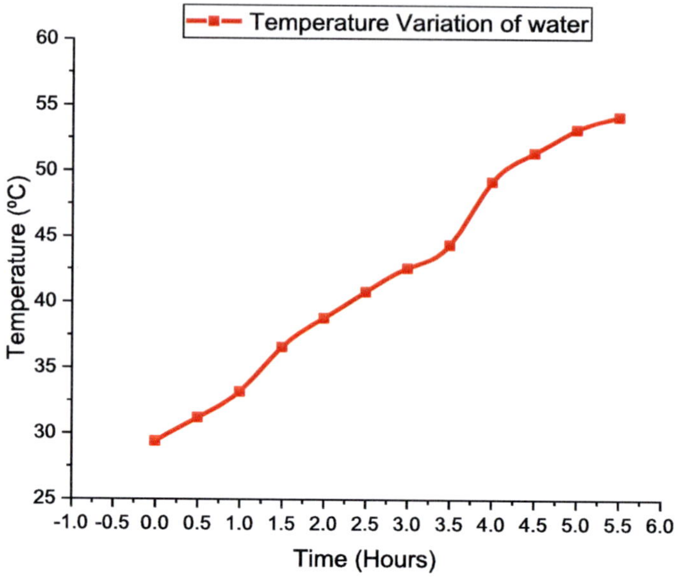


**Fig. 3** Temperature variation of solution versus time

## *3.2 Analysis of Flat-Plate Thermal Collector*

A flat-plate thermal collector delivers the thermal energy required for the regeneration process. Sun is used as a radiation source of the collector; the level of radiation emitted by the lamp is adjusted. Forced mode is used for the analysis of the collector; a pump circulates the water through the collector and storage tank, unlike thermo-siphon mode. Temperature, flow rate, and wind speed are displayed in the front panel of the thermal collector. The experiment is performed for an ambient temperature of 29 °C and a wind speed of 0.2 m/s. Storage tank valves are opened, and the required flow rate is maintained by adjusting the valves. Then, temperature reading and average radiation reading are noted, and the experiment continues till the required temperature reaches are shown in Fig. 3 (60 °C).

## *3.3 Dehumidification Setup*

The main duct with a vertical film dehumidifier is fixed with a duct section that provides the required air supply from the blower. The flow rate of air is measured using a U-tube manometer. The desiccant solution is pumped from the storage tank to vertical film setup (test unit) fabricated at Anna University Chennai are shown in Fig. 4, where the dehumidification process takes place. Desiccant flows into another storage tank, where it is cooled to require the temperature to complete the cycle. A

**Fig. 4** Experimental setup of dehumidifier unit

similar process takes place for the regeneration process, but the desiccant solution is heated using hot fluid from a thermal collector instead of cooling.

Temperature and humidity sensor are used to measure the inlet and outlet conditions of the dehumidifier (temperature and Humidity) [31–33]. Using this data, the enthalpy and specific humidity of air are determined, then dehumidification performance like mass removal rate, humidity effectiveness can be calculated.

## 3.4 *Theoretical Analysis Dehumidifier*

The water vapor from the air is absorbed by the dehumidifier as it undergoes heat and mass transfer. A mathematical model is used to determine the HMT of desiccant and air. Using effectiveness NTU ($\varepsilon$–NTU), the outlet condition of air from the dehumidifier is determined. Outlet condition of air (enthalpy and humidity ratio) is used to find the mass removal rate of water vapor from the air, and the effectiveness of humidifier and dehumidifier cooling output is noticed in Table 3, respectively.

## 3.5 *Experimental Analysis of Dehumidifier and Regeneration*

### 3.5.1 Influence of Temperature and Solution Concentration

The concentration of desiccant solution shows a positive impact on mass removal rate as the higher concentration ratio contributes to a high-dehumidification rate. The mass removal rate is 0.464 g/s for concentration of 37%, and 0.553 g/s at 45% for desiccant temperature of 18 °C shows an increase of 21.8%. Higher concentration partial vapor pressure of desiccant solution reduces, thus the pressure difference between

**Table 3** Dehumidification performance for 40% of desiccant

| S. No | Desiccant inlet temperature (°C) | Air inlet enthalpy (kJ/kg) | Air outlet enthalpy (kJ/kg) | Air outlet humidity (kg/kg of air) | Mass removal rate (g/sec) |
|---|---|---|---|---|---|
| 1 | 18 | 74 | 66.74 | 0.01565 | 0.06015 |
| 2 | 19 | 74 | 67.18 | 0.01573 | 0.5898 |
| 3 | 20 | 74 | 67.77 | 0.01584 | 0.5743 |
| 4 | 21 | 74 | 68.42 | 0.01588 | 0.5665 |
| 5 | 22 | 74 | 69.08 | 0.01602 | 0.5471 |
| 6 | 23 | 74 | 69.68 | 0.01615 | 0.5276 |
| 7 | 24 | 74 | 70.38 | 0.01625 | 0.5121 |

the air and desiccant solution increases it enhances the mass transfer potential, thus mass removal rate increases with increasing concentration. The temperature of the desiccant solution has a negative impact on mass dehumidification performance. As the temperature of the desiccant solution rises, the mass removal rate drops.

Figure 5 explicates the mass removal rate is 0.584 g/s by a temperature of 18 °C that reduce to 0.3354 g/s at 24 °C for a concentration of 45%, showing a 13.4% reduction in dehumidification performance. With a lower difference in pressure between air and desiccant solution, higher solution temperature creates higher vapor pressure on the solution side, resulting in less mass transfer in the dehumidification process. In comparing both effects of concentration and the temperature of desiccant, it is observed that desiccants with higher concentration at a lower temperature will have better dehumidification [30, 34].

### 3.5.2 Dehumidification Cooling Output

Dehumidifier cooling output depicts the amount of energy absorbed by the desiccant dehumidifier. The desiccant solution absorbs the moisture from the air, which reduces the latent heat of air, so the enthalpy of outlet air decreases. Thus, dehumidifier cooling output depends on the moisture absorption rate of the desiccant. Hence, high concentration and low-temperature dehumidifier cooling are more; maximum cooling output is 1.195 kW obtain at 18 °C and 45% concentration compared to the minimum output of 0.3654 kW at 24 °C and concentration of 35% are noticed in Fig. 6.

### 3.5.3 Effect of Desiccant Flow Rate

Figure 7 explains the dehumidification rate increase with desiccant flow rate but stagnates at the higher rate; after 0.1 kg/s (6 kg/min) of desiccant flow rate, the mass removed is significantly small. This could be due to the fact that raising the solution

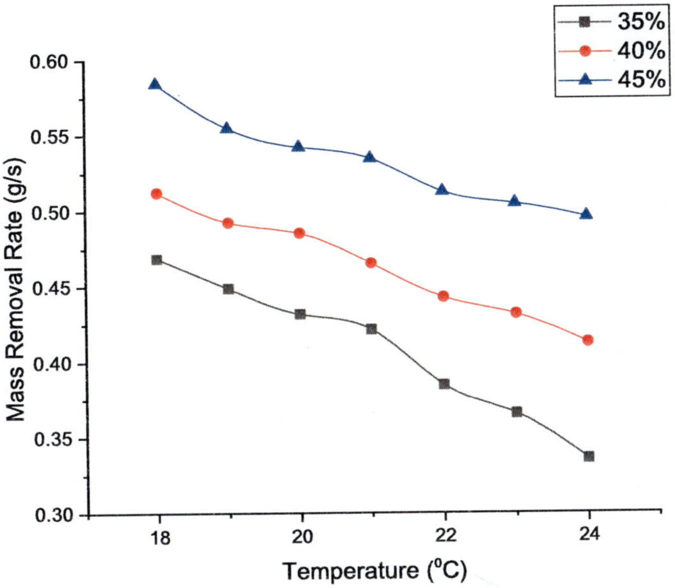


**Fig. 5** Mass removal rate versus temperature

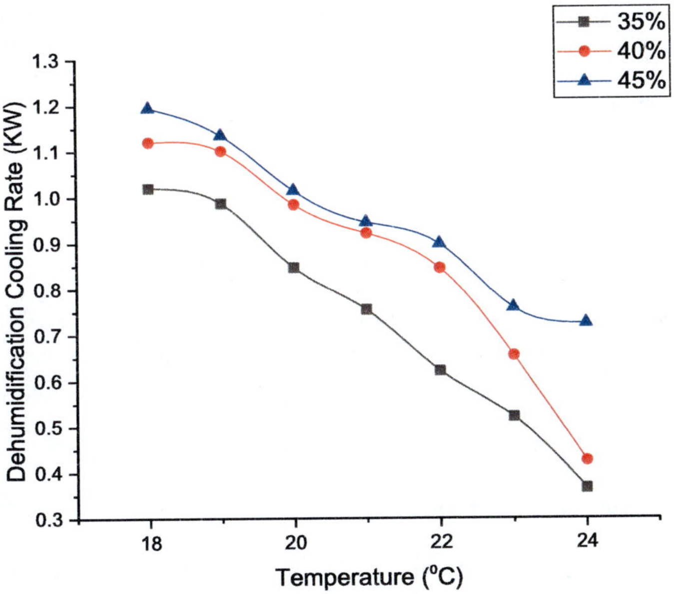


**Fig. 6** Dehumidification cooling output for different temperature

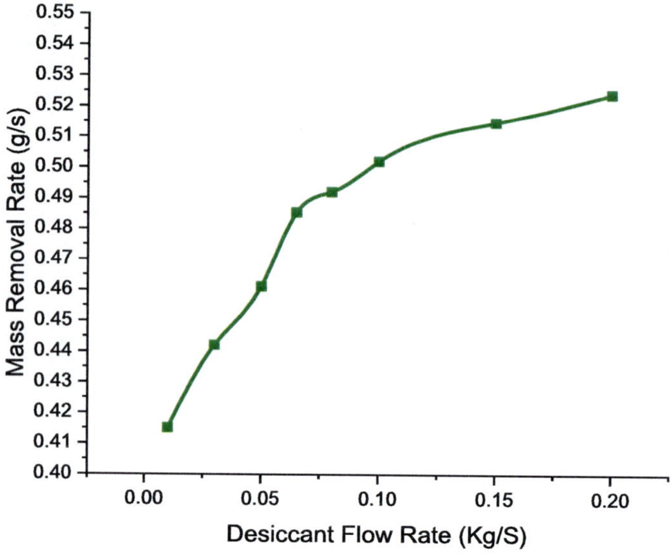


**Fig. 7** Mass removal rate versus desiccant flow rate

flow rate promotes proper connections between the air as well as the desiccant, which improves the heat transfer co-efficient. But a very high-desiccant flow rate may resist airflow through the control volume, and contact time between solution-air reduces, thus resulting in a decrease in dehumidification performance.

### 3.5.4 Effect of Air Flow Rate

Mass removal rate increases drastically as of 0.148 g/s to 0.512 g/s, as airflow rate varies from 30 to 450 $m^3$/hr, higher airflow rate increases the mass transfer co-efficient concerning the desiccant solution and air, so mass removal rate increases, but the very high-flow rate of air also reduces the contact time among air and desiccant solution that reduces the capability of absorption of water vapor by desiccant solution, shown in Fig. 8; as per result, the mass removal rate is less significant.

### 3.5.5 Regeneration Performance

Mass evaporate rate is 0.8874/s at 40 °C while it is 1.845/s at 60 °C, showing an increase of 52.75% as shown in Fig. 9. The higher temperature of the desiccant solution increases the vapor pressure of the solution. As a result of the large vapor pressure difference across air and solution, the mass transfer occurs, and the mass evaporation rate increases.

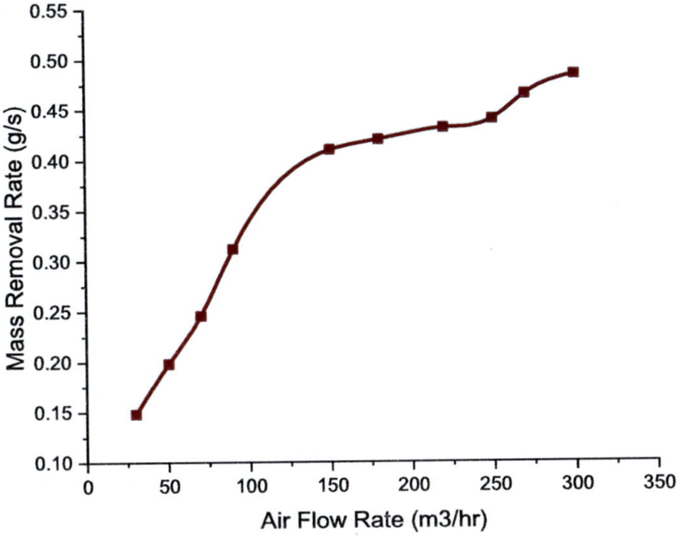


**Fig. 8** Mass removal rate versus air flow rate

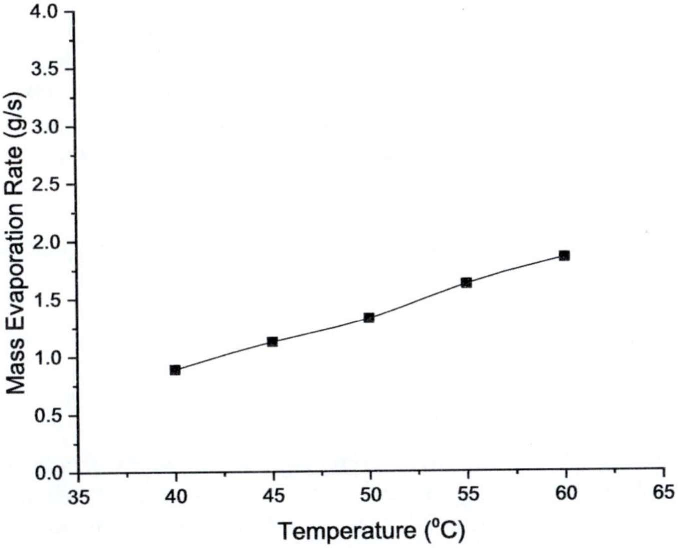


**Fig. 9** Regeneration rate at different temperature

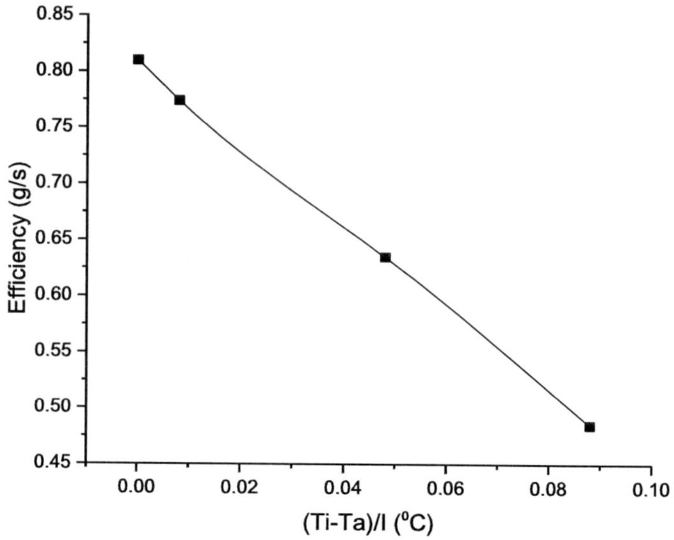


**Fig. 10** Solar thermal collector efficiency versus reduced temperature

### 3.5.6 Effect of Solar Thermal Collector

Solar thermal collector system is tested under forced mode condition pump is used for circulation of water, and input parameters are fixed. The efficiency of the collector varies from 81% when the temperature of the water is the same as ambient condition {28 °C} to 48.9% for water at 50 °C as shown in Fig. 10. Thus, the thermal collector area has to be increased to improve the incident energy falls on the collector consequently useful thermal energy also increases, so high temperature can be obtained in lesser time with higher efficiency.

## 4 Conclusion

Liquid desiccant dehumidifier with solar regeneration is experimentally analysis the desiccant solution and appropriateness of flat-plate collector till require temperature grasps. An analytical method is used to determine the dehumidification performance under different concentrations, temperature, and flow rate obtain the following results:

- A flat-plate thermal collector provides the heat energy required for regeneration of desiccant solution with an efficiency of 22.6%. It can be improved by increasing the area of the collector (>1.5 $m^2$) and incident radiation.

- Dehumidification performance increases with increasing concentration, thus mass removal rate increases by 20.8% in varying concentrations from 35 to 45% for the temperature of 18 °C.
- The temperature has a negative impact on dehumidification performance, the moisture removal decreases 11.4% as the temperature of the desiccant solution increases from 18 to 24 °C for the concentration of 45% because of the Indian location of the temperature range. Dehumidifier cooling output also shows the same trend in varying concentrations and temperatures. Thus, the system will have better performance at high concentrations and low temperatures.
- Humidity effectiveness of the system is less influenced by the desiccant concentration and temperature because equivalent humidity and surface tension also affect it.
- Increasing flow rate of air to more than 450 $m^3$/hr and desiccant flow rate of 0.1 kg/sec have less impact on the dehumidification performance of the proposed system for a desiccant solution at 45% and 18 °C.
- Regeneration performance increases with temperature, evaporation rate increases by 50% as temperature varies from 40 to 60 °C. The regeneration rate is better than the dehumidification rate.

## References


1. Nanda Kishore PVR, Dilip D (2013) Experimental analysis of a liquid desiccant dehumidifier using aqueous calcium chloride solution. Int J Inno Res Sci Eng Technol 2(2013):2319–8753
2. Seenivasan D, Selladurai V, Arjunan TV (2018) Experimental studies on the performance of dehumidifier using calcium chloride as a liquid desiccant. Int J Energy Technol Policy 14(2018):49–63
3. Chen X, Riffat S, Hongyu B, Zheng X, Reay D (2020) Recent progress in liquid desiccant dehumidification and air-conditioning: a review. Energy Built Environ 1(2020):106–130
4. Cho H-J, Cheon S-Y, Jeong J-W (2019) Experimental analysis of dehumidification performance of counter and cross-flow liquid desiccant dehumidifiers. Appl Therm Eng 150(2019):210–223
5. Rafique MM, Gandhidasan P, Bahaidarah HMS (2016) Liquid desiccant materials and dehumidifiers—a review. Renew Sustain Energy Rev 56(2016):179–195
6. Martin V, Goswami DY (2011) Effectiveness of heat and mass transfer processes in a packed bed liquid desiccant dehumidifier/regenerator. HAVC & R Res 6(1):21–39
7. Cho H-J, Jeong J-W (2018) Comparison of dehumidification performance of counter and cross-flow type liquid desiccant dehumidifiers. In: International refrigeration and air conditioning conference, pp. 2325, 1–10.
8. Emhofer J, Beladi B, Dudzinski P, Fleckl T, Kuhlmann HC (2017) Analysis of a cross-flow liquid-desiccant falling-film. Appl Therm Eng 124:91–102
9. Coca-Ortegon A, Prieto J, Coronas A (2016) Modelling and dynamic simulation of a hybrid liquid desiccant system regenerated with solar energy. Appl Therm Eng 97:109–117
10. Zhang L, Dang C, Hihara E (2010) Performance analysis of a no-frost hybrid air conditioning system with integrated liquid desiccant dehumidification. Int J Refrigeration 33(2010):116–124
11. Zhang Q et al (2020) Performance optimization of a heat pump driven liquid desiccant dehumidification system using exergy analysis. Energy 204(2020):117891

12. Mansuriya K et al (2020) Assessment of liquid desiccant dehumidification aided vapor-compression refrigeration system based on thermo-economic approach. Appl Therm Eng 164(2020):114542
13. Yinglin L, Xiaosong Z, Laizai T, Zhongbin Z, Wei W, Xueying X (2016) Performance analysis of a novel liquid desiccant-vapor compression hybrid air-conditioning system. Energy 109(2016):180–189
14. Sanaye S, Taheri M (2018) Modeling and multi-objective optimization of a modified hybrid liquid desiccant heat pump (LD-HP) system for hot and humid regions. Appl Therm Eng 129(2018):212–229
15. Liu X, Xie Y, Zhang T, Chen L, Cong L (2018) Experimental investigation of a counter-flow heat pump driven liquid desiccant dehumidification system. Energy Build 179:223–238
16. Yang Z, Lian Z, Tao R, Ni H, Zhong K (2019) Experimental study on the performance of the internally-heated ultrasonic atomization liquid desiccant regeneration system. Appl Therm Eng 163(2019):114211
17. Shen S, Cai W, Wang X, Wu Q, Yon H (2017) Investigation of liquid desiccant regenerator with heat recovery heat pipe system. Energy Build 146(2017):353–363
18. Chen Y, Luo Y, Yang H (2017) Energy saving potential of hybrid liquid desiccant and evaporative cooling air-conditioning system in Hong Kong. Energy Procedia 105:2125–2130
19. Ren H, Ma Z, Liu J, Gong X, Li W (2019) A review of heat and mass transfer improvement techniques for dehumidifiers and regenerators of liquid desiccant cooling systems. Appl Therm Eng 162(2019):114271
20. Lin J, Huang S, Wang R, Chua KJ (2019) On the dimensional analysis of a cross-flow flow-plate membrane liquid desiccant dehumidifier. Energy Procedia 158(2019):1467–1472
21. Vivekh P, Islam MR, Chua KJ (2020) Experimental performance evaluation of a composite superabsorbent polymer coated heat exchanger based air dehumidification system. Appl Energy 260(2020):114256
22. Emhofer J et al (2017) Analysis of a cross-flow liquid-desiccant falling-film. Appl Therm Eng 124:91–102
23. Thirumavalavan K, Sarukasan D (2022) Experimental investigation on multi-layered filament wound basalt/E-glass hybrid fiber composite tubes. Mater Res Express 9(4):045301
24. Wang C et al (2021) Study on heat transfer and dehumidification performance of desiccant coated microchannel heat exchanger. Appl Therm Eng 192(2021):116913
25. Guan B, Liu X, Zhang T (2021) Investigation of a compact hybrid liquid-desiccant air-conditioning system for return air dehumidification. Build Environ 187:107420
26. Bhowmik M, Muthukumar P, Anandalakshmi R (2021) Experimental study of coupled heat and mass transfer phenomena between air and desiccant in a solar-assisted thermal liquid desiccant system. Int J Therm Sci 162(2021):106795
27. Peng D et al (2020) Applicability, optimization, and performance comparison of self-preheated solar collector/regenerator. Solar Energy 198(2020):113–123
28. Unterberger V et al (2021) An adaptive short-term forecasting method for the energy yield of flat-plate solar collector systems. Appl Energy 293(2021):116891
29. Tao W et al (2021) Comparative study on the liquid desiccant dehumidification performance of lithium chloride and potassium formate. Renew Energy 167(2021):841–852
30. Abeens M et al (2021) An investigation on the effect of stacking fault energies on surface mechanical treatment process of AA6061 alloy and ETP-copper. Surf Topogr Metrol Prop 9(3):035010
31. Guan B, Liu X, Zhang T (2021) Exergy analysis on optimal desiccant solution flow rate in heat exchanger for air dehumidification using a liquid desiccant. Int J Refrig 128:129–138
32. Zhou X (2021) Thermal and energy performance of a solar-driven desiccant cooling system using an internally cooled desiccant wheel in various climate conditions. Appl Therm Eng 185:116077
33. Peng D, Xu S, Yang H (2020) Heat and mass transfer characteristics and dehumidification performance improvement of an evaporatively-cooled liquid dehumidifier. Appl Therm Eng 178(2020):115579

34. Sarukasan D, Thirumavalavan K, Prahadeeswaran M, Muruganandhan R (2021) Fabrication and mechanical characterization of jute-coir reinforced unsaturated polyester resin hybrid composites with various fiber size using compression moulding technique. Int J Recent Technol Eng 10(1). https://doi.org/10.35940/ijrte.a5934.0510121

# Experimental Analysis of Interfacial Evaporation Utilizing Solar and Electrically Driven Systems

**Balaji Kalaiarasu, R. Sriram, G. Kishore Madavan, S. S. Ajith Kumar, and S. Anish**

## 1 Introduction

Interfacial evaporation is a research domain that has gained tremendous traction in recent years. This process can be attributed to the fact that it can be seamlessly integrated into myriad thermal applications and provide optimistic results in an economical and eco-friendly manner. By definition, interfacial evaporation is a technique of producing steam through localized heating at the air–liquid interface. The advent of this idea stemmed from the fact that evaporation is a localized surface phenomenon. Fundamentally, the interfacial technique aims to enhance the evaporation rate and drastically reduce conductive thermal losses. This augments thermal applications like sterilization and desalination by decreasing cycle time substantially. A simple interfacial evaporation setup employs a permeable membrane dispersed on the water's surface to enhance the surface area to facilitate rapid steam generation. Additionally, a tailored insulator is positioned beneath the interfacial membrane to suppress thermal losses and facilitate water movement through capillary action. Hence, heat localization only materializes at the surface, and the bulk medium remains thermally undisturbed. Collectively, such a setup can obtain a water-steam conversion efficiency of over 80%.

Over the last decade, the central focus of improving the evaporation rate was from the standpoint of interfacial materials. In preliminary research, it was noticed that hydrophobic carbon fibers accelerate evaporation and hence were a prospective interfacial material [1]. Eventually, foam-based porous materials (3D nanoporous graphene) with low-thermal conductivity furnished a conversion efficiency of 80% [2]. It was also observed that a double-layered structure (DLS) of carbon foam supporting an exfoliated graphite layer was an appropriate substitute [3]. This opened fresh avenues for economical interfacial materials to be employed with structural

B. Kalaiarasu (✉) · R. Sriram · G. Kishore Madavan · S. S. A. Kumar · S. Anish
Department of Mechanical Engineering, Amrita Vishwa Vidyapeetham, Coimbatore, India
e-mail: k_balaji@cb.amrita.edu



N. L. Narasimhan et al. (eds.), *Recent Advances in Energy Technologies*, Lecture Notes in Mechanical Engineering, https://doi.org/10.1007/978-981-19-3467-4_3

modifications. One noteworthy observation was that carbon-based materials were exceedingly versatile, economical, and possessed favorable thermal absorptivity. Hence, they were primarily utilized for solar interfacial evaporation setups [4]. Eventually, more sophisticated materials were studied to assimilate them into existing thermal applications.

In parallel, research was also administered for superficial materials to improve interfacial efficacy. In terms of solar absorbers, $CuInSe_2$ (CIS) was an empirical material that could be produced economically using non-vacuum methods [5]. Similarly, for applications involving electricity generation, fabrics with good light affinity such as dye-sensitized materials (DSSC) and organic semiconductors are rational solutions [6]. Chemically, photothermal modifications in carbon compounds can be effectively cascaded with existing interfacial systems [7, 8]. Even the materials used in interfacial equipment need to be appropriately selected; as we noticed, overlooked thermal losses could impede the effect of localized heating [9]. Concurrently, the evaporation coefficient also heightens with water activity and pressure [10]. It is to be noted that a uniform temperature gradient is established below the liquid–vapor interface [11]. Hence, we remark that additional evaporation and apparatus parameters must also be contemplated when developing interfacial systems for thermal applications. Furthermore, this parallel research effectively incorporates interfacial evaporation in distinctive energy extensive applications such as individual-house power generation [12].

In subsequent years, it was documented that a thin coating of $TiNO_X$ absorber furnished a conversion efficiency of 87% under low-optical concentration [13]. This was an important conclusion because it promoted thin layered absorbers as interfacial membranes. Eventually, research performed on thin-film polypyrrole coating over polypropylene membrane successfully enriched the vapor generation efficiency of an interfacial solar-still application [14]. A new "airing" methodology was also introduced to improve the thin-film interfacial membranes, which further stimulated the evaporation efficiency using nanoparticles [15]. Additional research aimed to utilize hybrid materials and formulate alternate interfacial methodologies to enhance the evaporation rate and efficiency. All the data from these experiments were consolidated to gain a robust understanding of the significance of materials utilized in interfacial systems.

In a recent development, a biochar-based interfacial sterilization system was designed, which had a cycle time of just 8.5 min and a sterilization efficiency of 99.9% [16]. This was achieved by capitalizing on the kinetic advantage of carbon-based interfacial materials. Similarly, plasmonic nanocomposites also yielded a vapor generation efficiency of 87% for a brackish-water desalination application [17], and a reasonable interfacial membrane alternative was to utilize TiNO [18]. Hence, it was speculated that solar-based interfacial evaporation had the transformative potential to be incorporated into building thermal absorbers, evaporators, insulators, and concentrators in the near future [19]. A vast spectrum of interfacial materials was identified through supplementary research, including semiconductors and composite materials that could be extensively utilized [20]. In coherence to this, distinctive applications for interfacial evaporation were also introduced, such as optical enhancements, environmental energy enhancement, and photothermal membrane distillation [21].

There were some remarkable advancements in the possibility of achieving interfacial evaporation without a solid membrane. Hence, gold nanoparticles were dispersed on the water's surface to achieve accelerated evaporation using a process called plasmon resonance [22]. One major issue of this approach was that the absorption potential of gold nanoparticles was wavelength-specific [23]. Further membrane designs were assessed to overcome this, such as a 3D porous evaporator that provided high-quality steam at 100 °C under one sun illumination [24]. Subsequent research was also conducted on enhancing the solar still with external Fresnel lenses to maximize heat input. One of the experiments also yielded a productivity increase of 638%, which was notable [25]. Irrespective of these advancements, solar interfacial evaporation also had its share of deficiencies in terms of low-thermal stability, structural complexity, and insufficient scalability [26]. Hence, some significant developments aimed to overcome this issue in recent years. A radical new methodology was recently initiated to achieve interfacial evaporation using an electric disk heater. This approach overcame most shortcomings of solar-based systems and yielded favorable results. When incorporated into a sterilization system, it produced a conversion efficiency of 90% and a thermal response of just 20 s. This was one of the most efficient small-scale systems established [27].

One of the fundamental reasons for utilizing the idea of interfacial evaporation is to overcome the drawbacks of volumetric heating, which is predominantly used in most thermal applications. Volumetric heating is generally coupled with considerable thermal losses that harm system efficiency. This was one of the pivotal research gaps effectively bridged by solar interfacial evaporation systems. However, from our literature survey, we infer that solar-based systems rely on a diffuse source of energy and have a discontinuous operation cycle. This is one of the significant shortcomings of solar-based systems. Though electrical systems are enlisted and experimented with, they are still in the nascent stages of research and cannot yield substantial results for large-scale systems. Hence, the current research gap is the lack of experimentation of electrical methodology and the presence of a factual comparative study between solar and electrically driven interfacial evaporation approaches. Our present paper conducts systematic experiments on both processes and ascertains the most efficient and resilient interfacial system to be incorporated in practical thermal applications.

## 2 Theoretical Calculations

We performed a parametric analysis between solar-based interfacial evaporation and volumetric heating through this analytical calculation and demonstrated the kinetic advantage of interfacial evaporation over volumetric heating. This is pivotal because volumetric heating is principally employed in most thermal applications. The given estimation is conducted for both heating and cooling cycles individually. The base

**Table 1** Abbreviations of variables used in theoretical calculations for volumetric heating and interfacial evaporation

| Abbreviation | Connotation | Units |
|---|---|---|
| $\rho_{w,1}$ | Density of water at 30 °C | kg/m$^3$ |
| $h_{w,2}$ | Enthalpy of water at 121 °C | kJ/kg |
| $h_{w,1}$ | Enthalpy of water at 20 °C | kJ/kg |
| $P_{w,1}$ | Pressure of water at 20 °C | kPa |
| $P_{w,2}$ | Pressure of water at 121 °C | kPa |
| $\rho_{\text{air},1}$ | Density of air at 20 °C | kg/m$^3$ |
| $V_s$ | Volume of steam | m$^3$ |
| $C_{v,\text{air}}$ | Isobaric specific heat | kJ/kg-K |
| $\Delta T_{1-2}$ | Temperature change of air from 20 to 121 °C | K |
| $P_{s,2}$ | Pressure of steam at 121 °C | kPa |
| $P_{\text{out}}$ | Power of heat flux in a cooling phase | kW |
| $T_{121}$ | Time interval of the cooling phase | s |
| $\rho_{s,2}$ | Density of steam at 121 °C | kg/m$^3$ |
| $h_{s,2}$ | Enthalpy of steam at 121 °C | kJ/kg |
| $V_w$ | Volume of water | m$^3$ |
| $\Delta E$ | Energy change from 20 to 121 °C | kJ |

temperature is taken as room temperature, and the maximum temperature is 121 °C [16]. The justification for opting this temperature is that the steam at 121 °C has sufficient potential to desalinate or sterilize effectively. In this theoretical derivation, we extensively employ the energy balance equation because we wish to ascertain the relationship between conversion time of solar-based and volumetric evaporation systems (Table 1).

**Assumptions made**

1. During the heating phase, the system is assumed to have no thermal losses
2. The total volume of bulk water does not alter during the transformation phase
3. The ratio of the volume of steam and water is taken as constant
4. No liquid–vapor mixed phases are present at an elevated temperature of 121 °C

## 2.1 *Derivation*

**Interfacial Evaporation (Heating Phase)**

$$\begin{aligned}\Delta E &= \rho_{\text{air},1} V_s C_{v,\text{air}} \Delta T_{1-2} + \rho_{s,2} V_S \left(h_{s,2} - h_{w,1}\right) - V_s\left(P_{s,2} - P_{w,1}\right)\\ &\Rightarrow (1.2041)(V_s)(72.518) + (1.145)(V_S)(2623.485) - (V_s)(203.37)\\ &\Rightarrow 2887.8292 V_s \end{aligned} \tag{1}$$

**Volumetric Heating (Heating Phase)**

$$\begin{aligned}\Delta E &= \text{Eq.}(1) + \rho_{1w} V_w \left(h_{w,2} - h_{w,1}\right) - V_w \left(P_{w,2} - P_{w,1}\right) \\ &\Rightarrow 2887.8292 V_s + (948.2) V_w (44.29) - V_w (203.37) \\ &\Rightarrow 2887.8292 V_s + 424,327.8 V_w \end{aligned} \tag{2}$$

(Dividing Eq. (2) by Eq. (1))

$$\frac{\Delta E_{H,V,121}}{\Delta E_{H,I,121}} = \frac{(2887.829 V_s + 424,327.8 V_w)}{2887.8292 V_s}$$

$$\frac{\Delta E_{H,V,121}}{\Delta E_{H,I,121}} = \frac{T_{H,V,121}}{T_{H,I,121}} \approx 1 + 148\left(\frac{V_w}{V_s}\right)$$

**Interfacial Evaporation (Cooling Phase)**

$$\begin{aligned}P_{\text{out}} T_{121} &= \rho_{\text{air},1} V_s C_{v,\text{air}} \Delta T + \rho_{s,2} V_s \left(h_{s,2} - h_{w,1}\right) - V_s \left(P_{s,2} - P_{w,1}\right) \\ &\Rightarrow (1.204)(V_s)(15.078) + (1.145)(V_s)(2288.23) - V_s(104.29) \\ &\Rightarrow 18.155 V_s + 2620.02 V_s - 104.29 V_s \Delta 2533.885 V_s \end{aligned} \tag{3}$$

**Volumetric Heating (Cooling Phase)**

$$\begin{aligned}P_{\text{out}} T_{121} &= \text{Eq.}(3) + \rho_{w,1} (V_w) \left(h_{w2} - h_{w,1}\right) - V_w \left(P_{w,2} - P_{w1}\right) \\ &\Rightarrow 2533.855 V_s + (998.2)(V_w) \\ &\Rightarrow 2533.855 V_s + 87,879.728 V_w - 104.29 V_w \end{aligned} \tag{4}$$

(Dividing Eq. (4) by Eq. (3))

$$\frac{T_{C,V,121}}{T_{C,I,121}} = \frac{(2533.855 V_s + 87,879.728 V_w - 104.29 V_w)}{2533.885 V_s}$$

$$\frac{T_{V,121}}{T_{I,121}} \approx 1 + 32\left(\frac{V_w}{V_s}\right)$$

The primary inference from this theoretical analysis is that interfacial evaporation is 148 times faster than volumetric heating during the heating phase and 32 times faster during the cooling phase. This instantaneous heating and gradual cooling progression are best suited for practical applications like sterilization and desalination. This is the crucial kinetic advantage that interfacial evaporation possesses over volumetric heating, and hence, the research in this domain is gaining much traction.

# 3 Experimental Setup

The experimental investigation aims to curate a solar-based and electrically driven interfacial evaporation set up to facilitate a comparative study. The prime emphasis is to gage parameters independent of apparatus geometry, such as surface temperature, mass change, and conversion efficiency. Other performance characteristics are also calculated from measured data to support comparative analysis. Furthermore, all these experiments are conducted in a non-idealistic environment. The aim of performing this experiment in an innate configuration over a simulated one is to attain realistic results.

## *3.1 Solar-Based Interfacial Evaporation Setup*

A 10 ml beaker with two layers of external insulation (PTFE + foam) is utilized to establish this setup. An ester-based polyurethane foam operates as a floater for the interfacial layer and a thermal insulator. Commercially available spunlace fabric is used as the interfacial material. For the current experiment, a 5 cm × 5 cm spunlace fabric is wound around polyurethane foam with dimensions of 2.5 cm × 3 cm × 0.8 cm to create an interfacial coating. This layer is designed in such a way that the flanks of the spunlace are in perpetual contact with the bulk medium of water underneath. The interfacial setup is then dispersed on the water's surface, and further, experimentation is performed under natural solar irradiance. The experimental readings for the evaporation cycle are measured using a sensitive digital thermometer and a weighing scale with an accuracy of 0.01 g. One prominent compliance from this setup is that only commercially available materials that are inexpensive and eco-friendly have been utilized for the experimentation as shown in (Fig. 1).

The insulated beaker is initially filled with 44.7 g of water, and the interfacial membrane is dispersed on the water surface. This setup is then exposed to solar irradiance. The best time for executing this experiment is between 12 p.m. and 2 p.m. at

**Fig. 1** Materials used for solar-based interfacial evaporation setup: spunlace fabric, insulated beaker, ester-based polyurethane foam

noon, as this is the time when solar irradiance is at maxima. In the case of solar-based systems, the evaporation commences after 3–4 min and then accelerates. This experiment gages four significant parameters: time, surface temperature, bulk temperature, and mass loss. The values of these parameters are tabulated every 5 min for a net duration of 30 min. The experiment is conducted where external factors like wind do not enhance the evaporation rate. Supplementary performance metrics such as evaporation rate, solar thermal conversion efficiency, solar vapor generation efficiency, thermal-vapor generation efficiency, and thermal loss efficiency are calculated using the tabulated data. Plotting these parameters as a function of time gives us a holistic picture of the performance of the solar-based interfacial evaporation systems.

## 3.2 *Electrically Driven Interfacial Evaporation Setup*

The electrically driven setup is analogous in comparison with the solar-based interfacial design. In this case, instead of confiding in solar irradiance, thermal energy is supplied via an external heating source. We utilized a 200 W disk heater made of mild steel as the heating source for our experiment. This was clasped on the water surface by a floater made of PVC sheets and wooden sticks. A 2-ply tissue paper of dimension 60 mm × 120 mm was utilized as the interfacial material. All these materials were encapsulated in a fixed volume beaker. Furthermore, the disk heater was engaged to an external AC source via a rheostat of range 1–100 Ω. The rheostat is utilized to experiment with different loading conditions. The experimental readings are measured using an IR thermometer and a digital weighing scale. All the apparatus have been procured individually to meet experiment specifications as shown in (Fig. 2).

Initially, the beaker is filled with 500 ml of water. In this experiment, the disk heater and the floater are considered isolated modules suspended on the water's surface. Similar to the solar-based interfacial system, the interfacial membrane is wound around the heater-floater set up such that the flanks of the 2-ply tissue are in explicit contact with the bulk medium of water. Initially, the rheostat is set at 20

**Fig. 2** All materials used for electrically driven interfacial evaporation setup and the corresponding disc heater CAD model

Ω load, and a buffer time is provided for the heater to attain constant temperature. In the case of an electrically driven system, evaporation spontaneously occurs as the thermal response is less than the 40 s. Three notable parameters are measured: mass loss, surface temperature, and bulk temperature. The values are tabulated every 5 min with 1 min as settling time for every reading for a net duration of 30 min. The experiment is reiterated with 50 Ω rheostat resistance, and similar parameters are measured. We need to point out that this investigation is administered where external factors do not augment the evaporation rate.

## 4 Results and Discussion

Two distinctive interfacial evaporation setups (i.e., solar-based and electrically driven) are utilized to perform systematic experiments. Using the empirical data, the performance characteristics of both systems are studied exclusively. The implication of this study is to interpret the behavior of interfacial parameters such as evaporation rate and conversion efficiencies as a function of time. The experiment is conducted under solar irradiance for solar-based systems, and for electrically driven systems, the investigation is performed using an external heating source under different loading conditions. The results give us a holistic picture of which approach would be more effective when integrated into traditional thermal applications.

### *4.1 Solar-Based Interfacial Evaporation*

In this investigation, three significant parameters (i.e., surface temperature, bulk temperature, and mass loss) are tabulated for 30 min. The experimental study is performed under direct solar irradiance between 12 and 2 p.m. at noon. The interfacial material used is spunlace fabric, and the floater material is ester-based polystyrene foam. This experiment is conducted using a small-scale setup, and any external factors which could augment the evaporation rate have been neglected. The obtained data have been shown (Table 2).

Using this data, further performance metrics, namely solar thermal conversion efficiency ($\eta_{st}$), solar vapor generation efficiency ($\eta_{sv}$), thermal-vapor generation efficiency ($\eta_{tv}$), and thermal loss efficiency ($\eta_{tl}$), are evaluated. The previous interfacial research has cited the analytical formulae for these parameters, using which the data have been shown (Table 3) [20].

**Empirical Plots for Solar-Based Interfacial Evaporation**

All the graphs have been plotted for a time interval of 30 min. From Fig. 3, we can infer that the mass loss keeps increasing over a period; this promptly implies that the evaporation rate also increases over the given time frame. This occurs due

**Table 2** Experimental tabulation of parameters measured for solar interfacial evaporation

| Time (min) | Mass loss (g) | Surface temperature (°C) | Bulk temperature (°C) |
|---|---|---|---|
| 0 | 0 | 32.7 | 32.7 |
| 5 | 0.43 | 34 | 32 |
| 10 | 0.64 | 36.5 | 35.6 |
| 15 | 0.94 | 37 | 36.2 |
| 20 | 1.15 | 37.2 | 36.4 |
| 25 | 1.44 | 37.3 | 36.9 |
| 30 | 1.63 | 37.7 | 36.9 |

**Table 3** Performance characteristics of solar interfacial evaporation

| Time (min) | Solar thermal $\eta_{st}$ (%) | Solar vapor $\eta_{sv}$ (%) | Thermal-vapor $\eta_{tv}$ (%) | Thermal $\eta_{tl}$ (%) |
|---|---|---|---|---|
| 0 | 0 | 0 | 0 | 100 |
| 5 | 83.65 | 58.32 | 41.32 | 58.68 |
| 10 | 81.98 | 59.65 | 42.36 | 57.64 |
| 15 | 80.65 | 60.21 | 44.79 | 55.21 |
| 20 | 79.58 | 61.23 | 45.66 | 54.34 |
| 25 | 78.66 | 62.35 | 46.78 | 53.22 |
| 30 | 76.54 | 63.45 | 47.89 | 52.11 |

to effective localized heating at the air–liquid interface. In Fig. 4, all the performance metrics are compiled into a single graph to aid comparative study. From this plot, we can observe that solar thermal conversion efficiency, solar vapor generation efficiency, and thermal-vapor generation efficiency show an upward trend. They reached a maximum at the end of the 30-min timeframe and yielded an efficiency of 76.5%, 63.45%, 47.89% and 52.11%, respectively. Furthermore, the only plot showing a downward trend is thermal loss efficiency. This is because the external insulation prevents undesirable thermal losses. This effect is more pronounced when the temperature of the system increases. From these plots, we can infer that the efficacy of interfacial evaporation only keeps rising over a period of time; this happens because of the continuous increase in surface temperature. Figure 5 describes the variation of surface and bulk temperature as a function of time. We can observe an abrupt rise in surface temperature between 5 and 10 min; this can be associated with the maximum thermal response at this position. Furthermore, the bulk temperature does not increase drastically like surface temperature or fluctuate significantly. This can be attributed to the fact that ester-based polystyrene foam, used as a thermal insulator, is effective.

We can observe that this experiment achieved high efficiency from a small-scale setup under natural conditions. Though some marginal errors are present due to

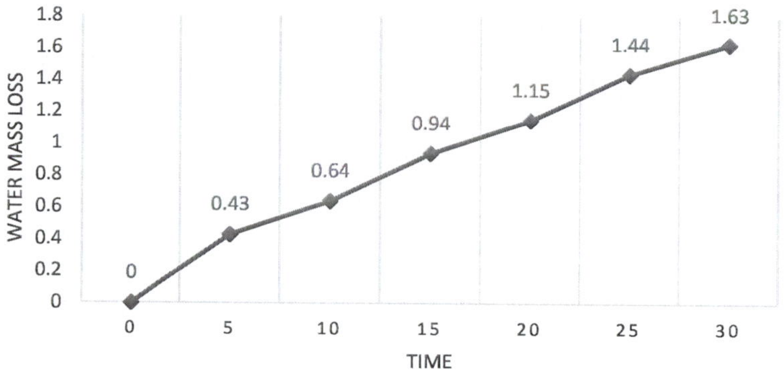


**Fig. 3** Water mass loss as a function of time

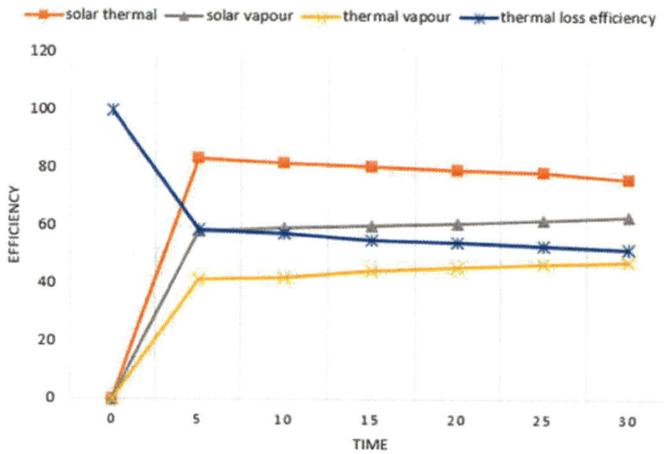


**Fig. 4** Performance characteristics of the system as a function of time

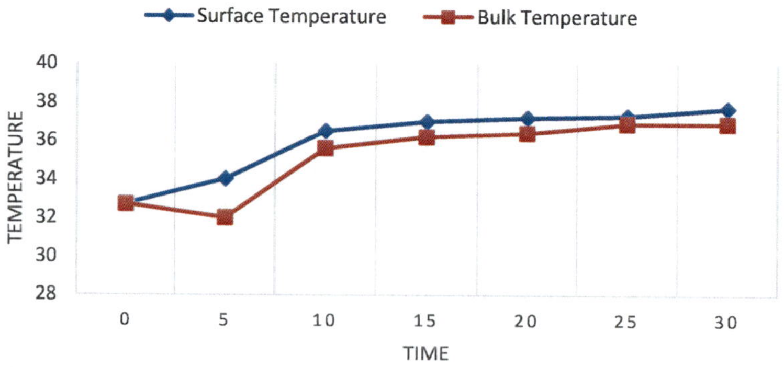


**Fig. 5** System temperature as a function of time

measurement inaccuracies, the overall performance signifies an effective alternative to traditional evaporation methods. Our previous theoretical calculation also ascertained how solar-based interfacial evaporation is around 148 times faster than volumetric heating. Furthermore, the above experiment is conducted with white spunlace fabric, which is an economical material that is commercially available. Utilizing more complex materials such as carbon-based compounds or other effective porous solar absorbers can yield conversion efficiencies higher than 85%.

## 4.2 *Electrically Driven Interfacial Evaporation*

The experimental setup is very similar to solar interfacial evaporation; the only change is the heating source; an external disk heater is utilized over solar irradiance. This experiment obtains two primary parameters, namely water mass loss and conversion efficiency, under different loading conditions (20 Ω , 50 Ω). Furthermore, supporting data such as steam temperature, surface temperature, and bulk temperature are also measured and tabulated. In this experiment, a 2-ply tissue is used as an interfacial material, and a dedicated PVC floater is designed to support the disk heater. It is to be noted that even this experiment is conducted under natural conditions, and any external factors which enhance evaporation rate have been neglected. The obtained parameters are plotted as a function of time, and these graphs provide us with a clear picture of the performance characteristics of the system (Tables 4 and 5).

**Empirical Plots for Electrically Driven Interfacial Evaporation Systems**

All the experimental graphs have been plotted over a 30-min time frame. From Fig. 6, we can discern that the mass loss trend as a function of time for electrical systems is significantly higher than solar-based interfacial systems. This can be attributed to the fact that electrical systems have a quick thermal response time of 40 s. This parameter also reflects on the accelerated evaporation rate of the system. From Fig. 7, we can observe the conversion efficiency for both loading conditions of 20 Ω and 50

**Table 4** Experimental data for 20 Ω loading condition

| Time (min) | Mass change (g) | Surface temperature (°C) | Thermal energy (J) | Conversion $\eta$ (%) |
|---|---|---|---|---|
| 0 | 0 | 30 | 0 | 0 |
| 5 | 15 | 82 | 5715.28 | 59.35 |
| 10 | 31.2 | 86 | 12,089.43 | 65.37 |
| 15 | 51 | 86.6 | 21,260.79 | 76.65 |
| 20 | 68.9 | 88.5 | 32,214.67 | 87.11 |
| 25 | 86.1 | 88.3 | 40,433.75 | 87.47 |
| 30 | 101 | 83.9 | 47,943.73 | 86.43 |

**Table 5** Experimental data for 50 Ω loading condition

| Time (min) | Mass change (g) | Surface temperature (°C) | Thermal energy (J) | Conversion $\eta$ (%) |
|---|---|---|---|---|
| 0 | 0 | 31 | 0 | 0 |
| 5 | 6.5 | 79.2 | 5200.57 | 53.94 |
| 10 | 19.4 | 81.7 | 12,202.15 | 63.8 |
| 15 | 33.44 | 81.4 | 19,622.17 | 67.84 |
| 20 | 46 | 83 | 26,779.95 | 69.44 |
| 25 | 57.4 | 82.7 | 32,428.84 | 67.27 |
| 30 | 69.7 | 80.4 | 37,190.73 | 64.29 |

Ω. In both cases, we see a sharp upward trend during the first 5 min and then a more gradual increase over the remaining time frame. The sharp increase is correlated with the enhanced evaporation rate due to the quick thermal response. The 20 Ω loading condition provides a higher conversion efficiency because, according to Ohm's law, at a constant voltage lower the resistance, the higher is the current. Hence, effective localized heating occurs when a larger current is supplied to the disk heater. These characteristics can be further explained from the context of surface and bulk temperature. From Fig. 8, we can infer that the interfacial surface temperature also rises steeply during the first 5-min due to the quick thermal response, which results in accelerated evaporation. Figure 9 also follows a similar trend initially and gives us an idea of the steam temperature. We can infer from both these plots that localized heating is the major driving factor in enhancing the evaporation rate and conversion efficiency (Table 6).

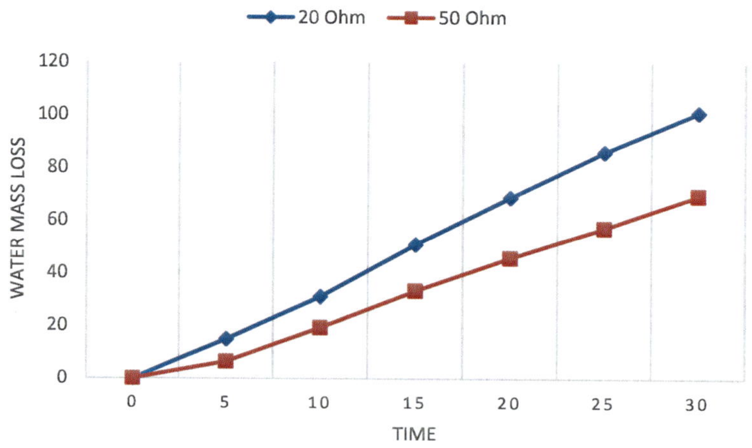


**Fig. 6** Water mass loss as a function of time for different loading capacity

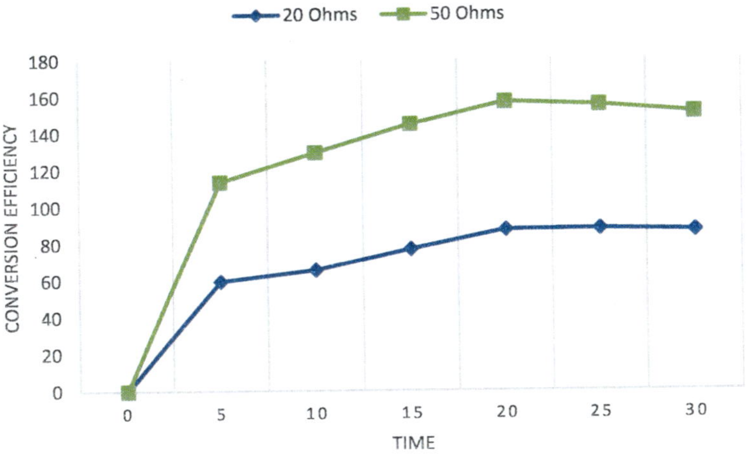


**Fig. 7** Conversion efficiency as a function of time for different loading capacity

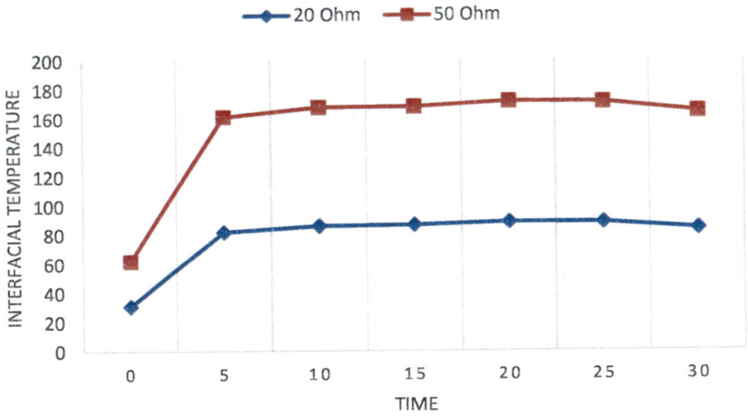


**Fig. 8** Interfacial temperature variation as a function of time for different loading capacity

## *4.3 Limitations of Interfacial Evaporation*

Though interfacial evaporation has numerous advantages, some notable shortcomings are that it is still an arcane research domain in its nascent stages. Solar-based interfacial systems rely on a diffuse energy source; hence extracting substantial steam to drive energy extensive applications would require utilizing an interfacial membrane with a vast surface area, which is not economical. Furthermore, complications in conversion efficiency also arise due to salt accumulation in interfacial material in the case of desalination applications. An electrically driven system also possesses little scalability, as designing a heater with a large surface area is

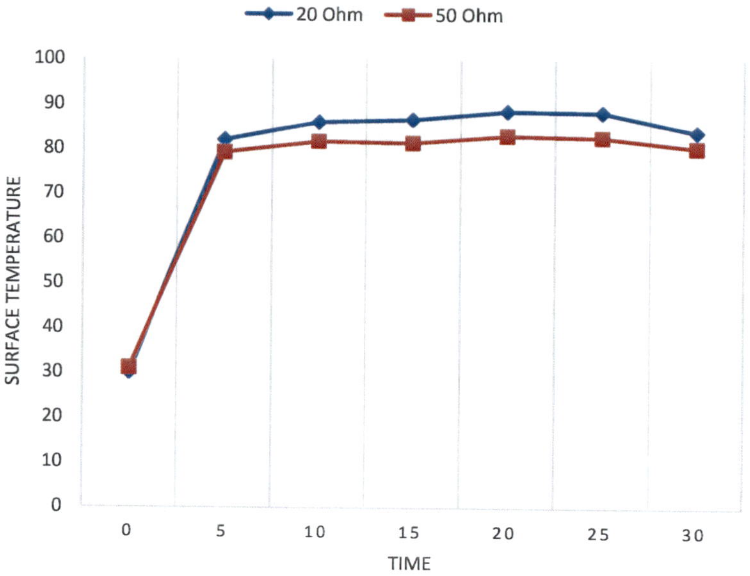


**Fig. 9** Surface temperature variation as a function of time for different loading capacity

**Table 6** Consolidated experimental data for interfacial systems

| Parameters | Solar-driven | Electrically driven | Units |
|---|---|---|---|
| Avg. evaporation rate | 0.0013 | 0.077 | g/cm$^2$ min |
| Peak surface temp | 38.2 | 83.9 | °C |
| Peak conversion $\eta$ | 47.89 | 86.43 | % |

not currently feasible. Furthermore, interfacial membrane and floater material must possess good porosity and not char under high-heater temperatures. Due to the above shortcomings, most of the research have been inclined to find effective interfacial materials. Most interfacial materials that yield high-conversion efficiency are structurally modified. Hence, they are limited to laboratory applications and have not been assimilated into practical scaled-up apparatus for real-time usage. However, when compared to the vast potential of this approach, active research is still ongoing to explore this domain.

## 5 Conclusion

This paper aimed to expose the expansive potential of interfacial evaporation. In summary, we demonstrated the kinetic benefits of interfacial evaporation over volumetric heating through an analytical calculation. We discerned that interfacial methodology is 148 times faster than volumetric heating. Furthermore, we also elucidated the two most prominent interfacial evaporation approaches, namely electrically driven and solar-based interfacial evaporation systems. Propelling further, we performed an experimental comparative study between these two approaches under natural conditions and observed the performance characteristics. The maximum thermal-vapor conversion efficiency achieved by the solar interfacial system was 47.89%, whereas, for electrically driven systems, it was 86.43% for 20 Ω loading conditions and 64.29% for 50 Ω loading conditions. Utilizing experimental data, we concluded that electrically driven interfacial systems have tremendous potential to be incorporated into existing thermal applications as it possesses 38.5% higher vapor generation efficiency and has a thermal response under 50 s. Primarily, economical and eco-friendly materials were utilized for experimentation as one of the fundamental goals of this research was to supplement the scalability and feasibility of interfacial systems, and interfacial evaporation possesses the immense potential to foster further research in this domain.

**Acknowledgements** We acknowledge our professors, Dr. K. Balaji and Dr. Sudip Kumar Batabyal, for giving us valuable guidance and insights into the prime concepts and basics of this experimental-oriented project.

## References

1. Zhang L, Tang B, Wu J, Li R, Wang P (2015) Hydrophobic light-to-heat conversion membranes with self-healing ability for interfacial solar heating. Adv Mater 27:4889–4894. https://doi.org/10.1002/adma.201502362
2. Ito Y, Tanabe Y, Han J, Fujita T, Tanigaki K, Chen M (2015) Multifunctional porous graphene for high-efficiency steam generation by heat localization. Adv Mater 27(29):4302–4307. https://doi.org/10.1002/adma.201501832py
3. Ghasemi H, Ni G, Marconnet AM, Loomis J, Yerci S, Miljkovic N, Chen G (April 21, 2014) Solar steam generation by heat localization. Nat Commun 5(1).https://doi.org/10.1038/ncomms5449
4. Hou Y, Wang Q, Wang S, Wang M, Chen X, Hou X (2021) Hydrophilic carbon nanotube membrane enhanced interfacial evaporation for desalination. Chin Chem Lett. ISSN 1001-8417. https://doi.org/10.1016/j.cclet.2021.09.007
5. Sun J, Batabyal SK, Tran PD, Wong LH (2014) Electrodeposition of single-phase $CuInSe_2$ for solar energy harvesting: role of different acidic additives. J Alloys Compd 591:127–131. https://doi.org/10.1016/j.jallcom.2013.12.178
6. Sreekala CSNOA, Indiramma J, Kumar KBSP, Sreelatha KS, Roy MS (2013) Functionalized multi-walled carbon nanotubes for enhanced photocurrent in dye-sensitized solar cells. J Nanostructure Chem 3. https://doi.org/10.1186/2193-8865-3-19

7. Devi R, Prabhavathi G, Yamuna R, Ramakrishnan S, Kothurkar NK (2014) Synthesis, characterization and photoluminescence properties of graphene oxide functionalized with azo molecules. J Chem Sci 126. https://doi.org/10.1007/s12039-013-0536-1
8. He W, Zhou L, Wang M, Cao Y, Chen X, Hou X (2021) Structure development of carbon-based solar-driven water evaporation systems. Sci Bull 66(14):1472–1483. ISSN 2095-9273. https://doi.org/10.1016/j.scib.2021.02.014
9. Ponnusamy S, Gangadharan SSK, Balaji K (2020) An exergy analysis for overall hidden losses of energy in solar water heater. In: Thermal science. Society of Thermal Engineers of Serbia, pp 343–343. http://thermalscience.vinca.rs/online-first/4283\
10. Bénet J-C, Ouoba S, Ouedraogo F, Cherblanc F (2021) Experimental study of water evaporation rate, at the surface of aqueous solution, under the effect of a discontinuity of chemical potential—effect of water activity and air pressure. Exp Therm Fluid Sci 121. https://doi.org/10.1016/j.expthermflusci.2020.110233
11. Guo RF, Zhang L, Mo DM, Wu CM, Li YR (2020) Measurement of temperature profile near the evaporating interface in an annular pool with radial temperature gradients at low pressures. Exp Therm Fluid Sci 119. https://doi.org/10.1016/j.expthermflusci.2020.110221
12. Bhargav MMSRS, Ratna Kishore V, Anbuudayasankar SP, Balaji K (2016) Power generation by high headwater in a building using micro-hydro turbine—a greener approach. Environ Sci Pollut Res 23:9381–9390. https://doi.org/10.1007/s11356-015-5317-6
13. Chang C, Tao P, Fu B, Xu J, Song C, Wu J, Shang W, Deng T (2019) Three-dimensional porous solar-driven interfacial evaporator for high-efficiency steam generation under low solar flux. ACS Omega 4(2):3546–3555.https://doi.org/10.1021/ACSomega.8b03573
14. Huang X, Yu Y-H, de Llergo OL, Marquez SM, Cheng Z (2017) Facile polypyrrole thin film coating on polypropylene membrane for efficient solar-driven interfacial water evaporation. RSC Adv 7(16):9495–9499. https://doi.org/10.1039/c6ra26286d
15. Peng G, Deng S, Sharshir SW, Ma D, Kabeel AE, Yang N (2020) High efficient solar evaporation by airing multifunctional textile. Int J Heat Mass Transf 147:118866. https://doi.org/10.1016/j.ijheatmasstransfer.2019.118866
16. Li J, Du M, Lv G, Zhou L, Li X, Bertoluzzi L, Zhu J (2018) Interfacial solar steam generation enables fast-responsive, energy-efficient, and low-cost off-grid sterilization. Adv Mater:1805159.https://doi.org/10.1002/adma.201805159
17. Xu Z, Rao N, Tang C-Y, Law W-C (2020) Seawater desalination by interfacial solar vapor generation method using plasmonic heating nanocomposites. Micromachines 11(9):867. https://doi.org/10.3390/mi11090867
18. Chang C, Liu M, Pei L, Chen G, Wang Z, Ji Y (2021) Porous TiNO solar-driven interfacial evaporator for high-efficiency seawater desalination. AIP Adv 11:045228. https://doi.org/10.1063/5.0047390
19. Tao P, Ni G, Song C et al (2018) Solar-driven interfacial evaporation. Nat Energy 3:1031–1041. https://doi.org/10.1038/s41560-018-0260-7
20. Zhu L, Gao M, Peh CK, Ho GW (2019) Recent progress in solar-driven interfacial water evaporation: advanced designs and applications. Nano Energy 57:507–518. ISSN 2211-2855. https://doi.org/10.1016/j.nanoen.2018.12.046
21. Kashyap VG (March 25, 2020) Solar heat localization: concept and emerging applications. J Mater Chem A 8.https://doi.org/10.1039/D0TA01004A
22. Chang C, Tao P, Fu B, Xu J, Song C, Wu J, Shang W, Deng T (2019) Three-dimensional porous solar-driven interfacial evaporator for high-efficiency steam generation under low solar flux. ACS Omega 4(2). https://doi.org/10.1021/ACSomega.8b03573
23. Juhász L, Moldován K, Herman P, Erdélyi Z, Fábián I, Kalmár J, Cserháti C (2020) Synthesis and stabilization of support-free mesoporous gold nanoparticles. Nanomaterials 10(6):1107. https://doi.org/10.3390/nano10061107
24. Liu X, Liu H, Xiaoqiang Y, Zhou L, Zhu J (2019) Solar thermal utilizations revived by advanced solar evaporation. Curr Opin Chem Eng 25:26–34. https://doi.org/10.1016/j.coche.2019.07.004
25. Johnson A, Mu L, Park YH, Valles DJ, Wang H, Xu P, Kota K, Kuravi S (2019) A thermal model for predicting the performance of a solar still with Fresnel lens. Water 11:1860. https://doi.org/10.3390/w11091860

26. Xu J, Wang Z, Chang C, Song C, Wu J, Shang W, Tao P, Deng T (2019) Electrically driven interfacial evaporation for high-efficiency steam generation and sterilization. ACS Omega 4(15). https://doi.org/10.1021/ACSomega.9b02475
27. Tao F, Valenzuela Garcia A, Xiao T, Zhang Y, Yin Y, Chen X (2020) Interfacial solar vapor generation: introducing students to experimental procedures and analysis for efficiently harvesting energy and generating vapor at the air-water interface. J Chem Educ 97:1093–1100.https://doi.org/10.1021/acs.jchemed.9b00643

# Effect of Perforated Tube Insert on Thermal Behavior of Flat-Plate Solar Water Collector

Elumalai Vengadesan and Ramalingam Senthil

## *Nomenclature*

| | |
|---|---|
| $A_c$ | Collector area ($m^2$) |
| $A_p$ | Absorber area ($m^2$) |
| $C_{pw}$ | Specific heat of water (kJ/kg K) |
| $H$ | Convective heat transfer coefficient (W/$m^2$ K) |
| $I$ | Solar radiation (W/$m^2$) |
| $\dot{m}$ | Mass flow rate of water (kg/s) |
| $T_a$ | Ambient temperature (K) |
| $T_f$ | Mean temperature of water (K) |
| $T_i$ | Inlet temperature of water (K) |
| $T_o$ | Outlet temperature of water (K) |
| $T_s$ | Sun temperature (K) |
| $T_w$ | Absorber tube wall temperature (K) |
| $Q_{conv}$ | Convective heat transfer (W) |
| $Q_{water}$ | Heat transfer to water (W) |

## *Greek Symbol*

| | |
|---|---|
| $\eta_{th}$ | Instantaneous thermal efficiency (%) |
| $\eta_{ex}$ | Exergy efficiency (%) |

E. Vengadesan (✉) · R. Senthil
Department of Mechanical Engineering, SRM Institute of Science and Technology, Kattankulathur, Chennai, India
e-mail: venkatmech66@gmail.com

R. Senthil
e-mail: senthilr@srmist.edu.in



N. L. Narasimhan et al. (eds.), *Recent Advances in Energy Technologies*, Lecture Notes in Mechanical Engineering, https://doi.org/10.1007/978-981-19-3467-4_4

## *Abbreviations*

| | |
|---|---|
| FPSWC | Flat-plate solar water collector |
| HTF | Heat transfer fluid |
| HTC | Heat transfer coefficient |
| SWH | Solar water heater |
| TT | Twisted tape |

## 1 Introduction

The energy need of developing countries like India has been increasing from past decades due to the country's economic growth. Fossil fuel has a major contribution still which is causing environmental pollution and global climate change. Renewable energy usage has been increasing, and specifically, solar energy has advantages like clean and cost-free energy [1]. The solar water heaters (SWHs) contribute more to the thermal applications, an eco-friendly system that requires lower operational and maintenance costs [2]. The flat-plate solar water collector (FPSWC) is a simple and well-known SWH applicable in many industrial and building heating applications. The heat output of FPSWCs is lower due to more heat loss to the surrounding and lesser heat conduction by the water. The different techniques such as reducing heat loss, increasing heat absorption by water, and increasing energy capture are the way to improve the thermal behavior of FPSWCs [3]. Various enhancement methods like improved riser tube cross-section profile, enhanced HTF's thermal conductivity, and using turbulence creator (passive insert devices) inside the riser tubes are familiar. The inclusion of nanofluid combined with passive insert devices enhanced the collector's thermal performance to a greater level. However, passive insert devices significantly contribute to improving the collector's thermal performance [4–9]. In addition, nanofluids have chemical and physical instability, higher viscosity, higher cost, and more pump power requirements. But, the investment of the tube insert devices is very cheaper.

The passive method is identified as an effective enhancement method with lesser investment on different tube heat exchangers and FPSWCs. The FPSWCs with different types of passive insert devices are investigated to make the enhanced fluid flow absorb maximum heat from the absorber, leading to reduced absorber temperature and heat loss. The FPSWC with twisted tape (TT) inserts increased heat transfer due to more heat transfer area and enhanced turbulence intensity of the flow. The overall efficiency was increased by 5% [10]. The contribution of TT and wire coil insert on decreasing the collector's absorber temperature was examined [11]. The enhanced HTC at tube inside flow decreased absorber temperature by a maximum of 5.4 °C. The selection insert was based on pressure drop, where the TT insert was suitable for the harp-type FPSWC and the wire coil insert was suitable for serpentine-type FPSWC due to the chance of minimum pressure drop. The thermal efficiency was increased to 31%, and useful power collection was up to 12%. The increased HTF flow rate had increased the thermal efficiency further [12]. Riser tube with TT, louvered strip, and laser-peened absorber of FPSWC was investigated, and collector reached the maximum outlet temperature with louvered strip insert. The heat absorption and collector's efficiencies were 678.13 W and 33.24% which had 7 and 10% higher efficiency than collector with TT inserted absorber and laser-peened absorber tubes. The HTC was also higher than the other two configurations by 440 W/m$^2$ K than TT inserted absorber tubes and 547 W/m$^2$ K than laser-peened absorber tubes. They suggested that the laser-peened absorber tube with any type of insert device can improve the better heat transfer [13].

The tube heat transfer had increased when plain, perforated, and dimpled TT inserts were used as passive heat transfer enhancing devices with a considerable increase in pressure drop. A secondary flow that occurred through the fluid flow improved fluid mixing at the core and surrounding region which tends to increase heat transfer due to convection. The perforated tape has more swirl flow which enhanced secondary flow development and fluid mixing. The dimpled tape decreased the gap between the tube wall and flow, decreasing the laminar boundary layer formation. The thermal and hydraulic performance was higher for the dimpled insert with water [14]. The rate of heat transfer was enhanced by 18–70%, with a pressure drop increase of 87–132% using TT. The collector's thermal performance increased by 30% and reduced the thermal loss [15].

V-tough solar collector inserted with plain helix TT, helix TT with square, and V-cut was examined for augmentation in heat transfer. The heat transfer rate of the V-tough collector was higher by 8.86% than FPSWC. The V-cut TT inserts resulted in higher heat transfer than the other two configurations. A secondary and swirl flow generation for the V-cut insert was higher which increased the core and hot region fluid mixing [16].

A numerical simulation was conducted to study the effect of vortex generators (rectangle and delta-winglet) on the thermal output of FPSWC. The rectangle vortex generators were created in more corners and main vortex than delta-winglet which increased the riser tube's heat transfer and collector's thermal efficiency. Though the development of secondary flow enhanced the thermal response of the collector, more pressure drop was observed than conventional system [17].

The reaction of thermal output of FPSWC with rod and tube enhancers was experimentally investigated, the heat transfer due to convection was enhanced, and the increase of pressure drop was lesser. The rod insert has a superior thermal response than other configurations as the area of heat transfer is improved [18]. The thermal response of the collector with rod and tube enhancers was higher. The difference in HTC of the collector with plain tubes and rod enhancers was 60 W/m$^2$ K [19]. The peak efficiency (74%) was earned at a higher flow rate, and free convection contributed more to total heat transfer [20]. The exergy efficiency was also higher using enhancers with more useful exergy [21]. Although the enhancers upgraded the thermal output, the pressure drop had increased by 1.081 times higher than conventional collector when using rod insert and 1.044 times using tube enhancer [22].

The porous copper block-inserted riser tube was tested, and the heat transfer is enhanced due to improved conduction and convection. The fluid mixing was enhanced, which intensified the heat transfer due to convection. The heat transfer due to conduction was also improved due to the presence of a porous structure. The temperature difference of 8.2 °C was observed with a notable pressure drop of about 153% [23]. SWH with dimpled absorber tubes was simulated to find the contribution to an intensification of heat transfer. Due to a dimple shape, the swirl flow was developed and flow acceleration was observed. The swirls created near the boundary region improved fluid mixing rate and physical contact between the HTF, and tube was increased, resulting in heat transfer intensification. The thermal efficiency was enhanced by 32.3% compared to smooth tubes [24]. The effect of mass flow is significant on the collector's thermal efficiency as useful energy is higher in more HTF flow. The increasing flow rate increased the HTC and efficiency significantly [25].

The combined effect of more surface area and swirl motion intensified the heat transfer as the water has more contact with the tube with improved fluid mixing. The boundary layer due to laminar flow was disturbed and resulted in higher convective HTC [12]. From the review of insert devices' contribution to solar thermal energy systems, swirls created due to passive devices improve the hot and cold fluid region merging. Inserts with additional cuts created a secondary flow further and added the convective HTC. Hence, absorber temperature reduced considerably which leads to reduced heat loss [26]. The flow deviation from the center to the other region makes the fluid strike over the inner tube surface, decreasing the chance of the creation of a laminar boundary layer. The effective mixing enhances the heat transfer locally which results in more total heat exchanged [13]. The thermal output of the solar water collector is being increased when the water has more contact with the absorber for an extended time [27]. The passive insert devices increased the heat transfer area, do not allow the fluid to flow freely, disturb the flow to have more surface contact with the inner tube wall, and get mixed with other fluid zones.

Passive insert devices increase the chance of more area and extended flow time that leads to more heat absorption from the system. A secondary flow is created, and swirl flow generation is possible if a cut or hole is provided over the insert device. The perforations made on the insert can give an additional flow gap, increasing the chance of enhanced fluid velocity toward the inner wall of the tube and fluid mixing. Further, the passive devices improve the HTF flow by creating turbulence, which can act as an extended surface if they have frictional contact with the absorbers. This experimental study investigates the contribution of a plain and perforated oval-shaped tube insert on the thermal functioning of the FPSWC by comparing it with a conventional collector. The tube insert makes frictional contact with the absorber tube by its oval shape at the top and bottom portion. Further, the perforations are provided over two sides of the oval-shaped tube for its entire length to improve HTF flow nature by creating turbulence and more surface area improvement with the absorber.

The present article structure is arranged as follows.

- The introduction section has information on the need for solar energy, the importance of passive enhancing devices on intensification of heat output, and the study's objective.
- Section 2 contains the experimental materials and methods.
- The thermal energy and exergy analysis and their calculation are provided in Sect. 3.
- Section 4 has the discussion and findings of measured performance parameters.
- The conclusion is provided in Section 5 based on the observed results and findings.

## 2 Experimentation

The experimental setup consists of an insulated storage tank, electrical pump, flow meter, control valves, collector, thermocouple wires, data logger, and computer as presented in Fig. 1. The FPSWC consists of two riser tubes connected with a header tube. The single glass and glass wool insulation are provided, and the other collector's specifications are provided in Table 1. The glass cover's thickness and transmissivity, insulation material and properties, absorber material, and thermal conductivity are taken from the earlier studies [28, 29]. Collector type-I has only plain riser tubes, type-II has a plain oval-shaped tube insert into the riser tubes, and type-III has a perforated oval-shaped tube insert in the riser tubes. The details of oval-shaped tube inserts have appeared in Fig. 2. The riser tube and oval-shaped insert profiles are presented in Fig. 3.

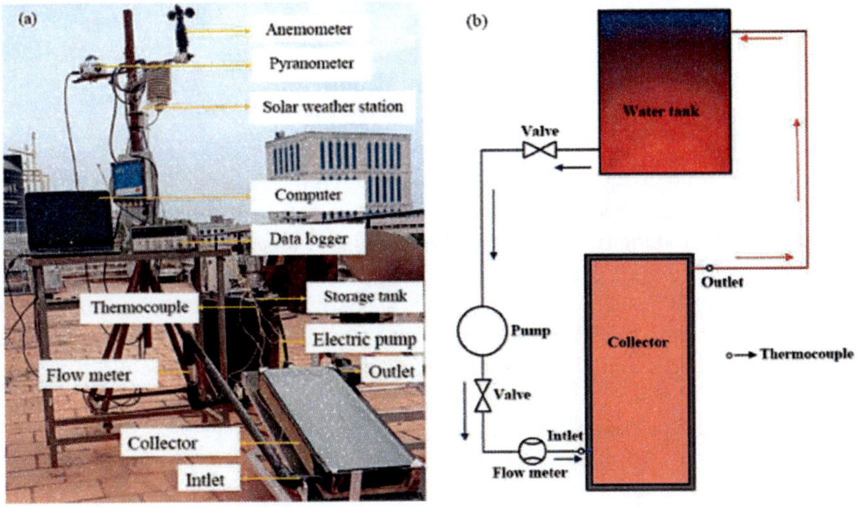


**Fig. 1** Testing setup of FPSWC, **a** experimental photograph, **b** schematic diagram

**Table 1** Dimensions and materials of FPSWC [28, 29]

| Items | Dimensions and materials |
|---|---|
| Area of the collector | $0.185 \text{ m}^2$ ($0.26 \text{ m} \times 0.71 \text{ m}$) |
| Riser tube diameter | 12.7 mm |
| Header tube diameter | 22 mm |
| Number of tubes | 2 |
| The thickness of the glass cover | 4 mm |
| Transmissivity of the glass cover | 0.95 |
| Absorber | Copper sheet (absorptivity-0.92) |
| Thermal conductivity of absorber | 387.6 W/m K |
| Insulation | 0.040 W/m K (glass wool) |

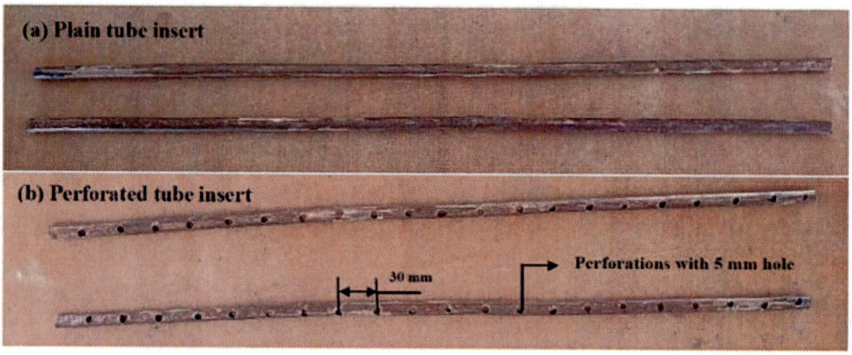


**Fig. 2** Details of the tube inserts

The temperatures measurements are done with a K-type thermocouple and data logger (Agilent-34970A). An average three-point temperature has been taken as the temperature of the absorber plate. The solar weather station is utilized for measuring solar radiation, air velocity, and ambient temperature. The uncertainty of instruments used for measurements is considered from the earlier studies and is displayed in Table 2 [28, 29]. The experimental test is conducted at 0.0167 kg/s (laminar flow condition) for several continuous days between 8 a.m. and 5 p.m., and observations are measured at 30 min intervals. A similar rate of mass flow is followed for all three collectors. The best readings with almost similar solar radiation are considered for the performance analysis of three types of collectors.

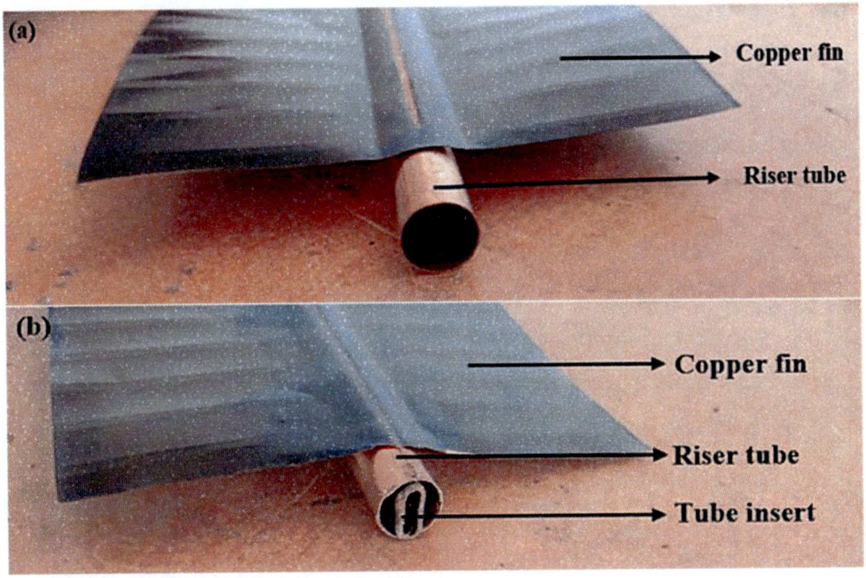


**Fig. 3** Riser tube, **a** without insert, **b** with insert

**Table 2** Uncertainty and measurement range of instruments [28, 29]

| Instrument | Range of measurement | Uncertainty |
|---|---|---|
| Pyranometer | 0–4000 W/m$^2$ | ±20 W/m$^2$ |
| Anemometer | 0–25 m/s | ±0.01 m/s |
| K-type thermocouple | −200–1000 °C | ±0.5 °C |
| Flow meter | 0–0.0555 kg/s | ±0.0011 kg/s |

## 3 Thermal Analysis

The heat absorbed by water ($Q_{\text{water}}$) is calculated from Eq. (1) [13].

$$\dot{Q}_{\text{water}} = \dot{m}_w C_{\text{pw}}(T_{\text{o}} - T_{\text{i}}) \tag{1}$$

where $\dot{m}$, $C_{\text{pw}}$, $T_{\text{i}}$, and $T_{\text{o}}$ are mass flow rate, the specific heat of water, inlet, and outlet temperature of the water, respectively.

The heat absorbed by water is considered equal to convective heat transfer of water by Eq. (2).

$$\dot{Q}_{\text{water}} = Q_{\text{conv}} \tag{2}$$

The convective heat transfer ($Q_{\text{conv}}$) is calculated by Eq. (3).

$$Q_{\text{conv}} = hA_p\left(T_{\text{w}} - T_f\right) \tag{3}$$

Here, $A_p$, $T_{\text{w}}$, and $T_f$ are absorber area, tube wall temperature, and mean temperature of the water, respectively.

Mean temperature of water ($T_f$):

$$T_f = \frac{T_{\text{o}} + T_{\text{i}}}{2} \tag{4}$$

Absorber tube wall temperature ($T_{\text{w}}$):

$$T_{\text{w}} = \frac{T_1 + T_2 + T_3 + T_4 + \cdots + T_n}{n} \tag{5}$$

The HTC ($h$) is calculated by Eq. (6)

$$h = \frac{\dot{m}_w C_{\text{pw}}(T_{\text{o}} - T_{\text{i}})}{(T_{\text{w}} - T_f)A_p} \tag{6}$$

The instantaneous thermal efficiency ($\eta_{\text{th}}$) is calculated as follows (Eq. 7) [13, 27]

$$\eta_{\text{th}} = \frac{\dot{m}_w C_{\text{pw}}(T_{\text{o}} - T_{\text{i}})}{I A_{\text{c}}} \tag{7}$$

where $I$ and $A_{\text{c}}$ are solar radiation intensity and collector area.

The exergy efficiency ($\eta_{\text{ex}}$) is calculated from Eq. (8) [27]

$$\eta_{\text{ex}} = \frac{\dot{m}_w C_{\text{pw}}\left[(T_{\text{o}} - T_{\text{i}}) - T_{\text{a}}\left(\ln \frac{T_{\text{o}}}{T_{\text{i}}}\right)\right]}{A_{\text{c}} I\left[1 - \frac{T_{\text{a}}}{T_{\text{s}}}\right]} \tag{8}$$

where $T_{\text{a}}$ and $T_{\text{s}}$ are the ambient and the sun temperature. The sun temperature is considered as 4500 K [30].

## 4 Results and Discussion

The real-time experiment is conducted for several days at a similar flow rate. The best observations for each configuration have been taken for the thermal performance calculation. The average radiation intensity varies from 689 to 718 W/m$^2$, and ambient temperature is observed from 28 to 44 °C during the considered days. The peak solar radiation of 945, 952, and 965 W/m$^2$ is observed during the first, second, and third day. An experiment day for types-I, II, and III is considered as the first, second, and third day. Figure 4 shows the temperature of ambient and radiation intensity during testing of types-I, II, and III.

### 4.1 *Outlet Water Temperature*

The inlet, outlet water, and absorber temperature are measured with 30 min intervals. The inlet water temperature is allowed in a closed cycle where the inlet temperature is increased with solar radiation. The temperature of the absorber steeply increased when solar radiation is increasing. An average of three absorber's point temperatures has taken as absorber plate temperature. The steep increase in absorber temperature is minimized when the water gets heated. If the temperature of riser tube is higher, the temperature difference is minimal due to more thermal loss. If the temperature difference increases, the heat absorption is more, whereas the absorber temperature is decreased. The peak absorber temperature of types-I, II, and III is observed as 63 °C, 62.5 °C, and 61.5 °C, respectively. The temperature difference of absorber for the plain riser tube and a plain tube-inserted riser tube is low when compared between plain and perforated inserted riser tubes. The perforated insert reduced absorber temperature better than the riser tube with plain tube insert and without any insert. Figure 5 shows the temperature variations of absorbers of three different configurations.

The outlet water temperature of type-III is greater than types-I and II. The peak water outlet temperatures are 60 °C, 56.2 °C, and 57.4 °C for riser tubes with perforated tube insert, plain tube insert, and without any insert, respectively. The temperature of outlet water is maintained at more than 50 °C from 12:30 to 5 p.m. because of higher heat absorbed by water. The water flow through the plain riser tube flows freely without any disturbance, where the water contact and residence time are lower. For tube-inserted riser tube, the frictional engagement of the oval-shaped insert reduces space for free HTF flow and makes the water contact more with the insert device and inner wall of the riser tube. The outlet water temperature difference between plain and tube-inserted riser tubes showed that the insert device acts as a flow enhancer and an extended surface for heat transfer. The perforated insert directs the water flow at both sides through the perforations provided at both sides of the oval-shaped tube insert. Hence, the increased flow residence and fluid mixing are possible with the perforated tube insert. The enhanced fluid flow mixing and more contact at the riser tube inner wall make the water absorb more heat and attain a higher temperature than the other two types. The chance of lower temperature at the center of the HTF flow is reduced when perforations are given in the tube insert. The comparison of outlet temperature of water for three collectors is presented in Fig. 6. The maximum temperature difference of 2 °C is reached using a perforated insert. Though the temperature difference was measured lesser, the outlet water temperature is higher for both insert devices. Figure 7 compares the temperature difference for three types.

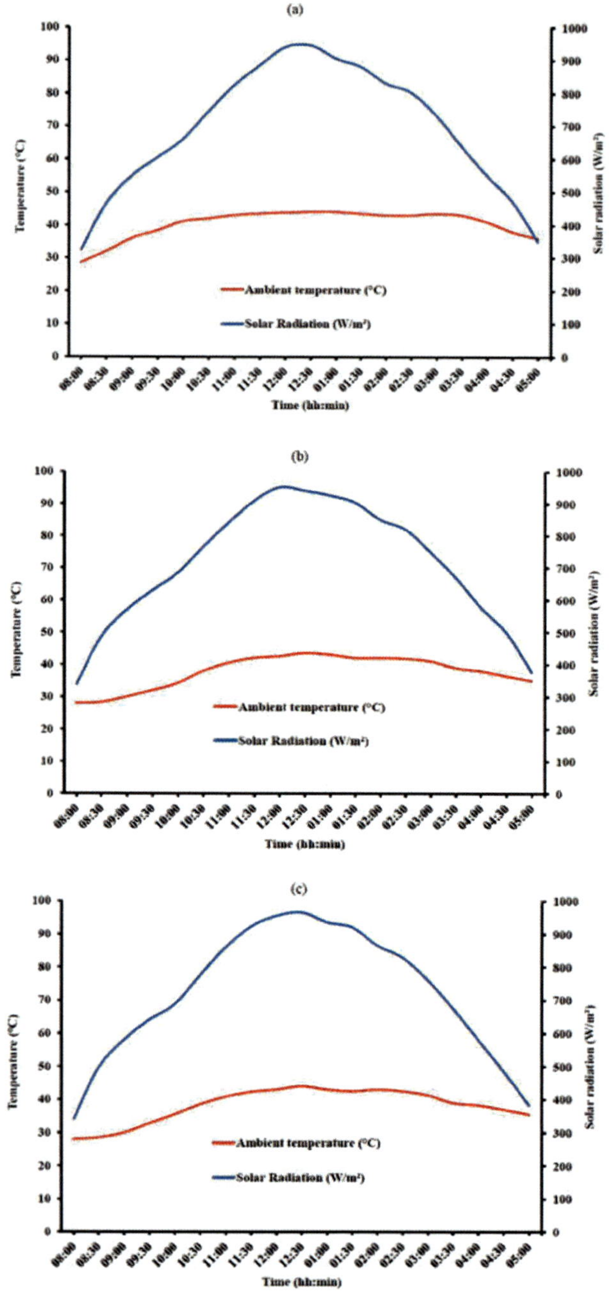


**Fig. 4** Ambient temperature and radiation. **a** Type-I, **b** type-II, **c** type-III

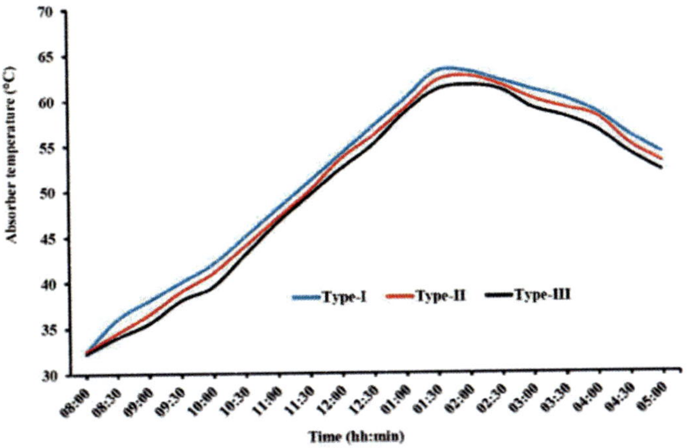


**Fig. 5** Comparison of absorber temperature

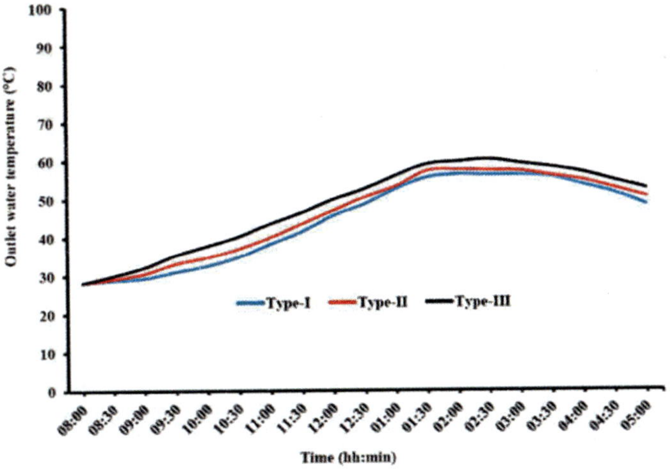


**Fig. 6** Comparison of outlet water temperature

## *4.2 Heat Transfer Coefficient (h)*

The convective HTC is an important factor as convective mode heat transfer is dominant between the absorber tube wall and water. The plain riser tube results in lower HTC due to less contact at the riser tube inner wall and lower residence time. The water temperature at the center is lower, and the hot fluid region is not mixed with the cold region. The heat transfer between the hot region and the cold region is slower as no physical interaction between the two regions. For a plain oval tube-inserted

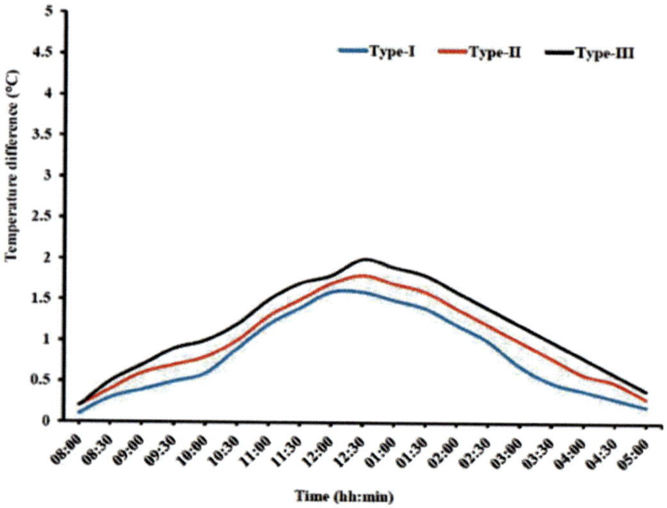


**Fig. 7** Comparison of temperature difference

collector, the free space of the HTF flow is disturbed by the insert. The insert is having frictional contact with the riser tube wall, which means the insert device also acts as an extended surface. Hence, the fluid flows through the insert device and the other between the riser and the device. The thin water flow is possible instead of bulk flow which can increase heat transfer to HTF. The perforations provided at both sides of the oval-shaped insert make the flow toward the riser tube wall continuously along its length. This perforated tube insert device reduces the chance of lower temperature at the center flow as the perforation directs the flow at two sides. Hence, mixing the hot and cold fluid region has enhanced the heat transfer rate to the water.

The HTC of riser tube without inserts, plain oval-shaped tube insert, and perforated oval-shaped insert is 2.35, 3.98, and 5.79 W/m$^2$ K, respectively. The perforated insert has more than twice the times of higher HTC of the riser tube without an insert. The perforation made on the insert device increased the convective HTC significantly. The average HTC of types-I, II, and III is 1.39, 2.89, and 4.18 W/m$^2$ K, respectively. The variations and comparisons of HTC for three configurations are presented in Fig. 8.

### 4.3 *Instantaneous Thermal Efficiency*

The instantaneous thermal efficiency is calculated by taking temperature differences at 30 min time intervals. The temperature difference of type-III is more significant than the other two types with a lower difference. The maximum instantaneous thermal efficiencies are 64.4%, 72.4%, and 78.4%, respectively, for types-I, II, and III. The

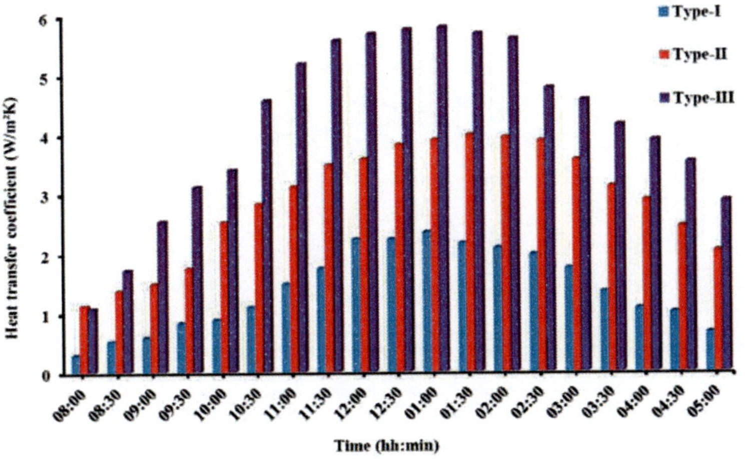


**Fig. 8** Comparison of convective HTC

plain insert and perforated insert have higher efficiencies by 12.4% and 21.7% than the collector with plain riser tubes. The average instantaneous thermal efficiencies are 41.2%, 49.8%, and 57.7%, respectively, for types-I, II, and III. The peak, and average efficiencies of type-III are higher by 8.3% and 15.9% than type-II.

The plain and perforated insert tube inserts have 20% and 40% enhancement in average efficiency than using plain riser tube. The plain riser tube results in more absorber temperature and less outlet temperature as heat exchange is poor. The hot and cold fluid heat exchange is lower for plain risers due to uniform flow throughout its length. The laminar boundary layer development inside the absorber tube wall restricts the heat transfer due to convection. The water flow in a plain tube is uniform, and the flow at the center continues its pass along its length with lower temperature. The riser tube with a plain tube insert splits the flow into the insert device and the space between the tube insert and the riser tube. The free water flow is disturbed, and thin water flow through the outer pace of the insert and riser tube makes the fluid get more surface contact, leading to increased HTC. The perforation made on the insert directs the water flow toward the riser tube wall and flow; it splits the flow into two directions. The temperature of the core water is enhanced due to better fluid mixing. The frictional contact of both inserts increased the tube heat transfer further by their extended surface functionality. The instantaneous thermal efficiency of the collector for three configurations is presented in Fig. 9.

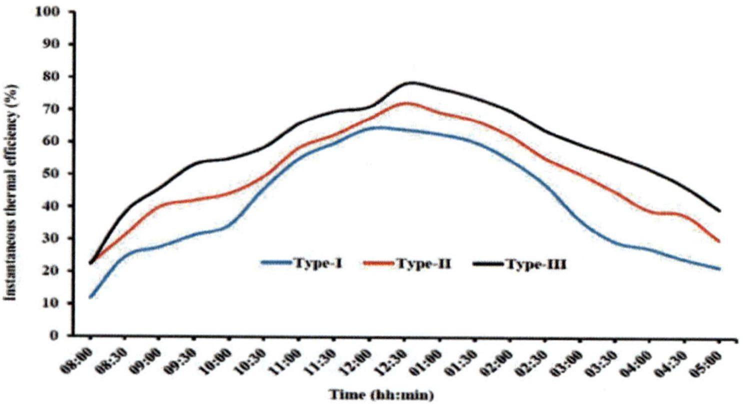


**Fig. 9** Instantaneous thermal efficiency

## 4.4 Exergy Efficiency

The useful energy absorbed by the riser tube with tube insert is a higher riser tube. The enhanced convective HTC of the riser tube with insert devices improved the thermal response of the FPSWC significantly. The maximum exergy efficiencies of types-I, II, and III are 2.5%, 3.3%, and 3.9%, respectively. The average exergy efficiencies are 1.1%, 1.6%, and 2.1% for types-I, II, and III, respectively. The collector with a perforated tube insert has almost double the average exergy efficiency of FPSWC with plain riser tubes. The perforation provided at both sides of the inert device deviates the water toward the inner wall of the riser and observed better fluid mixing between the center fluid and surrounding fluid regions that increased the HTC. Hence, the useful energy absorption is higher for the collector with passive insert devices, resulting in more exergy efficiency than plain riser tubes. The variation of exergy efficiency is shown in Fig. 10.

Uncertainty in the results of performance parameters of type-III is calculated as 1.22% (HTC), 3.30% (thermal efficiency), and 3.31% (exergy efficiency) by using the following formulae.

$$\frac{\Delta h}{h} = \left[\left(\frac{\Delta \dot{m}}{\dot{m}}\right)^2 + \left(\frac{\Delta T_o}{T_o}\right)^2 + \left(\frac{\Delta T_{in}}{T_{in}}\right)^2 + \left(\frac{\Delta T_f}{T_f}\right)^2 + \left(\frac{\Delta T_p}{T_p}\right)^2 + \left(\frac{\Delta A_p}{A_p}\right)^2\right]^{0.5} \tag{9}$$

$$\frac{\Delta \eta_{th}}{\eta_{th}} = \left[\left(\frac{\Delta Q_u}{Q_u}\right)^2 + \left(\frac{\Delta Q_{in}}{Q_{in}}\right)^2\right]^{0.5}. \tag{10}$$

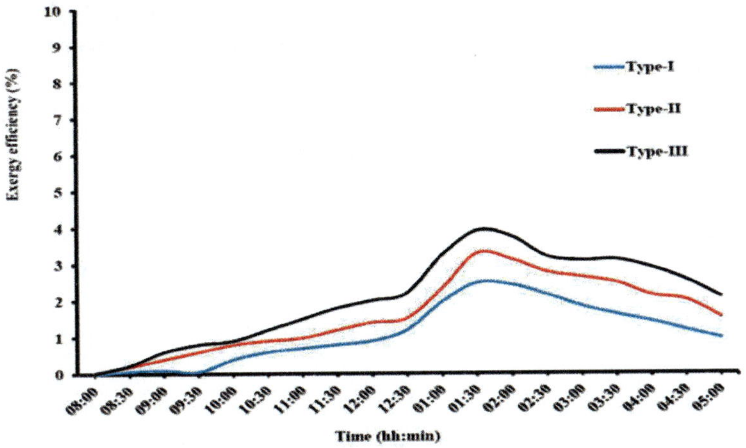


**Fig. 10** Exergy efficiency

$$\frac{\Delta\eta_{\text{ex}}}{\eta_{\text{ex}}} = \left[\left(\frac{\Delta\dot{m}}{\dot{m}}\right)^2 + \left(\frac{\Delta T_{\text{o}}}{T_{\text{o}}}\right)^2 + \left(\frac{\Delta T_{\text{in}}}{T_{\text{in}}}\right)^2 + \left(\frac{\Delta T_{\text{a}}}{T_{\text{a}}}\right)^2 + \left(\frac{\Delta A_c}{A_c}\right)^2 + \left(\frac{\Delta I}{I}\right)^2\right]^{0.5} \tag{11}$$

## 5 Conclusion

The experimental test on FPSWCs with plain riser tube without insert device, plain tube insert, and perforated tube insert. A similar flow rate is maintained for all experimental tests. HTC, instantaneous thermal, and exergy efficiency of FPSWCs are calculated from the observations. The findings of this study lead to the following conclusions.

- A higher outlet water temperature is observed when the riser tube is inserted with a perforated tube insert. The enhancement of fluid heat transfer and mixing improved the collector's thermal behavior. The maximum water temperature (60 °C) is observed for type-III.
- Absorber temperature is reduced when the heat absorption rate is increased, which can also reduce the overall thermal loss of the collector.
- The perforations of the insert devices enhanced the heat gained by water which has the convection HTC of 5.79 W/m$^2$ K. The average HTC of type-III is 4.18 W/m$^2$ K which is higher than a plain riser tube without an insert and with plain tube inserts.
- The perforated inserted collector's maximum and average instantaneous thermal efficiencies are 21.7% and 40% higher than FPSWC with plain riser tubes and 8.3% and 15.9% higher than FPSWC with plain tube inserts.

- The maximum exergy efficiency with a perforated tube insert is 3.9%. The average exergy efficiency for the perforated and plain tube-inserted collectors is 90.9% and 45.5% higher than plain riser tubes without the insert.

The effect of tube insert profiles with perforations has to be investigated further to find the optimal performance. Numerical simulation is required to know the physics behind the thermal output enhancement.

**Acknowledgements** The authors are grateful to the SRM Institute of Science and Technology, Kattankulathur Campus, Chennai, for supplying the required research infrastructure.

## References

1. Sharma P, Jamwal A, Sharma N, Agrawal R (2021) Opportunities and issues with clean renewable energy development in India: a review. In: Sikarwar BS, Sundén B, Wang Q (eds) Advances in fluid and thermal engineering. Lecture notes in mechanical engineering. Springer, Singapore. https://doi.org/10.1007/978-981-16-0159-0_47
2. Chadge RB, Sunheriya N, Mahatme C, Giri JP (2021) Experimental and numerical investigation of flat plate solar water heater. In: Sharma BP, Rao GS, Gupta S, Gupta P, Prasad A (eds) Advances in engineering materials. Lecture notes in mechanical engineering. Springer, Singapore. https://doi.org/10.1007/978-981-33-6029-7_4
3. Gorjian S, Ebadi H, Calise F, Shukla A, Ingrao C (2020) A review on recent advancements in performance enhancement techniques for low-temperature solar collectors. Energy Convers Manag 222:113246. https://doi.org/10.1016/j.enconman.2020.113246
4. Bellos E, Tzivanidis C, Tsimpoukis D (2018) Enhancing the performance of parabolic trough collectors using nanofluids and turbulators. Renew Sust Energ Rev 91:358–375. https://doi.org/10.1016/j.rser.2018.03.091
5. Munuswamy DB, Devarajan Y, Babu MN, Ramalingam S (2020) Experimental investigation on lowering the environmental hazards and improving the performance patterns of solar flat plate collectors by employing the internal longitudinal fins and nano additives. Environ Sci Pollut Res 27:45390–45404. https://doi.org/10.1007/s11356-020-10311-3
6. Saffarian MR, Moravej M, Doranehgard MH (2020) Heat transfer enhancement in a flat plate solar collector with different flow path shapes using nanofluid. Renew Energy 146:2316–2329. https://doi.org/10.1016/j.renene.2019.08.081
7. Sundar LS, Kirubeil A, Punnaiah V, Singh MK, Sousa ACM (2018) Effectiveness analysis of solar flat plate collector with $Al_2O_3$ water nanofluids and with longitudinal strip inserts. Int J Heat Mass Transf 127:422–435. https://doi.org/10.1016/j.ijheatmasstransfer.2018.08.025
8. Sundar LS, Sintie YT, Said Z, Singh MK, Punnaiah V, Sousa ACM (2020) Energy, efficiency, economic impact, and heat transfer aspects of solar flat plate collector with $Al_2O_3$ nanofluids and wire coil with core rod inserts. Sustain Energy Technol Assess 40:100772. https://doi.org/10.1016/j.seta.2020.100772
9. Sheikholeslami M, Farshad SA, Said Z (2021) Analyzing entropy and thermal behavior of nanomaterial through solar collector involving new tapes. Int Commun Heat Mass Transf 123:105190. https://doi.org/10.1016/j.icheatmasstransfer.2021.105190
10. Behura AK, Kumar A, Todkari VC, Dwivedi G, Gupta HK (2021) Analysis of thermal efficiency of solar flat plate collector using twisted tape. In: Ramgopal M, Rout SK, Sarangi SK (eds) Advances in air conditioning and refrigeration. Lecture notes in mechanical engineering. Springer, Singapore. https://doi.org/10.1007/978-981-15-6360-7_9

11. García A, Herrero-Martin R, Solano JP, Pérez-García J (2018) The role of insert devices on enhancing heat transfer in a flat-plate solar water collector. Appl Therm Eng 132:479–489. https://doi.org/10.1016/j.applthermaleng.2017.12.090
12. García A, Martin RH, Pérez-García J (2013) Experimental study of heat transfer enhancement in a flat-plate solar water collector with wire-coil inserts. Appl Therm Eng 61(2):461–468. https://doi.org/10.1016/j.applthermaleng.2013.07.048
13. Abeens M, Meikandan M, Sheriff J, Murunganadhan R (2020) Experimental analysis of convective heat transfer on tubes using twisted tape inserts, louvered strip inserts and surface treated tube. Int J Ambient Energy 41(5):540–546. https://doi.org/10.1080/01430750.2018.1476263
14. Dagdevir T, Ozceyhan V (2021) An experimental study on heat transfer enhancement and flow characteristics of a tube with plain, perforated, and dimpled twisted tape inserts. Int J Therm Sci 159:106564. https://doi.org/10.1016/j.ijthermalsci.2020.106564
15. Kumar A, Prasad BN (2000) Investigation of twisted tape inserted solar water heaters—heat transfer, friction factor and thermal performance results. Renew Energy 19(3):379–398. https://doi.org/10.1016/S0960-1481(99)00061-0
16. Saravanan A, Jaisankar S (2019) Heat transfer augmentation techniques in forced flow V-trough solar collector equipped with V-cut and square cut twisted tape. Int J Therm Sci 140:59–70. https://doi.org/10.1016/j.ijthermalsci.2019.02.030
17. Silva FASD, Dezan DJ, Pantaleão AV, Salviano LO (2019) Longitudinal vortex generator applied to heat transfer enhancement of a flat plate solar water heater. Appl Therm Eng 158:113790. https://doi.org/10.1016/j.applthermaleng.2019.113790
18. Balaji K, Iniyan S, Goic R (2018) Thermal performance of solar water heater using velocity enhancer. Renew Energy 115:887–895. https://doi.org/10.1016/j.renene.2017.09.014
19. Balaji K, Ganesh Kumar P, Sakthivadivel D, Vigneswaran VS, Iniyan S (2019) Experimental investigation on flat plate solar collector using frictionally engaged thermal performance enhancer in the absorber tube. Renew Energy 142:62–72. https://doi.org/10.1016/j.renene.2019.04.078
20. Balaji K, Idrish Khan A, Ganesh Kumar P, Iniyan S, Goic R (2019) Experimental analysis on free convection effect using two different thermal performance enhancers in absorber tube of a forced circulation flat plate solar water heater. Sol Energy 185:445–454. https://doi.org/10.1016/j.solener.2019.04.089
21. Balaji K, Iniyan S, Swami MV (2018) Exergy, economic and environmental analysis of forced circulation flat plate solar collector using heat transfer enhancer in riser tube. J Clean Prod 171:1118–1127. https://doi.org/10.1016/j.jclepro.2017.10.093
22. Balaji K, Iniyan S, Swami MV (2017) Experimental investigation on heat transfer and pumping power of forced circulation flat plate solar collector using heat transfer enhancer in absorber tube. Appl Therm Eng 112:237–247. https://doi.org/10.1016/j.applthermaleng.2016.09.074
23. Nima MA, Ali AM (2017) Effect of metal foam insertion on thermal performance of flat-plate water solar collector under Iraqi climate conditions. Arab J Sci Eng 42:4863–4884. https://doi.org/10.1007/s13369-017-2683-z
24. Manoram RB, Moorthy RS, Ragunathan R (2021) Investigation on influence of dimpled surfaces on heat transfer enhancement and friction factor in solar water heater. J Therm Anal Calorim 145:541–558. https://doi.org/10.1007/s10973-020-09746-0
25. Bharti A, Sharma B, Paswan MK (2021) CFD and thermal analysis of the flat plate collector—solar water heater under steady-state conditions. In: Das LM, Kumar N, Lather RS, Bhatia P (eds) Emerging trends in mechanical engineering. Lecture notes in mechanical engineering. Springer, Singapore. https://doi.org/10.1007/978-981-15-8304-9_15
26. Rashidi S, Kashefi MH, Hormozi F (2018) Potential applications of inserts in solar thermal energy systems—a review to identify the gaps and frontier challenges. Sol Energy 171:929–952. https://doi.org/10.1016/j.solener.2018.07.017
27. Pathak PK, Chandra P, Raj G (2019) Comparative analysis of modified and convectional dual purpose solar collector: energy and exergy analysis. Energy Sources A: Recov Util Environ Eff. https://doi.org/10.1080/15567036.2019.1692974

28. Vengadesan E, Senthil R (2022) Experimental thermal performance and enviroeconomic analysis of serpentine flow channeled flat plate solar water collector. Environ Sci Pollut Res 29:17241–17259. https://doi.org/10.1007/s11356-021-16985-7
29. Vengadesan E, Senthil R (2022) Experimental study on the thermal performance of a flat plate solar water collector with a bifunctional flow insert. Sustain Energy Technol Assess 50:101829. https://doi.org/10.1016/j.seta.2021.101829
30. Petela R (1964) Exergy of heat radiation. J Heat Transfer 86:187–192. https://doi.org/10.1115/1.3687092

# Effect of Parking Direction and Radiation Shields on the Indoor Cabin Environment of a Stationary Passenger Car: Experimental Study

**N. Lakshmi Narasimhan, M. Praveen Kumar, B. Sathish, R. Sathish Kumar, and V. Sivaraj**

## Nomenclature

| | |
|---|---|
| AC | Air conditioning |
| $Q_{body}$ | Heat transfer from car body, W |
| $Q_{conv,cabin}$ | Convective heat transfer in cabin, W |
| $Q_{conv,skin}$ | Convective heat transfer from car skin, W |
| $Q_{diff,rad}$ | Diffuse radiation, W |
| $Q_{duty,ac}$ | Heat duty of the AC system, W |
| $Q_{rad}$ | Incident radiation, W |
| $Q_{rad,front}$ | Radiation entering front windshield, W |
| $Q_{rad,rear}$ | Radiation entering rear windshield, W |
| $Q_{refl}$ | Reflected radiation, W |
| $T_{bs}$ | Back seat temperature, °C |
| $T_{cabin,b}$ | Bottom cabin air temperature, °C |
| $T_{cabin,c}$ | Central cabin air temperature, °C |
| $T_{DBc}$ | Front dashboard central temperature, °C |
| $T_{DBl}$ | Front dashboard left temperature, °C |
| $T_{DBr}$ | Front dashboard right temperature, °C |
| $T_{deck}$ | Rear deck temperature, °C |
| $T_{fwsi}$ | Front windshield inner temperature, °C |
| $T_{fwso}$ | Front windshield outer temperature, °C |
| $T_{lfsf}$ | Left side front seat front temperature, °C |
| $T_{lfsr}$ | Left side front seat rear temperature, °C |
| $T_{rfsf}$ | Right side front seat front temperature, °C |
| $T_{rfsr}$ | Right side front seat rear temperature, °C |

N. L. Narasimhan (✉) · M. P. Kumar · B. Sathish · R. S. Kumar · V. Sivaraj
Department of Mechanical Engineering, Sri Sivasubramaniya Nadar College of Engineering, OMR, Kalavakkam, Tamil Nadu 603110, India
e-mail: lakshminaras74@gmail.com



N. L. Narasimhan et al. (eds.), *Recent Advances in Energy Technologies*, Lecture Notes in Mechanical Engineering, https://doi.org/10.1007/978-981-19-3467-4_5

| | |
|---|---|
| $T_{rwsi}$ | Rear windshield inner temperature, °C |
| $T_{rwso}$ | Rear windshield outer temperature, °C |
| $T_{skin}$ | Cabin skin temperature, °C |

## 1 Introduction

There has been a significant rise in the demand for air-conditioned passenger cars especially in countries like India and China. Given the soaring need for thermal comfort, mobile air conditioning (a/c) has become more of a necessity than luxury in the tropical countries. On the other hand, operating air-conditioning (a/c) systems in vehicles leads to additional fuel consumption and engine emissions as well. Therefore, development of efficient a/c systems with superior performance and optimal fuel consumption is very critical in the mobility industry. The performance of a mobile a/c system is strongly influenced by the total heat gain of the vehicle. The heat gain is predominantly by the cabin materials when the vehicle was under parking or driving; depending on the intensity of the incident solar radiation and engine heat infiltrated via conduction. A larger heat gain translates into a larger heat load/heat duty to the vehicle's a/c system and has to be essentially handled by it ensuring passenger thermal comfort. In a real situation, the heat infiltration into the cabin space is highly dynamic in nature resulting in a complex heating pattern that makes it difficult for an estimation of the heat duty a priori. From a design point of view, a prior knowledge on the actual heating pattern of cabin space and the materials inside is vital for the heating, ventilation, and air-conditioning (HVAC) engineers. There have been many models developed and controlled experiments carried out in the past investigating the thermal response of passenger cars under running conditions [1–4]. Studies have shown that a 20% increase in the a/c cooling capacity is needed to decrease the cabin temperature of a passenger car by about 4 °C [4]. On the other hand, in tropical countries, the warming inside the cabin space could be quite significant under parked conditions too due to absorption of short-wave solar radiation by the cabin materials. As a consequence, heat duty of mobile a/c systems increases resulting in a prolonged cooldown time and poor passenger thermal comfort. Therefore, it is of specific interest to study in detail, the dynamic response of a passenger car's cabin environment under parked conditions. The present work aims to investigate the same for a stationary passenger car conducting real-time experiments under an open sunshine environment for four different cases described later.

As mentioned earlier, the net heat load of a car a/c system depends strongly on the active and passive heats received by the cabin space and thermal masses present inside [1]. Figure 1 shows as a representative case, the thermal load distribution for a passenger car. At starting conditions, the initial heat load handled by the a/c system would be very high, depending on the intensity of the heat gained during the entire parking duration [2]. For example, in metro cities, there could be numerous occasions when passenger cars have to be parked under open sunshine for a longer duration having all of its windows closed. During the period, the heat gained would be significant by the cabin structure, dashboard, seats/cover surfaces exposed to solar radiation, evaporating coils of the a/c system, steering wheel, and so on. In addition, a local greenhouse effect could prevail resulting in a gradual increase of cabin air temperature over a period of time. Whereas the hot air could be vented off from the cabin by partial or full opening of the windows for a short interval, the heat gained by the materials cannot be dissipated off without operating an air conditioner. Eventually, the cooldown gets prolonged, and the a/c system is severely loaded during the initial stages thus resulting in (i) excess thermal load, (ii) excess power requirement, and (iii) additional fuel consumption. The excess thermal load leads to poor performance of the a/c system and passenger discomfort as well. In tropical countries like India, where many cities experience long and hot summers and short and warm winters, passenger cars need to handle such excess heat loads quite often. An understanding on the cabin temperature variation and heating pattern is essential for proper sizing and rating of the vehicle's a/c system. While parameters like speed of the car, cool air flow rate, number of passengers, and so on influence the cabin air temperature for a moving car, direct radiation and parasitic heat infiltration via conduction are those that influence the cabin environment for a stationary (parked) car [4]. The present work attempts to study the cabin temperature variation for a stationary passenger car, having the vehicle parked in four different directions, viz. north, south, east, and west in an open arena. Temperatures were measured at several locations within the cabin as well as the cabin top surface, throughout the day for several days. Experiments were also carried out in the later part of the study, to eliminate the direct and diffuse radiations entering through the windows for investigating their influence on the cabin environment. The experiments were carried out systematically under four different cases (Case-1 to 4), the results of which are discussed later.

There have been many studies in the past carried out by different authors on passenger car cabin environment. A CFD model coupling the convection and radiation effects together with the airflow movement, heating, and air conditioning, and external climate was presented by Fujita et al. to investigate the response of cabin environment of a passenger car under different conditions [1]. The accuracy of the model was verified by carrying out experiments on a four-seater Sedan car using a

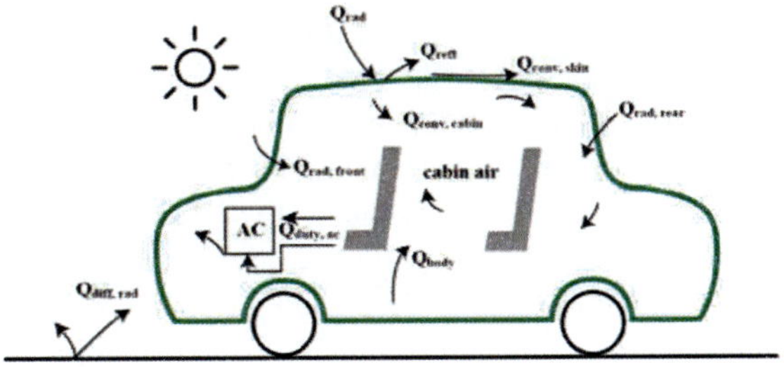


**Fig. 1** Thermal load distribution for a passenger car cabin

chassis dynamometer and model human body placed inside the cabin space. A parameter called thermal sensation vote (TSV) was introduced to correlate the temperature interval for human perception against the accuracy of the simulation model for different air flow rates. The air velocities as well as air temperatures were found to vary across knee and shoulder locations in both passenger and driver seats under heating as well as air-conditioning modes. Overall, their model was found to predict the cabin's thermal environment with a reasonable accuracy. Mezrhab and Bouzidi proposed a method to reduce the air-conditioning power requirements of a passenger car [2]. They demonstrated using a numerical simulation model that the influence of radiation infiltered and radiative properties of the car body materials was significant on the cabin's thermal environment. It was predicted that the dashboard of a stationary car exposed to direct sunshine (at Paris) for about 2 h could reach about 100 °C at 2 p.m. during summer (with peak solar radiation of about 650 W) affecting the overall performance of the indoor a/c system. On the other hand, colour of the car body was also found to have a significant influence on the cabin temperature and power drawn by the a/c system. Car body painted white resulted in lower cabin temperatures and reduced a/c power consumption compared to that painted black. Rugh et al. [3] introduced different techniques such as solar reflective glasses, solar reflective paints onto the car body, solar-powered ventilation to reduce cabin temperatures, and air-conditioning power requirements of a passenger car. With only the solar reflective coating over the cabin roof, the possible reduction in car roof skin temperature was found to be about 6.7 °C. However, the reflective coating did not influence the cabin side temperatures. On the other hand, providing only solar-powered ventilation had resulted in a decrease in the cabin air and seat temperatures by about 5–6 °C. They found that the solar reflective paint together with solar reflective glasses yielded a reduction in the cabin air, seat, windshield, and dashboard by about 9.7, 8.7, 19.7, and 14.6 °C, respectively. Further reduction by about 1–3 °C with the later technique was achieved operating a solar-powered ventilation in parallel. Through simulation, the estimated reduction in a/c power consumption was about 30%. Dadour et al. [5] presented a greenhouse model to estimate the cabin temperature rise for a parked car. Field experiments were carried out at Floreat Park, Western Australia, during summer to study the actual temperature rise within the cabin space. A set of experiments with seven passenger cars of black and white colours was carried out parking the vehicles facing east. The effect of ventilation and solar radiation infiltered on the cabin temperature was their prime focus. The cabin temperature was found to rise by 20 °C in summer when all windows were closed, whereas with windows lowered up to 2.5 cm could reduce the cabin temperature only by about 3 °C. The colour of the vehicle was indeed found to have an effect on the cabin temperature rise with black vehicle remaining at 5 °C higher than that of white vehicle under the given conditions. The cabin side heat transfer and the dynamics behind are quite complex. For a given vehicle, it requires a dynamic control of the fresh air inducted and conditioned air circulated, to realize an optimal power input to the a/c system. Fuzzy controllers have been proposed and demonstrated for its successful deployment in passenger cars to achieve faster cooling with optimal ventilation and power requirements [6]. Realizing the computational complexities involved in the estimation of car a/c loads, attempts have been made by few authors to estimate vehicle a/c loads and proposed a relatively simpler approach adopting quasi-steady lumped heat model [7, 8]. Such simpler models were found to be less intensive in computation even when actual driving cycles were integrated into the models. Pokorny et al. [9] developed a virtual testing stand (VTS)—a sophisticated code using MATLAB combining its features to predict the indoor cabin thermal environment of passenger cars. Qi et al. [10] have shown that the cabin thermal comfort of a passenger car could become better when the vehicle was parked under shade with windows opened in full. Their results showed that a 4–12 °C temperature difference could be achieved between the car cabin and ambient by parking the vehicle under a shade. They concluded the operation of air conditioners as an obvious option for achieving a desired thermal comfort under driving conditions of a passenger car.

It is evident from the literature survey that many mathematical models have been presented in the past to map the cabin's thermal environment of passenger cars under driving conditions. Only few studies focused on parked vehicles. As per the literature, the influence of heat gained during parked conditions was found to be significant on the power drawn by the air-conditioning systems of passenger cars as well as thermal comfort of passengers. To the author's knowledge, there are no studies available in the open literature that investigates the effect of parking directions on a car cabin environment. The present work attempts to address the gap and appears to be the first of its kind carried out for a tropical metropolitan city, Chennai, located in South India.

## 2 Experimentation

Real-time experiments were carried out on a white Maruti Suzuki Swift passenger car (specifications mentioned in Table 1) parking the vehicle in an open ground at our institution's campus (12° 45′ N 80° 11′ E) located near Chennai, South India. All the experiments were conducted between 10.30 a.m. and 4.00 p.m. during Feb–Mar 2017 and temperatures measured at 17 different locations (16 locations for the vehicle and one for ambient) using K-type thermocouples (uncertainty ±0.5 °C) at every 20 min time interval. All the thermocouples were calibrated before commencing the experiments, and their precisions were compared to yield confidence levels as high as 95% or more on the measurements. Table 2 lists out the different locations, viz. the dashboard, seats, rear deck, windshields (inner and outer surfaces), and cabin skin surface (facing the sky), where the temperatures were measured. Two sets of digital indicators were used to display the measured temperatures. A pyranometer was employed to measure the solar insolation on the days experiments were performed.

**Table 1** Technical specifications of the Maruti Suzuki Swift passenger car studied

| | |
|---|---|
| Engine | DDiS diesel engine |
| Capacity | 1248 cc |
| Number of cylinders | 4 |
| Bore × stroke | 69.6 mm × 82 mm |
| Compression ratio | 17.6:1 |
| Maximum power | 75 ps @ 4000 rpm |
| Emission | BSIV norms, India |

**Table 2** Location of the temperature sensors

| Location | Nomenclature | Location | Nomenclature |
|---|---|---|---|
| Front windshield inner | $T_{fwsi}$ | Central cabin air | $T_{cabin,c}$ |
| Front windshield outer | $T_{fwso}$ | Bottom cabin air | $T_{cabin,b}$ |
| Rear windshield inner | $T_{rwsi}$ | Left side front seat front surface | $T_{lfsf}$ |
| Rear windshield outer | $T_{rwso}$ | Left side front seat rear surface | $T_{lfsr}$ |
| Front dashboard (left end) | $T_{DBl}$ | Right side front seat front surface | $T_{rfsf}$ |
| Front dashboard (centre) | $T_{DBc}$ | Right side front seat rear surface | $T_{rfsr}$ |
| Front dashboard (right end) | $T_{DBr}$ | Back seat centre | $T_{bs}$ |
| Rear deck | $T_{deck}$ | Cabin top skin surface | $T_{skin}$ |

Experiments were carried out individually for the four different cases mentioned below to investigate the effect of parking direction and incident solar radiation on the cabin's thermal response.

**Case-1**: Car parked in four different directions with all windows closed (referred to as base line case to identify the most influencing direction of heating).

**Case-2**: Experiment conducted with all windows kept opened.

**Case-3**: Experiment with all windows closed and dashboard provided with a radiation shield.

**Case-4**: Experiment with radiation shields all over the windows and windshields (both front and rear).

While Case-1 was performed to identify the most influential parking direction that yields a higher cabin temperature, Case-2 was performed to find out the effect of 'open windows' on the cabin heating. Both Cases-3 and 4 were separately studied to investigate the effect of radiation incident on the dashboard and that entering through the windows/windshields, respectively. Cases-2 to 4 were both performed parking the car in the most influential direction identified from the initial Case-1 studies. The following section describes the results obtained from the experimental studies (Fig. 2).

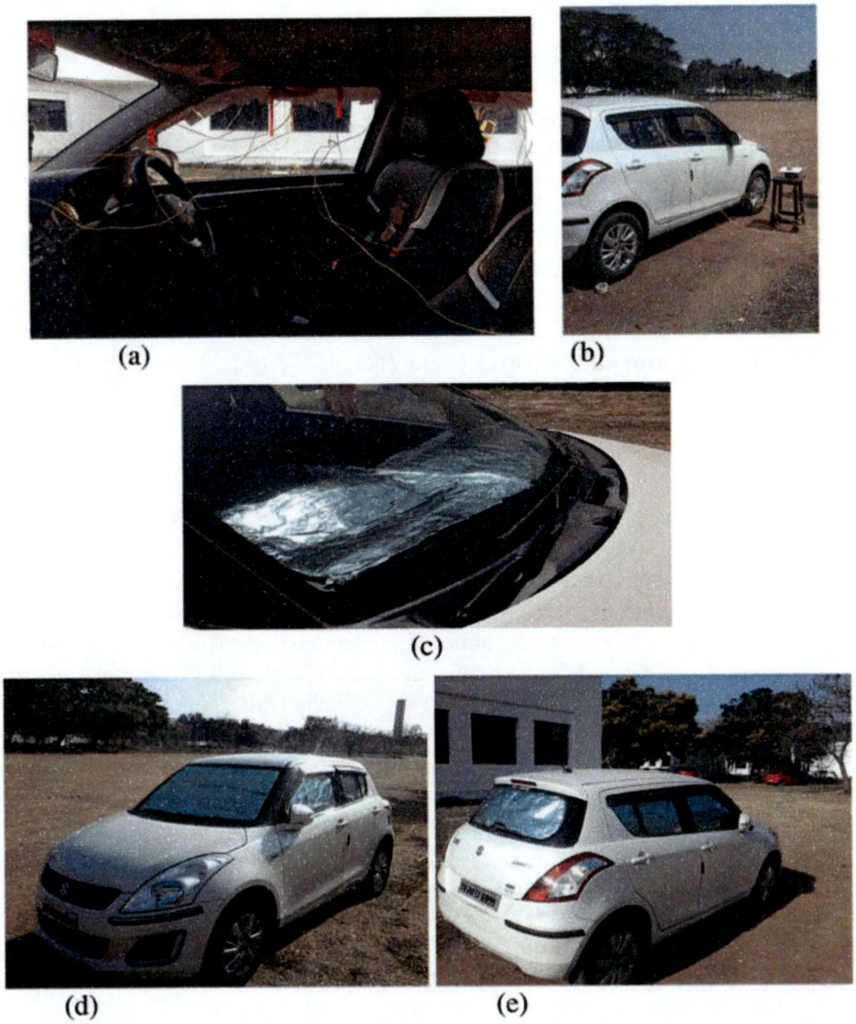

**Fig. 2** Pictorial view of the experimental setup **a** thermocouple locations within the cabin, **b** outside view of the car during Case-1 experiments, **c** view of the radiation shield on the dashboard (Case-3), **d** and **e** respective front and rear views of the radiation shielded car prepared for Case-4 studies

# 3 Results and Discussion

## *3.1 Baseline Studies (Case-1)*

Considering that a majority of passenger cars in metro cities are parked having the windows closed (for safety reasons), the initial set of experiments on the vehicle was performed with all windows closed. Four different individual experiments were performed during Feb 9–13, 2017 (except Feb 12, 2017) parking the vehicle in east, west, north, and south directions, respectively, to identify the most influencing direction that results in higher cabin temperatures. These experiments have been taken as a baseline (Case-1) to proceed with further experiments (Cases-2 to 4) described in the later part of the discussion. In order to understand the heating pattern (heat gain), temperatures at different locations, viz. inside the cabin space, on the windshield surfaces (inner and outer), seats (front and rear), dashboard, rear deck, and cabin top skin surface have been measured real time throughout the day. Discussions of the results obtained are presented below, and all the experiments reported were for the days when the peak solar insolation was between 900 and 1000 W/m$^2$.

**Windshield Temperatures (Front and Rear)**: Windshield temperatures were measured during the individual experiments carried out parking the vehicle facing north, south, east, and west directions. Figure 3 shows the temporal variation of the inner front windshield temperatures ($T_{\mathrm{fwsi}}$) studied for the four different parking directions. It can be observed that the heating pattern of windshield was different for different parking directions studied. Whereas the temperatures reached about 72 °C when the car was facing west, it was only about 50 °C when facing north, 65 and 67 °C with east and south, respectively. Compared to other directions, parking the car facing east had resulted in a rapid rise in inner windshield temperature to around 65 °C during the morning hours (10.00 a.m.). However, the temperatures were observed to settle around 65 °C for about two hours (till 12.00 p.m.). As the trace of the sun moved from overhead to west, a gradual fall in the inner windshield temperature could be observed with the east parking. On the other hand, the rise in temperature was gradual between 10.00 a.m. and 12.00 p.m. when the car was parked facing west. Unlike the parking directions north/south/east, parking west had resulted in relatively higher windshield temperature (65 °C) even during 12.00 and 3.00 p.m. The temperature started to fall gradually only after 3.00 p.m. for the west parking case. While rest of the parking directions showed a steep fall in the windshield temperatures beyond 2.00 p.m., parking west had the temperature maintained relatively higher. The lowest temperature of the windshield was obtained (45–50 °C) when the car was parked facing north. The drop in windshield's temperature was steep after noon when the car faced north/south/east, due to significant reduction in the incident radiation from mid-noon onwards. With west parking, the incident radiation from the west after mid-noon maintained the inner surface temperature of the windshield close to about 68 to 70 °C. In the case of south parking, the rise in temperature was higher than that observed for north or west parking. At around 12.30 p.m., the windshield temperatures reached about 65 °C irrespective of the parking direction being east or west or south. It can be observed that, on an average, both west and south parking directions resulted in higher windshield temperatures and hence not recommended as a suitable parking direction. At any point of time during the day, parking the car facing north is highly recommended. Similarly, beyond 1 p.m., parking the car facing east appears to be a better option as compared to west or south. It can be suggested from the trend of the curves that parking the car facing east or north is beneficial compared to parking west or south in Chennai. From this plot, it is

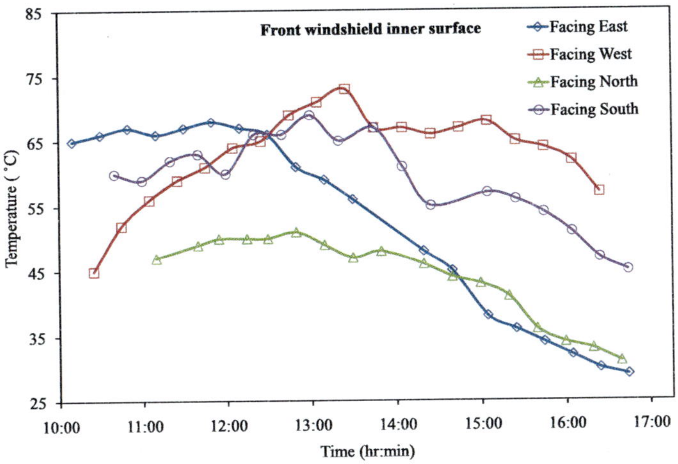


**Fig. 3** Time-wise variation of front windshield inner surface temperature

evident that the direction of parking influences the heating pattern of the windshield quite significantly. Similar to inner surface of the front windshield, the outer surface temperature of the front windshield ($T_{fwso}$) was also measured for the four parking directions. Figure 4 shows as a representative case, the measured temperatures across the inner and outer surfaces of front windshield when the car was parked facing east. The maximum temperature difference across the windshield was found to be between 20 and 25 °C. The inner surface of the windshield remained at a relatively higher temperature than that of the outer surface. Though heating at the external surface of the windshield happened due to partial absorption of the incident solar radiation, a simultaneous cooling due to natural wind movement kept the exterior at a relatively lower temperature than that of the interior as shown in Fig. 4.

The temperature rise within the cabin space was essentially due to heat transfer by all the three modes, viz. conduction, convection, and radiation. Most of the radiation received was through the front windshield made of toughened glass with emissivity around 0.7 to 0.8. While warming of the windshield happened at both inner and outer surfaces, a simultaneous cooling at the outer surface occurred due to forced convection of the ambient wind. On the other hand, the interior of the windshield experiences a local natural convective environment with relatively low values of heat transfer coefficient within the cabin space. Therefore, the interior front windshield temperature was obviously higher than that of the exterior as shown in Fig. 4.

Figure 5 shows the rear windshield temperature history for the different parking directions studied. Comparing with Fig. 3, it can be observed that the heating pattern of the windshields in the front and rear is quite different depending on the orientation of parking. The temperature of the inner surface of rear windshield ($T_{rwsi}$) reached as high as 65 °C within few minutes of parking in the west direction. On the other hand, it took about till 12.00 noon to reach up to 55–65 °C when the car was facing

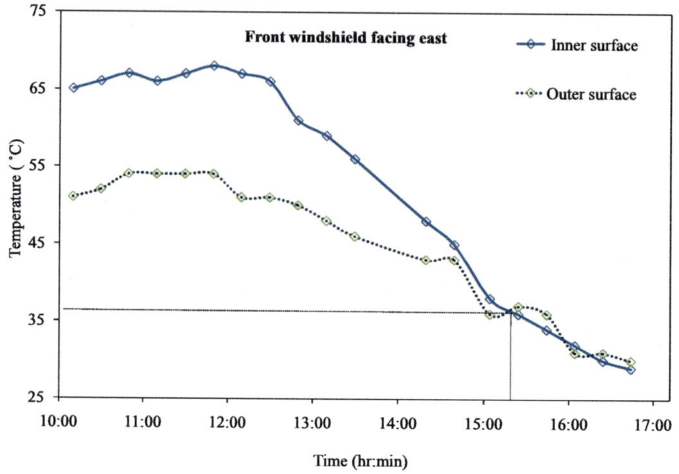


**Fig. 4** Time-wise temperature variation across front windshield while facing east

east and north, respectively. It is to be noted that the rear side of the car faced east and continued to receive the sunshine till noon when parked facing west. The reverse was the situation when the car was parked facing east as heating of rear windshield (facing west) occurred post noon due to incident solar radiation from the west. As shown in Fig. 5, the rear windshield temperature kept increasing up to 70 °C till 3.00 p.m. when the vehicle was parked facing east. Parking the vehicle south (rear windshield facing north) did not cause any significant warming of the rear windshield with temperatures staying moderately between 35 and 45 °C throughout the day.

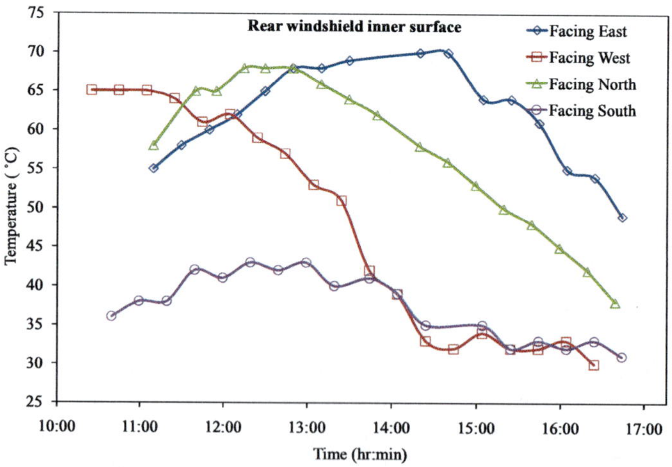


**Fig. 5** Time-wise variation of the rear windshield inner surface temperature ($T_{rwsi}$)

**Dashboard and Rear Deck Temperature**: Dashboard forms the important part within the vehicle's cabin that intercepts and absorbs a considerable portion of the incident solar radiation entering the vehicle. Being made of a special polymer material and coloured black as well, the dashboard itself was found to absorb more heat from the direct sunshine. The dashboard thus became a passive local hot source resulting in an additional thermal load for the air-conditioning system. Similarly, the rear deck though covered with felt-like material (coloured dark ash) has been found to gain heat through incident radiation depending on the direction of parking. It is to be noted that a large portion of heat entering the cabin space is retained by the dashboard, rear deck as well as the seats. Figures 6, 7, and 8 show the dashboard and rear deck temperature histories for different parking directions. It can be seen that the dashboard reached up to 80 °C at around 1.00 p.m. when the car was parked facing west (Fig. 6). The dashboard temperatures increased gradually, and on an average, the temperature remained higher (>60 °C) till 3 p.m. for all orientations except north. Figure 7 shows the temperatures at different locations on the dashboard (left, centre, and right) when the vehicle was parked facing east. It is evident from the figure that the entire dashboard remained at nearly uniform temperature throughout. A gradual reduction in the dashboard temperature was observed as the sun (incident radiation) traced from east to west (Fig. 7). A similar trend could be observed for the rear deck when the car was parked facing east or west. The rear deck peak temperature was close to 60 °C and occurred at 12.30 p.m. and 1.00 p.m., respectively, when the car was parked west and east. It was observed that there existed a similarity in the heating pattern of the rear deck between east and west parking conditions. Whereas rear deck temperatures reached up to 60 °C between 10.00 a.m. and 1.00 p.m. when the vehicle was parked facing west, a similar heating pattern was observed between 1.00 p.m. and 4.00 p.m. when the vehicle was parked facing east (Fig. 8).

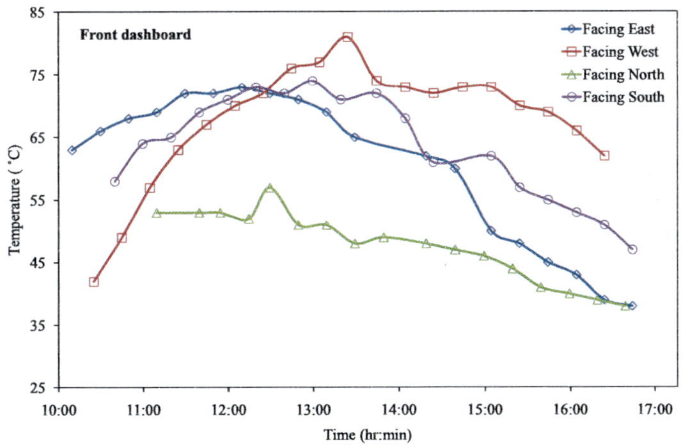


**Fig. 6** Time-wise temperature variation of front dashboard

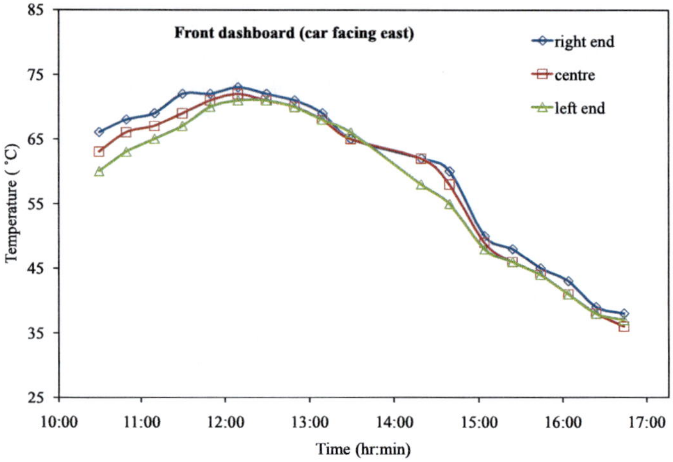


**Fig. 7** Time-wise temperature variation across front dashboard while facing east

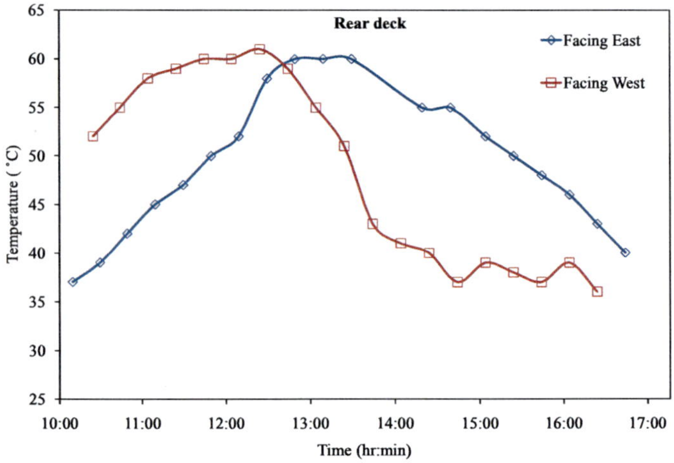


**Fig. 8** Time-wise variation of the rear deck temperature

**Cabin Air Temperature**: Figures 9 and 10 show the effect of heat gains on the cabin air temperatures measured at centre ($T_{\text{cabin,c}}$) and bottom ($T_{\text{cabin,b}}$) of the cabin, respectively. The cabin air temperatures reached well above 50°C till 12.00 p.m. and 1.30 p.m. when the car was parked facing east and west, respectively. The cabin temperature was very much oscillating when the car was parked facing south with a temperature difference about 5 to 8 °C. It can be seen that parking east, north, or south had resulted in lower central cabin air temperatures after 12.00 p.m. compared to that when car was parked facing west (Fig. 9). In the absence of forced convection within the cabin space, the air remained quiescent with the heat transfer taken care predominantly by natural convection. Due to density difference, the low-temperature (high density) air remained at the bottom of the cabin, while warmer (low density) air settled at the top within the cabin space. It can be observed from Figs. 9 and 10 that the temperature at the bottom of the cabin essentially remained lower than the centre by about 10 °C. Thus, it is evident that an appreciable temperature gradient exists within the cabin space due to heat gain from the dashboard, seats, side walls, and rear deck.

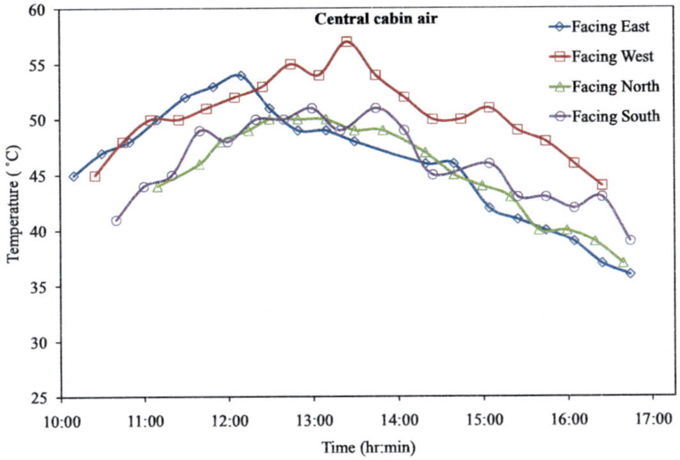


**Fig. 9** Time-wise variation of central cabin air temperature

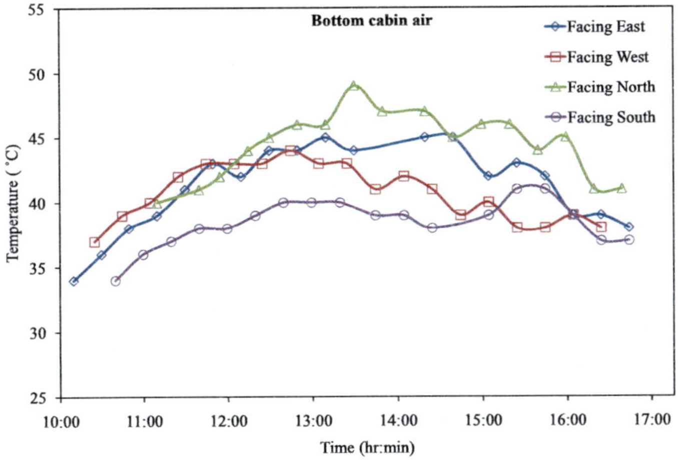


**Fig. 10** Time-wise variation of the bottom cabin air temperature

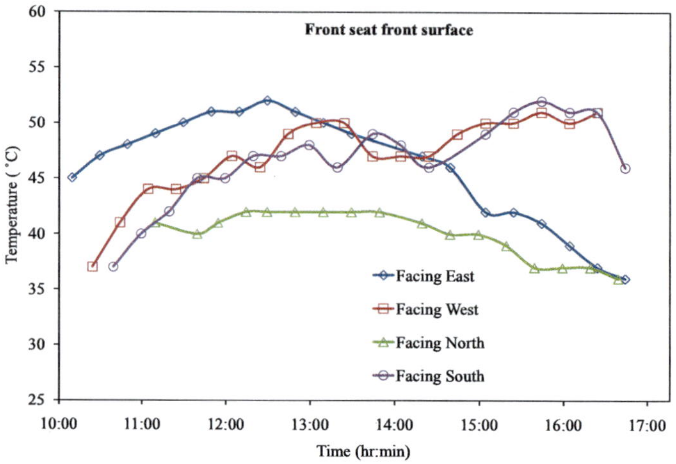


**Fig. 11** Time-wise variation of the front seat front surface temperature

**Front Seat Temperature**: Figures 11 and 12 show the temperatures measured across the front seat of the car. It can be observed that the seat temperature rose to about 45 °C and above for most of the period irrespective of the directions except north. The heat gained was minimal when the car was parked facing north, and therefore, the temperature was found to be below 40 °C for quite a longer period (Fig. 11). Figure 12 shows the temperature gradient across the front and rear surfaces of the front seat when the car was parked west. The temperature difference across the seat was found to remain at about 5 °C until 1.00 p.m. with the temperature of the rear surface higher than that of the front. Beyond 2.00 p.m., the front surface temperature rose rapidly due to direct sunshine from the west resulting in a 10 °C temperature gradient across the front seat.

**Skin Temperature ($T_{skin}$)**: A substantial portion of the solar radiation incident on the vehicle is intercepted by its skin surface facing the sky. While part of the radiation is absorbed, the rest of gets reflected back depending of the quality of paint applied and its colour. Larger the heat absorbed by the top surface of the car, larger would be the parasitic heat load diffused into the cabin space. Present-day cars are painted with special paints that reflect back the incident radiation minimizing the internal parasitic thermal gain of the cabin space. However, the thermal gain due to absorption cannot be avoided due to limitations of the paint material and colour characteristics. The top skin temperature was measured during the experiments and found to be around 50 °C on an average despite the outside ambient convective cooling when the car was facing east/west/north (Fig. 13). The measured skin temperature was relatively lower (40–45 °C) as shown in Fig. 13 when the car was parked facing south.

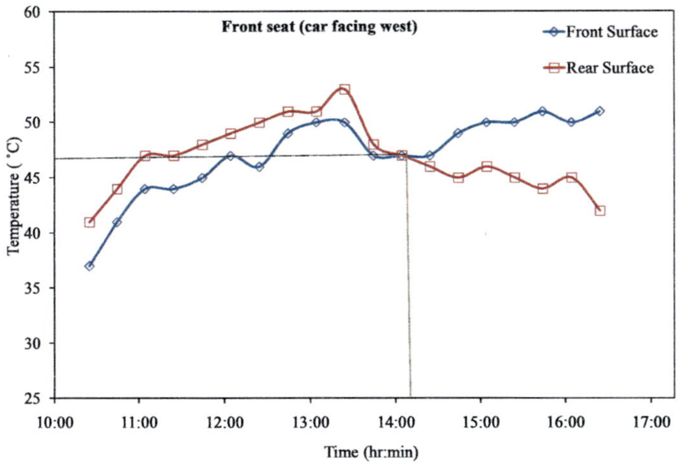


**Fig. 12** Time-wise temperature variation across the front seat while facing west

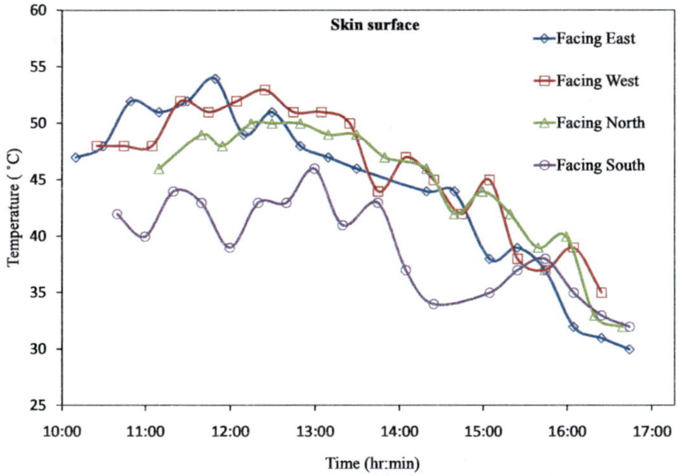


**Fig. 13** Time-wise variation of top skin surface temperature

## 3.2 *Case-2 (Studies with All Windows Opened)*

The base line study (Case-1) was carried out with all windows closed to investigate the actual scenario of a parked vehicle in metro cities. The results show that the temperatures of cabin space as well as materials inside the cabin temperatures stay reasonably higher (between 40 and 50 °C) for quite a long duration irrespective of the direction of parking. Thus, the situation leads to passenger discomfort as well as overloading of the air-conditioning system. One of the easiest methods of handling the issue is to allow natural air into the cabin space by keeping the windows wide open. Despite the fact that safety gets compromised with wide open windows for stationed cars, it can be expected to improve passenger comfort as well as reduce parasitic thermal loads with a natural flow of air through the window openings. Therefore, experiments similar to Case-1 were carried out in Case-2 with an exception that all the windows were kept fully opened unlike in Case-1. The base line experiments revealed that parking the vehicle west had resulted in a higher cabin air temperature. Hence, Case-2 experiments were carried out parking the car facing west with all windows kept open from 10.30 a.m. to 4.00 p.m. (conducted on 24/02/17). Similar to Case-1, the temperatures were measured at every 20 min time interval for Case-2 experiment. As a representative case, Figs. 14 and 15, respectively, show the measured dashboard and cabin air temperatures comparing the two Cases-1 and 2. It can be seen that allowing the ambient air through the side windows had resulted in lowering the dashboard temperature quite significantly in spite of the heating happening by solar radiation. The air movement within the cabin space enhanced the local convective heat transfer mechanism quite appreciably. While the dashboard temperatures reached as high as 80 °C with all windows closed (Case-1), it was only around 50 °C when the windows were fully opened (Case-2). Thus, it is evident that opening of windows has a significant effect on the heating pattern of parked vehicles. It is worth to mention at this point that, keeping the windows opened throughout the day may not be practically viable for stationed cars due to reasons including safety and personal as well. Quite interestingly, with open window condition, the central cabin air temperature remained appreciably lower and closer to the ambient throughout the entire day (30–35 °C) as shown in Fig. 14. On a comparative scale, the cabin air temperature was higher by about 15–20 °C when the windows remained closed. Case-2 study revealed that the opening of the windows during parking had resulted in lowering the temperatures at all locations inside the cabin space by about 15–20 °C as compared to closed windows (Case-1). As the windows cannot be kept opened in parked passenger cars, there is a need for minimizing thermal gains that results in higher temperatures as shown in Case-1.

### 3.3 *Case-3 (Dashboard with Radiation Shield)*

It is evident from the previous discussions (Figs. 6, 7, and 14) that heat gained by the dashboard is quite significant resulting in temperatures as high as 80 °C. The dashboard thus becomes a pain point for designers as it is impossible to avoid its heating due to direct solar radiation unless it is prevented from receiving the same. Therefore, experiments were conducted covering the entire dashboard surface with a radiative shield made of polished alumina as shown in Fig. 2c to minimize the effects of direct radiation. Similar to Case-2, Case-3 experiments were also conducted parking the vehicle west and temperatures measured at every 20 min between 10.30 and 4.00 p.m. on 17/03/2017. It can be observed from Fig. 16 that the radiation shield on the dashboard had resulted in a phenomenal drop in its temperature (by about

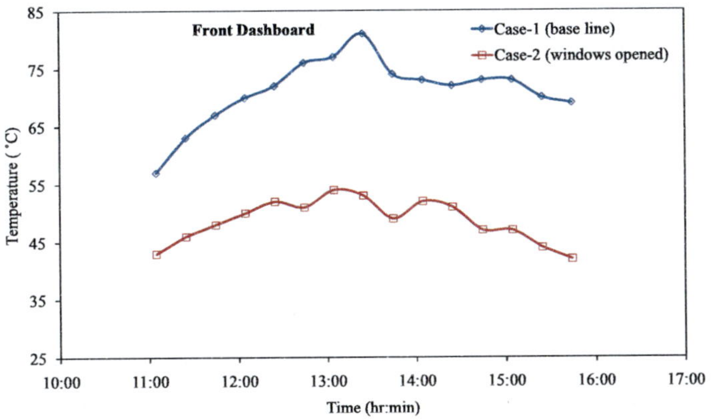


**Fig. 14** Effect of window opening on dashboard temperature

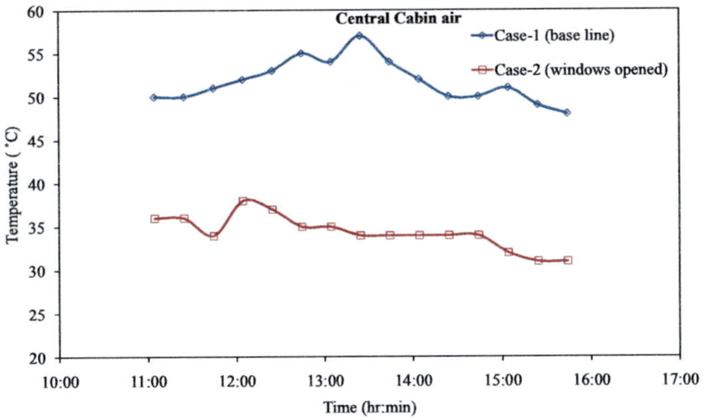


**Fig. 15** Effect of window opening on cabin air temperature

25 °C) when the vehicle was parked facing west. While the maximum temperature reached was about 80 °C without shielding (Case-1), it was only about 55 °C when the dashboard was shielded with a radiation shield. Similarly, the cabin air temperature got reduced by about 5–8 °C as shown in Fig. 17 due to the presence of radiation shield on the dashboard.

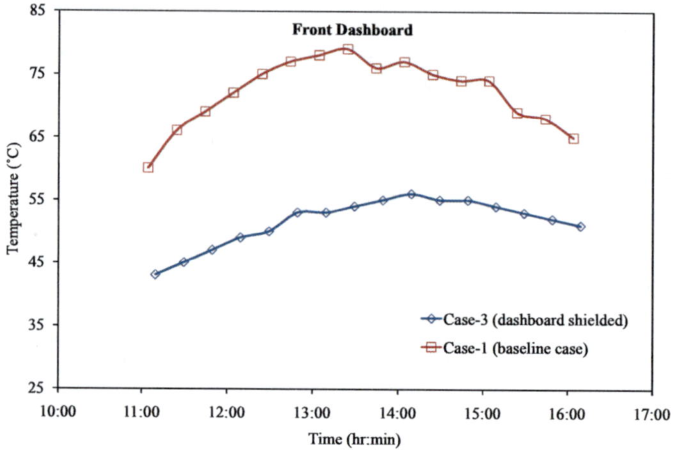


**Fig. 16** Effect of dashboard radiation shield on dashboard temperature

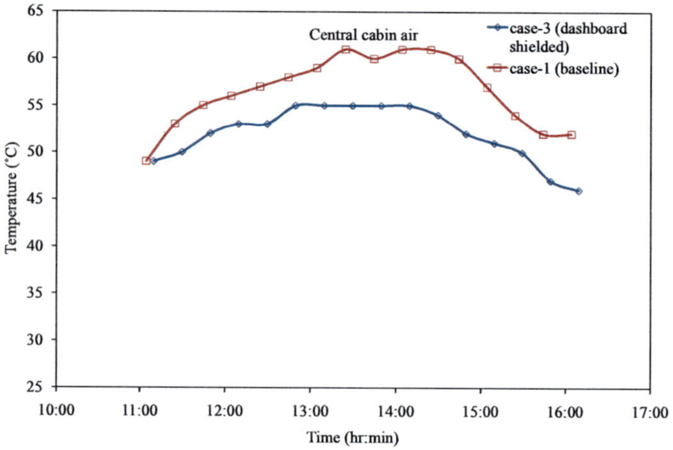


**Fig. 17** Effect of dashboard radiation shield on cabin air temperature

## *3.4 Case-4 (All Windows and Front/Rear Windshields Radiation Shielded)*

Noting the significant reduction in cabin temperatures obtained with radiation shield on the dashboard alone (Case-3), further experiments were conducted with complete shielding of all the windows and front/rear windshields (Case-4), thereby eliminating the influence of radiation on the cabin environment. Figures 2d, e show the vehicle with all windows and front and rear windshields completely covered with radiation shields made of polished alumina. Thus, the effect of radiation on the cabin side

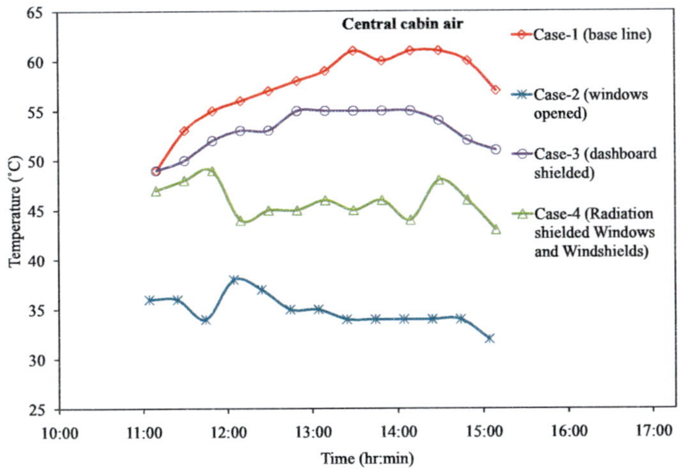


**Fig. 18** Comparison of cabin air temperature for the four different cases studied

was almost eliminated ensuring the modes of heat transfer to be dominated by both conduction and natural convection. Similar to Cases-2 and 3, Case-4 experiments were performed parking the vehicle west with temperatures measured at every 20 min time interval between 10.00 and 4.00 p.m. on 25/03/2017.

Figure 18 shows the comparison of central cabin air temperature ($T_{\text{cabin,c}}$) history between the four cases studied. It can be observed that the complete radiation shielding of windshields and windows (Case-4) had resulted in a significant drop in the cabin air temperature by about 15 to 20 °C as compared to Cases-1, 3, and 5. Yet, the best reduction in cabin side temperatures was achieved only with Case-2, where all the windows were kept open. Thus, it is evident from Figs. 3, 4, 5, 6, 7, 8, 9, 10, 11, 12, 13, 14, 15, 16, 17, and 18 that the direction of parking and heat gain of the dashboard and seat materials influences the cabin environment of a passenger car quite significantly. The higher cabin air temperature measured under parked conditions obviates the abundant scope for addressing the thermal load issues of passenger car HVAC systems in order to achieve fuel economy, thermal comfort, and emission reduction.

## 4 Conclusion

The thermal response of a stationary passenger car cabin (Maruti Suzuki Swift) under open parking condition at a location near Chennai, India, was studied experimentally. As a baseline study (Case-1), the influence of parking direction on the cabin environment was investigated parking the vehicle in four different directions. Based on the results, it was found that the warming of the cabin space was relatively less when the vehicle was parked facing north direction, whereas parking the vehicle facing west had resulted in a higher heating of the cabin space resulting in higher temperatures. The study was further extended parking the car in west direction, to understand the effect of full opening of the car windows (Case-2), radiation shield on dashboard alone (Case-3), and radiation shield on all the four windows and front/rear windshields (Case-4) on the dynamics of cabin air and dashboard temperatures. The following conclusions have been drawn based on the actual experimental results obtained:

(1) Heat gain was essentially high in dashboard, rear deck, and seats resulting in higher temperatures within the car's cabin space.
(2) Under closed window parking, dashboard temperature reached as high as 85 °C, whereas cabin air temperature and windshields reached about 60–65 °C.
(3) Indeed, the extent of heating within the cabin space was found to vary significantly with the direction of parking. Cabin air, dashboard, seats, and windshields were all found to be influenced by the direction of parking of the vehicle.
(4) For the location studied, parking the car west has been found to result in higher temperatures inside the cabin compared to rest of the three directions. Parking the vehicle north was found to be favourable resulting in relatively lower temperatures. Hence for the location Chennai (India), parking direction west is not recommended, while north is recommended to avoid high parasitic heat gains.
(5) Dashboard was found to have a uniform temperature distribution along its surface.
(6) Similar to dashboard, rear deck also showed a significant heat gain and similar heating pattern.
(7) Cabin air temperature was found to be non-uniform. A temperature gradient of about 10 °C prevailed within the cabin space from top to bottom.
(8) The temperature of seats reached as high as 45 °C with a temperature gradient of about 5 °C across its front and rear sides.
(9) Irrespective of the direction, the car's outer skin temperature reached about 50 °C during the summer period considered for the study.
(10) Keeping all the windows wide open minimized the thermal gain of the cabin space maintaining the cabin air temperatures close to ambient (30 to 35 °C) throughout the day. There was significant drop in the dashboard temperatures (by about 30 °C) due to wide open windows. Yet, the maximum dashboard temperatures were still higher (about 50 to 55 °C) with wide open windows. Therefore, radiation shielding of the dashboards during parking is highly recommended to keep temperatures reasonably close to ambient.
(11) With a proper thermal management like the radiation shielding of windows, windshields, and dashboard, the maximum temperatures of the cabin space and components inside could be brought down to less than 55 °C resulting in a reduction of unnecessary heat gain. It is hoped that the present experimental results will be useful in understanding the real effect of parking direction on

the thermal gains of a passenger car and serve as a tool for further R&D in line with the scope of the work.

**Conflict of Interest** The authors wish to state that there is no conflict of interest with any individual or any firm/organization as regards the present work.

## References

1. Fujita A, Kanemaru J, Nakagawa H, Ozeki Y (2001) Numerical simulation method to predict the thermal environment inside a car cabin. JSAE Rev 22:39–47
2. Mezrhab A, Bouzidi M (2006) Computation of thermal comfort inside a passenger car compartment. Appl Therm Eng 26:1697–1704
3. Rugh J, Chaney L, Lustbader J, Meyer J (2007) Reduction in vehicle temperatures and fuel use from cabin ventilation, solar-reflective paint, and a new solar-reflective glazing. SAE Technical Paper 2007-01-1194
4. Sanaye S, Dehghandokht M (2011) Thermal modelling for prediction of automobile cabin air temperature. Int J Autom Eng 1(3):152–164
5. Dadour IR, Almanjahie I, Fowkes ND, Keady G, Vijayan K (2011) Temperature variations in a parked vehicle. Forensic Sci Int 207:205–211
6. Sanaye S, Dehghandokht M, Fartaj A (2012) Temperature control of a cabin in an automobile using thermal modeling and fuzzy controller. Appl Energy 97:860–868
7. Fayazbakhsh M, Bahrami M (2013) Comprehensive modeling of vehicle air conditioning loads using heat balance method. SAE Technical Paper 2013-01-1507
8. Lee H, Hwang Y, Song I, Jang K (2015) Transient thermal model of passenger car's cabin and implementation to saturation cycle with alternative working fluids. Energy 90:1859–1868
9. Pokorny J, Fiser J, Jicha M (2014) Virtual testing stand for evaluation of car cabin indoor environment. Adv Eng Softw 76:48–55
10. Qi C, Helian Y, Liu J, Zhang L (2017) Experiment study on the thermal comfort inside a car passenger compartment. Procedia Engineering 205:3607–3614

# Solar Photovoltaics

# An Extended Boost Topology of Z-Source Inverter Suitable for PV Interfacing

**R. Ramaprabha and E. Oliviya Joselin Komagal**

## 1 Introduction

In today's world, a rise in power demand along with the decline of fossil resources has paved a way for photovoltaics. A crucial role is performed by the inverters in the extraction of the alternating current (ac) output from the direct current (dc) output of PV systems. The broad categories of conventional inverters are voltage source (VSI) and current source (CSI), but their output range is still insufficient. Therefore, they require an extra buck/boost dc–dc converter for some applications. In addition, the occurrence of short circuit or open circuit conditions in these inverters directly affects the input sources. Z-source inverter (ZSI) is the new topology introduced into the power category [1], which rules out the restrictions of conventional inverters. It can buck/boost the voltage, so it does not need any extra dc–dc converter and mainly can raise the voltage to the desired level using shoot-through (ST) condition, which is banned in conventional VSIs.

## 2 Z-Source Inverter (ZSI)

Power-conditioning units in renewable energy sources, especially photovoltaic [2], necessitate a concept of power conversion from dc to ac using ZSI that is a very favorable one-stage buck-boost inverter topology. ZSI has the potential to allow the inverter's dc-link to be shorted, which is known as "shoot-through". This distinct shoot-through characteristic renders the boosting. ZSI depicted in Fig. 1 consists of capacitors $C_1$ and $C_2$ and inductors $L_1$ and $L_2$ affixed in X shape that constitute the

R. Ramaprabha (✉) · E. O. J. Komagal
Department of Electrical and Electronics Engineering, Sri Sivasubramaniya Nadar College of Engineering, Kalavakkam, Tamilnadu 603110, India
e-mail: ramaprabhar@ssn.edu.in

 
N. L. Narasimhan et al. (eds.), *Recent Advances in Energy Technologies*, Lecture Notes in Mechanical Engineering, https://doi.org/10.1007/978-981-19-3467-4_6

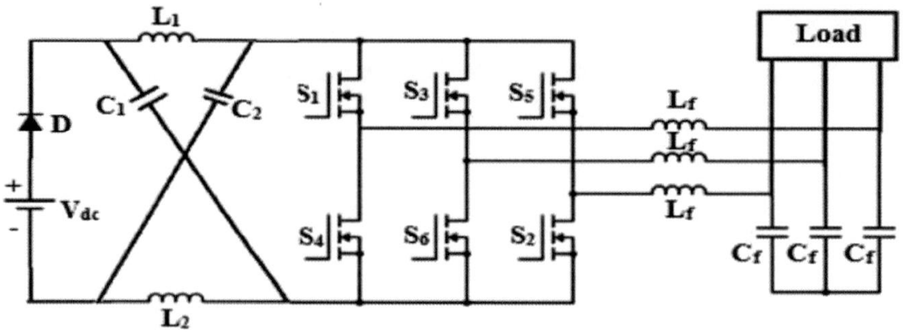


**Fig. 1** ZSI connected to a three-phase load

two-port impedance network [3] of the inverter. This network is surrounded by a dc source with a diode on the left end whereas by an inverter bridge on its right. The special impedance network makes the ZSI a buck-boost-type inverter. Moreover, it provides the inverter with privileges that lack in conventional types of inverters.

Usually, the three-phase VSI operation is well depicted with six active and two zero (non-active) vectors, where input voltage appears across the load when any one state of the vectors is active. The non-active "zero" state is produced when the top or bottom three devices are turned on together and so the output voltage is zero. Besides these, in ZSI, one extra zero state is attained in both devices by turning on a similar phase either in one leg or in a combination of two or all three legs. Therefore, there exists an extra seventh vector to generate an extra zero state. This additional zero state is labeled as the ST state, which is undesirable in VSI, but here, this distinct zero state contributes to the boosting nature of the inverter.

There are two modes of operation, which are described as follows:

## 2.1 Non-shoot-Through (nST) Mode

Here, the devices in the same leg will not conduct simultaneously. This mode consists of active and zero states (shown in Table 1). The inverter bridge operates in any one of the six vectors during the active state and conceivably be regarded as a current source. Since the structure is symmetrical, similar current values flow through both the inductors. The dc input is built over the capacitors and the inductors in this mode. The capacitor is charged, and the inductor passes the energy to the load.

The bridge operates in one of the two zero vectors during the zero state. Here, the top three or bottom three devices are shorted with the bridge as forming an open circuit. Again, the input voltage is built over the impedance network, but with no current passing to the load. Figure 2 shows the circuit during active and zero states, respectively.

**Table 1** Switching states of ZSI

| Modes | States | $S_1$ | $S_4$ | $S_3$ | $S_6$ | $S_5$ | $S_2$ |
|---|---|---|---|---|---|---|---|
| Active | $A_1$ | 1 | 0 | 0 | 1 | 0 | 1 |
| | $A_2$ | 1 | 0 | 1 | 0 | 0 | 1 |
| | $A_3$ | 0 | 1 | 1 | 0 | 0 | 1 |
| | $A_4$ | 0 | 1 | 1 | 0 | 1 | 0 |
| | $A_5$ | 0 | 1 | 0 | 1 | 1 | 0 |
| | $A_6$ | 1 | 0 | 0 | 1 | 1 | 0 |
| Zero | $Z_1$ | 1 | 0 | 1 | 0 | 1 | 0 |
| | $Z_2$ | 0 | 1 | 0 | 1 | 0 | 1 |
| Shoot-through | $ST_1$ | 1 | 1 | $\overline{S_3}$ | $\overline{S_3}$ | $\overline{S_5}$ | $\overline{S_5}$ |
| | $ST_2$ | $\overline{S_1}$ | $\overline{S_1}$ | 1 | $\overline{S_3}$ | $\overline{S_5}$ | $\overline{S_5}$ |
| | $ST_3$ | $\overline{S_1}$ | $\overline{S_1}$ | $\overline{S_3}$ | $\overline{S_3}$ | 1 | 1 |
| | $ST_4$ | 1 | 1 | 1 | 1 | $\overline{S_5}$ | $\overline{S_5}$ |
| | $ST_5$ | $\overline{S_1}$ | $\overline{S_1}$ | 1 | 1 | 1 | 1 |
| | $ST_6$ | 1 | 1 | $\overline{S_3}$ | $\overline{S_3}$ | 1 | 1 |
| | $ST_7$ | 1 | 1 | 1 | 1 | 1 | 1 |

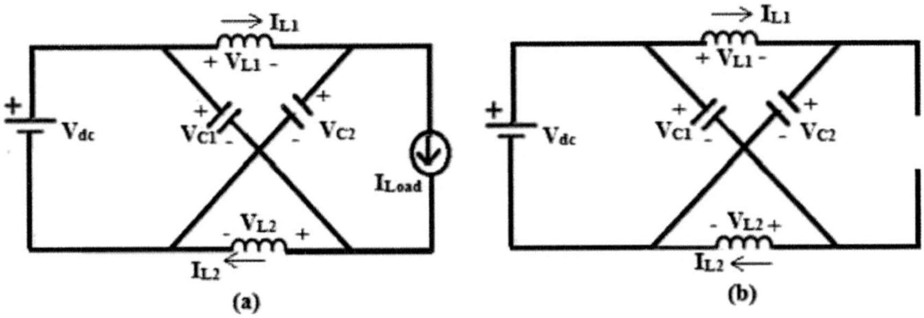


**Fig. 2** **a** Active state **b** zero state

## 2.2 *Shoot-Through (ST) Mode*

Here, devices in the same leg conduct simultaneously. The bridge operates in one of the seven ST vectors (given in Table 1) in this state. In this state, the bridge circuit is perceived to be shorted from the input link of the inverter. Like the zero-state operation, no voltage is found across the load during this mode, but the dc value of the capacitor is raised to a specific level based on the shoot-through duty ratio. Also, the capacitors are connected in parallel with inductors. Thus, the charge is transferred from the capacitors to inductors causing the inductor currents to build up. This makes the diode reverse biased and hence turned off segregating the source and the dc-link as represented in Fig. 3. Therefore, this mode helps to enhance the

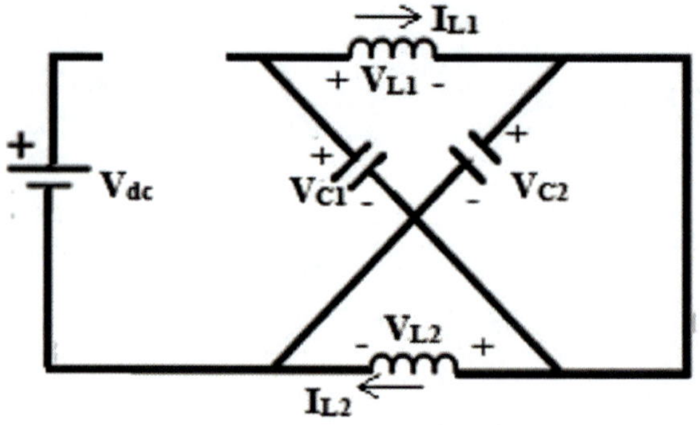


**Fig. 3** ZSI during ST state

voltage at times when a PV module fails to supply the appropriate voltage or in times of any voltage sags owing to temperature fluctuations and deviating solar irradiance. The boosting capacity of the inverter is connected to the ST interval, which can be altered according to the requirement.

## 3 Quasi Z-Source Inverter (qZSI)

With a slight change in the topology from the conventional ZSI, a new structure is obtained namely quasi Z-source inverter (qZSI) [4], in accordance with the qualities of ZSI, such as deduction of power conditioning, inversion, and buck-boost in one stage with better performance. Besides these, the lesser component ratings and the continuous input current in qZSI make it much more beneficial and superior. Figure 4 shows the quasi ZSI. The output voltage ($V_{ac}$) for both ZSI and qZSI is given by the equation,

$$V_{ac} = BM\frac{V_{dc}}{2} \tag{1}$$

where $M$ stands for modulation index and $V_{dc}$ stands for dc input voltage. Using the boost factor ($B$), the duty ratio of shoot-through ($D_{sh}$) can be calculated as given by the equation,

$$B = \frac{1}{1 - D_{sh}} \tag{2}$$

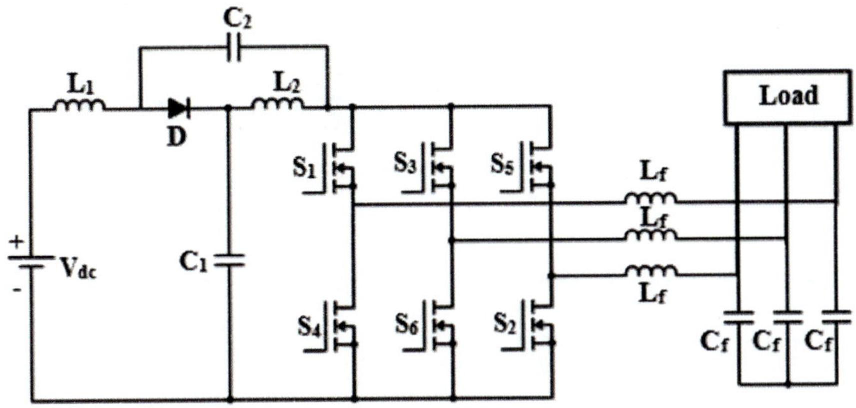


**Fig. 4** Quasi Z-source inverter

## 4 Diode-Assisted Extended Boost Quasi Z-Source Inverter (DAEBQZSI)

A DAEBQZSI is a derivation of extended boost ZSI [5], shown in Fig. 5.

DAEBQZSI incorporates an added impedance network apart from conventional ZSI or qZSI topology. Inductor ($L_3$), capacitor ($C_3$), and two diodes ($D_1$ and $D_2$) are present in the additional network; as an outcome, the impedance network is a set $L_1 - L_3$, $C_1 - C_3$, and $D_1 - D_3$ that enhances the boosting function. This network is positioned connecting the source and the inverter bridge. The operating modes are explained below:

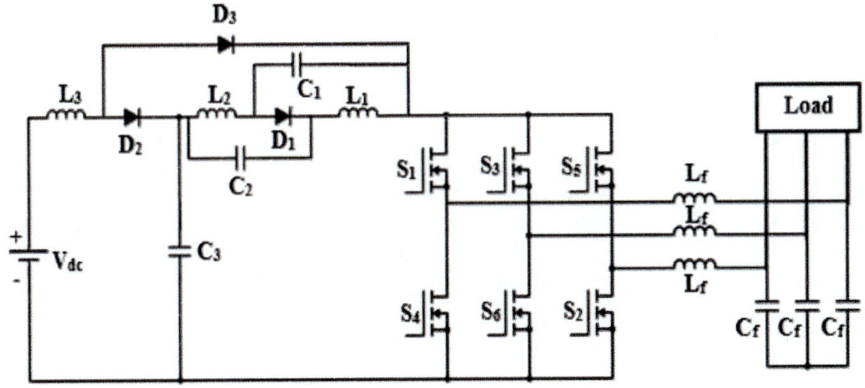


**Fig. 5** DAEBQZSI connected to a three-phase load

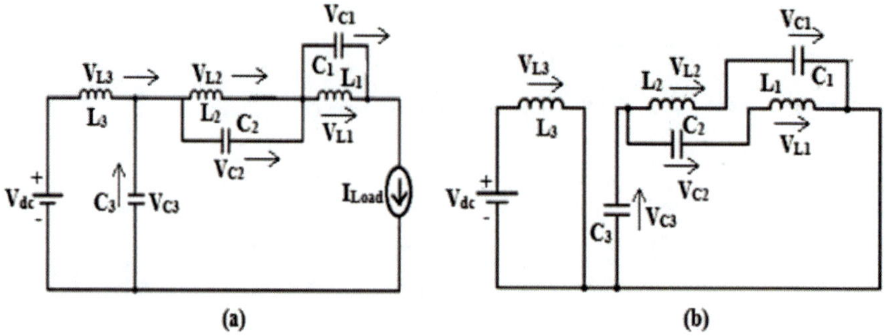


**Fig. 6** DAEBQZSI operation **a** mode-1 **b** mode-2

### 4.1 Mode-1 (nST Mode)

Figure 6a shows the mode-1 operation of DAEBQZSI which is an nST mode, where the inverter is regarded as an open switch. In this mode, $D_1$ and $D_2$ are in ON condition, while $D_3$ is OFF. Consequently, the inductors get discharged, whereas the capacitors are charged.

### 4.2 Mode-2 (ST Mode)

The structure depicted in Fig. 6b shows the mode-2 operation of DAEBQZSI, which is the ST mode, where the inverter is regarded as a closed switch. In this mode, $D_3$ is in ON condition, while $D_1$ and $D_2$ are OFF. Consequently, the inductors get charged by the capacitors, thereby reducing their stored energy. The output voltage of ac from the DAEBQZSI is expressed as,

$$V_{\text{ac}} = BM\frac{V_{\text{dc}}}{2} \tag{3}$$

where $B$ can be obtained from the equation below,

$$B = \frac{1}{(1 - 2D_{\text{sh}})(1 - D_{\text{sh}})} \tag{4}$$

## 5 Simulation Results

The DAEBQZSI is simulated along with qZSI in this section, where pulse generation is accomplished using simple boost control [6]. The parameters of simulation and its value are listed in Table 2.

The output voltage waveforms of qZSI in nST mode without filter and with filter are depicted in Fig. 7. Shoot-through pulses are depicted via Fig. 8a. The output voltage ($V_o$) waveforms of qZSI in ST mode without filter and with filter are depicted in Figs. 8b, c. The capacitor voltages $V_{C1}$ and $V_{C2}$ of qZSI are shown in Fig. 9.

The output voltage waveforms of DAEBQZSI in nST mode without and with filter are presented in Fig. 10. The output voltage waveforms of DAEBQZSI in ST mode without and with filter are as shown in Fig. 11.

The capacitor voltages $V_{C1,}$ $V_{C2}$, and $V_{C3}$ of DAEBQZSI are shown in Fig. 12.

PV array sizing for the system described above is 9 × 1. PV modeling [7–9] is done for a single panel and 9 × 1 array using the mathematical modeling equation with their parameters listed in Table 3, and their PV and PI characteristics are shown in Fig. 13.

Since the diode-assisted topology is found to have a higher boosting capability, the boosted output voltage after interfacing with the PV array is obtained. For a PV

**Table 2** Simulation parameters

| Parameters | Value |
|---|---|
| Voltage at input ($V_{dc}$) | 150 V |
| Inductors | 250 μH each |
| Capacitors | 400 μF each |
| Switching frequency ($f_s$) | 10 kHz |
| Modulation index ($M$) | 0.642 |
| Duty ratio of shoot-through ($D_{sh}$) | 0.358 |
| Filter inductors ($L_f$) | 21.3 mH each |
| Filter capacitors ($C_f$) | 450 μF each |
| $R$ load | 9.7 Ω |

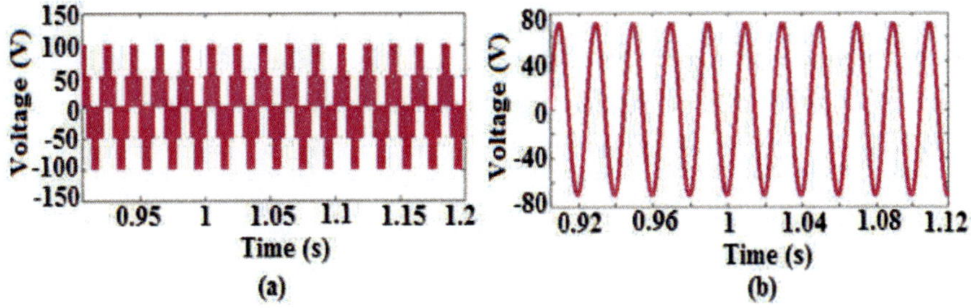


**Fig. 7** Output voltage of qZSI in nST mode **a** without filter **b** with filter

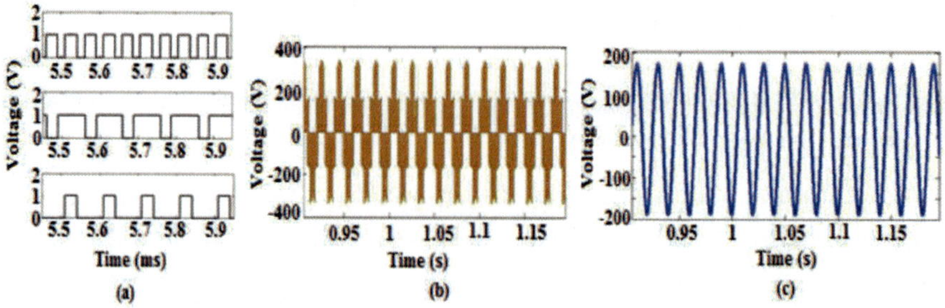


**Fig. 8** **a** Shoot-through pulses; output voltage of qZSI in ST mode **b** without filter **c** with filter

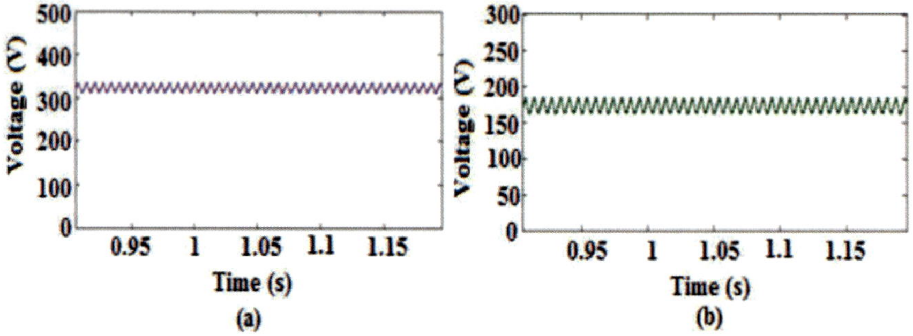


**Fig. 9** Capacitor voltage of qZSI **a** $V_{C1}$ **b** $V_{C2}$

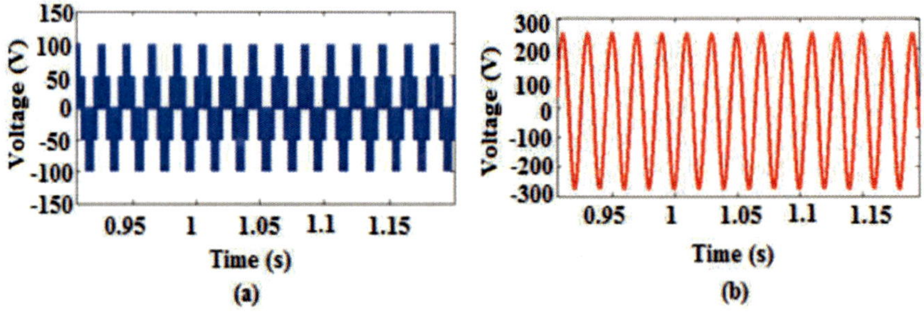


**Fig. 10** Output voltage of DAEBQZSI in nST mode: **a** unfiltered **b** filtered

input voltage ($V_{mp}$) of 149 V, the filtered inverter output voltage is observed as 258 V (Fig. 14).

The output voltages for the shoot-through mode of the qZSI and the DAEBQZSI are compared and presented in Table 4.

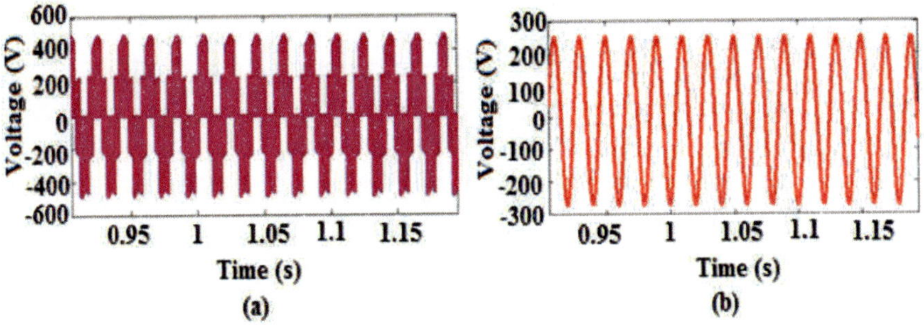


**Fig. 11** Output voltage of DAEBQZSI in ST mode: **a** unfiltered **b** filtered

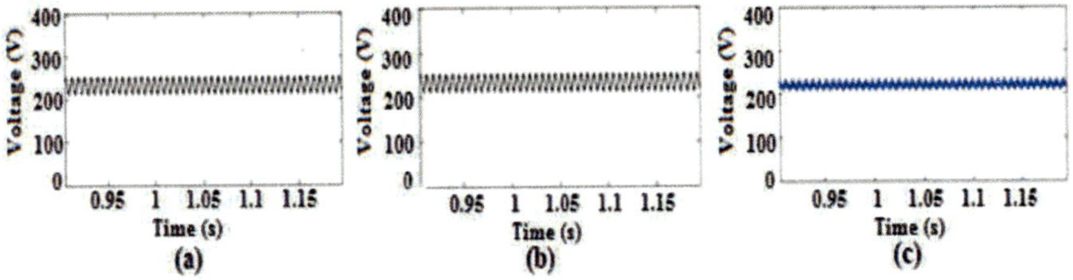


**Fig. 12** Capacitor voltage of DAEBQZSI **a** $V_{C1}$ **b** $V_{C2}$ **c** $V_{C3}$

**Table 3** PV array parameters

| Parameters | Value |
|---|---|
| Rated power ($P_{max}$) | 330 W |
| Max power voltage ($V_{mp}$) | 149.04 V |
| Max power current ($I_{mp}$) | 2.25 A |
| Voltage across open circuit ($V_{oc}$) | 191.16 V |
| Current through short circuit ($I_{sc}$) | 2.55 A |
| PV array size | 9 × 1 |

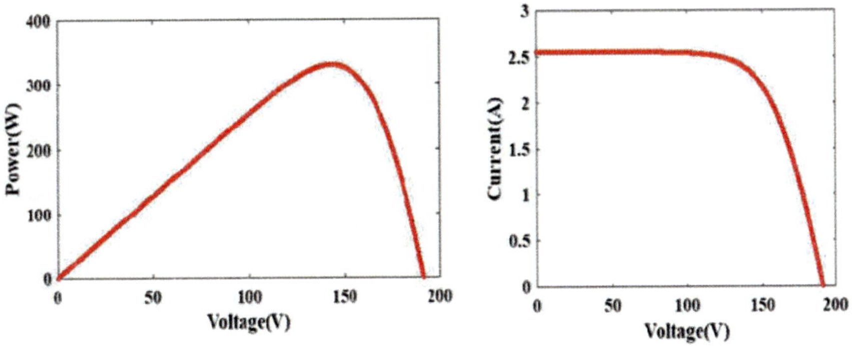


**Fig. 13** Characteristics of 9 × 1 PV array

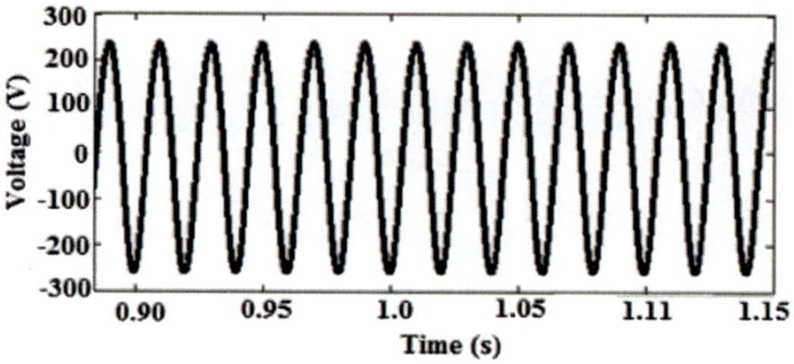


**Fig. 14** PV-interfaced inverter output voltage

**Table 4** Comparison table

| Parameters | qZSI | | DAEBQZSI | |
|---|---|---|---|---|
| Input dc voltage | 150 V | | 150 V | |
| Boost factor | $B = \frac{1}{1-D_{Sh}}$ | | $B = \frac{1}{(1-2D_{Sh})(1-D_{Sh})}$ | |
| | 3.52 | | 5.48 | |
| Output ac voltage | $V_{ac} = BM\frac{V_{dc}}{2}$ | | $V_{ac} = BM\frac{V_{dc}}{2}$ | |
| | Theoretical | Simulated | Theoretical | Simulated |
| | 169.5 V | 171.1 V | 263.8 V | 260 V |

## 6 Experimentation

The hardware setup of DAEBQZSI is depicted in Fig. 15. The gating pulses are generated using STM32F103C8T6 processor. MOSFET IRF460, diodes 1N4008, optocoupler TLP250 with rated values of passive components listed in Table 2 are used for realizing the experimental setup. The generated pulses are shown in Fig. 16. The output voltage without a filter in both modes is shown in Fig. 17, and the inverter voltage after the filter is shown in Fig. 18.

## 7 Results and Discussion

It can be viewed from Figs. 10, 11, 17, and 18 and Table 4 that the experimentation results closely follow that of the simulation results as shown in Table 5.

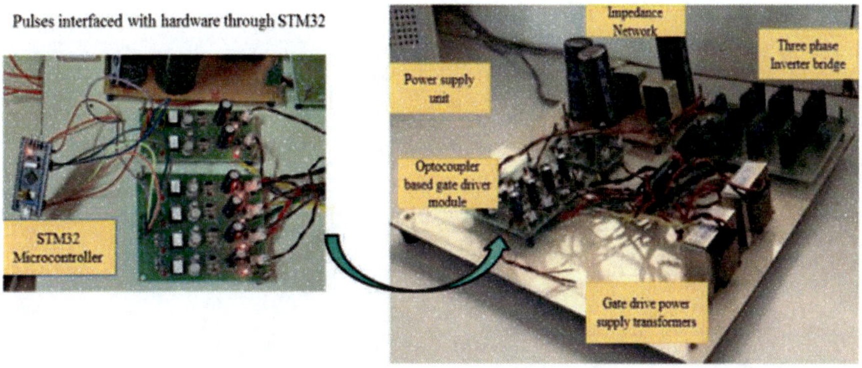


**Fig. 15** Experimental setup of DAEBQZSI

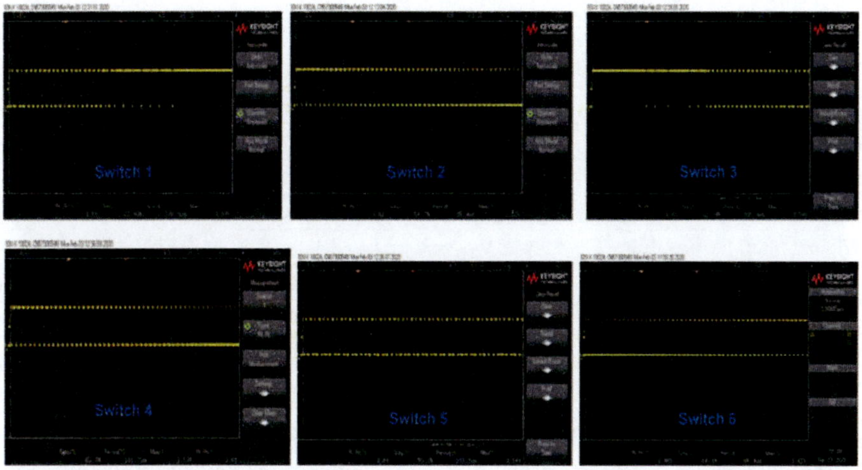


**Fig. 16** Gating pulses generated

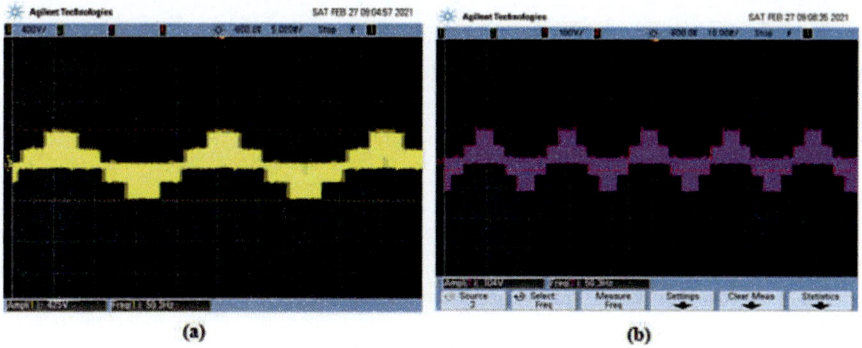


**Fig. 17** Output voltage without filter **a** ST mode **b** nST mode

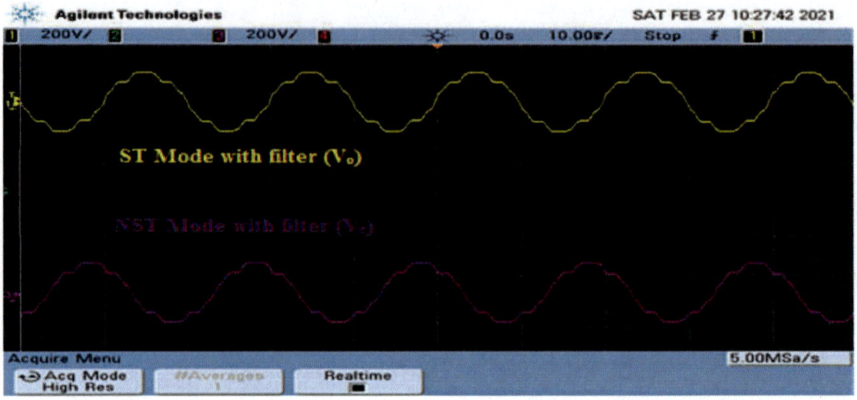


**Fig. 18** Output voltage with filter

**Table 5** Simulation and experimentation results

| Parameters | qZSI | | DAEBQZSI | |
|---|---|---|---|---|
| Input dc voltage | 150 V | | 150 V | |
| Mode | nST | | ST | |
| Output ac voltage | Simulated | Experimentation | Simulated | Experimentation |
| Without filter | 100 | 98 V | 420 | 418 V |
| With filter | 230 V | 210 V | 260 V | 258 V |

## 8 Conclusion

Both qZSI and DAEBQZSI have been investigated for both the nST and ST modes. The output voltage waveforms of shoot-through modes are examined completely, and it is concluded that the diode-assisted topology has the enhanced boosting capability which makes it suitable for interfacing it with PV module.

**Acknowledgements** The authors are pleased to thank the SSNCE management, Chennai, because of supplying the required facilities to incubate this idea through internal funding.

## References

1. Hanif Basu M, Malabika Gaughan K (2011) Understanding the operation of a Z-source inverter for photovoltaic application with a design example. Power Electronics, IET 4:278–287
2. Adle R, Renge M, Muley S, Shobhane P (2017) Photovoltaic based series z-source inverter fed induction motor drive with improved shoot through technique. Energy Procedia 117:329–335
3. Rajakarunaand S, Jayawickrama RL (2005) Designing impedance network of z-source inverters. In: Proceedings of 7th international power engineering conference (IPEC), vol 2, pp 962–967

4. Cintron JG, Li Y, Jiang S, Peng FZ (2011) Quasi-Z-source inverter with energy storage for photovoltaic power generation systems. In: Proceedings of 26th annual IEEE applied power electronics conference and exposition, pp 401–406
5. Gajanayake CJ, Luo FL, Gooi HB, So PL, Siow LK (2010) Extended boost z-source inverters. IEEE Trans Power Electron 25(10):2642–2652
6. Loh PC, Vilathgamuwa DM, Lai YS, Tin CG, Li Y (2005) Pulse-width modulation of z-source inverters. IEEE Trans Power Electron 20(6):1346–1355
7. Chowdhury S, Taylor GA, Chowdhury SP, Saha AK, Song YH (2007) Modelling simulation and performance analysis of a PV array in an embedded environment. In: Proceedings 42nd international universities power engineering conference (UPEC), pp 781–785
8. Pandiarajan N, Ramaprabha R, Ranganath M (2011) Application of circuit model for photovoltaic energy conversion system. Int J Adv Eng Technol 2(4):118–127
9. www.mathworks.com

# 31-Level RCC Multilevel Inverter for Photovoltaic System Interfacing

**R. Ramaprabha and N. Ramya Krishnan**

## 1 Introduction

Multilevel inverters (MLI) are persistent to acquire important aspects due to their high efficiency for higher voltage capability, with low electromagnetic inference (EMI) and reduced switching losses. For this reason, it is approved in power electronics applications, as they meet the increasing demand for utilities. Power electronic devices play an important role in various forms of renewable energy sources, especially solar energy which is foresighted as the best energy source. Although the photovoltaic (PV) system generates dc power, a large part of our utility applications required an ac power source. For this conversion from dc into ac, inverters are used. Sunlight is converted into electricity by the PV system, which in turn fed into the inverter to get the required power with lower THD. Previously, inverters are used only for light loads when the grid gets off or encountered any problem. Over the last decade, the technological advancement in the electrical field and power conversion has led to the implementation of MLI and with constant proper scrutiny has become a well-known developed technique in many areas of power applications [1, 2].

Since industries are needed to keep updating their drives, so the drive performance relies on THD in the inverters output voltage. For this, a superior alternative is deploying an MLI because it smoothens the output waveform of the staircase which consecutively decreases the effect of harmonic distortions, thereby its spectral quality improves by means of the number of levels. The system performance is improved when a multilevel inverter with larger power ratings is employed which significantly reduces the harmonics.

Though the MLI has some advantages, the drawback of the classical type should also be noted. There is an increase in the components like capacitors, inductors,

R. Ramaprabha (✉) · N. R. Krishnan
Department of Electrical and Electronics Engineering, Sri Sivasubramaniya Nadar College of Engineering, Kalavakkam, Tamilnadu 603110, India
e-mail: ramaprabhar@ssn.edu.in



N. L. Narasimhan et al. (eds.), *Recent Advances in Energy Technologies*, Lecture Notes in Mechanical Engineering, https://doi.org/10.1007/978-981-19-3467-4_7

dc sources, and power devices when we increase the number of levels [3, 4]. To overcome this drawback, a distinctive division of inverters known as the reduced component count (RCC) inverters is studied in the literature, which generates the required levels of output voltage with the help of minimum devices [5–7]. This paperwork has examined one type of RCC inverter, the performance is analyzed with distinct modulation techniques, and the best modulation technique is stated here.

## 2 The Basic Ladder-Type MLI

The RCC primary inverter schematic structure is observed as shown in Fig. 1. The inverter circuit is constructed with unidirectional switches—six ($K_1, K_2, K_3, K_4, S_x$, $S_y$), bidirectional switches—two ($S_1$, $T_1$), with dc supplies—four.

The functioning of MLI should be carefully operated such that we will be able to obtain our desired levels of output voltage. Here, the power semiconductor switches utilized are MOSFET, which we can control according to our load or required consumer utilities. The devices are initiated such that their dc source magnitudes are properly cascaded [8]. To obtain 0 V across the load, switches such as $K_1$, $K_3$, and $S_x$ are turned on, when the switches $S_1$, $K_4$, and $S_y$ are given rise to a voltage of $V_1$. During the operation of switches like $K_1$, $K_4$, and $S_y$ are set in motion, we get a voltage level of $2V_1$ across the load by rightly cascading the sources on the left-hand side. When the source voltage from both ends is cascaded suitably by turning on the switches $S_1$, $T_1$, and $S_y$, we obtain a voltage level of $V_1 + V_2$. When the devices $K_1$, $T_1$, and $S_y$ are set on, the right of two dc sources shall be cascaded along a single voltage source on the left to obtain a level of $2V_1 + V_2$. To get a voltage of $2V_1 + 2V_2$, switches such as $K_1$, $K_3$, and $S_y$ must be turned on. The detailed operation is mentioned in [9]. The symmetrical and asymmetrical cases differ in their magnitudes, and their respective switch ratios are ($1V_1 = 1V_2 = V$) and ($1V_1 = 3V_2 = V$). The basic structure operation is illustrated in Fig. 2.

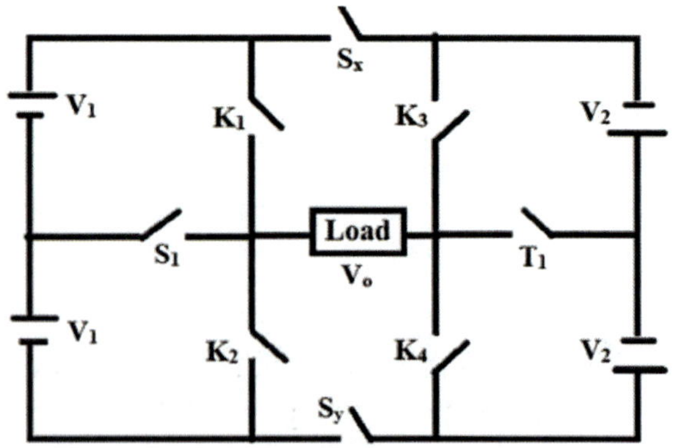


**Fig. 1** Root configuration

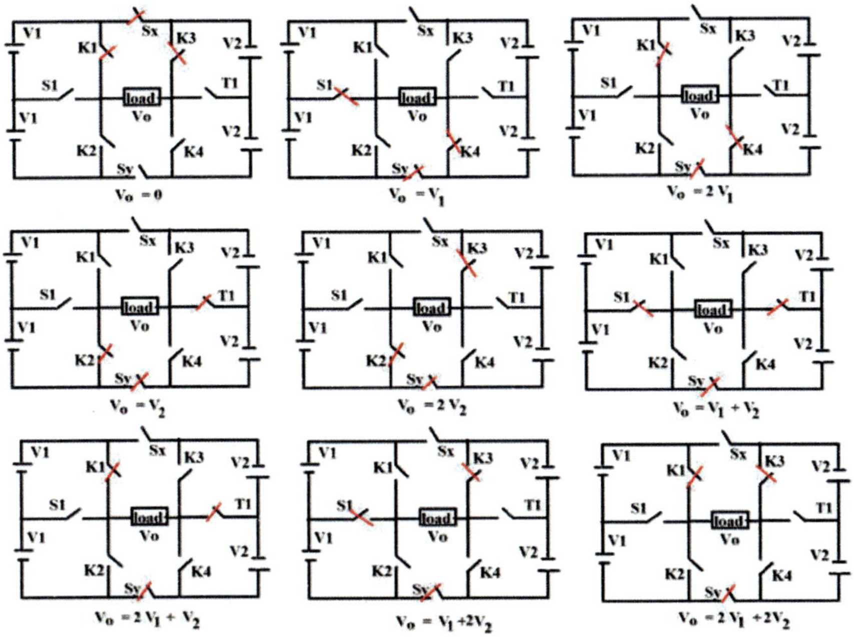


**Fig. 2** Basic structure operation

In Table 1, the voltage generated for individual switching states is given. The switch $S_y$ is required to be turned on for generation of positive level, whereas $S_x$ needs to be turned on to generate the negative level. The basic configuration can generate 9 and 17 levels, and their switching patterns are given in Tables 2 and 3.

The basic structure in symmetric mode generates only 9 distinct output voltages. The basic structure can be operated, in asymmetric mode, where the magnitudes are not in symmetry to attain a larger number of voltage levels [10, 11]. When the 9-level MLI is operated with dc sources ratio 1:3, i.e., ($1V_1 = 3V_2 =$ V), we can build up a 17-level MLI structure. As given in Table 3, the switching sequence of the MLI changes even though modes of operation remain the same. As listed in the tables, the switching of the MLI is represented as (ON—✓ or 1, OFF—✗ or 0).

## 3 31-Level Ladder-Type MLI

The RCC for a 31-level RCC-MLI structure is considered as displayed in Fig. 3. The 31-level structure has unidirectional switches—six ($K_1$, $K_2$, $K_3$, $K_4$, $S_x$, $S_y$, $S_1$, $S_2$, $T_1$, $T_2$), bidirectional switches—four ($S_1$, $S_2$, $T_1$, $T_2$), and dc sources— six. The operation is similar to the basic structure, and only the switching states changes

**Table 1** Basic switching patterns [9]

| Modes | $K_1$ | $K_2$ | $K_3$ | $K_4$ | $S_1$ | $T_1$ | $S_x$ | $S_y$ | Voltage ($V_o$) |
|---|---|---|---|---|---|---|---|---|---|
| i | ✓ | ✗ | ✓ | ✗ | ✗ | ✗ | ✓ | ✗ | 0 |
| ii | ✗ | ✗ | ✗ | ✓ | ✓ | ✗ | ✗ | ✓ | $V_1$ |
| iii | ✗ | ✗ | ✓ | ✗ | ✓ | ✗ | ✓ | ✗ | $-V_1$ |
| iv | ✗ | ✓ | ✗ | ✗ | ✗ | ✓ | ✗ | ✓ | $V_2$ |
| v | ✓ | ✗ | ✗ | ✗ | ✗ | ✓ | ✓ | ✗ | $-V_2$ |
| vi | ✓ | ✗ | ✗ | ✓ | ✗ | ✗ | ✗ | ✓ | $2V_1$ |
| vii | ✗ | ✓ | ✓ | ✗ | ✗ | ✗ | ✓ | ✗ | $-2V_1$ |
| viii | ✗ | ✓ | ✓ | ✗ | ✗ | ✗ | ✗ | ✓ | $2V_2$ |
| ix | ✓ | ✗ | ✗ | ✓ | ✗ | ✗ | ✓ | ✗ | $-2V_2$ |
| x | ✗ | ✗ | ✗ | ✗ | ✓ | ✓ | ✗ | ✓ | $V_1 + V_2$ |
| xi | ✗ | ✗ | ✗ | ✗ | ✓ | ✓ | ✓ | ✗ | $-(V_1 + V_2)$ |
| xii | ✓ | ✗ | ✗ | ✗ | ✗ | ✓ | ✗ | ✓ | $2V_1 + V_2$ |
| xiii | ✗ | ✓ | ✗ | ✗ | ✗ | ✓ | ✓ | ✗ | $-(2V_1 + V_2)$ |
| xiv | ✗ | ✗ | ✓ | ✗ | ✓ | ✗ | ✗ | ✓ | $V_1 + 2V_2$ |
| xv | ✗ | ✗ | ✗ | ✓ | ✓ | ✗ | ✓ | ✗ | $-(V_1 + 2V_2)$ |
| xvi | ✓ | ✗ | ✓ | ✗ | ✗ | ✗ | ✗ | ✓ | $2V_1 + 2V_2$ |
| xvii | ✗ | ✓ | ✗ | ✓ | ✗ | ✗ | ✓ | ✗ | $-(2V_1 + 2V_2)$ |

**Table 2** Switching patterns (symmetric mode)—9 level [9]

| Modes | $K_1$ | $K_2$ | $K_3$ | $K_4$ | $S_1$ | $T_1$ | $S_x$ | $S_y$ | Voltage ($V_o$) |
|---|---|---|---|---|---|---|---|---|---|
| i | ✓ | ✗ | ✓ | ✗ | ✗ | ✗ | ✓ | ✗ | 0 |
| ii | ✗ | ✗ | ✗ | ✓ | ✓ | ✗ | ✗ | ✓ | V |
| iii | ✗ | ✗ | ✓ | ✗ | ✓ | ✗ | ✓ | ✗ | −V |
| iv | ✓ | ✗ | ✗ | ✓ | ✗ | ✗ | ✗ | ✓ | 2 V |
| v | ✗ | ✓ | ✓ | ✗ | ✗ | ✗ | ✓ | ✗ | −2 V |
| vi | ✓ | ✗ | ✗ | ✗ | ✗ | ✓ | ✗ | ✓ | 3 V |
| vii | ✗ | ✓ | ✗ | ✗ | ✗ | ✓ | ✓ | ✗ | −3 V |
| viii | ✓ | ✗ | ✓ | ✗ | ✗ | ✗ | ✗ | ✓ | 4 V |
| ix | ✗ | ✓ | ✗ | ✓ | ✗ | ✗ | ✓ | ✗ | −4 V |

accordingly. The output levels range from 0 to ±3 V and ±5 V to ±15 V (+V, −V, + 2 V, −2 V, etc.). In asymmetric mode, this structure can generate thirty-one distinct voltage levels at the outputs that include 15 positive and 15 negative levels.

To be precise, when the 17-level MLI is further cascaded with magnitudes of dc sources ratio of 1:4, i.e., ($1V_1 = 4V_2 = V$), we can build up a 31-level MLI structure. The modes of operation remain the same with a difference in the switching patterns [12, 13]. Table 4 represents the switching states in asymmetrical mode.

**Table 3** Switching patterns (asymmetrical mode)—17 level [9]

| Modes | $K_1$ | $K_2$ | $K_3$ | $K_4$ | $S_1$ | $T_1$ | $S_x$ | $S_y$ | Voltage ($V_o$) |
|---|---|---|---|---|---|---|---|---|---|
| i | ✓ | ✗ | ✓ | ✗ | ✗ | ✗ | ✓ | ✗ | 0 |
| ii | ✗ | ✗ | ✗ | ✓ | ✓ | ✗ | ✗ | ✓ | V |
| iii | ✗ | ✗ | ✓ | ✗ | ✓ | ✗ | ✓ | ✗ | −V |
| iv | ✓ | ✗ | ✗ | ✓ | ✗ | ✗ | ✗ | ✓ | 2 V |
| v | ✗ | ✓ | ✓ | ✗ | ✗ | ✗ | ✓ | ✗ | −2 V |
| vi | ✗ | ✓ | ✗ | ✗ | ✗ | ✓ | ✗ | ✓ | 3 V |
| vii | ✓ | ✗ | ✗ | ✗ | ✗ | ✓ | ✓ | ✗ | −3 V |
| viii | ✗ | ✗ | ✗ | ✗ | ✓ | ✓ | ✗ | ✓ | 4 V |
| ix | ✗ | ✗ | ✗ | ✗ | ✓ | ✓ | ✓ | ✗ | −4 V |
| x | ✓ | ✗ | ✗ | ✗ | ✗ | ✓ | ✗ | ✓ | 5 V |
| xi | ✗ | ✓ | ✗ | ✗ | ✗ | ✓ | ✓ | ✗ | −5 V |
| xii | ✗ | ✓ | ✓ | ✗ | ✗ | ✗ | ✗ | ✓ | 6 V |
| xiii | ✓ | ✗ | ✗ | ✓ | ✗ | ✗ | ✓ | ✗ | −6 V |
| xiv | ✗ | ✗ | ✓ | ✗ | ✓ | ✗ | ✗ | ✓ | 7 V |
| xv | ✗ | ✗ | ✗ | ✓ | ✓ | ✗ | ✓ | ✗ | −7 V |
| xvi | ✓ | ✗ | ✓ | ✗ | ✗ | ✗ | ✗ | ✓ | 8 V |
| xvii | ✗ | ✓ | ✗ | ✓ | ✗ | ✗ | ✓ | ✗ | −8 V |

**Fig. 3** Structure of 31-level MLI

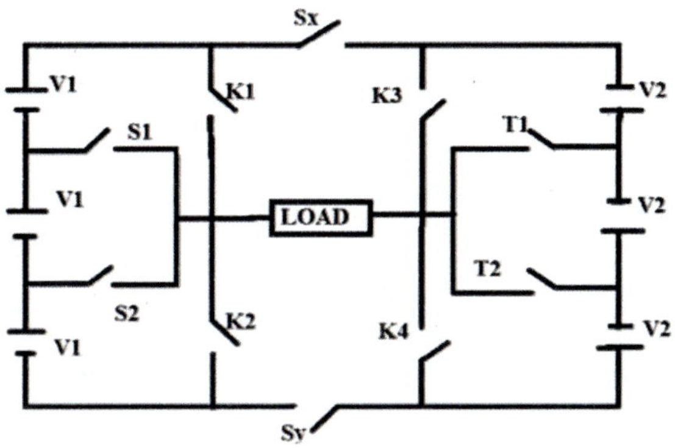


## 4 Switching Angle Calculation

The switching angle of fundamental frequency calculation is illustrated in this section [14]. It must be carefully analyzed so to examine the performance of the inverter. The RCC-MLI uses either MOSFET/IGBT switches with on and off condition so as to generate the appropriate voltage output of staircase with minimum total harmonic distortion, which is expected. The distinct strategies used in RCC-type MLIs can

**Table 4** Asymmetrical mode switching patterns—31 level [9]

| Modes | $K_1$ | $K_2$ | $K_3$ | $K_4$ | $S_x$ | $S_y$ | $S_1$ | $S_2$ | $T_1$ | $T_2$ | Voltage ($V_o$) |
|---|---|---|---|---|---|---|---|---|---|---|---|
| i | ✗ | ✓ | ✗ | ✓ | ✗ | ✓ | ✗ | ✗ | ✗ | ✗ | 0 |
| ii | ✗ | ✗ | ✗ | ✓ | ✗ | ✓ | ✗ | ✓ | ✗ | ✗ | +V |
| iii | ✗ | ✗ | ✗ | ✓ | ✗ | ✓ | ✓ | ✗ | ✗ | ✗ | +2 V |
| iv | ✓ | ✗ | ✗ | ✓ | ✗ | ✓ | ✗ | ✗ | ✗ | ✗ | +3 V |
| v | ✗ | ✓ | ✗ | ✗ | ✗ | ✓ | ✗ | ✗ | ✗ | ✓ | +4 V |
| vi | ✗ | ✗ | ✗ | ✗ | ✗ | ✓ | ✗ | ✓ | ✗ | ✓ | +5 V |
| vii | ✗ | ✗ | ✗ | ✗ | ✗ | ✓ | ✓ | ✗ | ✗ | ✓ | +6 V |
| viii | ✓ | ✗ | ✗ | ✗ | ✗ | ✓ | ✗ | ✗ | ✗ | ✓ | +7 V |
| ix | ✗ | ✓ | ✗ | ✗ | ✗ | ✓ | ✗ | ✗ | ✓ | ✗ | +8 V |
| x | ✗ | ✗ | ✗ | ✗ | ✗ | ✓ | ✗ | ✓ | ✓ | ✗ | +9 V |
| xi | ✗ | ✗ | ✗ | ✗ | ✗ | ✓ | ✓ | ✗ | ✓ | ✗ | +10 V |
| xii | ✓ | ✗ | ✗ | ✗ | ✗ | ✓ | ✗ | ✗ | ✓ | ✗ | +11 V |
| xiii | ✗ | ✓ | ✓ | ✗ | ✗ | ✓ | ✗ | ✗ | ✗ | ✗ | +12 V |
| xiv | ✗ | ✗ | ✓ | ✗ | ✗ | ✓ | ✗ | ✓ | ✗ | ✗ | +13 V |
| xv | ✗ | ✗ | ✓ | ✗ | ✗ | ✓ | ✓ | ✗ | ✗ | ✗ | +14 V |
| xvi | ✓ | ✗ | ✓ | ✗ | ✗ | ✓ | ✗ | ✗ | ✗ | ✗ | +15 V |
| xvii | ✓ | ✗ | ✓ | ✗ | ✓ | ✗ | ✗ | ✗ | ✗ | ✗ | 0 |
| xviii | ✗ | ✗ | ✓ | ✗ | ✓ | ✗ | ✓ | ✗ | ✗ | ✗ | −V |
| xix | ✗ | ✗ | ✓ | ✗ | ✓ | ✗ | ✗ | ✓ | ✗ | ✗ | −2 V |
| xx | ✗ | ✓ | ✓ | ✗ | ✓ | ✗ | ✗ | ✗ | ✗ | ✗ | −3 V |
| xxi | ✓ | ✗ | ✗ | ✗ | ✓ | ✗ | ✗ | ✗ | ✓ | ✗ | −4 V |
| xxii | ✗ | ✗ | ✗ | ✗ | ✓ | ✗ | ✓ | ✗ | ✓ | ✗ | −5 V |
| xxiii | ✗ | ✗ | ✗ | ✗ | ✓ | ✗ | ✗ | ✓ | ✓ | ✗ | −6 V |
| xxiv | ✗ | ✓ | ✗ | ✗ | ✓ | ✗ | ✗ | ✗ | ✓ | ✗ | −7 V |
| xxv | ✓ | ✗ | ✗ | ✗ | ✓ | ✗ | ✗ | ✗ | ✗ | ✓ | −8 V |
| xxvi | ✗ | ✗ | ✗ | ✗ | ✓ | ✗ | ✓ | ✗ | ✗ | ✓ | −9 V |
| xxvii | ✗ | ✗ | ✗ | ✗ | ✓ | ✗ | ✗ | ✓ | ✗ | ✓ | −10 V |
| xxviii | ✗ | ✓ | ✗ | ✗ | ✓ | ✗ | ✗ | ✗ | ✗ | ✓ | −11 V |
| xxix | ✓ | ✗ | ✗ | ✓ | ✓ | ✗ | ✗ | ✗ | ✗ | ✗ | −12 V |
| xxx | ✗ | ✗ | ✗ | ✓ | ✓ | ✗ | ✓ | ✗ | ✗ | ✗ | −13 V |
| xxxi | ✗ | ✗ | ✗ | ✓ | ✓ | ✗ | ✗ | ✓ | ✗ | ✗ | −14 V |
| xxxii | ✗ | ✓ | ✗ | ✓ | ✓ | ✗ | ✗ | ✗ | ✗ | ✗ | −15 V |

be categorized in accordance with the value of the switching frequency. Here, half-height fundamental-type PWM is used for all the three types of MLIs with explaining only HH-PWM in detail.

### 4.1 *Half-Height PWM Method*

Illustration of the method of half-height (HH) implied that the switching angle is set when the fundament value rises to the HH of the level. Using the given formula below, the main switching angles are determined by,

$$\theta_i = \sin^{-1}\left(\frac{2i-1}{N-1}\right); \quad i = 1, 2, \ldots \left(\frac{N-1}{2}\right) \tag{1}$$

where $N$—number of output levels. For each type of RCC-MLI, the switching angles using HH-PWM are calculated as follows:

The switching angles are $\theta_1 = 7.18°$; $\theta_2 = 22.02°$; $\theta_3 = 36.68°$; $\theta_4 = 61.04°$ for 9-level inverter. For 17-level inverter, the switching angles are $\theta_1 = 3.58°$; $\theta_2 = 10.80°$; $\theta_3 = 18.20°$; $\theta_4 = 25.94°$; $\theta_5 = 34.22°$; $\theta_6 = 43.43°$; $\theta_7 = 54.34°$; $\theta_8 = 69.63°$. For 31-level inverter, the switching angles are $\theta_1 = 1.91°$; $\theta_2 = 5.73°$; $\theta_3 = 9.59°$; $\theta_4 = 13.49°$; $\theta_5 = 17.45°$; $\theta_6 = 21.51°$; $\theta_7 = 25.67°$; $\theta_8 = 30°$; $\theta_9 = 34.51°$; $\theta_{10} = 39.29°$; $\theta_{11} = 44.42°$; $\theta_{12} = 50.05°$; $\theta_{13} = 56.44°$; $\theta_{14} = 64.15°$; and $\theta_{15} = 75.16°$.

## 5 Simulated Results

The purpose of this segment is to present MATLAB simulations of RCC-type inverter for 9, 17, and 31 levels in both symmetric and asymmetric modes using the HH-PWM type. The proposed inverter type is analyzed only with the HH-PWM technique, which results in the best outcomes. From the interpreted results, it is endowed that the HH-PWM type signifies nine distinct VO levels, and its waveform is set forth in Fig. 4a.

Here, the frequency of 50 Hz with FFT analysis for a 9-level inverter is presented in Fig. 4b, where the THD is 9.37% with the fundamental voltage of 69 V using the HH-PWM method. The THD can be varied with different schemes with fundamental frequencies [15]. For a 9-level inverter, the switch ratio is 1.125.

When further investigated the basic structure, it can also generate 17 unique levels with change in magnitude [16]. It is analyzed with the HH-PWM type, and the result signifies seventeen different levels of voltage with a frequency of 50 Hz at the output which is displayed in Fig. 5a.

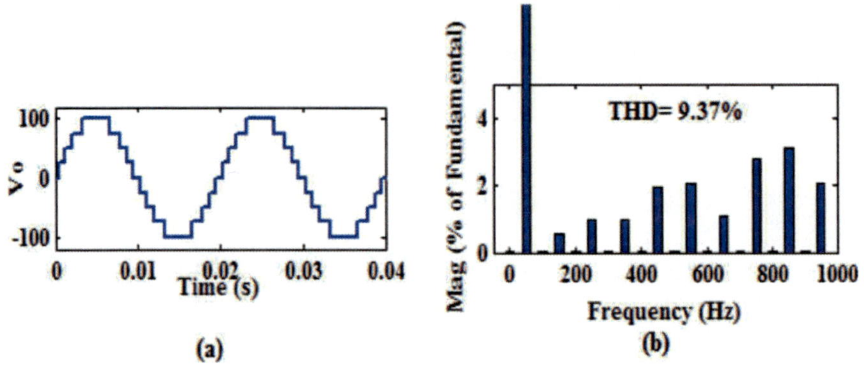


**Fig. 4** 9-level inverter **a** generated output (HH-PWM) **b** FFT analysis

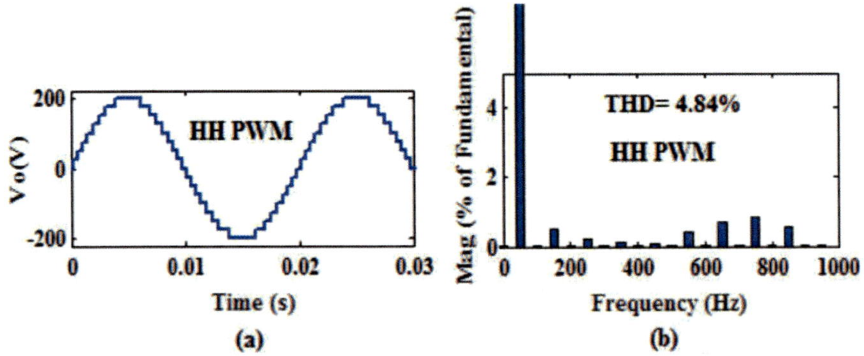


**Fig. 5** **a** Generated waveforms for 17-level inverter; **b** FFT analysis of 17-level inverter

Similarly, for a frequency of 50 Hz, the FFT analysis is done for 17-level RCC-MLI which is displayed in Fig. 5b. The THD percentage of the 17-level RCC-MLI is 4.84% with the fundamental voltage of 181 V using the HH-PWM method. The THD can be varied with different schemes with fundamental frequencies. For 17-level inverter, switch ratio is 2.125. With further study in basic structure, the inverter is cascaded to generate unique 31 levels. It is analyzed with the half-height-PWM type, and the result signifies thirty-one different voltage levels of voltage with a frequency of 50 Hz at the output which is shown in Fig. 6a.

For 31-level RCC-MLI, the FFT analysis with a frequency of 50 Hz is unveiled in Fig. 6b. The THD percentage of the 31-level inverter is 2.63% with the fundamental voltage of 289 V using the HH-PWM method. The THD can be varied with different schemes with fundamental frequencies. For a 31-level, MLI switch ratio is 3.1.

The 31-level inverter is interfaced with a photovoltaic (PV) array. PV array is simulated using [17]. In Fig. 3, $V_1$ and $V_2$ are replaced by $2 \times 1$ and $8 \times 1$ PV

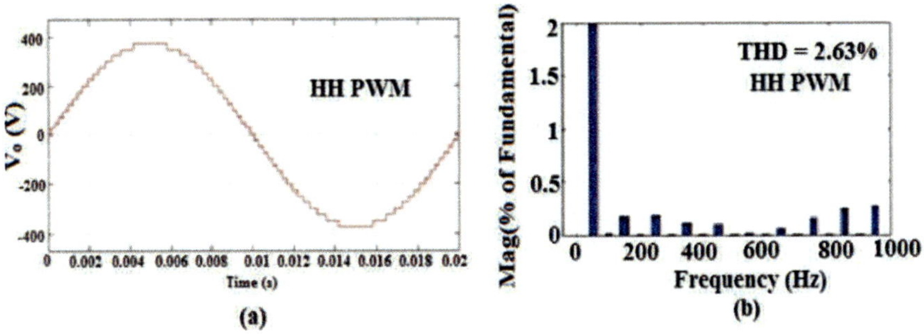


**Fig. 6** 31-level inverter **a** generated output; **b** FFT analysis

arrays, respectively, to meet voltage ratings and shown in Fig. 7. The characteristics of PV array were simulated, and the resultant final output, both are shown in Fig. 8.

The harmonic content (THD %) of a 31-level inverter, when interfaced with a PV array, varies with the PWM schemes. The 31-level inverter with HH-PWM gives a THD of 2.59%, which is observed to be well improved compared to the other various PWM schemes, so asymmetric RCC-MLI is chosen as the best choice.

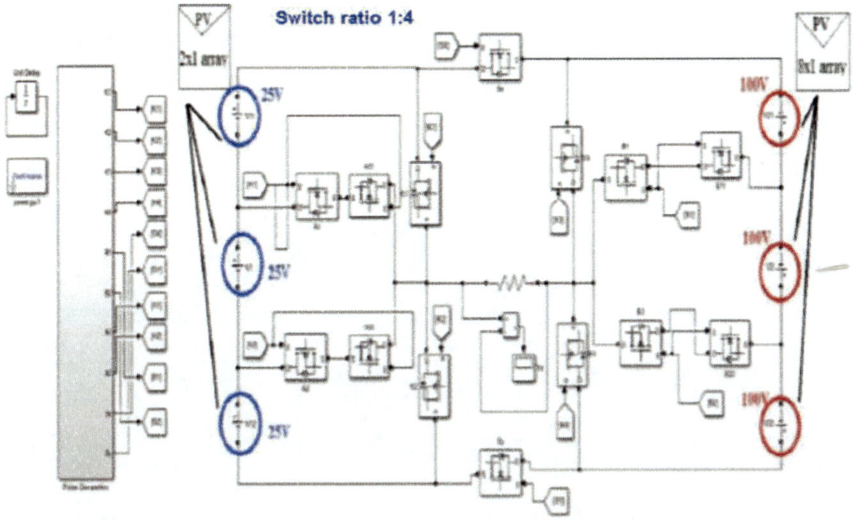


**Fig. 7** MATLAB diagram—PV interfacing of 31-level RCC inverter

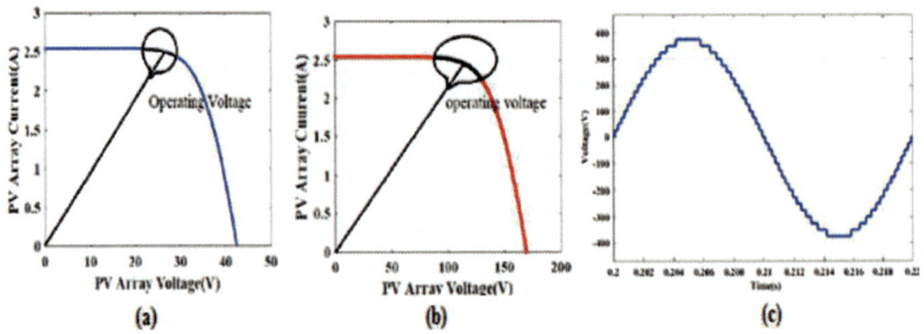


**Fig. 8** Simulated characteristics of **a** 2 × 1 PV array; **b** 8 × 1 PV array; **c** output voltage

## 6 Experimental Results

The hardware setup of 31-level MLI is represented in Fig. 9. The experimental setup consists of MLI circuit, DC supply unit, opto-coupler (TLP250), and Arduino processor. MLI circuit is realized using IRFP250 MOSFETs. The code is generated to generate the gating pulses to trigger the switches with the Arduino UNO as shown in Fig. 10. The waveform of output voltage and its measured THD are illustrated in Fig. 11.

The measured THD percentage of the 31-level inverter is 2.8% using the HH-PWM method, which closely agrees with the simulation result shown in Fig. 7.

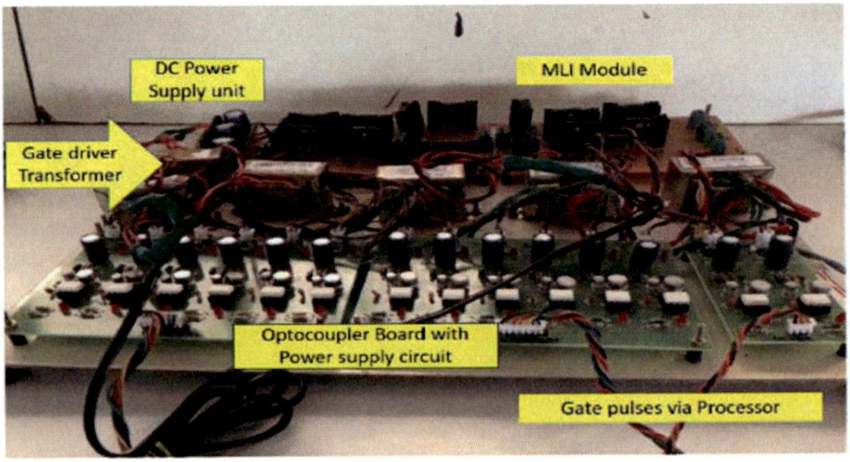


**Fig. 9** Hardware prototype

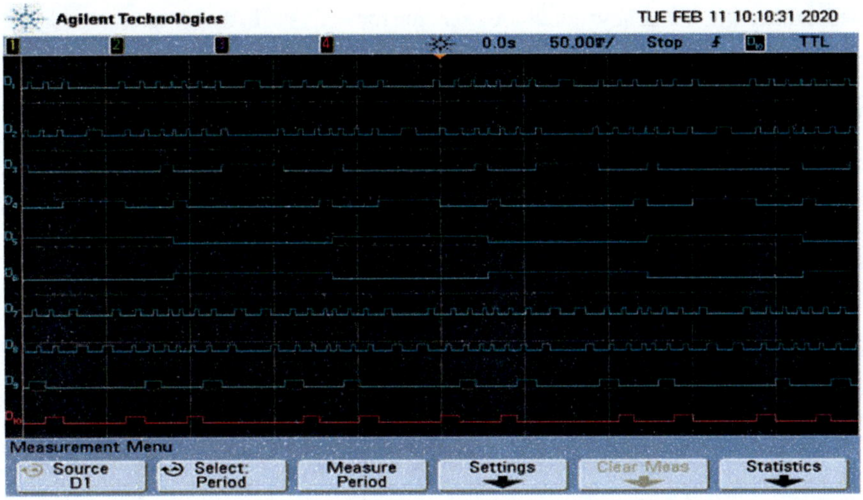


**Fig. 10** Switching pulses

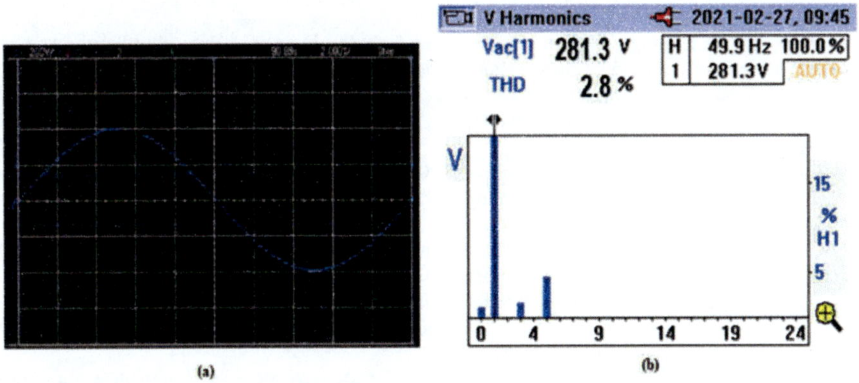


**Fig. 11** Measured voltage and %THD

## 7 Conclusion

The RCC inverter type has an elite function over the conventional type in terms of cost, components, reliability, dc supplies. The advantage of this paper is using lesser components with which it can develop a higher number of levels with half-height fundamental frequency PWM scheme, where it can be used in our daily utilities. The ultimate work of this paper is to work with a reduced component count which has several advantages over classical inverters. The generalized configuration of the RCC-type MLI and its symmetric and asymmetric operations are discussed briefly. The 9-level, 17-level, and 31-level MLI were built and simulated in MATLAB with the fundamental frequency modulation scheme for analysis. In the structure, levels are increased by adding and changing the dc supplies magnitude. It is inferred from the above results that the half-height-PWM method has lesser distortion when compared to other PWM techniques. The results were verified through hardware testing of the circuit.

**Acknowledgements** The authors are pleased to thank the SSNCE management, Chennai, for supplying the required facilities to incubate this idea through internal funding.

## References

1. Laxmi Srikanya B, Venkata Ramana G (2018) An advanced seven level multilevel inverter technique for grid-connected system applications with reduced switching devices. Int J Adv Res Sci Eng 7(2):83–94
2. Singh AV, Singh RS (2018) A comparative study of multilevel inverter topologies. Int Res J Eng Technol 5(3):1009–1014
3. Kim HJ, Jung DW, Sul SK (2000) A new discontinuous PWM strategy of neutral-point clamped inverter. In: Conference record—IAS annual meeting (IEEE Industry Applications Society), vol 3, pp 2017–2023
4. Chang-xin M, Li-ping S, Tai-xu W, Cheng-bao C (2009) Flying capacitor multilevel inverters with novel PWM method. Procedia Earth Planet Sci 1:1554–1560
5. Moses Stephen P, Ponraj M, Sathesh N, Balamurugan K, Karthikeyan K (2018) Modified cascaded H—bridge multilevel inverter for household appliances. Int Res J Eng Technol 5(3):3089–3093
6. Ruiz-Caballero D, Mussa SA, Arancibia S (2011) Symmetrical hybrid multilevel inverter with 'N' cells connected in parallel employing multi-state switching. In: Proceedings in XI Brazilian power electronics conference, pp 776–781
7. Chlaihawi A, Sabbar A, Jedi H (2020) A high-performance multilevel inverter with reduced power electronic devices. Int J Power Electron Drive Syst 11:1883–1889
8. Irusapparajan G, Periyaazhagar D, Prabaharan N, Rini ann Jerin A (2019) Experimental verification of trinary DC source cascaded H-bridge multilevel inverter using unipolar pulse width modulation. Automatika 60(1):19–27
9. Ramaprabha R, Malathy S, Ramya Krishnan N (2019) A 17-level asymmetric multilevel inverter with fundamental frequency PWM schemes. In: Proceedings of 2019 IEEE 2nd international conference on power and embedded drive control, ICPEDC2019, SSN College of Engineering, India
10. Oskuee MRJ, Karimi M, Ravadanegh SN, Gharehpetian GB (2015) An innovative scheme of symmetric multilevel voltage source inverter with lower number of circuit devices. IEEE Trans Ind Electron 62(11)
11. Laboure M, Cuniere A, Meynard TA, Forest F, Sarraute E (2009) A theoretical approach to intercell transformers, application to interleaved converters. IEEE Trans Power Electron 23(1):464–474
12. Thakre K, Mohanty KB, Chatterjee A (2018) Reduction of circuit devices in symmetrical voltage source multilevel inverter based on series connection of basic unit cells. Alexandria Eng J 57(4):2703–2712
13. Siddique MD et al (2018) Asymmetrical multilevel inverter topology with reduced number of components. In: 2018 IEEE international conference on power electronics, drives and energy systems (PEDES), pp 1–5

14. Babaei E, Kangarlu MF, Mazgar FN (2012) Symmetric and asymmetric multilevel inverter topologies with reduced switching devices. Electric Power Syst Res 86:122–130
15. Veenstra M, Rufer A (2005) Control of a hybrid asymmetric multilevel inverter for competitive medium-voltage industrial drives. IEEE Trans Ind Appl 41(2):655–664
16. Alishah RS, Hosseini SH, Babaei E, Sabahi M (2016) A new general multilevel converter topology based on cascaded connection of submultilevel units with reduced switching components, DC sources, and blocked voltage by switches. IEEE Trans Ind Electron 63:7157–7164
17. www.mathworks.com

# Synthesis and Characterization of NiO/CuO/$Fe_2O_3$ Multi-Layered Structure for Photovoltaic Applications

**R. Abhishek and Siddharth Joshi**

## 1 Introduction

Currently, due to the increase in human population, there is a huge demand for energy resources that are essential for human beings. Due to this increase in demand, fossils fuel are depleting, and on the other hand, one must keep the environment safe too. Therefore, the renewable energy, such as wind, wave and tidal energies and also from the solar radiation and heat energy are in huge demand. To fulfil these requirements, it is required to develop a PV technology at a minimal cost with improved efficiency [1, 2]. These PV cells are based on the principle of separation of electron–hole pairs by the electric field induced due to Schottky contact. Similarly, for several decade, silicon-based photovoltaic's is ruling the field of photovoltaic's due to its highest power conversion efficiency, but at the same time, these silicon-based semiconductors are very costly because of their high cost of production. To overcome these problems, metal oxide (MO)-based semiconductors stood as an alternative against the silicon-based PV devices due to their low cost of production, safe environment during deposition, tuneable properties based on the film thickness and the deposition conditions [3] and different structures such as nanorods [4], nano-flowers [5], nano-pillars [6], nano-tubes [7] etc., hence they are referred as next generation solar cells (NGSC). These MO-based thin films can be used in various fields such as gas sensors, perovskite-based solar cells, DSSC's, supercapacitors, optical coatings and many more, because of their ability to improve the device characteristics even being as an amorphous or crystalline in nature. These MO thin films have found their application as barrier layers in organic PV, perovskite SC and many more due to this the device lifetime is enhanced and also the use of expensive semiconducting materials can be mitigated [8]. Based on the parameter of charge extraction these buffer layers

R. Abhishek · S. Joshi (✉)
Centre for Nanotechnology, Department of Mechanical Engineering, The National Institute of Engineering, Mysuru, India
e-mail: sjoshi@nie.ac.in

 
N. L. Narasimhan et al. (eds.), *Recent Advances in Energy Technologies*, Lecture Notes in Mechanical Engineering, https://doi.org/10.1007/978-981-19-3467-4_8

can be differentiated as hole transport layers (HTL) and electron transport layers (ETL). The common HTL are $MoO_3$, NiO, $V_2O_5$, etc., and ETL such as ZnO. $TiO_2$, $SnO_x$, $Fe_2O_3$, etc. In the solar cell device, the purpose of using HTL is for the efficient transfer of holes to anode and blockage at the cathode interface and vice versa. At the same time, metal oxides can be grown by the wide range of deposition techniques, following the top-down or bottom-up approach.

The present work focuses on a cost-effective method for the synthesis of the multi-layered structure of NiO/CuO/$Fe_2O_3$ for pn-junction application. To study the thin film formation of such stacking by exploring the XRD, SEM and UV–VIS spectroscopy. The special feature of the article is that the implementation of $Fe_2O_3$ as an ETL in combination with the CuO absorber layer, wherein the efficiency of the transported electron to the terminal is probably more effective due to the low band gap of the $Fe_2O_3$ and also increased movement and production of the electron in the ETL which in turn has efficacy in the efficiency of the device. A couple of studies on multi-layered PV systems have been stated where Joshi et al., demonstrated structural, surface morphological and optical characterization by using state-of-the-art techniques such as XRD, SEM and UV–Vis spectroscopy of NiO/CuO/ZnO multi-layered structure synthesized by the wet chemical process. In the study, Joshi et al., synthesized NiO by CBD technique having poly-crystalline nature and porous structure with a transmittance of 95% in the visible and near infra-red region. The CuO layer was synthesized by hydrothermal bath deposition technique where the synthesized CuO layer uniformly covered the porous NiO. The CuO layer deposited on NiO/ITO substrate exhibited poly-crystalline nature with the absorbance of around 83% making an efficient absorber layer. Finally, the poly-crystalline ZnO layer synthesized on top of NiO/CuO by hydrothermal bath deposition technique exhibited a rod-like structure with different orientations throughout the material and absorbance of 98% making an efficient n-type layer for PV cell application [9]. Similarly, Dimopoulos et al. studied the efficiency of the p-CuO/n-ZnO multi-layered structure using the varied CuO layer thickness. In this study, he obtained maximum efficiency of 0.08% at 500 nm as an optimized CuO layer thickness [10]. Bouhjar et al. conducted a study on perovskite-based solar cells photovoltaic performance, where the compact $Fe_2O_3$ has its influence on the photovoltaic performance. In this study, the author has replaced $TiO_2$ with $Fe_2O_3$ as an ETL and got the power conversion efficiency of 5.7%. The resulted power conversion efficiency was the result of enhanced surface coverage of $Fe_2O_3$ on $CH_3NH_3PbI_3$ based perovskite solar cells which in turn had its impact on the efficacy of charge separation [11]. Manikandan et al. conducted a study on $Fe_2O_3$ based DSSC's by two different morphologies of nanorods and nanoparticles. The XRD study revealed the formation of poly-crystalline nature of $Fe_2O_3$ with crystallite size was varying in the range of 9.4–26.6 nm. The increase in annealing temperature increased crystallite size and reduced the bandgap. The power conversion efficiency obtained was 0.43% for nanorods and 0.29% for nanoparticles [12].

The present article focussed on the synthesis and characterization of NiO/CuO/$Fe_2O_3$ multi-layered thin film and studying its structural, optical and surface morphological properties towards the application of photovoltaic cells.

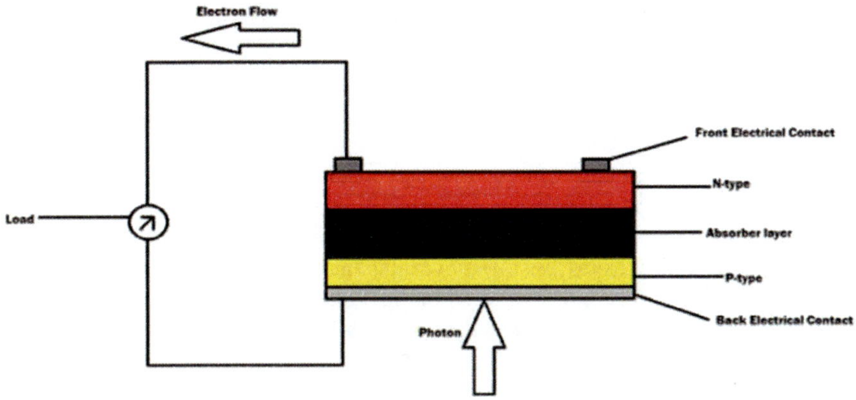


**Fig. 1** Schematic representation of device structure of NiO/Cu/$Fe_2O_3$

The individual layers and the multi-layered structure characterization have been performed by exploring the various surface science tools like XRD, SEM and UV–Vis Spectroscopy (Fig. 1).

## 2 Experimental Part

All the chemicals purchased were analytically reagent grade and were used without any further purification. The chemicals used are deionized water, nickel sulphate hexahydrate, potassium persulphate, ammonia, copper acetate monohydrate, ethanol, copper nitrate, hexamine, glucose and ammonium ferrous sulphate. Nickel oxide (NiO) thin film was deposited on indium tin oxide (ITO) coated glass substrates through the process of chemical bath deposition technique as referred by Xia et al. [13], for the deposition of NiO thin film, 0.1 M of 80 ml Nickel Sulphate and 0.25 M of 60 ml potassium persulphate solution is mixed in a 200 ml beaker. Sequentially, the ITO-coated glass substrates are placed in vertical fashion inside the beaker in which precursor solution is present. Afterwards, to the beaker containing the precursor solution 20 ml of 25% ammonia solution is added and allowed for constant stirring at 20 min at a speed of 300 rpm at room temperature. After 20 min, the substrates were removed from the beaker and washed with deionized water and acetone to remove the excess particles which are adhered on the top surface. Then, the as prepared NiO thin film was annealed at 300 °C for 1.5 h [13].

In the second stage, the cupric oxide (CuO) thin film was deposited using the hydrothermal bath deposition technique. For stability or to enhance the surface interface interaction, initially, a seed layer of CuO has been deposited on top of NiO thin film. Whereas, the seed layer solution was prepared by dissolving 10 mM copper acetate monohydrate in ethanol, followed by the dipping of the substrate in the seed layer solution for 3 min, later annealed at 100 °C on a hot plate for 5 min and followed by the second annealing inside the oven at 200 °C for 2.5 h. Once the seed layer is deposited, 25 mM copper nitrate and 25 mM hexamine are dissolved in Deionized water by stirring for 1 h at room temperature. The substrate must be placed at the bottom of the beaker and the solution is constantly heated at 95 °C for 3–4 h. After 4 h the substrates were taken out of the beaker and washed with deionized water followed by annealing at 100 °C for 30 min [14] to have a better thermodynamical stability and optimized structure as well as to remove the residuals too.

Later, the Iron (III) Oxide ($Fe_2O_3$) thin film is further deposited on ITO/NiO/CuO bilayer through the process of hydrothermal method and Dr. Blade method. For the preparation of $Fe_2O_3$ on top of the ITO/NiO/CuO multilayer, hydrothermal bath deposition technique is used. 3.75 g of glucose anhydrous is dissolved in 20 ml of deionized water, and 0.75 g of ammonium ferrous sulphate is dissolved in 10 ml of deionized water, and both the solutions are added to the stainless-steel autoclave and kept for the hydrothermal method for 24 h at 180 °C. Here, the carbohydrate to metal molar ratio should be maintained in the range of 5:1 to define the lateral shell thickness of the synthesized hollow spheres. After 24 h, the black-coloured $Fe_3O_4$ will be formed. This $Fe_3O_4$ is filtered off and washed with deionized water and ethanol and dried in a vacuum oven at 60 °C for 5 h. The dried $Fe_3O_4$ is annealed at 550 °C for 2 to 5 h and finally forms $Fe_2O_3$ powder [15]. The formed $Fe_2O_3$ powder is dispersed in Triton X binder solution and by using Dr. Blade technique, the $Fe_2O_3$ thin film is deposited and annealed at 500 °C for the formation of ITO/NiO/CuO/$Fe_2O_3$ multi-layer structure.

The individual layers of NiO, CuO and $Fe_2O_3$ and the multi-layered structure of NiO/CuO/$Fe_2O_3$ have been explored using the surface science techniques like X-ray diffractometer (XRD, Proto, Canada) with Cu kα radiation, $\lambda = 0.154$ nm, operated at 30 kV and 20 mA. The multi-layered surface morphological analysis has been done using scanning electron microscope (SEM, Zeiss (EVO LS 15), 15 kV acceleration voltage, Germany).

## 3 Results and Discussion

The SEM images of NiO thin film on top of ITO glass substrate synthesized using chemical bath deposition at room temperature are shown in Fig. 2. The SEM images reveal the formation of the porous nature of NiO with the random distribution of clusters all over the material over ITO-coated glass substrates with uniform coverage of NiO thin film. The overgrowth of clusters is probably due to the nucleation and coalescence process.

Figure 3 represents the SEM images of the CuO layer deposited upon NiO-coated ITO substrate by hydrothermal method. From Fig. 3a the formation of flower bud-like structures can be visualized with the grain size of 86 nm and from Fig. 3b can be visualized the irregular and inhomogeneous distribution of CuO particles all over the porous NiO thin film.

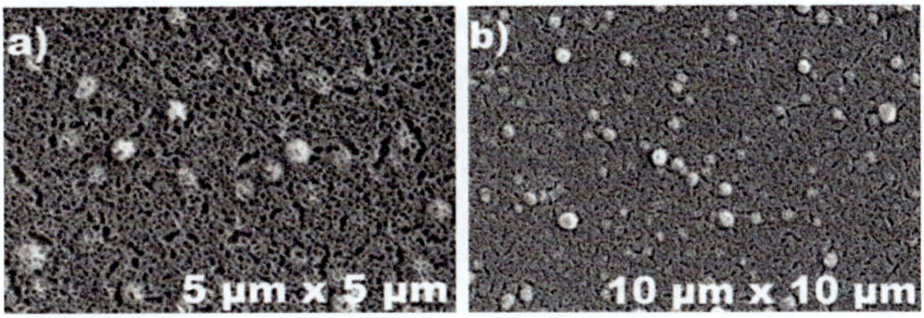


**Fig. 2** SEM images of NiO thin film on top of ITO glass plate, deposited using the chemical bath deposition technique

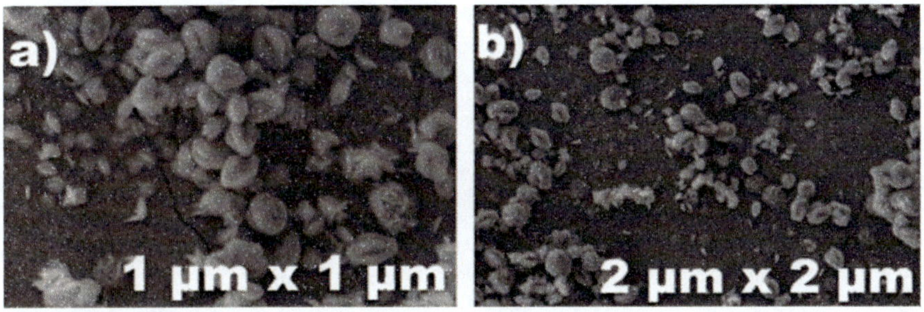


**Fig. 3** SEM images of CuO thin film deposited on NiO/ITO substrate using the chemical bath deposition method

This indicates the irregular surface coverage, and this can alter the properties measured at different places of the CuO-covered film.

Figure 4 represents the SEM images at two different magnifications of $Fe_2O_3$ thin film upon the CuO/NiO/ITO substrate, synthesized by Dr. Blade's technique. Figure 4a, b shows the formation of big sticky grains of CuO, where small particles of $Fe_2O_3$ got stuck with them because of the use of the binder.

This has made the entire structure sturdy. The use of a binder can affects the mechanical/electrical properties too (further studies required).

Figure 5 represents the XRD theta-2theta scan of the NiO thin film on top of ITO substrates after annealing at 300 °C for 1.5 h. From the XRD scan, the peaks at 37.21°, 44.01° and 63.25° corresponding to (111), (200) and (220) peaks, respectively, indicating the poly-crystalline nature of the synthesized thin film. The hump-like structure between 10° to 40° in 2θ range probably indicates the partial amorphous nature of the film.

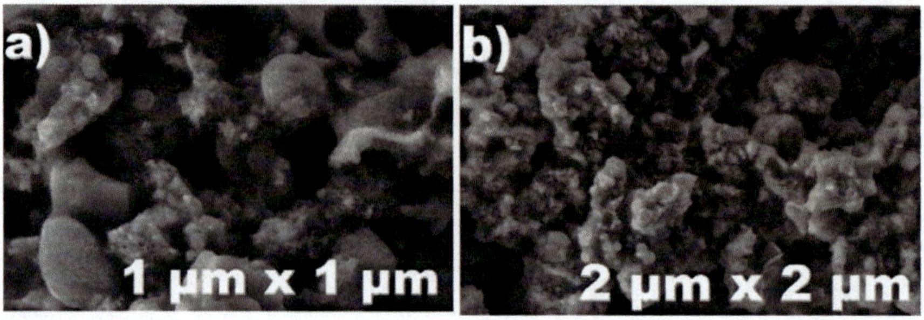


**Fig. 4** SEM images of $Fe_2O_3$ deposited on top of CuO/NiO/ITO substrate using the Dr. Blade technique

**Fig. 5** XRD scan of NiO thin film deposited on ITO substrate by CBD technique

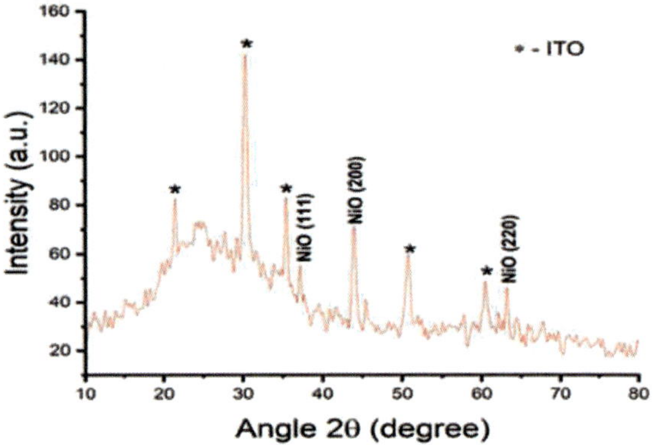


The crystallite size was calculated using the Debye Scherer formula and d spacing by Bragg's law shown in Eqs. 1 and 2, respectively. For the most pronounced peaks, the sizes of crystallites were varying in the range of 14–24 nm, and the d spacing was found to be 0.2 nm. The crystallite sizes and d spacing of the most pronounced peaks are tabulated in Table 1.

$$\text{Size of the crystallite} = \frac{k\lambda}{\beta \cos\theta} \quad (1)$$

$$n\lambda = 2d \sin\theta \quad (2)$$

Figure 6 depicts the theta-2theta XRD scan of CuO thin film deposited using low-temperature hydrothermal method upon ITO-coated glass substrate after annealing at 200 °C for 30 min. From the XRD scan (−110), (110), (002) and (111)/(200) were phases of dominant peaks at 29.93°, 32.11°, 35.26° and 38.59°, respectively [9]. Amongst all the peaks, the peak at 35.26° corresponding to (002) phase of

**Table 1** Crystallite size and d spacing of most pronounced NiO peaks

| Peak | Position, 2θ (degrees) | Full width half maximum (FWHM) (degrees) | D spacing (nm) | Crystallite size (nm) |
|---|---|---|---|---|
| NiO (111) | 37.21 | 0.341 | 0.24 | 24 |
| NiO (200) | 44.01 | 0.576 | 0.20 | 15 |
| NiO (220) | 63.25 | 0.415 | 0.14 | 23 |

the crystallite is having the highest intensity indicating the maximum numbers of crystallites growth along the surface normal direction.

The crystallite size of all the peaks mentioned in Table 2 is varying in the range from 9 to 94 nm (using the Debye Scherrer formula). The peaks obtained in the XRD scan are matching with the work done by Joshi et al., indicating the formation of poly-crystalline nature of CuO thin film [9]. The hump-like structure between 10° to 40° is probably due to the presence of the partial amorphous nature of CuO. Both amorphous and poly-crystalline matrixes are embedded in the synthesized CuO thin film. The crystallite size and d spacing have been tabulated in Table 2 for CuO peaks.

Figure 7 shows the theta-2theta XRD scan of $Fe_2O_3$ thin film deposited on top of ITO-coated glass substrate by Dr. Blade technique after annealing at 550 °C. It is observed that the strong $Fe_2O_3$ peaks with high intensities at 33.46°, 35.94°, 49.89° and 54.39° correspond to (104), (110) and (024) and (116) phase, respectively [9].

The high-intensity peak at 33.46° corresponding to (104) phase of the crystallite indicated the growth of the crystal in the surface normal direction. The high-intensity peaks are appearing in $Fe_2O_3$ due to the regular arrangement of crystallite all over the material with crystallite size in the range of 3–48 nm. The crystallite size and d spacing for all the peaks have been tabulated in Table 3.

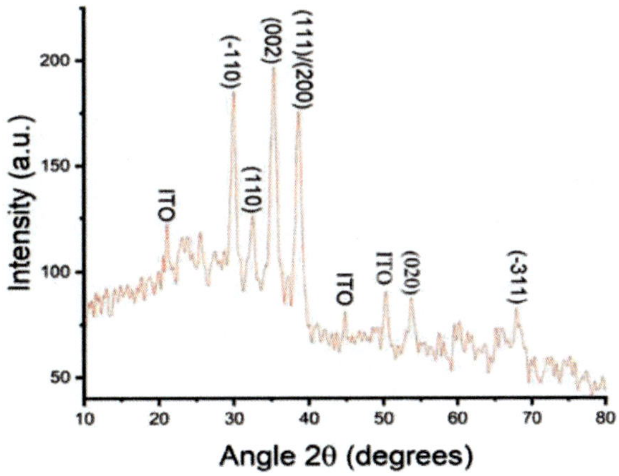


**Fig. 6** XRD scan of CuO deposited on ITO substrate by hydrothermal method

**Table 2** Crystallite size and d spacing of most pronounced CuO peaks

| Peak | Position, 2θ (degrees) | Full width half maximum (FWHM) (degrees) | D spacing (nm) | Crystallite size (nm) |
|---|---|---|---|---|
| CuO (−110) | 29.92 | 0.6198 | 0.3 | 13 |
| CuO (110) | 32.11 | 0.0933 | 0.28 | 88 |
| CuO (002) | 35.26 | 0.8846 | 0.25 | 9 |
| CuO (111)/(200) | 38.59 | 0.8378 | 0.23 | 10 |
| CuO (020) | 53.75 | 0.6022 | 0.1 | 14 |
| CuO (−311) | 67.81 | 0.1014 | 0.14 | 94 |

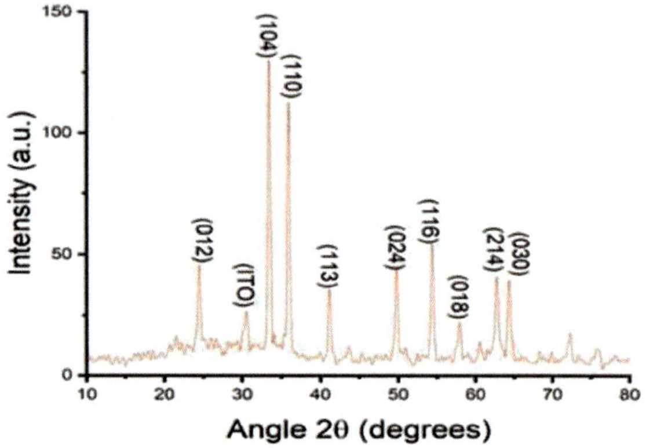


**Fig. 7** XRD scan of $Fe_2O_3$ deposited on ITO substrate by Dr. Blade method

**Table 3** Crystallite size and d spacing of most pronounced $Fe_2O_3$ peaks

| Peak | Position (degrees) | Full width half maximum (FWHM) (degrees) | D spacing (nm) | Crystallite size (nm) |
|---|---|---|---|---|
| $Fe_2O_3$ (012) | 24.47 | 0.2115 | 0.4 | 3 |
| $Fe_2O_3$ (104) | 33.46 | 0.2870 | 0.3 | 28 |
| $Fe_2O_3$ (110) | 35.94 | 0.1751 | 0.25 | 48 |
| $Fe_2O_3$ (113) | 41.17 | 0.3111 | 0.22 | 28 |
| $Fe_2O_3$ (024) | 49.89 | 0.3670 | 0.18 | 24 |
| $Fe_2O_3$ (116) | 54.39 | 0.2808 | 0.17 | 32 |
| $Fe_2O_3$ (018) | 57.91 | 0.3411 | 0.16 | 27 |
| $Fe_2O_3$ (214) | 62.74 | 0.3479 | 0.15 | 27 |
| $Fe_2O_3$ (030) | 64.39 | 0.4462 | 0.14 | 22 |

Figure 8 represents the XRD scan of the multi-layered structure of ITO/NiO/CuO/$Fe_2O_3$ in the range of 10°–80° for θ–2θ scan. The CuO and $Fe_2O_3$ peaks are named and respectively. From the XRD scan, it is clear that the formed multi-layered structure was poly-crystalline with the presence of dominant $Fe_2O_3$ peaks and CuO peaks of very low intensity. The dominance of $Fe_2O_3$ peaks could be due to the presence of $Fe_2O_3$ on the top surface.

For the optical characterization, the UV–Vis spectroscopy technique has been employed to study not only the individual layers but also the combinations of multi-layered systems too. Figure 9 represents the absorbance and transmittance curve of NiO deposited on ITO-coated glass substrate by the chemical bath deposition technique.

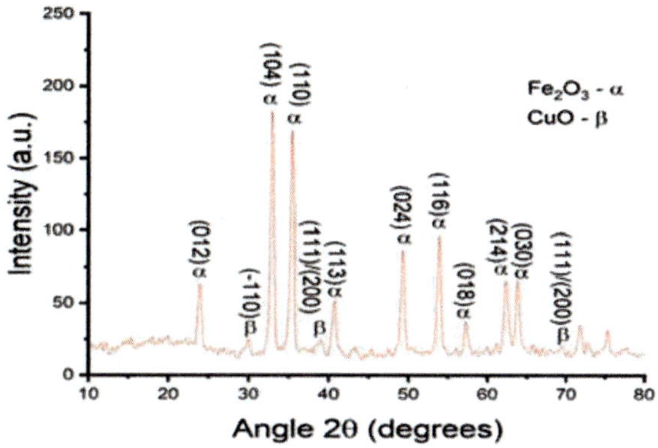


Fig. 8 XRD scan of ITO/NiO/CuO/$Fe_2O_3$ multi-layered structure

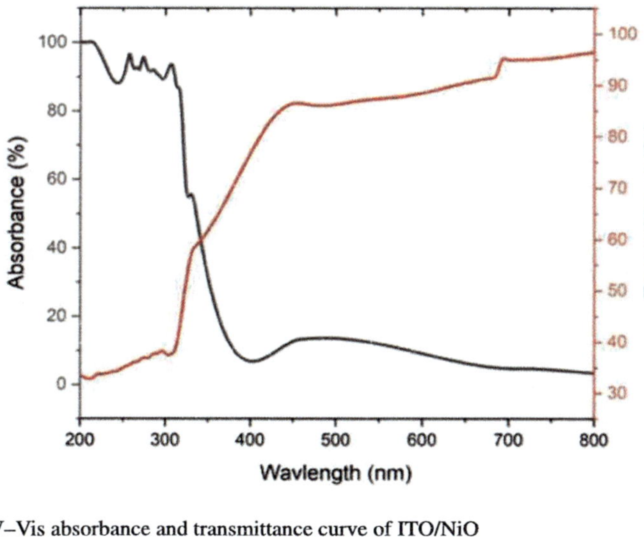


Fig. 9 UV–Vis absorbance and transmittance curve of ITO/NiO

Under the absorbance curve, one can notice the broad absorption peak at 290 nm, as well as the small and broad peak, cantered at 550 nm, which indicates the presence of broadly two absorption regions. In the transmittance spectra, due to the wide bandgap of 3.67 eV (not shown here), the transmittance is constantly increasing with an increase in wavelength. In transmittance spectra, a peak at 450 nm is also observed. The maximum transmittance obtained is around 96%. Due to this high transmittance nature, NiO is favourable as a light-transmitting material and also due to its p-type nature acting as a hole transporting layer too.

Figure 10 represents the absorbance and transmittance curve of the CuO layer deposited over the NiO/ITO substrate using a low-temperature hydrothermal deposition technique. From the absorbance curve, one can see that except for the UV region, i.e. from 200 to 294 nm, the absorption curve has two absorption peaks centred around 350 nm as well as at 568 nm. Further, due to lower energy at a high wavelength, the absorption curve decreases to around 72%.

The second peak has the highest absorption at 568 nm as well as it has the broad distribution of the incident light absorption, indicating that the maximum photon falling and getting absorbed of visible wavelength between 400 and 600 nm and ultimately, the maximum electron–hole pair generation in the given visible range. In the transmittance curve, the peak is observed at 291 nm indicating the maximum transmittance in the UV region, and the transmittance decreases till the end of the visible region.

Figure 11 represents the absorbance and transmittance curve of $Fe_2O_3$ deposited on top of CuO/NiO/ITO by Dr. Blade's technique. The transmittance curve of the multi-layered ITO/NiO/CuO/$Fe_2O_3$ structure indicates low transmittance in the

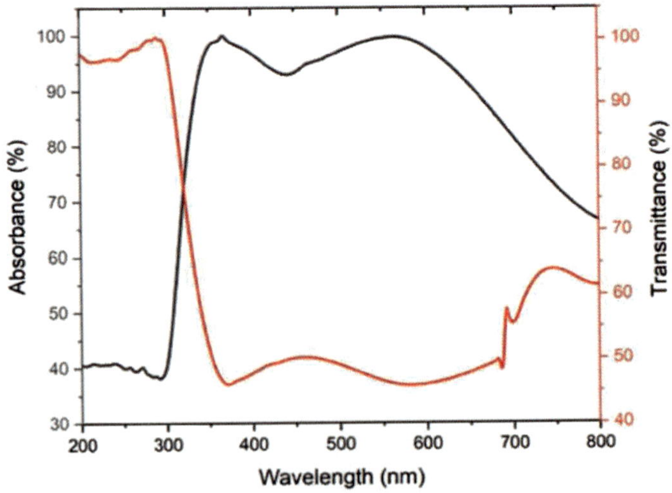


**Fig. 10** UV–Vis absorbance and transmittance curve of CuO on NiO/ITO substrate

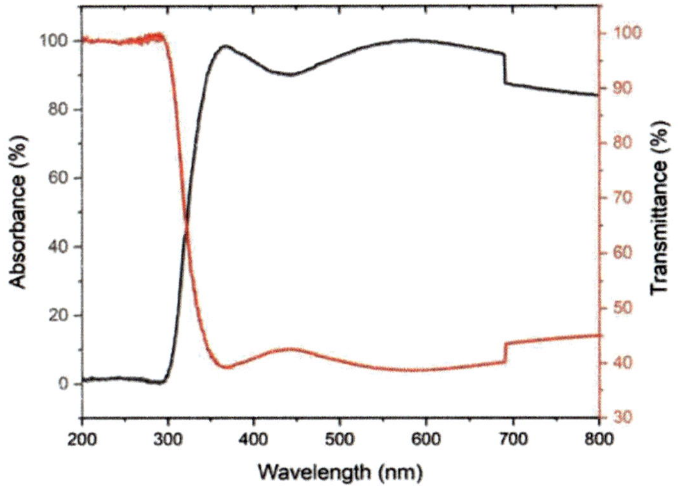


**Fig. 11** UV–Vis absorbance and transmittance curve of ITO/NiO/CuO/$Fe_2O_3$

visible region, which is the primary region of visible light absorption and indicates that the entire multi-layer structure ITO/NiO/CuO/$Fe_2O_3$ is suitable for the photovoltaic application.

The absorbance curve of the multi-layered structure has a peak at 369 nm at the edge of UV and visible band, afterwards, the curve fluctuates a little, which could be due to the multi-layer system and varied between 90 and 100%.

## 4 Conclusion

The simple cost-effective, wet chemical process has been performed for the synthesis of inorganic-based NiO, CuO and $Fe_2O_3$ semiconducting materials which has the potential applications in the field of Photovoltaic devices. The core focus of present work is on synthesis and characterization of multi-layered NiO/CuO/$Fe_2O_3$ structure for photovoltaic applications. In the present work, the synthesis of all the individual layers of NiO, CuO and $Fe_2O_3$ was successfully carried out on top of the modified substrate. The XRD analysis of NiO thin film on top of ITO-coated glass substrate shows the poly-crystalline nature with the highest surface coverage with porous structure with uniform and highest surface coverage. At the same time, the UV–Vis measurement of NiO film reveals the gradual increase in transmittance in the visible region makes it a perfect layer for light transmitting with minimum absorption.

The CuO thin film deposited on top of NiO/ITO substrate using the hydrothermal technique, shows the poly-crystalline nature with the growth of crystallite along the surface normal direction. The crystallite appears to be flower-like structure.

The UV–Vis spectroscopy study reveals the maximum absorption of around 95% for CuO/NiO/ITO in the visible region. Finally, the $Fe_2O_3$ was coated on top of CuO/NiO/ITO by Dr. Blade's technique. The synthesized $Fe_2O_3$ thin film is poly-crystalline with the random arrangement of crystallites all over the material indicating the formation of the various miller planes along the surface's normal direction. Whereas, the SEM micrograph of $Fe_2O_3$ indicates the formation of small spherical particles grains, which agglomerated with flower buds of CuO in the presence of a binder, both CuO and $Fe_2O_3$ particles are making a rigid structure, where small particles of $Fe_2O_3$ are nicely distributed within the flower buds of CuO. At the same time, UV–Vis spectroscopy study of multilayer structure ITO/NiO/CuO/$Fe_2O_3$ shows almost constant absorption, which could be due to the contribution of three individual layers like NiO, CuO and $Fe_2O_3$.

Overall, all the measurements performed and analysed indicated the overall poly-crystalline nature of the three-metal oxide thin films prepared using solution-processable methods, whereas the optical properties using UV–Vis indicate the perfect suitability of this entire multi-layered structure, which could be utilized for the photovoltaic/sensors applications.

**Acknowledgements** The authors like to thank UGC DAE and NIE Mysuru for providing the financial support for purchasing chemicals for the above work. One of the authors like to thank Dr. Siddharth Joshi for providing an opportunity to develop master thesis work during this pandemic situation as well as for his constant support throughout the project work.

## References

1. Mureddu M, Caldarelli G, Chessa A, Scala A, Damiano A (2015) Green power grids: how energy from renewable sources affects networks and markets. PLOS ONE 10:1–15
2. Shukla AK, Sudhakar K, Baredar P (2016) Design simulation and economic analysis of stand-alone rooftop solar PV system in India. Solar Energy 136:437–449
3. Calnan S (2014) Applications of oxide coatings in photovoltaic devices. Coatings 4:62–202
4. Bazrafshan H, AlipourTesieh Z, Dabirnia S, ShajarehTouba R, Manghabati H, Nasernejad B (2017) Synthesis of novel $\alpha$-$Fe_2O_3$ nanorods without surfactant and its electrochemical performance. Powder Technol 308:266–272
5. Nishino F, Jeem M, Zhang L, Okamoto K, Okabe S, Watanabe S (2017) Formation of CuO nano-flowered surfaces via submerged photo-synthesis of crystallites and their antimicrobial activity. Scientific Reports 7:1–11
6. Welegergs G, Akoba R, Sacky J, Nuru ZY (2020) Structural and optical properties of copper oxide (CuO) Nanocoatings as a selective solar absorber. Mater Today: Proc 5:1–5
7. Wu G, Wang C, Zhang X, Yang H, Qi Z, Li W (1998) Lithium insertion into CuO/carbon nanotubes. J Power Sour 75:175–179
8. Pérez-Tomás A, Mingorance A, Tanenbaum D, Lira-Cantú M (2018) Metal oxides in photovoltaics: all-oxide, ferric, and perovskite solar cells. Future Semiconductor Oxides Next-Generation Solar Cells 8:267–356
9. Joshi S, Mudigere M, Krishnamurthy L, Shekar G (2014) Growth and morphological studies of NiO/CuO/ZnO based nanostructured thin films for photovoltaic applications. Chem Papers 68:1–9

10. Dimopoulos T, Peic A, Müllner P, Neuschitzer M, Resel R, Abermann S, Brückl H (2013) Photovoltaic properties of thin film heterojunction with cupric oxide absorber. J Renew Sustain Energy 5:1–12
11. FerielBouhjar M, Mollar S, Ullah B, Marí B, Bessaïs B (2018) Influence of a compact $Fe_2O_3$ layer on the photovoltaic performance of perovskite-based solar cells. J Electrochem Soc 165:30–38
12. Manikandan A, Saravanan A, Antony SA, Bououdina M (2015) One-pot low-temperature synthesis and characterization studies of nanocrystalline $\alpha$-$Fe_2O_3$ based dye-sensitized solar cells. J Nanosci Nanotechnol 15:4358–4366
13. H X. Xia J, Tu P, Zhang J, Wang XL, Zhang WK, Huang H (2008) Electrochromic properties of porous NiO thin films prepared by a chemical bath deposition. Solar Energy Mater Solar Cells 92:628–633
14. Senthilkumar V, Kim YS, Chandrasekaran S, Rajagopalan B, Kim EJ, Chung JS (2015) Comparative super capacitance performance of CuO nanostructures for energy storage device applications. RSC Adv 5:20545–20553
15. Tian W, Wang X, Zhi C, Zhai T, Liu D, Zhang C, Bando Y (2013) Ni $(OH)_2$ Nanosheet @ $Fe_2O_3$ nanowire hybrid composite arrays for high-performance supercapacitor electrodes. Nano Energy 2:754–763

# Binary Decision Diagram Model Based Reliability Prediction of PV Powered Quasi Z-Source Inverter

**S. Madhumitha, R. Sudiksha, R. Seyezhai, D. Umarani, and R. Sujatha**

## 1 Introduction

Owing to the spike in power requirement of the globe there has been an increasing shift toward the greener and more efficient photovoltaic (PV) power fed grids. In conventional PV systems string inverters, central inverters or module inverters are generally employed, but they have a major shortcoming like a lower output voltage than the DC voltage thus resulting in a dismal spectral quality. Therefore, a PV quasi Z-source inverter with (PVQZSI) is proposed in this work. The major gain in this system is that it provides step up or step-down operation with direct to alternating current, i.e., single-stage power inversion by adding an impedance network and allowing short circuit across the switches [1].

A PV system comprises of major components, namely, the solar array and inverter. Failure of one of these components will result in the failure of the whole PV system [2]. Owing to this reason, the selection and implementation of power converter circuit is a crucial step in the design phase which may increase in the cost of the entire system. Nowadays, power electronic converters act as interface for renewable energy systems, electric vehicles, and various other applications. Sensitive commercial, military, and space applications demand high reliability in each part to prevent any major disasters. So, it is very essential to study the reliability modeling of the power converter using mathematical approaches.

Reliability of any system can be best described as the capability of that system to be eligible to fulfill the required functionality expected from it at specified time

S. Madhumitha
Department of CSE, Sri Sivasubramaniya Nadar College of Engineering, Chennai, India

R. Sudiksha · R. Seyezhai (✉) · D. Umarani
Department of EEE, Sri Sivasubramaniya Nadar College of Engineering, Chennai, India
e-mail: seyezhair@ssn.edu.in

R. Sujatha
Department of Mathematics, Shiv Nadar University, Chennai, India



N. L. Narasimhan et al. (eds.), *Recent Advances in Energy Technologies*, Lecture Notes in Mechanical Engineering, https://doi.org/10.1007/978-981-19-3467-4_9

with constraints. The reliability modeling majorly throws light upon the relational algebraic dependency between the system and various components. The system's characteristics determine the choice of reliability modeling method, such as fault tree, reliability block diagram, petri net model, network reliability model, or markov's model [3]. This is a quantitative analytic tool to finally obtain the failure rate of the entire system. In this paper, the fault tree model is considered for establishing the reliability model of PVQZSI system to find the fault probability.

The quasi Z-source inverter discussed in this paper includes an H-bridge consisting of MOSFET switches and an impedance network. The failure of the inverter depends on failure of MOSFET switches, inductors, capacitors, diode, wirings, etc. Hence, finding the ability of the components to perform the assigned function under the given environment conditions for a specific period of time would help to know the performance of the PV system. This paper focuses on reliability analysis of PV-based QZSI using part count model, fault tree model, and binary decision diagram. The electrical and the principle diagram of PV based QZSI system are elucidated. The fault tree analysis model (FTA) is developed for the PVQZSI. The binary decision diagram (BDD) is derived from the FTA for reliability analysis. The proposed method is proved to be suitable and practically viable for the proposed system reliability analysis for the estimation of failure probability of the topmost event.

## 2 Single-Phase Quasi Z-source Inverter

In the class of impedance source inverters, QZSI is a transformerless DC-AC converter topology. It involves a specific impedance network between PV source and the inverter H-bridge to perform DC-DC and DC-AC conversion. This inverter works under shoot-through stage to boost the input voltage by LC or impedance network. During the other stage, the boosted input voltage is converted to AC. So, the main advantage of QZSI is the single stage power conversion with high efficiency. It possess the advantages of conventional impedance source inverter (ZSI) such as of lower component ratings, improved reliability and continuous input current from the PV source [4]. It has a variable voltage gain that is required for any PV system as its output varies with varying temperature and irradiation. The shoot-through states and the index of modulation are used to control the output voltage of PVQZSI. Different modulation strategies are found to be reported in the literature for the PVQZSI [5]. This paper implements simple boost control for generating the shoot through and PWM pulses. Simulation of the QZSI is implemented in PSIM. The circuit diagram of the proposed QZSI is shown in Fig. 1.

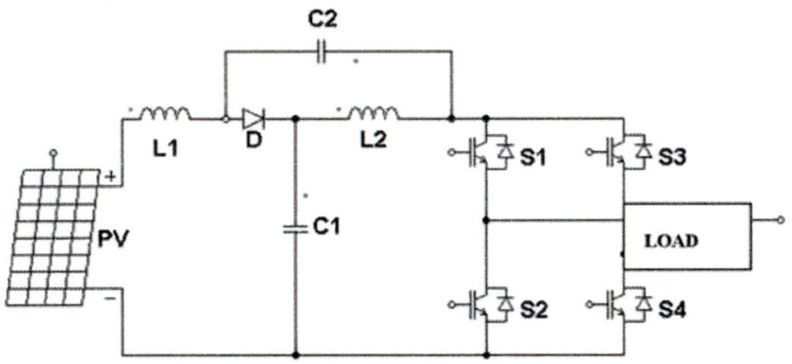


**Fig. 1** Single-phase quasi Z-source inverter with PV source

## 2.1 *Operating Modes of QZSI*

The QZSI operates in two states such as non-shoot-through state and shoot-through state [5, 6].

**Non-Shoot-through mode**

As shown in Fig. 2, the diode is short circuited and the switching pulses applied are similar to VSI. The input PV voltage which is boosted will be available as DC link voltage to the H-bridge. In this stage, boosted DC is converted to AC.

**Shoot-through mode**

As depicted in Fig. 3, the switches in the same H-bridge leg are turned on together for quick short interval. The resulting short circuit will not affect the source due to the presence of quasiimpedance network. Now, the input voltage is boosted, and DC link voltage depends on the boost factor which depends on shoot through for the given modulation index.

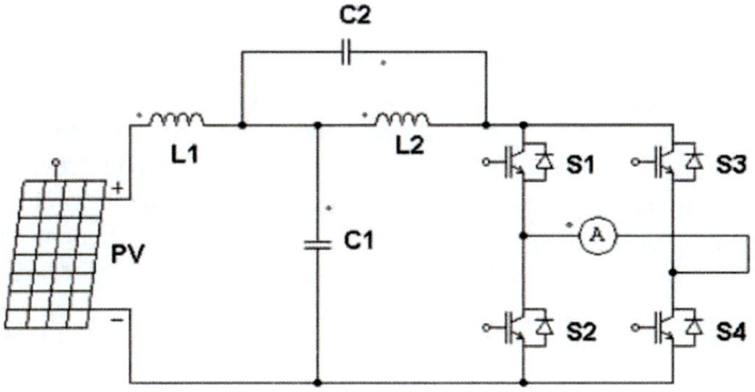


**Fig. 2** Non-shoot-through mode of operation of QZSI

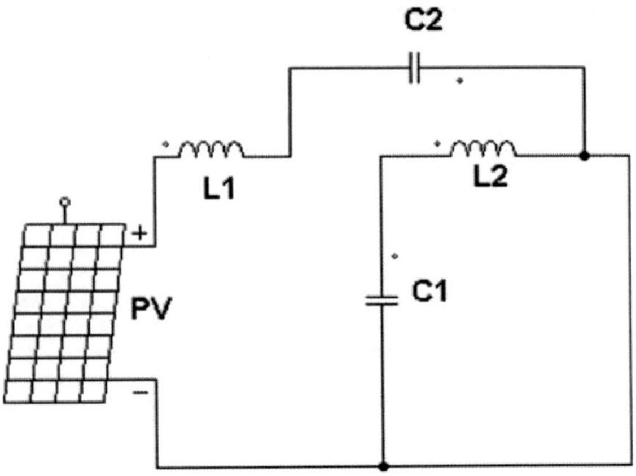


**Fig. 3** Shoot-through mode of operation of QZSI

**Table 1** Design parameters of PVQZSI

| Parameter | Rating |
|---|---|
| PV input voltage | 30.5 V |
| Switching frequency | 5 kHz |
| Boost factor | 2 |
| Shoot-through duty ratio | 0.8 |
| Inductor $L_1, L_2$ | 3 mH |
| Capacitor $C_1, C_2$ | 2.2 mF |

### 2.2 *Design Parameters for PVQZSI*

The design parameters of PVQZSI are shown in Table 1.

### 2.3 *Simple Boost Control Technique for QZSI*

To vary the output voltage of PVQZSI, the simple boost control (SBC) technique is implemented. The shoot-through pulses are generated by the upper and lower shoot-through envelopes compared with the triangular carrier. The shoot-through envelopes are positioned in required bias to control the shoot-through duty ratio. It has uniform distribution of shoot-through states to make output ripple free. Figure 4. shows the pulse pattern generated by simple boost control, and Fig. 5. shows the output voltage of QZSI.

The applied input voltage of 30.5 V is boosted and converted to an AC peak voltage of 61 V.

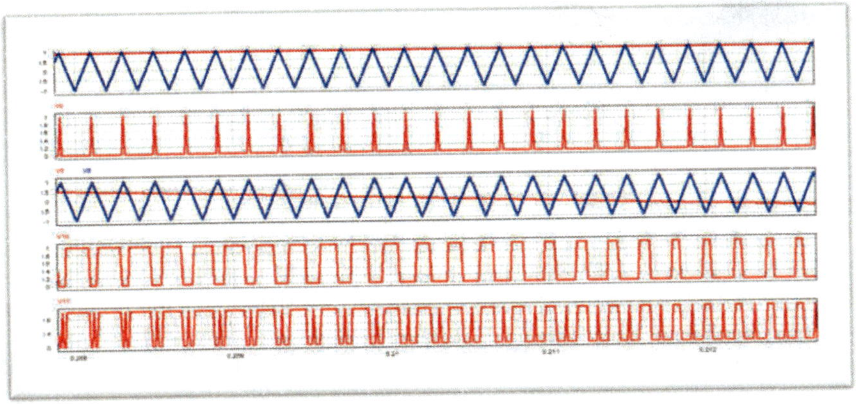

**Fig. 4** Gating pulse generation for simple boost control method

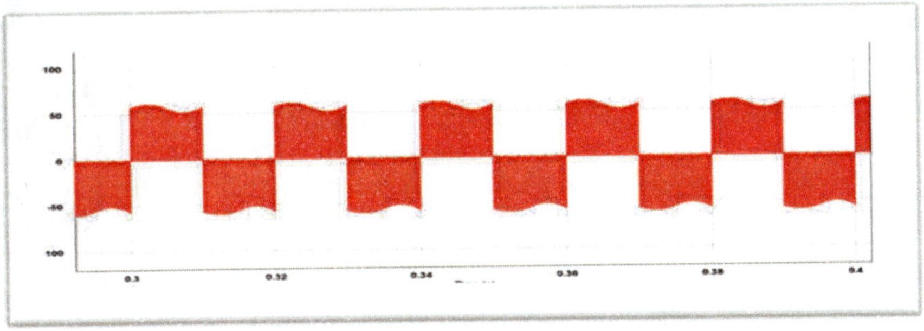

**Fig. 5** Output voltage waveform of QZSI

## 3 Reliability of PV System

The commercial success of PV systems majorly depends on the reliability. Evaluating the reliability of any PV system is of importance when considering its maintenance cost. Owing to the application in emerging areas, the need for enhancing the efficiency of a PV system is on the rise [7, 8]. Reliability methods help in estimating the lifetime of the components/systems which may be essential either at the initial design stage or while operating the system to predict the failure and repair rate. Several reliability assessment methods of power electronic converters are investigated in the literature [9, 10]. They are classified as part count, combinatorial, and state-space models. Part count model calculates the failure rate of each component in the system. If a component fails, the entire system is assumed to fail as the failure rate is assumed to be constant. This method is simple and applicable for small systems, whereas combinatorial and state-space models deal with system level failures. Combinatorial models like fault tree and reliability block diagram deals with simple redundant systems. The associated failure metrics discussed in this section assists in computing the system failure rate.

### 3.1 Failure Metrics

The important failure metrics are mean time to repair (MTTR), mean time to failure (MTTF), and mean time between failures (MTBF) for lifetime prediction of a system [11]. They provide numerical value for quantifying the failure rate and the expected lifetime. The numerical value can be expressed in hours.

MTBF is a significant reliability metric that is best defined as the number of failures of the system for a span of over million hours. This measure reflects the system's lifetime and is very crucial for evaluating the productivity measure of the system.

Mean time to repair (MTTR) proves to be another indispensable metric that gives the average amount of time needed to repair the system to bring it back to a functional state. This could range from any hardware to software component. This parameter also considers the replacement time along with the time window that is required to slate the installation of the new or replacement component.

Mean time to failure (MTTF) is a measure similar to MTTR but applied only for non-repairable components. This measure gives the average of the exact time frame that the component takes before it fails.

Failure in time (FIT) is an estimation of the expected failure rate of the component for a time period of one million hours. This metric is largely used in the electronics and semi-conductor industry. This metric can be equated to that of MTBF. The estimation can be scheduled for any one component over million hours or for a million components over an hour.

MTBF is given by

$$\text{MTBF} = 1 + \text{FR}_1 + \text{FR}_2 + \text{FR}_3 + \ldots. + \text{FR}_n \tag{1}$$

where FR represents the failure rate of $n$th component.

## 4 Part Count Model of PVQZSI

The part count method is used in the design phase. The components in the system like capacitors, resistors, and inductors are fixed with the condition that the design parameters are not changed in later stages of development. The time to failure of parts is assumed to be distributed exponentially. The failure rate of the system is the summation of failure rates of the components which are in series connection. If the system has non-series elements, failure rate is estimated either by considering only series elements by adding the failure rates of individual element [12]. The failure rates of the components are obtained from the database Military Handbook MIL-HBK 217. For a system having n subsystems, then the failure rate $\lambda$ is given by,

$$\lambda = \sum_{i=1}^{n} \lambda_i \tag{2}$$

where $\lambda_i$ is the failure rate of the $i$th subsystem.

The MTTF of the QZSI is given by,

$$\text{MTTF} = \frac{1}{\lambda_i} \tag{3}$$

If the equipment consisting of various parts, operating in various environments, then Eq. (1) is applied to each part of the equipment in that environment. Summing failure rates of all conditions will represent the overall equipment failure rate [13]. A PV-based QZSI is divided into three parts, namely, PV array, H-bridge, and impedance network. The part count model of PV QZSI is shown in Fig. 6.

The failure rate of solar panel is 2.96 failures per $10^6$ h and charge controller is 0.1750 failures per hours. The failure rate of components like inductor, capacitor, and diode are estimated from MIL HB 217 [14] and it is depicted in Tables 2, 3, 4 and 5.

From the $\lambda^P$ values of all components, the total failure for QZSI is computed as

$$\lambda_{p(\text{System})} = \lambda_{p(\text{Solar})} + 4\lambda_{p(\text{MOSFET})} + 2\lambda_{p(\text{Capacitor})} + 2\lambda_{p(\text{Inductor})} + \lambda_{p(\text{Diode})} \tag{4}$$

$$\lambda_{p(\text{System})} = 4.4928 \text{ failures}/10^6\text{h}$$

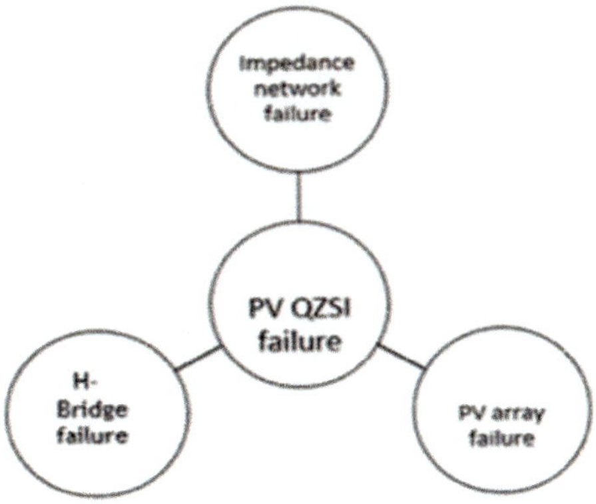


**Fig. 6** Part count model for PVQZSI

**Table 2** Failure rate calculation for Inductor

| Factor | Values |
|---|---|
| $\lambda b$ | 0.00053 |
| $\pi c$ | 1 |
| $\pi E$ | 1 |
| $\pi Q$ | 20 |
| $\lambda p = \lambda b * \pi c * \pi E * \pi Q$ | 0.0106 failures per $10^6$ h |

**Table 3** Failure rate calculation for capacitor

| Factor | Values |
|---|---|
| $\lambda b$ | 0.0428226 |
| $\pi c$ | 0.30 |
| $\pi E$ | 1 |
| $\pi Q$ | 10 |
| $\pi cv$ | 1.3627 |
| $\lambda p = \lambda b * \pi c * \pi E * \pi Q * \pi cv$ | 0.1750 failures per $10^6$ h |

**Table 4** Failure rate calculation for diode

| Factor | Values |
|---|---|
| $\lambda b$ | 0.0010 |
| $\pi T$ | 1.92006 |
| $\pi C$ | 1 |
| $\pi Q$ | 5.5 |
| $\pi E$ | 1 |
| $\pi S$ | 0.054 |
| $\lambda P = \lambda b * \pi T * \pi C * \pi Q * \pi E * \pi S$ | 0.00057 failures per $10^6$ h |

**Table 5** Failure rate calculation for MOSFET switch

| Factor | Values |
|---|---|
| $\lambda b$ | 0.012 |
| $\pi T$ | 1.1 at 30 °C |
| $\pi A$ | 4 |
| $\pi Q$ | 5.5 |
| $\pi E$ | 1 |
| $\lambda P = \lambda b * \pi T * \pi A * \pi Q * \pi E$ | 0.2904 failures per $10^6$ h |

MTBF is given by

$$\text{MTBF} = \frac{1}{\lambda_{p(\text{System})}} = 222578.3475\ \text{h} \tag{5}$$

**Table 6** Failure rate of each component of PVQZSI

| Components | Failure rate |
|---|---|
| Solar panel (q1) | 2.96 failures per $10^6$ h |
| Charge controller (q2) | 0.1750 failures per $10^6$ h |
| Inductor L1 (q3) | 0.0106 failures per $10^6$ h |
| Inductor L2 (q4) | 0.0106 failures per $10^6$ h |
| Capacitor C1 (q5) | 0.1750 failures per $10^6$ h |
| Capacitor C2 (q6) | 0.1750 failures per $10^6$ h |
| Diode D (q7) | 0.00057 failures per $10^6$ h |
| MOSFET Switches (q8,q9,q10,q11) | 0.2904 failures per $10^6$ h |

So, the failure rate of PV-based QZSI by part count method is 4.4928 failures/$10^6$ h.

## 5 Binary Decision Diagram of PVQZSI

Binary decision diagram is one of the methods of mathematical modeling used to represent fault tree or the logical combinations resulting in failure of top event in the form of boolean expression. They are effectively used to encode the boolean representation of the failure rate equations. A minimal cut set is defined as the minimal component failures that would further end up in the failure of the entire system [15]. BDD diagrams help to efficiently isolate these cut sets and thus calculate the failure rate by giving an expression for the same.

### *5.1 Fault Tree*

Fault tree analysis is one of the most primary techniques that help in calculating the reliability of any complex system. This involves a top down approach where the entire system is broken down into various modules which are then analyzed for calculating the failure rate. The fault tree structure is a deductive method in which one can obtain the logical relationships between the failure events of different submodules in the complex system. Here, fault refers to the event of failure of a particular submodule that in turn being computed for the entire system [16, 17]. This is regarded as the most widely used reliability assessment tool in the industrial sector. Fault trees make the debugging process much simpler and are easy to interpret. In this work, the event of failure of a module is related with probability of unity. This structure employs basic level gates such as AND, OR, XOR, and XNOR . for the logical relation. In this work, single-phase QZSI with lesser number of components and being a simpler system AND and OR gates are employed to derive the logical condition for the failure of the entire system. Fault tree diagram corresponding to QZSI is developed using fault tree analyzer software and it is depicted in Fig. 7. The binary decision diagram constructed from the above fault tree is given in Fig. 8.

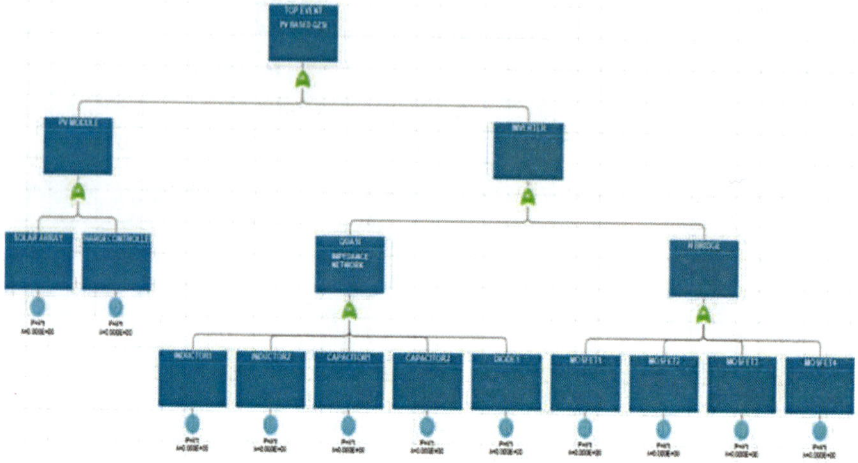


**Fig. 7** Fault tree diagram for QZSI

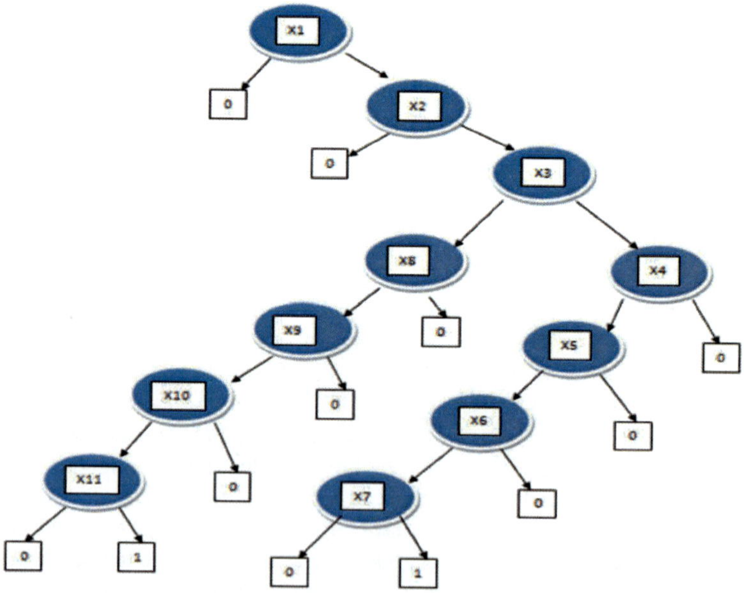


**Fig. 8** Binary decision diagram for QZSI

**Table 7** Labels for events in BDD

| Label | Event | Label | Event | Label | Event |
|---|---|---|---|---|---|
| X1 | Event of system failure | X5 | Inductor L2 failure | X9 | Event of switch S2 failing |
| X2 | Impedance network failure | X6 | Capacitor C1 failure | X10 | Event of switch S3 failing |
| X3 | H-bridge failure | X7 | Diode failure | X11 | Event of switch S4 failing |
| X4 | Inductor L1 failure | X8 | Event of switch S1 failing | | |

In Fig. 7, the PVQZSI is divided into various subsystems and eventually subcomponents. The failure of PV based QZSI is considered to be the prime event and followed by dividing it into subsystems. The PV module subsystem encompasses the solar array and the charge controller circuitry. The inverter subsystem covers the most important quasi impedance network and the H-bridge. These are then further decomposed into their constituents, namely, the capacitors, inductors, and MOSFET switches that forms the basic events of the fault tree.

## 5.2 Constructing BDD from Fault Tree

While developing the BDD from fault tree diagram the entire system is divided into subcomponents as specified in the fault tree. Then each event of failure of any component is labeled as X1, X2, etc., and the possible flow of events that would occur after the failure is sketched out in the flow of the BDD. This helps to deduce the working of the system in the event of failure of any one or certain combination of components. The labels and events for BDD of QZSI which is constructed from fault tree is listed in Table 7.

The Boolean equation deduced from the BDD are

$$\begin{aligned} &q1 + (1 - q1)q2 + (1 - q1)(1 - q2)q3q8 + (1 - q1)(1 - q2) \\ &\quad q3(1 - q8)q9 + (1 - q1)(1 - q2)q3(1 - q8)(1 - q9) \\ &\quad q10 + (1 - q1)(1 - q2)q3(1 - q8)(1 - q9)(1 - q10) \\ &\quad q11 + (1 - q1)(1 - q2)(1 - q3)q4 + (1 - q1)(1 - q2)(1 - q3)(1 - q4) \\ &\quad q5 + (1 - q1)(1 - q2)(1 - q3)(1 - q4)(1 - q5)q6 + (1 - q1) \\ &\quad (1 - q2)(1 - q3)(1 - q4)(1 - q5)(1 - q6)q7 = 2.081095\frac{\text{failures}}{10^6\text{h}} \end{aligned} \tag{6}$$

Thus, 2.081095 failures/$10^6$ h gives the failure rate of the system after substituting all the failure rates of the components listed above in this equation.

# 6 Hardware Setup of the PVQZSI

A hardware prototype of the quasi Z-Source inverter has been built with DC source, impedance network, H-bridge, and gating circuit. The gating pattern has been developed using MATLAB- Xilinx interface and implemented in FPGA processor SPARTAN 3E. The pulse generation circuit is shown in Fig. 9.

The pulse has been generated by using pulse width amplitude modulation technique and the pattern is shown in Fig. 10.

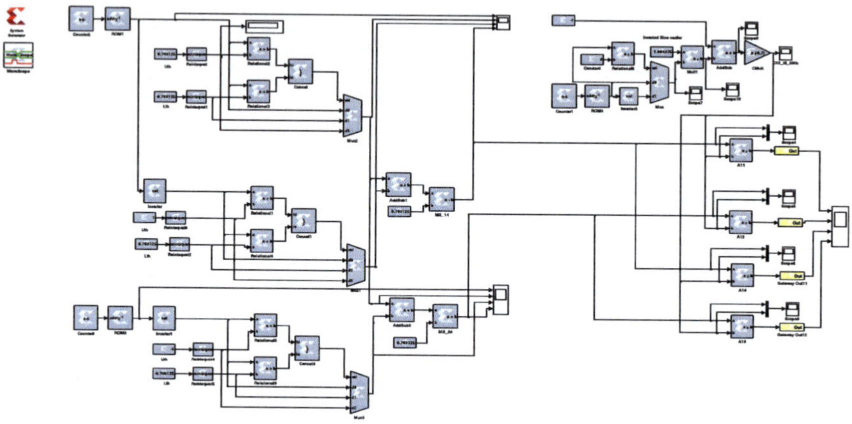

**Fig. 9** Pulse generating circuit in MATLAB-Xilinx interface

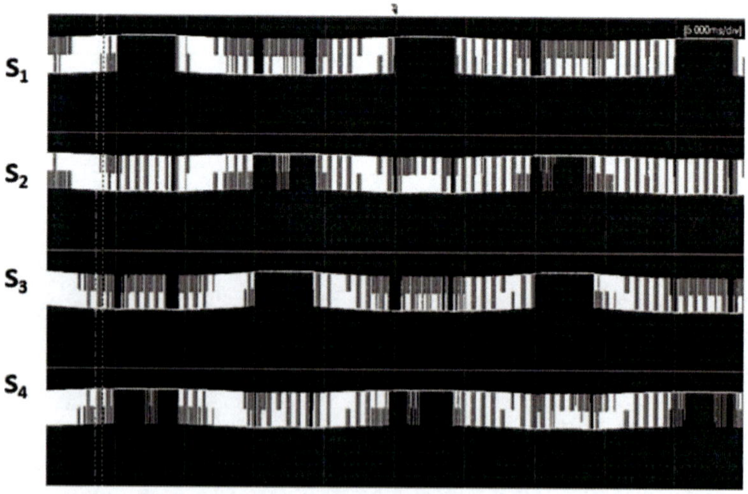


**Fig. 10** Pulse pattern for QZSI

The impedance network shown in Fig. 11 comprises of two inductors $L_1$ and $L_2$ of 3 mH each, a fast recovery diode and capacitors $C_1$ and $C_2$ of 2.2 mF each. The gate pulses for the IGBT switches $S_1$, $S_2$, $S_3$, and $S_4$ are applied through opto-coupler circuit which in turn is applied with input pulses from SPARTAN 3E Processor. Fig. 12 shows the complete hardware setup of QZSI. For 30 V of input voltage applied from the DC source, an output RMS voltage of 56 V is obtained.

**Fig.11** Quasi impedance network and H-bridge of QZSI

**Fig. 12** Output voltage of PV QZSI

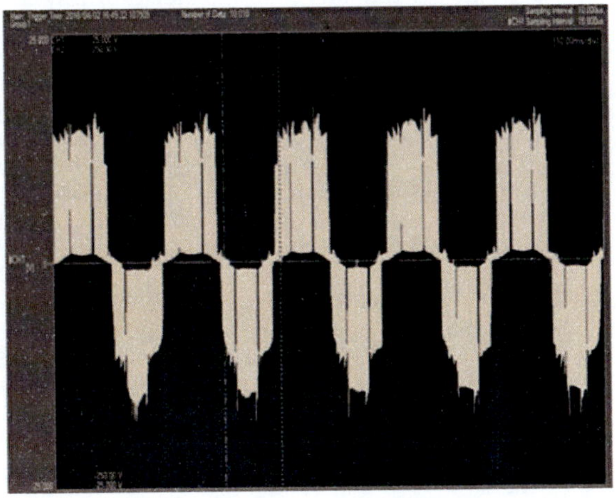

## 7 Conclusions

This paper emphasize on the importance of reliability estimation of photovoltaic quasi Z-Source Inverter. The proposed system is mathematically modeled using part count model and binary decision diagram. The BDD is derived from the fault tree structure of the PVQZSI. These models are used to estimate the possible failures in PVQZSI for every $10^6$ h. The failure rate of individual components were calculated using MIL HB-217. The failure rate of the PVQZSI is estimated as $4.4928/10^6$ h and $2.0810/10^6$ h from the part count method and the binary decision diagram, respectively. As the part count method is based on series structure of the subsystems, the expected failure rate is more than the binary decision diagram which provides reliability estimation based on logical combination of events. Therefore, the PVQZSI system reliability can be more accurately predicted using the combinatorial model. A hardware prototype of the PVQZSI has been built. It provides an output voltage 56 V for the input voltage of 30 V DC.

**Acknowledgements** The authors would like to sincerely thank the SSN Trust and Management for funding the research work.

## References

1. Anderson J, Peng FZ (2008) A class of quasi-Z-source inverters. In: IEEE industry applications society annual meeting, pp 1–7
2. Preiser K (2003) Photovoltaic systems. Handbook of photovoltaic science and engineering, pp 753–798
3. Gupta N, Garg R, Kumar P (2017) Sensitivity and reliability models of a PV system connected to grid. Renew Sustain Energy Rev 69:188–196
4. Ge B, Abu-Rub H, Peng FZ, Lei Q, De Almeida AT, Ferreira FJ, Sun D, Liu Y (2012) An energy-stored quasi-Z-source inverter for application to photovoltaic power system. IEEE Trans Industr Electron 60(10):4468–4481
5. Yuan L, Joel A, Peng FZ, Liu D (2009) Quasi Z-source inverter for photovoltaic power generation system. In: 24th annual IEEE applied power electronics conference, pp 918–924
6. Park JH, Kimy HG, Nho EC, Chun TW (2010) Power conditioning system for a grid connected pv power generation using a quasi-Z-source inverter. J Power Electron 10(1):79–84
7. Maish AB, Atcitty C, Hester S, Greenberg D, Osborn D, Collier D, Brine M (1997) Photovoltaic system reliability. In: Twenty sixth IEEE photovoltaic specialists conference, pp 1049–1054
8. Spagnuolo G, Weidong X, Cecati C (2015) Monitoring, diagnosis, prognosis, and techniques for increasing the lifetime/reliability of photovoltaic systems. IEEE Trans Indus Electron 62(11):7226–7227
9. Petrone G, Giovanni S, Remus T, Mummadi V, Massimo V (2018) Reliability issues in photovoltaic power processing systems. IEEE Trans Industr Electron 55(7):2569–2580
10. Wohlgemuth J, Daniel C, Paul M, Jay M, Andy N (2006) Long term reliability of photovoltaic modules. In: IEEE 4th world conference on photovoltaic energy conference, pp 2050–2053
11. Patrick D, Andre K (2012) Practical reliability in engineering, 5th edn. John Wiley and Sons
12. Yang Y, Wang H, Sangwongwanich A, Blaabjerg F (2012) Design for reliability of power electronic systems. In: IEEE IECON 2012–38th annual conference on IEEE industrial electronics society, pp 33–44

13. Song Y, Wang B (2013) Survey on reliability of power electronic systems. IEEE Trans Power Electron 28(1):591–604
14. Pecht G, Nash F (1994) Predicting the reliability of electronic equipment [and prolog]. Proc IEEE 82(7):992–1004
15. Shaoyong Y, Bryant A, Mawby P, Dawei X, Ran L, Tavner P (2011) An industry-based survey of reliability in power electronic converters. IEEE Trans Ind Appl 47(3):1441–1451
16. Shaoyong Y, Xiang D, Bryant A, Mawby P, Ran L, Tavner P (2010) Condition monitoring for device reliability in power electronic converters: a review. IEEE Trans Power Electron 25(11):2734–2752
17. MIL-HDBK-217F (1991) Military handbook: reliability prediction of electronic equipment. Department of Defense, Washington DC

# Hybrid Boost Converter for AC and DC Loads Using PV System

T. Porselvi, Preethi Kannan, S. R. Y. Aouthithiye Barathwaj, C. S. Sai Ganesh, and R. Kothai

## 1 Introduction

In the energy sector, the day-by-day increase in demand for electricity has become one of the primary problems these days. Nevertheless, non-renewable resources are not enough in quantity and accessibility to meet current energy consumption [1]. While considering the future accessibility of traditional power generating sources like renewable energies, conventional energy production systems must be utilized to satisfy the energy demand. One renewable technique of solving the present energy dilemma is to draw energy from the entering solar energy radiation, which is free to all worldwide [2, 3]. In the next several years, the world's electricity generation is rising and will continue to rise. The greenhouse effects and environmental issues resulting from conventional power (coal, oil, and natural gas) may limit its usage to meet the demand, whereas photovoltaic (PV) power generation is clean. Moreover, photovoltaic energy production has increased throughout the years and will continue more quickly than before [4, 5]. The PV runs at its maximum power at a unique max point because of a photovoltaic array's (PV) nonlinear intensity feature. However, due to the extreme voltage and current waveforms, a variable PV array changes with radiation and temperature, the maximum power point (MPP) shifts under ambient environment circumstances [6–8]. The PV array when operated on the highest available power boosts its efficiency leading to a reduction in running expenses.

The pulse width modulation generator is utilized to provide the duty cycle. A DC–DC boost converter with a switch is employed for opening and closing operations [9]. In boost converter, the inductor store energy when the switch is closed and discharge it through the diode when the switch is opened [10–12]. The voltage at load side merely remains constant with the help of output capacitor connected across

T. Porselvi (✉) · P. Kannan · S. R. Y. A. Barathwaj · C. S. S. Ganesh · R. Kothai
Department of EEE, Sri Sairam Engineering College, Chennai-44, Tamil Nadu, India
e-mail: porselvi.eee@sairam.edu.in



N. L. Narasimhan et al. (eds.), *Recent Advances in Energy Technologies*, Lecture Notes in Mechanical Engineering, https://doi.org/10.1007/978-981-19-3467-4_10

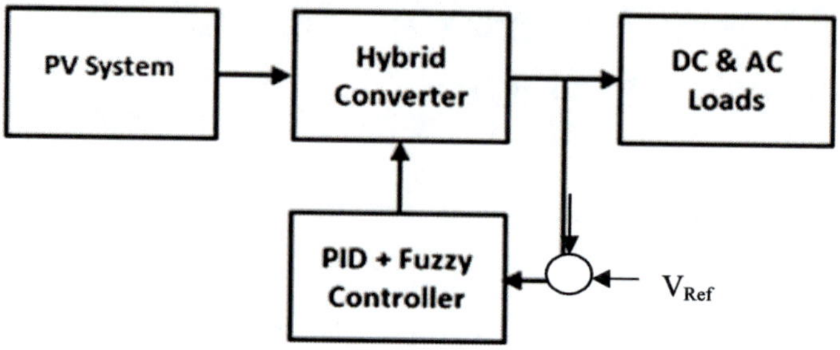


**Fig. 1** The proposed system as a block diagram

the load resistance. Whenever the switch is closed, the inductor current reaches its maximum value. On the other hand, it reaches minimum value when the switch is in open condition [13, 14]. The controller, which is coupled to the boost converter's output, provides feedback [15]. The solar panel converts the solar energy into electrical energy of DC voltage, and the DC–DC boost converter is used to improve the magnitude of input voltage. The PI controller maintains a steady output voltage which is used to regulate the output voltage [16]. The output voltage is regulated at a constant desired value.

## 2 Proposed System

In the system, the PV source voltage is regulated with fuzzy PID controller and hence keeps the PV output power at high value with the help of DC–DC converter. Figure 1 shows the proposed block diagram which includes the DC–DC hybrid converter. By using this converter, increased voltage gain is obtained with AC and DC loads. The single-phase inverter which converts DC to AC is utilized to connect AC electrical load. Multiple systems can be added to the bus bar and thus connecting DC loads and AC loads. The output voltage is compared with reference and the error is given to the controller. The error is processed in the controller (PI/PID/PID + Fuzzy), and the control action is performed to regulate the voltage at the reference value, by changing the duty cycle of the converter [17, 18].

## 3 The PV System

### *3.1 PV Source*

The system uses a PV source whose output voltage is boosted and its power is increased by utilizing a system with a controller coupled to a DC–DC converter that

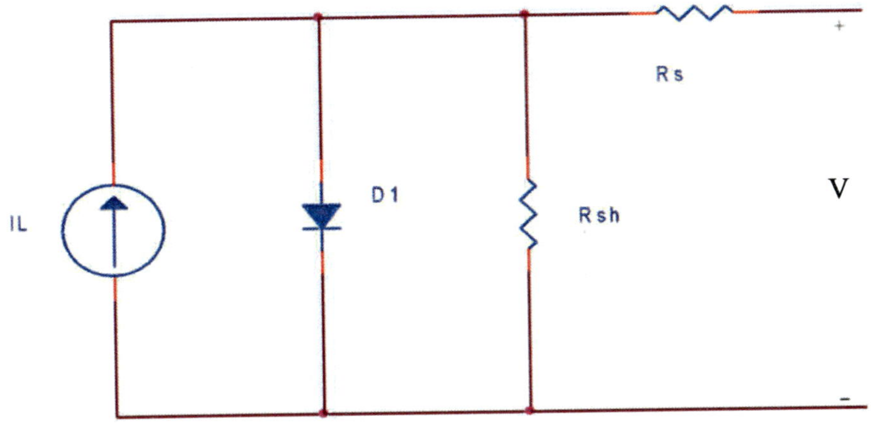


**Fig. 2** PV model (single diode)

performs the maximum power point tracking. The maximum power point tracking (MPPT) varies with insulating temperature and level because of the variable nature of photovoltaic panels. In general, MPPT refers to a solitary point on the *V–I* or *V–P* curve where the complete PV system functions at peak efficiency and provides maximum output power. At 25 °C, 250 W/m$^2$ irradiation is obtained. Figure 2 shows the widely used PV solar cell equivalent circuit [19, 20]. The solar light determines the intensity of the source at any present moment. The solar cell cannot sustain a fixed current, as and when the load resistance increases.

A solar cell's characteristic equation is given by [20]:

$$I = I_L - I_o\left(\exp\left(\frac{q}{nkT}(V + IR_s)\right) - 1\right) - \frac{(V + IR_s)}{R_{sh}}$$

Where $I_L$, is the generated current, $I_o$ is the reverse saturated current of diode, $n$, $k$, $q$ and $T$ are, respectively, diode ideality factor, Boltzmann's constant, elementary charge, and absolute temperature. Figure 3 depicts the standard *I–V* and *P–V* characteristics of solar cell. $I_{mp}$ and $V_{mp}$ are, respectively, the current and voltage of the solar cell corresponding to maximum power. $I_{sc}$ and $V_{oc}$ are short circuit current and open circuit voltage of the cell, respectively [20].

Current (*I*)–Voltage (*V*) and Power (*P*)–Voltage (*V*) characteristics of a photovoltaic system are important because they have been used to analyze various approaches and methodologies, such as maximal power point tracking (MPPT).

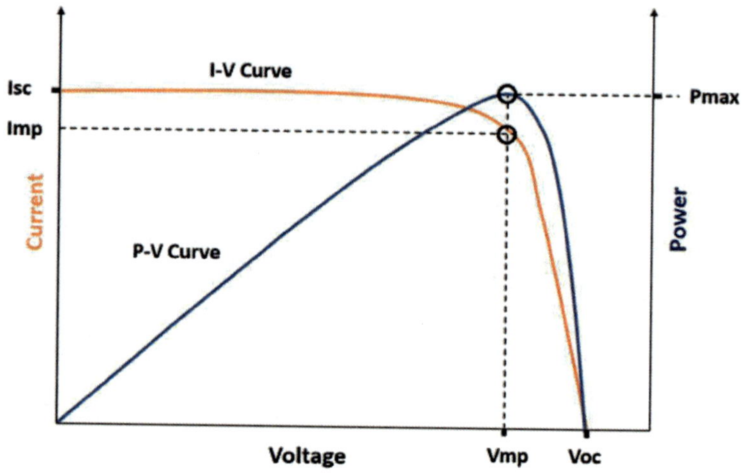


**Fig. 3** PV characteristics of solar cell

## 3.2 *Hybrid DC–DC Converter*

The PV system is used to supply both DC and AC loads using the hybrid boost converter as proposed by Olive Ray et al., that requires a lesser number of switches for providing DC and AC outputs. This converter has increased reliability due to the inherent presence of protection against shoot-through fault in the inverter stage [21]. The voltage at the output of the converter is regulated using the controllers. The converter is consisting of a DC-link capacitor, inductor as well as power semiconductor components. The hybrid boost converter has conventional boost converter with diode D, inductor $L_1$, and Capacitor $C_2$, with the switch being replaced by a bridge circuit with $S_1$, $S_4$ in its first leg and $S_3$, and $S_4$ in its second leg. The AC load is connected between the mid points of two legs through the filter capacitor $C_1$. The DC load is connected across $C_2$.

**Mode 1:**

The converter system which is connected with the PV system is operated in shoot-through mode. During the mode 1, the switches of any particular leg (either $S_1$, $S_4$ or $S_3$, $S_2$) are turned on. The proposed conversion pulse width is determined by the interval of the shoot-through operation. At the same period, the diode 'D' is reverse biased.

**Mode 2:**

During this mode, the diode is in ON condition, the voltage across the load is a sum of input and inductance voltages. The inductor energy stored is dissipated and the charge is stored in the capacitor $C_2$.

The design equations for $L_1$ and $C_2$ are given by (1) and (2)

$$L_1 = V_S \frac{k}{f \Delta I} \quad (1)$$

$$C_2 = \frac{I_o k}{f \Delta v} \quad (2)$$

where $k$ is the duty cycle, $\Delta I$ and $\Delta v$ are, respectively, the ripple current of $L_1$ and ripple voltage of $C_2$, $I_o$ is DC load current.

For supplying AC load, the bridge is operated in inverter mode and the load is fed with the AC output.

## 3.3 Control Method

Controllers are used to control the performance of the hybrid boost system model. The PID closed-loop control is used for converter operation control in the proposed system. The proposed system acts in a way to maintain the voltage gain at a high value. A PID controller typically has three components: proportional, integral, and derivative as shown in Fig. 4.

In case of PI controller, the derivative part is missing, the error is processed, and the duty signal is changed according to the error value.

To obtain the desired output voltage, PID, and fuzzy logic controllers are included in the control design, as shown in Fig. 5.

The fuzzy controller is used in the proposed model to deal with conversion and the function of the hybrid converter that increases the input DC supply. This 3 * 3 law-based fuzzy logic technique is implemented. In this scheme, the gains of the

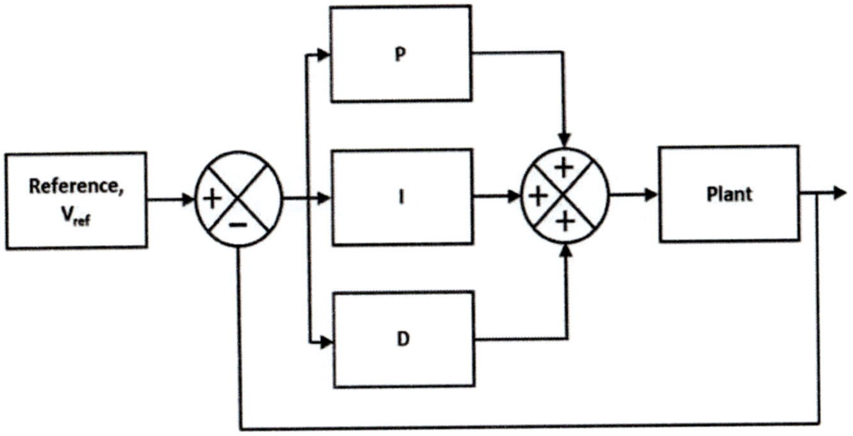


**Fig. 4** PID controller

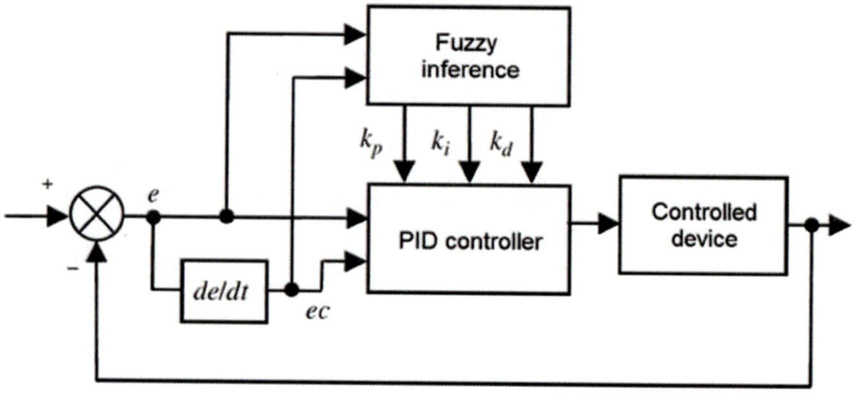


**Fig. 5** Control system block diagram

controller, $K_p$, $K_i$, and $K_d$ are found using the fuzzy concepts. The error between the reference and actual is fed to the controllers, and three outputs $K_p$, $K_i$, and $K_d$ are included in the fuzzy PID input. The settlement time and the control of the fuzzy PID are reduced.

## 4 Simulation of the Proposed PV System

The proposed PV system with hybrid converter is simulated in MATLAB. The simulink model is shown in Fig. 6a. High voltage gain is achieved in both direct and alternating loads. The disturbances in the load voltage and current are reduced using the proposed control method of the fuzzy controller. The power from the PV array is fed to the load through a DC–DC hybrid converter in the proposed system, the hybrid converter consists of an inverter arrangement for AC load. The fuzzy with PID controller is used to make a DC bus bar voltage at constant. Figure 6b depicts the simulink model of the PV system. The simulation parameters are listed in Table 1.

The whole system is simulated with PI, PID, and PID + Fuzzy controllers. The output voltage waveforms for both DC and AC loads with three controllers are obtained.

## 5 Results and Discussion

The system is simulated with three controllers PID, PI, and PID + Fuzzy and the results are obtained. Figures 7 and 8, respectively, show the voltage outputs of both

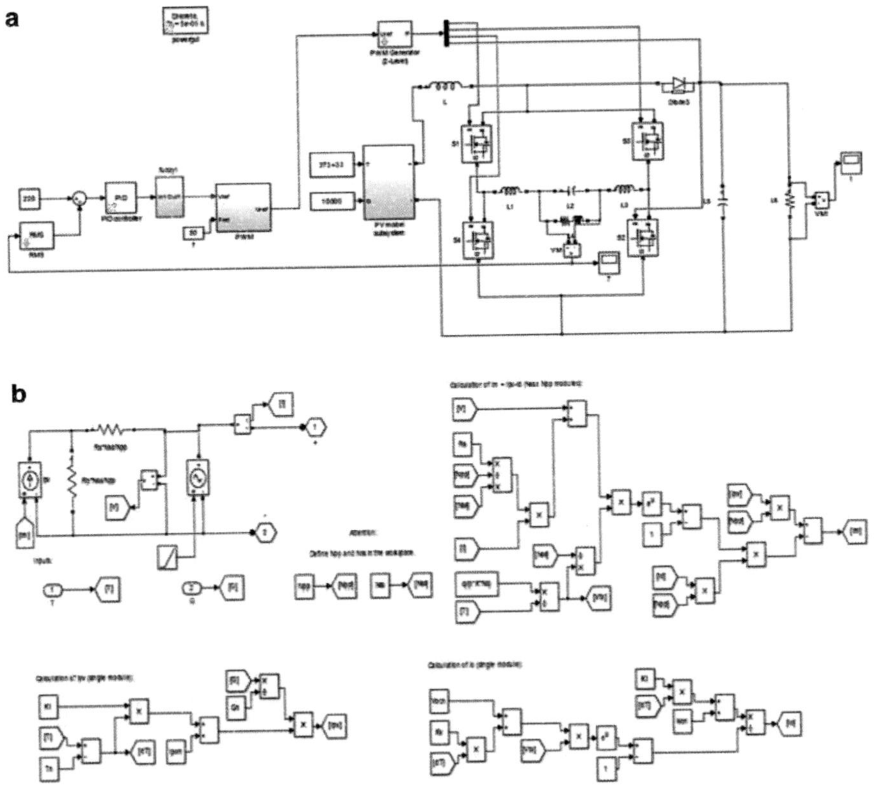


**Fig. 6** **a** Simulink model of the proposed system, **b** Simulink model of PV system

**Table 1** Simulation parameters of the proposed system

| Parameter | Unit |
|---|---|
| Input DC (PV output) | 48 V |
| PV temperature | 33 °C |
| Irradiance | 10,000 W/m$^2$ |
| R load | 100 ohms |
| Boost inductance | 10 mH |
| Filter capacitance at DC load | 2000 Micro Farad |
| Filter capacitance at AC load | 75 Micro Farad |
| Switching frequency | 1 kHz |
| Output DC voltage | 480 V |
| Output AC voltage | 220 V (*p–p*) |
| $K_p$ | 0.01 |
| $K_i$ | 0.1 |
| $K_d$ | 0.01 |

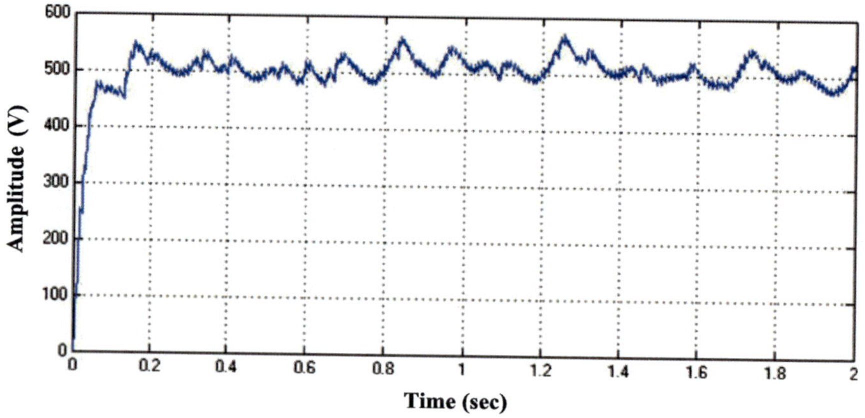


**Fig. 7** DC load voltage with PID controller

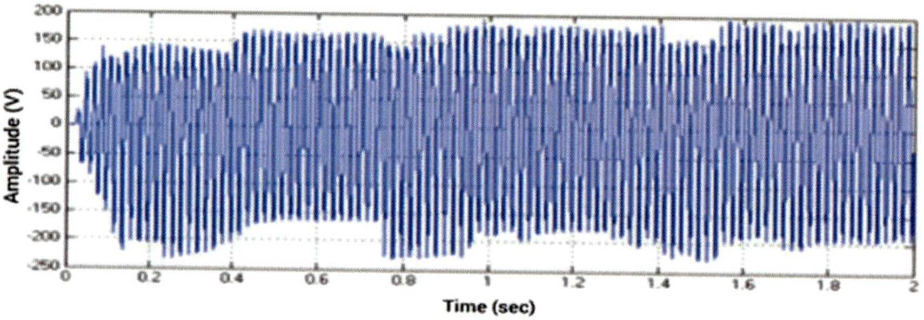


**Fig. 8** AC load voltage with PID controller

DC and AC loads with PID controller. It is observed that the settling time for DC output is 1.1 s and the steady-state error is approximately 80 V.

The voltage outputs for both DC and AC loads with PI controller are shown in Figs. 9 and 10, respectively. It is observed that the settling time for DC output is 2 s, and the steady-state error is approximately 20 V. The peak overshoot is 1000 V.

Figures 11 and 12 show the fuzzy + PID control-based outputs for both AC and DC loads. It is observed that the settling time for DC output is 0.2 s, and the steady-state error is approximately 20 V. The peak overshoot is 880 V.

Table 1 shows the system simulation parameters, and Table 2 gives the comparison of time domain specifications of output voltage for the three controllers.

$$\text{The efficiency of the system} = \frac{\text{Pout}}{\text{Pin}}$$

$$\text{Pout} = \frac{480^2}{100} = 2304\ \text{W}$$

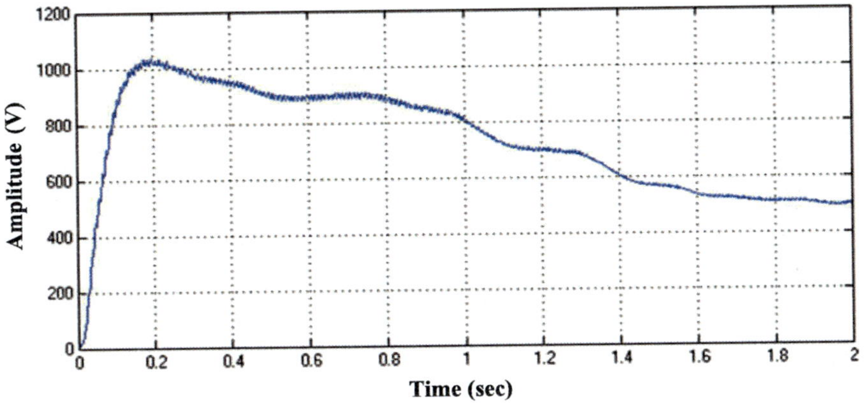


**Fig. 9** DC load voltage with PI controller

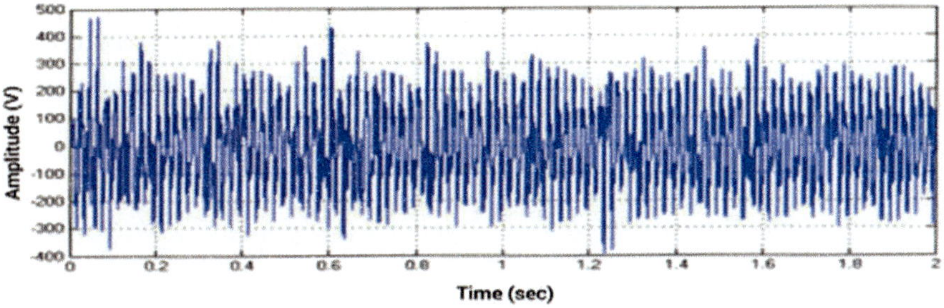


**Fig. 10** AC load voltage with PI controller

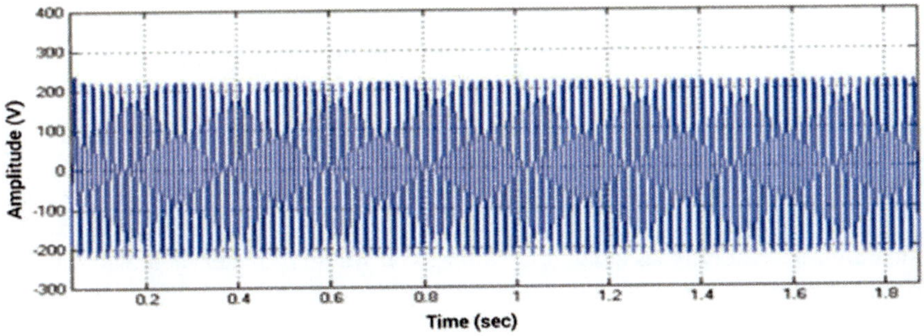


**Fig. 11** AC load voltage with PID and fuzzy

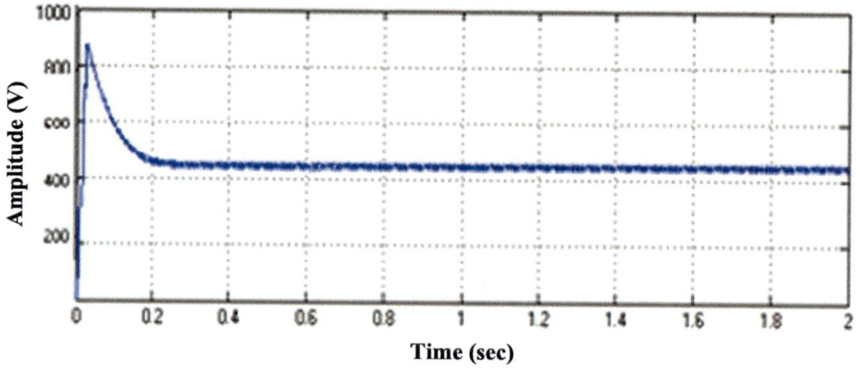


**Fig. 12** DC load voltage with PID and fuzzy

**Table 2** Comparison of time domain specifications for three controllers

| Parameter | PI controller | PID controller | PID + Fuzzy controller |
|---|---|---|---|
| Settling Time(sec) | ~2.0 | 1.1 | 0.2 |
| Peak overshoot(V) | 1000 | 560 | 880 |
| Steady-state error(V) | 20 | 80 | 20 |

$$\text{Pin} = 48 \times 53 = 2544\ \text{W}$$

$$\eta = \frac{2304}{2544} \times 100 = 90.56\%$$

It is found that the proposed system's efficiency is 90.56%, which is an acceptable value.

It is found from Table 2 that the fuzzy + PID controller provides the satisfying results with comparable steady-state error to PI controller, less settling time. The maximum overshoot is less than that with PI and slightly greater than that with the PID controller.

## 6 Conclusion

The AC and DC loads are interfaced with the sustainable power source of the PV system with the help of hybrid boost converter to maximize the generated PV power by using the PID and fuzzy controllers. The system consists of a single-phase inverter which converts DC to AC for AC load. The load voltage and current are regulated with the help of the control action. The suggested converter achieves a high voltage gain that the input voltage of 48 V is converted to output voltage of 220 V (*p*–*p*) for AC load and 480 V for DC. The proposed system is simulated, and the results are validated using MATLAB/Simulink for three types of controllers. It is found that the fuzzy PID controller provides the satisfying results for both AC and DC loads with comparable steady-state error to PI controller, less settling time. The maximum overshoot is less than that with PI and slightly greater than that with the PID controller.

## References


1. Walekar VR, Murkute SV (2018) Speed control of BLDC motor using PI & Fuzzy approach: a comparative study. In: International conference on information, communication, engineering and technology (ICICET). https://doi.org/10.1109/icicet.2018.8533723
2. Mamadapur A, UndeMahadev G (2019) Speed control of BLDC motor using neural network
3. Chaudhary V, Bhargava A, Bhasin S (2019) Modeling and simulation of grid-connected hybrid power systems integrated with solar PV/Wind and controlled by voltage regulator. In: International conference on communication and electronics systems ICCES, pp 1316–1320, IEEE
4. Berrazouane S, Mohammedi K (2014) Parameter optimization via cuckoo optimization algorithm of fuzzy controller for energy management of a hybrid power system. Energy Conver Manage 78:652–660
5. Wang S, Janabi A, Wang B (2020) Generalized optimal SVPWM for the switched-capacitor voltage boost converter, ECCE, pp 2708–2711. IEEE
6. Manie N, Pattanaik B (2019) Zeta DC–DC converter based on MPPT technique for BLDC application. Int J MC Square Sci Res 11(2):1–2
7. Soumya K, Aruna D (2021) Effective design and implementation of three port hybrid converter for hybrid power sources. Int J Phys Conf Series 1717(1):012051
8. Ishaque K, Salam Z, Amjad M, Mekhilef S (2012) An improved particle swarm optimization (PSO)—based MPPT for PV with reduced steady-state oscillation. IEEE Trans Power Electron 27(8):3627–3638
9. Chaohui Z (2008) Emerging technology: photo-voltage generation status and trends. J Shanghai Dianji University 11(2):104–109
10. Boutasseta N (2012) PSO-PI based control of photovoltaic arrays. Int J Comput Appl 17:36–40
11. Gules R, Pacheco JDP, Hey HL, Imhoff J (2008) A maximum power point tracking system with parallel connection for PV standalone applications. IEEE Trans Indus Electron 55(7):2674–2683
12. Kwon JM, Kwon BH, Nam KH (2009) Grid connected photovoltaic multi-string PCS with PV current variation reduction control. IEEE Trans Industr Electron 56(11):4381–4388
13. Meena R (2014) Simulation study of boost converter with various control techniques. Int J Sci Res 3(9):74–79
14. Nagarajan R, Saravanan M (2014) Performance analysis of a novel reduced switch cascaded multilevel inverter. J Power Electron 14(1):48–60
15. Nagarajan R, Sathishkumar S, Balasubramani K, Boobalan C, Naveen S, Sridhar N (2016) Chopper fed speed control of DC motor using PI controller. IOSR J Electr Electron Eng (IOSR-JEEE) 11(3Ver. I):65–69
16. Nagarajan R, Saravanan M (2012) Comparison of PWM control techniques for cascaded multilevel inverter. Int Rev Autom Control 5(6):815–828

17. Anil G, Murugan N, Ubaid M (2013) PI controller based MPPT for a PV system. IOSR J Electr Electron Eng 6(5):10–15
18. Raj A, Gopinath A (2015) Proportional plus integral (PI) control for maximum power point tracking in photovoltaic systems. Int Res J Eng Technol (IRJET) 02(06):408–412
19. Porselvi T, Krithika K, Sai Ganesh C (2021) PV system based induction motor with landsman converter using IFOC controller. In: 2021 2nd global conference for advancement in technology (GCAT), pp 1–4. https://doi.org/10.1109/GCAT52182.2021.9587724
20. Bhadra SN, Kastha D, Banerjee S (2008) Wind electrical systems. Oxford University Press
21. Ray O, Mishra S (2014) Boost-derived hybrid converter with simultaneous DC and AC outputs. In: IEEE Trans Ind Appl 50(2):1082–1093. https://doi.org/10.1109/TIA.2013.2271874

# Energy Storage and Management

# Design and Analysis of Cumulative Regenerative System for a 42 kWh Battery Pack

**M. C. Sabareesh, R. Karthik Raja, Arockia Selvakumar Arockia Doss, and Michael Short**

## *Abbreviations*

| | |
|---|---|
| HV | High Voltage System |
| EV | Electric Vehicle |
| V–I Curve | Voltage–Current Curve |
| P–I Curve | Power–Current Curve |
| $I_{sc}$ | Short circuit Current |
| $V_{oc}$ | Open circuit voltage |
| ICE | Internal Combustion Engine |
| Li-ion | Lithium-Ion Battery |
| CC | Constant Current |
| CV | Constant Voltage |
| AC | Alternating Current |
| DC | Direct Current |
| mF | Milli Farad |

## *Annotations*

| | |
|---|---|
| $V_{tot}$ | Total Voltage of Battery |

M. C. Sabareesh · A. S. A. Doss (✉)
School of Mechanical Engineering, Vellore Institute of Technology, Chennai, India
e-mail: arockia.selvakumar@vit.ac.in

R. K. Raja
School of Electrical Engineering, Vellore Institute of Technology, Chennai, India

M. Short
School of Computing, Engineering and Digital Technologies, Teesside University, Middlesbrough, UK


N. L. Narasimhan et al. (eds.), *Recent Advances in Energy Technologies*, Lecture Notes in Mechanical Engineering, https://doi.org/10.1007/978-981-19-3467-4_11

| | |
|---|---|
| $Ns$ | Cell modulus in Series |
| $Nsp$ | Series connected packs |
| $N_p$ | Cells modulus in parallel |
| $N_{pp}$ | Parallel connected packs |
| $V_{max}$ | Maximum Cell Voltage |
| $V_{min}$ | Minimum Cell Voltage |
| $C_d$ | Discharge Capacity |
| $\omega_m$ | Maximum disc speed [rad/sec] |
| $J$ | Total inertia referred to braked shaft [kgm$^2$] |
| $T_b$ | Braking time [sec] |
| $E_{out}$ | Output Energy |
| $E_{in}$ | Input Energy |
| $H$ | Efficiency |
| $D_c$ | Cylinder Diameter |
| $D_R$ | Piston Rod Diameter |
| $P_c$ | Compression Chamber Pressure |
| $E_{susp}$ | Energy Produced By Suspension |
| $E_{bp}$ | Energy Produced by Battery Pack |
| $E_{pv}$ | Energy produced by a single PV panel |
| $N_{pv}$ | Total Number of Solar panels used |
| $E_{ss}$ | Total Energy generation by Solar system unit |
| $T_{wor}$ | Time taken without regeneration system (secs) |
| $T_{wr}$ | Time taken with regeneration system (secs) |
| $T_h$ | Time to convert seconds to hour |
| $D_{wor}$ | Distance covered without regeneration (km) |
| $D_{wr}$ | Distance covered with regeneration (km) |

## 1 Introduction

The present context gives us insight into the energy storage system that includes a lithium-ion battery pack [1, 2]. The energy-storage system comprises of the high voltage (HV) Li-ion battery pack, electrically coupled in parallel with the regenerative system [3, 4]. The featured aim of this paper is to characterize a multipurpose regenerative system, which results in the active charging of the battery pack mentioned in above paragraphs. A strategy to improve real-time energy consumption limit as developed by Zhu [5]. In the context of Regeneration, we have summed up three types of regenerative energy which is converted into electrical energy. These include combining the braking (includes frictional energy), suspension (includes damping energy) and solar energy into a closed-loop system for obtaining a better efficient performance of the vehicle stated by Takahiro Hirano [6, 7]. For developing this self-designed battery model is configured [8–10]. Hence, the idea, modeling, logic, results, and scope of development of this embodiment would be explained in the further sections of the context provided [11–14]. Calculation of opposing forces in a vehicle as stated by model specifications of microcar for cities [15]. In mechanical braking system when brake is applied, the kinetic energy is dissipated to the brake drums or brake pads where the energy is lost as heat energy due to friction [16, 17]. The regeneration braking concept uses the recovery mechanism that involves converting the kinetic energy into the form of reverse flux that is converted back from the motor to the battery pack for charging. Similarly, the suspension involves damping energy that is being lost via the spring-damper arrangement in cars [18, 19]. The regenerative suspension utilizes this energy being lost and hence improves the power supply efficiency of vehicles [20]. Along with both regenerative systems, the solar charging technology is involved that includes charging the car using solar panels mounted over the car [7, 21–24]. Finally, all the three are embedded into a cumulative closed-loop system and a comparative study is carried out.

## 2 Model Development

In order to create an optimum regenerative system, the auxiliary battery pack configuration is modeled. The modeled system is further studied, analyzed, and then it is modified for obtaining ideal regeneration characteristics that are performed through a series of simulations performed.

### 2.1 *Battery Cell Specification*

To model optimum battery pack for the study, lithium-ion cells are considered, and the cell configuration is shown in Table 1.

**Table 1** Li-ion cell configuration [25]

| Item | Specification |
|---|---|
| Discharge capacity | 4900 mAH |
| Max. voltage | 4.2 V |
| Nominal voltage | 3.6 V |
| Min. voltage | 2.4 V |
| Charging method | CC |
| Charging current | 2450 mA |
| Charging time | 3 h |
| Operating temperature | Charge: 0 to 45 °C |
| Cell weight | Discharge: −20 to 60 °C |
| Cell dimension | 69 g |

**Table 2** Battery pack configuration [25]

| Item | Specification |
|---|---|
| Module arrangement | 5p cells |
| Module voltage | 4.2 V |
| Module discharge capacity | 24.5 Ah |
| Pack configuration | 24 s modules |
| Battery pack configuration | 4s5p packs |
| Battery pack energy | 42,336 W-h |

## 2.2 *Battery Pack Configuration*

According to the design calculations and cell configuration, the battery pack assembly is designed. The design is a 24 module assembly comprising of 120 cells overall in a single pack. The designed battery pack specifications are shown in Table 2:

## 2.3 *Battery Pack Calculation*

Energy calculations:

$$\text{Total Voltage of Battery Pack} = V_{\text{max}} \times N_s \times N_{\text{sp}} \tag{1}$$

$$\text{Maximum Energy of Battery Pack} = V_{\text{tot}} \times C_d \times N_p \times N_{\text{pp}} \tag{2}$$

$$\text{Energy of Battery Pack} = V_{\text{nom}} \times C_d \times N_p \times N_{\text{pp}} \tag{3}$$

where $V_{\text{tot}}$ = Total Voltage of Battery, $N_s$ = Cell modulus in Series, $N_{\text{sp}}$ = Series connected packs, $N_p$ = Cells modulus in parallel, $N_{\text{pp}}$ = Parallel connected packs, $V_{\text{max}}$ = Maximum Cell Voltage, $V_{\text{min}}$ = Minimum Cell Voltage, $C_d$ = Discharge Capacity.

## 2.4 *Battery Pack Characterization and Arrangement*

Overall, battery pack assembly consists of 2400 cells arranged in series and parallel connection with each module comprising of five cells arranged in parallel.

The modules are placed within internal packs where each internal pack comprises of 24 modules. Finally, the whole structure of battery pack consists of 20 internal battery pack connected in a four series and with a five parallel row combination, hence creating a path for current to flow as shown in Fig. 1.

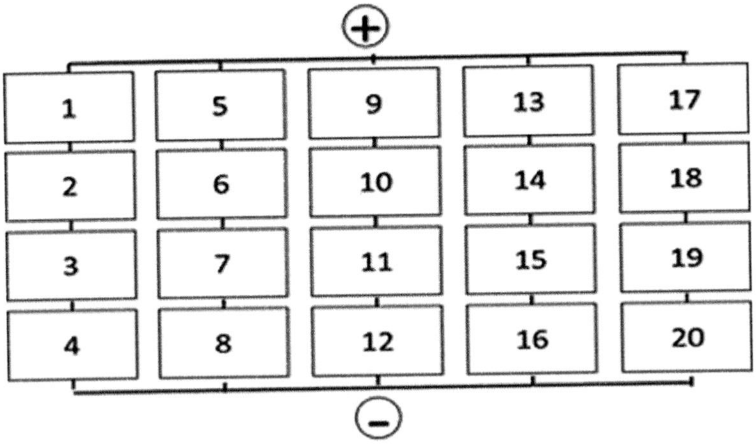


**Fig. 1** Battery packs arrangement

## 2.5 *Battery Pack Discharge Study*

With a fully assembled battery pack that is functional, the important parameter is to analyze the charging and discharging cycles of assembled battery. Because various factors like voltage overloading, lack of voltage, and faulty series–parallel connection may affect the battery life and even may result in battery failure permanently.

The battery pack discharge characterization is studied and analyzed using MATLAB with defined parameters set for the batteries as shown in Fig. 2. The nominal discharge characteristic curve is obtained as shown in Fig. 3. The first plot depicts the discharge current versus voltage plot for maximum current drawn scenario, i.e., for 245 Amps of maximum current drawn the battery discharges within 30 min. The second plot depicts the real-case scenario for different speed ranges during throttle where the current drawn varies and hence battery discharge rate also varies accordingly.

During the study constant voltage (CV), charging is used because usually Li-ion batteries are too sensitive for voltage overloading and charging and discharging are not uniform. So by constant voltage flowing, rate of current can be controlled properly. The discharging rate of every cell is not same rather it differs according to the load. After a cell reached its cut-off voltage furthermore extraction of current cannot be expected. Battery becomes open circuit. If we force the cell to operate beyond that point cell battery might not respond to the power supply.

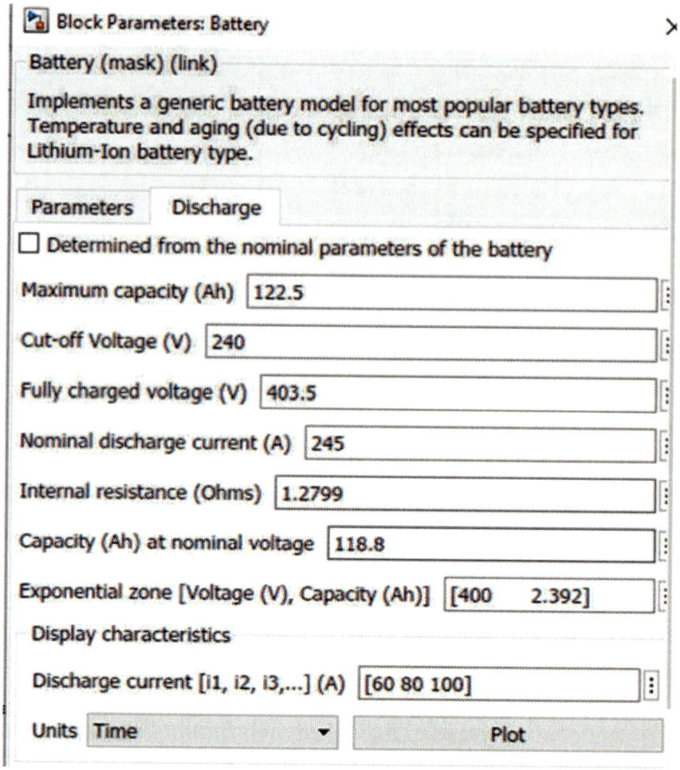


**Fig. 2** Battery parameters

## 3 Modeling Regeneration System

After the design, study and analysis of the vehicle battery pack with an additional source of recharging is called as regeneration. Regeneration converts the lost form of energy during braking and damping of suspension and also by solar charging that involves conversion of solar energy to electrical energy. A MATLAB-Simulink block is modeled as shown in Fig. 4 to configure the power requirement.

### *3.1 Theory and Working of Regenerative Braking System*

Braking is the phenomenon of reducing the speed of the moving vehicle and bringing the vehicle to a stationary position. In mechanical braking system, the rotation or revolution of wheel is reduced/stopped with the help of brake pads which rubs with the brake disk/drum and hence a frictional force comes into act. During this, the

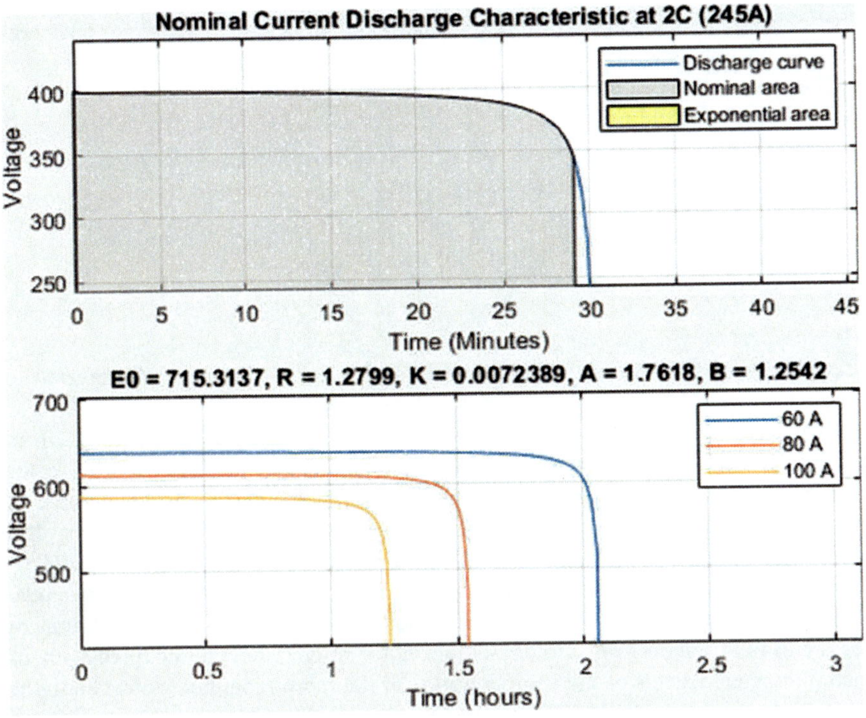


**Fig. 3** Battery pack discharge graph

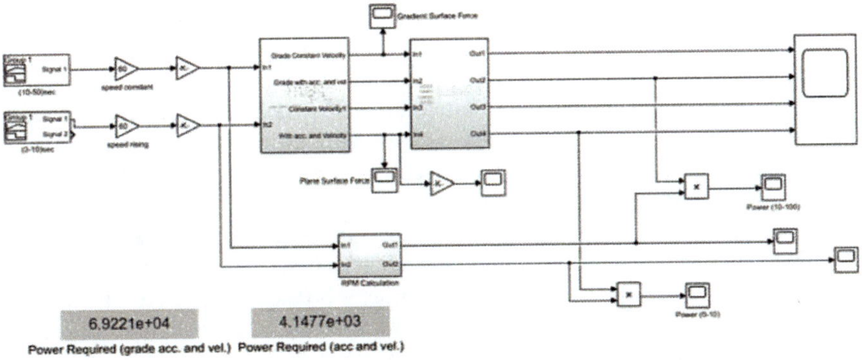


**Fig. 4** Simulink block diagram used for total power requirement in a vehicle system

kinetic energy of vehicle is converted to thermal energy lost to the brake pads and disks/drums.

The kinetic energy and power dissipation during normal braking on the braking pad/disk can be calculated as:

$$\text{KE} = J\frac{\omega_m^2}{2} \tag{4}$$

$$\text{Mean Power Dissipation} = \frac{\text{KE}}{T_b} \tag{5}$$

where $\omega_m$—Maximum disk speed [rad/sec] = 2.Π.N; $J$—Total inertia referred to braked shaft [kgm$^2$] = $\frac{\pi\rho L(D_4-d_4)}{32}$ [kgm$^2$]; $T_b$—Braking time [sec] = $\frac{V}{\propto}$.

Regenerative braking system is activated when vehicle starts to decelerate from its initial speed and power extraction from the battery pack is restricted by motor controller. Motor controller is a controlling device, which is present in the intersection of motor and battery pack. It is a system with capacitor bank (250 mF), converts DC supply to AC supply. During braking action, motor controller terminal switches produce inverse duty cycle. In which supply coming from battery Pack stopped. Later on, present energy stored in the capacitor bank supplies a mild amount of energy to field windings of motor. As the motor rotor is directly coupled with 14 inch (diameter) wheels of the vehicle. Rotor of the motor continues to rotate in the same direction due to inertia of the vehicle.

However, due to negative voltage terminal present in field windings and changing of induced flux in the field winding causes flux linkage production by the rotation. It opposes the change in flux linkage value; soon the rotation stopped causing a braking action, until the rotation of the wheel stops. Later on, those energies passed via motor controller to battery pack.

Hence, through this within regenerative braking system, vehicle kinetic energy is converted into electrical energy. By the reverse rotation of the motor and with variation in slip angle it leads to the production of reverse DC voltage. That is finally sent back to the energy storage device.

The EV regenerative system model for braking is modeled using MATLAB-Simulink [26]. The battery configuration is taken as input block for the design.

For the simulation, the braking action parameter is taken occurring for the car at a deceleration from 60 to 0 kmph in 50 s. According to that required torque waveform is plotted using the Simulink block diagram in Fig. 5.

After modeling the regenerative braking model, the performance/efficiency of the model is studied and finalized on how much the voltage/energy associated with the battery has increased during the braking action when the vehicle decelerates.

The change in voltage of the battery pack is studied throughout the regeneration period using the MATLAB-Simulink model from Fig. 5. The input to DC motor is taken as torque, a resistor and battery that are connected in series to the store and also controls the amount of current flow in the circuit. Using a deceleration waveform

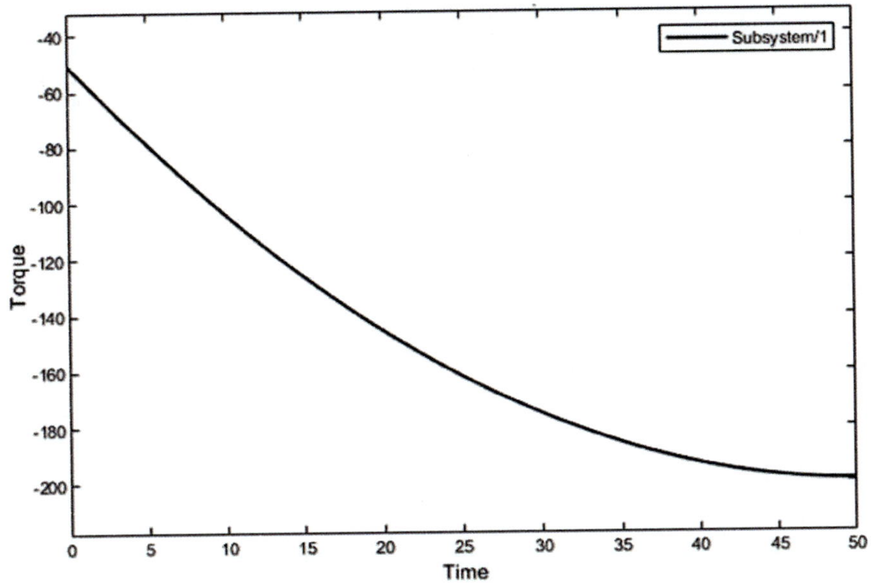


**Fig. 5** Torque versus time graph during deceleration from 60 to 0 kmph in 50 s

from 60 to 0 kmph in 50 secs, we have obtained the charging/regeneration action performed on the battery at Fig. 6.

Efficiency:

$$\text{For Single Brake Cycle} = \frac{E_2 - E_1}{E_1} \tag{6}$$

$$= \frac{V_2 - V_1}{V_1} (\text{When load is Constant})$$
$$= 0.8\% \tag{7}$$

For 20 brake cycles, $\eta = 16.43\%$.

Where $E_2$ = Final Energy; $E_1$ = Initial Energy; $V_1$ = Initial voltage; $V_2$ = Final Voltage; $\eta$ = Efficiency.

Table 3 shows the boundary conditions assumed for the study. For 20 braking cycles, the increase in efficiency is analyzed as 16.43%. The increase in energy/voltage is added back to full integration model that is explained in further sections.

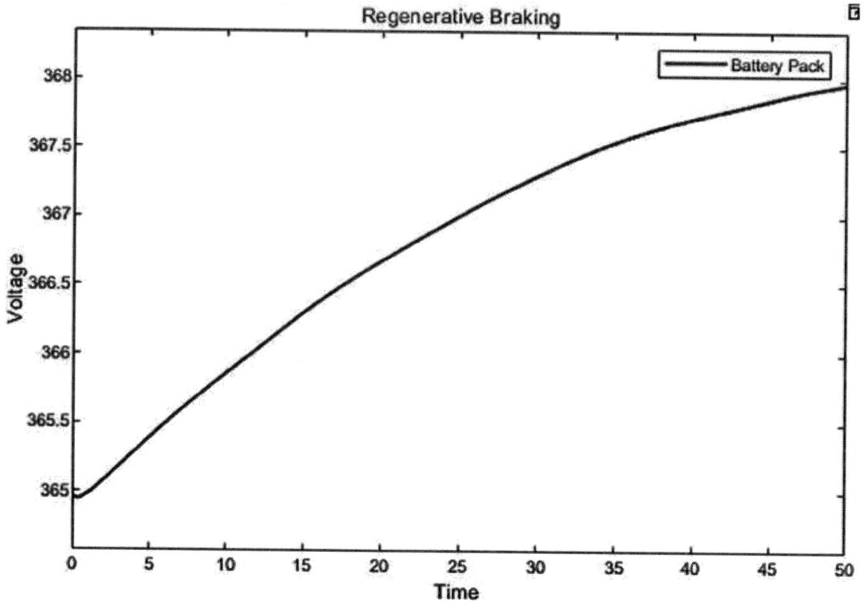


**Fig. 6** Voltage versus time during the time of regeneration

**Table 3** Boundary condition involved

| | |
|---|---|
| Deceleration from | 60 kmph–0 kmph |
| No of times brake applied | 20 Times in avg |

## 3.2 Theory and Working of Regenerative Suspension System

The suspension of a vehicle is the most important factor that gives the driver/passengers a comfortable ride and also protects the frame/components from taking loads (like torsional, vertical loading, etc.). The suspension basically comprises of linkages attached to the wheel to frame and also spring, shock absorber/damper. This assembly enables first of all constraining the degrees of freedom for the wheel and also results in the action of damping, i.e., also called as shock absorbing during turns or a bumpy road. The suspension spring mass system can be as a vibrating system undergoing compression.

The equation of the spring mass system and forced produced during damping can be given as:

The transfer functions for the unsprung mass system:

$$(M_1 S_2 + b_1 s + K_1)X_1(s) - (b_1 s + K_1)X_2(s) = U(s) \tag{8}$$

Damper force:

$$F_{\mathrm{Dc}} = \frac{\pi}{4} \times 2D_c \times (P_C - P_R)(\text{During Compression Stroke}) \tag{9}$$

$$F_{\mathrm{DR}} = \frac{\pi}{4} \times (D_C - D_R)^2 \times (P_C - P_R)(\text{During Rebound Stroke}) \tag{10}$$

Where $D_c$—Cylinder Diameter, $D_R$—Piston Rod Diameter, $P_R$—Rebound Chamber Pressure, $P_c$—Compression Chamber Pressure, $F_{\mathrm{Dc}}$—Damping force during compression, $F_{\mathrm{DR}}$—Damping force during rebound.

In IC engine vehicles, the role of dampers is to absorb and dissipate energy. However, in regenerative suspension at four linkage joints nearby wheels in place of suspensions a combination of linear motors with springs are used. At sudden or instantaneous bumps to avoid damage to the linear motors springs used. At irregular pathway situation, there will be sudden shift of vehicle CG, due to which up and down motion linear motion occurs. In the construction of these linear motors, stator made up of permanent magnets and rotor is a coil wound with the wires. At vibrating state, rotor moves up and down due to which flux changes rapidly. An opposition force is generated to stop change in flux linkage. Due to change in flux and EMF generated in the rotor winding alternative current is generated, later on it is regulated and rectified using Regulating and Rectifying unit. Further, the Control Box (02) provides the logic to ensure the efficient transfer of the suspension energy which is transferred to Battery pack (Main Storage unit). At each vibrations, the amount of energy produced by the linear motor is unique. So due to that, four controllers are placed in suspension control box for the purpose of monitoring and restoring energy.

The regenerative suspension model is designed using MATLAB-Simulink. In the Fig. 7, signal 1 indicates the vehicle acceleration and deceleration actions performed by the driver. Initially, it goes from rest to 60 kmph, and it will maintain the action for a time, then decelerates. During the whole course of action, 1541 J energy was dissipated.

The following block model labels are used as shown in Fig. 8

- Shows the input of vehicle drive cycle. How the vehicle accelerates/decelerates with in a particular amount of time.
- Converts vehicle linear motion into a vertical motion (in bumps).

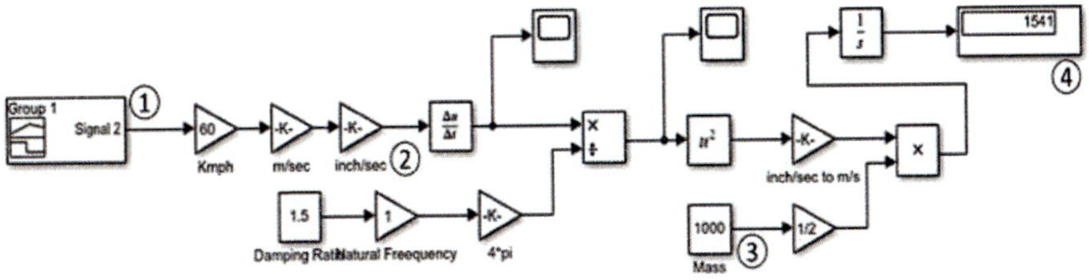


**Fig. 7** Simulink block diagram used in suspension energy calculation

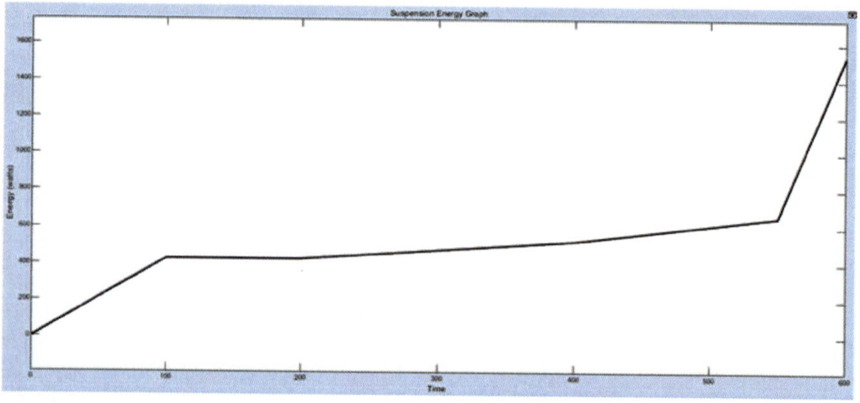


**Fig. 8** Energy versus time graph for regenerative suspension system (0–600 s)

- Vehicle mass is represented. That vertical movement/damping affect is considered as kinetic energy lost via suspension.
- Display overall acquired energy in total drive cycle (1541 J).

After modeling the regenerative suspension model in Simulink, the performance of the model is studied and analyzed. The voltage/energy increased due to damping action taking place. The required input as described in block models are taken to create the plot (energy vs. time).

From (0–100 s): Acceleration: The vehicle is accelerating steadily to achieve 60 kmph speed, and hence, it is assumed the road has certain bumps and vertical acceleration is produced. Hence, due to which increase in energy production by suspension systems is drawn.

At (100–560 s): Constant Speed Travel: The vehicle is trying to maintain constant speed and vertical acceleration produced is limited. Hence, due to this, there is not that much significant change in energy produced.

At (560–600 s): Deceleration: The driver implements sudden braking action due to which the suspension comes in action for absorbing the damping load. Hence, due to this sudden increase in energy production by suspension systems is drawn.

Efficiency calculation by graphical analysis:

$$E_{\text{susp}} = 1541\ \text{J};\ E_{\text{bp}} = 49392\ \text{J}$$

$$\text{Efficiency}(0 - 600\ \text{s}) = \frac{E_{\text{Susp}}}{E_{\text{bp}}}$$

$$\text{Efficiency}(\eta) = 3.11\% \tag{11}$$

Where $E_{\text{susp}}$ = Energy produced by suspension, $E_{\text{bp}}$ = Energy produced by battery pack, $\eta$ = Efficiency.

### 3.3 Solar Energy

The concept of charging using solar energy is currently the most common technology, it is being under research and development process. It is an electromagnetic radiation energy, which is converted into electrical energy via photovoltaic cells. Tiny charges known as photons hits the solar panels, by transferring its energy to one of the outer-orbit energy. Due to this phenomenon, EMF generates inside a PV cell further causes into the production of current.

The solar irradiance is a determinant factor in the prediction of energy generation from the solar panels. Solar irradiance can be termed as the power produced per unit area that is received from the sun in the form of electromagnetic radiation. The solar irradiance energy versus months period in different locations data is referred [10, 11]. The data obtained in this plot is crucial in determining the solar energy that is produced and therefore depicting regions with higher potential for this solar based technology.

The solar energy model is designed using MATLAB-Simulink. The solar irradiance of a specific region is taken and hence our photovoltaic conversion is assessed accordingly.

The following block model labels are used as shown in Fig. 9 (labeled from 1 to 3):

(1) Solar panel with open circuit voltage rating 30 V, short circuit current 8.3 A. By a constant block, (250-W h) solar irradiance energy passes through the solar panel.
(2) Variable resistor is used as a load to increase the value of current in the circuit.
(3) Display block is used to display the total energy produced by the solar panel in a day. Totally, 632 watts of energy is generated by a single solar array.

After modeling the solar energy production block model we need to analyze/simulate and study the model created. For simulation purpose, we have

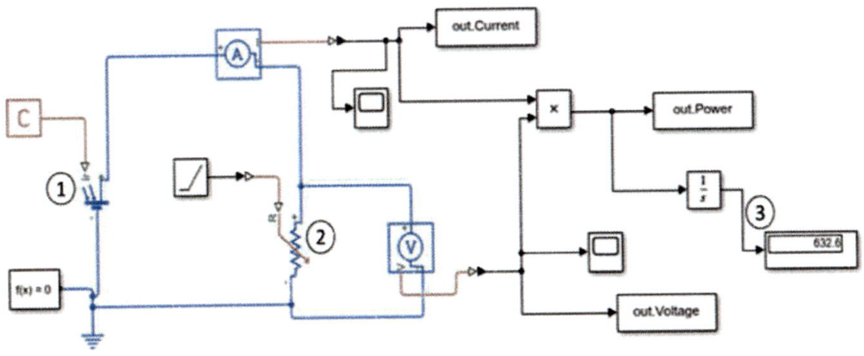


**Fig. 9** Simulink diagram for calculating solar energy

considered the solar irradiance data of a specific location, the location Chennai, Tamil Nadu being with an annual average solar irradiance value of 5.08 kWh/m$^2$/day.

The 2D basic view of solar panel arrangement is displayed in Fig. 10. The solar panels cover from top hatch to rear to the front portion of the car according to the assumed calculations. The complete assumption/ boundary condition and specifications for solar energy model analysis is described as shown in Tables 4 and 5:

With boundary conditions specified the output and graph of solar energy production is analyzed and efficiency is calculated.

In the V-I curve, it is noted that at the 23.45 V setting, current value 3.887 A passed in the circuit. So, by operating the panel at 23.45 V setting, maximum power point conversion (MPPT) occurs as shown in Fig. 11. In the PV curve, it is noted that at the 23.45 V setting of a solar panel. The expected power production is 90.81 watts from a single PV array as shown in Fig. 12.

This method is used to find out the maximum power production done by a PV array. Efficiency calculation as follows,

$$E_{pv} = 632\ W,\ N_{pv} = 16$$

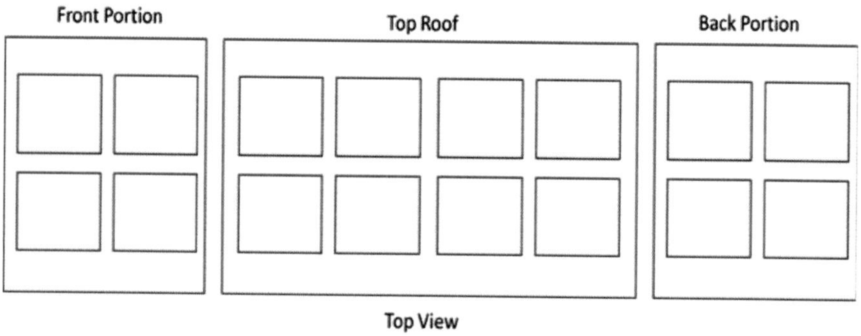


**Fig. 10** Top view arrangement of solar panel

**Table 4** Boundary conditions

| | |
|---|---|
| Total number of solar panels used | 16 |
| Total area of solar panels used | 9.6 m$^2$ |
| Solar panel configuration | 8s2p (8-Series, 2-Parallel) |
| Solar irradiance (Chennai, TN) | 5.08 kWh/m$^2$/day |

**Table 5** Panel specification

| | |
|---|---|
| $V_{oc}$ (Open circuit voltage) | 30 V |
| $I_{sc}$ (Short circuit current) | 8.3 A |
| $P_m$ (Maximum power) | 250 W |

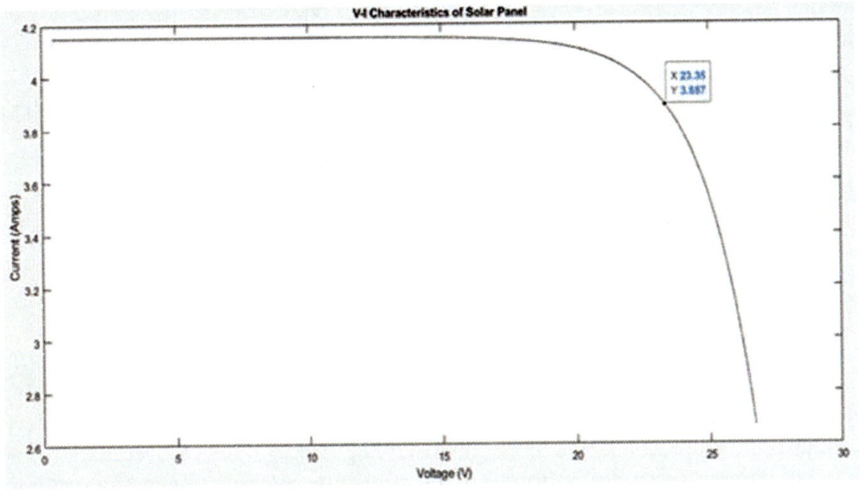


**Fig. 11** V–I curve of a solar array

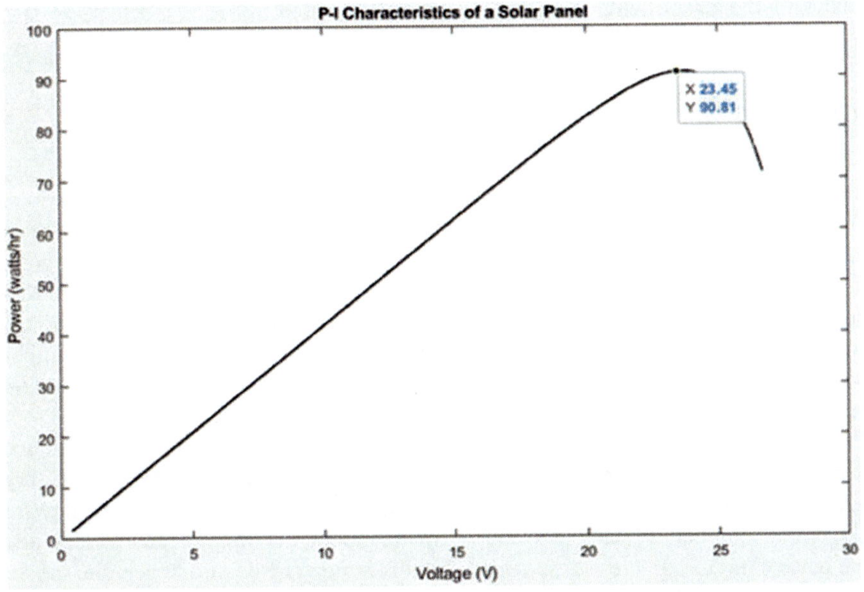


**Fig. 12** P–V curve of a solar array

$$E_{ss} = N_{pv} \times E_{pv} \quad (12)$$

$$\text{Efficiency } (\eta) = \frac{E_{ss}}{E_{pv}} = 20.47\% \quad (13)$$

where, $E_{pv}$ = Energy produced by a single PV panel, $N_{pv}$ = Total Number of Solar panels, $E_{ss}$ = Total Energy generation by Solar system unit, $\eta$ = Efficiency.

# 4 Modeling Regeneration System

In order to maintain smooth charging and discharging of our configured battery pack in order to prevent the cells from sustaining damage due to overcharging or deep charging, a defined monitoring process must be initiated. Hence, in order to monitor/survey the battery pack and internal cell characteristics through the process of discharging of battery to regeneration process a defined algorithm or model for analyzing the state of charge of the EV system is created.

## *4.1 Modeling SOC Analysis Block*

State of charge determines the energy left in the battery pack during the course of time. Hence, it is really important to use the available energy efficiently and calculation is performed in the further sections to measure the change in its endurance effect with or without using regenerative energy systems using this SOC Study. Hence, for this purpose, a MATLAB-Simulink Model is created to study the SOC characteristics as shown in Fig. 13. The two cases are studied, one without regeneration and the other when regeneration is embedded to the system whose characteristics are plotted and results are analyzed.

The above MATLAB-Simulink model created analyzes the battery performance during the scenario where the vehicle is operated with only the source of battery pack where the regeneration source it cutoff. This is done in order to analyze the bare performance of battery pack and vehicle. In the model, the input parameters include the battery pack characteristics, vehicle load parameters etc. And hence, the model is simulated in order to obtain the performance data (that includes SOC study and endurance calculation).

So, from the SOC versus time plot, it can be accessed that the full extent of endurance for overall 403.2 V of the battery pack is 233 km as shown in Fig. 14. But, by the course of action at reaching 20% of SOC, the voltage is cutoff and hence the real-time endurance would be less than 233 km.

The above MATLAB-Simulink model created analyses the battery performance during the regeneration scenario. The model represents the signal builder that

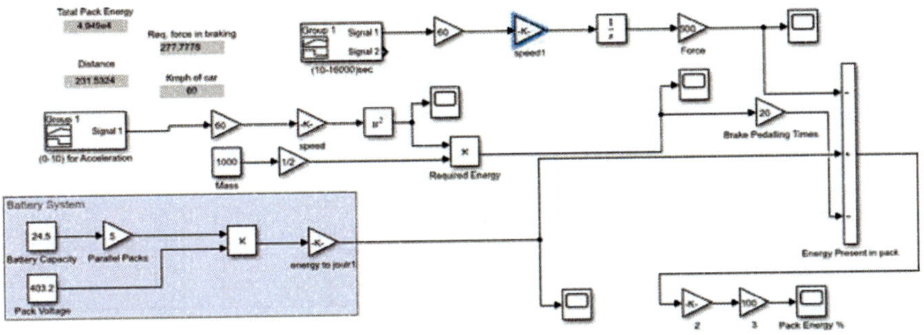


**Fig. 13** Simulink block diagram for SOC study without regeneration and solar charging

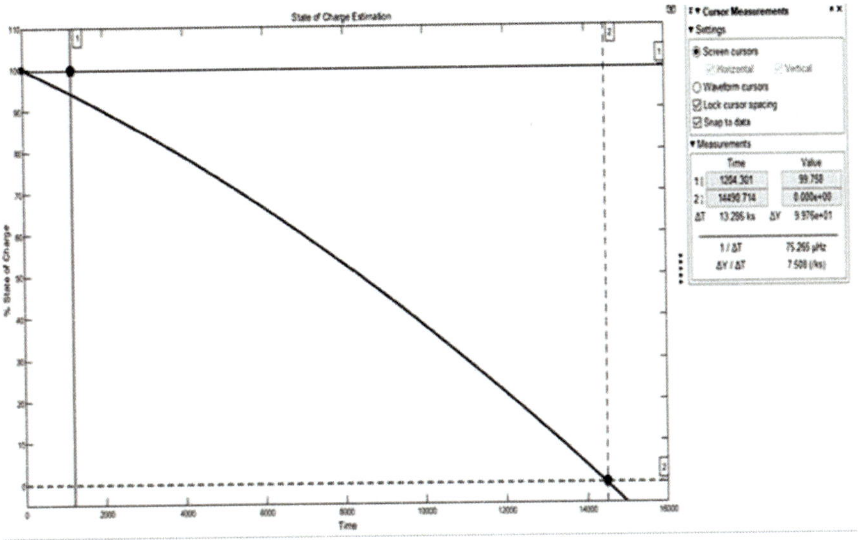


**Fig. 14** Percentage state of charge versus time without regeneration system 0–16,000 s

includes input such as the vehicle speed, battery pack characteristics, braking, solar, and damping energy data as shown in Fig. 15.

Hence, from the SOC versus time plot, we determined that using regeneration, the maximum endurance limit can reach up to 332 km as shown in Fig. 16. During a course of action, only 20% of state of charge is needed to be left over in a pack so, that it can be recharged properly. In the case of real-time drive, the vehicles endurance would be below 332 km in margin of 20% for recharging purpose.

From the above SOC, study performed for the two cases the efficiency of the vehicle with regeneration can be analyzed.

- Endurance limit without regeneration:

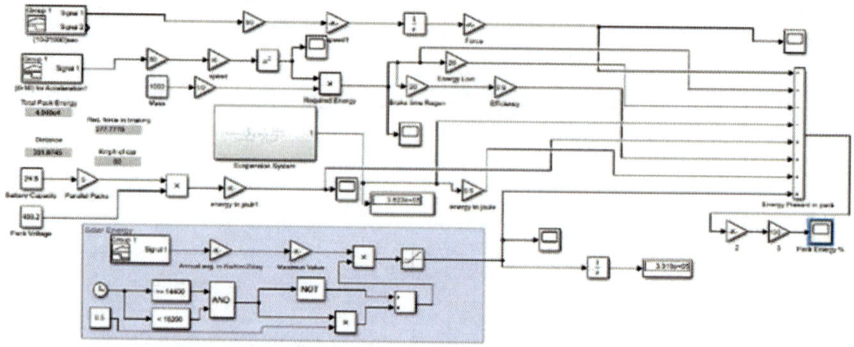

**Fig. 15** Simulink block diagram for state of charge calculation with regenerative braking, solar. and suspension energy

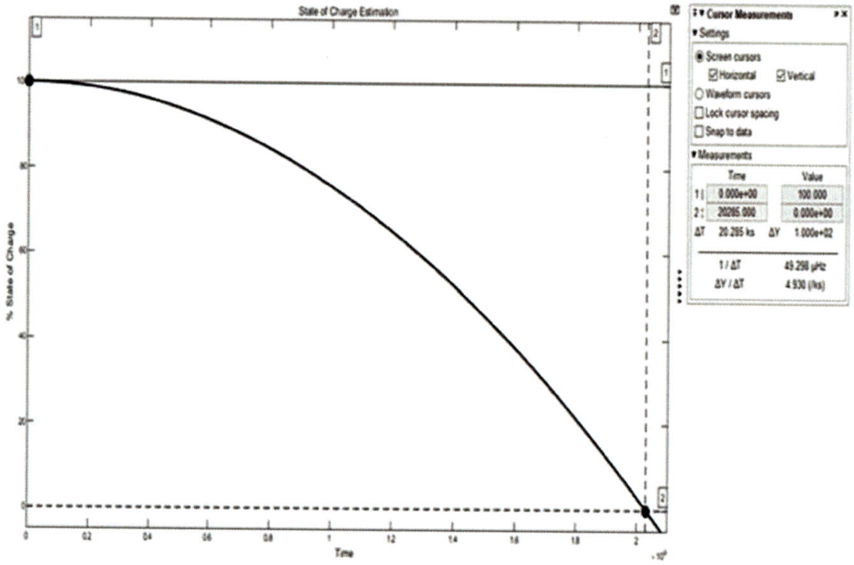


**Fig. 16** State of charge versus time without regeneration system 0–21,000 s

Analyzing for 20% SOC limit at 11,562 s the cutoff of the endurance can be calculated as:

$$D_{\mathrm{wor}} = \frac{T_{\mathrm{wor}}}{T_h} N_v$$

$$D_{\mathrm{wor}} = 193.2 \text{ Km}. \tag{14}$$

- Endurance Limit with Regeneration:

Analyzing for 20% SOC limit at 15,582 s for cutoff the endurance can be calculated as:

$$D_{wr} = \frac{T_{wr}}{T_h} N_v$$
$$D_{wr} = 270.4 \text{ Km.} \quad (15)$$

- Overall Efficiency:

  The overall % increase endurance can be estimated as:

$$\eta = \frac{D_{wr} - D_{wor}}{D_{wor}} \times 100$$
$$\eta = 39.95\% \quad (16)$$

Where $T_{wor}$ = Time taken without regeneration system (secs), $T_{wr}$ = Time taken with regeneration system (secs), $T_h$ = Time to convert seconds to hour, $D_{wor}$ = Distance covered without regeneration (km), $D_{wr}$ = Distance covered with regeneration (km), $\eta$ = Efficiency.

## 5 Modeling Regeneration System

### *5.1 Vehicle SubSystem Configuration*

In order to understand the integration of the overall regenerative system with the EV system frame, the vehicle subsystem is modeled and characterized to have a better understanding on how all the individual subsystems are integrated with regenerative system as shown in Fig. 17.

The vehicle subsystem model configured using MATLAB-Simulink.

Blocks specification:

The description of every block of the vehicle subsystem model is as follows:

(1) Control box: Part (1) implies embedded systems for the safety purpose of entire EV System. It main role is to isolate HV system in case of emergencies. It mainly deals with exterior parts except HV system batteries like crash, system failure, or short circuit, etc. Part (2) implies control system for controlling rechargeable systems by braking and solar energy.
(2) Battery management system (BMS): Part (3) implies control systems mainly designed to monitor battery parameters, regular battery health check-ups, prevents it from overcharging/over discharging cases, etc. BMS controls power delivery by the HV battery pack to propulsion system.
(3) Motor controller: Part (4) mainly focuses on power conversion. converts required power from the battery bank and feeds it to motor to acquire desired speed.

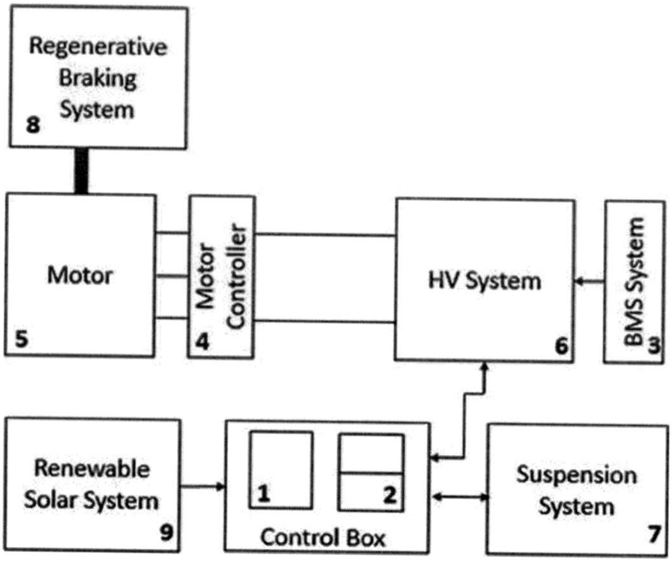


**Fig. 17** Block diagram of vehicle subsystem

(4) Motor: Part (5) is the required type of motor selected. According to it, motor controller is used. It is an electromechanical device, converts electrical energy to mechanical energy.
(5) HV system: Part (6) consists of Li-ion battery modules. It is the primary energy storage unit, with two accumulator isolation relay (AIR's) connecting positive and negative terminals.
(6) Suspension system: Part (7) converts suspension energy produced by jerks due to uneven road structures, into electrical energy via linear motors, which mounted on side portion of the wheels of vehicle. It feds the renewable energy back to one of the power storage unit.
(7) Regenerative braking system: Part (8) is used to obtain energy loss occurs by frictional force, converts that energy (as dissipated by heat) by electromagnetic conversion into electrical energy, which is again fed back to one of the power storage unit.
(8) Renewable solar system: Part (9) converts solar energy produced by sun into electrical energy via photovoltaic solar panels, which mounted on top portion of the vehicle. It feds the renewable energy back to one of the power storage unit.

### 5.2 *EV Control System Integration*

After the design and analysis of the individual regenerative system (including braking, suspension, and solar) and the vehicle subsystem, the main task is to integrate all these models into a single unified system and process with EV system and study the output characteristics of the integrated model [27].

The physical modeling of the overall vehicle system with the regeneration model is done using simulink, hence creating a full vehicle performance scenario integrated as one. A PID controller is used to control in speed, which the vehicle is to be driven, it will acts as accelerator in real-time case. The motor, battery, and driver inputs are provided for the designed model input. Suspension energy, solar energy, and regenerative braking energy blocks are directly add with battery component to increase to overall efficiency of vehicle. The model is simulated and the integrated performance is analyzed [28], obtained the required motor torque with respect to vehicle speed. It provides the dataset in selecting the motor with ideally configured motor ratings to drive the vehicle in a specific speed that is required.

Battery discharge rate with respect to time, as is observed that there is a significant raise in battery SOC at end. The raise in SOC can be analyzed due to the regeneration by suspension and braking and by solar energy production that is being added up to HV system [29].

## 6 Results and Discussions

The numerical and MATLAB-Simulink analysis were performed concurrently to obtain the best cell arrangement structure. Each cell type varies in their performance level, from that nickel manganese cobalt (NMC) cell chemistry chosen because of its rigid structure and enhanced safety features.

The battery pack discharge curve analysis reveals the period the pack can supply energy to the vehicle without having any regenerative features on it. By arranging selected cell series in various arrangements, to achieve desired torque and speed characteristics results in 4s5p configuration. As torque and speed are directly proportional to battery pack voltage (345.6 V) and current (155 A).

To model a vehicle with the best performance, both voltage and current level were considered to be at maximum value. Normally, the energy density of storage Li-ion is 100 times lesser than gasoline energy density. So, while designing, it was crucial to estimate the pack energy and pack weight because increase in battery pack energy simultaneously increases vehicle overall weight. Hence, results in a decrement in endurance range were obtained. So, two factors were considered while designing a model system: Pack energy as well as pack weight, by performing solar, suspension and regenerative braking Simulink analysis, it was concluded that using a battery pack of 42.3 kWh results in providing an efficient increase in endurance range by without having a plug-in connection.

The solar charging model proposed in the study was found to be satisfactory compared to existing with overall 10.1 kWh energy being produced with an increase in endurance by 24.8 miles and 20.4% increase in endurance range obtained rather than a study carried out on Prius PHV obtained 3.5 km. Increase in endurance range obtained due to increased surface area for solar cells in vehicle surface area, considering Indian terrain irradiance value emitted by sun annually, and full day time sunlight availability for recharging battery pack via photovoltaic systems.

The comparative study of battery pack with and without regeneration (including solar energy) yielded an overall increase in endurance from 121.2 miles without regeneration to 168.75 miles using the combined regenerative charging. The increase in battery pack energy being from 42.3 kWh to 57.75 kWh increasing the drive time about 1.26 h.

## 7 Conclusion

This paper provides a detailed overview over the design of a battery pack integrated with a regenerative braking, suspension, and solar charging that aims to improve the endurance efficiency for an EV. Based on the study, following concluding remarks are given:

- Compared to the existing battery pack model the designed and configured 4s5p model in the study was found to be satisfactory with overall battery pack capacity about 42.3 Kwh and overall usable energy to be 20.73 kWh for a constant speed of 60 kmph.
- The discharge study of the battery pack based on the MATLAB study resulted in overall endurance about 121.5 miles for a nominal voltage drawn about 162 V and overall drive time about 2 h.
- Compared to the existing regenerative braking model the designed model in the study was found to be satisfactory resulting in overall increase in the endurance by 39.5% and 16.7 kWh energy being added to battery pack. The braking parameters being accessed at 60 kmph velocity with 20 times/charge braking cycle.
- Compared to the existing regenerative suspension model the designed model in the study was found to be satisfactory resulting in overall increase in the endurance by 3.1% and 1.28 kWh energy being added to the battery pack. The suspension parameters being accessed at damping coefficient 1.5, 60 kmph velocity with a natural frequency of 1 Hz.
- The solar irradiance study based on the monthly average plot of solar irradiance data for states of different countries depicted India being an ideal place for harnessing solar energy with an annual solar irradiance average about 5.08–5.77 kWh/m$^2$/day and also location like Beijing (China) and Washington (US) with annual solar irradiance about 4.8 kWh/m$^2$/day and 3.7–3.85 kWh /m$^2$/day having the potential to harness solar energy technology.

- The solar charging model proposed in the study was found to be satisfactory. Further the proposed model is found to improve 20.4% endurance efficiency compared to existing battery packs. For the proposed solar panel configuration, 5 $m^2$ of vehicle roof and rear panel area, solar power of each panel with 250 watts based on annual solar irradiance data for Chennai, India is considered.
- The integral control system designed for the overall EV configuration was found to be satisfactory and SOC variation due to all the three, i.e., regenerative braking, regenerative suspension, and solar charging acceptable for the working of the configured battery pack.

Thus, the designed and proposed model was found to be acceptable and satisfactory to improve the EV endurance and to reduce the charging cycles for the battery pack. As for the technological advancements in the solar industry, soon solar panels can be sprayed on the whole vehicle surface area and an increase in their conversion efficiency can be expected shortly.

## References

1. Santos S, Aranha J, Cerri F, Nascimento T et al (2018) Design of an energy storage system with blended of Li-ion batteries for pure electric vehicle of high performance. SAE technical paper 2018-36-0180. https://doi.org/10.4271/2018-36-0180
2. Hu X, Stanton S (2014) A complete li-ion battery simulation model. SAE technical paper 2014-01-1842. https://doi.org/10.4271/2014-01-1842
3. Van Sterkenburg S, Rietveld E, Rieck F, Veenhuizen B, Bosma H (2011) Analysis of regenerative braking efficiency—a case study of two electric vehicles operating in the Rotterdam area. 2011 IEEE Vehicle power and propulsion conference, Chicago, IL, pp 1–6. https://doi.org/10.1109/VPPC.2011.6043109
4. Ulikov I, Karpukhin K (2018) Model analysis of efficiency and energy distribution in the powertrain of an electric vehicle equipped with a solar cell battery. SAE technical paper 2018-01-5026. https://doi.org/10.4271/2018-01-5026
5. Modeling of lithium-ion battery management system and regeneration control strategy for hybrid electric vehicles Zhengshu Zhu Michigan Technological Univ. 2013-01-0939. Published 04/08/2013 Copyright © 2013 SAE International. https://doi.org/10.4271/2013-01-0939
6. Cha S, Carter P, Bradow R (1983) Simulation of automobile brake wear dynamics and estimation of emissions. SAE technical paper 831036. https://doi.org/10.4271/831036
7. MATLAB and Simulink racing lounge: vehicle modelling. From File Exchange https://in.mathworks.com/matlabcentral/fileexchange/63823-matlab-and-simulink-racing-lounge-vehicle-modeling
8. Minarcin M (2012) Navistar, Inc. Eric Rask Argonne National Laboratory "Considerations in Estimating Battery Energy for Hybrid and Electric Vehicles" 2012-01-0660, Published 04/16/2012Copyright © 2012 SAE International. https://doi.org/10.4271/2012-01-0660
9. Lee T-K, Zoran S (2011) Filipi Univ. of Michigan. Impact of model-based lithium-ion battery control strategy on battery sizing and fuel economy in heavy-duty HEVs, 2011-01-2253 Published 09/13/2011 Copyright © 2011 SAE International. https://doi.org/10.4271/2011-01-2253
10. Go K, Hirano T, Miyoshi T, Sato D (2017) Development solar charging system of vehicle. SAE Int J Passeng Cars Electron Electr Syst 10(2):353–358. https://doi.org/10.4271/2017-01-1598
11. Synergy Enviro Engineers: Solar irradiance data: http://www.synergyenviron.com/tools/solar-irradiance/india/tamil-nadu/chennai

12. Taksale AS (2013) Modeling, analysis and control of passive and active suspension system for a quarter car. IOSR IEEE 8:973–4562
13. Scott Peck Thermo Analytics Inc. Theodore Olszanski and Sonya Zanardelli US Army TARDEC, WArren, Michigan Matt Pierce Thermo Analytics Inc. Validation of a thermal-electric Li-Ion battery model. Published 04/16/2012Copyright © 2012 SAE International. https://doi.org/10.4271/2012-01-0332
14. Ma Y, Teng H, Thelliez M (2010) Electro-thermal modeling of a lithium-ion battery system. SAE Int J Engines 3(2):306–317. https://doi.org/10.4271/2010-01-2204
15. Akula P, Jandhyala L, Herb F, Narayana A (2012) Development of energy management strategies and analysis with standard drive cycles for fuel cell electric vehicles. SAE technical paper 2012-01-1609. https://doi.org/10.4271/2012-01-1609
16. Deschamps E (2000) Fundamental physics behind new suspension concept for automobiles. SAE technical paper 2000-01-1647. https://doi.org/10.4271/2000-01-1647
17. Samadani E, Farhad S, Panchal S, Fraser R et al (2014) Modeling and evaluation of Li-Ion battery performance based on the electric vehicle field tests. SAE technical paper 2014-01-1848. https://doi.org/10.4271/2014-01-1848
18. Xu W, Zheng H, Liu Z (2013) The regenerative braking control strategy of four-wheel-drive electric vehicle based on power generation efficiency of motors. SAE technical paper 2013-01-0412. https://doi.org/10.4271/2013-01-0412
19. Shaik A, Tappa R, Kannan P, Bhaskar P et al (2019) Design considerations and analysis of electric microcar for cities. SAE technical paper 2019-28-0161. https://doi.org/10.4271/2019-28-0161
20. Yang C, Wei X, Fang Q, Dai H (2019) SOC estimation of battery pack considering cell inconsistency. SAE technical paper 2019-01-1309. https://doi.org/10.4271/2019-01-1309
21. Wendel G, Steiber J, Stecklein G (1994) Regenerative active suspension on rough terrain vehicles. SAE technical paper 940984. https://doi.org/10.4271/940984
22. Ahmad K, Alam M (2017) Design and simulated analysis of regenerative suspension system with hydraulic cylinder, motor and dynamo. SAE technical paper 2017-01-1284. https://doi.org/10.4271/2017-01-1284
23. Wendel G, Stecklein G (1991) A regenerative active suspension system. SAE technical paper 910659. https://doi.org/10.4271/910659
24. Gunti S, Gunti S, Shaik A, Rao S et al (2015) Development of range extended—solar power assisted electric vehicle. SAE technical paper 2015-26-0111. https://doi.org/10.4271/2015-26-0111
25. Battery Datasheet. https://batteryservice.bg/wpcontent/uploads/2018/12/INR21700-50E.pdf
26. Samadani SE, Fraser RA, Fowler M (2012) A review study of methods for lithium-ion battery health monitoring and remaining life estimation in hybrid electric vehicles. University of Waterloo. 2012-01-0125 Published 04/16/2012 Copyright © 2012 SAE International. https://doi.org/10.4271/2012-01-0125
27. Nagai H, Morita M, Satoh K (2016) Development of the Li-ion battery cell for hybrid vehicle. SAE technical paper 2016-01-1207. https://doi.org/10.4271/2016-01-1207
28. Hartley J, Day A, Campean I, McLellan R et al (2010) Braking system for a full electric vehicle with regenerative braking. SAE technical paper 2010-01-1680. https://doi.org/10.4271/2010-01-1680
29. Bao R, Griggs P, Baxter J (2018) Simulation based control strategy design of all wheel drive electric vehicle regenerative braking system. SAE technical paper 2018-01-0411. https://doi.org/10.4271/2018-01-0411

# Modeling Methodology of Flywheel Energy Storage System for Microgrid Applications

R. Ramaprabha, C. Karthik Rajan, R. Niranjan, and J. Kalpesh

## 1 Introduction

Environmental issues led to the decentralized power production, which also include the renewable energy generation. This results the great fascination toward the research in the microgrid/smart grid areas. Microgrid (MG) integrates the different renewable and other sources. The major issue of balancing energy generation from different sources and load demand is met by energy storage systems in the microgrid. The storage system must quickly respond to maintain the power balance [1–3]. In the literature, it is reported that the most appropriate technology of FESS is considered to increase the stability in microgrid [4–6]. This paper discusses the step-by-step procedure for modeling a PV-based FESS suitable for the microgrid is discussed.

A flywheel acts like a mechanical battery that stores energy in kinetic form. The flywheel works based on Newton's first law of motion applied to rotating systems, wherein the flywheel keeps rotating even after removal of the source transferring rotational energy. This rotation of the flywheel after the removal of the source is then utilized to harness energy when required by the system interconnected to it. FESS is utilized for short to medium duration high-power storage and discharge operation. They can help in smoothing out voltage and current transients due to intermittency in power generation and control transients that can occur during switching operations [7–15]. They can act as an uninterruptible power supply source during switching operations due to their quick response time, which makes it more suitable to microgrid.

The schematic of the complete FESS is depicted in Fig. 1. The power from the source is conditioned accordingly based on the motor rating using a power-conditioning unit (PCU). In this stage, electrical energy is converted to mechanical

R. Ramaprabha (✉) · C. Karthik Rajan · R. Niranjan · J. Kalpesh
Department of Electrical and Electronics Engineering, Sri Sivasubramaniya Nadar College of Engineering, Rajiv Gandhi Salai, Kalavakkam, Tamil Nadu 603110, India
e-mail: ramaprabhar@ssn.edu.in



N. L. Narasimhan et al. (eds.), *Recent Advances in Energy Technologies*, Lecture Notes in Mechanical Engineering, https://doi.org/10.1007/978-981-19-3467-4_12

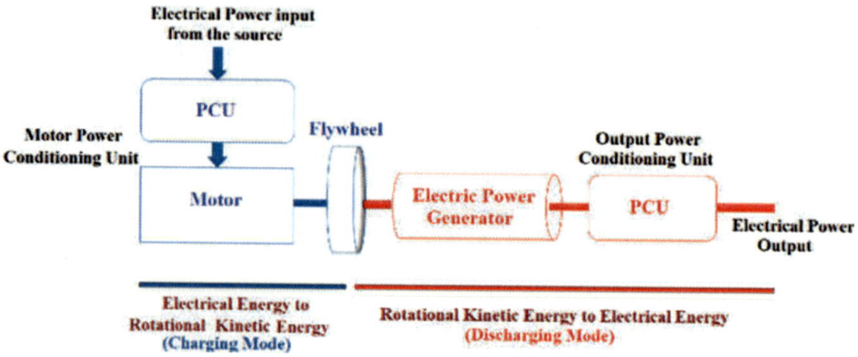


**Fig. 1** Line diagram of FESS

energy. The motor generates higher torque, which drives the flywheel at a higher rotational speed. Hence, the flywheel stores the energy kinetically, which is proportional to the square of its rotational speed and its moment of inertia (M.I). This energy can be used to operate an electric generator. In this stage, mechanical energy is converted back to electrical energy, and the desired power output can be obtained by using another PCU.

This paper aims to design and simulate a FESS for microgrid application with an appropriate power electronic interface. Moreover, the work focuses to test the system under different conditions, observe the characteristics & performance parameters of FESS and analyze its utility for microgrid applications. This paper results in the availability of a sophisticated FESS model suitable for given specifications for users. The following sections describe the detailed modeling of the same.

## 2 Overall Description of the System

A microgrid is an independently working mini-grid that can supply power to small loads. Figure 1 provides an overall indication for the system. In this paper, the utilization of a flywheel that can power a 1 kW system is considered. The system design depends on the flywheel and its storage capacity of energy. Based on the flywheel and its energy storage capacity, the system design is described. Here, a PV-based energy source for controlling the flywheel is taken. To drive the flywheel, a BLDC motor and a separately excited alternator are used. The excitation can be provided through another converter from the PV source or through suitable self-excitation methods with suitable converters for real-life implementation. FESS is designed and implemented on MATLAB/Simulink. The system consists of multiple subsystems interfaced with each other, which are PV array, buck-boost converter, three-phase inverter, BLDC motor, flywheel, and alternator.

The subsystems are connected together, and the performance of the system is studied and analyzed. The PV array based on the environmental conditions produces a DC output voltage and an output current. This output voltage is fed to the DC–DC buck-boost converter, which produces an output voltage of 48 V (which is the rated voltage of the BLDC motor). If the PV array output is higher than 48 V, buck operation is performed, and if the PV array output is lower than 48 V, boost operation is performed. A controller is used in this converter to get the required output voltage of 48 V. This is then fed to the inverter, which produces a three-phase output voltage with 48 V peak value and acts as input to the BLDC motor. The motor starts rotating with a speed and produces torque. The torque produced drives the flywheel whose shaft is associated with the motor. The flywheel is utilized to store energy in rotational kinetic form. The alternator with appropriate excitation provides electrical power output that can be used to supply the required load.

## 3 Design Components of System

The design components of the system are described in this section.

### *3.1 PV Array*

The procedure for obtaining the PV power rating [14] for the whole FESS is explained in detail below:

- A load of 1.2 kW that the microgrid must be capable of supporting is assumed. The source of the microgrid is a PV array. Hence, assumed load = **1.2 kW**
- Assuming 8 h operation of the system, Watt-hour capacity of the system = operating hours × total load = **8 x 1200 = 9600 Wh**
- A PV panel with a peak rating of 250 W is used in the simulation. Considering the operating factor (equal to 0.75), actual PV output power = operating factor × peak power = **0.75 x 250 = 187.5 W**
- Also considering the system efficiency to be roughly 81%, power available to the user after considering system efficiency = **187.5 x 0.81 = 152 W**
- In a day, energy output from 250 W per panel = actual PV output power × 8 h/day = **152 x 8 = 1216 Wh**
- Therefore, the number of PV panels required for the system is given by N = (Watt-hour rating)/(Energy output per panel) = **9600/1216 = 7.89 = 8 (rounded off)**

Hence, 8 panels of 250 Wp are required. They are connected in a 2 × 4 configuration for the simulation. The wiring diagram for the PV array is represented in Fig. 2. In Table 1, the PV array specifications are listed.

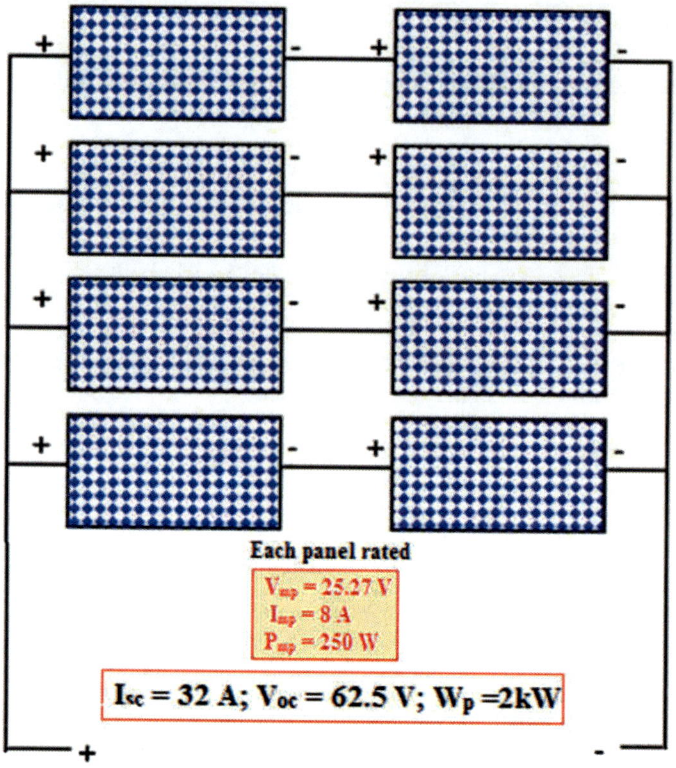


**Fig. 2** Wiring diagram of the PV array

**Table 1** PV array specifications

| Parameters | Specifications/ratings |
|---|---|
| Power rating of array | 2 kW |
| No. of panels and configuration | 8 panels, 2 × 4 |
| Maximum power rating of panel $P_m$ | 250 W |
| Maximum power voltage $V_m$ | 25.27 V |
| Voltage at open circuit $V_{oc}$ | 31.25 V |
| Maximum power current $I_m$ | 8 A |
| Current at short circuit $I_{sc}$ | 8.97 A |
| System maximum voltage | 1000 V |

## 3.2 Buck-Boost Converter

The desired voltage is obtained by using a buck-boost converter with an appropriate control scheme. The design procedure for a buck-boost converter is as follows:

**Table 2** Converter specifications

| Parameters | Specifications/ratings |
|---|---|
| $L$ | 12 μH |
| $C$ | 546 μF |
| Switching frequency | 25 kHz |
| Desired output voltage | 48 V |
| $K_p$ and $K_i$ | 0.001 and 0.04 |

- A maximum voltage of 48 V is required, which is the maximum DC voltage, the motor can withstand. Hence, $V_o = 48$ V
- The duty cycle of the converter is given by $D = V_o/(V_o + V_s)$, where $V_s$ is the source voltage from the PV panel.
- Inductor current $I = I_o/(1 - \mathrm{D})$, where $I_o$ is the expected output current from the converter which is fed to the inverter.
- $L$ is calculated by assuming the ripple current $\Delta i$ about 20% of the output or load current.
- $L$ is calculated using $\Delta i = (V_s\ \mathrm{D}T)/L$, where $T = 1/f_s$ and $f_s$ is switching frequency.
- Now, we compute ripple voltage value using $\Delta v = 0.04\ V_o$ or 4% of $V_o$.
- The capacitance value $C$ is calculated using the equation, $\Delta v = \frac{V_S D^2 T}{[C R_o (1-D)]}$, where $R_o$ is the output resistance or load resistance. Assume CCM of operation.

In Table 2, the designed values are tabulated.

### 3.3 Motor-Drive System

Based on the literature [9–12] study, in this work, a BLDC is taken as a motor, while a separately excited alternator is considered for generating the required power. So, in this work, the brushless DC motor is used in the FESS. The BLDC motor specifications are enumerated in Table 3.

**Table 3** BLDC motor specifications

| Parameters | Specifications/ratings |
|---|---|
| Rating | 1.5 kW |
| Voltage and current | 48 V DC and 31.25 A |
| Speed | 3000 rpm |
| Power | 2 hp |

### 3.4 Flywheel System

The flywheel is designed based on (1)–(6). The calculation steps are discussed below and consolidated in Table 3.

M.I of a disk/solid cylinder,

$$J = \frac{1}{2}mr^2 \tag{1}$$

M.I of a hollow cylinder,

$$J = \frac{1}{2}m(r^2 + R^2) \tag{2}$$

Energy stored in the flywheel,

$$E = \frac{1}{2}J\omega^2 \tag{3}$$

Torque,

$$T = J\frac{d\omega}{dt} \tag{4}$$

Mass,

$$m = \text{Density} \times \text{volume} \tag{5}$$

Volume of the cylinder,

$$V = \pi r^2 h \tag{6}$$

where $\omega$—Angular speed; $m$—Flywheel mass; $r$, $h$—Flywheel radius and thickness

- The first step is to decide the requirement of output power and time. In this paper, **1 kW power** for **2.5 s** is finalized.
- Hence, it is required about **2500 J of energy**. Now, assume a rotational velocity of **1200 RPM (125.7 rad/sec)**.
- Using the above values, the moment of inertia required to produce the required energy is calculated from (3), and this results in about **0.3125 kg/m$^2$**.
- With this value for a moment of inertia, we then calculate the dimensions of the disk for the flywheel after assuming the thickness and choosing a particular material of given density.
- The material we chose is steel, and the density is about **7850 kg/m$^3$** on average, the dimensions obtained are listed in Table 4.

**Table 4** Flywheel specifications

| Parameters | Specifications/ratings |
|---|---|
| Material | Steel |
| Mass of flywheel | 10 kg |
| Material density | 7850 kg/m$^3$ |
| Shape | Thin disk/cylindrical |
| Radius and thickness of flywheel | 0.25 m and 0.04 m |
| Hollow shaft diameter (inner, outer) | 0.043 m, 0.023 m |
| Shaft length on either side of flywheel | 0.035 m |
| Total moment of inertia | 0.3125 kg/m$^3$ |
| Speed of rotation | 1200 rpm |

### *3.5 Alternator*

In the FESS, a separately excited synchronous AC generator or alternator is used, because the machine to supply a three-phase load of 1 kW is required. Here, to generate about 1 kW of power from the flywheel whenever possible and to achieve at a phase voltage of about 230 V at a speed of 1500 RPM (as per specifications), the alternator with specifications listed in Table 5 is used.

## 4 MATLAB Implementation, Results, and Discussions

The components and subsystems (Fig. 3) are designed using MATLAB/Simulink [16]. The design specifications of the various components used in the FESS are incorporated into appropriate subsystems. These subsystems are then converted into blocks for an easy and better understanding of the overall system. These blocks are connected together as a whole as shown in Fig. 3, and the simulation is run to analyze

**Table 5** Alternator specifications

| Parameters | Specifications/ratings |
|---|---|
| Power rating | 1 kVA |
| Armature voltage output | 415 V |
| Armature current output | 4 A |
| Speed | 1500 rpm |
| Field voltage | 220 V |
| Field current | 0.5 A |
| Number of poles | 4 |
| Phase voltage | 230 V |

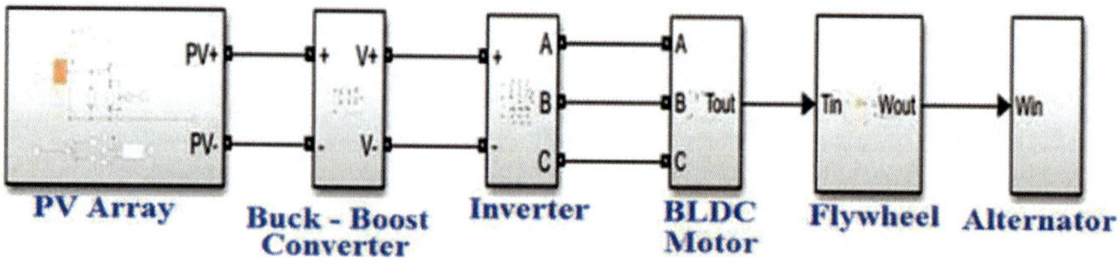


**Fig. 3** MATLAB circuit for FESS

the FESS performance. The required observations are made, and the waveforms are plotted.

## *4.1 MATLAB Implementation of PV Array*

The designed PV array (Table 1) is simulated, and the characteristics are plotted (Fig. 4). The attained maximum power point values are 77.2 V, 25.89 A, and 2 kW.

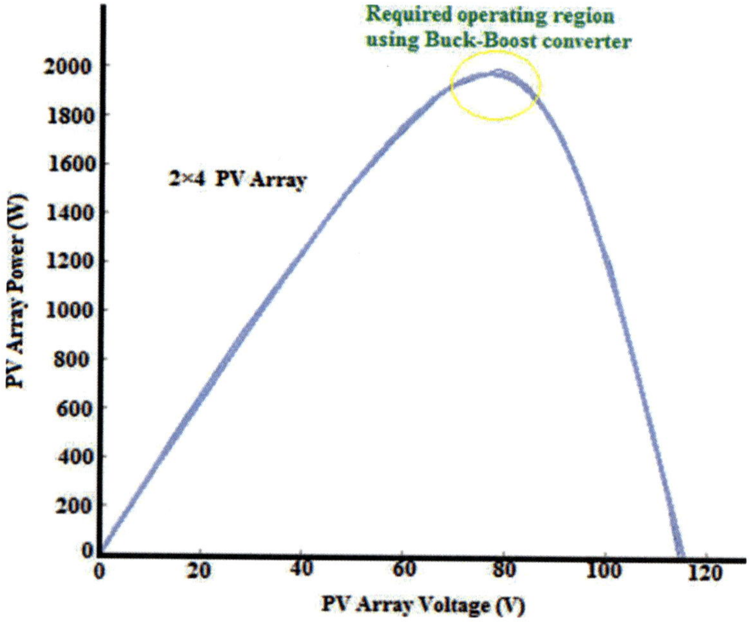


**Fig. 4** P–V characteristics of PV array

## 4.2 MATLAB Implementation of Three-Phase Inverter and Buck-Boost Converter

The design of buck-boost (DC–DC) converter with controller (Table 2) is simulated, and the characteristics with controller are plotted (Fig. 5). The three-phase inverter is simulated, and line-line voltage is plotted (Fig. 6).

It is observed that DC converter gives the regulated output voltage of 48 V, and hence, the inverter peak voltage follows that.

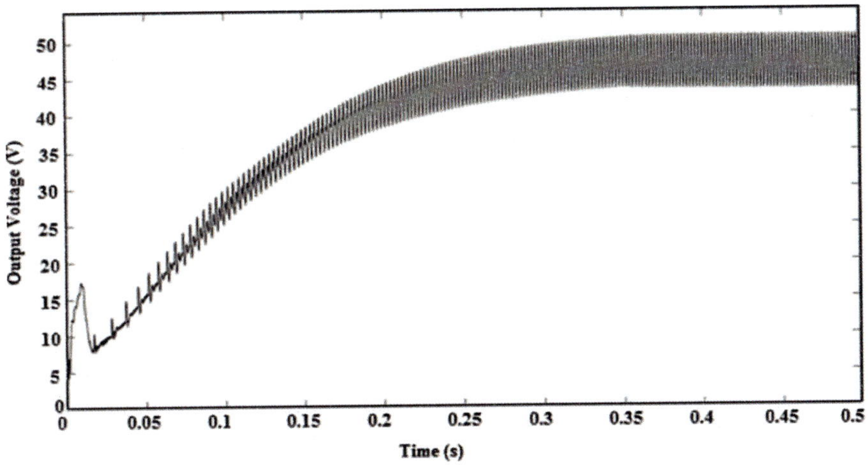


**Fig. 5** Output voltage of buck-boost converter

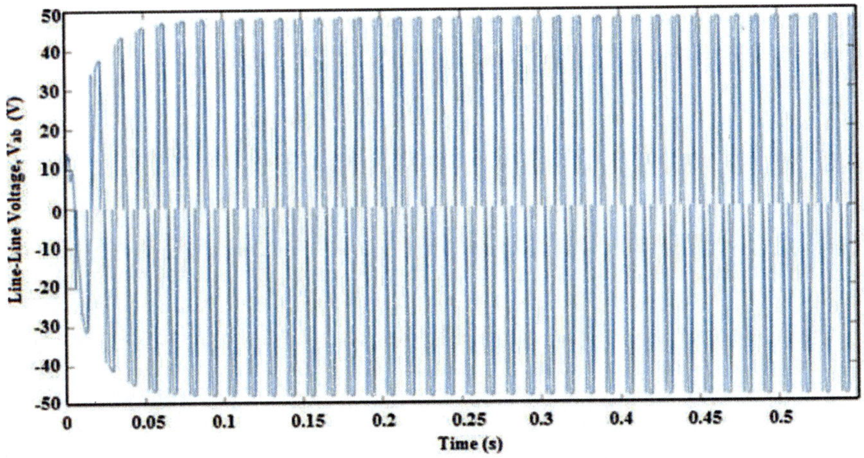


**Fig. 6** Line-line voltage

**Table 6** Truth table for switching for BLDC motor windings

| S. No. | $H_a$ | $H_b$ | $H_c$ | $S_1$ | $S_2$ | $S_3$ | $S_4$ | $S_5$ | $S_6$ |
|---|---|---|---|---|---|---|---|---|---|
| 1 | 1 | 0 | 0 | 0 | 1 | 1 | 0 | 0 | 1 |
| 2 | 1 | 1 | 0 | 1 | 0 | 1 | 0 | 0 | 1 |
| 3 | 0 | 1 | 0 | 1 | 0 | 1 | 1 | 0 | 0 |
| 4 | 0 | 1 | 1 | 1 | 1 | 0 | 1 | 0 | 0 |
| 5 | 0 | 0 | 1 | 0 | 1 | 0 | 0 | 1 | 0 |
| 6 | 1 | 0 | 1 | 0 | 1 | 1 | 0 | 1 | 0 |

### *4.3 MATLAB Implementation of BLDC Motor*

The BLDC motor is the driver of the entire FESS system. It produces speed and torque with the appropriate electrical input, and hence, it drives the flywheel as well as the alternator that is coupled to the motor. A torque input equal to the machine's rated torque (which is 4.774 Nm for the given machine) is given as a step input to the motor. Since the three-phase AC is given to the motor, the motor runs and produces a speed and torque output. Hall sensors are present in the BLDC motor, which indicates the rotor position, and based on this, the respective winding is charged. Accordingly, in the three-phase inverter, the gate pulses are generated. The manner in which the gate pulses are generated and subsequently the inverter switches being operated is based on a truth table. The logic of the truth table is shown in Table 6. Based on the specifications of the motor, the parameters are given in the Simulink model of BLDC drive, and waveforms are shown in Fig. 7.

Even though the BLDC motor rated speed is 3000 rpm and the alternator rated speed is 1500 rpm (Table 5), the flywheel rated speed is 1200 rpm. Hence, 1200 rpm of reference speed is applied to the entire system. A speed control loop with PI controller ($K_p = 16.6$ and $K_i = 0.01$) is used for speed control and maintains the speed at 1200 rpm.

### *4.4 MATLAB Implementation of Flywheel*

The flywheel rotates with the torque produced by the BLDC motor and drives to its rated speed. Hence, it acts as an energy storage device in the formation of rotational kinetic energy. Therefore, the speed of the flywheel will be the same as that of the BLDC motor since the shafts of both the machines are connected together. The angular speed is shown in Fig. 8.

In a microgrid, suppose the output of the PV array is below a threshold value, the switchover takes place to the conventional grid to prevent load fluctuation and high transients. In this simulation, let us consider disconnection from the microgrid in such a case. An ideal torque source is used, and rated torque is given, due to

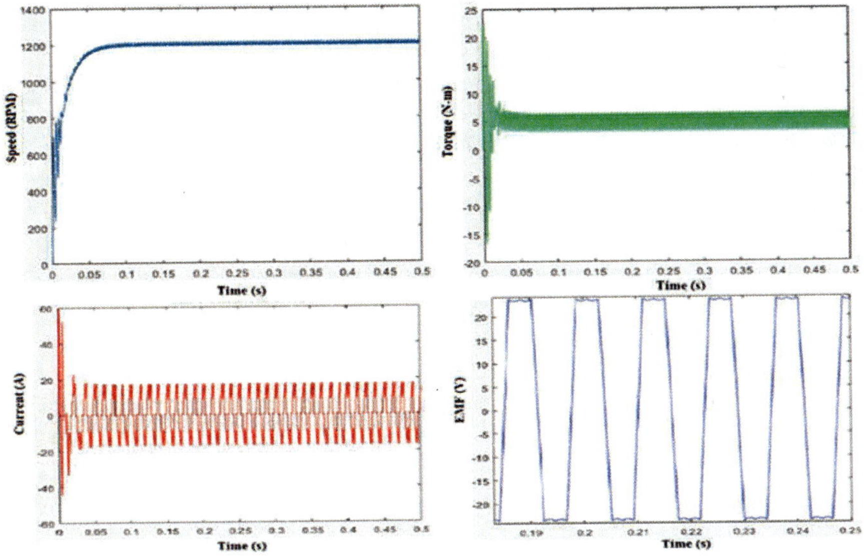


Fig. 7 Waveforms of BLDC motor

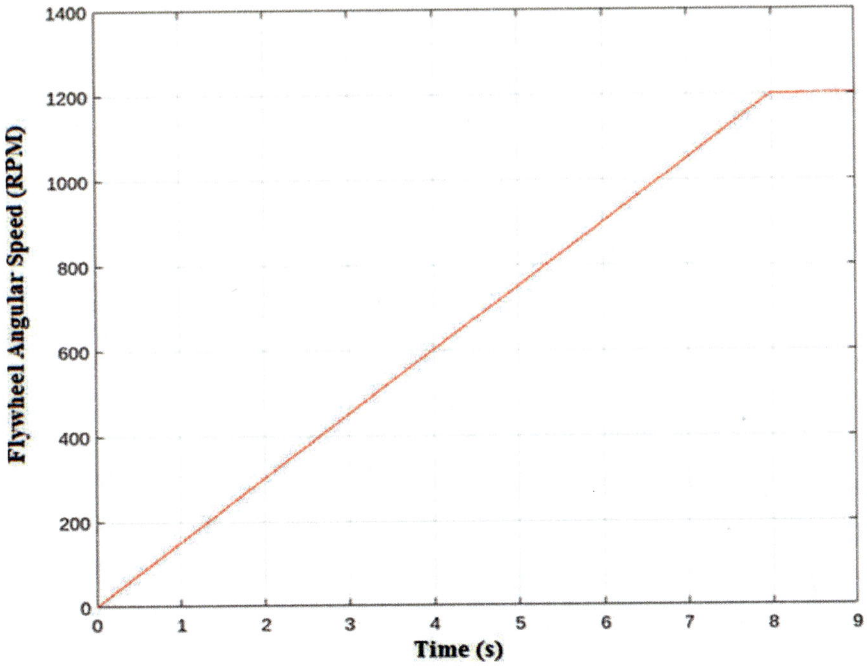


Fig. 8 Flywheel speed

which the speed is running at 1200 rpm without any fluctuations. To indicate the disconnection phenomenon for the simulation, let us assume the motor's operation stops at $t = 22.385$ s. Hence, the flywheel will stop rotating. The flywheel speed will gradually reduce to 0 if the torque given to the flywheel reduces linearly to 0. The waveform is shown in Fig. 9 and observed parameters are listed in Table 7.

The significance of this reading is that if disconnection from the microgrid happens, the flywheel with the given specifications will be able to rotate for 13.052 s more and operate the alternator, which will sustain the load for this time period. Therefore, the flywheel is capable of acting like a storage device in the form of rotational kinetic energy during this period.

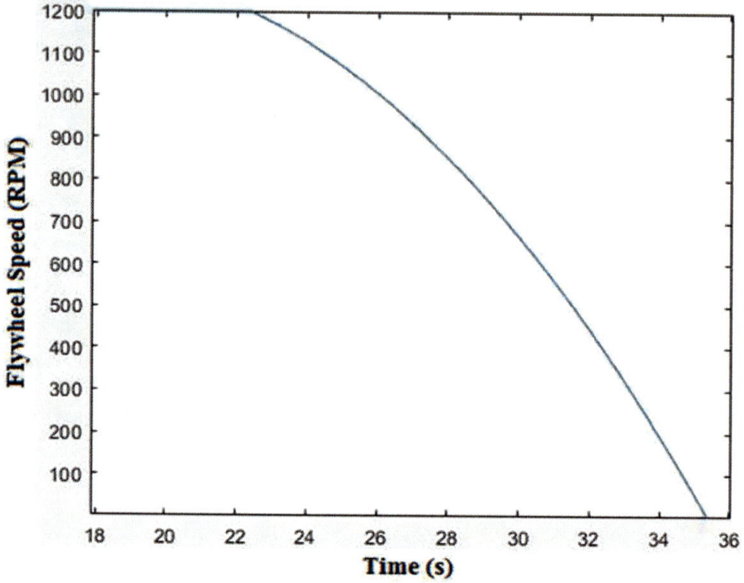


**Fig. 9** Flywheel speed after disconnection from the source

**Table 7** Parameters of flywheel at the time of disconnection from the source

| Initial torque input given to the rotating flywheel (N-m) parameters | Time at switchover (s) | Time at zero speed (s) |
|---|---|---|
| 4.774 | 22.385 | 35.437 |
| Time period for flywheels runs at same speed after disconnection | 35.437–22.385 = 13.052 s | |

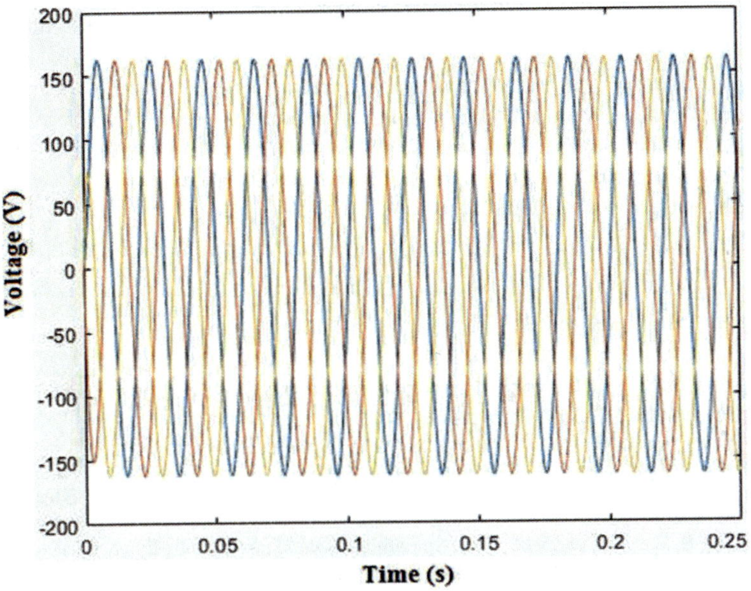


**Fig. 10** Output voltage of alternator

### 4.5 *MATLAB Implementation of Alternator*

The alternator is modeled based on Table 4. The flywheel drives the alternator, which produces electrical power output to a load of 1 kW. The three-phase voltage waveform is represented in Fig. 10. The terminal voltage is observed as 162. 5 V. The alternator supplies a constant electrical power output to the three-phase load. The electrical power is observed as 855.4 W.

The alternator is the machine of the FESS in which the load is to be connected. Hence, it is the machine whose output characteristics are of importance.

## 5 Conclusions

In this work, a solar PV-based FESS is designed, simulated, and evaluated for micro-grid application. An extensive study has been conducted to understand the progress behind the implementation of FESS for RES. The system proposed consists of a PV solar energy source that powers a FESS consisting of a flywheel driven by a BLDC motor that is speed controlled by a buck-boost converter and an inverter circuit. The output of the flywheel is then fed to an alternator to generate power in AC. The design and calculations for the ratings of the components are carried out, and the subsystems are interfaced and simulated on MATLAB-Simulink and the necessary characteristics are obtained. Based on the simulation, observations and the values obtained, it can be concluded that the flywheel can be utilized as a great ESS for microgrid. Under any abnormal conditions, the flywheel prevents voltage transients and sudden load variations. The flywheel can be designed to run for any amount of time based on the ratings of the system and the energy requirement that the flywheel has to supply during that time period. This work can be extended to systems of different ratings for implementation.

## References

1. Nikkhajoei H, Lasseter RH (2009) Distributed generation interface to the CERTS microgrid. IEEE Trans Power Del 24(3):1598–1608
2. Stimoniaris D, Tsiamitros D, Poulakis N, Kottas T, Kikis V, Dialynas E (2011) Investigation of smart grid topologies using pilot installations experimental results. In: Proceeding of 2011 2nd IEEE PES international conference and exhibition on innovative smart grid technologies (ISGT Europe), pp 1–8
3. Gayathri S, Kar IN, Senroy N (2016) Smoothing of wind power using flywheel energy storage system. IET Renew Power Gener 11
4. Samineni S, Johnson BK, Hess HL, Law JD (2006) Modeling and analysis of a flywheel energy storage system for voltage sag correction. IEEE Trans Ind Appl 42(1):42–52
5. Rajapakshe A, Madawala UK, Muthumani D (2008) A model for a fly-wheel driven by a grid-connected switch reluctance machine. In: IEEE International conference sustainable energy technology, pp 1025–1030
6. Ibrahim H, Ilinca A, Perron J (2008) Energy storage systems—characteristics and comparisons. Renew Sustain Energy Rev 12(5):1221–1250
7. Jin C et al (2017) Research on coordinated control strategy of flywheel energy storage array for island microgrid. In: 2017 IEEE conference on energy internet and energy system integration (EI2)
8. Hirofumi A, Sato H (2002) Control and performance of a doubly-fed induction machine intended for a flywheel energy storage system. IEEE Trans Power Electron 17(1):109–116
9. Mustafa A, Keith E, Pullen R (2017) A review of flywheel energy storage system technologies and their applications. Appl Sci 7(3):286
10. Khodadoost A et al (2017) Review of flywheel energy storage systems structures and applications in power systems and microgrids. Renew Sustain Energy Rev 69:9–18
11. Qian C et al (2015) Coordinated control for flywheel energy storage matrix systems for wind farm based on charging/discharging ratio consensus algorithms. IEEE Trans Smart Grid 7(3):1259–1267
12. Xing L, Erd N, Binder A (2016) Evaluation of flywheel energy storage systems for residential photovoltaic installations. In: 2016 International symposium on power electronics, electrical drives, automation and motion (SPEEDAM)
13. Pullen Keith R (2019) The status and future of flywheel energy storage. Joule 3(6):1394–1399
14. Pranav P et al (2017) Simplified implementation and control of a flywheel energy system for microgrid applications. In: 2017 IEEE global conference on signal and information processing (GlobalSIP)
15. Pandiarajan N, Ramaprabha R, Muthu R (2012) Application of circuit model for photovoltaic energy conversion system. Int J Photo Energy 14
16. www.mathworks.com.

# Effect of Flow Disturbers on the Performance of Sensible Energy Storage Device

R. S. Shriram, S. Gowtham, and A. S. Krishnan

## Nomenclature

| | |
|---|---|
| $E$ | Energy, J |
| ID | Inner diameter, mm |
| $\dot{m}$ | Mass flow rate, kg/s |
| OD | Outer diameter, mm |
| $P$ | Power, W |
| PEP | Performance enhancement parameter, dimensionless |
| $Q$ | Heat transfer rate, W |
| $T$ | Temperature, °C |
| $V$ | Volume flow rate, cc/s |

## Subscripts

| | |
|---|---|
| $f$ | Fluid |
| $i$ | Including flow disturber |
| in | Inlet |
| $m$ | Mean |
| $o$ | Without flow disturber |
| out | Outlet |
| $p$ | Pumping |
| $s$ | Solid |
| $R$ | Ratio |

R. S. Shriram · S. Gowtham · A. S. Krishnan (✉)
Department of Mechanical Engineering, Coimbatore Institute of Technology, Coimbatore 641014, India
e-mail: krishhttp@alumni.iitm.ac.in



N. L. Narasimhan et al. (eds.), *Recent Advances in Energy Technologies*, Lecture Notes in Mechanical Engineering, https://doi.org/10.1007/978-981-19-3467-4_13

| | |
|---|---|
| $T$ | Transferred |
| RT | Required to transfer |

## 1 Introduction

Advances in energy technologies are never ending ventures due to ever increasing demand and constraints of supply of energy from clean sources. The supply of energy from clean resources is time dependent, and generally not in phase with the time varying demand. Hence, it is indispensable to store energy. There exists multitude of ways in which energy can be stored and each has its own pros and cons. The choice of the storage system, apart from application, depends on constraints of availability of space, time, cost, etc. Akinyele and Rayudu [1] presented a comprehensive review of the plethora of energy storage technologies. The authors have discussed various forms of energy storage like pumped hydro, super capacitors, thermal energy storage, etc., thus encompassing storage in mechanical, electrical, and magnetic forms. The authors have presented a comparison of characteristics of these systems in terms of power rating, discharge time and losses, etc., along with their respective economical aspects. A similar study was presented by Guney and Tepe [2]. The authors have attempted at assessment of different energy storage system based on various parameters. Prominent among them were impact on the environment, feasibility of implementation, and application variations. Given the need for research in energy storage, it is increasingly becoming necessary for networking among different fields for both fundamental research as well as application. For instance, a combination of different types of energy storage can be envisaged for a system. Parra et al. [3] reported a review of energy storage system for communities from an interdisciplinary view point. The authors have also discussed on the opportunities and challenges associated with such solutions.

As mentioned earlier, energy can be stored in the form of heat, and such systems are generally called thermal energy storage systems [TES]. Research on TES has been well documented. Sarbu and Sebarchievici [4] presented a comprehensive review of TES. Different types of both sensible and latent heat storage systems have been discussed, which include packed bed, chemical storage, cold thermal storage, etc. The author(s) have explained on calculation of thermal storage capacity, performance, and economics of TES. Li et al. [5] reported their research on review on TES technologies. They also carried out experiments and simulations on adsorption TES for residential applications. In a more recent article, Alva et al. [6] discussed in breadth and depth on TES. The authors have addressed various aspects which include thermophysical properties, mechanisms, performance, and operations of these systems. TES is classified as sensible heat storage (SHS) and latent heat storage (LHS). While it is possible to store large amount of heat in LHS, SHS is relatively simpler. Burcu et al. [7] presented a comprehensive review of SHS. The authors discussed in length and depth on the application of SHS in solar energy utilization for industries. The authors have examined systems and materials for SHS and pointed out that SHS can help provide cost-effective solar thermal energy utilization. The pebble bed and cored brick regenerators (CBR) are two popular SHS systems. CBRs have significantly lesser pressure drop than the former [8]. They also do not have the constraint of fluidization. Consequently, they are among the front runners as SHS devices.

Any system has scope for improvement, and the CBR is no exception. It is basically a system with internal flows and there exist multiple, in both active and passive ways of heat transfer enhancement in internal flows. Liu and Sakr [9] reported a comprehensive review of heat transfer enhancement techniques in tubular heat exchangers. Introduction of a flow disturber is one of the methods, and this has been explored by Ramakrishnan et al. [8] for experimentally analyzing performance improvement of a single-hole CBR. The authors explored the effect of placing two flow disturbers at five different locations and arrived at best among them based on the mean temperature, exergy efficiencies, etc. On a natural curiosity, in this study, the number of flow disturbers has been increased to five to explore all possible combinations of flow disturbers placements. The novelty present work lies in to develop a wholistic parameter that would not only include heat transfer rates and pressure drop for comparison, but also facilitate in comparison of a given system against a reference. The forthcoming sections present on evolving this parameter and utilize to perceive a panoramic view of effect of location and number of flow disturbers.

# 2 Experimentation

## *2.1 Experimental Arrangement*

The experimental arrangement has been described in detail in [7]. Hence, the salient features alone are presented in this section. The schematic of the experimental arrangement is given in Fig. 1. A CBR made of aluminum forms the heart of the system.

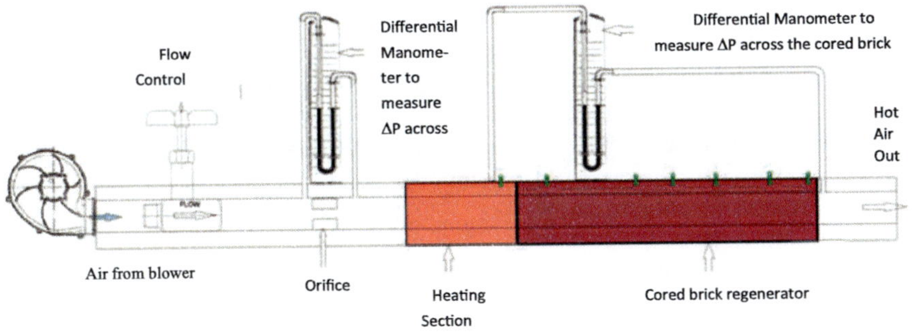


**Fig. 1** Experimental arrangement

**Table 1** Specifications of the CBR

| | |
|---|---|
| Material | Aluminum AISI 5051 |
| Length | 455 mm |
| ID/OD | 26/40 mm |

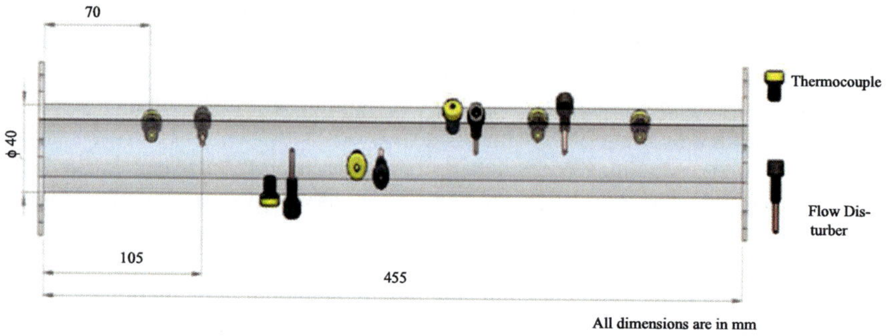


**Fig. 2** Thermocouples and flow disturbers in CBR

An electrical heater is used to heat air that is supplied to the CBR using blower with provision to control the flow rate. The material and dimensions of the CBR are given below in Table 1. The CBR has provisions for measurement of temperature at six locations using K type thermocouples and placement of five flow disturbers. The thermocouples and flow disturbers are placed as shown in Fig. 2.

In Fig. 2, the thermocouples are represented by a yellow color small lug nut and flow disturbers by lengthier nail. The diameter of the flow disturber is 1.5 mm and protrude 13 mm radically inward to the CBR. The CBR has five locations along its length, as indicated below in Table 2. Thermocouples in the CBR are placed 35 mm upstream of the flow disturber.

The experimental arrangement has flow control valves for varying the flows, and a couple of U-tube manometers, one each for measurement of flow and pressure drop across the CBR. A data acquisition system is used to record the temperature history of the CBR and that of the fluid at inlet and exit. The temperature along the cored brick was measured at six locations, the orientations of which are shown in Fig. 2. Two more thermocouples are used to measure the inlet and exit air temperature of cored brick. The temperatures were recorded at a regular interval of 30 s by a data acquisition system. The primary objective being to experimentally analyze the effect of position and number of flow disturbers on the heat transfer and pressure drop, a possibility matrix for the same was constructed. This is given below in Table 3.

**Table 2** Provision for location of the flow disturbers

| Position | 1 | 2 | 3 | 4 | 5 |
|---|---|---|---|---|---|
| Distance from inlet, mm | 105 | 170 | 235 | 300 | 365 |

**Table 3** Possibility matrix

| No. of disturbers | Various possible locations |
|---|---|
| One (5 cases) | 1,2,3,4,5 |
| Two (10 cases) | (1,2), (1,3), (1,4), (1,5), (2,3), (2,4), (2,5), (3,4), (3,5), (4,5) |
| Three (10 cases) | (1,2,3), (1,2,4), (1,2,5), (1,3,4), (1,3,5), (1,4,5), (2,3,4), (2,3,5), (2,4,5), (3,4,5) |
| Four (5 cases) | (1,2,3,4), (1,2,3,5), (1,3,4,5), (2,3,4,5), (2,4,5,1), |
| Five (1 case) | (1,2,3,4,5) |

Experiments were conducted for all the above possibilities and for the base case of without flow disturber.

## 2.2 *Experimentation*

The charging cycle—Air from the variable speed blower is heated using the nichrome heater and then allowed to pass through the aluminum CBR, thus heating the regenerator. Velocity is measured from by the head difference of the differential u-tube water manometer across an orifice. Pressure drop across the CBR is measured by means of a differential u-tube water manometer. Charging is performed typically for a time 15 min.

The discharge cycle - The stored heat from the CBR is retrieved by passing ambient air through it, but in the opposite direction to that done during the charging cycle. The same flow rates are used as that of the charging cycle. Discharging cycle is also carried out for 15 min.

During both the charging and the discharging processes, the temperatures of the CBR along its length, inlet, and exit temperatures of the working fluid and the pressure drop across the CBR were recorded. The experiments were carried out for a varied number and location of flow disturbers, and flow rate ranging from 96 cc/s to 214 cc/s.

## 2.3 *Analysis*

The parameters considered for performance assessment of the CBR are the mean temperature, instantaneous heat transfer rates, energy transferred to the CBR, exit temperature of the fluid during discharge and PEP. But for temperatures, all the others were derived quantities, and the following assumptions were made in their estimation:

(1) The material is homogenous in composition throughout the cored brick.
(2) There is no phase change occurring in the cored brick throughout the process.
(3) The outer wall of the cored brick is well insulated that the heat lost to the surroundings is limited to a minimum.

(4) Heat transfer by the radiation is negligible.

The pressure drop across the regenerator was estimated by the standard expression as

$$\Delta p = \rho g \Delta h \tag{1}$$

The instantaneous heat lost by the hot fluid during the charging process of the regenerator was estimated using

$$Q_f = mc_p(T_i - T_o) \tag{2}$$

In Eq. (2), the temperature varies on a temporal basis as well. Hence, the inlet and outlet temperatures were taken at times which were marginally separated depending upon the velocity of the fluid. The energy transferred during the charging cycle, over a period, say 15 min is given by

$$E_T = \int_{t=0}^{15\,\text{min}} Q_f \mathrm{d}t \tag{3}$$

The energy transferred over a period was evaluated numerically. Likewise, a pumping power and hence, a pumping energy is required to effectuate this energy transfer. This is given as energy required to transfer, given by

$$E_{RT} = \int_{t=0}^{15\,\text{min}} P_P \mathrm{d}t \tag{4}$$

It is expected that any change in the fluid flow pattern could result in a change in heat transfer rate and pressure drop, or in other terms, energy transferred to energy required for this transfer. Consequently, a ratio of these two parameters will serve as a good comparison between two or more types of flow alteration modes. Hence, a parameter "energy ratio" is introduced, which is defined as

$$E_R = \frac{E_T}{E_{RT}} \tag{5}$$

While it would suffice to use $E_R$ for comparison, PEP allows for easier visualization of the increase or decrease in performance. Further, it is worthwhile to compare all the cases with flow disturber(s) against the base case of without them, it is more useful to define a performance enhancement parameter (PEP) defined as below:

$$\text{PEP} = \frac{E_{Ri}}{E_{Ro}} \tag{6}$$

where the additional subscripts *i* and *o* represent the energy ratios for cases with and without flow disturber, respectively.

## 3 Results and Discussion

There are many ways in which the performance of a regenerator can be assessed, for a given flow rate and inlet temperature of the working fluid. In the present study, the mean temperature of the CBR after charging, instantaneous heat transfer rates, energy stored, and PEP. These effects of number and locations of flow disturbers on these performance parameters are discussed in the ensuing paragraphs.

As far the charging process is concerned, the temperature of the CBR at the end of charging is a vital parameter. The mean temperature of the CBR has been considered for this purpose. Figure 3 depicts variation of mean temperature with time for the charging cycle.

In this figure, only five different possibilities of disturber placements that yielded higher temperatures are given. The results indicate that increased number of disturbers need not necessarily result in higher mean temperatures of the CBR. For the flow rate considered in Fig. 3, a lone disturber at location number 3 resulted in the highest mean temperature. This could possibly be attributed to mixing of the fluid. The next parameter to be considered is the heat transfer rate. Figure 4 presents the variation of instantaneous heat transfer during charging. It is noticed that trends of variation of heat transfer are quite similar, the magnitudes are not. For the flow rate considered, placing disturbers at (2, 5) and (3, 4) help improving the instantaneous heat transfer compared to others.

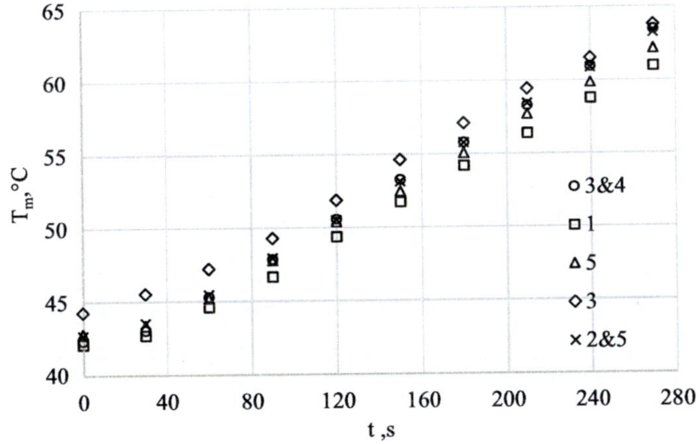


**Fig. 3** Mean temperature history

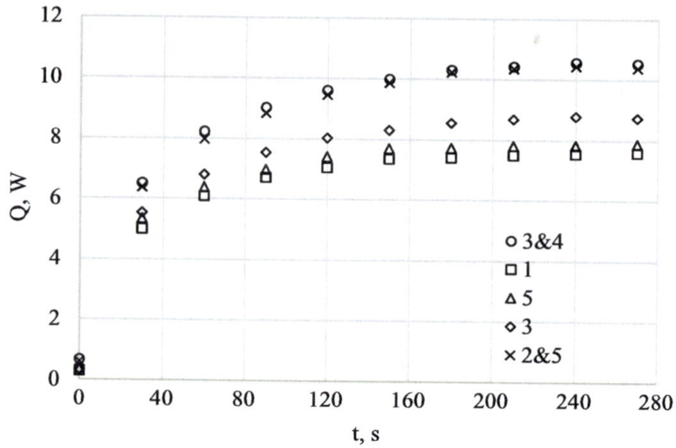


**Fig. 4** Variation of instantaneous heat transfer

A larger instantaneous heat transfer rate is expected to result in an increased amount of energy stored. The energy stored during the charging time for different flow rates for the top five cases is presented in Fig. 5.

It could be seen from the Fig. 5 that the positioning the disturbers at (3, 4) consistently leads to better performance of the CBR in most of the cases, followed by disturbers at (2, 5). The sole purpose of storing the heat is to retrieve at a later stage. In the present work, the discharging process was carried out soon after the charging.

The Fig. 6 indicates that the top five positions are shared again by the same group of disturber locations. It is also seen that the drop-in temperature over a period of 4.5 min is about 5 °C.

It is quite natural that introducing a disturbance in the flow causes the fluid to mix and hence results in an increased heat transfer. This could, however, result in

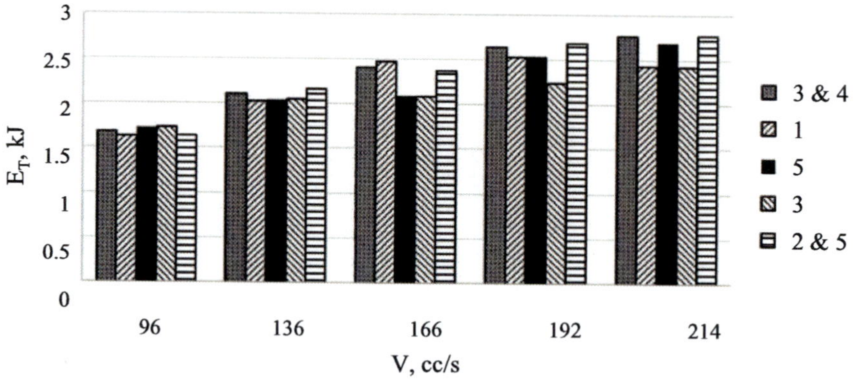


**Fig. 5** Effect of flow disturber position on energy stored

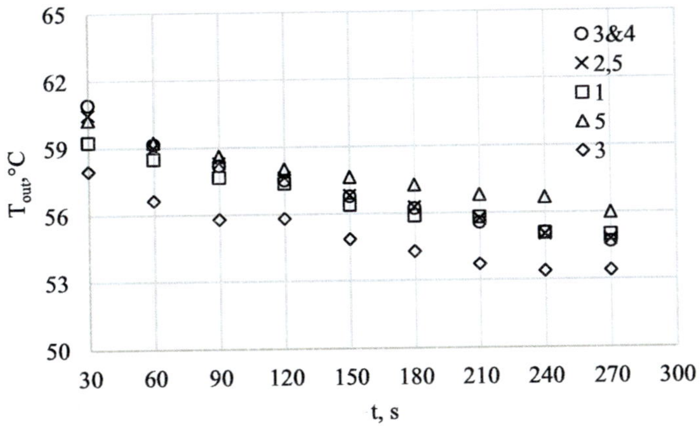


**Fig. 6** Variation of fluid temperature at exit

increased pressure drop and hence an increased pumping power too. So, it is essential to have a quantitative information on the increased pumping energy that needs to be spent for augmenting the energy transfer.

In the present work, PEP has been used as defined earlier for comparing the performance of the cases with flow disturber to the base case of without them. PEP was estimated for all the cases, and Fig. 6 represents the PEP obtained for different cases of number of flow disturbers and their locations for various flow rates considered in the present study. PEP of unity represents the base case of CBR without any flow disturber.

From Figs. 3, 4, 5, and 6, the performance of the CBR appears to increase with increase in the number of disturbers from 1 to 2. Figure 7, however, gives a holistic view of the results of all the cases analyzed. Tableau was used to analyze the data of about 150 experimental runs to holistically visualize various performance parameters. It could be seen that as the number of disturbers are increased beyond 2, the performance starts decreasing and in some cases are even less than the base case of CBR without any flow disturber. It is obvious that, irrespective of the performance parameter considered, whether it is temperature, heat transfer rate, pressure, or energy stored, it is significantly affected by presence of flow disturber—its quantum and the placement.

It is a well-known fact that boundary layer represents a resistance to heat transfer. The heat transfer coefficient decreases as the boundary layer develops. Further, it is within the boundary layer, where the temperature gradient and hence the heat transfer occurs predominantly. Hence, the heat transfer rate would increase, if it the fluid out of the boundary layer is also brought into participation. However, it must be taken care that the process of mixing is not obtained at an expensive increase in pressure drop. This is precisely carried out by the flow disturbers, that helps in mixing and hence an increased heat transfer, and slender enough not to result in a significant increase in pressure drop. The flow disturber could preferably be placed when the

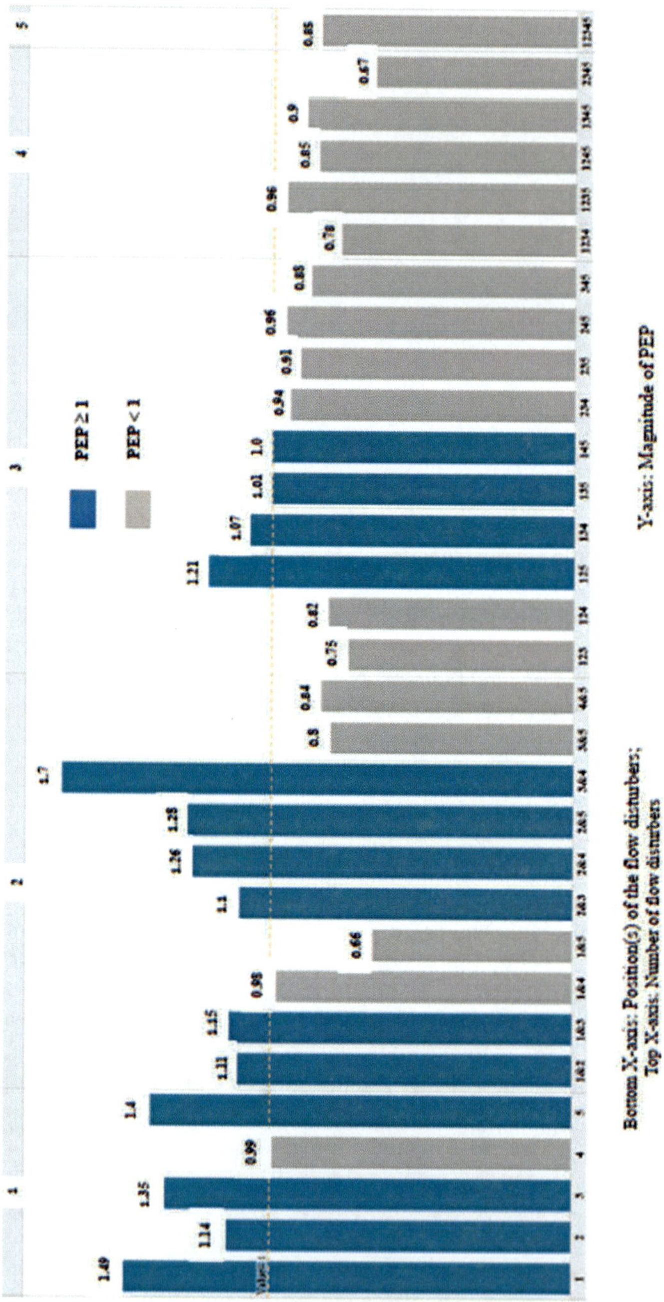


**Fig. 7** Performance enhancement parameter (PEP) for all cases in the present study

flow is just developed and hence facilitate a break in the boundary layer, resulting in a fresh developing flow. In congruence, the number of flow disturbers per unit length could play major role in inducing fresh mixing of the fluid. Hence, the observation that an increased number of disturbers does result in increased pressure drop, but not in a proportionate increase in heat transfer rate cannot be extrapolated to longer dimensions.

## 4 Conclusions

A study of effect of number and locations of flow disturbers on the performance of a single-hole cored brick regenerator was carried out. The following conclusions could be made from the study:

(1) The effect of flow disturbers and the location of the disturbers play a significant role in the heat transfer rate and pressure drop. The variations in heat transfer rate and the pressure drop were estimated to be 25% and 45%, respectively.
(2) An increased heat transfer to the CBR results in an increased temperature of CBR. Whether or not this requires a proportionate increase in pressure drop depends on the way of deployment of flow disturbers. Hence PEP, by the way, it has been defined could be used as a single performance assessment parameter.
(3) The introduction of flow disturbers can result in an increased energy transfer for a relatively less increase in the pumping energy requirement. This is evident from a maximum increase of up to 70% in PEP among the cases investigated.

While the above conclusions have been made from this research, the research itself is limited in terms of the following aspects.

(1) The flow rate considered falls within an order of magnitude. Increasing the range of flow may bring out interesting results.
(2) The study has been made only a single-hole cored brick regenerator. In general, cored brick heaters have multiple holes, and hence, alternate options of fixing the flow disturbers need to be explored.
(3) Sensible heat storage system has been used to study the effect of flow disturbers. Extending the same for latent heat storage system is expected to enhance the applicability of the present research.
(4) The study is limited to a parametric analysis, and hence, design and optimization are the among the next obvious scope for future work.

Considering the quantum of ongoing research in the field of thermal energy, along with healthy scope for its applications like building heating, steel industries, and high-enthalpy testing facilities, it is hoped that the above challenges are overcome in due course of time.

## References

1. Akinyele DO, Rayudu RK (2014) Review of energy storage technologies for sustainable power networks. Sustain Energy Technol Assessments 8:74–91
2. Guney MS, Tepe Y (2017) Classification and assessment of energy storage systems. Renew Sustain Energy Rev 75(C):1187–1197
3. Parra D, Swierczynski M, Stroe DI, Norman SA, Abdon A, Worlitschek J, O'Doherty T, Rodrigues L, Gillott M, Zhang X, Bauer C, Patel MK (2017) An interdisciplinary review of energy storage for communities: challenges and perspectives. Renew Sustain Energy Rev 79:730–749
4. Sarbu I, Sebarchievici C (2018) A comprehensive review of thermal energy storage. Sustain 10(1):191–222
5. Li G (2013) Review of thermal energy storage technologies and experimental investigation of adsorption thermal energy storage for residential application. Statew Agric L
6. Alva G, Lin Y, Fang G (2018) An overview of thermal energy storage systems. Energy 144(C):341–378
7. Burcu K, Ana IF, Halime P (2020) Review of sensible thermal energy storage for industrial solar applications and sustainability aspects. Solar Energy 209:135–169
8. Ramakrishnan P, Krishnan AS, Gowtham S (2018) Experimental studies on performance augmentation of single hole cored brick sensible heat storage system using turbulence inducers. J Therm Anal Calorim 136(1):345–354
9. Liu S, Sakr M (2013) A comprehensive review on passive heat transfer enhancements in pipe exchangers. Renew Sustain Energy Rev 19:64–81

# Pitch Controller for Isolated Wind-Diesel System with Super Conducting Magnetic Energy Storage Unit Based on Fractional-Order Fuzzy PID Controller

**K. Kalpana and M. Mohamed Thameem Ansari**

## 1 Introduction

Dominance by a single utility firm for electrical power transmission, distribution, and generation has an impact on the system's economic and dependable functioning in traditional power systems. The power division has undertaken various measures to address these issues since its founding, namely deregulation. Independent enterprises such as Transmission Companies (TRANSCOs), generation companies (GENCOs), Independent System Operators (ISOs), and Distribution Companies (DISCOs) are the mainstay of the deregulated power paradigm (DISCOs). In an open power market scenario, any DISCO control region has the ability to make a contract with GENCO in any control region at competitive costs for power transactions. ISO will operate as a supervisor and manage power transactions between various DISCOs and GENCOs to ensure that the power system runs smoothly. For deregulated power systems with reheated and non-reheated power generating units, as well as non-renewable energy resources, researchers have developed a number of optimal load frequency control (LFC) schemes under a range of contracts. Traditional energy resources are the primary source of electrical power generation; however, these resources are rapidly decreasing and are also sensitive to global climate change [1]. Electrical energy consumption is always increasing in the physical world due to population increase and industrial development. Currently, the greater percentage of electrical energy demand is fulfilled by fossil fuel generating units. Traditional power plants, however, are unable to meet load demand due to gradual depletion of fossil fuels and an increase in fuel costs. Renewable energy sources (RES) offer a potential future in this situation because they reduce reliance on fossil fuels and provide clean energy. Wind energy,

K. Kalpana (✉) · M. M. T. Ansari
Department of Electrical Engineering, Annamalai University, Annamalainagar, India
e-mail: cmksambathkalpana@gmail.com

M. M. T. Ansari
e-mail: ansariaueee@gmail.com



N. L. Narasimhan et al. (eds.), *Recent Advances in Energy Technologies*, Lecture Notes in Mechanical Engineering, https://doi.org/10.1007/978-981-19-3467-4_14

solar radiation, ocean energy, biomass, hydropower, geothermal resources, hydrogen, and biofuels are some of the clean RES that are used to generate heat and electricity. Wind energy and biomass units have gained popularity among these renewable energy sources due to their environmentally friendly characteristics and rapid technological development. It also has a positive impact on the country's financial and social development. When these renewable energy sources are combined with traditional sources, both the power quality and reliability of the produced electricity improve [2].

The development of small-scale hybrid generating methods for residential use has been aided by a growing awareness of the impact of traditional energy generation methods on sustainability, the advancement of increasingly affordable small-scale distributed generation technology, and then public policies that frequently incorporate renewable resources in the matrix of energy generation. Renewable energy will be used to minimize local demand on the public grid while remaining linked to the grid as a backup in the event that renewable energy is unavailable. Due to the improvement of regulatory schemes in the small consumer market that allow for hourly price separation, small residential customers now have the option of using a hybrid generating method, as well as the ability to export or import energy to/from the grid based on hourly prices and energy resource availability [3]. There will be numerous challenges in transitioning from the current fossil-based economy to a sustainable one. Many sustainable energy resources, such as wind and biomass, have irregular availability as well as seasonality, time, location, and day-to-day variability, necessitating the use of storage methods and a combination of resources [4]. For a limited time, the biomass will be stored. On a larger scale, the unit commitment issue will address the combination of various sources, thus organizing power production from various sources based on demand, at a local level [5]. As a result, the distributed generation concept (i.e., a small isolated power system with or without grid connection) is used to regulate and support system power in large process industries, commercial areas, and rural applications. The following is the format of the paper: The current work is reported in Sect. 2, and the diesel-wind system with super conducting magnetic energy storage unit is discussed in Sect. 3, FOPID's design is discussed in Sect. 4. Section 5 contains the results, while Sect. 6 has the conclusion.

In light of the aforementioned, the following is the work's objectives:

- Create and implement a new fractional-order PID controller for multi-source isolated power systems.
- To see how well the FOPID controller performs with a random load pattern and a greater step load perturbation. The purpose was to examine how a superconducting magnetic energy storage (SMES) unit might influence the same instantaneous performance.

•Use sensitivity analysis to assess the proposed controller's resiliency.

- To test the robustness of the fractional-order Fuzzy PID controller in a variety of power system operating scenarios.

## 2 Related Works

In Ireland, an established technique [6] performs research on the outcomes for stakeholders and the government involved in wind turbine and small-to-medium firm investments. The goal of a research method is to examine and collect quantitative numerical data from a range of sources, such as electrical energy and power meters, firm sustainability records, and historical electricity bills. This study benefits all stakeholders, including the federal government, which promotes wind energy as a key component of its energy policy because it cuts greenhouse gas emissions.

Grid-connected wind turbines and Doubly Fed Induction Generators (DFIG) have substantial challenges in terms of fault-ride-through (FRT) output power and performance fluctuation [7]. Another approach includes the output power and performance variance of fault-ride-through (FRT). To address these issues, setting the SMES and Fault-Current Limiter (FCL) optimal values with the conventional section of the superconducting coil (SC) is suggested. In addition to DFIG's output power, the SC will act as a short circuit current limiter to improve FRT's potential.

Storage methods are utilized to lessen the output power changes of wind farms since wind speed and energy are created stochastically. A variable wind turbine speed-driven Permanent Magnet Synchronous Generator's output power is smoothed down more by SMES unit [8]. The PMSG is connected to the grid via a back-to-back converter. The optimal proportional-integral controller parameters are determined using the gray wolf optimizer, which are then utilized to regulate the discharge and charge SMES unit operations. EMTDC/PSCAD software is used to simulate and modify PMSG's wind turbine process.

To solve a load frequency problem, another way is to integrate two area power systems, each with a wind and water power plant. HVDC and SMES lines have been added to enhance the system's performance [9]. The PID controller is designed in all circumstances, and the differential evolution (DE) approach is used to fine-tune the controller settings, HVDC, and SMES connection.

According to one existing strategy, the SMES method is utilized to regulate the fluctuation problem of an interconnected wind and solar farm, as well as the traditional power way of delivering system load. [10]. To reduce fossil fuel usage, rooftop solar systems, wind farms, and solar farms are utilized in conjunction with traditional power systems. When renewable plants produce more energy, the SMES unit stores it and releases it when renewable power is unavailable owing to atmospheric conditions or during high load times.

Another approach looks into the impact of the LVRT mechanism on the system's transient stability, and a SMES scheme for the interconnected wind farm power system is developed [11]. To achieve the LVRT, the SMES approach is employed; the rotor's crowbar is changed based on the law of influence on the equivalent system's mechanical power to create a superior performance of transient stability in the system.

The LVRT standards have been advocated in the literature in view of the expanding use of renewable energy in power networks. Every LVRT technique often utilized in the DFIG, on the other hand, has a habit of changing its running state, which has a secondary effect on the system's transient stability [12]. The authors describe a SMES technique that uses location data to improve grid-connected wind farm features while also improving system performance.

Salama et al. [13] offer a new technique for increasing the LVRT capacity of a 5 * 1.5 MW DFIG-based wind farm. To establish a parallel connection with the rotor side of each DFIG, a single SMES unit is being used. In addition, the SMES rotor side and energy storage side converters use two LVRT techniques for transient voltage and inductance mimicking control. Both asymmetrical and symmetrical grid fault examples are employed for performance evaluations and extended simulations.

The SMES impact for controlling the voltage of a system's electrical power is investigated, which is related to the variable power of a wind farm. The SMES system consists of a converter transformer, a two-quadrant DC–DC chopper, a power conditioning unit, a vacuum/cryostat vessel, and a large inductance coil to store the coil in a superconducting state. A Fuzzy logic controller (FLC) with two input variables, wind speed and SMES current variation, is used to improve control performance [14].

The effect of different wind gusts on the operation of a wind energy conversion system and the use of a modern controller for the SMES unit decrease the use of DFIG-based grid-connected technologies. A DC-DC chopper controls the voltage converter, which includes the SMES coil and the investigated approach, as well as fuzzy logic controllers and hysteresis current. The findings show that the suggested SMES controller is effective and simple to use in both new and existing wind energy systems [15].

Another solution employs photovoltaic electricity to inject energy into a different unbalanced three-phase radial distribution approach that is fed from SCIG's WECS [16]. The SMES unit is a type of energy storage that is used to compensate for the intermittent power provided by renewable energy sources. In this scenario, FLC is employed to handle the discharging and charging of SMES.

It is proposed that a SMES unit be utilized to level the DFIG-based power variation of a wind power producing system. The DC-chopper is controlled by an FLC. The superconducting magnets, rotor side converter (RSC), and grid side converter (GSC) will handle the rotor's power flow and the DFIG's power grid separately [17]. Based on the system's needs, the SMES unit will provide DFIG output power stabilization.

Many research papers have discussed the use of SMES units in power systems [18]. Many researchers are drawn to superconducting magnetic technology to conduct research on its applications in a variety of fields. The early study of application in power systems in transmission line stabilization is reported in 1987. In addition, SMES researchers have been tempted to the smart grid, for example, to improve the system's reliability.

Another way was to manage the voltage and frequency of an isolated hybrid wind-diesel power system at the same time using a static VAR compensator (SVC) and a SMES system [19]. The system frequency is evaluated using the center of inertia (COI) technique, which is largely used to regulate the SMES unit.

The evaluation and design of a GdBCO magnet for hybrid energy storage (HES) in a kW-class dynamic voltage restorer (DVR) were described to use an existing technique [20–25]. A single battery energy storage (BES) unit and a single SMES unit will be integrated with the HES-based DVR concept. [25–31] A new continuous disk winding (CDW) technique is offered for enhancing the magnet's crucial current while ignoring the interior soldering joints.

## 3 Isolated Hybrid Power System with SMES

It is critical to consider both the environment and the generation of electrical energy problems concurrently, so that the increasing demand for industrialization for electricity is met with minimal environmental degradation. In the proposed method, biomass-based diesel is used in a hybrid system as an alternative source of diesel. It is made from animal waste, such as poultry waste and poultry waste, as well as other waste products, such as sludge, municipal solid waste, wastewater treatment, and water-related oils.

It is also a renewable fuel derived from renewable resources. It has gained popularity due to its sustainability, lower contribution to the carbon cycle, and, in some cases, low amount of greenhouse gases.

The generation of wind power is increasing all over the world. The wind does not blow 24 hours a day. As a result, those plants are combined with reliable power resources, such as a set of diesel engines together to operate.

Diesel hybrid systems can be classified based on their wind penetration levels. When wind penetration is minimal, the diesel generator continues to run indefinitely while wind power sustains the diesel generator output.

Because low wind penetration levels do not necessitate sophisticated control systems, the hybrid arrangement is simple, resulting in up to a 20% reduction in fuel consumption.

In medium wind penetration, the diesel generator must also run continuously. When the available wind power is high, the secondary loads are also powered to ensure that the diesel generator is properly loaded; however, when the wind power is high but the load is low, the secondary loads are not powered. Figure 1 shows a wind-diesel system in isolation.

The turbines are not performing at maximum capacity whenever the wind speed is too high as well as the system load is low. All that is required to manage this type of load dispatching is a basic control system. In the event of high wind penetration, the diesel generator can be turned off, and auxiliary components are required to maintain voltage and frequency. This function is handled by the pitch angle controller and super conducting magnetic energy storage in this project. It is impractical to keep the diesel generator running when wind power generation alone can meet the load requirement. The SMES system is utilized as a storage device. Its primary function is to store electrical energy in a magnetic field created by direct current flowing through it.

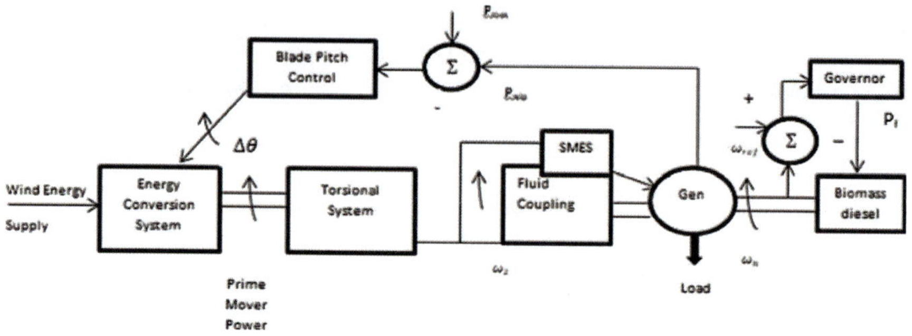


**Fig. 1** Isolated wind-diesel system is depicted as a conceptual model

## 3.1 Modeling of Wind Speed

The base wind, ramp, gust, and random noise are all calculated in wind speed modeling. A wind turbine generator's power output is proportional to the wind speed. ($V_W$).

(1) expresses the wind model as follows:

$$V_W = V_{WB} + V_{WG} + V_{WR} + V_{WN} \tag{1}$$

The model of base wind (2) is

$$V_{WB} = K_B \tag{2}$$

where $K_B$ indicates a wind speed model constant component. $V_{WB}$ is a constant wind component and $V_{WG}$, $V_{WR}$, and $V_{WN}$ are wind components of gust disturbance, ramp disturbance, and random noise.

The model of the gust wind is given in (3)

$$V_{WG} = \left\{ \begin{array}{ccc} 0 & \text{for} & t < T_{\text{gust}S} \\ V_{\cos} & \text{for} & T_{\text{gust}S} < t < T_{\text{gust}S} + T_{\text{gust}} \\ 0 & \text{for} & t < T_{\text{gust}S} + T_{\text{gust}} \end{array} \right\} \tag{3}$$

where $T_{\text{gust}S}$, $T_{\text{gust}E}$, and $T_{\text{gust}}$ are a gust start time, end time, and gust period.

The ramp wind model is given in (4)

$$v_{WR} = \left\{ \begin{array}{ccc} 0 & \text{for} & t < T_{\text{ramp}S} \\ v_{\text{ramp}} & \text{for} & T_{\text{ramp}S} < t < T_{\text{ramp}E} \\ 0 & \text{for} & t > T_{\text{ramp}E} \end{array} \right\} \tag{4}$$

where $T_{\mathrm{ramp}S}$ and $T_{\mathrm{ramp}E}$ is a ramp start and end time.

Thus, the noise wind model is given in (5)

$$V_{WN} = 2\sum_{i=1}^{N}[S_V(\Omega_i)\Delta\Omega]^{1/2}\cos(\Omega_i t + \emptyset_i) \tag{5}$$

The dynamics of the wind power model are analyzed by considering these four components.

## 3.2 System Modeling

Pitch angle control is a technique used to maximize turbine output when there are strong winds. The blades are positioned at an appropriate angle for performance using pitch servo motors, which might be hydraulic or electrical. Blade pitch angle adjustments in the range of 0–45 degrees are normally conducted with rotational speeds of around 5–10% under normal working conditions.

In the isolated hybrid power system, a pitch-controlled wind turbine generator is coupled in parallel with a biogas-powered diesel engine generator through an alternating current bus bar. An isolated biomass diesel-wind hybrid power system transfer function model is shown in Fig. 2. It is made of controllers that control the pitch of the blades.

The equation for power flow is given as (6):

$$\Delta P_{\mathrm{TOT}} = \Delta P_{\mathrm{DG}} + \Delta P_{\mathrm{WTG}} - \Delta P_{\mathrm{Es}} - \Delta P_L \tag{6}$$

where $\Delta P_{\mathrm{DG}}$ the output power deviation in diesel generator is $\Delta P_{\mathrm{WTG}}$ which is the output power variation in wind turbine, $\Delta P_{\mathrm{Es}}$ indicates deviated power of the energy storage unit (SMES), and $\Delta P_{\mathrm{Load}}$ represents the power deviation in load.

## 3.3 Modeling of SMES Unit

The SMES unit is faster acting and has nearly no self-discharge when compared to the CES system. Depending on the needs of the power system, energy is discharged or absorbed from the SMES coil. Because the coil is held at a low temperature when cryogenic coolers are utilized, the ohmic loss is nearly zero. The transfer function design of the SMES unit is shown in Fig. 3.

The SMES system can deliver or absorb massive amounts of electricity and discharge and charge very quickly. Also, the important feature of SMES consists of longer life that has millions of discharge and charge cycles.

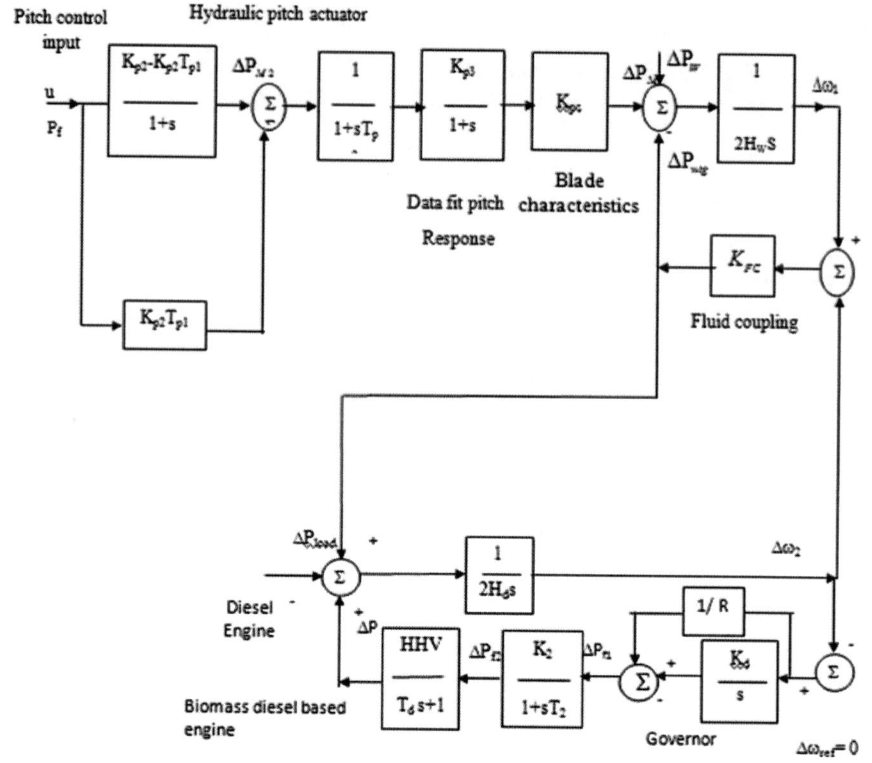


**Fig. 2** Isolated biomass diesel—wind hybrid power system transfer function model

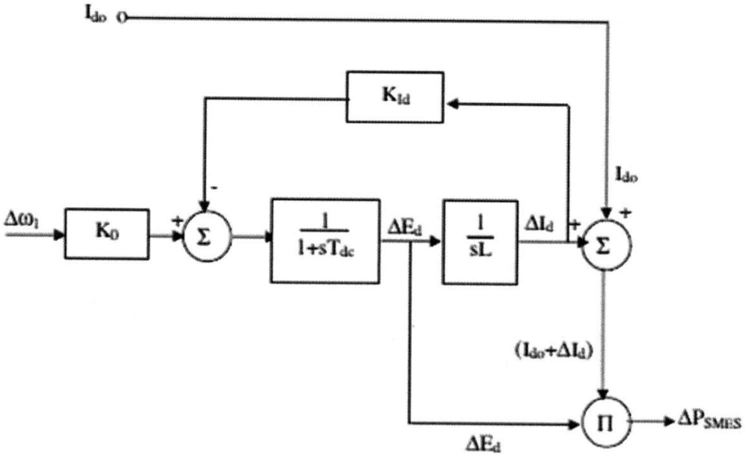


**Fig. 3** Transfer function design of SMES unit [32]

# 4 Fractional-Order Fuzzy PID Controller

Fuzzy logic controllers are linguistic systems based on the knowledge that is capable to manage uncertain nonlinear methods. However, the controller of PID is an effective and simple linear controller, which is suitable to manage uncertain linear methods. But, tuning the PID parameters is a difficult operation. Currently, great attention is received for fractional-order PID (FOPID) controllers. The FOPID controller has more level of freedom than the simple PID controllers. Thus, it will reach good results. Hence, the combination of fractional PID controllers and fuzzy logic will suitable to design an effective controller for a system of nonlinear with a significant number of uncertainties. The FOFPID structure is depicted in Fig. 4.

The controller's structure of FOFPID consists of six adjustable parameters, i.e., $K_d$, $K_e$, $K_i$, $K_p$, $\lambda$, and $\mu$. Where $K_d$ and $K_e$ are known as factors of input scaling, while $K_i$ and $K_p$ are known as factors of output scaling. The controller is FOPID if $\lambda$ and $\mu$ are set to 1. Then, the structure uses a Fuzzy logic interface which will decide depending on two inputs: error rate (fractional rate) and error.

## 4.1 *Fractional Controller*

To decode fractional-order control systems, fractional-order differential equations are used. In fractional calculus, the derivatives and integrals can be any real number. Fractional-order controllers are supported by the concept of fractional calculus. The fractional calculus-based FOPID controller is an enhancement to the classic PID controller. In general, approximating fractional systems with integer-order approximation can lead to significant disparities between the mathematical model and the real system. The fundamental rationale for utilizing integer-order models was the unavailability of solution methods for fractional-order differential equations.

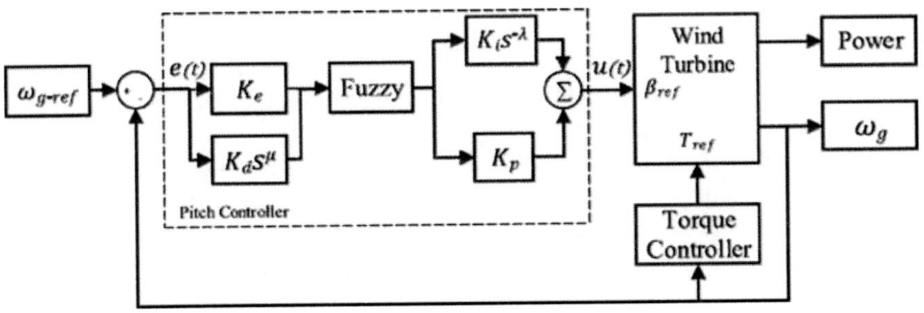


**Fig. 4** FOFPID structure

PID controllers are among the most commonly used industrial controllers, and efforts to improve their quality and reliability are ongoing as a result. Using fractional-order controllers with non-integer derivation components is one way to improve PID controllers.

The integrals and derivatives in fractional calculus will be real numbers. To improve PID controllers, fractional-order controllers with a non-integer derivation portion are used. The transfer function is expressed as

$$C(s) = K_p + \frac{K_I}{S^\lambda} + K_d\ S^\mu \tag{7}$$

The control signal or output response provided by a controller of PID in the time domain is given as

$$u(t) = K_p e(t) + \int_0^t \frac{K_I}{S^\lambda} e(t) + K_d \frac{d}{dt} \mathrm{e(t)} \tag{8}$$

where $u(t)$ represents the control signal and $e(t)$ indicates the generated error signal. A conventional PID is formed when $\lambda = 1$ and $\mu = 1$. The proposed controller is setup as a fractional-order PID controller if $\lambda$ and $\mu$ are set as an arbitrary number between 0 and 1.

Changes in process and controller settings are also less noticeable with FOPID. The parameters of FOPID controllers can be modified instead of the traditional two, enabling for more design criteria to be accomplished through the $\lambda$ and $\mu$ adjustment to improve design flexibility.

## 5 Results and Discussion

The fractional Fuzzy logic controller, together with a SMES unit is designed for an isolated biomass-based diesel-wind hybrid power system. As a comparison to the fractional-order Fuzzy logic controller, the conventional PI and then fuzzy logic controller are used as benchmarks.

The FOPID controller's control strategies are implemented using Simulink/MATLAB. As a result, the suggested FOFPID method is compared to Fuzzy logic and PI controllers in order to show the model's capabilities and efficacy. A time-domain analysis is done on the test system to demonstrate the effectiveness of the load frequency controller design. The system is subjected to a sudden shift in step load to compute the efficiency of the recommended technique, which would interrupt the observed power system's usual performance. The suggested hybrid power system's SMES influence on the environment is valued.

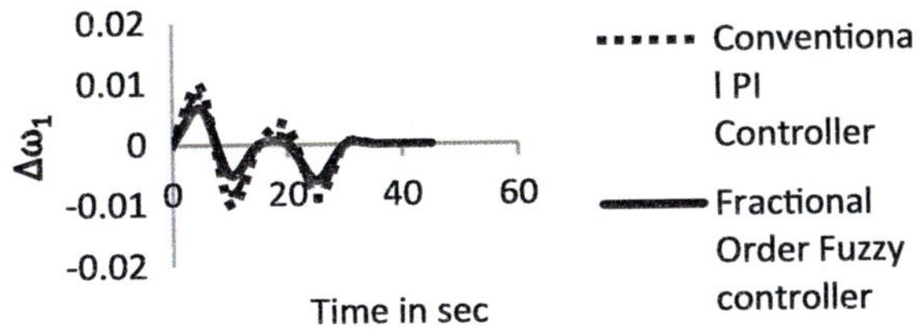


**Fig. 5** Biomass diesel frequency deviation

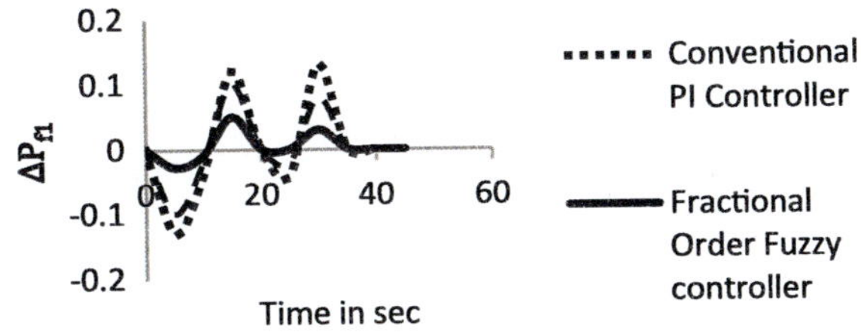


**Fig. 6** Biomass diesel power deviation

Figure 5 depicts a change in frequency deviation using the proposed FOFPID controller, as well as the traditional PI controller and the Fuzzy logic controller. It is clear from these contrasting curves that the frequency deviation for load changes is rapidly driven back to zero.

Figure 6 depicts the power deviation caused by the proposed FOFPID controller, as well as the conventional PI controller and Fuzzy logic controller. When compared to other techniques, this figure yielded a fast settling period and less overshoot.

A comparison of frequency and power variations using the proposed FOFPID controller, a normal PI controller, and a Fuzzy logic controller is shown in Fig. 7.

The simulation results show that the suggested controller delivers a significant performance improvement well over Fuzzy logic controller and the digital PI controller.

## 6 Conclusions

For wind and biomass gas-fueled diesel power generators that will operate on an independent electric power system, a thorough numerical technique is created. For the biomass-based wind-diesel isolated power system, changes in frequency deviations are achieved by employing a fractional Fuzzy controller that monitors the turbine and then varies the pitch point of the cutting edges in a similar manner. Incorporating SMES devices improves the power quality of the hybrid technique. As a result, while regulating the restrictions, the suggested method successfully decreases the frequency and power deviation of a hybrid system.

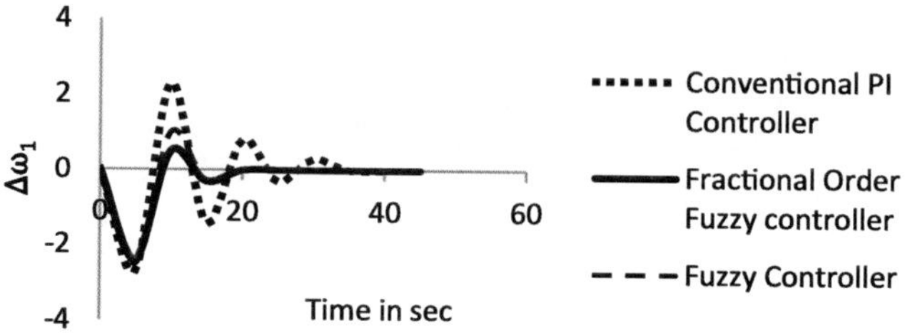


**(a) Wind generator frequency deviation**

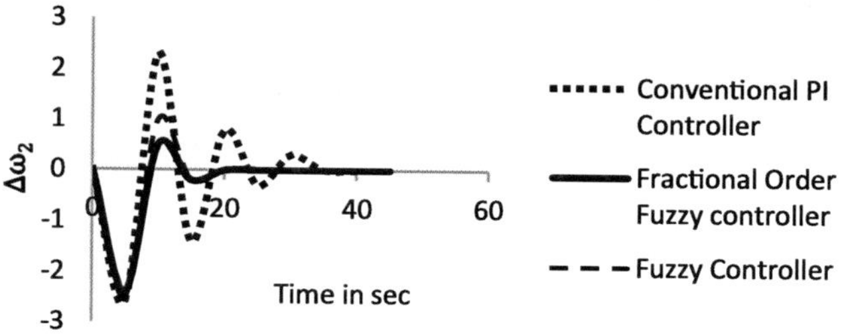


**(b) Diesel generator frequency deviation**

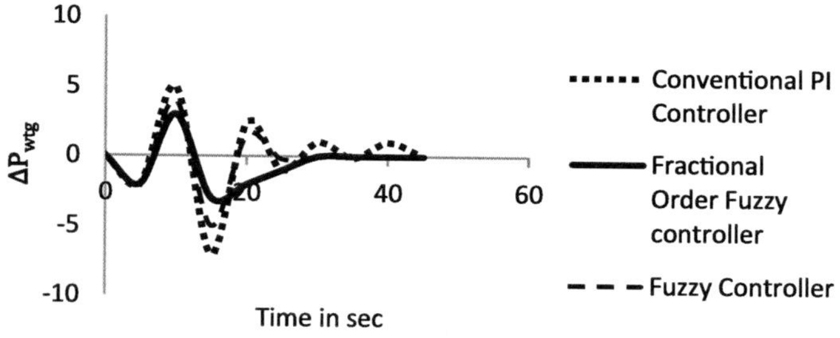


**(c) Wind generator power deviation**

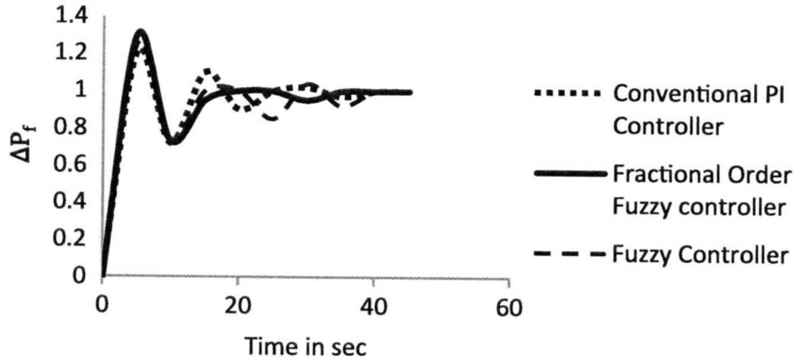


**(d) Diesel generator power deviation**

**Fig. 7** **a** Wind generator frequency deviation, **b** diesel generator frequency deviation, **c** wind generator power deviation, and **d** diesel generator power deviation

## Appendix

Inertia constant of wind system ($H_w$) = 3.52 s, Inertia constant of Diesel system ($H_d$) = 8.7 s, MGWS = 12 m/s, MRWS = 10 m/s, $V_{WB}$ = 7 m/s, $K_{fc}$ = 16:2 pu kW/Hz, $K_{hp2}$ = 1.25, $K_d$ = 16.5 pu Kw/Hz, $K_{hp3}$ = 1.40, $T_{hp1}$ = 0.60 s, $T_{hp2}$ = 0.041 s, $P_{max}$ = 0.6, $P_{load}$ = 1.0 pu, $K_{pc}$ = 1.080, $T_{gust}$ $S$ −5 s, $T_{gust}$ $E$ −15 s, $T_{ramp}$ $S$ −30 s, $T_{ramp}$ $E$ −40 s, $T_{noise}$ $S$ −0 s, $T_{noise}$ $E$ −80 s.

## References

1. Kealy T (2017) Stakeholder outcomes in a wind turbine investment; is the Irish energy policy effective in reducing GHG emissions by promoting small-scale embedded turbines in SME's. Renewable Energy 101:1157–1168
2. Said SM, Aly MM, Hartmann B (2018) Application of SMES for voltage control of power systems with high wind power penetration. In: 2018 International conference on innovative trends in computer engineering (ITCE), pp 461–466. IEEE
3. Choudhary R, Dahiva AK (2018) Stabilization of synchronous generator based PV and wind farm system using SMES. In: 2018 Second international conference on electronics, communication and aerospace technology (ICECA), pp 1031–1035. IEEE
4. Ngamroo I (2016) Optimization of SMES-FCL for augmenting FRT performance and smoothing output power of grid-connected DFIG wind turbine. IEEE Trans Appl Supercond 26(7):1–5
5. Yunus AS, Abu-Siada A, Masoum MA, El-Naggar MF, Jin JX (2020) Enhancement of DFIG LVRT capability during extreme short-wind gust events using SMES technology. IEEE Access 8:47264–47271
6. Qais MH, Hasanien HM, Alghuwainem S, Elgendy MA (2019) Output power smoothing of grid-tied PMSG-based variable speed wind turbine using optimal controlled SMES. In: 2019 54th international universities power engineering conference (UPEC), pp 1–6. IEEE
7. Behera SP, Biswal A (2019) PID controller in two area multi unit power systems with SMES and HVDC link. In: 2019 Global conference for advancement in technology (GCAT), pp 1–5. IEEE
8. Jiang H, Zhang C (2018) Coordinated control scheme for superconducting magnetic energy storage unit to boost transient stability of wind farm connected power system. In: 2018 IEEE international conference on applied superconductivity and electromagnetic devices (ASEMD), pp 1–2. IEEE
9. Yunus AS, Saini M (2016) Overview of SMES units application on smart grid systems. In: 2016 International seminar on intelligent technology and its applications (ISITIA), pp 465–470. IEEE
10. Xiao XY, Yang RH, Zheng ZX, Wang Y (2018) Cooperative rotor-side SMES and transient control for improving the LVRT capability of grid-connected DFIG-based wind farm. IEEE Trans Appl Supercond 29(2):1–5
11. Zargar MY, Lone SA (2017) Voltage and frequency control of a hybrid wind-diesel system using SVC and predictively controlled SMES. In: 2017 6th international conference on computer applications in electrical engineering-recent advances (CERA), pp 25–30. IEEE
12. Mukherjee P, Rao VV (2018) Fuzzy logic controlled superconducting magnetic energy storage for leveling power fluctuation of grid connected wind generator. In: 2018 International conference on power energy, environment and intelligent control (PEEIC), pp 665–669. IEEE
13. Salama HS, Aly MM, Vokony I (2019) Voltage/Frequency control of isolated unbalanced radial distribution system fed from intermittent wind/PV power using fuzzy logic controlled-SMES.

In: 2019 International conference on innovative trends in computer engineering (ITCE), pp 414–419. IEEE
14. Zheng ZX, Chen XY, Xiao XY, Huang CJ (2017) Design and evaluation of a mini-size SMES magnet for hybrid energy storage application in a kW-class dynamic voltage restorer. IEEE Trans Appl Supercond 27(7):1–11
15. Jiang H, Zhang C (2019) A method of boosting transient stability of wind farm connected power system using s magnetic energy storage unit. IEEE Trans Appl Supercond 29(2):1–5
16. Salama HS, Abdel-Akher M, Aly MM (2016) Development energy management strategy of SMES-based Microgrid for stable islanding transition. In: 2016 Eighteenth international middle east power systems conference (MEPCON), pp 413–418. IEEE
17. Aly MM, Salama HS, Abdel-Akher M (2016) Power control of fluctuating wind/PV generations in an isolated Microgrid based on superconducting magnetic energy storage. In: 2016 Eighteenth international middle east power systems conference (MEPCON), pp 419–424. IEEE
18. Xie Q, Zheng Z, Xiao X, Huang C, Zheng J, Ren J (2020) Enhancing the HVRT capability of DFIG-based wind farms using cooperative rotor-side SMES considering blocking fault of LCC-HVDC system. CSEE J Power Energy Syst
19. Yunus AS, Abu-Siada A, Djalal MR, Jin JX (2020) Optimal design of SMES and PSS for power system stability based on ant colony optimization. In: 2020 IEEE international conference on applied superconductivity and electromagnetic devices (ASEMD), pp 1–2. IEEE
20. Kondratowicz-Kucewicz B (2017) The energy and the magnetic field in HTS superconducting magnetic energy storage model. In: 2017 International conference on electromagnetic devices and processes in environment protection with seminar applications of superconductors (ELMECO & AoS), pp 1–4. IEEE
21. Elshiekh ME, Mansour DEA, Zhang M, Yuan W, Wang H, Xie M (2018) New technique for using SMES to limit fault currents in wind farm power systems. IEEE Trans Appl Supercond 28(4):1–5
22. Jin JX, Chen XY, Liu XD, Chen ZH, Peng JG (2018) Influence of flux diverter on energy storage property of small SMES magnet wound by 100-m-class GdBCO tape. IEEE Trans Appl Supercond 28(4):1–5
23. Ganesan S, Subramanian S, Chidambaram IA, Samuel Manoharan J (2017) Flower pollination algorithm based decentralized load-frequency controller for a two-area interconnected restructured power system considering gas turbine unit. Middle East J Sci Res 25(6):1308–1314
24. Musarrat MN, Islam MR, Muttaqi KM, Sutanto D (2018) Enhanced frequency support from a PMSG-based wind energy conversion system integrated with a high temperature SMES in standalone power supply systems. IEEE Trans Appl Supercond 29(2):1–6
25. Chen L, Li G, Chen H, Tao Y, Tian X, Liu X, ... Tang Y (2018) Combined use of a resistive SFCL and dc-link regulation of a SMES for FRT enhancement of a DFIG wind turbine under different faults. IEEE Trans Appl Supercond 29(2):1–8
26. Pahasa J, Ngamroo I (2021) Two-Stage optimization based on SOC control of SMES installed in hybrid wind/PV system for stabilizing voltage and power fluctuations. IEEE Trans Appl Superconductivity
27. Lin X, Lei Y (2017) Coordinated control strategies for SMES-battery hybrid energy storage systems. IEEE Access 5:23452–23465
28. Huang C, Xiao XY, Zheng Z, Wang Y (2019) Cooperative control of SFCL and SMES for protecting PMSG-based WTG s under grid faults. IEEE Trans Appl Supercond 29(2):1–6
29. Xiao XY, Yang RH, Chen XY, Zheng ZX (2018) Integrated DFIG protection with a modified SMES-FCL under symmetrical and asymmetrical faults. IEEE Trans Appl Supercond 28(4):1–6
30. Molla EM, Kuo CC (2020) Voltage sag enhancement of grid connected hybrid PV-wind power system using battery and SMES based dynamic voltage restorer. IEEE Access 8:130003–130013
31. Zhang Z, Miyajima R, Sato Y, Miyagi D, Tsuda M, Makida Y, ... Hirano N (2016) Characteristics of compensation for fluctuating output power of a solar power generator in a hybrid energy storage system using a Bi2223 SMES coil cooled by thermosiphon with liquid hydrogen. IEEE Trans Appl Supercondu 26(4):1–5

32. Ansari MMT, Velusami S (2010) Dual mode linguistic hedge fuzzy logic controller for an isolated wind–diesel hybrid power system with superconducting magnetic energy storage unit. Energy Convers Manage 51(1):169–181

# Fluid Flow and Heat Exchange

# Influence of Particle Size on Turbulent Flow Using Mono and Hybrid Nanofluids in a Heat Exchanger—An Experimental Investigation

**Ratchagaraja Dhairiyasamy and Mohan Govindasamy**

## *Nomenclature*

| | |
|---|---|
| $\mu_{bf}$ | Dynamic viscosity of the base fluid |
| $\mu_{nf}$ | Dynamic viscosity of the nanofluids |
| $a$ | Radius of the spherical particle |
| $A$ | Tube surface area |
| $C_p$ | Specific heat |
| $D$ | Density of the particle |
| $d$ | Density of the fluid |
| $d$ | Hydraulic diameter of the tubes |
| $g$ | Acceleration due to gravity |
| HTC | Heat transfer coefficient |
| K | Constant |
| $k$ | Thermal conductivity of the fluid |
| $k_f$ | Thermal conductivity of the base fluid (W/mK) |
| $k_{nf}$ | Thermal conductivity of nanofluids (W/mK) |
| $l$ | Length of the flattened tube of a car radiator |
| Nu | Nusselt number |
| P | Perimeter of the tubes |
| Re | Reynolds number |
| $T$ | Average temperature |
| $T_w$ | Wall temperature |
| $w$ | Width of the flattened tube of a car radiator |
| $\Delta P$ | Pressure drop |
| $\mu c$ | Viscosity of the suspension |

R. Dhairiyasamy (✉) · M. Govindasamy
Department of Mechanical Engineering, University College of Engineering, Tamil Nadu, Villupuram 605103, India
e-mail: ratchagaraja@gmail.com



N. L. Narasimhan et al. (eds.), *Recent Advances in Energy Technologies*, Lecture Notes in Mechanical Engineering, https://doi.org/10.1007/978-981-19-3467-4_15

$\varphi$ Concentration in %

## Subscripts

| | |
|---|---|
| bf | Base fluid |
| $\varphi$ | Particle size |
| nf | Nanofluid |

## 1 Introduction

Increased thermal loads as a method of growing vehicle technology need increased cooling levels. Due to the limited flow area inside the radiator tubes, the enhancement of heat transfer is important which is based on thermal properties of flowing fluid. Using nanofluids to increase heat transfer in cooling systems is the recent development that can help reduce power consumption while enhancing vehicle performance. However, the industry's interest in this heat transfer technology has waned due to higher pressure drops, friction factor, and sedimentation. In modern years, developments in nanomaterial technology have supported the fabrication of appropriate particles with a range of nanometre diameters to solve these difficulties. Various nanofluids were utilized as coolants in the car radiator, with an emphasis on their heat transfer properties. Once nanofluids are being used in car radiators, several thermophysical properties must be assessed to determine their overall cooling effectiveness. It should be examined how particle size affects heat transmission in laminar, transitional, and turbulent flows. Before employing nanofluids in an automobile radiator, all parameters such as nanoparticle properties, nanofluid production techniques, stability studies, thermophysical properties, Nusselt number (Nu), and heat transfer characteristics, pressure drop and friction factor should be researched. These characteristics must result in a more usable output than conventional coolants used in car radiators, such as DI water and ethylene glycol, while also being less expensive.

Li et al. [1] added 50 nm Cu particles to EG at concentrations of 1.0, 2.0, and 3.8 wt%. They discovered that increase in ultra sonification time increases the viscosity. According to Mahbubul et al. [2], longer ultrasonication minimizes nanoparticle sedimentation and so enhances nanofluid stability. They generated alumina (DI water/$Al_2O_3$) nanofluids at a concentration of 0.5 vol% for five hours of sonication and processed them for 30 days in a stationary condition. Akhgar et al. [3] studied the thermal conductivity of DI water/ethylene glycol/$TiO_2$-MWCNT hybrid nanofluid with a 50:50 volume fraction. They discovered that altering the volume fraction and temperature enhances the nanofluid thermal conductivity. [4–6] summarize the few published efforts on nanofluid production and thermophysical property augmentation, suggesting that nanofluids can be employed as a possible base fluid substitute.

Goudarzi et al. [7] showed that employing (EG/$Al_2O_3$) as nanofluid along with inserts of wire coil increases heat transfer by 9% in automobile radiators. Hatami et al. [8] computationally modelled a flat tube to optimize the cooling using nanofluid in a radiator and discovered that EG-$TiO_2$ has a structure of platelet and a higher volume of nanoparticles provides significant cooling efficiency for the engine. Selvam et al. [9] discovered that employing nanofluid-based graphene nanoplatelets as a coolant enhanced the overall heat transfer coefficient (HTC). At 35 °C, the average increase in OHTC concentration is 0.114%, whereas, at 45 °C, the increase in OHTC concentration is 0.81% at 0.5 vol%, flow rate of 62.5 g/s, and air speed of 5 m/s. References [10, 11] summarize a few pieces of literature on the usage of nanofluids in car radiators, demonstrating that nanofluids may be utilized as radiator coolants.

Akilu et al. [12] examined thermal properties of SiC nanofluids based on EG and PG and discovered that increasing density enhanced electrical and thermal conductivity. Huminic and colleagues [13] examined the thermal-physical characteristics of SiC nanofluids based on water, indicating that increasing nanoparticle weight and temperature thermal conductivity improves with. The concentration of nanoparticles in SiC/water nanofluids increases according to their size. References [14–17] summarize numerous articles on the use of hybrid nanofluids in heat exchangers, demonstrating that hybrid nanofluids can replace mono nanofluids. Asadi et al. [18] evaluated the heat transmission capability of an $Al_2O_3$-MWCNT hybrid nanofluid observing a slight change in dynamic viscosity at all tested stable concentrations between 0.5 and 1% at 50 °C. Sundar et al. [19] studied the turbulent heat transfer of nanodiamond-nickel hybrid nanofluids flowing in a pipeline (ND-Ni) and friction factor. At 0.3% volume concentration results in 35.43% increase Nu, with a friction factor loss of 1.12 times that of pure water at Reynolds number (Re) of 22,000. [20–29] summarize various types of literature on the influence of particle size on heat transmission in heat exchangers, indicating that nanoparticle size is critical for improving heat transfer comparing mono and hybrid nanofluids based on turbulent flow. Consequently, the increase in heat transfer in an automobile radiator using mono and hybrid nanofluids was explored, and friction factor and Nu correlations were generated.

## 2 Preparation and Characterization of Nanofluids

### *2.1 Preparation of Nanoparticles*

Silicon carbide (SiC), aluminium oxide ($Al_2O_3$) nanoparticles, and ethylene glycol (EG) were acquired in India from Sigma Aldrich. Deionized water (DI water) was given for this experiment by Astrra Chemicals, Chennai. The SiC with particle size of 110 nm is further reduced by milling it in a ball mill. A ball mill (Retsch-PM400) is employed to reduce the size of SiC. Milling decreased the crystallinity of SiC-L by 25%, making it more amorphous. The sun wheel, which is placed at the bottom

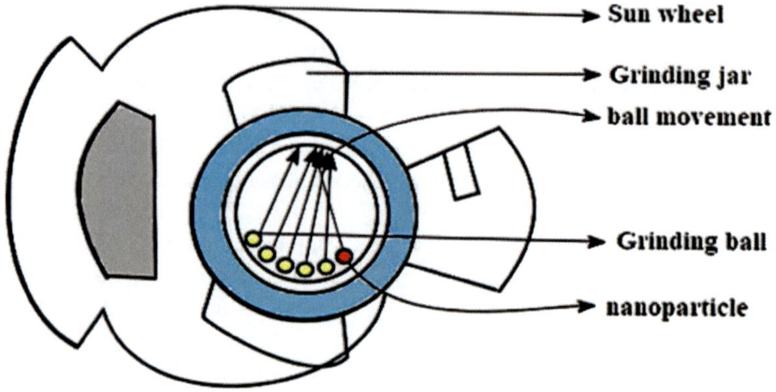


**Fig. 1** Planetary ball mill

of the planetary ball mill, revolves at a 1:−2 ratio concerning the grinding jars, as seen in Fig. 1. The planetary carrier rotates at 200 revolutions per minute, while the sun wheel rotates at 300 revolutions per minute and grinds in 52:59:59 s. Due to the rotation of grinding jars, grinding balls are susceptible to Coriolis forces. Due to the impact and frictional forces caused by the speed differential between the balls and the grinding jar. These pressures result in a considerable collapse of the nanoparticles, leading to a decrease in thickness [30]. The SiC has an EDS wt% of 52.3 Si and 36.12 C, with a Fe content of 11.58% after milling.

The morphology of SiC nanofluids is examined using scanning electron microscopy (SEM) and energy dispersive spectroscopy (EDS) for various sizes. Mono nanofluids comprise nanoparticles of a single substance in the base fluid irrespective of size and shape [31]. According to the findings, the resulting SiC-L is angular with massive agglomerates. Figures 2 and 3 illustrate the SEM images of milled SiC-S with sub-angular flakes.

In enlarged Fig. 3, the SEM distribution of SiC nanoparticles, 24 nm in diameter, spherical with particle size, is displayed. To test the particle size effect on thermophysical characteristics of nanofluids, the produced SiC is milled to lower the size. Figures 4 and 5 illustrate the SEM of hybrid nanofluids with particle size of 90 nm.

Hybrid nanofluids, regardless of particle size distributions or geometries, include different nanoparticles. $Al_2O_3$/SiC-L is spherical and includes massive agglomerates, whereas $Al_2O_3$/SiC-S is spherical and contains more particles.

## 2.2 *Preparation and Characterization of Nanofluids*

Four different nanofluids were utilized in this experiment: (i) $Al_2O_3$ doped with SiC-L, (ii) $Al_2O_3$ doped with SiC-S, (iii) SiC-L, and (iv) SiC-S at a concentration of 0.1% volume. The two-step preparation of nanofluids is in this experimental investigation.

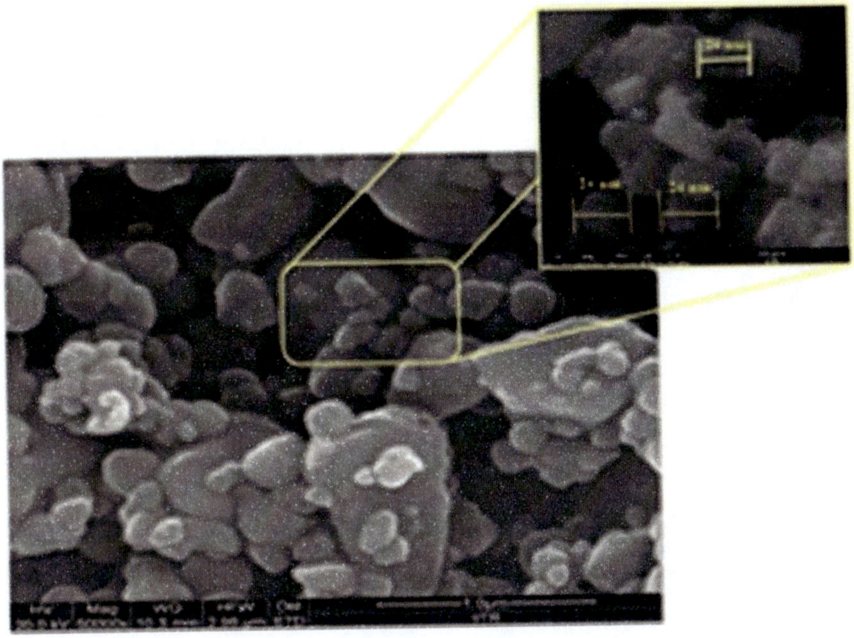

**Fig. 2** SEM image of silicon carbide nanoparticle of large size (SiC-L)

First, nanoparticles are dissolved in base fluid made up of (50:50 EG/DI water) at a concentration of 0.1% volume. The following four hours are spent sonication [32], followed by continuous stirring to prevent agglomeration. The procedure is repeated with the hybrid nanofluids $Al_2O_3$/SiC-L and $Al_2O_3$/SiC-S, as well as with the mono nanofluids SiC-L and SiC-S.

In Fig. 6, absorption spectra of mono and hybrid nanofluids are shown for 0.1 volume concentration, due to the massive number of molecules in the most concentrated nanofluid. Because the number of molecules increases with concentration, the absorbance value is frequently quite high. By comparing mono and hybrid nanofluids, the highest value of $Al_2O_3$/SiC-S at 0.1% was established. This is because the milling process introduces countless additional molecules into the system in addition to SiC-S. After 70 days of evaluating their stability with a 14-day interval, the nanofluids created showed no sedimentation [33], indicating that they may be used in place of conventional coolants in automobile radiators.

## 2.3 *Stability Study*

Nanofluid stability is critical to address because nanoparticle sedimentation may significantly alter the nanofluids thermal conductivity. The produced concentrations

**Fig. 3** SEM image of silicon carbide nanoparticle of small size (SiC-S)

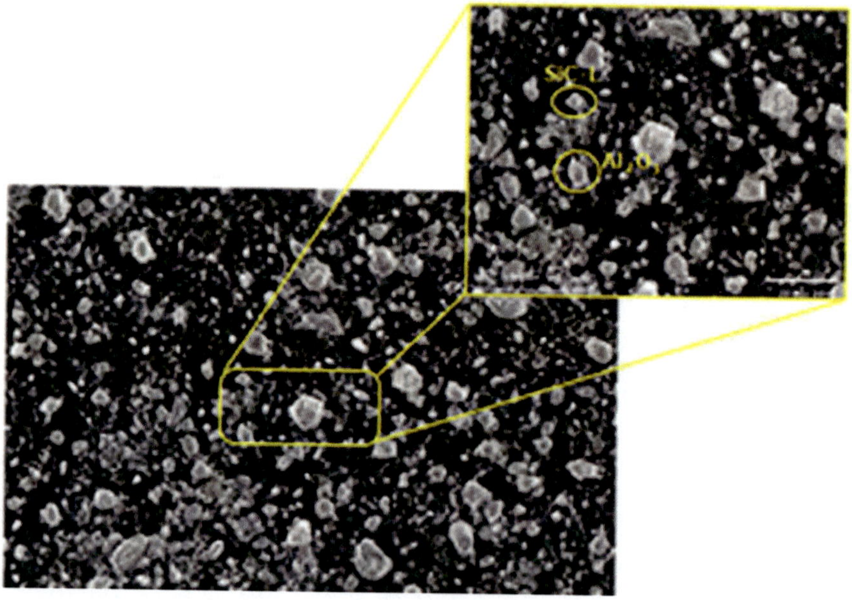


**Fig. 4** SEM image of $Al_2O_3$-SiC-L

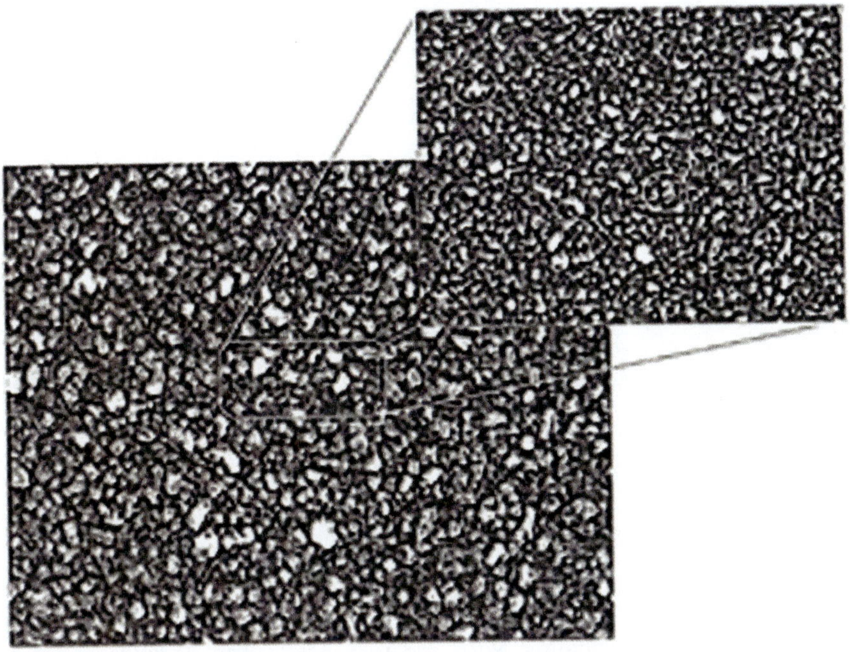

**Fig. 5** SEM image of $Al_2O_3$-SiC-S

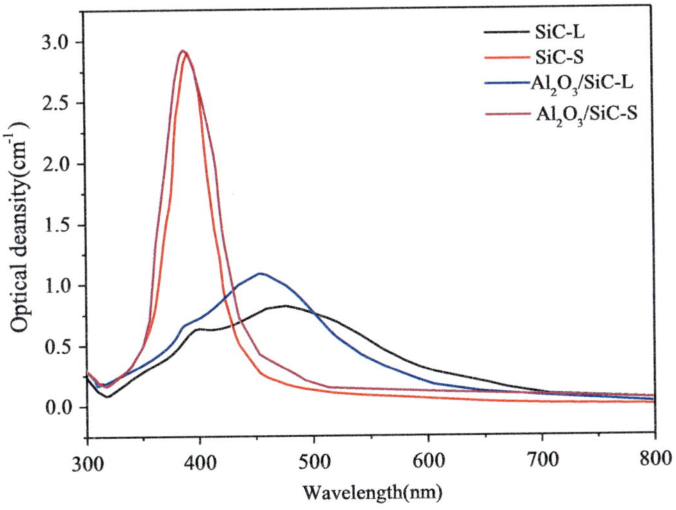


**Fig. 6** UV absorption spectra

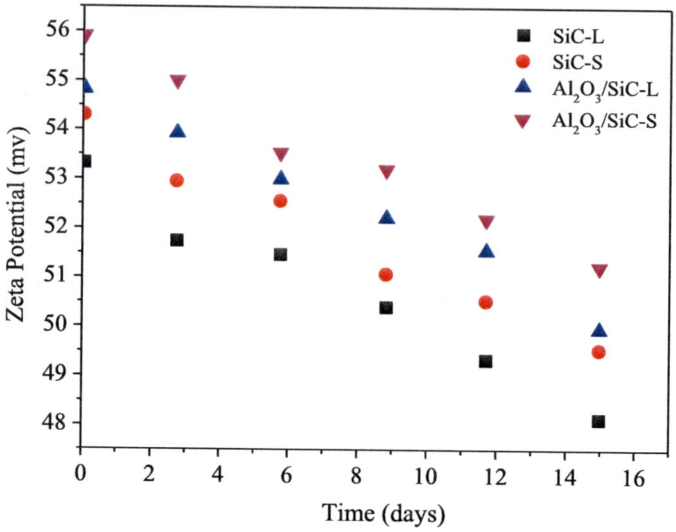


**Fig. 7** Stability analysis

are examined using a (Microtrac MRB, USA) Nanotrac Wave II with a range of –200 mV to +200 mV and repeatability of ±1 mV based on zeta potential analysis.

Figure 7 displays the nanofluids zeta potential over time in days, demonstrating that the prepared nanofluids have a zeta potential of between 49 and 57 mv, suggesting outstanding stability of the created nanofluids after days, as good zeta stability surpasses 30 mv according to the literature.

The hybrid nanofluid containing smaller nanoparticles is more stable than the mono and large-sized nanofluids. Although the zeta potential decreases gradually as the number of days grows owing to particle agglomeration, it stays steady beyond 47 mv, confirming the stability of the formed nanofluid. As depicted in Fig. 8, the prepared mono and hybrid nanofluids are sedimentation monitored for two days before testing, leading to increased stability.

## 2.4 *Measurement of Thermophysical Properties*

The viscosity is determined using a viscometer with a range accuracy of 1.0% and a repeatability of 0.2%. The nanofluids viscosity is determined between 35 and 75 °C at 10-degree intervals. Batchelor (1972) established a relationship for determining the viscosity by the Brownian motion of spherical particles as Eq. (1)

$$\mu_{nf} = \mu_{bf}\left(1 + 2.5\varphi + 6.5\varphi^2\right) \tag{1}$$

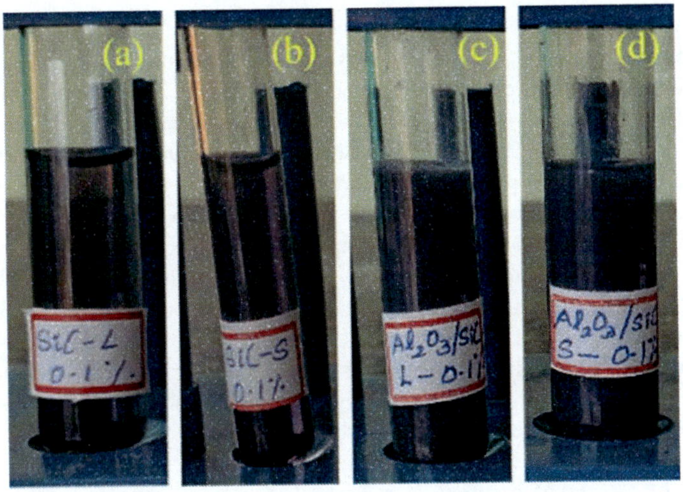


**Fig. 8** Prepared nanofluids **a** SiC-L, **b** SiC-S, **c** $Al_2O_3$/SiC-L, and **d** $Al_2O_3$/SiC-S

The viscosity of the hybrid nanofluids are calculated by Eq. (2)

$$\mu_{hnf} = \mu_{nf} \Big/ (1 - \varphi_{s1})^{2.5}(1 - \varphi_{s2})^{2.5} \tag{2}$$

Thermal conductivity is determined with a precision of 5 and a range of 0.2–2 W/mK. Equation (3) calculates the thermal conductivity using the Maxwell model.

$$\mathrm{k_{nf}} = \mathrm{k_f}\left[1 + 3(\alpha - 1)\phi \Big/ (\alpha + 2) - (\alpha - 1)\phi\right] \text{ where } \alpha = \mathrm{k_p} \Big/ \mathrm{k_f} \tag{3}$$

Equation (4) gives the thermal conductivity of hybrid nanofluids where $n$ is 3 in the case of spherical particles.

$$\frac{k_{hnf}}{k_{bf}} = \frac{k_{s2} + (n-1)k_{bf} - (n-1)\varnothing_2(k_{bf} - k_{s2})}{k_{s2} + (n-1)k_{bf} + \varnothing_2(k_{bf} - k_{s2})} \tag{4}$$

## 3 Experimental Setup and Procedure

The experimental setup is schematically depicted in Fig. 9. The radiator is constructed of vertical aluminium tubes and is equipped with fan. The design consists of a hot liquid and cooling air circuits. A geyser equipped with thermostat, pump, and flow regulator to control the flow of hot fluid. In the cold air circuit, axial fan is used to distribute the cold air. In the input and exit of the radiator, temperature sensors are attached and connected to data acquisition system. A digital anemometer is used to determine the air velocity. Two points are used to regulate the feed temperature, and the data gathering software is used to operate the PID power control unit. At the input, a heater is attached to warm the fluid before it entering the radiator. Additionally, the heater has facility for manual control from 0 to 100%. Automatic bleeding will be supplied by a safety group that controls pressure. Due to the inability of water to be squeezed during expansion, an expansion vessel containing a gas gap compensates for this expansion. The water level corresponds to the pressure at which hot water enters the gas space. A gate valve regulates the radiator's fluid flow. The cooling fluid is circulated throughout the system through a circulating pump. The pump is linked to a recycling line to reach the needed flow rate, with the excess fluid being returned to the tank via a gate valve. The heater raises the temperature of the nanofluid to 75 °C, the pump activates, allowing the fluid to flow through the radiator, and the fan absorbs and dissipates the heat generated by the hot fluid.

The radiator's inlet and exhaust temperatures are recorded. The radiator is capable of 13, 26, 39, 52, and 65 g/s flow rates. Air at 1, 2, 3, 4, and 5 m/s is utilized to cool the tubes that circulate hot fluid. The temperature of the air entering and exiting the radiator is monitored. Temperature sensors are used to monitor the coolant temperature as it flows through the system. The heat transmission properties of nanofluids are examined in a radiator with various flow rates. By introducing nanoparticles at concentrations of 0.1, 0.2, and 0.3%, the particle size effect on the base fluid is

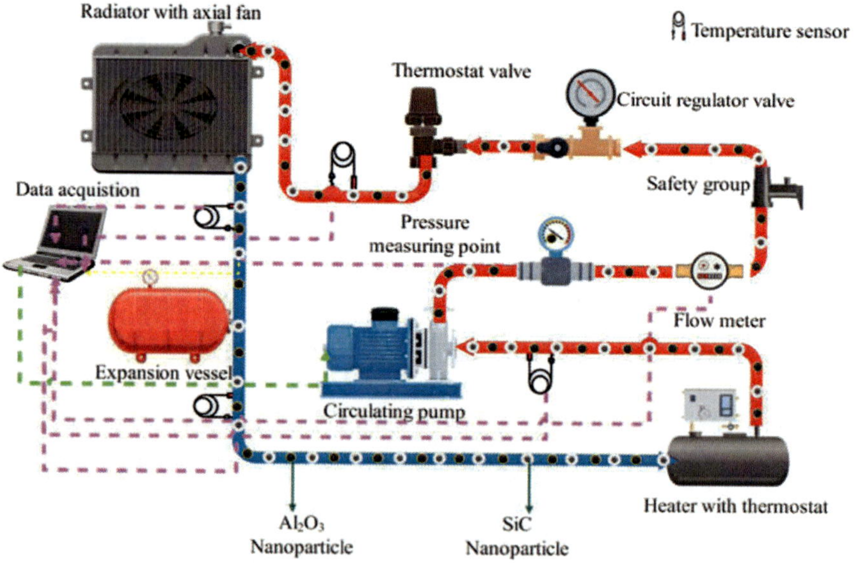


**Fig. 9** Experimental setup [Reprinted from Ref. [31], Copyright (2020), with permission from Elsevier]

**Table 1** Measurements of the radiator

| Parameters | Magnitude (cm) |
|---|---|
| Length | 34 |
| Width | 1.7 |
| Height | 31 |
| Fin size | 0.01 × 1.1 |
| Tubes | 42 No's |
| Tube length | 32 |
| Tube thickness | 0.007 |
| Hydraulic diameter of tube | 0.373 |

**Table 2** Experimental factors

| Parameters | Variation in parameters |
|---|---|
| $Al_20_3$/SiC-L (vol%) | 0.1 and 0.2 |
| $Al_20_3$/SiC-S (vol%) | 0.1 and 0.2 |
| Mass flow (g/s) | 13, 26. 39.52 and 65 |
| Experimental factors | $T_{in}$, $T_{out}$, $T_a$, $T_b$ and $T_w$ |
| Nanofluid inlet temperature (°C) | 75 |
| Air velocity (m/s) | 1, 2, 3, 4 and 5 |

examined. The heat transfer parameters of 8000 < Re < 20,000 at 50 and 60 °C are determined by adjusting the flow rate. Table 1 provides the radiator dimension.

The test parameters for the experiments are specified in Table 2.

## 4 Data Reduction

The following equations analyse the heat transmission properties that are critical for determining the particle size effect. HTC is determined by Eq. (5)

$$h = \frac{mC_p(T_{in} - T_{out})}{A[T_b - T_w]} \tag{5}$$

The specific heat is estimated using Eq. (6).

$$(C_P)_{nf} = \frac{Q_{nf}}{m_{nf}\Delta T} \tag{6}$$

Equation 7 gives the hydraulic diameter of the tubes

$$d = \frac{4A}{P} \tag{7}$$

A: Cross section of tube in $m^2$.

The velocity of the fluid is calculated using Eq. (8)

$$v = {}^{V}\!\Big/_{A.\eta_{\text{tube}}} \tag{8}$$

Equation 9 gives the correlation for Re

$$\text{Re} = \frac{\rho.v.d}{\mu} \tag{9}$$

Equation 10 gives the Nu

$$\text{Nu} = {}^{hd}\!\Big/_{k} \tag{10}$$

Equation 11 gives the friction factor, "$f_{nf}$" for laminar flow

$$f_{nf} = {}^{64}\!\Big/_{\text{Re}} \tag{11}$$

Friction factor is estimated using Eq. (12)

$$f = \left(\Delta p/\left[\rho v^2/2\right]\right) \times (d/l) \tag{12}$$

The pressure loss is calculated by Eq. (13)

$$\Delta P = \lambda \times {}^{L}\!\Big/_{d} \times {}^{\rho}\!\Big/_{2} \times w^2 \tag{13}$$

λ: friction coefficient of pipe based on pipe roughness.

## 4.1 *Uncertainty Analysis*

Due to measurement inaccuracies, all estimated numbers to approximate the Nu number and frictional factor are subject to uncertainty. However, uncertainty analysis is performed on calculation errors, not on probable fluid characteristics' inaccuracies.

**Nn.**

The uncertainties in the Nu number can be calculated from Eq. (14)

$$\begin{aligned} &{}^{\Delta Nu}\!\Big/_{Nu} \\ &= \left[\left({}^{\Delta I}\!/_{I}\right)^2 + \left({}^{\Delta V}\!/_{V}\right)^2 + \left({}^{\Delta T_w}\!/_{T_w}\right)^2 + \left({}^{\Delta T_f}\!/_{VT_f}\right)^2 + \left({}^{\Delta L}\!/_{L}\right)^2\right]^{0.5} \end{aligned} \tag{14}$$

**Table 3** Analysis values of measured and derived quantities

| Parameters (measured) | Quantities (%) | Parameters (derived) | Quantities (%) |
|---|---|---|---|
| Thermal conductivity | ± 4 | Density | ± 0.3 |
| Water mass flow rate | ± 1 | Re | ± 0.2 |
| Nanofluid mass flow rate | ± 0.2 | HTC | ± 3.5 |

**Friction Factor**

The uncertainties in friction factor and Re number can be calculated from Eqs. (15) to (16)

$$\Delta f / f = \left[ \left( \Delta(\Delta P) / P \right)^2 + \left( 3\Delta D / D \right)^2 + \left( 2\Delta \mathrm{Re} / \mathrm{Re} \right)^2 + \left( \Delta L / L \right)^2 \right]^{0.5} \quad (15)$$

$$\Delta \mathrm{Re} / \mathrm{Re} = \left[ \left( \Delta m / m \right)^2 + \left( \Delta D / D \right)^2 \right]^{0.5} \quad (16)$$

Uncertainty results of different parameters are given in Table 3.

## 5 Results and Discussion

### 5.1 Thermophysical Properties

The thermal conductivity of the base fluid, mono nanofluids, and hybrid nanofluids as a function of temperature is depicted in Fig. 10. $Al_2O_3$/SiC-S has better thermal conductivity than other nanofluids and base fluid. At 50 °C, maximum thermal conductivity of $Al_2O_3$/SiC-S was 0.723 W/mK, which was 4.8% more than base fluid. With temperature, amount of micro convectional heat transfer increases, improving thermal conductivity. Thermal conductivity was found to be 1.2% greater in $Al_2O_3$/SiC-S than in $Al_2O_3$/SiC-L, suggesting that tiny nanoparticles have considerable impact in enhancement of conductivity.

Furthermore, SiC-L and SiC-S have thermal conductivities that are 2.8 and 3.6% higher than the base fluid, respectively, implying that SiC-S has 3% higher thermal conductivity than SiC-L due to the reduction size of the particles. Smaller nanoparticles move and collide at a faster pace than larger nanoparticles, which enhances heat conductivity. Due to the reduced aggregation of various nanoparticles, hybrid nanofluids have higher thermal conductivity than mono nanofluids. Additionally, the

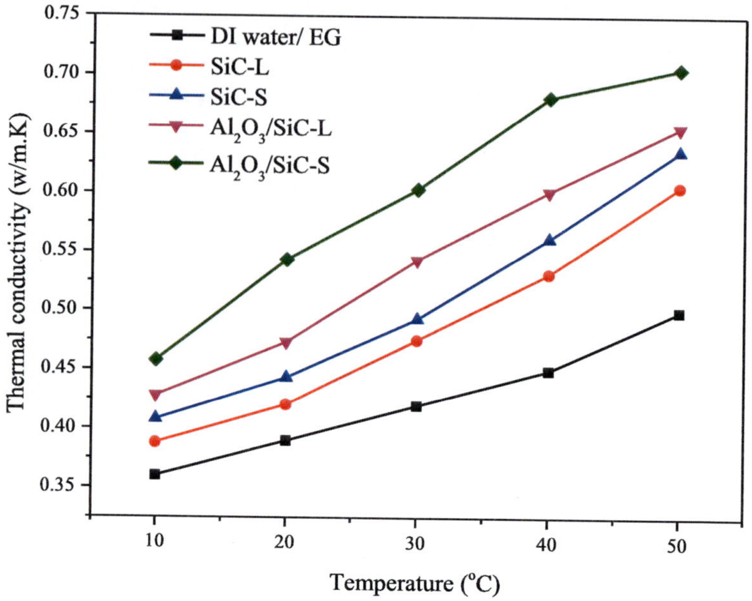


**Fig. 10** Thermal conductivity with temperature

quasi-convection of SiC-S with $Al_2O_3$ is greater than that of SiC-L with $Al_2O_3$, showing that small-sized nanoparticles are preferable for heat transfer physiognomies. In mono and hybrid nanofluids, particle size has considerable impact on thermal conductivity.

In Fig. 11, the viscosities of the base fluid, mono nanofluids, and hybrid nanofluids are shown against their respective temperatures. Due to the presence of various nanoparticles, the $Al_2O_3$/SiC-S mixture had the greatest viscosity of all the mixtures. The weak van der Waals forces of larger nanoparticles, the interaction between $Al_2O_3$ SiC-S is weaker than that between $Al_2O_3$ and SiC-L. Temperature increases the velocity and resistance of the molecule, which results in a decrease in viscosity. $Al_2O_3$/SiC-S produces a maximum enhancement of 42% over base fluid due to a reduction in the shearing force of the flowing fluid. The nanoparticles in $Al_2O_3$/SiC-S were smaller than those in $Al_2O_3$/SiC-L, resulting in a 7% increase in size.

As the density of the fluid layer lowers and the temperature rises, the kinetic energy tends to increase, making the viscosity dependent on the fluid layer thickness. When mono nanofluids SiC-L and SiC-S are compared, the aggregation of nanoparticles raises the viscosity of hybrid nanofluids. Viscosity of SiC-S is 3.1% higher than base fluid and 9% lower than $Al_2O_3$/SiC-S, indicating that the same molecular structures interact in the viscosity in the case of single- and multi-phase nanofluids.

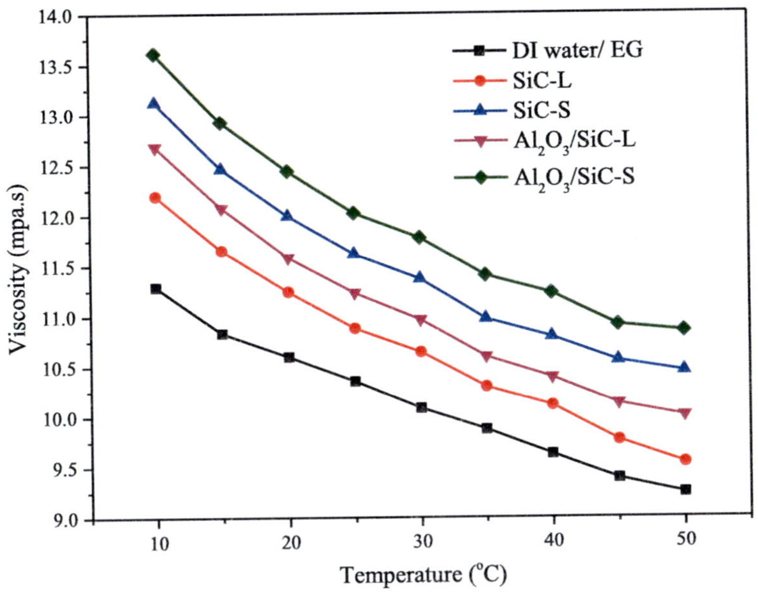


**Fig. 11** Viscosity with temperature

## 5.2 *Heat Transfer in Turbulent Condition*

To study the increased heat transmission, studies in a car radiator were done to determine particle size influence on heat transfer in turbulent regimes. The particle size effect is at 0.1% concentration by increasing the flow level of circulating pump.

**HTC influence on Re**

Figure 12a and b shows the HTCs influence on Re at 50 and 60 °C for 0.1% volume concentration and base fluid variations. $Al_2O_3$/SiC-S has a maximum heat transmission coefficient of 13–49% when compared to other nanofluids. The low viscosity and increased conductivity, the HTC at 60 °C is more than 50 °C. The $Al_2O_3$/SiC-S has a higher HTC than other materials, and the temperature of the nanofluid inlet increases from 50 to 60 °C. Similarly, elevated temperatures enhanced the HTCs of $Al_2O_3$/SiC-L, SiC-S, and SiC-L. The thermal conductivity is lowered by clustering and agglomeration, which is further reduced by turbulent movement.

The HTC rises by 24% when the particle size of $Al_2O_3$/SiC-S was lowered from 110 to 24 nm. The heat transfer along the fluid's streamlines was accomplished by the conduction of nanofluid molecules, which was helped by the nanoparticles' thermophysical characteristics. Increased Re results in a thin boundary layer and increased convection, which improves heat transmission.

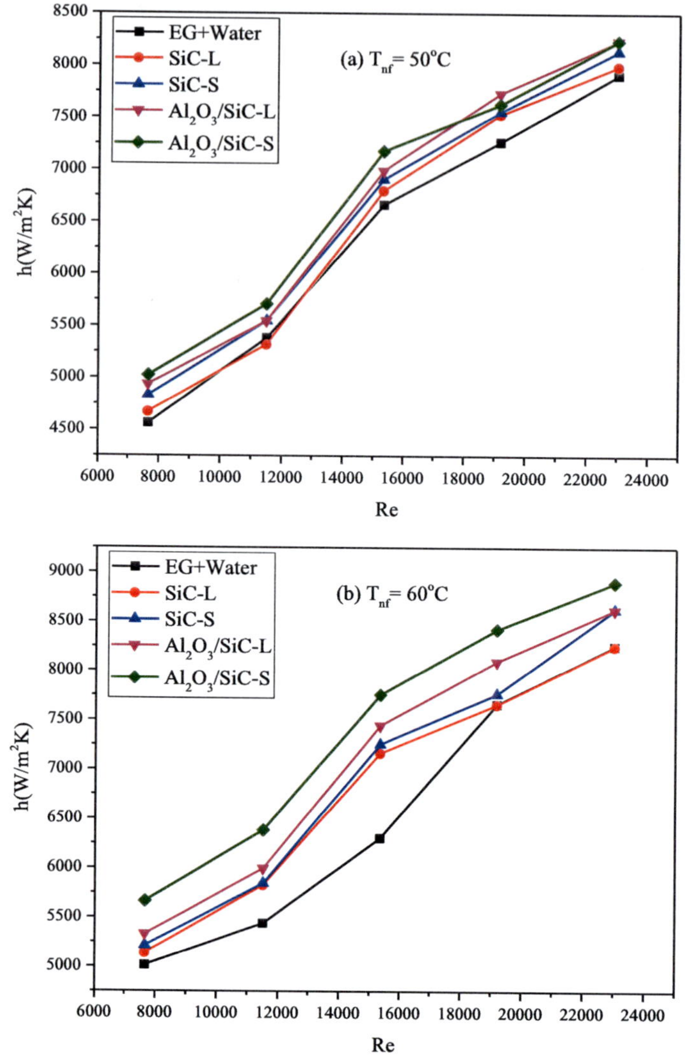


**Fig. 12** Variation of HTC with Re number for (**a**) 50 °C (**b**) 60 °C

## Variation of Nu with Re

Figure 13a and b shows the relationship between Nu and Re for 0.10% mono and hybrid nanofluids at 50 and 60 °C, respectively. The Nu grows as the size of the nanoparticles decreases, reaching a maximum value for $Al_2O_3$/SiC-S. The typical rise in Nu is between 20 and 75%, whereas Re increases by between 7000 and 22,000. This increase is more significant in turbulent flow than in mono nanofluids, with $Al_2O_3$/SiC-S exhibiting the most significant boost. By increasing the temperature of the nanofluids to 60 °C, the average enhancement for $Al_2O_3$/SiC-S was 26–68%, accompanied by a drop in viscosity and an increase in thermal conductivity. $Al_2O_3$/SiC-S showed a maximum increase in Nu of 42%, which was 14% greater than $Al_2O_3$/SiC-L. The Nu grows when the random mobility and collision of the nanoparticles increase due to the turbulent flow. The thermal conductivity and Nu of all the nanofluids were improved with the inlet temperature. The temperature fluctuations reduce the viscosity of $Al_2O_3$/SiC-S more than $Al_2O_3$/SiC-L or the mono nanofluids SiC-S and SiC-L. In comparison with large-scale nanoparticles, the turbulent movement of tiny nanofluids is caused by the presence of numerous molecules.

**Variation of friction factor with Re**

Figure 14 illustrates the fluctuation of the friction factor with Re in turbulent flow at 50 °C and 60 °C (a and b). SiC-L has higher friction factor than SiC-S and hybrid nanofluids ($Al_2O_3$/SIC-L and $Al_2O_3$/SIC-S), increasing by an average of 57%. The friction factor decreases with Re and viscosity. The friction factor reduces when the temperature is increased from 50 °C to 60 °C due to the reduced viscosity of nanofluids at higher temperatures. Despite the temperature increase from 50 °C to 60 °C, viscosity and friction factor of the nanofluids were reduced by 41% and 17%, respectively. The big nanoparticles in SiC-L, which has a higher viscosity, it has a friction factor that is 28% more than SiC-S and hybrid nanofluids.

The pressure drop of the radiator reduces as the temperature escalates from 50 to 60 °C. The viscosity of SiC-nanofluid L is reduced by 22%, and the average friction factor decreases by 26% with the temperature. When compared to other mono and hybrid nanofluids, SiC-L exhibits a maximum increase in turbulent flow of 54%.

**Pressure drop influence on Re**

The comparison of pressure drops for mono and hybrid nanofluids with Re number at 50 and 60 °C for 0.1 vol% is shown in Fig. 15a and b. It is deduced that the variance in pressure drop for both nanofluids is nearly identical. Due to the increased density of hybrid nanofluid, the pressure drop is somewhat greater than that of mono nanofluid. As mass flow rate rises, the pressure drop of mono nanofluid grows from 3.12 to 59 kPa, whereas the pressure drop climbs from 232 to 57 kPa.

**Correlations of nanofluids under turbulent flow**

For turbulent flow, a correlation equation is developed based on the regression analysis and the experimental values in Eq. (17) with a range of $7000 < Re < 22{,}000$, $= \varphi = 0.1\%$, and $13.35 < Pr < 20.36$ for the Nu.

$$NU_{Reg} = 0.5643Re^{0.7956}Pr^{0.4}(1 + \varphi)^{0.4639} \quad (17)$$

When the experimental values and the regression equation derived and shown in Fig. 16 are compared, the observed values and the predicted values have an average

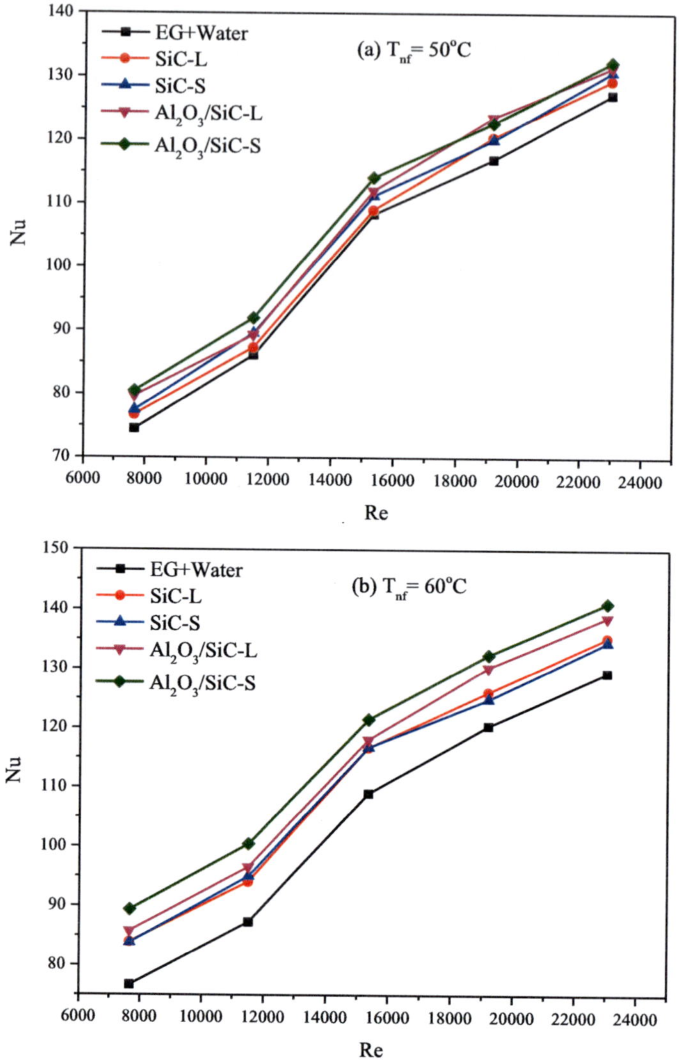


**Fig. 13** Nu influence on Re at (**a**) 50 °C (**b**) 60 °C

deviation of 5.34% and standard deviation of 4.74%. From, turbulent flow experiments, a new regression equation for predicting friction factor has been developed and given by Eq. (18) for base fluid, mono nanofluid, and hybrid nanofluid with ranges of $7000 < Re < 22{,}000$, $\varphi = 0.1\%$.

$$f_{Reg} = 29.93 Re^{-0.3946}(1 + \varphi)^{0.1943} \tag{18}$$

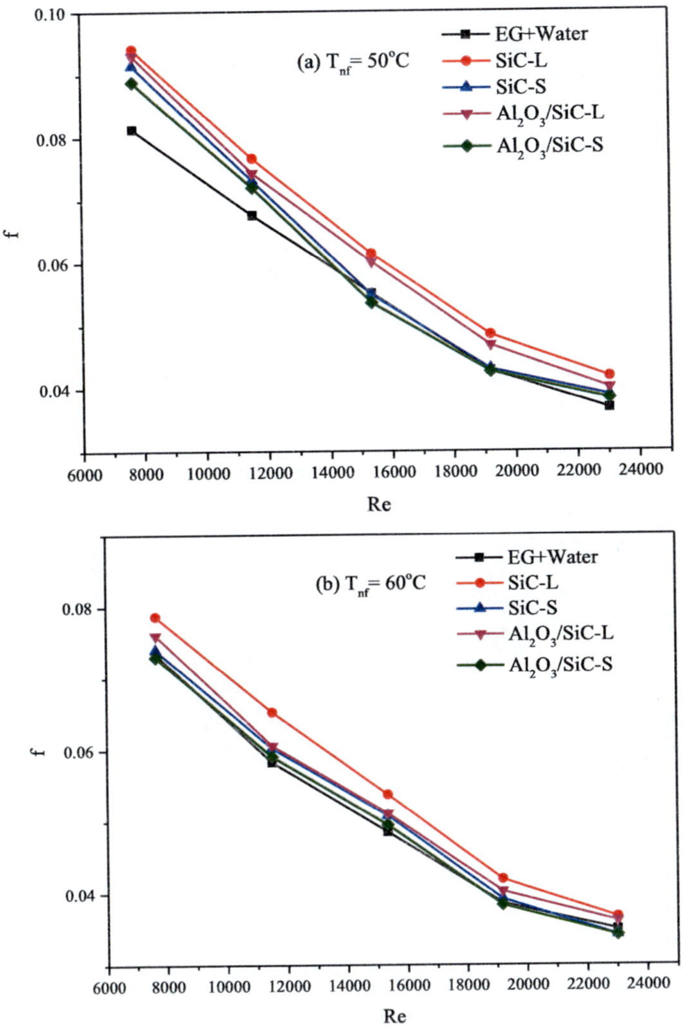


**Fig. 14** Friction factor influence on Re at (**a**) 50 °C (**b**) 60 °C

Figure 17 shows a comparison of the experimental regression equation-derived friction factor values [31]. The experimental and the regression friction factor have a standard deviation of 5.43% with an average deviation of 5.17% for turbulent flow.

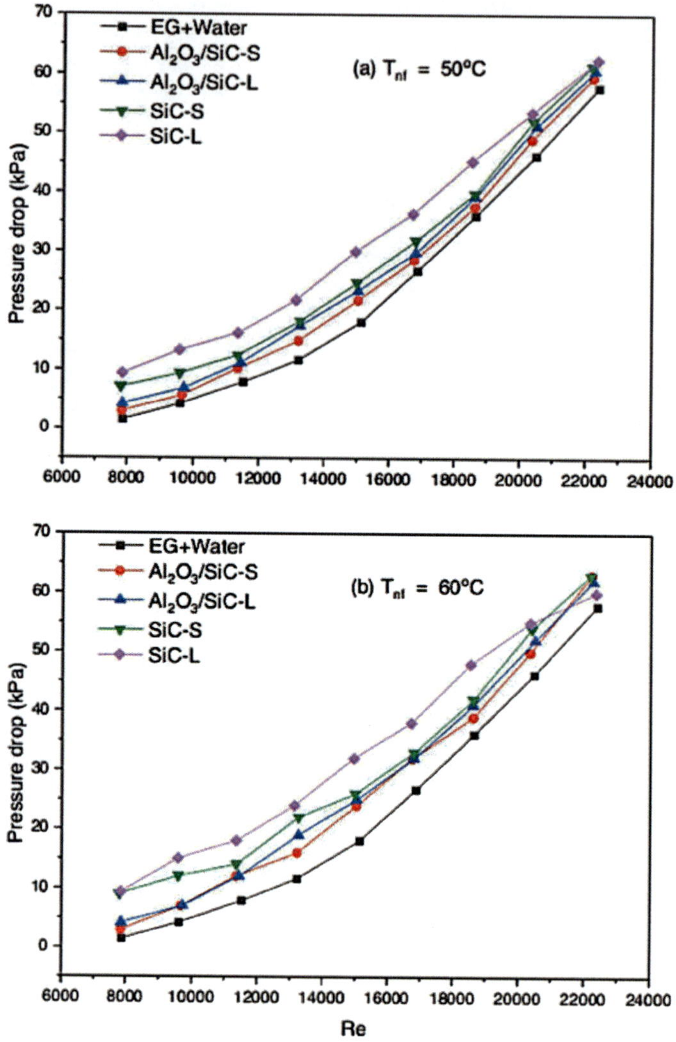


**Fig. 15** Pressure drop impact on Re at (**a**) 50 °C (**b**) 60 °C

## 6 Conclusions

The particle size effect in nanofluids is analysed using the findings, and correlations between the friction factor and Nu are generated. The key thermophysical parameters of heat conductivity and viscosity and the findings are examined in terms of particle size. The following outcomes were obtained from tests and analyses.

At 50 °C, $Al_2O_3$/SiC-S has a maximum thermal conductivity of 0.723 W/mK, which is 4.8% more than that of the base fluid. Micro-convection is enhanced as the

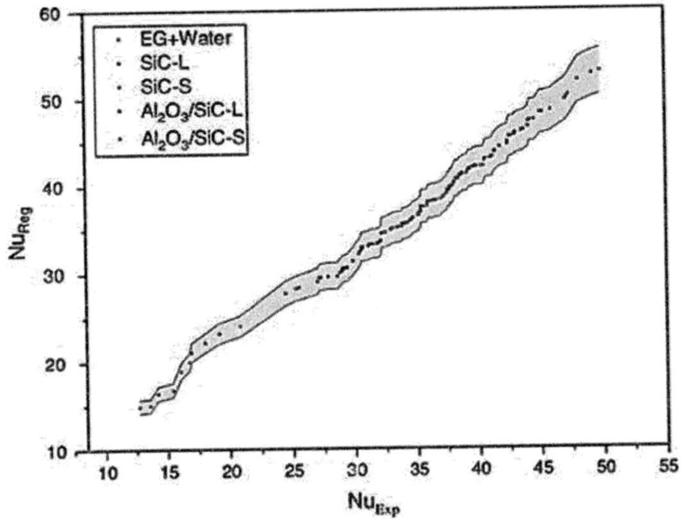


**Fig. 16** Exp versus pred Nu for turbulent flow

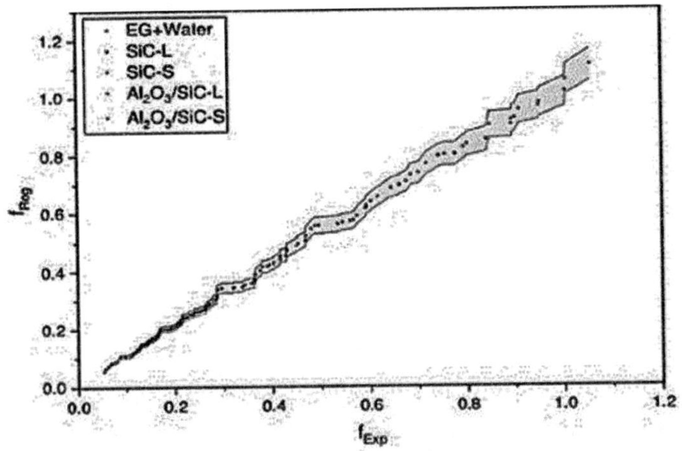


**Fig. 17** Exp versus pred friction factor for turbulent flow

temperature changes, hence enhancing thermal conductivity. Thermal conductivity was found to be 1.2% greater in $Al_2O_3$/SiC-S than in $Al_2O_3$/SiC-L, demonstrating that tiny nanoparticles have substantial effect heat transfer.

Furthermore, the thermal conductivities of SiC-L and SiC-S are 2.8% and 3.6% higher than the base fluid, respectively, implying that SiC-S is 3% more conductive than SiC-L. SiC-S has a viscosity that is 3.1% more than water and 9% less than $Al_2O_3$/SiC-S, showing that the same molecular structures are involved in the viscosity. According to the findings, hybrid nanofluids outperform mono nanofluids. When $Al_2O_3$/SiC-S and mono nanofluids are compared, a maximum boost of 2.2% is recorded.

$Al_2O_3$/SiC-HTC increases by 4.8% in comparison with other fluids. $Al_2O_3$/SiC-S rises by an average of 2.0–5.1% as temperature raised from 50 to 60 °C. This was owing to lower viscosity and enhanced thermal conductivity. The viscosity reduces by 3.5%, and the friction factor decreases by 14% as temperature elevated 50 °C to 60 °C. When the particle size of $Al_2O_3$/SiC-S was lowered, the HTC rises by 2.6%, allowing fluid to flow uninterruptedly along the streamline. $Al_2O_3$/SiC-S increased the Nu by 4.4%, 1.6% more than $Al_2O_3$/SiC-L, suggesting that tiny nanoparticles had a greater effect on big nanoparticles.

As the inlet temperature elevated from 50 to 60 °C, viscosity of SiC-L lowers by 2.4%, and the average friction factor reduces by 2.8%. In comparison with conventional coolants, including smaller hybrid nanoparticles of $Al_2O_3$/SiC-S into the base fluid enhances heat transmission and may allow for a smaller radiator. Hybrid nanoparticles have significant effect on temperature profiles than mono nanoparticles, and milling decreases their size, allowing for increased heat transmission in the automobile system.

## References

1. Li F, Li L, Zhong G, Zhai Y, Li Z (2019) Effects of ultrasonic time, size of aggregates and temperature on the stability and viscosity of Cu-ethylene glycol (EG) nanofluids. Int J Heat Mass Transf 129:278–286
2. Mahbubul IM, Elcioglu EB, Amalina MA, Saidur R (2019) Stability, thermophysical properties and performance assessment of alumina–water nanofluid with emphasis on ultrasonication and storage period. Powder Technol 345:668–675
3. Akhgar A, Toghraie D (2018) An experimental study on the stability and thermal conductivity of water-ethylene glycol/$TiO_2$-MWCNTs hybrid nanofluid: developing a new correlation. Powder Technol 338:806–818
4. Shah SNA, Shahabuddin S, Sabri MFM, Salleh MFM, Ali MA, Hayat N, Saidur R (2020) Experimental investigation on stability, thermal conductivity and rheological properties of rGO/ethylene glycol-based nanofluids. Int J Heat Mass Transf 150:118981
5. Li L, Zhai Y, Jin Y, Wang J, Wang H, Ma M (2020) Stability, thermal performance and artificial neural network modelling of viscosity and thermal conductivity of $Al_2O_3$-ethylene glycol nanofluids. Powder Technol 363:360–368
6. Xian HW, Sidik NAC, Saidur R (2020) Impact of different surfactants and ultrasonication time on the stability and thermophysical properties of hybrid nanofluids. Int Commun Heat Mass Transfer 110:104389
7. Goudarzi K, Jamali H (2017) Heat transfer enhancement of $Al_2O_3$-EG nanofluid in a car radiator with wire coil inserts. Appl Therm Eng 118:510–517
8. Hatami M, Jafaryar M, Zhou J, Jing D (2017) Investigation of engines radiator heat recovery using different shapes of nanoparticles in $H_2O/(CH_2OH)_2$ based nanofluids. Int J Hydrogen Energy 42(16):10891–10900
9. Selvam C, Solaimalai Raja R, Mohan Lal D, Harish S (2017) Overall heat transfer coefficient improvement of an automobile radiator with graphene-based suspensions. Int J Heat Mass Transf 115:580–588

10. Asadi A, Alarifi IM, Ali V, Nguyen HM (2019) An experimental investigation on the effects of ultrasonication time on stability and thermal conductivity of MWCNT-water nanofluid: finding the optimum ultrasonication time. Ultrason Sonochem 58:104639
11. Elsaid AM (2019) Experimental study on the heat transfer performance and friction factor characteristics of $CO_3O_4$ and $Al_2O_3$ based $H_2O/(CH_2OH)_2$ nanofluids in a vehicle engine radiator. Int Commun Heat Mass Transfer 108:104263
12. Akilu S, Baheta AT, Kadirgama K, Padmanabhan E, Sharma KV (2019) Viscosity, electrical and thermal conductivities of ethylene and propylene glycol-based β-SiC nanofluids. J Mol Liq 284:780–792
13. Huminic G, Huminic A, Fleaca C, Dumitrache F, Morjan I (2017) Thermo-physical properties of water based SiC nanofluids for heat transfer applications. Int Commun Heat Mass Transfer 84:94–101
14. Zufar M, Gunnasegaran P, Kumar HM, Ng KC (2020) Numerical and experimental investigations of hybrid nanofluids on pulsating heat pipe performance. Int J Heat Mass Transf 146:118887
15. Anitha S, Thomas T, Parthiban V, Pichumani M (2019) What dominates heat transfer performance of hybrid nanofluid in single pass shell and tube heat exchanger? Adv Powder Technol 30(12):3107–3117
16. Kumar V, Sarkar J (2019) Numerical and experimental investigations on heat transfer and pressure drop characteristics of $Al_2O_3$-$TiO_2$ hybrid nanofluid in minichannel heat sink with different mixture ratio. Powder Technol 345:717–727
17. Kaska SA, Khalefa RA, Hussein AM (2019) Hybrid nanofluid to enhance heat transfer under turbulent flow in a flat tube. Case Studies in Thermal Engineering 13:4–13
18. Asadi A, Asadi M, Rezaniakolaei A, Aistrup L, Afrand M (2018) Heat transfer efficiency of $Al2O_3$-MWCNT/thermal oil hybrid nanofluid as a cooling fluid in thermal and energy management applications: An experimental and theoretical investigation. Int J Heat Mass Transf 117:474–486
19. Sundar LS, Singh MK, Sousa ACM (2018) Turbulent heat transfer and friction factor of nanodiamond-nickel hybrid nanofluids flow in a tube with longitudinal strip inserts. Int J Heat Mass Transf 121:390–401
20. Bhattad A, Sarkar J, Ghosh P (2019) Experimentation on effect of particle ratio on hydrothermal performance of plate heat exchanger using hybrid nanofluid. Appl Therm Eng 162:114309
21. Jai S, Krishnan S, Nagarajan PK (2019) Influence of stability and particle shape effects for an entropy generation based optimized selection of magnesia nanofluid for convective heat flow applications. Appl Surf Sci 489:560–575
22. Ambreen T, Kim M (2018) Effects of variable particle sizes on hydrothermal characteristics of nanofluids in a microchannel. Int J Heat Mass Transf 120:490–498
23. Maheshwary PB, Handa CC, Nemade KR (2017) A comprehensive study of effect of concentration, particle size and particle shape on thermal conductivity of titania/water based nanofluid. Appl Therm Eng 119:79–88
24. Nikkam N, Ghanbarpour M, Khodabandeh R, Toprak MS (2017) The effect of particle size and base liquid on thermo-physical properties of ethylene and diethylene glycol based copper micro and nanofluids. Int Commun Heat Mass Transfer 86:143–149
25. Rabbani M, Mohseni E, Rao M (2016) Effect of particle size and viscosity on thermal conductivity enhancement of graphene oxide nanofluid. Int Commun Heat Mass Transfer 76:308–315
26. Hossein M, Darvanjooghi K, Esfahany MN (2016) Experimental investigation of the effect of nanoparticle size on thermal conductivity of in-situ prepared silica—ethanol nanofluid. Int Commun Heat Mass Transfer 77:148–154
27. Teng T, Hung Y, Teng T, Mo H, Hsu H (2010) The effect of alumina/water nanofluid particle size on thermal conductivity. Appl Therm Eng 30(14–15):2213–2218
28. Anoop KB, Sundararajan T, Das SK (2009) Effect of particle size on the convective heat transfer in nanofluid in the developing region. Int J Heat Mass Transf 52(9–10):2189–2195
29. Graham A (1981) On the viscosity of suspensions of solid spheres. Appl Sci Res 37:275–286

30. Cardenas Contreras EM, Oliveira GA, Bandarra Filho EP (2019) Experimental analysis of the thermohydraulic performance of graphene and silver nanofluids in automotive cooling systems. Int J Heat Mass Transf 132:375–387
31. Ramalingam S, Dhairiyasamy R, Govindasamy M, Muthaiah VMR (2020) Consequence of nanoparticles size on heat transfer characteristics of a radiator. Powder Technol 367:213–224
32. Said Z, El Haj AM, Hachicha AA, Bellos E, Abdelkareem MA, Alazaizeh DZ, Yousef BAA (2019) Enhancing the performance of automotive radiators using nanofluids. Renew Sustain Energy Rev 112:183–194
33. Ardekani AM, Kalantar V, Heyhat MM (2019) Experimental study on heat transfer enhancement of nanofluid flow through helical tubes. Adv Powder Technol 30(9):1815–1822

# Experimental Studies on the Two-Phase Pressure Drop Through Minichannel

S. B. Charthankar and A. T. Autee

## *Nomenclature*

| | |
|---|---|
| $A$ | Cross section ($m^2$) |
| $C$ | Chisholm parameter |
| $d$ | Characteristics diameter (m) |
| $\Delta p$ | Pressure drop (Pa) |
| Fr | Froude number |
| $g$ | Gravitational acceleration ($m/s^2$) |
| $j$ | Superficial velocity (m/s) |
| $m$ | Mass flow rate (kg/s) |
| Re | Reynolds number |
| We | Weber number |
| $G$ | Mass flux ($kg/m^2$ s) |
| $D$ | Diameter of the tube (mm) |
| $E$ | A parameter in Friedel correlation for two-phase pressure drop |
| $F$ | A parameter in Friedel correlation for two-phase pressure drop |
| $X$ | Mass fraction of gas in a two-phase mixture (vapor quality) |
| $H$ | A parameter in Friedel correlation |
| $dp/dz$ | Pressure drop (Pa) |

## *Greek symbols*

| | |
|---|---|
| $\chi$ | Lockhart–Martinelli parameter |

S. B. Charthankar (✉) · A. T. Autee
Mechanical Engineering Department, Maharashtra Institute of Technology, Aurangabad, Maharashtra, India
e-mail: shantanubc@gmail.com


N. L. Narasimhan et al. (eds.), *Recent Advances in Energy Technologies*, Lecture Notes in Mechanical Engineering, https://doi.org/10.1007/978-981-19-3467-4_16

| | |
|---|---|
| $\phi$ | Two-phase frictional multiplier |
| $f$ | Friction factor |
| $\mu$ | Dynamic viscosity (Ns/m$^2$) |
| $\nu$ | Kinematic viscosity (m$^2$ /s) |
| $\rho$ | Density (kg/m$^3$) |
| $\sigma$ | Surface tension (N/m) |
| $\Omega$ | Correction factor |

## *Subscripts*

| | |
|---|---|
| a | Acceleration |
| v | Gas phase only |
| vo | All mass is assumed as gas only |
| H | Hydrostatic |
| l | Liquid |
| lo | All mass is assumed as liquid |
| HOM | Homogeneous |
| TP | Two-phase |
| TPF | Two-phase frictional |

## 1 Introduction

Mini and microchannels with a two-phase flow are widely used in contemporary and advanced technological applications, such as biomedical equipment, computer chip cooling, cooling systems of data centers, process plants, genetic. Various researchers have proposed the classification of the mini and microchannels. According to Kandlikar classification, micro and minichannels sizes ranged from 10 to 200 μm and 200 μm to 3 mm, respectively, [1] while as proposed by Mehendale et. al. the micro and meso-channel categorization varies from 1 μm to 100 μm and 100 μm to 1 mm, respectively [2]. Several studies on two-phase pressure drops have been conducted in recent years, and numerous forecasting approaches have been offered by various researchers, still due to different variables such as channel size, channel geometry, surface roughness, improper methods of data reduction there are differences in the data, giving scope for developing new database [3]. An experimental study of 1.48–25.83 mm diameter circular tubes was presented by Lockhart and Martinelli. For adiabatic blends of air with water, kerosene, benzene, along with different lubricants, friction pressure gradients owing to two phases were assessed using two-phase multipliers. The actual two-phase frictional pressure change is related to the single-phase pressure change by two-phase multipliers. This multiplier which is usually mentioned as the Lockhart–Martinelli parameter (X) would depend on the proportion of liquid and vapor phase pressure gradients [4]. Akers utilized their experimental data for plain tube as well as micro fin extruded aluminum microchannels (rectangular port with Dh of 2.64 mm and 1.56 mm, respectively) which fit well. The two-phase friction factor is calculated using a comparable all-liquid flow that has the identical frictional pressure drop as the two-phase flow [5]. For fully rough surfaces, Chisholm improved friction factor which is independent of Reynolds number [6]. Friedel established a two-phase multiplier correlation by performing the experiments, using tubes as small as 4 mm, for flow in horizontal and vertical direction [7]. Jung and Radermacher developed the correlation by using a copper tube that closely matched with the Lockhart–Martinelli parameter, Xtt. Experimentally studied horizontal annular flow boiling of pure and mixed refrigerants was used to develop this correlation. The refrigerant mixture's pressure drop was found to be independent of the mixture but shows a good fit with the Lockhart–Martinelli parameter, Xtt [8]. Souza proposed a two-phase multiplier correlation by testing R-12 and R-134a in a 10.9 mm diameter smooth copper tube. The total two-phase flow pressure decrease was split into frictional and accelerational components, with the multiplier determined using the Froude rate effect [9]. Mishima and Hibiki conducted the study using glass and aluminum tubes to measure the pressure drop of air–water two-phase flow for upright capillaries of size varying between 1 and 4 mm diameter. By neglecting pressure loss caused by fluid acceleration, frictional pressure drop was determined. The findings then matched with Lockhart and Martinelli's correlation. As a function of inner diameter, the Chisholm parameter was updated with a new Chisholm coefficient C, which reproduced frictional pressure drop. The new equation predicted pressure drop for all the experimentation well within the error limit of $+12\%$ to $-12\%$ except for ammonia vapor flow. For ammonia—vapor flow, the error was $+25\%$ to $-25\%$ [10]. Circular tubes of diameter 2.46 and 2.92 mm, and a rectangular channel of dimensions 4.06 mm $\times$ 1.7 mm during evaporation, were used by Tran et al. to perform experiments wherein R134a, R12, and R113 with mass fluxes varying between 50 and 800 kg/s m$^2$ were used. Experiments concluded that most of the large tube correlation under-predict annular region (high mass flux and qualities) pressure drop results [11]. The pressure drop study for R-134a, R-22, and R-404a through the copper tube of diameter 6.5–3.25 mm was made by Zhang and Webb. It was reported that, especially for highly reduced pressure, Friedel correlation could not satisfactorily predict the pressure drop [12]. To identify differentiation parameters for microchannels and minichannels, Chung and Kawaji conducted experiments in circular channels of 530, 250, 100, and 50 $\mu$m diameter, using a blend of gaseous nitrogen with liquid water to find out how the diameter of circular tubes are affected by the two-phase flow. The experimental result data, void percent, and two-phase frictional pressure drop data demonstrated the effect of channel diameter on two-phase flow. The two-phase flow characteristics were distinct for channel sizes between 250 and 100 $\mu$m diameter. Traditional correlations no longer apply because the two-phase flow characteristics in channels with $D < 100$ $\mu$m differ from those in minichannels. The current findings are confined to similar gas–liquid couplings since the effects of surface tension and liquid property in nondiabetic two-phase flow in microchannels are still mainly unexplored [13]. The Chen factor (homogeneous) was used by Chen et al. to improve the homogeneous correlation [14]. Müller Steinhagen used a huge database of air and oil, cryogenic, vapor and water, air water and liquid combinations, and numerous refrigerants to compare the two-phase pressure drop prediction model [15]. By conducting studies on two-phase flow patterns, void percent, and pressure drop for circular microchannels of 100 m diameter, Kawahara et al. explained the concern of friction multiplier factors in single-phase and two-phase based on pressure drop evaluation [16]. The impact of tube diameter was investigated by Venkatesan, on two-phase frictional pressure loss with air and water flowing through thin tubing with inner diameters of 0.6, 1.2, 2.6, and 3.4 mm [17]. By comparing various earlier macro and minichannel correlations with experimental data for circular tubes of 3–8 mm diameter, Autee et al. proposed a novel interrelationship for the prediction of adiabatic two-phase pressure decrease [18]. Hamad et al. led an experimental and quantitative evaluation of the influence of quality and velocity of the mixture of liquid and gas on pressure drop of liquid–gas flow in horizontal pipes of several sizes and found that for constant water flow, the increased gas flow rate increases the friction pressure drop [19]. In various process industries, Giri and Autee emphasized the need for improvement in design and analysis for the secure and efficient functioning of different thermal machines by using dependable and precise prediction of pressure drop [20]. Lee and Lee studied the adiabatic air and water two-phase flow pressure drop in rectangular mini and macro channels kept in horizontal orientation [21]. By linking the multiplication factor of the two-phase to the effects caused by viscosity, inertia, and surface tension, Charthankar et al. suggested a new-found connection for two-phase pressure decrease in mini and macro channels in a horizontal orientation [22].

In any system with a two-phase flow, calculating and predicting pressure drop is significant. The design of compact heat exchangers, components of HVAC systems, thermal power plants, petrochemical plants, etc., greatly depends on a wide-ranging two-phase pressure drop knowledge because it alters the system's functioning. The pressure drops predicted by using available methods may differ by large.

The present work represents an investigational and analytic study of pressure drops of two-phase across the 1.00, 1.50, 2.00, 2.50, and 3.00 mm diameter circular acrylic tubes in horizontal orientation under adiabatic conditions. The setup is devised and utilized to study the two-phase pressure drop phenomenon. The predicted values are compared to the measured experimental data using the correlations chosen for comparison in this study to analyze the predictive potential of some of the correlations available from the open literature. The results of experimental investigations into Lockhart–Martinelli, Friedel, and Chisholm techniques employed to estimate the two-phase pressure drop during flow in minichannels are presented in the current study. Finally, for tube diameters of 1.00, 1.50, 2.00, 2.50, and 3.00 mm, a generalized correlation is created to estimate the current experimental data.

## 2 Experimentation

The experimental setup is designed to conduct experiments using an air–water mixture. Figure 1 represents the experimental setup schematically.

The experiment setup is designed as well as developed to conduct experiments for the two-phase flow of adiabatic air–water flowing through circular minichannels. The circular acrylic tubes with inner diameter 1.00, 1.50, 2.00, 2.50, and 3.00 mm are used during the investigation. The test sections were aligned horizontally. For this setup, two separate plyboard portions were made. The mixing chamber, bypass valves, and water pump are placed on the first portion, while the test section is fitted on the second section. A centrifugal pump is used to pump water into the test section, from a tank of 100-L capacity. The control valve and the bypass valve are used to control and regulate water and airflow rate. An air compressor is used to store compressed air into a storage tank which is further supplied into the test segment. A standardized pressure gauge connected to the air supply line measures pressure. Air and water rotameters of range 10–100 LPM and 3–30 LPM, respectively, are used so that in the mixing chamber, air and water flowing at different rates are combined, resulting in a two-phase flow. The mixture is released into the atmosphere through a small tube in the test portion. Differential pressure transducers (DPT) with a measuring range of 0–2.5 bar are put in to monitor the frictional pressure drop of the air–water mixture. The static pressure was recorded with a DPT of range 0–2.5 bar. A suitably designed data gathering system is used to collect the experimental data. Acquired experimental data is then stored on a computer by using customized software. 1.00,

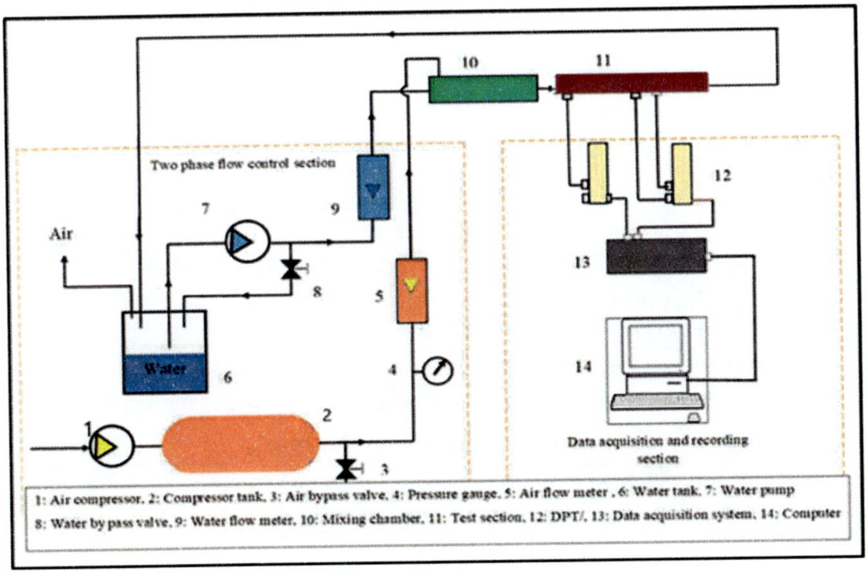


**Fig. 1** Schematic layout of the experimental setup

**Table 1** Uncertainty of the measuring instruments

| Measuring instrument | Uncertainty (%) |
|---|---|
| Air Rota meter (10 to 100 LPM) | 4.52 |
| Water Rota meter (3 to 30 LPM) | 3.12 |
| DPT 1 (0–2.5 bar) | 5.16 |
| DPT 2 (0–2.5 bar) | 5.51 |

1.50, 2.00, 2.50, and 3.00 mm diameters tubes are used during the investigation. The experimental results were recorded by repeating the experimental procedure for each tube. The air–water flowing at different flow rates is used for measuring frictional pressure drop in this study. The uncertainty of the measuring instruments is shown in Table 1.

The experimental setup and instrumentation validation are carried out by performing single-phase pressure drop tests. Experiments were carried out on a 300 mm long test segment with an interior diameter of 8.0 mm. The experimental pressure drop was recorded for water flowing through the test section. Figure 2 compares the measured friction factor f to the Blasius correlation predicted values. The experimental results are found to be very close to the Blasius correlation ($f = 0.079/Re^{0.25}$) prediction. Table 2 shows different existing correlations used for the comparison of experimental data.

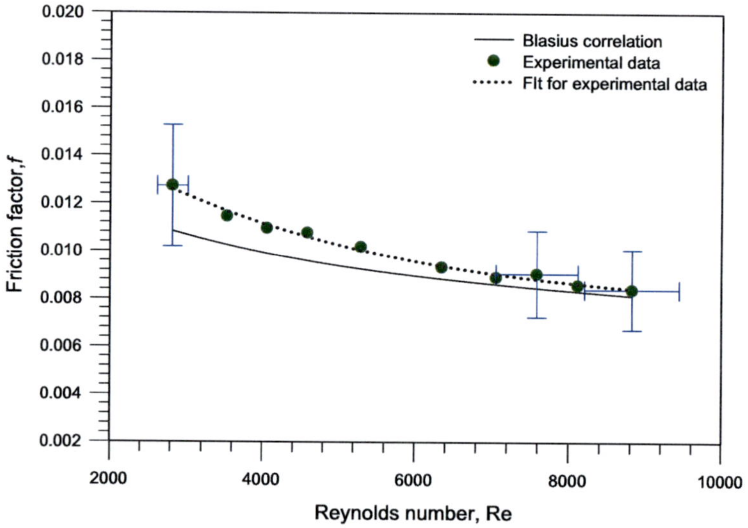


**Fig. 2** Friction factor versus Blasius correlation

**Table 2** Various existing two-phase pressure drop correlations

| Authors | Correlations |
|---|---|
| Lockhart–Martinelli [1949] | $\left(\frac{dp}{dz}\right) = \varphi_L^2\left(\frac{dp}{dz}\right)_l = \varphi_G^2\left(\frac{dp}{dz}\right)_g$, $\left(\frac{dp}{dz}\right)_l = \frac{2f_l G^2(1-x)^2}{\rho_l D}, \left(\frac{dp}{dz}\right)_g = \frac{2f_g G^2 x^2}{\rho_g D}, \varphi_L^2 = 1 + \frac{C}{X} + \frac{1}{X^2}$, $\varphi_G^2 = 1 + CX + X^2, X^2 = \frac{(dp/dz)_l}{(dp/dz)_g}, Re_L = \frac{[G(1-x)]D}{\mu_l}, Re_G = \frac{(Gx)D}{\mu_g}$ |
| Chisholm [1972] | $\varphi_L^2 = 1 + (X^2 - 1)\left[Bx^{0.875}(1-x)^{0.875} + x^{1.75}\right]$, $X^2 = \frac{\left(\frac{dp}{dz}\right)_{vo}}{\left(\frac{dp}{dz}\right)_{lo}}$ |
| Friedel [1979] | $\left(\frac{dp}{dz}\right)_{TP} = \varphi_L^2\left(\frac{dp}{dz}\right)_{lo}, \varphi_L^2 = E + \frac{3.24FH}{F_r^{0.045}W_e^{0.035}}$, $F_r = \frac{G^2}{gD\rho_H^2}, W_e = \frac{G^2 D}{\sigma\rho_H}, E = (1-x)^2 + x^2\frac{\rho_l f_{vo}}{\rho_{vo} f_l}$, $F = x^{0.78}(1-x)^{0.224}, H = \left(\frac{\rho_l}{\rho_v}\right)^{0.91}\left(\frac{\rho_v}{\rho_l}\right)^{0.19}\left(1 - \frac{\rho_v}{\rho_l}\right)^{0.7}$ |

## 3 Results and Discussions

The frictional pressure drop for air and water two-phase flow across horizontal tubes with inner diameters of 1.00, 1.50, 2.00, 2.50, and 3.00 mm is calculated using statistical analysis of the experimental results. Figures 3, 4, 5, 6, 7 show a frictional pressure drop evaluation computed applying available correlations and investigational data. The experimental pressure drop is indicated on the abscissa, whereas the projected pressure drop is shown on the ordinate. Dotted lines in these diagrams indicate the ±50% error band.

For a 1.00 mm tube, Fig. 3 illustrates the Lockhart–Martinelli, Friedel, and Chisholm correlations in comparison with experimental two-phase pressure drop data. The predictions from Friedel and Chisholm equation show that 100% of data points are outside the 50% error range and are overpredicted, while the Lockhart–Martinelli method prediction reveals that 89% of record points are inside the 50% error band.

Figure 4 compares two-phase pressure drop data from experiments with the Lockhart–Martinelli, Friedel, and Chisholm correlations for a 1.50 mm tube. According to the Lockhart–Martinelli correlation prediction, resulting in 53% of points within the 50% error band, whereas 100% of Friedel correlation data points are outside the 50% error band and are overpredicted. However, for the Chisholm correlation, 66% of data points are outside the 50% error zone and are overpredicted.

Figure 5 shows experimental friction pressure drop comparison using Lockhart–Martinelli correlation, Friedel correlation, and Chisholm correlation for 2.00 mm, tube. The prediction made by using Lockhart–Martinelli correlation shows that 100%

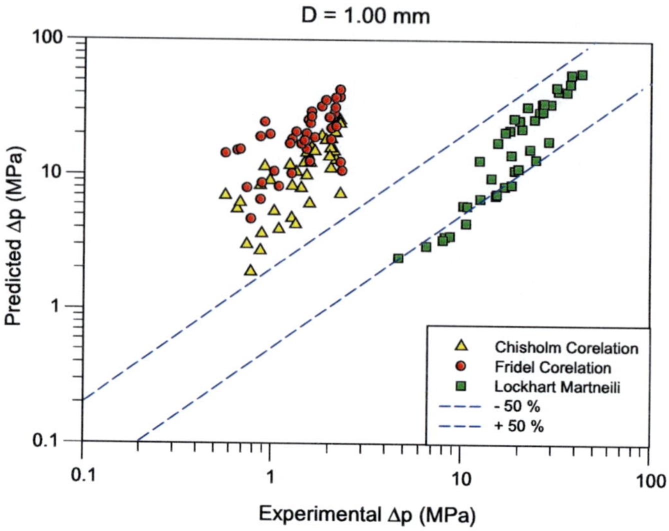


Fig. 3 Predicted $\Delta p$ versus experimental $\Delta p$ for $D = 1.00$ mm

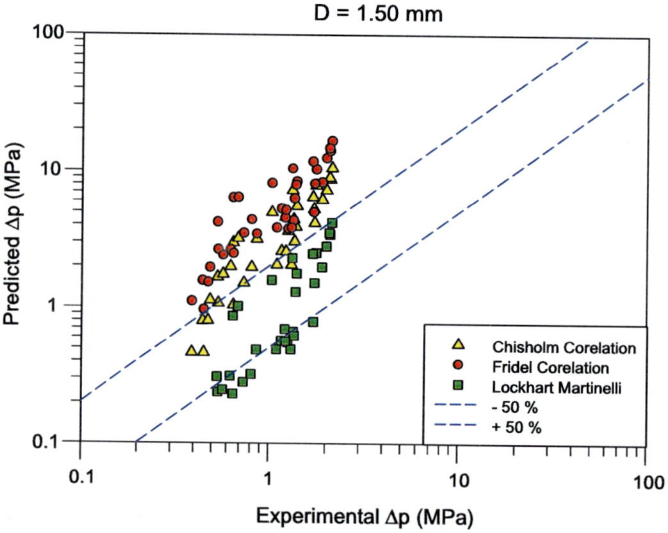


Fig. 4 Predicted $\Delta p$ versus experimental $\Delta p$ for $D = 1.50$ mm

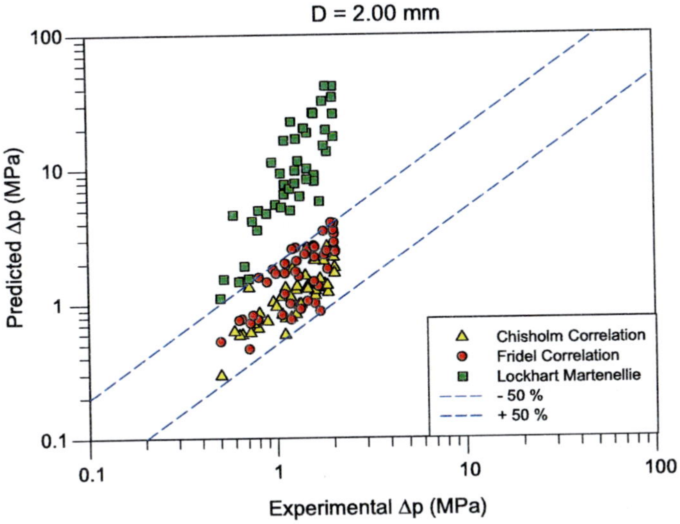


**Fig. 5** Predicted $\Delta p$ versus experimental $\Delta p$ for $D = 2.00$ mm

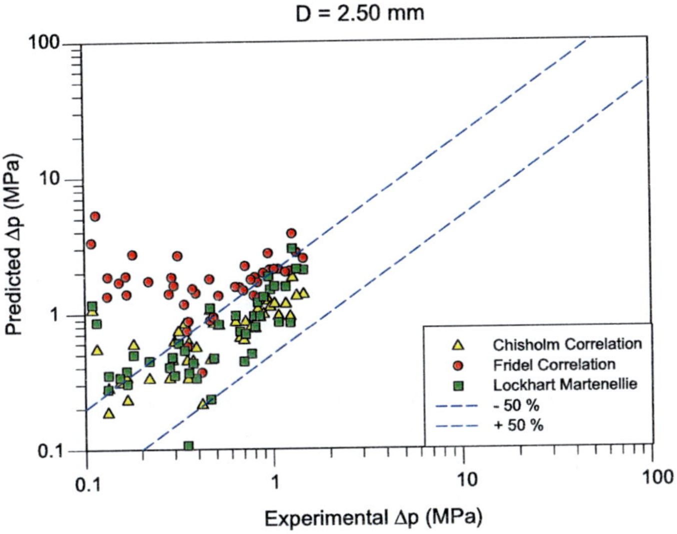


**Fig. 6** Predicted $\Delta p$ versus experimental $\Delta p$ for $D = 2.50$ mm

of data points are overpredicted, while 100% data points for Friedel correlation and Chisholm correlation are within ± 50% error band.

For 2.50 mm tubes, Fig. 6 presents a comparison of two-phase friction pressure drop data using the Lockhart–Martinelli, Friedel, and Chisholm correlations.

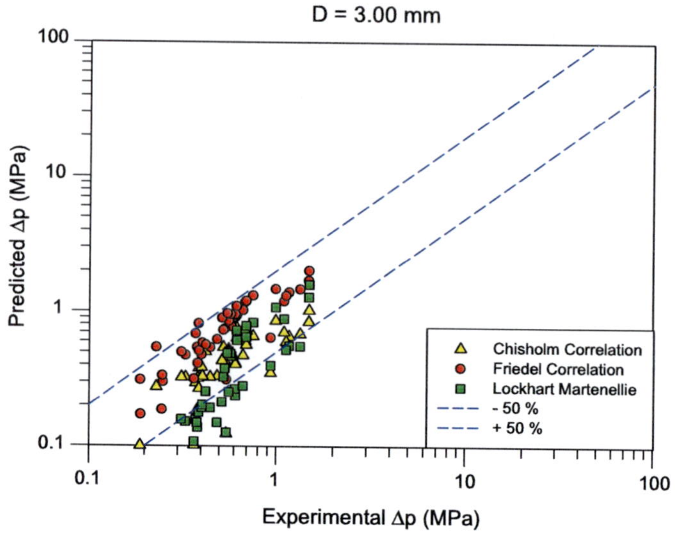


**Fig. 7** Predicted $\Delta p$ versus experimental $\Delta p$ for $D = 3.00$ mm

According to the Lockhart–Martinelli correlation, 80% of the variables are in the 50% error range, but the Friedel correlation reveals that 85% of the recorded points are in the 50% error group. Chisholm correlation is overstated since almost 75% of the variables are beyond the error range of 50%.

Figure 7 displays the Lockhart–Martinelli correlation, Friedel correlation, and Chisholm correlation for a 3.00 mm tube in comparison with experimental frictional pressure drop data, Friedel correlation, and Chisholm correlation. The Lockhart–Martinelli correlation prediction reveals that 37% of data sets are inside a 50% error range and rest of the observed points are underpredicted, but Friedel and Chisholm correlation predictions show that 96% of measured values are within a 50% error band.

The total mass flux spans from 307 to 7883 kg/m$^2$ s, and the two-phase flow quality ranges from 0.0005 to 0.1217. In the current investigation raised mass fluctuation, flow pressure drops are rather considerable, notwithstanding a slight decrease in flow quality. When the aggregate mass flux through the system is low but the flow quality is high, the pressure decrease observed is minor. It means that the pressure drop in the segment is primarily determined by the rate of water flow. The Reynolds number connects inertia forces with viscous forces. The interaction between the fluids intensifies as the mass instability of liquid and air increases, and inertial force gains dominance. As the diameter of the tubes decreases, the frictional pressure loss increases. According to the available literature, the surface tension force is a significant characteristic for macro channels and dominates below 5.0 mm. The two-phase flow literature for small diameter tubes generally agrees that as channel diameter decreases, surface tension has a greater impact on pressure drop. Surface tension has an impact on Weber and Bond numbers.

According to the overall assessment of experimental results, the increasing flow rate resulted in a rise in pressure drop. The presence of some uncertainties, such as fluid density, viscosity, surface tension, test section size, and so on, can explain the mean absolute error. Additionally, whenever the tube's scale is reduced, the pressure drop increases.

As previously stated, the predicting capacities of various prevailing equations and methods are evaluated by assessing them to the current trial data. The existing correlations used in the comparison were unable to effectively predict this experimental database. Therefore, a different connection is needed. The equation is expressed applying the constant $C$ established by Lockhart–Martinelli, which is simple and suitable for use. Hence, a new correlation is proposed for the constant $C$. The pressure drop multiplier depends on constant $C$. The constant C is anticipated to be dependent on the quality of two-phase flow ($X$), and dimensionless numbers Re, Fr, We, and Eu. The expression obtained for $C$ is as shown in the following equation.

$$C = 60 \times \left(\text{Eu}^{-1.2}\right) \times \left(\text{Re}^{-0.78}\right) \times \left(\text{Fr}^{-0.01}\right) \times \left(\text{We}^{0.61}\right) \times \left(\text{X}^{0.02}\right) \tag{1}$$

The recently built equation shows a nice fit with the investigational data. It is noted that nearly all the data is lying inside the ±35% error band as shown in Fig. 8.

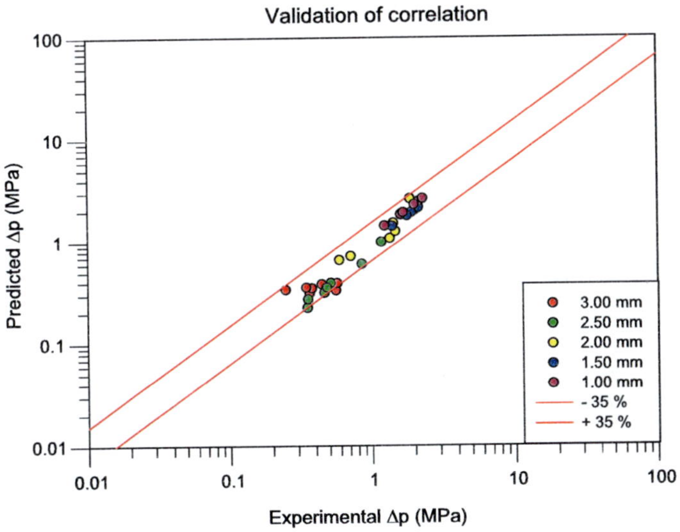


**Fig. 8** Predicted $\Delta p$ versus experimental $\Delta p$ for new correlation

## 4 Conclusions

Current experimental studies investigate the adiabatic flow of water and air through horizontal minichannels with different inner diameters. Three current interrelationships are investigated and evaluated with experimental data recorded in this study. The newly developed equation is compared to the experimental data. It appears to be a decent fit for the data. The conclusions of the current work are as follows:

1. The experimental setup was validated and found to be suitable for performing two-phase flow studies.
2. Increased frictional pressure drop is caused by an increase in flow rate and a decrease in tube internal diameter.
3. The previous correlations did not adequately predict the experimental frictional pressure drop.
4. The new correlation is constructed in such a way that it agrees well with experimental data. Most of the data was found to be within the 35% error range.

## References

1. Kandlikar SG (2003) Microchannels and minichannels—history, terminology, classification, and current research needs. In: First international conference on microchannels and minichannels, New York
2. Mehendale SS, Jacobi AM, Shah RK (2000) Fluid flow and heat transfer at micro- and meso-scales with application to heat exchanger design. Appl Mech Rev 53(7):175–193
3. Ribatski G (2013) A critical overview on the recent literature concerning flow boiling and two-phase flows inside microscale channels. Exp Heat Transf 26:198–246. https://doi.org/10.1080/08916152.2012.737189
4. Lockhart R, Martinelli TC (1949) Proposed correlation of data for equal two-phase, two-component flow in pipes. Chem Eng Prog 45:39–48
5. Akers WW, Deans HA, Crosser OK (1960) Condensation heat transfer within horizontal tube. Chem Eng Prog S Ser 55(29):171–176
6. Chisholm D (1972) Pressure gradients due to friction during the flow of evaporating two-phase mixtures in smooth tubes and channels. International Journal Heat Mass Transfer 16:347–358
7. Friedel L (1979) Improved friction pressure drop correlations for horizontal and vertical two-phase flow. In: Proceedings of European two-phase flow group meet, Ispra, Italy
8. Jung DS, Radermacher R (1989) Prediction of pressure drop during horizontal annular flow boiling of pure and mixed refrigerants. Int J Heat Mass Transf 32(12):2435–2446
9. Souza AL, Chato JC, Jabardo JMS, Wattelet JP, Panek JS, Christoffersen BR, Rhines N (1992) Pressure drop during two-phase flow of refrigerants in horizontal smooth tubes. Air Conditioning and Refrigeration Centre, College of Engineering. University of Illinois at Urbana-Champaign
10. Mishima K, Hibiki T (1996) Some characteristics of air-water two-phase flow in small diameter vertical tubes. Int J Multiph Flow 22(4):703–712
11. Tran TN, Chyu M-C, Wambsganss MW, France DM (2000) Two-phase pressure drop of refrigerants during flow boiling in small channels: an experimental investigation and correlation development. Int J Multiph Flow 26(11):1739–1754
12. Zhang M, Webb RL (2001) Correlation of two-phase friction for refrigerants in small-diameter tubes. Exp Thermal Fluid Sci 25(3–4):131–139

13. Chung PM-Y, Kawaji M (2004) The effect of channel diameter on adiabatic two-phase flow characteristics in microchannels. Int J Multiph Flow 30(7–8):735–761
14. Chen IY, Yang IY, Yang SS (2002) An empirical correlation for two-phase frictional performance in small diameter tubes. Int J Heat Mass Transfer 45:3667–3671
15. Muller-Steinhagen H, Heck K (1986) A simple friction pressure drop correlation for two-phase flow in pipes. Chem Eng Process 20:291–308
16. Kawahara A, Chung PM-Y, Kawaji M (2002) Investigation of two-phase flow pattern, void fraction and pressure drop in a microchannel. Int J Multiph Flow 28:1411–1435
17. Venkatesan M, Das SK, Balakrishnan AR (2011) Effect of diameter on two-phase pressure drop in narrow tubes. Exp Thermal Fluid Sci 35:531–541
18. Autee AT, Rao SS, Puli R et al (2012) Experimental study on two-phase pressure drop of air-water in small diameter tubes at horizontal orientation. Thermal Science (online-first). Issue 00
19. Hamad FA, Faraji F, Santim CGS et al (2016) Investigation of pressure drops in horizontal pipes with totally different diameters. International Journal of Multiphase Flow 120–129
20. Giri SV, Autee AT (2016) Experimental study on two-phase flow pressure drop by small diameter bends. Preservatives in Science 8:621–625
21. Lee HJ, Lee SY (2001) Pressure drop correlations for two-phase flow within horizontal rectangular channels with small heights. Int J Multiph Flow 27:783–796
22. Charthankar SB, Autee AT, Shrivastava R, Rao SS (2019) Experimental investigations on two-phase pressure drop in small diameter at horizontal orientation. In: Proceeding of the 25th national and 3rd international ISHMT-ASTFE heat and mass transfer conference (IHMTC-2019) pp 515–520

# Comparison of Hydrodynamic Characteristics of Porous and Solid Square Cylinder at Low Reynolds Number Using CFD

S. Seralathan, V. Hariram, M. Sathishkumar, and K. L. Vasudev

## Nomenclature

| | |
|---|---|
| Re | Reynolds number |
| $U$ | Velocity (m/s) |
| $D$ | Characteristic length (M) |
| $\nu$ | Kinematic viscosity ($m^2$/s) |
| Pr | Prandtl number |
| Da | Darcy number |
| $\theta$ | Temperature (K) |
| $x$ | Position (m) |
| $k$ | Permeability ($m^2$) |
| $\varepsilon$ | Porosity |
| $t$ | Time flow (s) |
| $C_l$ | Coefficient of Lift |
| $C_d$ | Coefficient of Drag |

S. Seralathan
Department of Aeronautical Engineering, Hindustan Institute of Technology and Science, Chennai, India

V. Hariram
Department of Mechanical Engineering, Hindustan Institute of Technology and Science, Chennai, India

M. Sathishkumar · K. L. Vasudev (✉)
Department of Mechanical Engineering, Amrita School of Engineering, Amrita Vishwa Vidyapeetham, Chennai, India
e-mail: kl_vasudev@ch.amrita.edu



N. L. Narasimhan et al. (eds.), *Recent Advances in Energy Technologies*, Lecture Notes in Mechanical Engineering, https://doi.org/10.1007/978-981-19-3467-4_17

## 1 Introduction

Flow past bluff bodies have evolved into a diverse field and deeply interesting study with many application areas, for example, cooling of electronics systems, flow inside nuclear reactor around nuclear rods, heat exchange inside a building, design of skyscrapers, flow past submarines, etc. Many times, it is required to look for an alternate design to replace these bluff bodies with shapes that are better in flow characteristics and heat transfer performance. One such design alternative is the use of porous material in place of complete solid. This is because a porous media offers larger contact with fluids which in return, results in better flow mixing enhancement along with heat transfer. The challenging part is to design a porous structure with equivalent strength and other mechanical characteristics as that of a solid structure.

Many studies have been carried out in the past by considering conditions like porosity, Darcy (Da) and Reynolds numbers (Re), flow and geometric characteristics. Hadim [1] studied steady-state heat transfer characteristics of a channel, by placing partially and fully layers of porous material above the heat sources and placing non-porous material wherever else in the channel. He concluded that the variation in convection characteristics in the channel with partial and fully covered porous material is negligible. Fu et al. [2] explored characteristics of steady-state flow and thermal convection around the block, where the heat source is placed at only one wall of the channel. Sung et al. [3] investigated where to place and what should be the size of the block with the porosity to be mounted over the source of heat onto a single wall of the channel to improve the heat convection characteristics. Recent studies [4–6] demonstrated the heat exchange inside a channel with more than one source of heat fixed to the wall of the channel and covered by a block with porosity. Huang et al. [6] concluded that a significant increase in heat exchange can be achieved by employing blocks with porosity at multiple locations on the wall of the channel.

Jue et al. [7] employed a semi-implicit projection finite element method for vortex flow. He found a solution to the 2D unsteady flow passing through a square cylinder with porosity and concluded that the vortex shedding in the flow occurs late and for a shorter duration. Chen et al. [8] studied numerically the interface stress jump conditions in the flow around a cylinder with porosity and a square cross section and concluded that above-mentioned condition causes flow instability. Dhiman et al. [9] carried out a two-dimensional numerical study to find the influence of Prandtl number and cross-buoyancy on heat exchange and flow characteristics of an isothermal cylinder with a square cross section in a channel. They concluded that the effect of the Richardson number on the drag coefficient is minimal when compared with that on the lift coefficient. Wu et al. [10] numerically studied the characteristics of a heated porous cylinder with a square cross section in unsteady flow. The parameters studied are Da, Re and porosity on heat exchange performance.

Nazari et al. [11] investigated the use of the lattice Boltzmann method (LBM) as a potential solution to solve fluid dynamics problems in porous media. They performed 2D channel simulations to validate their results. They proposed two correlations for a random and regular arrangement of obstacle between Re, Nusselt number

(Nu), blockage ratio, power index and porosity. Liu et al. [12] developed a two-dimensional flow field and employed as temperature field to study the heat exchange through porous media. They concluded that the proposed model is outperforming the Lattice Bhatnagar-Gross-Krook method. Mahdhaoui et al. [13] studied 2D laminar flow passing through a porous cylinder with a square cross section. The porous medium is described by the classic Darcy-Brinkman-Forcheimer model. Similarly, the forced convection model has been used for the flow and heat transfers in the channel. Casanova et. al. [14] analyzed active mixing mechanism to mix two fluids with different densities and flowing with Re = 1 using computational fluid dynamics (CFD). The mechanism involves a cylinder with a square cross section located at the center of a straight channel. He also investigated the cost involved in mixing and found that the proposed method is 1.7 times less expensive.

From the literature, it is evident that CFD is used intensively in all most all the works, where numerical studies were carried out. The reason behind using CFD to analyze the flow field around the cylinder in the channel is that CFD considers the local fairing effects. The recent advancements in CFD helped a lot in the study of these types of problems and help to replace few experiments at the preliminary stage of the design. Technically conducting experiments are possible, but highly expensive and time taking. In many cases, the basic economics of design does not allow us that freedom [15]. Hence, CFD is preferred in this study, to analyze the flow characteristics of a porous cylinder with a square cross section. It is also evident from the literature that all the attempts so far were done using 2D simulations. In the present work, we could run 3D simulations due to the recent advancements in CFD. The computations were carried out in ANSYS Fluent. Meshing and 3D modeling were carried out in ICEM CFD.

## 2 Numerical Methodology

Flow considered in the present study has been assumed to be three-dimensional, laminar and unsteady. Mass and momentum conservation equations considered in the present study are given as,

$$\nabla \cdot \vec{V} = 0 \tag{1}$$

$$\frac{\partial \vec{V}}{\partial t} + \vec{V} \cdot \nabla \vec{V} = -\frac{\nabla P}{\rho} + \nu \nabla^2 \vec{V} \tag{2}$$

where $\tilde{V}$ is the velocity vector, and $\rho$ are density and $\upsilon$ kinematic viscosity of the fluid, respectively, and $P$ is the pressure. Fluid properties corresponding to water have been considered in the present study. Finite volume-based commercial computational fluid

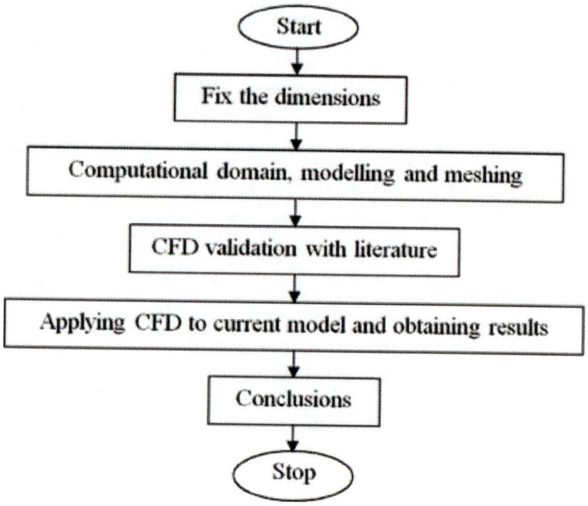


**Fig. 1** Flow chart of the current work

dynamics (CFD) package ANSYS Fluent 15.0 has been used in the present study. The detailed plan of action has been reported in Fig. 1 in the form of a flowchart.

## 2.1 Geometric Modeling

The three-dimensional computational domain has been created using ANSYS ICEM CFD software. The velocity inlet is located ahead of the cylinder at a distance of 7.5 times that of the length of the cylinder ($D$), where $D = 25$ mm has been considered in this work. The pressure outlet is located behind the cylinder at a distance of 20 times ($D$), making the total length of the domain to be 27.5 times ($D$). The width of the computational domain is 6 times ($D$) and similarly, the height of the domain is 15 times ($D$). The detailed view of the computational domain in both side view and top view is presented in Fig. 2.

The constants and solver parameters used in this study are reported in Table 1.

## 2.2 Boundary Conditions

The following conditions are set on the fluid domain boundaries:

(a) On the upstream inlet, velocity perpendicular to the boundary is specified.
(b) The downstream outlet is imposed with gauge pressure (i.e., $P = 0$), and the k and ω gradients are set to 0.
(c) All other boundaries of the domain are set as '0' shear stress.
(d) The surface of the cylinder is specified as no-slip.

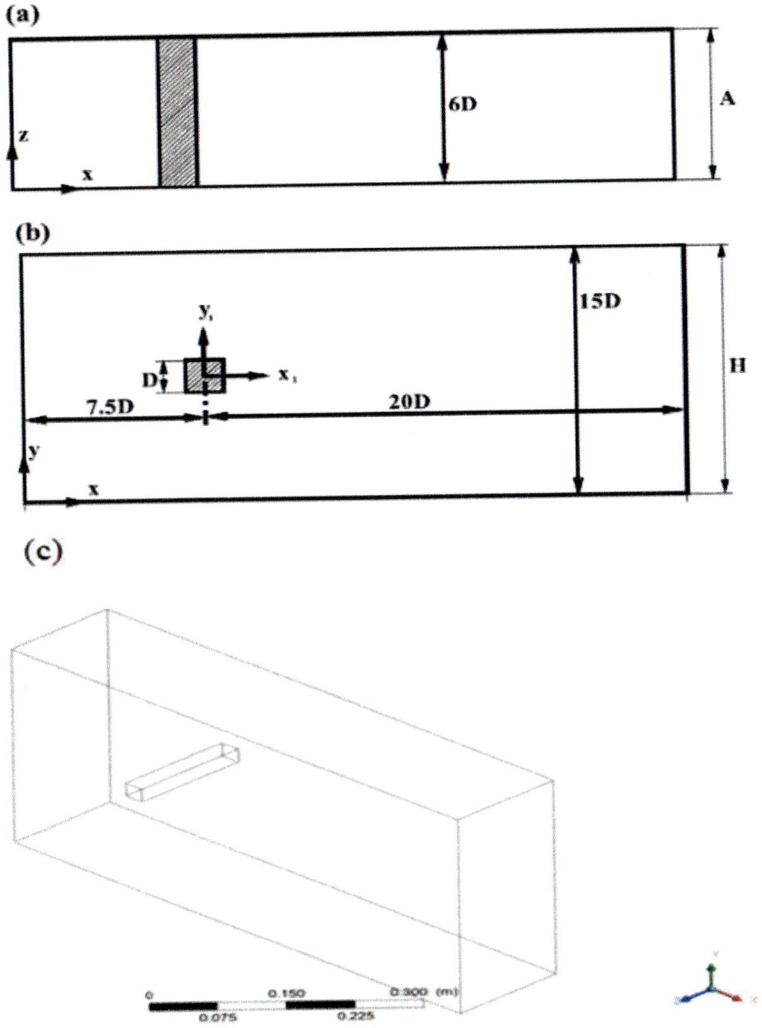


**Fig. 2** Computational domain: (a) side(x-z) view, (b) top(x-y) view, (c) 3d view

## 3 Results and Discussions

### *3.1 CFD Validation*

To validate the CFD methodology, initially, a circular cylinder of diameter $D = 1$ m has been modeled as reported in Silva et al. [16]. The computational domain has been modeled with the following dimensions ($L \times B \times H = 20\text{D} \times 7.5\text{D} \times 7.5\text{D}$).

**Table 1** Constants and solver parameters used in the study

| | |
|---|---|
| Discretization | |
| Pressure | Body force weighted |
| Momentum; turbulent kinetic energy and dissipation rate | Second order upwind |
| *Fluid properties* | |
| Density | 1000 kg/m$^3$ |
| Kinematic viscosity | $10^{-6}$ m$^2$/s |
| *Roughness parameters* | |
| Roughness constant ($C_S$) | 0.5 |
| Roughness height ($K_S$) | 0 |
| $\varepsilon = \text{Porosity}\begin{cases} 1, & \textit{outside the porous cylinder} \\ 0 < \varepsilon < 1, & \textit{inside the porous cylinder} \end{cases}$ | |
| *Turbulence criteria at inlet* | |
| Turbulent intensity ($T$) | 5% |
| Turbulent length scale ($l_T$) | 0.001$L$ |
| *Under relaxation factors* | |
| Pressure | 0.3 |
| Density | 1 |
| Body forces | 1 |
| Momentum | 0.7 |
| Turbulent viscosity | 1 |
| Turbulent kinetic energy | 0.8 |
| Turbulent dissipation rate | 0.8 |

A structured mesh with 407,850 elements and 431,782 nodes has been generated inside the computational domain. Figure 3 shows the cylinder with circular cross section modeled and the generated mesh around the cylinder and in the computational domain. For validating purposes, the conditions are mentioned in [16]; we recreated with Re = 80, and the coefficient of drag and lift was compared. Figure 4 represents the plot from Silva et al. [16] for the coefficient of lift. Figure 4 reports the lift coefficient from the current simulation. As can be noticed on the ordinate, the range in which the lift coefficient in both the figures fluctuates within the same range. Similarly, Fig. 5 shows the plot from Silva et al. [16] for the coefficient of drag and Fig. 5 presents the drag coefficient from the simulation. The drag coefficient is converging to ~1.4 in both cases as can be read from the figures. Similarly, the lift coefficient is converging close to 0.0009.

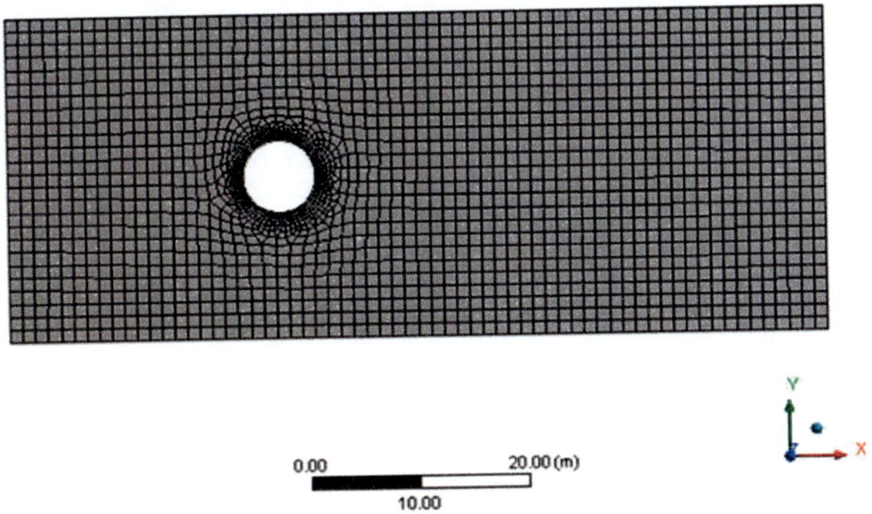


Fig. 3 Mesh inside the computational domain with circular cylinder

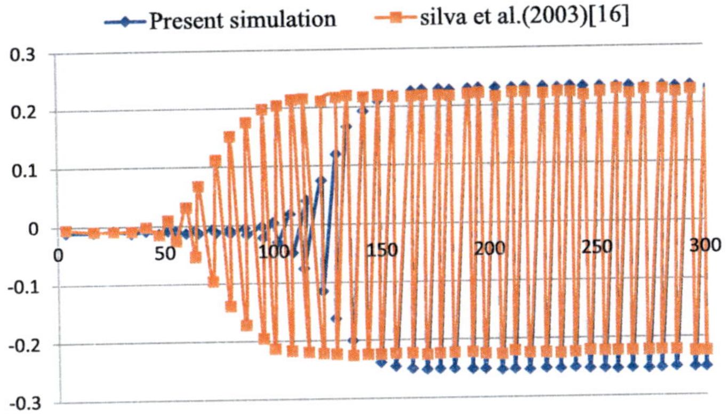


Fig. 4 Comparison of coefficient of lift with Silva et al. [16]

### 3.2 *Comparison of Drag Coefficients of Porous and Non-Porous Cylinders with a Square Cross Section*

The computations for both porous and non-porous cylinders with square cross sections have been carried out for different Re, namely 170, 185, 200 and 250. Similarly, discretization of the computational domain and the boundary conditions as mentioned in Sect. 2 is used. The number of nodes and elements in the domain is specified as mentioned in Sect. 3.1. Grid independence study for a solid square

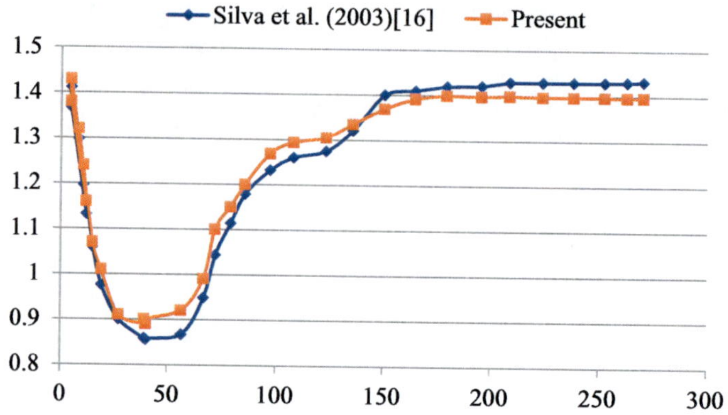


**Fig. 5** Comparison of coefficient of drag with Silva et al. [16]

cylinder has been carried out, and the results are reported in Table 2. Initially, approximately 140,000 elements were considered and further with a rise of $\sqrt{2}$ times in the number of elements is carried out. From Table 3, it is evident that with 407,850 elements, the difference in the drag value is less than 1% with 586,720 elements. Hence, we conclude the grid independence study for our problem with 407,850 elements

The drag coefficients of solid and porous square cylinders for different Re are reported in Fig. 6. Similarly, in Fig. 7, lift coefficients of the two cylinders are presented. It is evident from Fig. 6 that for both the cylinders, the drag coefficient is decreasing with an increase in Re. As can be noticed from the figure, drag coefficient

**Table 2** Grid independence study of solid square cylinder

| No of elements | *Re* | Coefficient of drag, $C_D$ | Error % | Coefficient of lift, $C_L$ | Error % |
|---|---|---|---|---|---|
| 144196 | 200 | 0.10235 | 20.6155 | 0.0005 | 0.3 |
| 203950 | 200 | 0.12345 | 12.5476 | 0.00065 | 0.153846 |
| 288450 | 200 | 0.13894 | 9.5509 | 0.00075 | 0.186667 |
| 407850 | 200 | 0.15221 | 0.0854 | 0.00089 | 0.011236 |
| 586720 | 200 | 0.15234 | | 0.0009 | |

**Table 3** Comparison of lift and drag coefficients of solid and porous square cylinder for different Reynolds numbers

| *Re* | Solid, $C_L$ | Porous $C_L$ | Difference % | Solid, $C_D$ | Porous $C_D$ | Difference % |
|---|---|---|---|---|---|---|
| 170 | 0.0009 | 0.00075 | 20 | 0.18565 | 0.09152 | 50.70 |
| 185 | 0.00089 | 0.00074 | 20 | 0.17125 | 0.08725 | 49.05 |
| 200 | 0.0008 | 0.00066 | 21 | 0.15199 | 0.07995 | 47.39 |
| 250 | 0.0007 | 0.00058 | 20 | 0.13952 | 0.07543 | 45.93 |

values of porous media are 0.09152, 0.08725, 0.07995 and 0.07543 for Re 170, 185, 200 and 250, respectively. Similarly, for non-porous cylinder, the values are 0.18565, 0.17125, 0.15199 and 0.13952, respectively. The reduction in the drag values is 50%, 48%, 47% and 49%, respectively. The values for porous media are almost reduced by 50%. This proves the ability of porous media as a potential replacement for the solid cylinder.

Figure 8 presents the velocity contours of both porous and non-porous cylinders with a square cross sections for Re = 170, 185, 200 and 250, respectively. Figure 8a, c, e, g shows the velocity contours of solid cylinder with square cross section for Re = 170, 185, 200 and 250, respectively. Similarly, Fig. 8b, d, f, h presents the velocity contours of the cylinder with porous media. It is clear from the figure that relatively higher velocity values are observed in the top and bottom sides of the cylinder, and relatively, lower values of velocity values are observed behind the cylinder for both cases. From Fig. 8, it is evident that in the wake of the porous cylinder, the magnitude

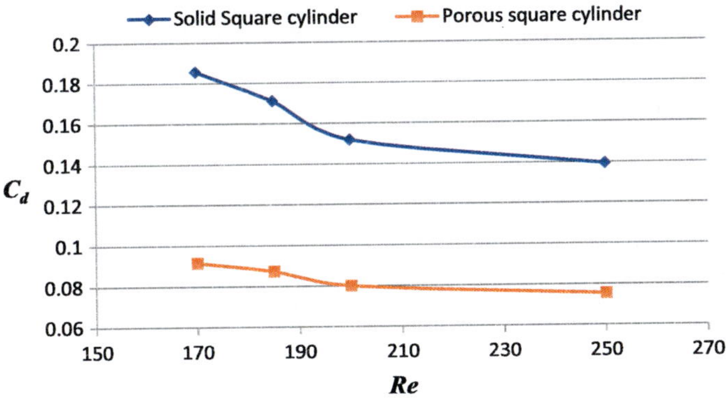


**Fig. 6** Comparison of coefficient of drag of solid and porous square cylinders

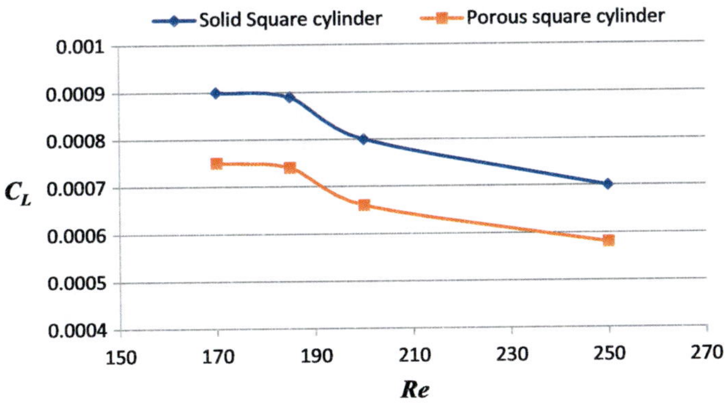


**Fig. 7** Comparison of coefficient of lift of solid and porous square cylinders

of the velocities is higher in comparison with a solid cylinder. This proves that porous cylinder offers better mixing of fluid flow. Figure 9 shows the pressure contours for both cylinders. The maximum value of pressure is noticed at the stagnation points in both the cylinder this should happen as the magnitude of velocity at the stagnation point is ‘0’ the pressure should be maximum.

## 4 Conclusion

A comparison of flow characteristics of square cylinders with and without porous media is presented in this work. It is observed that cylinder with porous media outperforms the cylinder with non-porous media both in flow mixing and minimum drag criteria. This is happening because the porous media enhances the ability to pass the fluid through the media. The drag coefficient of both the cylinders reduced with an increase in *Re*. This is evident because the range of *Re* considered in this work fall under laminar flow characteristics. The reduction in drag between the two cylinders is noticed as close to 50% across the range of the *Re* considered in the current work.

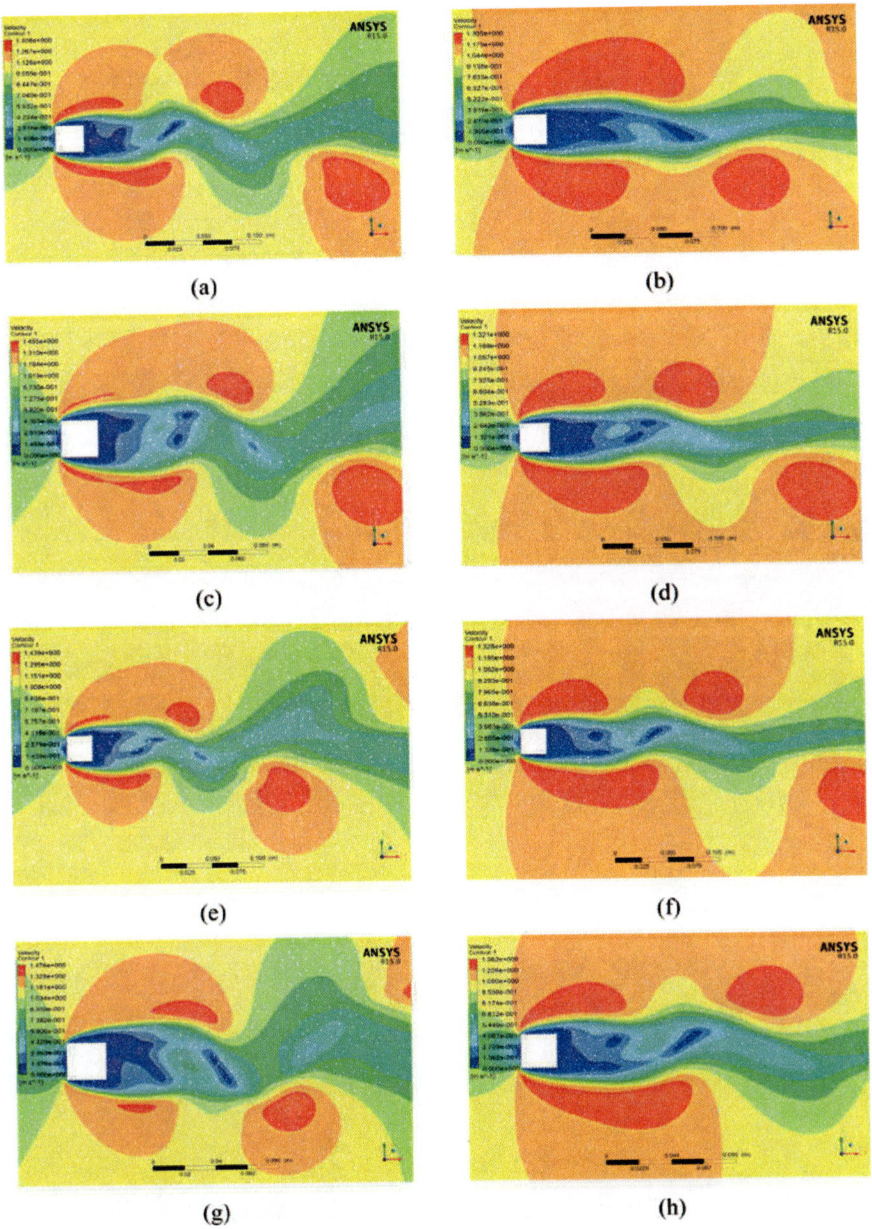

**Fig. 8** Velocity contours at Re = 170, 185, 200, 250 respectively (a), (c), (e), (g) for solid cylinder; (b), (d), (f) and (h) for porous cylinder

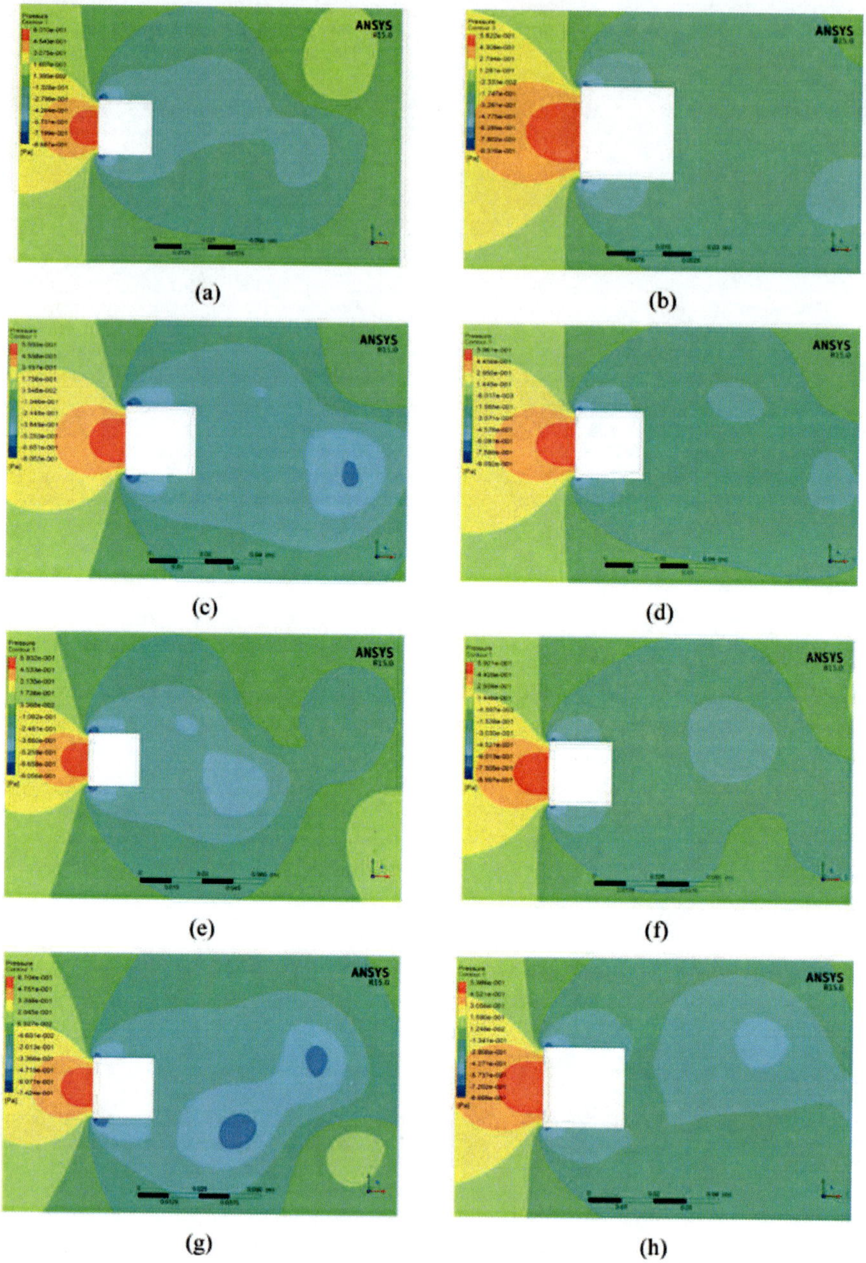


**Fig. 9** Pressure contours at Re = 170, 185, 200, 250 respectively (a), (c), (e), (g) for solid cylinder; (b), (d), (f) and (h) for porous cylinder

## References


1. Hadim A (1994) Forced convection in porous channel with localized heat sources. J Heat Transfer 116:465–472
2. Fu WS, Huang HC, Liou WY (1996) Thermal enhancement in laminar channel flow with a porous block. Int J Heat Mass Transfer 39(10):2165–2175
3. Sung HJ, Kim SY, Hyun JM (1995) Forced convection from an isolated heat source in a channel with porous medium. Int J Heat Fluid Flow 16:527–535
4. Huang PC, Vafai K (1994) Analysis of forced-convection enhancement in a channel using porous blocks. J Thermophys Heat Transfer 8:563–573
5. Chikh S, Boumedien A, Bouhadef K, Lauriat G (1998) Analysis of fluid and heat transfer in a channel with intermittent heated porous blocks. Heat Mass Transfer 33:405–413
6. Huang PC, Yang CF, Hwang JJ, Chiu MT (2005) Enhancement of forced-convection cooling of multiple heated blocks in a channel using porous covers. Int J Heat Mass Transfer 48:647–664
7. Jue TC (2004) Numerical analysis of vortex shedding behind a porous square cylinder. Int J Numer Methods Heat Fluid Flow 14:649–663
8. Chen XB, Yu P, Winoto SH, Low HT (2008) Numerical analysis for the flow past a porous square cylinder based on the stress-jump interfacial-conditions. Int J Numer Methods Heat Fluid Flow 18(5):635–655
9. Dhiman AK, Chhabra RP, Eswaran V (2008) Steady mixed convection across a confined square cylinder. Int Commun Heat Mass Transfer 35:47–55
10. Wu H-W, Wang R-H (2010) Convective heat transfer over a heated square porous cylinder in a channel. Int J Heat Mass Transf 53(9–10):1927–1937
11. Nazari M, Mohebbi R, Kayhani MH (2014) Power-law fluid flow and heat transfer in a channel with a built-in porous square cylinder: Lattice Boltzmann simulation. J Nonnewton Fluid Mech 204:38–49
12. Liu Q, He Y-L, Li Q, Tao W-Q (2014) A multiple-relaxation-time lattice Boltzmann model for convection heat transfer in porous media. Int J Heat Mass Transf 73:761–775
13. Mahdhaoui H, Chesneau X, Laatar AH (2017) Numerical simulation of flow through a porous square cylinder. Energy Procedia 139:785–790
14. Ortega-Casanova J (2017) CFD study on mixing enhancement in a channel at a low Reynolds number by pitching a square cylinder. Comput Fluids 145:141–152
15. Vasudev KL, Sharma R, Bhattacharyya SK (2016) A modular and integrated optimization model for UVs. Defense Sci J 66(1):71–80
16. Lima E Silva ALF, Silveira-Neto A, Damasceno JJR (2003) Numerical simulation of two-dimensional flows over a circular cylinder using the immersed boundary method. J Comput Phys 189(2), 351–370

# Energy Interactions in Built Environment

# Effect of Building Orientation, Window Glazing, and Shading Techniques on Energy Performance and Occupant Comfort for a Building in Warm-Humid Climate

S. Karthick, Adithya Sivakumar, S. Bhanu Chander, S. Bharathwaj, and B. V. V. Nagendra Kumar

## Nomenclature

| | |
|---|---|
| $\varepsilon$ | Emissivity |
| $\rho$ | Reflectivity |
| $U$ | Overall heat transfer coefficient (W/m$^2$ K) |
| $\dot{Q}_{\text{HG}}$ | Average Heat Gain Rate (kW) |
| $\dot{Q}_{\text{SR}}$ | Average Solar Radiation Rate (kW) |
| $\dot{Q}_{\text{CL}}$ | Ideal supply air maximum cooling load (kW) |

## 1 Introduction

The last two decades have witnessed an increase in global surface temperatures, which has led to a surge in energy demand especially in developing countries [1]. The data presented in Energy statistics 2020 reports that, in India, 32.6% of electricity is consumed by the domestic and commercial sectors [2]. These primarily include residential, commercial, and public buildings which are the second largest electricity consumer. The three primary applications that consume energy in buildings are the basic electric devices (Tube lights, Refrigerators, Entertainment units, etc.), domestic geyser, and HVAC (Heating, Ventilating, and Air-Conditioning) systems for space heating/cooling and ventilation [3].

The major portion of energy consumption in commercial buildings arise from the usage of HVAC systems [4]. This shows the importance given to maintaining a

S. Karthick (✉) · A. Sivakumar · S. Bhanu Chander · S. Bharathwaj · B. V. V. N. Kumar
Department of Mechanical Engineering, Amrita School of Engineering, Amrita Vishwa Vidyapeetham, Coimbatore, India
e-mail: s_karthick@cb.amrita.edu



N. L. Narasimhan et al. (eds.), *Recent Advances in Energy Technologies*, Lecture Notes in Mechanical Engineering, https://doi.org/10.1007/978-981-19-3467-4_18

thermally comfortable environment. The American Society of Heating, Refrigerating and Air conditioning Engineers (ASHRAE) defines thermal comfort in indoor spaces as the state of mind which expresses satisfaction with the environment [5]. Thermal comfort in a confined space depends on ambient factors (air temperature, humidity, and air velocity) and personal factors (physical activity and clothing). To achieve thermal comfort in a confined space focusing on these parameters is necessary [6].

Predicted Mean Vote (PMV) and Percentage of People Dissatisfied (PPD) are indexes, which consider both environmental and personal factors for expressing the state of thermal comfort condition inside a confined space. PMV shows on a seven-point scale from $-3$ to $+3$, the thermal sensation of a large group of people corresponding to the categories "cold," "cool," "slightly cool," "neutral," "slightly warm," "warm," and "hot." The percentage of people thermally discontented is established by PPD which is determined from PMV [5]. The acceptable ranges for PMV and PPD according to specifications of ASHRAE standard 55–2020 are between $-0.5$ and $+0.5$ and below 20%, respectively. The PPD and PMV are interrelated under the assumption that people voting $-3$, $-2$, $+2$, or $+3$ are dissatisfied.

In today's world, there is an increasing realization on the rapid depletion of non-renewable energy sources used for electricity generation. The environmental impact due to burning coal is colossal in the radio-ecological, human health, and global warming viewpoints. Thus, there is a need for developing sustainable environments by creating new building designs or modifying the existing buildings, to have reduced building energy consumption and carbon dioxide emission, and equally achieve required thermal comfort levels [7].

To mitigate these concerns, two types of cooling systems namely active and passive systems are available [3]. The active cooling systems require an external power source to provide thermal comfort in the confined space. In some cases, efficient energy management is achieved using active systems in smart buildings, by means of various algorithms, which are developed or modified to obtain net zero energy consumption [8–11]. The energy required to power active systems is not completely green. Moreover, shortage of conventional energy sources along with the rising energy cost adds burden to the usage of active cooling methods.

An alternate way to achieve thermal comfort in a confined space is to use passive cooling systems [12]. Adopting these systems in a confined space reduces the temperature without the need for electric power. These systems use renewable energy sources like wind and sun to provide cooling and heating, respectively [13]. This reduces the cooling load of a confined space and diminishes the capacity of the installed HVAC unit also reducing the period of usage [14]. Achieving thermal comfort using passive cooling techniques does not require external power input, thereby making it sustainable. Considering the advantages, this study will focus on passive systems. Multiple passive cooling systems have been tried and tested over the years, and some of them have proven successful in reducing cooling load. Passive systems such as building orientation [15], window glazing [16], thermal insulation [13, 17, 18], shading devices [19], greening systems [20], evaporative cooling [21], and much more are majorly used. The passive techniques analyzed in this paper are discussed below.

One of the primary passive design strategies, namely building orientation, has a sizeable impact on the energy performance. It is defined as layout of the site on the horizontal plane such that heat gain into the indoor is minimum. Conventionally, 0° is assigned to North, 90° to East, 180° to South, and 270° to West. The performance of passive strategies is examined by analyzing average incident solar radiation for different building orientations by rotating it between 0° and 360° [22]. Though building envelop is the major contributor to indoor heat gain, orientation of facades also plays an important role. A simulation study performed in tropical climate showed building with windows facing south is likely to give better performance [23].

Windows dissipate large amount of energy to the indoor in comparison with other building fabrics. This might be due the large difference in the U-value between windows and other building elements. To prove windows transfer a large proportion of energy in a building, a review conducted in a house with a WWR of 30 indicates that the total energy loss from the building envelope that occurs through windows is about 60%. Furthermore, in commercial buildings, the extensive usage of glass enclosures has caused the rise in heat gain and decline in thermal comfort of indoor space [24]. Hence, there is a necessity to search for alternate window glass materials for reducing the cooling load requirements of the buildings. Glazing technique is one of the solutions to this problem, in which a layer of glass or transparent material is coated on the window of a confined space. Glazing can be classified as static and dynamic. Static glazing refers to glazing which has fixed properties (thermal and optical) while in dynamic glazing, the properties vary within a range for fixed glazing thickness [25]. This study extensively deals with static glazing techniques.

The performance of static glazing is measured in terms of thermal and optical properties like Overall heat transfer coefficient (U), Solar Heat Gain Coefficient (SHGC), Visible Transmittance (VT) [25, 26], and Light to Solar Gain (LSG). The rate of heat transfer through window surface is determined using overall heat transfer coefficient. Glazing with lower U will be more energy efficient. SHGC is the portion of solar radiation passing through a window either transmitted directly and/or absorbed which is released as heat to the indoors. Lower SHGC of the glazing is more favorable as it transmits less solar heat and has greater shading ability. The amount of visible light transmitted through the window is an optical property known as VT. Window glazing with high VT allows more daylight in a space which can make up for electric lighting and its accompanying cooling loads [26, 27]. LSG is the ratio of VT to SHGC. Larger LSG of a glazing will reduce the solar heat gain with minimal reduction in VT.

Additionally, it is important to implement measures that can be adopted to reduce heat gain/loss through windows without changing the amount of daylight harvested. Daylighting is an essential aspect of indoor environmental comfort which refers to illumination of building using natural light from the sun. It is measured in terms of illuminance (lux). To estimate the range in which daylight can be beneficial, "Useful Daylight Illuminance" (UDI) parameter is used. UDI is defined by Energy Conservation Building Code (ECBC, 2017) [28] as illumination of daylight in a work plane which is preferable by occupants. Thus, providing an equipoise between the maximum daylight harvested and discomforts due to glare and overheating is the key to maintaining visually comfortable space [29]. Window shading is one of the suitable approaches to control undesirable ingress of heat to the indoor space [30] and maintain required visual comfort levels. There are two major types of shading techniques namely external and internal shading. If shading system is installed in the exterior of the building, the solar radiation is efficiently blocked before entering the indoor through fenestrations. Contrarily, the usage of interior shading devices during summer period absorbs heat and emits it to the indoor space, which increases cooling energy requirement [31].

In the literature, several investigations have been conducted to assess the performance of window glazing and shading systems. An investigation was carried out on the impact of window glazing and shading of high-rise building in temperate climate of Netherland using DesignBuilder. The results showed glazing and shading reduces annual energy consumption by 7.48% and 3.13%, respectively [32]. A study performed using DesignBuilder emphasized the energy saving and economics of including external shading in a hotel building in hot-humid climate. The proposed shading system engendered a reduction of 20.5% to the annual building energy consumption [33]. In addition to energy aspect, studies focusing on the indoor comfort levels have been reported. The simulation performed for a building with large areas of glass facade in hot-humid climate region of Taiwan predicted the impact of eight glazing types, four glazing areas, and five types of overhangs in terms of discomfort and overheating. The outcomes of PMV-based comfort control illustrated the aforementioned systems effectively contribute toward thermal comfort [34]. According to a study that evaluated the impact on thermal comfort using solar shading in tropical climate of Uganda, usage of shading showed considerable influence during periods of high temperature in the year, by diminishing the risk of overheating up to 52% [35]. Even though the results suggest significant reduction in solar heat gain using shading, it is not effective in maintaining thermal comfort requirement in tropical climate. A multi-objective investigation on windows and shading system for an office building for different climate regions of China has been attempted [36]. The results highlight the importance of windows and shading system in improving indoor thermal comfort. From the above study, it is apparent that window glazing and shading system are effective in reducing the energy consumption of the building and improving thermal comfort levels.

The present work aims to address the research gap in investigating the combined effect of building orientation, window glazing, and shading on indoor comfort (of a computer laboratory application) for tropical climate zone according to ASHRAE standard 55–2020. This is a simulation-based study that determines the reduced capacity of Air-Conditioning Unit (ACU), occupant comfort, and annual cooling energy savings obtained in implementing these passive strategies. The paper is structured as follows: Sect. 2 describes the assumptions, methodology, and simulation procedure underlying in the study. Section 3 provides detailed results and discussion of passive cooling techniques and its impact on thermal comfort. Section 4 comes up with conclusions and offers prospective ideas.

## 2 Methodology

The simulation is carried out using DesignBuilder Version 6.1.8.021 and energy simulation software—EnergyPlus version 8.9. The former was used for modeling the building, material selections, and Computational Fluid Dynamics (CFD) simulations. To assess the passive cooling techniques, the dynamic simulation model of the building was carried out within EnergyPlus simulation environment. The assumptions made in the theoretical study include the following: Windows are positioned at the middle of the wall and facades in the air-conditioned building remain closed.

A building of 6 m × 7 m × 3.5 m dimensions is modeled as shown in Fig. 1a). The location of the study is Chennai, Meenambakkam (13.0827° latitude, 80.2707° longitude). The building was modeled as a computer laboratory with six occupants, computer systems, and desks (Fig. 1b). Additionally, four LED lights and ACUs were modeled.

The external walls and roof (Fig. 2) are modeled such that U-value complies with ECBC guidelines. The external walls consist of Polyurethane foam (50 mm), red brick (230 mm), and cement plaster (10 mm) on either side. The roof consists of Reinforced Concrete slab (150 mm), Polyurethane foam (50 mm) covered by cement plaster (10 mm) on either side of roof structure. The U of external walls and roof surface is 0.4 W/m2 K and 0.23 W/$m^2$ K, respectively. Window to Wall Ratio (WWR) represents the fenestration area as a proportion of external wall area. The WWR for the designed model is 30%, which is within the limits prescribed by ECBC. The occupancy and air-conditioning schedule are chosen from 8:00 AM to 6:00 PM. The occupant's internal heat source is 105 W/person. The rated power for each lighting and computer equipment is 15 W and 200 W, respectively. ACU used in the study is four-pipe fan coil unit air-cooled chiller which maintains the cooling set point temperature for the indoor space at 25 °C with supply air louver angle of 20°. In the air-conditioned space, air infiltration rate is 0.0106 $m^3$/s.

Climate characteristics of the chosen location from EnergyPlus weather data file are shown in Fig. 3. The graph depicts the outdoor Dry Bulb Temperature (DBT), Relative Humidity (RH), Wind speed, and Direct Solar Radiation Rate—SRR

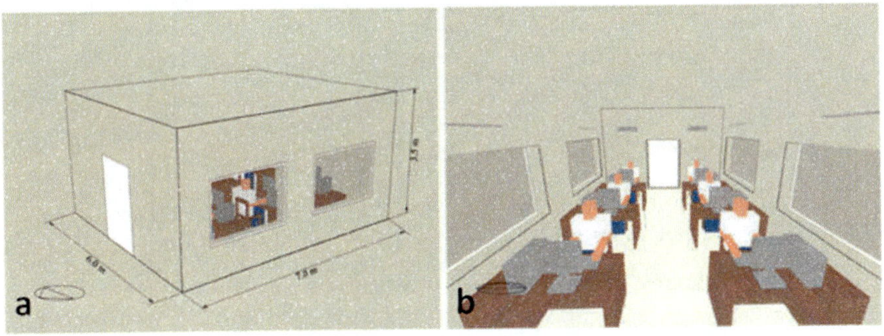


**Fig. 1** **a** Exterior model of reference building, **b** Interior model of reference building

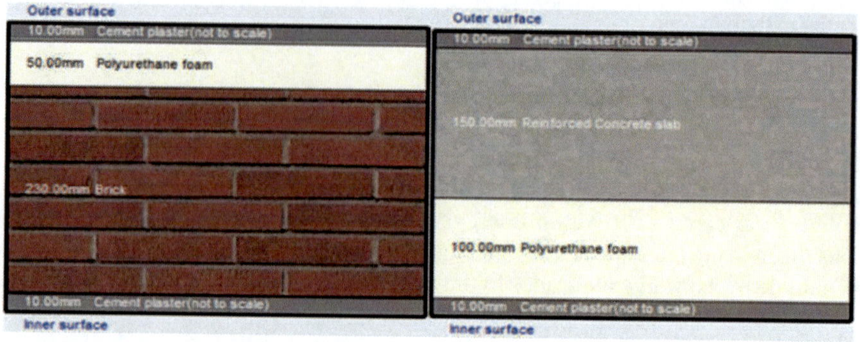


**Fig. 2** Cross-section of external wall and roof according to ECBC compliance

(measured in hecto-Watt per area—$hW/m^2$) through the year for Chennai region. The monthly average DBT from the climate data indicates the months of April, May, and June to have the highest temperatures throughout the year.

Therefore, subsequent EnergyPlus simulations in the study are performed for these three summer months to determine the optimal choice of window glazing and shading systems. Building orientation is analyzed for a year, as it is required to minimize incident solar radiation round the year. To assess the chosen passive cooling systems, a reference building is established. This building is orientated at 0° and does not include any passive systems, while ACUs are installed. The parameters of interest for this study include the following: (a) thermal properties comprising average window Heat Gain Rate—HGR ($\dot{Q}_{HG}$), average window SRR ($\dot{Q}_{SR}$), and ideal supply air maximum cooling load ($\dot{Q}_{CL}$), (b) visual property—UDI, and (c) annual cooling energy savings percent. Moreover, Simple Payback Period (SPP) of passive techniques is estimated by dividing the total initial investment with annual cost savings. The indoor visual comfort is ensured by estimating the potential daylit time within the UDI range of 100–2000 lx according to ECBC. The UDI should meet

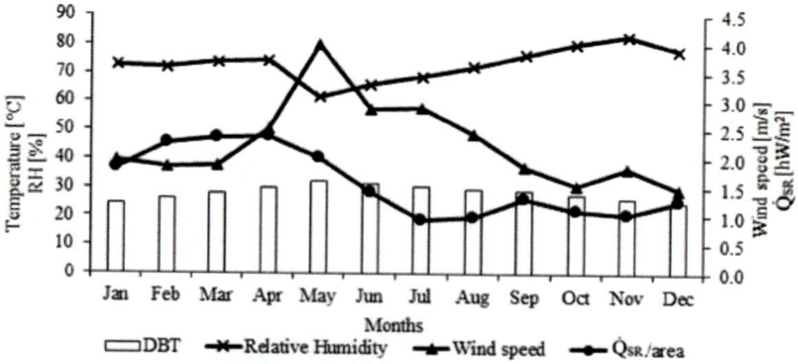


**Fig. 3** Monthly average outdoor air-dry bulb temperature, relative humidity, wind speed, and direct SRR/area for Chennai, Meenambakkam location

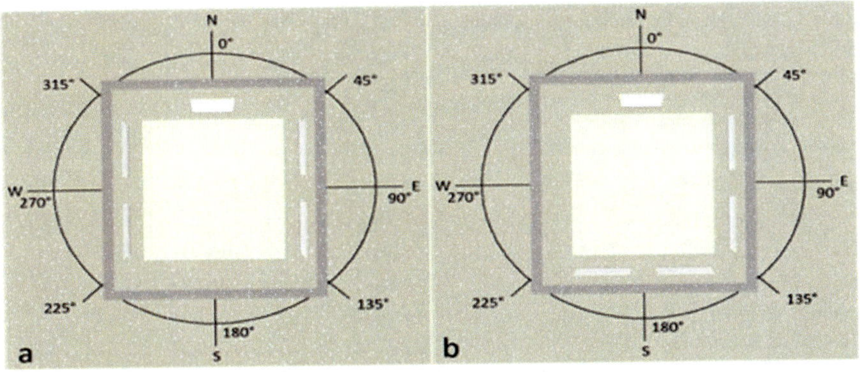


**Fig. 4** **a** Orientation of building for windows positioned on opposite walls. **b** Orientation of building for windows positioned on adjacent walls

90% of the potential daylight time (considered from 9:00 AM IST to 4:00 PM IST), which sums up to 2920 h in a year. Besides UDI, the daylight discomfort glare also has an effect on visual comfort of the occupants especially in computer laboratory. From the EnergyPlus documentation, maximum allowable discomfort glare index recommended for laboratory is 22.

Firstly, building orientation is analyzed for every 45° by placing the windows on opposite and adjacent walls (Fig. 4a, b, respectively). The chosen building orientation is referred to as Combined system 1 (C1). Secondly, impact of three types of windows glazing such as Single glazing—SG (3, 6, 12 mm), Double-glazed air-filled—DGa (3/6, 3/13, 6/6, 6/13 mm), and Double-glazed low emissivity argon-filled—DGε (6/13 mm for $\varepsilon = 0.1$, 0.2, and 0.4) in reference building is simulated. For example, DGa 3/6 mm represents two 3-mm-thickness glass panes separated by 6 mm air gap.

In Table 1, the properties of window glazing are listed. Furthermore, the outcomes of using different glazing on the chosen building orientation are evaluated. The combination of optimal orientation and glazing choice in a building is represented as Combined system 2 (C2). Subsequently, the performance of three shading systems, which are overhang, louver, and side-fin of different projection lengths and louver angles are investigated for the reference building. Moreover, the thermal and the visual performance of shading types on C2 (building orientation and window glazing) is analyzed. The collective system of ideal orientation, glazing, and external shading solutions is indicated as Combined system 3 (C3). Following this, internal shading using horizontal blind slats of low, medium, and high reflectivity is reviewed on C3 (orientation, glazing, and external shading) for glare control in indoor. The collective inclusion of the building orientation, glazing, external shading, and internal shadings along with ACUs is referred as passive building.

Figure 5 illustrates the approach used in studying the passive cooling systems. On performing CFD simulations, the thermal comfort parameters (air temperature, air velocity, PMV, and PPD) are compared in the reference and passive building at a distance of 1.4 m from the walls and 0.6 m above the floor as per ASHARE

**Table 1** Thermal and visual properties of window glazing types

| Type | Glazing | Thickness (mm) | U-Value (W/m$^2$K) | SHGC | VT | LSG |
|---|---|---|---|---|---|---|
| R | Reference building (Perfectly clear glass) | 3 | 3.12 | 0.99 | 0.99 | 1.00 |
| G1 | Single glazed | 3 | 5.89 | 0.86 | 0.90 | 1.04 |
| G2 | | 6 | 5.78 | 0.82 | 0.88 | 1.08 |
| G3 | | 12 | 5.56 | 0.74 | 0.84 | 1.14 |
| G4 | Double-glazed air-filled (DGa) | 3/6 | 3.16 | 0.76 | 0.81 | 1.07 |
| G5 | | 3/13 | 2.72 | 0.76 | 0.81 | 1.06 |
| G6 | | 6/6 | 3.09 | 0.70 | 0.78 | 1.12 |
| G7 | | 6/13 | 2.67 | 0.70 | 0.78 | 1.12 |
| G8 | Double-glazed low emissivity argon (DGε) | 6/13 ($\varepsilon = 0.1$) | 1.49 | 0.57 | 0.75 | 1.31 |
| G9 | | 6/13 ($\varepsilon = 0.2$) | 1.69 | 0.64 | 0.72 | 1.14 |
| G10 | | 6/13 ($\varepsilon = 0.4$) | 2.22 | 0.69 | 0.76 | 1.10 |

standard 55–2020. The thermal comfort simulations are performed at 4:00 PM on May 28 due to highest outdoor air DBT (41.1 °C) and ideal cooling load requirement (4.69 kW). The following CFD simulation conditions prevail for both reference and passive building used in the study:

- Air flow in the building is steady and incompressible.
- The air flow model used is K—epsilon turbulence model.
- Upwind discretization scheme is chosen.
- The convergence criteria of $10^{-6}$ are used to achieve accuracy.

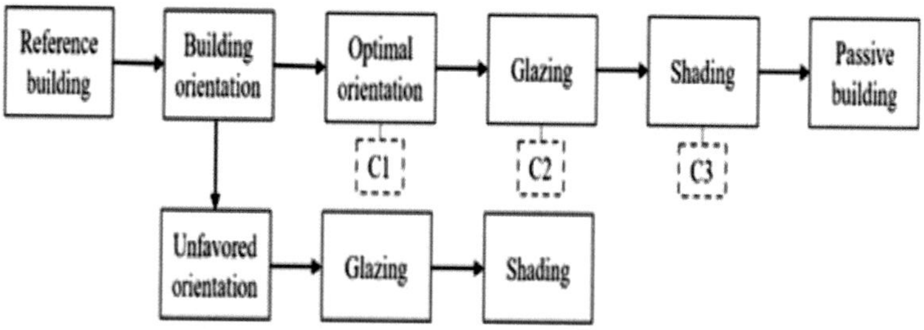


**Fig. 5** Framework for passive design strategies

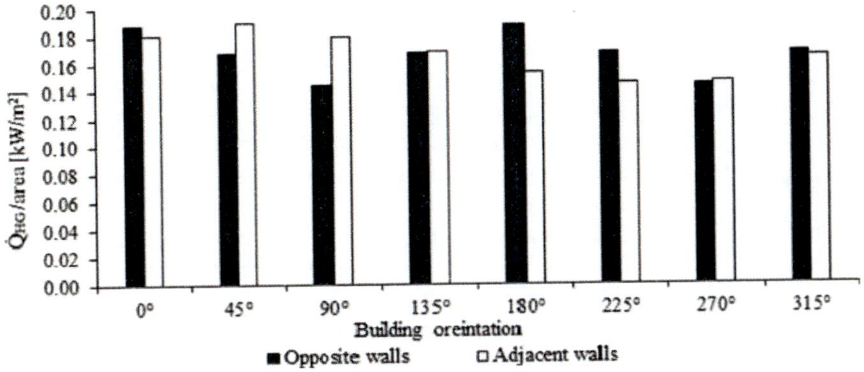


**Fig. 6** Total HGR per surface area through windows for different window orientations

## 3 Results and Inference

The annual cooling energy savings reported in the following results are calculated considering, 6.35 INR/unit of electricity as per Tamil Nadu Generation and Distribution Corporation Limited (TANGEDCO).

### *3.1 Building Orientation*

Most ideal building orientation is determined by studying the total $\dot{Q}_{HG}$ per surface area through building envelope in a year as shown in Fig. 6, for windows positioned on opposite and adjacent walls. The least $\dot{Q}_{HG}$ per surface area of 1262.5 kWh/m$^2$ is transmitted into the building orientated at 90° with windows on opposite walls. Building orientated at 45° having windows on the adjacent walls transmits maximum $\dot{Q}_{HG}$ per surface area of 1654.6 kWh/m$^2$, which is the least recommended. Orientating the building at 90° with windows on opposite walls reduces $\dot{Q}_{HG}$ per surface area by 23.7% when compared to 45° orientation with windows on adjacent walls. Thus, 90° orientation of building with windows on North and South walls (referred as C1) is the suitable choice. This result coincides with the building orientation recommended by ECBC (2017).

### *3.2 Window Glazing*

Figure 7 illustrates $\dot{Q}_{CL}$ and annual cooling energy savings of glazing types implemented in reference building. Perfectly clear glass used in the reference building is included to compare the performance of other glazing types. G8 (DG$\varepsilon$ 6/13 mm, ε = 0.1) shows superior performance over other glazing types, which is indicated by the reduction of $\dot{Q}_{CL}$ in April, May, and June by 12.7%, 13.8%, and 13.8%, respectively, when compared to reference building with no passive cooling strategies. To support the thermal performance results with economic perspective, the annual energy savings is illustrated in Fig. 6. The graph confirms G8 has the highest energy savings of 19.1%. Therefore, G8 is the preferable glazing type for reference building.

Furthermore, the performance of glazing types is analyzed on C1 (Fig. 8). The graphical result depicts that G8 shows lowest $\dot{Q}_{HG}$ through window of 0.92, 1.02, and 1.15 kW while G1 (SG 3 mm) has the highest $\dot{Q}_{HG}$ of 1.54, 1.80, and 1.97 kW in summer months. In addition to $\dot{Q}_{HG}$, $\dot{Q}_{SR}$ through window is also examined. Among the three heat transfer modes, radiation is found to be the primary contributor to heat infiltrated through windows. From the different types of glazing and thicknesses reviewed, $\dot{Q}_{SR}$ influences $\dot{Q}_{HG}$ through window with highest contribution of 81.1% in G1 during June and minimum contribution of 65.1% in G6 (DGa 6/6 mm) for May (Fig. 9).

$\dot{Q}_{CL}$ for G8 is the lowest through all three summer months with a maximum difference of 1.12 kW in April, relative to the reference building and 25.5% in May with respect to C1. Also, the annual energy savings of G8 compared to the reference building escalates to 24.3% (Fig. 10). To validate the above results, the properties of glazing in Table 1 show G8 has the lowest U-value (1.49 W/m$^2$ K), SHGC (0.57), and highest LSG value (1.3). Thus, for the reference building and C1, G8 proves to be the optimal choice of glazing system with SPP of 9.9 years and 7.8 years, respectively. Daylight simulation results indicate that significant percentage of annual potential daylight time is above the UDI threshold (2000 lx) for all glazing types; hence, visual comfort is achieved by implementing shading system.

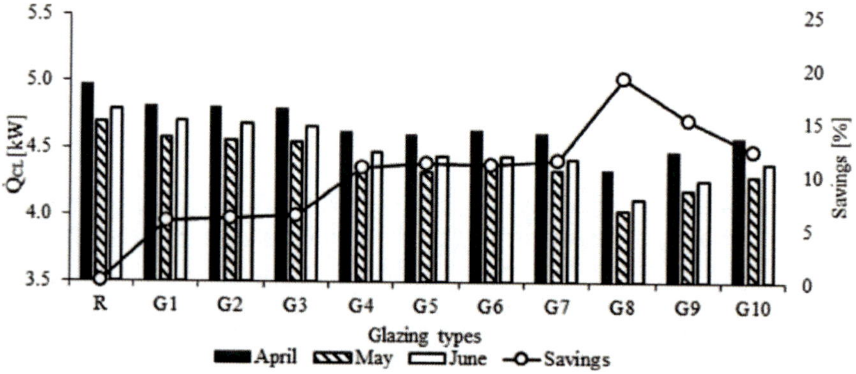


**Fig. 7** Ideal supply air maximum cooling load during summer months and annual cooling energy savings of glazing types implemented in the reference building

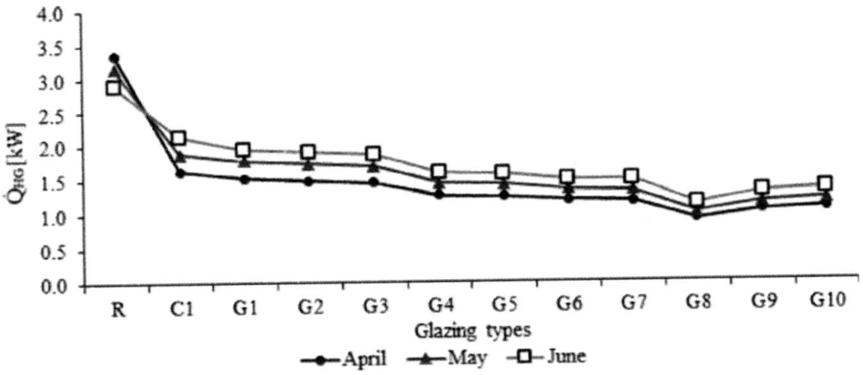


**Fig. 8** Average window HGR of glazing types implemented in C1 for summer months

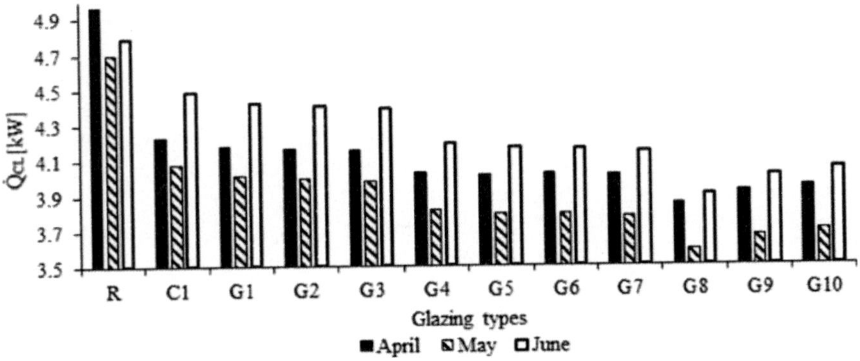


**Fig. 9** Ideal supply air maximum cooling load of glazing types implemented in C1 during summer months at setpoint temperatures 25 °C

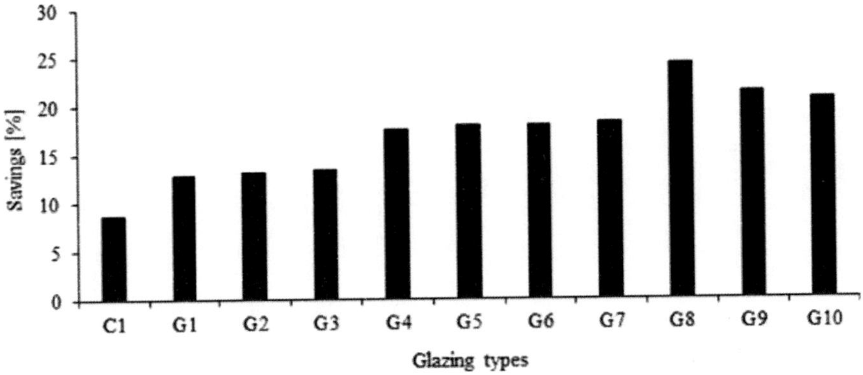


**Fig. 10** Annual cooling energy savings of glazing types implemented in C1 for summer months

### 3.3 Window Shading

**External shading devices.** Parameters that influence performance of external shading techniques are presented in Table 2 based on simulations. The annual potential daylight time within UDI for external shading types in the reference and C2 is shown in Table 3. Analyzing shading system in the reference building, out of 18 types, 5 have more than 90% of daylight time within UDI range.

Figure 11 depicts S12 (Louver 1 m 25°) has least $\dot{Q}_{CL}$ requirement at 3.89 kW, 3.61 kW and 3.84 kW during three summer months and has highest annual energy savings of 32.5% (Fig. 12). Thus, S12 is a suitable external shading for reference building with SPP of 14.58 years.

Furthermore, performance of shading types in C2 are examined. As in Table 3, among the shading cases, S3 and S7 have more than 90% potential daylit time in UDI range and are extensively studied. S7 is highly recommended shading type due to least $\dot{Q}_{HG}$ through window at 0.52, 0.58, and 0.54 kW during summer months, whereas, S3 (Overhang 1.5 m) has the highest $\dot{Q}_{HG}$.

Relative to the reference building, highest $\dot{Q}_{CL}$ reduction of 1.29 kW is observed in S7 during April and May (Fig. 13), whereas substantial increase in annual energy saving of 35.6% is obtained (Fig. 14). When compared to C2, S7 shows largest $\dot{Q}_{CL}$ drop of 6.9% during June. Therefore, S7 is an optimal choice of shading for building with C2 and an SPP of 12.01 years.

**Internal shading devices.** The glare index for C3 is above the threshold of 22, during 92% of the daylight time in a year with maximum discomfort glare index of 26.8. It is observed that usage of horizontal slatted blinds with reflectivity of 0.2, 0.5, and 0.8 (low, medium, and high, respectively) reduces the discomfort glare index below 22.

In Fig. 15, compared to C3, high reflectivity slats show negligible difference in $\dot{Q}_{CL}$ while a minimal decline in annual energy savings of 0.24% is observed. The low and medium reflectivity slats demonstrate a rise in $\dot{Q}_{CL}$ with respect to

**Table 2** Parameters used in simulation of basic external shading systems

| Shading system | Parameter | Value |
|---|---|---|
| Overhang | Projection length | 0.5 m/1 m/1.5 m |
| | Vertical offset from window top | 0 m |
| | Horizontal window overlap | 0.2 m |
| Side-fin | Projection length | 0.5 m/1 m/1.5 m |
| | Horizontal offset from window left or right | 0 m |
| Louver | Projection length | 0.5 m/1 m/1.5 m |
| | Distance from window | 0.1 m |
| | Louver angle | 15°/25°/35°/45° |

**Table 3** Annual potential daylight time within UDI range (100–2000 lx) for external shading types during summer months

| Type | Shading Type | Projection length (m) | Angle (°) | Annual potential daylight time in UDI range (%) | |
|---|---|---|---|---|---|
| | | | | Shading in reference building | Shading in C2 |
| R | Reference building | | | 04.34 | 59.45 |
| S1 | Overhang | 0.5 | 0 | 11.27 | 71.85 |
| S2 | | 1 | | 17.50 | 82.91 |
| S3 | | 1.5 | | 99.32 | 100.0 |
| S4 | Side-fins | 0.5 | 0 | 04.52 | 83.36 |
| S5 | | 1 | | 12.60 | 67.74 |
| S6 | | 1.5 | | 13.05 | 66.47 |
| S7 | Louver | 0.5 | 15 | 23.22 | 100.0 |
| S8 | | | 25 | 95.86 | 84.04 |
| S9 | | | 35 | 94.21 | 75.03 |
| S10 | | | 45 | 84.04 | 37.97 |
| S11 | | 1 | 15 | 94.48 | 67.63 |
| S12 | | | 25 | 91.54 | 64.93 |
| S13 | | | 35 | 81.19 | 25.65 |
| S14 | | | 45 | 74.10 | 19.76 |
| S15 | | 1.5 | 15 | 87.05 | 28.56 |
| S16 | | | 25 | 87.91 | 54.18 |
| S17 | | | 35 | 72.95 | 15.58 |
| S18 | | | 45 | 66.47 | 15.41 |

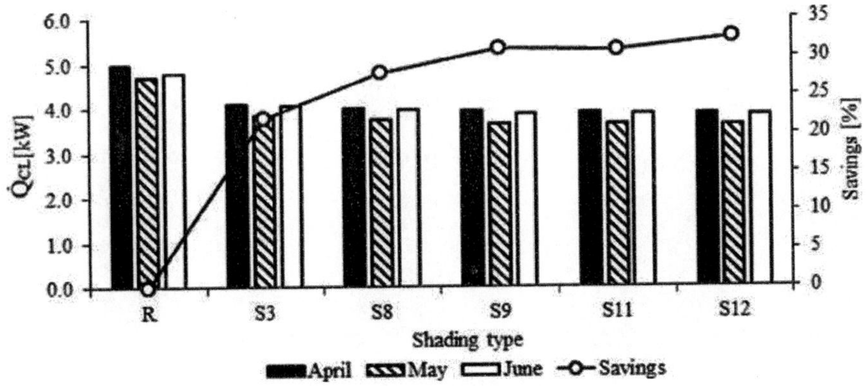


**Fig. 11** Ideal supply air maximum cooling load during summer months and annual cooling energy savings of shading types implemented in the reference building

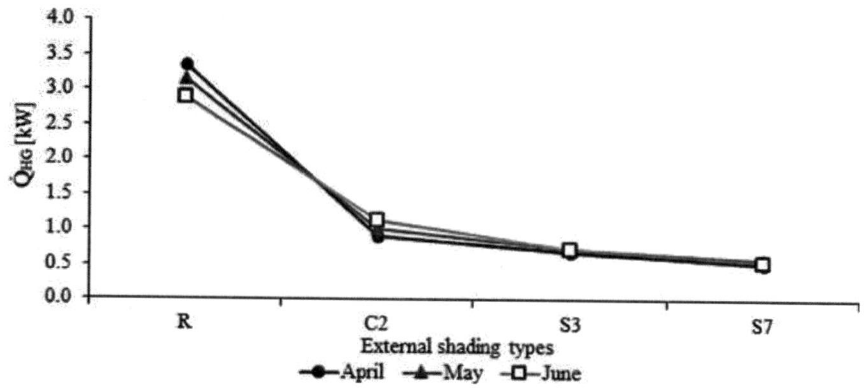


**Fig. 12** Average window HGR of external shading types implemented in the C2 for summer months

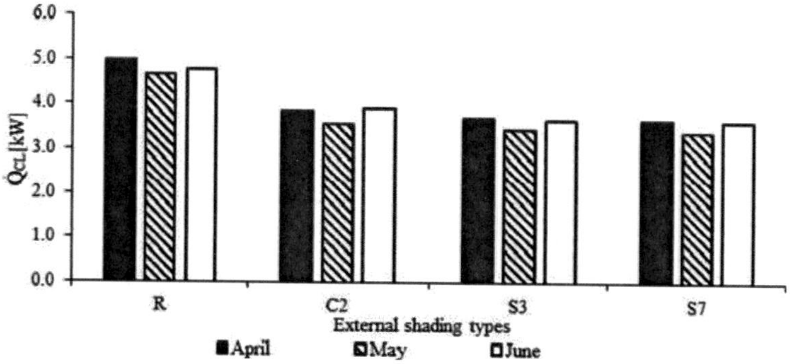


**Fig. 13** Ideal supply air maximum cooling load of external shading types during summer months at setpoint temperature 25 °C

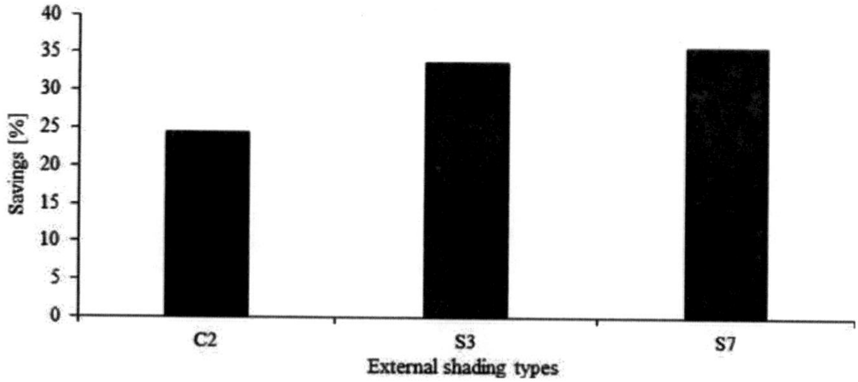


**Fig. 14** Annual cooling energy savings of external shading types implemented in C1 for summer months

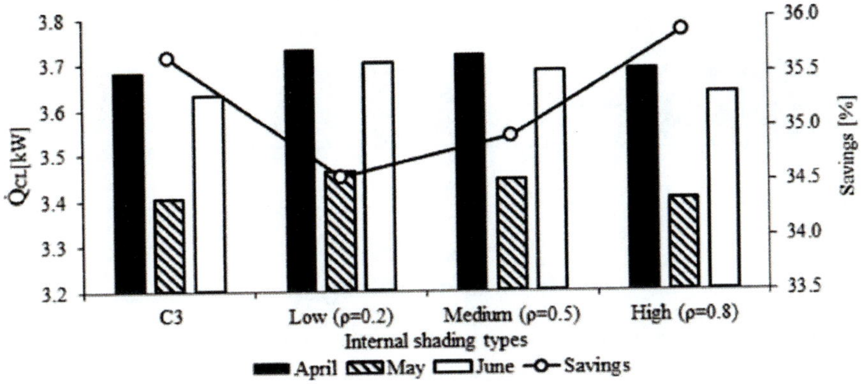


**Fig. 15** Ideal supply air maximum cooling load and annual cooling energy saving of internal shading types implemented in C3 during summer months

C3. Consequently, annual energy savings report a decline of 1.1 and 0.73% for low and medium reflectivity slats, respectively, making them undesirable. Hence, high reflectivity slats are suitable internal shading system with an SPP of 12.91 years.

## 3.4 Annual Cooling Energy Savings in Passive Building

The passive building equipped with the aforementioned passive systems (90° orientation, G8, S7, and horizontal slatted blind with high reflectivity) has a total annual cooling energy savings of 35.9% relative to the reference building. Figure 16 depicts the contribution of each passive cooling technique on energy savings. It is inferred that window glazing has enormous effect on the cooling energy by saving 15.6% relative to C1. Consequently, shading system provides savings of 11.3% on cooling energy with respect to C2. Orientating the building such that there is minimal $\dot{Q}_{HG}$ per surface area saves 8.6% of cooling energy relative to reference building while the internal shadings have least influence on cooling energy savings (0.2%) relative to C3.

## 3.5 CFD Analysis

The input boundary conditions for CFD module of DesignBuilder at 4:00 PM are imported from EnergyPlus simulation and represented in Table 4. The contour scales for thermal comfort parameters are presented in Table 5.

**Indoor air temperature.** Figure 17 illustrates the difference in air temperature distribution in reference and passive building. In the reference building except for

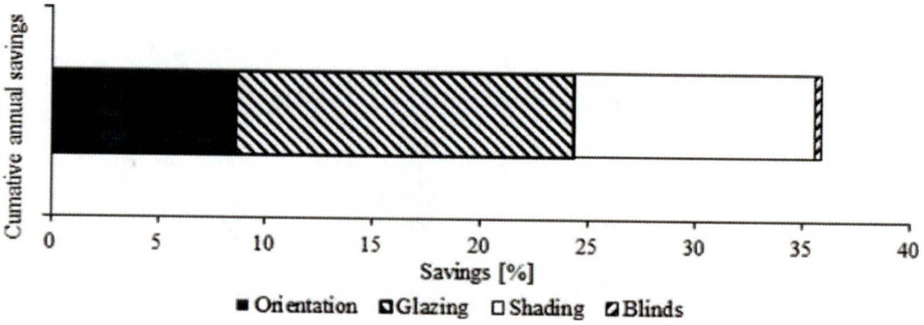


**Fig. 16** Contribution of passive techniques in annual cooling energy savings for passive building

regions near the computer units where the temperature reaches 27 °C, the indoor temperature is in the range from 24.4 to 26.3 °C. Whereas, in passive building, the temperature is in range of 22.6–25.3 °C for most regions of the indoor space. Thus, usage of passive systems results in observable decrease in the indoor air temperature.

**Indoor air velocity.** Air velocity distribution in reference and passive building is represented in Fig. 18. Majority of occupied space in reference building has air velocity in range of 0.14–0.23 m/s while, in passive building, the range varies from 0.05–0.14 m/s. This is explained by the fact, the reference building requires high volume flow rate of supply air (0.0946 $m^3/s$) to decrease the indoor temperature to 25 °C compared to its passive counterpart (0.0521 $m^3/s$). Since an air velocity below 0.25 m/s is recommended by ASHRAE for thermal comfort, both the reference and passive buildings meet the required indoor air velocity levels.

**PMV index.** The thermal comfort parameter PMV is shown in Fig. 19, for both the building types. Overall, in reference building, the PMV value has a wide range from + 0.27 to + 1. For passive building, the PMV value in most of the occupied region is in the range of -0.09 and + 0.27. For the entire indoor space, the average PMV of reference and passive building are + 0.63 and + 0.16, respectively. The average PMV value in case of passive building is closer to 0, providing occupants with neutral sensation (neither warm nor cold). Hence, passive building shows predominant reduction in the average PMV value by 0.46 and thereby improving the indoor thermal comfort levels.

**PPD index.** The parameter which specifies the occupant dissatisfaction, the PPD, is studied for both building types as in Fig. 20. For the reference building, PPD value has a wide range varying from 6 to 25%.

In case of passive building, expect for regions very close to the ACU, the PPD stays uniform around 5%. For reference and passive building models, the average value of PPD for the entire indoor space is found to be 13.5% and 5.6%, respectively. This indicates passive building has significantly contributed to the reduction of average PPD in the indoor by 7.9%.

**Table 4** CFD boundary conditions for reference and passive building model

| Boundary condition | Reference building values | Passive building values |
|---|---|---|
| | *Inside surface temperature (°C)* | |
| North wall | 28.90 | 25.69 |
| East wall | 28.79 | 25.90 |
| South wall | 28.86 | 25.66 |
| West wall | 28.56 | 25.85 |
| Roof | 30.42 | 26.20 |
| Floor | 25.68 | 22.01 |
| North wall windows | – | 29.23 |
| East wall windows | 36.69 | – |
| South wall windows | – | 29.23 |
| West wall windows | 27.28 | – |
| | *Heat gain rate (W)* | |
| Computer | 200 | |
| Human | 105 | |
| Lights | 15 | |
| | *Supply air condition/ACU* | |
| Velocity | 1.586 m/s | 0.87 m/s |
| Temperature | 18 °C | |
| Volume flow rate | 0.0946 $m^3/s$ | 0.0521 $m^3/s$ |
| | *Occupant parameters* | |
| Clothing | 0.7 clo | |
| Metabolic rate | 1.1 met | |
| | *Relative humidity (%)* | |
| RH | 66.39 | 67.42 |

## 4 Conclusion

During summer period, the space cooling is of utmost importance, especially in warm and humid climate zones. In the present study, passive cooling techniques are implemented in computer laboratory to reduce the cooling energy consumption and sustain indoor comfort levels. Results reveal that orientating the building in North–South considerably reduces the annual Heat Gain Rate by 23.1%. Usage of double glazed low-ε argon filled window in building gives peak energy savings of 15.6%. Introducing external shading (louver of 0.5 m projection with 15° inclination) in the

**Table 5** CFD plot contour scale for air temperature, air velocity, PMV, and PPD

| Color | Temperature (°C) | Velocity (m/s) | PMV | PPD (%) |
|---|---|---|---|---|
| | 19.00 | 0.00 | −1.00 | 0.00 |
| | 19.91 | 0.05 | −0.82 | 2.27 |
| | 20.82 | 0.09 | −0.64 | 4.55 |
| | 21.73 | 0.14 | −0.46 | 6.82 |
| | 22.64 | 0.18 | −0.27 | 9.09 |
| | 23.55 | 0.23 | −0.09 | 11.36 |
| | 24.45 | 0.27 | 0.09 | 13.64 |
| | 25.36 | 0.32 | 0.27 | 15.91 |
| | 26.27 | 0.36 | 0.46 | 18.18 |
| | 27.18 | 0.41 | 0.64 | 20.45 |
| | 28.09 | 0.46 | 0.82 | 22.73 |
| | 29.00 | 0.50 | 1.00 | 25.00 |

building offers highest reduction in ideal maximum cooling load of 0.27 kW during June. Even though external shading helps to maintain indoor visual comfort levels by providing the required illuminance for more than 90% of potential daylit time, it fails to control the discomfort glare. Inclusion of internal blinds with high reflectivity slats provides substantial reduction in glare index while its impact on cooling energy is negligible. From simulation results for the passive building, following conclusions can be drawn:

- For peak summer month of May, introducing the passive systems in building reduces the cooling load by 1.29 kW. This diminishes the cooling capacity of the Air-Conditioning Units to 0.95 TR as compared to 1.31 TR without passive system.

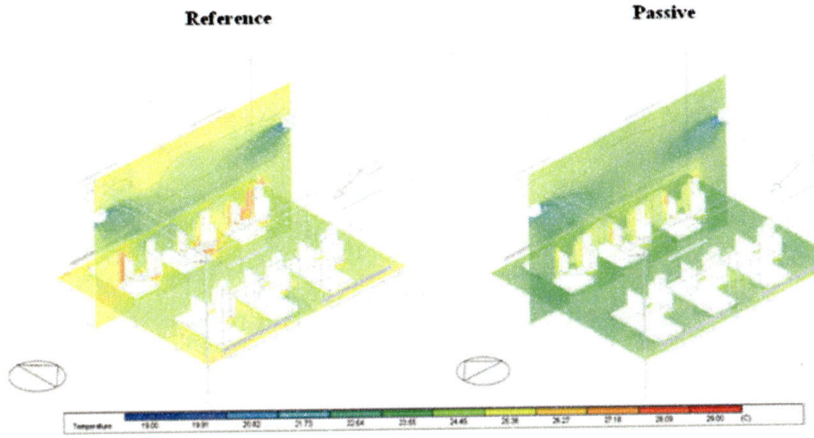


**Fig. 17** Indoor air temperature distribution in reference and passive building

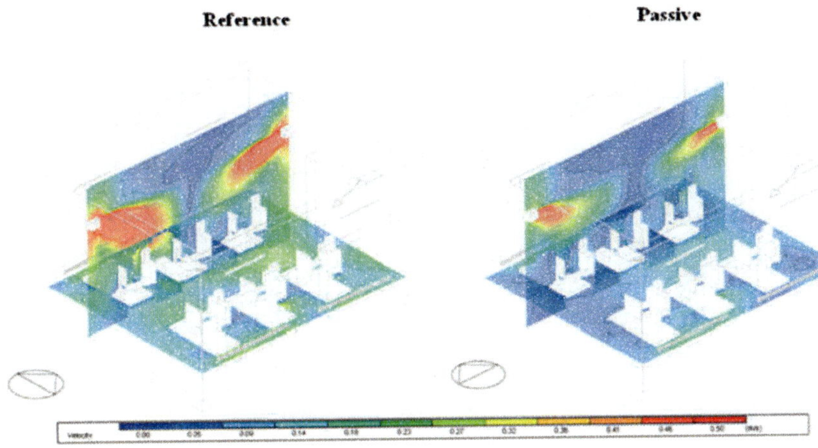


**Fig. 18** Indoor velocity distribution in reference and passive building

- The annual energy consumption for cooling the laboratory reduces from 1646.8 kWh to 1055.6 kWh, thereby providing annual cooling energy savings of 35.9%.
- The inclusion of passive systems is found to reduce Predicted Mean Vote value by 0.46 while, Percentage People Dissatisfied index is decreased by 7.9%.
- The overall Simple Payback Period of collective system is calculated to be 12.91 years.

Finally, installing passive systems establishes substantial reduction in the transfer of heat through the building, thereby creating a drop in energy required to cool the space and providing the ideal indoor comfort conditions. Future work can consider

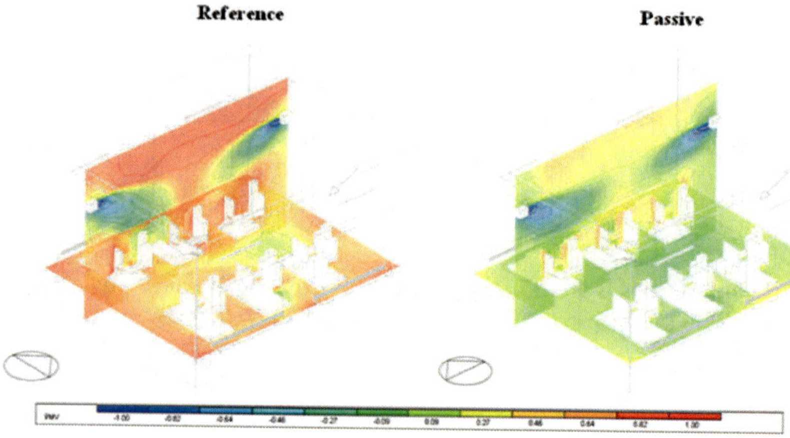


**Fig. 19** Variation of PMV index in reference and passive building

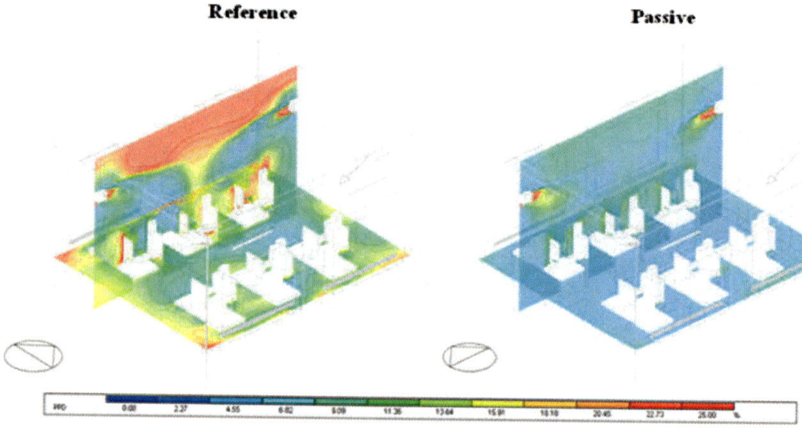


**Fig. 20** Variation of PPD index in reference and passive building

evaluation of changing the relevant building envelop parameters, assessment of artificial lighting, and Indoor Air Quality. Correspondingly, inclusion of radiant barriers and indoor greening to the present study can be considered.

## References

1. Akpinar-Ferrand E, Singh A (2010) Modeling increased demand of energy for air conditioners and consequent $CO_2$ emissions to minimize health risks due to climate change in India. Environ Sci Policy 13(8):702–712
2. Energy Statistics (2020) National Statistics Office, Ministry of Statistics and Programme

Implementation, New Delhi, India
3. Chetan V, Nagaraj K, Kulkarni PS, Modi SK, Kempaiah UN (2020) Review of passive cooling methods for buildings. J Phys: Conf Ser 1473:012054
4. Kumara GD, Naik M, Shankara RCR (2018) Impact of green building concept on sustainability of office buildings-a case study. Pollut Res 37:160–165
5. ANSI/ASHRAE thermal environmental conditions for human occupancy standard 55-2020
6. Alwetaishi MS (2016) Impact of building function on thermal comfort: a review paper. Am J Eng Applied Sci 9:928–945
7. Attia S (2018) Net zero energy buildings (NZEB): concepts, frameworks and roadmap for project analysis and implementation. Butterworth-Heinemann
8. Madathil D, Pandi VR, Ilango K, Nair MG (2017) Differential evolution based energy management system for zero net energy building. In: International conference on technological advancements in power and energy (TAP energy) IEEE, pp 1–5
9. Ramesh MV, Devidas AR, Athira K, Rangan V (2017) Using CPS enabled microgrid system for optimal power utilization and supply strategy. Energy and Buildings 145:32–43
10. Devidas AR, George SR, Ramesh MV (2015) A system for energy conservation through personalized learning mechanism. In: International conference on smart cities and green ICT systems (SMARTGREENS) IEEE, pp 1–7
11. Anjana MS, Athira K, Devidas AR, Ramesh MV (2016) A smart positioning system for personalized energy management in buildings. In: International conference on wireless communications, signal processing and networking (WiSPNET) IEEE, pp 742–747
12. Bansal V (2016) A review on performance analysis of passive cooling and ventilation system. International Journal of Engineering Technology, Management and Applied Sciences 4:355–365
13. Taleb HM (2014) Using passive cooling strategies to improve thermal performance and reduce energy consumption of residential buildings in UAE buildings. Frontiers of Architectural Research 3(2):154–165
14. Magaji M, Muhammad SA (2017) Analysing the performance of passive cooling system in buildings: designing natural solution to summer cooling loads and Architectural Interventions. American Journal of Engineering Research (AJER) 6(10):272–280
15. Kumar TS, Chandrasekar J, Moorthy SK, Sakthikala A, Bharath SA (2016) Optimization of building envelope to reduce air conditioning. Indian J Sci Technol 9:4
16. Gorantla K, Shaik S, Setty ABTP (2017) Effect of different double glazing window combinations on heat gain in buildings for passive cooling in various climatic regions of India. Materials Today: Proceedings 4(2):1910–1916
17. Lakshan RR, Rosini AM, Sathiyan K, Sathyan D, Mini KM, Gangadharan D (2020) Comparison of different dosages of PCM incorporated wallpanels. Conference Series: Materials Science and Engineering 872(1):012119
18. Raj A, Sathyan D, Balaji K, Mini KM (2021) Heat transfer simulation across a building insulated with foam concrete wall cladding. Materials Today: Proceedings 42:1442–1446
19. Al-Masrani SM, Al-Obaidi KM, Zalin NA, Isma MA (2018) Design optimisation of solar shading systems for tropical office buildings: challenges and future trends. Sol Energy 170:849–872
20. Raji B, Tenpierik MJ, van den Dobbelsteen A (2015) The impact of greening systems on building energy performance: a literature review. Renew Sustain Energy Rev 45:610–623
21. Boukhanouf R, Amer O, Ibrahim H, Calautit J (2018) Design and performance analysis of a regenerative evaporative cooler for cooling of buildings in arid climates. Build Environ 142:1–10
22. Haase M, Amato A (2009) An investigation of the potential for natural ventilation and building orientation to achieve thermal comfort in warm and humid climates. Sol Energy 83(3):389–399
23. Mangkuto RA, Rohmah M, Asri AD (2016) Design optimisation for window size, orientation, and wall reflectance with regard to various daylight metrics and lighting energy demand: a case study of buildings in the tropics. Appl Energy 164:211–219

24. Gustavsen A, Jelle BP, Arasteh D, Kohler C (2007) State-of-the-art highly insulating window frames—research and market review (technical report). Norwegian university of science and technology SINTEF Building and Infrastructure. https://doi.org/10.2172/941673
25. Bhamare DK, Rathod MK, Banerjee J (2019) Passive cooling techniques for building and their applicability in different climatic zones—the state of art. Energy and Buildings 198:467–490
26. Atef Faggal A, Mohamed Moustafa A, Yasser Arafat M (2019) Effect of different windows' glazing types on energy consumption of a residential building in a hot-arid climate "case study: residential building in New Cairo City." Journal of Engineering Sciences 47(5):706–719
27. Alhagla K, Mansour A, Elbassuoni R (2017) Optimizing windows for enhancing daylighting performance and energy saving. Alex Eng J 58(1):283–290
28. Energy Conservation Building Code (ECBC) (2017) Bureau of energy efficiency. New Delhi., India, pp 1–2
29. Bakmohammadi P, Noorzai E (2020) Optimization of the design of the primary school classrooms in terms of energy and daylight performance considering occupants' thermal and visual comfort. Energy Rep 6:1590–1607
30. Dutta A, Samanta A, Neogi S (2017) Influence of orientation and the impact of external window shading on building thermal performance in tropical climate. Energy and Buildings 139:680–689
31. Kim G, Lim HS, Lim TS, Schaefer L, Kim JT (2012) Comparative advantage of an exterior shading device in thermal performance for residential buildings. Energy and Buildings 46:105–111
32. Raji B, Tenpierik MJ, Van Den Dobbelsteen A (2016) An assessment of energy-saving solutions for the envelope design of high-rise buildings in temperate climates: a case study in the Netherlands. Energy and Buildings 124:210–221
33. Alhuwayil WK, Mujeebu MA, Algarny AMM (2019) Impact of external shading strategy on energy performance of multi-story hotel building in hot-humid climate. Energy 169:1166–1174
34. Hwang RL, Shu SY (2011) Building envelope regulations on thermal comfort in glass facade buildings and energy-saving potential for PMV-based comfort control. Build Environ 46(4):824–834
35. Hashemi A, Khatami N (2017) Effects of solar shading on thermal comfort in low-income tropical housing. Energy Procedia 111:235–244
36. Zhao J, Du Y (2020) Multi-objective optimization design for windows and shading configuration considering energy consumption and thermal comfort: a case study for office building in different climatic regions of China. Sol Energy 206:997–1017

# Performance Assessment of a Solar/Gas Driven $NH_3/LiNO_3$ Absorption Cooling System for Malls

**Carlos Amaris, Andres Rodriguez, Alexis Sagastume, and Mahmoud Bourouis**

## *Nomenclature*

| | |
|---|---|
| $A_c$ | Area of collectors ($m^2$) |
| $C_p$ | Specific heat of water (kJ/kg K) |
| $h$ | Specific enthalpy (kJ/kg) |
| $I_s$ | Solar irradiation ($W/m^2$) |
| $\dot{m}$ | Mass flow rate (kg/s) |
| $P$ | Pressure (kPa) |
| $\dot{Q}$ | Heat transfer flow (kW) |
| $T$ | Temperature (°C) |
| $\upsilon$ | Specific volume ($m^3/kg$) |
| $\dot{W}$ | Power (kW) |
| $X$ | Concentration of $LiNO_3$ in solution |
| COP | Coefficient of performance |
| SCOP | Solar Coefficient of performance |

## *Subscripts*

| | |
|---|---|
| $a$ | Absorber |
| Amb | Environment |
| $p$ | Pump |
| $c$ | Solar collector |

C. Amaris · A. Rodriguez · A. Sagastume
Deparment of Energy, Universidad de La Costa, 080008 Barranquilla, Colombia

M. Bourouis (✉)
Department of Mechanical Engineering, Universitat Rovira i Virgili, 43007 Tarragona, Spain
e-mail: mahmoud.bourouis@urv.cat



N. L. Narasimhan et al. (eds.), *Recent Advances in Energy Technologies*, Lecture Notes in Mechanical Engineering, https://doi.org/10.1007/978-981-19-3467-4_19

| | |
|---|---|
| cond | Condenser |
| $e$ | Evaporator |
| $d$ | Desorber |
| $s$ | Solar |
| shx | Heat exchanger |
| $u$ | Useful |

## *Greek letters*

| | |
|---|---|
| $\varepsilon$ | Effectiveness |
| $\eta_o$ | Solar collector efficiency |
| $\eta_{GN}$ | Gas burner overall efficiency |

# 1 Introduction

The building sector requires around one third of the total energy consumed all over the world [1], whereas air-conditioning systems of buildings in developed countries consume about half of the total energy input to buildings [2]. Also, it is predicted that the energy requirements in buildings would increase by half around 2030 together with the $CO_2$ equivalent emissions [3, 4].

Malls in tropical climates require to provide cooling to large, occupied volumes to satisfy the comfort of visitors. Consequently, shopping malls consume large amounts of electrical energy to power the air conditioning units [3]. The mechanical refrigerant compression units are the most common systems used for air cooling of large spaces in malls which makes them responsible of high amounts of electricity for their operation [3].

Absorption heat pump is considered as a promising alternative to lower the electricity consumption of HVAC systems, and might eventually play a key role to move to sustainable cooling technologies [5]. Absorption heat pumps require very small fractions of the electrical energy, while the main power source can come from heat sources related to renewable energies, residual heat, or natural gas [6–9]. The key processes and development needs in the absorption technology rely on the absorber and generator due to the complex transport processes occurring there [10, 11].

The performance of technologies for solar cooling relies mainly on climate conditions [12]. Tropical regions generally show favourable conditions for solar cooling applications [13]. Moreover, the performance of solar thermal panels surpasses that of the photovoltaic panels, which makes them technically attractive for solar air conditioning in buildings [12]. A comparative study of a conventional compression chiller driven by photovoltaic panels and an absorption chiller assisted by solar thermal energy is presented by Alrwashdeh et al. [14]. This study evidence shows that both configurations are cost-effective. Moreover, absorption refrigeration systems take advantage given the option to be powered by various available thermal heat sources [5, 9]. For instance, Dispenza et al. [9] studied the economic feasibility of an absorption chiller assisted by waste heat. The authors evidenced that an investment time of return below 4 years could be obtained. Gomri [15] showed that a 10 kW solar/gas driven water/LiBr absorption chiller could provide a COP up to 0.82, while the solar energy could cover up to 58% of its total energy input in Algeria. Al-Falahi et al. [16] evidenced that parabolic collectors are preferable to collectors of evacuated tubes to assisted a water/LiBr absorption cooling in terms of performance. Moreover, it has been demonstrated that solar-assisted absorption systems for air conditioning are feasible in buildings for commercial use in Saudi Arabia [17]. Also, the positive operational feasibility of a solar-driven double effect absorption system for air conditioning of buildings for residential use in the United Arab has been presented as well [18]. Recently, Rodriguez et al. [19] showed that the use of solar/gas assisted $H_2O/LiBr$ absorption system could lower $CO_2$ equivalent emissions between 17 and 76% if contrasted to the emissions of mechanical compression chillers in malls, whereas the internal rate of return could vary between 40 and 54.6% depending on the thermal load provided by the chillers in malls of Barranquilla, Colombia.

Solar collector dimensions play a key role in technical and economic feasibility studies related to solar cooling absorption systems. It varies together with the payback periods considering external conditions, the building cooling needs, working fluids used and the cooling system operating conditions [12, 13]. In terms of fluids used in these systems, the ammonia/lithium nitrate ($NH_3/LiNO_3$) mixture appears as a promising working fluid to water/LiBr because there is no risk of crystallization at high ambient temperature and the heat dissipation can be carried out by dry air [20–22].

The large solar irradiation and availability of natural gas in Colombia are not exploited to power thermal cooling systems. The main reasons are related to the lack of knowledge about these technologies and their potential for solar cooling in the region [23, 24]. Particularly, Barranquilla (capital of the Atlantic department with high solar potential) has tropical weather generating thermal discomfort [25], which results in high demand for air conditioning [26]. The electrical energy needs for air conditioning in the city malls are around 8 $kWh/m^2$, which is quite higher than that in other regions of Colombia [27]. Absorption cooling technologies are recognized as key contributors to reduce the electrical energy demand in malls; therefore, this study aims at assessing the technical potential of solar/gas assisted absorption chillers using the working pair $NH_3/LiNO_3$ in three malls in Barranquilla. To identify proper control actions for absorption chillers in the malls, the influence of environmental variables, condensation temperature and chilled water temperature on their performance is estimated.

## 2 Methodology

This section shows the environmental conditions of the city of Barranquilla, Colombia, the details of the malls used in this study, the configuration of the chiller simulated and its thermodynamic model.

### 2.1 *Environmental Variables and Shopping Malls Details*

The altitude and landscape define the climate zones in Colombia. Barranquilla is one of the cities with the highest economic growth in Colombia, with a population over 1.274 million people. Its climate is considered as tropical savanna, with exterior temperatures ranging from 24 to 35 °C and a relative humidity above 68% [28, 29]. Figure 1 presents the solar radiation during each month of the year. The data is available every hour.

The highest peak irradiation is reached during the first and second month of the year and the lowest peak irradiation is reached in October. Those months with lowest solar irradiation correspond to the months of rainfall in the year. The solar irradiation in Barranquilla varies between 600 Wh/m$^2$ and nearly 1000 Wh/m$^2$. The maximum irradiation during a day is reached from 11:00 to 15:00 h and the sunlight time per day is on averages around 7.2 h. The irradiation per day is on average around 846.4 Wh/m$^2$ [25].

Three malls of the city (M1, M2 and M3) were chosen for the present study. These malls are located in the Riomar area. Table 1 presents the air-conditioned areas and the roof available areas for solar collectors of the shopping malls.

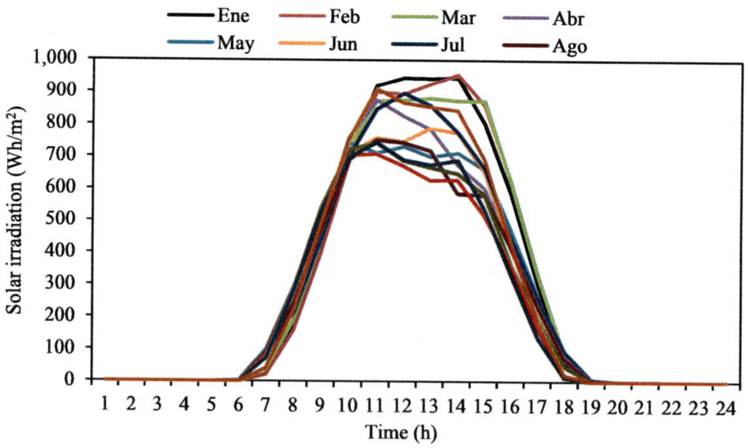


**Fig. 1** Available solar irradiation data for Barranquilla [25]

**Table 1** Area details of the malls

| Mall | Area for air conditioning ($A_{CA}$) ($m^2$) | Available roof space for solar panels ($A_{SC}$) ($m^2$) | Ratio ($A_{CA}/A_{SC}$) |
|---|---|---|---|
| M1 | 28,390 | 4043 | 7 |
| M2 | 21,000 | 2391 | 9 |
| M3 | 19,800 | 1214 | 16 |

The area ratio for M3 is above that of M1 and M2. Then, for M3 is expected the lowest potential for solar cooling.

A power analyser FLUKE 435 II was used to record the energy required to run the conventional compression chillers in the malls. Data were collected from 7:00 to 16:30 h, every one minute. The power data for M2 were collected in March 2019. The power data for M1 and M3 were collected in October 2019.

## 2.2 *Absorption Cooling System Characteristics*

Figure 2 presents a basic scheme of the $NH_3/LiNO_3$ absorption chiller.

The system consists of evacuated tube solar collectors and a natural gas burner coupled to an absorption chiller. The description of the chiller is well presented by Amaris et al. [10]. The operation characteristics of absorption systems are similar to that of mechanical compression chillers, but the pressurization of the refrigerant to the condition required for its condensation is conducted by the refrigerant vapour absorber, solution pump and refrigerant vapour generator pack. The solution internal heat exchanger is used to preheat the flow of $NH_3/LiNO_3$ coming from the absorber. Streams 11, 12 correspond to pressurized water which is employed to transport the energy required to the generator.

Two scenarios were evaluated to study the operation of the $NH_3/LiNO_3$ absorption chillers:

1. Baseline case: Cooling effect provided by conventional compression chillers.
2. Solar/Gas case: Cooling effect produced by solar/gas assisted absorption chillers.

## 2.3 *Thermodynamic Model*

The solar energy flow collected $\left(\dot{Q}_s\right)$, is estimated as [30]:

$$\dot{Q}_s = A_c I_s \tag{1}$$

where $A_c$ is the solar panel area, and $I_s$ is the solar irradiation. The available solar irradiation data for Barranquilla shows constant values per hour. The solar panel

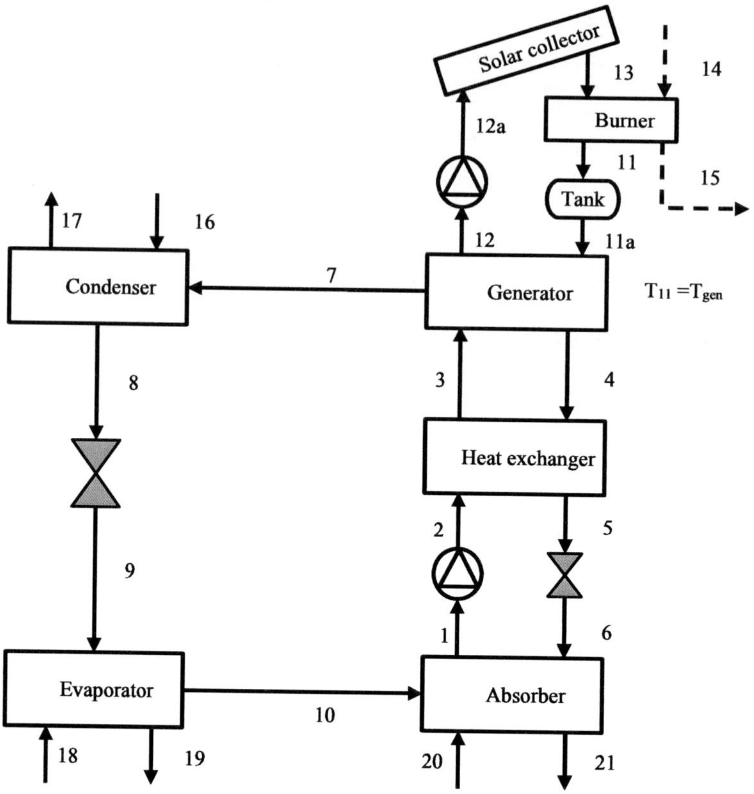


**Fig. 2** Absorption cooling system configuration

thermal efficiency is expressed as:

$$\eta_c = \frac{\dot{Q}_u}{\dot{Q}_s} = \frac{\dot{m}_{11} c_p (T_{12} - T_{13})}{A_c I_s} \tag{2}$$

where $(\dot{Q}_u)$ is the heat flow gained by the pressurized water. The evacuated tube collector efficiency is estimated as [31, 32]:

$$\eta_c = 0.82 - \frac{2.198 \cdot \Delta T_{avg}}{I_s} \tag{3}$$

The thermodynamic model of the system in Fig. 2 relies on energy and mass balances in each main component of the system. The main considerations are presented as follows:

- The refrigerant leaving the generator is saturated vapour.

- There is saturated liquid at the condenser outlet.
- Adiabatic flow in pipes.
- Gas burner with efficiency ($\eta_{GN}$) of 0.85 [15].
- The solution pump with efficiency of 0.6.

The software used in this study was the EES. Thermodynamic and transport properties of the $NH_3/LiNO_3$ were approached from [33–35]. Enthalpies were obtained according to the methodologies reported in [36, 37]. Table 2 presents the energy and mass energy balances for each main element of the absorption system.

The coefficient of performance (COP) was estimated as in Eq. 34. The solar coefficient of performance (SCOP) was estimated as in Eq. 35:

$$\mathrm{COP} = \frac{\dot{Q}_e}{\dot{Q}_{gen} + \dot{W}_p} \tag{34}$$

**Table 2** Energy and mass balances

| Components | Mass balance | Eq. | Energy balance and effectiveness | Eq. |
|---|---|---|---|---|
| Pump | $\dot{m}_1 = \dot{m}_2$ | 4 | $\dot{W}_p = n_p \cdot \dot{W}_e$ | 14 |
| | | | $\dot{W}_p = \dot{m}_1 \cdot \upsilon_1 \cdot (P_2 - P_1)$ | 15 |
| Heat Exchanger | $\dot{m}_2 = \dot{m}_3$ | 5 | $\dot{Q}_{shx} = \dot{m}_2 \cdot (h_3 - h_2)$ | 16 |
| | $\dot{m}_4 = \dot{m}_5$ | 6 | $\dot{Q}_{shx} = \dot{m}_4 \cdot (h_4 - h_5)$ | 17 |
| | | | $\varepsilon = \frac{T_4 - T_5}{T_4 - T_2}$ | 18 |
| Generator | $\dot{m}_3 = \dot{m}_7 + \dot{m}_4$ | 7 | $\dot{Q}_{gen} = \dot{Q}_u + \dot{Q}_{GN}$ | 19 |
| | $\dot{m}_3 X_3 = \dot{m}_7 + \dot{m}_4 X_4$ | 8 | $\dot{Q}_{GN} = \dot{m}_{GN} \cdot LHV \cdot \eta_{GN}$ d | 20 |
| | | | $\dot{Q}_{gen} = \dot{m}_7 \cdot h_7 - \dot{m}_3 \cdot h_3 + \dot{m}_4 \cdot h_4$ | 21 |
| | | | $\varepsilon = \frac{T_4 - T_3}{T_{11a} - T_3}$ | 22 |
| Absorber | $\dot{m}_{10} = \dot{m}_1 - \dot{m}_6$ | 9 | $\dot{Q}_a = \dot{m}_{10} \cdot h_{10} + \dot{m}_6 \cdot h_6 - \dot{m}_1 \cdot h_1$ | 23 |
| | | | $\dot{Q}_a = \dot{m}_{20} \cdot C_p \cdot (T_{21} - T_{20})$ | 24 |
| | $\dot{m}_6 X_6 + \dot{m}_{10} = \dot{m}_1 X_1$ | | | |
| | | | $\varepsilon = \frac{T_6 - T_1}{T_6 - T_{20}}$ | 25 |
| Condenser | $\dot{m}_7 = \dot{m}_8$ | 10 | $\dot{Q}_{cond} = \dot{m}_7 \cdot h_7 - \dot{m}_8 \cdot h_8$ | 26 |
| | | | $\dot{Q}_{cond} = \dot{m}_{16} \cdot C_p \cdot (T_{17} - T_{16})$ | 27 |
| | | | $\varepsilon = \frac{T_7 - T_8}{T_7 - T_{16}}$ | 28 |
| Evaporator | $\dot{m}_9 = \dot{m}_{10}$ | 11 | $\dot{Q}_e = \dot{m}_{10} \cdot h_{10} - \dot{m}_9 \cdot h_9$ | 29 |
| | | | $\dot{Q}_e = \dot{m}_{18} \cdot C_p \cdot (T_{18} - T_{19})$ | 30 |
| | | | $\varepsilon = \frac{T_{18} - T_{19}}{T_{18} - T_{19}}$ | 31 |
| Valve 1 | $\dot{m}_8 = \dot{m}_9$ | 12 | $h_8 = h_9$ | 32 |
| Valve 2 | $\dot{m}_5 = \dot{m}_6$ | 13 | $h_5 = h_6$ | 33 |

$$\text{SCOP} = \frac{\dot{Q}_e}{\dot{Q}_s} \tag{35}$$

where $\dot{Q}_e$ is the evaporator heat flow, $\dot{Q}_{gen}$ refers to the energy input to the vapour generator and $\dot{W}_P$ refers to the pump power input. Equation 36 represents the relation between the cooling effect produced by the absorption units and that produced by the mechanical compression chillers:

$$\text{Shareofcooling}(\%) = \frac{\dot{Q}_e}{\dot{Q}_{ecomp}} \cdot 100 \tag{36}$$

# 3 Results and Discussion

This section shows the performance assessment of the solar/gas assisted absorption chiller.

## 3.1 *Sensitivity Analysis*

The development of the simulation model was performed taking into account the external mass flow and temperature of the streams for absorption chillers from the Thermax series with water/LiBr for a minimum cooling capacity of 352 kW at nominal operating conditions: a temperature of 6.7 °C for the chilled water, a temperature of 29.4 °C for the heat dissipation, and a temperature of 90.6 °C for the pressurized water employed as heat transport medium to the generator [38]. The mass flow and temperatures of the external streams at the inlet of each component were defined as proposed in [38]. Then, the solution flow rate leaving the absorber for the absorption chiller but using the $NH_3/LiNO_3$ was estimated as 4.791 kg/s. This value is 2.88 times higher than that of the system with water/LiBr because of the differences in the transport properties values such as thermal conductivity and viscosity of the mixtures. For higher cooling capacities, the solution flow required at the absorber outlet rises almost linearly with the cooling effect of the chillers with $NH_3/LiNO_3$ and water/LiBr, as shown in Table 3. Reasonable effectiveness values were set for the evaporator, condenser, generator, absorber and solution heat exchanger (see Table 4).

If evacuated tube collectors (ETC) are integrated to the $NH_3/LiNO_3$ absorption chiller, a collector area around 1105 m$^2$ is needed. This value is higher than the area of 840 m$^2$ needed to drive a chiller with water/LiBr at the same conditions because of higher COP of the water/LiBr chiller (around 0.78 at nominal conditions) [19]. Moreover, if flat plate collectors are used, the area required is around 1.68 times

**Table 3** Cooling capacity versus solution mass flow at the absorber outlet

| Cooling effect (kW) | Water/LiBr mass flow (kg/s) | $NH_3/LiNO_3$ mass flow (kg/s) |
|---|---|---|
| 352 | 1.66 | 4.791 |
| 422 | 1.90 | 5.744 |
| 492 | 2.30 | 6.696 |
| 563 | 2.60 | 7.663 |
| 633 | 2.90 | 8.615 |
| 738 | 3.40 | 10.040 |

**Table 4** Main components effectiveness values [39–41]

| Component | Effectiveness for water/LiBr chillers | Effectiveness for $NH_3/LiNO_3$ chillers |
|---|---|---|
| Evaporator | 0.815 | 0.850 |
| Absorber | 0.500 | 0.456 |
| Solution heat exchanger | 0.640 | 0.800 |
| Generator | 0.302 | 0.500 |
| Condenser | 0.676 | 0.679 |

higher than that using the ETC. Given the higher efficiency of ETC, the capital cost of the collectors [31] decreases around 110,232 USD if compared to that of flat plate collectors.

Figure 3a shows the effect of the chilled water temperature and heat dissipation temperature on the heat flow in the evaporator and generator for Is = 500 W/m$^2$. Moreover, Fig. 3b shows the effect of the chilled water temperature and heat dissipation temperature on the COP and SCOP for the same Is.

Figure 3 shows that the chiller could reach a cooling capacity of 250 kW at a chiller water temperature of 15 °C and a heat dissipation temperature of 28 °C. Then, it decreases to 183 kW by decreasing the temperature of the outlet chilled water to 8 °C and increasing the temperature of heat dissipation medium to 34 °C. These variations also have negative effects on the COP and SCOP of the system under study. In this case, the COP and SCOP were up to 0.687 and 0.453, respectively.

Results show that the drop in the cooling capacity is more pronounced than the drop in the thermal energy input to the generator when the chilled water temperature is decreased and the heat dissipation temperature is increased. The higher the heat dissipation temperature, the lower the potential for heat transfer in the absorber and condenser. Moreover, the lower the chiller water temperature, the lower the potential for heat transport in this component. Therefore, it is evident that for Is below 500 W/m$^2$, the absorption chiller could not reach the nominal capacity of 352 kW.

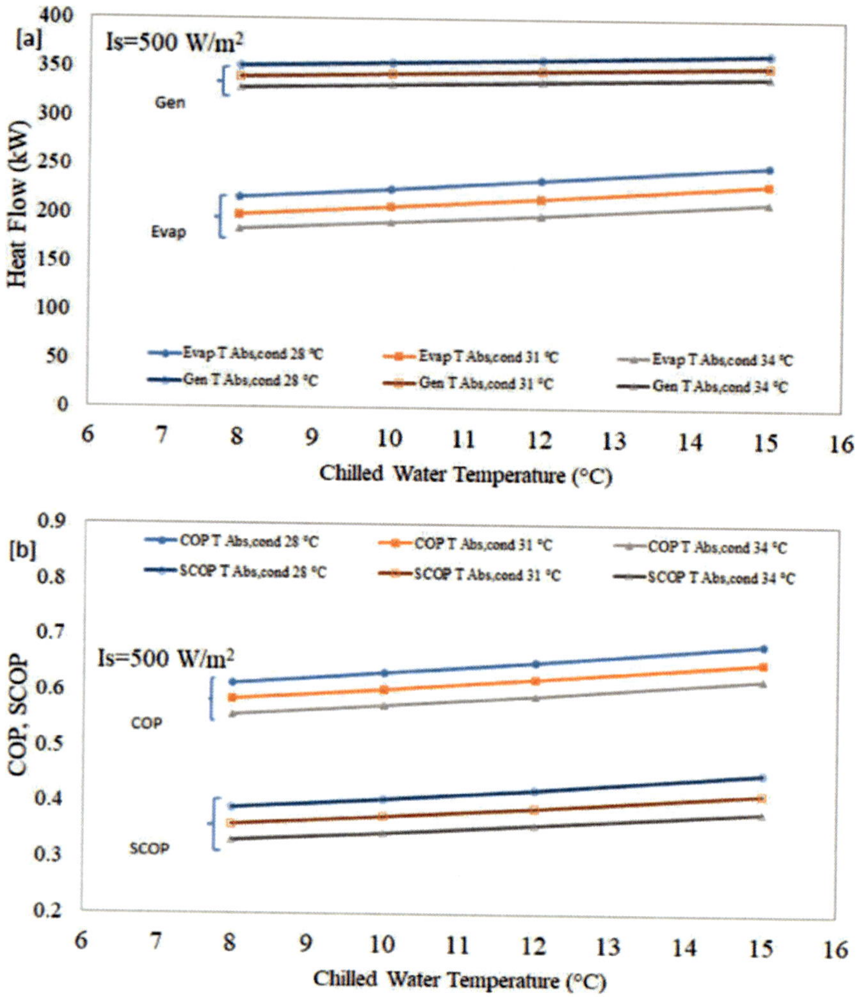


**Fig. 3** Chilled water temperature and heat dissipation temperature versus (**a**) Heat flow in evaporator and generator and (**b**) COP and SCOP, for Is = 500 W/m$^2$

Figure 4 shows the performance of the chiller for Is = 700 W/m$^2$. Results indicate that the chiller approaches the nominal capacity at the lowest heat dissipation temperature (28 °C) and the highest chilled water temperature (15 °C) considered, while the COP and SCOP are up to 0.658 and 0.443, respectively. Moreover, the lowest cooling capacity obtained at the studied conditions is 270 kW, while the minimum COP and SCOP are 0.557 and 0.349, respectively.

Additionally, it is noted that at the operation conditions studied, there is no risk of crystallization of the $NH_3/LiNO_3$ solution. Results obtained can be employed

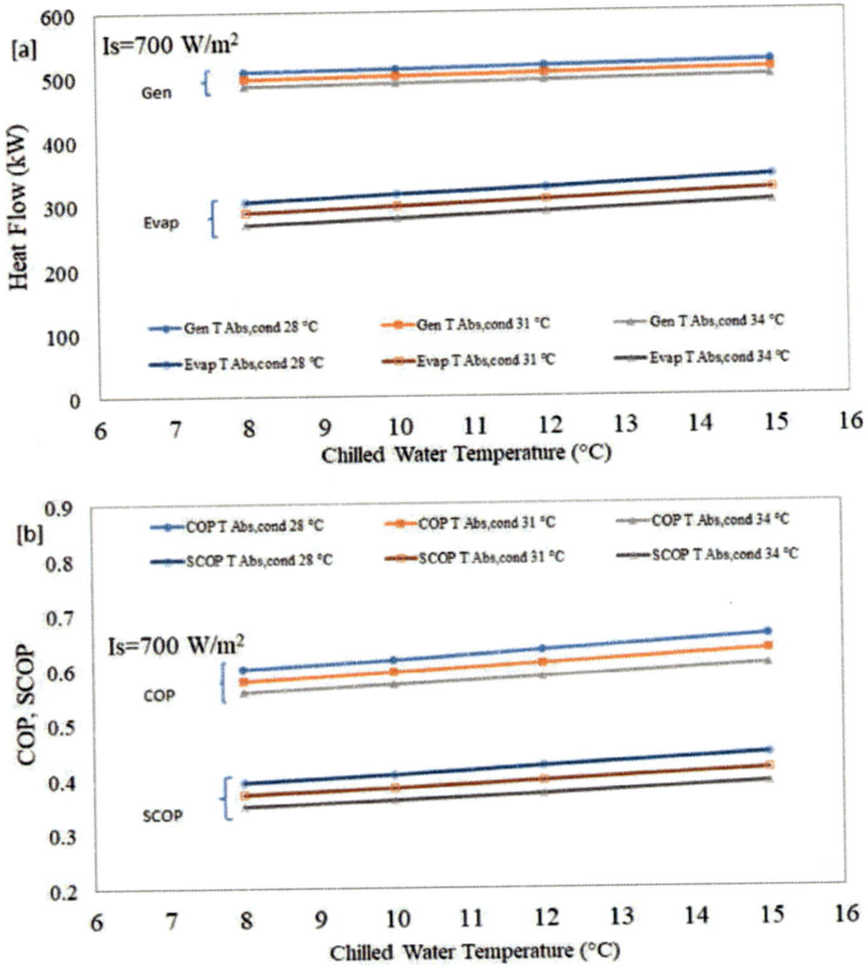


**Fig. 4** Chilled water temperature and heat dissipation temperature versus (**a**) Heat flow in evaporator and generator and (**b**) COP and SCOP, for Is = 700 W/m$^2$

to establish proper control actions for the $NH_3/LiNO_3$ absorption chillers in Barranquilla.

## 3.2 *Energy Consumption of Mechanical Chillers*

Figure 5 shows the power measurements of the conventional mechanical chillers in the malls. The data measured is to some extent stable during the chiller's operations. Malls in Barranquilla are usually opened around 13 h per day. The M3 showed power

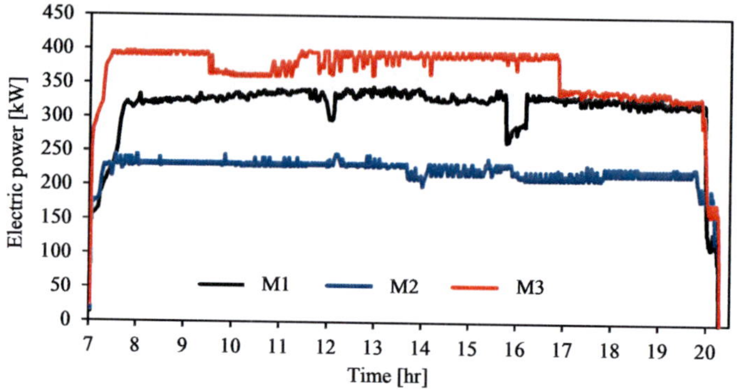


**Fig. 5** Power consumption measurements of chillers

data ranging from 330 to 400 kW for most of the measurement day. Then, it dropped from 5 p.m. because of the shutdown of some backup systems. Moreover, in M1, the power input to the chillers changed between 267 and 330 kW for the measurement day. In M2, the power requirements of the chillers varied between 200 and 240 kW. In each mall, changes in the power input were registered due to the normal operation of the malls, i.e. environmental conditions, people entering and leaving the mall, etc. A $COP_{comp}$ of three was contemplated to estimate the total cooling load provided by the conventional chillers in the malls [13, 42]. Considering the above, the cooling effects produced by the chillers are up to 1001 kW, 735 kW and 1187 kW, for M1, M2 and M3, respectively. Results of the estimated thermal loads are used in the following sections as the cooling loads objectives for the absorption cooling system configurations investigated.

It is worth to highlight that low cooling loads are estimated when the chillers were turned on at 7 h and when the chillers were shut down at 20:30 h. It can be noted that the cooling load covered increased just after the chillers are turned on, from around 20 to 320 kW in M1, from around 20 to 230 kW in M2 and from around 20 to 400 kW in M3. It can also be noted that the cooling load covered dropped before the chillers were shut down.

### *3.3 Share of Cooling Demand from the Solar/gas Assisted Chillers*

This section presents the main simulation outputs of the absorption chillers for malls 1, 2 and 3. In this case study, the chilled water temperature and heat dissipation temperature were fixed to 12 °C and 29.4 °C, respectively. A low heating value of 47.04 MJ/kg was assumed for the natural gas [39]. The chillers are expected to operate above their nominal because the temperature of the chilled water is adjusted to 12 °C (the nominal temperature is 6 °C). Table 5 presents the chiller units used in the malls and the solar panel area required for each case.

The number of absorption chillers required in each mall were estimated considering the maximum cooling loads provided by the mechanical chillers and the area available on the roof of the malls. In the case of M1, the mechanical chillers could provide a cooling effect up to 1001 kW, therefore, three absorption chillers of 352 kW working at nominal conditions could be assumed to reach the cooling target of 1001 kW. However, they could operate beyond the nominal conditions because a temperature of 12 °C was set for the chilled water. In this case, two chillers assisted by natural gas and solar energy can be used in M1. For M3, the number of chillers was only one because of the limited mall area available for solar collectors. Given the relatively steady operation of the total energy input, the COP and SCOP of the solar/gas assisted absorption chillers in M1, M2 and M3 are around 0.64 and 0.41, respectively.

Figure 6 displays the share of the cooling covered by the solar/gas assisted absorption chillers in M1. Results show that between 9:00 a.m. and 2:00 p.m., the solar collectors provide from 55 to 65% of the energy required to power the chillers in the malls. The remaining energy is obtained from the combustion of the gas in the burner. Moreover, between 7:00 a.m. and 9:00 a.m., and between 2:00 p.m. and 8:00 p.m. most of the energy required to power the absorption chillers come from the natural gas.

Figure 6 also shows that the solar/gas assisted chillers could provide around 83% of the total cooling target for most of the time at the operating conditions given. An additional absorption chiller unit would exceed the cooling target for more than 280 kW.

Figure 7 displays the share of the cooling provided by the solar/gas assisted chillers in M2. Results evidence that between 8:00 a.m. and 3:00 p.m., the solar collectors could provide from 66 to 92% of the energy required to drive the absorption chillers in the mall. Moreover, most of the energy required to power the absorption chillers come from the natural gas before 8:00 a.m., and after 3:00 p.m. It can also be noted that the solar/gas assisted chillers could provide 100% of the total cooling target for most of the day at the operating conditions given.

Figure 8 displays the share of the cooling provided by the solar/gas assisted chillers in M3. Results evidence that between 8:00 a.m. and 3:00 p.m., the solar collectors could provide only from 20 to 28% of the energy required to drive the chillers in the mall. With only one solar/gas chiller assisted absorption chiller, the share of cooling covered varies between 30 and 35% for most of the day at the operating conditions given. The share of cooling could be increased by adding more chillers, but given

**Table 5** Absorption chillers and solar panel area required in the malls

| Mall | Chiller units | Area of collectors ($m^2$) |
|---|---|---|
| M1 | 2 | 2210 |
| M2 | 2 | 2210 |
| M3 | 1 | 1105 |

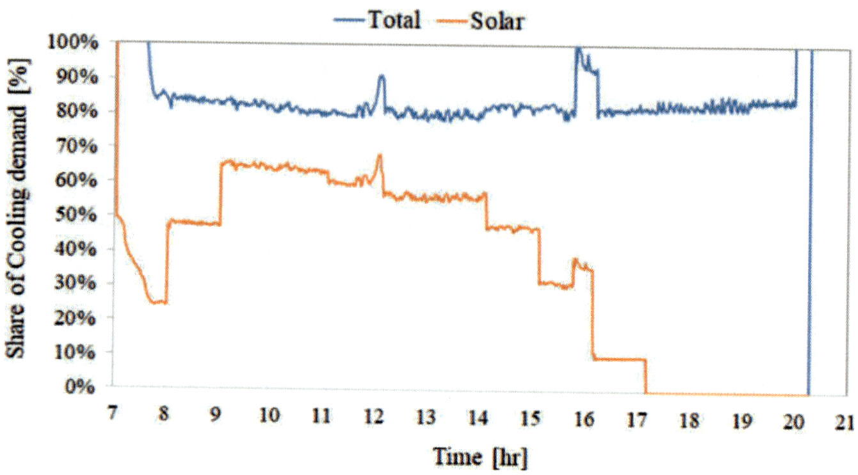


**Fig. 6** Share of cooling covered by the solar/gas assisted chillers for M1

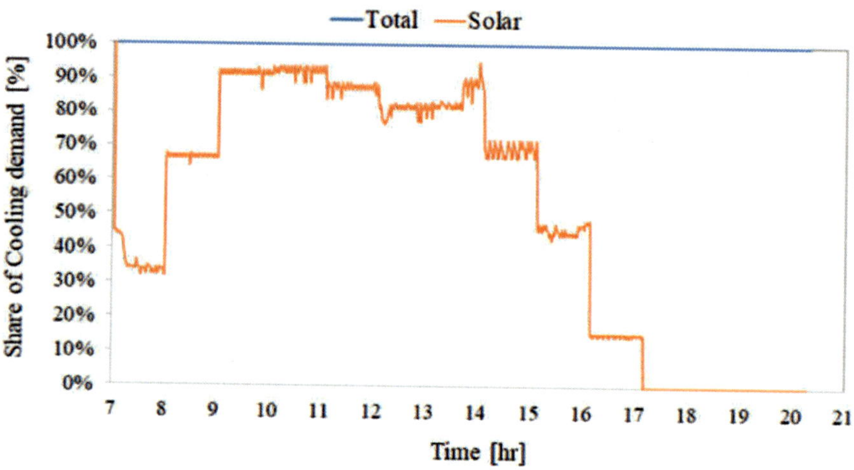


**Fig. 7** Share of cooling covered by the solar/gas assisted chillers for M2

the limitations in the area available for the solar collectors in mall 3 (1214 $m^2$), it is not possible to add them if the objective is to exploit the solar energy to partially drive the absorption chillers (1105 $m^2$ per chiller).

Results from these case studies indicate that a solar/gas assisted chiller with $NH_3/LiNO_3$ is technically feasible in tropical weather, whereas crystallization issues are not a concern. Therefore, the heat dissipation can be by dry cooling considering dry-air temperatures up to 35 °C as is the case of the city of Barranquilla. Additionally, the area available for solar collectors is a limiting factor if only solar energy is

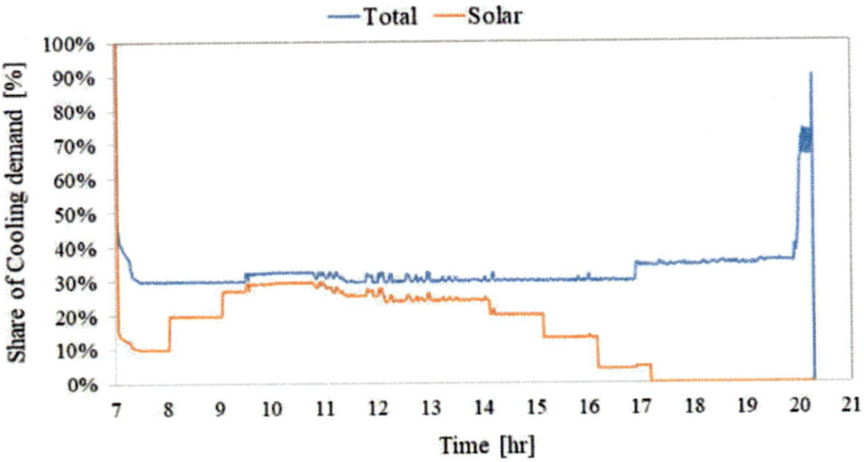


**Fig. 8** Share of cooling covered by the solar/gas assisted chillers for M3

considered to drive absorption chillers for malls, especially for those with a vertical architecture.

## 4 Conclusions

In the present investigation, a performance study of an $NH_3/LiNO_3$ absorption chiller using natural gas and solar energy as energy inputs was conducted. A thermodynamic model was employed to simulate the absorption chillers operation for three malls located in Barranquilla city. The main conclusions are presented as follows:

The sensitivity analysis allowed characterizing and identifying the control strategies to operate a solar-assisted $NH_3/LiNO_3$ absorption chiller in the city of Barranquilla. Solar energy was found not enough to power absorption chillers at full load conditions given the limitations in solar collector area. Therefore, additional energy input from natural gas was found required. The maximum COP and SCOP obtained from the sensitivity study were 0.687 and 0.453, respectively, at the highest chilled water temperature and lowest heat dissipation temperature considered.

The solar energy input could cover up to 65%, 92% and 28% of the share of cooling only during the pick solar hours in malls one, two and three, respectively, considering the number of chiller units used in each mall. In general, the solar/gas assisted chillers could cover up to 83%, 100% and 35% of the peak cooling requirements in malls one, two and three, respectively. The area available for solar collectors in mall three did not allow using more than one solar/gas assisted absorption chiller to fully cover the cooling requirements. The maximum COP and SCOP values of the solar/GAS assisted chillers were up to 0.64 and 0.41, respectively.

The available area for the installation of solar collectors is a limiting factor for solar absorption cooling applications in malls, especially for those with a vertical architecture.

**Acknowledgements** The authors acknowledge the Universidad de la Costa for supporting the present research through the INDEX CONV-13-2018.

## References

1. International Energy Agency—IEA (2010) Energy technology perspectives 2010. www.iea.org/etp
2. Pérez-Lombard L, Ortiz J, Pout C (2008) A review on buildings energy consumption information. Energy Build. 40:394–398. https://doi.org/10.1016/J.ENBUILD.2007.03.007
3. Li H, Li X (2018) Benchmarking energy performance for cooling in large commercial buildings. Energy Build. 176:179–193. https://doi.org/10.1016/j.enbuild.2018.07.039
4. Hassan JS, Zin RM, Majid MZA, Balubaid S, Hainin MR (2014) Building energy consumption in Malaysia: an overview. J Teknol 70:33–38. https://doi.org/10.11113/jt.v70.3574
5. Ravi Kumar K, Krishna Chaitanya NVV, Sendhil Kumar N (2021) Solar thermal energy technologies and its applications for process heating and power generation—a review. https://doi.org/10.1016/j.jclepro.2020.125296
6. Nikbakhti R, Wang X, Hussein AK, Iranmanesh A (2020) Absorption cooling systems—review of various techniques for energy performance enhancement. Alexandria Eng J 59:707–738. https://doi.org/10.1016/J.AEJ.2020.01.036
7. Amaris C, Miranda BC, Balbis-Morejón M (2020) Experimental thermal performance and modelling of a waste heat recovery unit in an energy cogeneration system. Therm Sci Eng Prog 20. https://doi.org/10.1016/j.tsep.2020.100684
8. Yousefzadeh M, Lenzen M, Tyedmers EK, Hassan Ali SM (2020) An integrated combined power and cooling strategy for small islands. J Clean Prod 276:122840. https://doi.org/10.1016/j.jclepro.2020.122840
9. Dispenza A, La Rocca V, Messineo A, Morale M, Panno D (2013) Absorption equipment for energy savings: a case study in Sicily. Sustain Energy Technol Assessments 3:17–26. https://doi.org/10.1016/j.seta.2013.05.002
10. Amaris C, Vallès M, Bourouis M (2018) Vapour absorption enhancement using passive techniques for absorption cooling/heating technologies: a review. Appl Energy 231:826–853. https://doi.org/10.1016/j.apenergy.2018.09.071
11. Amaris C, Bourouis M (2021) Boiling process assessment for absorption heat pumps: a review. Int J Heat Mass Transf 179:121723. https://doi.org/10.1016/J.IJHEATMASSTRANSFER.2021.121723
12. Allouhi A, Kousksou T, Jamil A, Bruel P, Mourad Y, Zeraouli Y (2015) Solar driven cooling systems: an updated review. Renew Sustain Energy Rev 44:159–181. https://doi.org/10.1016/J.RSER.2014.12.014
13. Bellos E, Tzivanidis C (2017) Energetic and financial analysis of solar cooling systems with single effect absorption chiller in various climates. Appl Therm Eng 126:809–821. https://doi.org/10.1016/j.applthermaleng.2017.08.005
14. Alrwashdeh SS, Ammari H (2019) Life cycle cost analysis of two different refrigeration systems powered by solar energy. Case Stud Therm Eng 16. https://doi.org/10.1016/j.csite.2019.100559
15. Gomri R (2013) Simulation study on the performance of solar/natural gas absorption cooling chillers. Energy Convers Manag 65:675–681. https://doi.org/10.1016/J.ENCONMAN.2011.10.030

16. Al-Falahi A, Alobaid F, Epple B (2020) Design and thermo-economic comparisons of large scale solar absorption air conditioning cycles. Case Stud Therm Eng 22:100763. https://doi.org/10.1016/j.csite.2020.100763
17. Al-Ugla AA, El-Shaarawi MAI, Said SAM, Al-Qutub AM (2016) Techno-economic analysis of solar-assisted air-conditioning systems for commercial buildings in Saudi Arabia. Renew Sustain Energy Rev 54:1301–1310. https://doi.org/10.1016/j.rser.2015.10.047
18. Ghaith FA, Razzaq HH (2017) Performance of solar powered cooling system using parabolic trough collector in UAE. Sustain Energy Technol Assessments 23:21–32. https://doi.org/10.1016/j.seta.2017.08.005
19. Rodríguez-Toscano A, Amaris C, Sagastume-Gutiérrez A, Bourouis M (2022) Technical, environmental, and economic evaluation of a solar/gas driven absorption chiller for shopping malls in the Caribbean region of Colombia. Case Stud Therm Eng 30:101743. https://doi.org/10.1016/J.CSITE.2021.101743
20. Amaris C, Alvarez ME, Vallès M, Bourouis M (2020) Performance assessment of an NH3/LiNO3 bubble plate absorber applying a semi-empirical model and artificial neural networks. Energies 13. https://doi.org/10.3390/en13174313
21. Zapata A, Amaris C, Sagastume A, Rodríguez A (2021) CFD modelling of the ammonia vapour absorption in a tubular bubble absorber with $NH_3/LiNO_3$. Case Stud Therm Eng 27:101311. https://doi.org/10.1016/J.CSITE.2021.101311
22. Oronel C, Amaris C, Vallès M, Bourouis M (2010) Experiments on the characteristics of saturated boiling heat transfer in a plate heat exchanger for ammonia/lithium nitrate and ammonia/(lithium nitratewater). In: Proceedings of 2010 3rd international conference on thermal issues in emerging technologies, theory and applications, ThETA3 2010. pp 217–225. https://doi.org/10.1109/THETA.2010.5766401
23. García A, Bajar costos: reto de la refrigeración solar | ACR Latinoamérica.
24. EPM, Tarifario del mes|Gases del Caribe.
25. IDEAM, Boletín Climatológico Mensual
26. Murcia Ruiz JF (2010) Cambio climático en temperatura, precipitación y humedad relatica para Colombia usando modelos meterolígicos de alta resolución. Ideam. 1–91
27. De Minas M, Energía Y, De U, Minero P, Caracterización E, De Los E, Residencial S, Terciario CY (2007) República de Colombia
28. Weather Underground, Barranquilla, Colombia weather conditions|weather underground
29. Castro AO (2010) Análisis del potencial energético solar en la Región Caribe para el diseño de un sistema fotovoltaico. INGE CUC. 6:95–102
30. Kalogirou SA (2004) Solar thermal collectors and applications. Prog Energy Combust Sci 30:231–295. https://doi.org/10.1016/J.PECS.2004.02.001
31. Bellos E, Tzivanidis C, Symeou C, Antonopoulos KA (2017) Energetic, exergetic and financial evaluation of a solar driven absorption chiller—a dynamic approach. Energy Convers Manag 137:34–48. https://doi.org/10.1016/j.enconman.2017.01.041
32. Asadi J, Amani P, Amani M, Kasaeian A, Bahiraei M (2018) Thermo-economic analysis and multi-objective optimization of absorption cooling system driven by various solar collectors. Energy Convers Manag 173:715–727. https://doi.org/10.1016/j.enconman.2018.08.013
33. Libotean S, Martín A, Salavera D, Valles M, Esteve X, Coronas A (2008) Densities, viscosities, and heat capacities of ammonia + lithium nitrate and ammonia + lithium nitrate + water solutions between (293.15 and 353.15) K. J Chem Eng Data 53:2383–2388. https://doi.org/10.1021/je8003035
34. Libotean S, Salavera D, Valles M, Esteve X, Coronas A (2007) Vapor-liquid equilibrium of ammonia + lithium nitrate + water and ammonia + lithium nitrate solutions from (293.15 to 353.15) K. J Chem Eng Data 52:1050–1055. https://doi.org/10.1021/je7000045
35. Cuenca Y, Salavera D, Vernet A, Teja AS, Vallès M (2014) Thermal conductivity of ammonia + lithium nitrate and ammonia + lithium nitrate + water solutions over a wide range of concentrations and temperatures. Int J Refrig 38:333–340. https://doi.org/10.1016/j.ijrefrig.2013.08.010

36. Haltenberger W (1939) Enthalpy-concentration charts from vapor pressure data. Ind Eng Chem 31:783–786. https://doi.org/10.1021/ie50354a032
37. McNeely LA (1979) Thermodynamic properties of aqueous solutions of lithium bromide. ASHRAE Trans 85:413–434
38. Absorsistem (2014) Plantas enfriadoras de agua por ciclo de absorción, accionadas por agua caliente
39. Elberry MF, Elsayed AA, Teamah MA, Abdel-Rahman AA, Elsafty AF (2018) Performance improvement of power plants using absorption cooling system. Alexandria Eng J 57:2679–2686. https://doi.org/10.1016/j.aej.2017.10.004
40. Khan Z, Khan ZA (2019) Thermodynamic performance of a novel shell-and-tube heat exchanger incorporating paraffin as thermal storage solution for domestic and commercial applications. Appl Therm Eng 160. https://doi.org/10.1016/j.applthermaleng.2019.114007
41. Herold K, Radermacher R, Klein S (2016). Applications of absorption chillers and heat pumps. https://doi.org/10.1201/b19625-14
42. Jiang J, Gao W, Wei X, Li Y, Kuroki S (2019) Reliability and cost analysis of the redundant design of a combined cooling, heating and power (CCHP) system. Energy Convers Manag 199:111988. https://doi.org/10.1016/j.enconman.2019.111988

# Energy Conservation

# A Method for Recovering the Waste Heat to Achieve Overall Energy Conservation in Aluminum Casting Industries

**R. Aakash Kumar, K. Prabhuram, V. Subrammaniyan, and M. Thenarasu**

## 1 Introduction

There is an imminent need for unconventional methods to curtail energy wastage. One of the industries that consume a large amount of energy for functioning is the casting industry. A huge amount of energy is required as input in foundries, wherein around 70% [1] of the total energy in casting is spent in melting the solid metal. A method that results in even minuscule reductions in the local usage of energy will prove to be an enormous saving of energy in the global scenario. One such method is discussed here. Metal casting is one of the most important processes of manufacturing, and it is obligatory in the production of components in various fields such as agriculture, automotive, aerospace, industrial and domestic products [2]. The idea to make sure the process consumes less amount of energy is to preheat the next batch of raw materials scrap by absorbing the heat emitted by the solidifying casting. This paper proposes an industrially feasible method to conserve energy in foundries to reduce carbon emission from fossil fuels by reducing energy wastage [3]. The real-life consequences of other such Industrial energy conservation have been compiled in [4] when experimented in Basque Country. About the human exploitation and the extended usage of fossil fuels, the reduction of the emission of carbon and its components is indispensable. An effective analysis of the environmental impact by conservation in resources consumed during traditional manufacturing processes such as sand casting is mentioned in [5]. Such an energy conservation process if

R. A. Kumar · M. Thenarasu (✉)
Department of Mechanical Engineering, Amrita School of Engineering, Vishwa Vidyapeetham, Coimbatore, Amrita, India
e-mail: m_thenarasu@cb.amrita.edu

K. Prabhuram
Tata Consultancy Services, Bangalore, India

V. Subrammaniyan
Department of Human Resources, Indian Institute of Management, Tiruchirappalli, India

 
N. L. Narasimhan et al. (eds.), *Recent Advances in Energy Technologies*, Lecture Notes in Mechanical Engineering, https://doi.org/10.1007/978-981-19-3467-4_20

implemented in large industrial applications could be very beneficial in mitigating the rampant use of resources.

## 2 Analytical Model

Some other methods for scrap preheating using recovered heat include preheating steel scrap using exhaust gas from steelmaking furnaces [6]. Preheating of scrap is also found to be more effective in countries with optimal sunlight availability [7]. This method is very eco-friendly but in places where there is very minimal sunlight available, conservation of dissipating heat holds good rendering it unusable. Design of risers in such a way that they can be used as pipes, etc., [8]. Scrap preheating technology by the rotary kiln was investigated as a research activity of the Japanese national project "Shinseiko Process Forum" for developing a high energy efficient scrap melting process for a total scrap recycling system [9].

This method differs from the others as it focuses on preheating the solid raw material by extracting energy from solidifying molten metal. The idea is to embed aluminum shots in the sand casting, after the aluminum shot absorbs some of the sensible and latent heat that is conducted to the sand mold from the molten metal, it is then used to preheat the solid metal. This idea of using the energy from solidifying metal for scrap preheating is an innovative and novel one. In the sand mold, the aluminum shots are added in such a way that they are placed near the cavity to absorb most of the heat emitted as the molten metal solidifies. The aluminum shots are taken out of the mold after the casting process is over and then sent onto an insulated box with the solid raw material, where the heat gets transferred from the shots to the raw material and preheating occurs. Thus, the preheating process reduces the amount of energy required to melt it. The step by step process to prepare the setup is,

1. Along with the molding sand, aluminum—shots are mixed.
2. The molding sand-shots mixture is then filled around the cuboidal pattern.
3. They are then rammed and compacted around the pattern to form a mold.
4. The pattern is removed carefully to obtain a perfectly finished mold.
5. At the required distances, the K-type thermocouples are placed. (10 cm and 20 cm from the mold cavity).

The molten aluminum is decanted inside the mold cavity created by the pattern. The molten aluminum is decanted inside the mound cavity created by the pattern. The first batch is processed in the traditional method of sand casting. The procedure for the second batch and onwards is depicted in Fig. 1. The additional steps done to implement the energy conservation method are indicated in color, wherein we mix the aluminum shots along with the molding sand before pouring the molten metal into the cuboidal pattern. The same aluminum shots which now have absorbed the energy released by the metal in the solidification process are separated from the molding sand once the solidification process is completed. The aluminum shots are

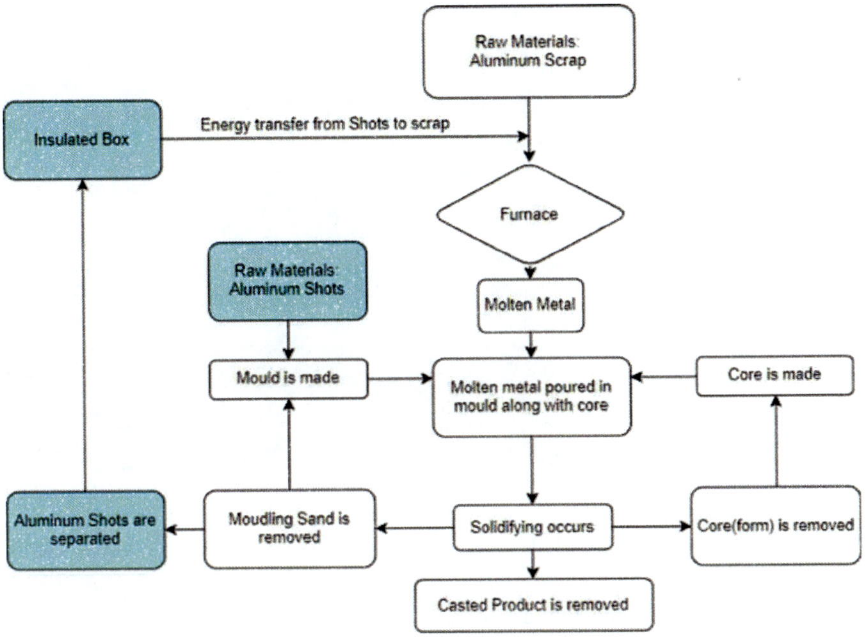


**Fig. 1** Flowchart depicting the procedure for the second batch and onwards

separated from the sand from the first batch by using sieve. The aluminum shots are placed along with the raw material for the next batch in a well-insulated box for preheating the next batch of raw materials.

## 2.1 *Experimental Setup*

The setup of the experiment consists of standard foundry components and the pattern. Cope and drag of dimensions 31 × 31 × 9 cm and a cuboidal pattern of dimensions 7.5 × 7.3 × 5 cm with the cast metal being aluminum (LM-25 grade) are utilized for the experiment.

An electric muffle furnace is used to melt the aluminum metal. The thermocouples were positioned in the drag (Fig. 2) so that the temperature variation in the metal cast and the temperature acquired by the aluminum shots could be observed and compared. The K-type thermocouple (NiCr–NiAl) used here has a cap of 1473 K, having a stabilizing time of 2 min, a working temperature range of 223–1023 K and a precision range of ±1 K. A digital thermometer was utilized.

**Fig. 2** Position of the thermocouple and temperature display

## 3 Results and Discussion

The hot molten metal is decanted into the mold cavity where it solidifies to acquire its shape of it. Due to the phase change taking place during solidification, energy is lost in the form of heat to the sand [10] and cut-wire shot mixture surrounding it [11]. This temperature change data is collected from the thermocouples and recorded as shown in Fig. 3. This data is then used to plot a temperature–time graph (Fig. 4), from which a temperature rise history of the aluminum shots is observed.

Now, the aluminum shots are separated from the mold and mixed with the next batch of scrap. The temperature changes of both the shots and the raw material scrap are noted periodically with the help of two thermocouples with one in the mixture and the other one placed in the solid metal. The data obtained is used to construct a

**Fig. 3** Arrangement of the thermocouple inside the aluminum shots and the connection to the temperature display

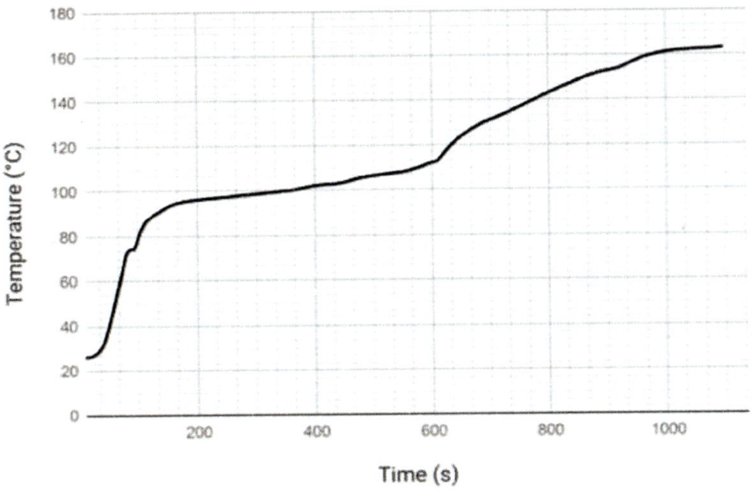


**Fig. 4** Temperature rise of the shots placed in the sand mold

temperature–time graph (Fig. 5) from which it is observed that when the temperature of the cut-wire shots decreases, the temperature of the solid metal increases. The heat energy emitted from the wire shots is transferred by conduction to the solid raw material which is at proximity. This reduces the electrical energy required to heat the raw material scrap from the initial temperature thereby conserving energy.

This phenomenon is governed by the heat conduction equation given by Fourier's law of heat conduction in Eq. (1) where $k$ is the thermal conductivity of aluminum LM25 and $T$ is the temperature in kelvin.

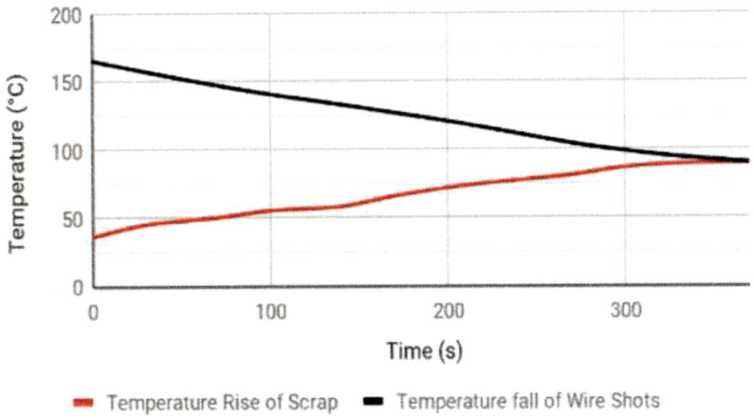


**Fig. 5** Temperature progression trend of shots and scrap metal

**Table 1** Properties of aluminum LM24

| Density (ρ) | Specific heat capacity (c) | Volume (V) | Mass to be melted (ρ × V) |
|---|---|---|---|
| 2680 kg/m$^3$ | 0.91 kJ/kg K | 390 × 10$^{-6}$ m$^3$ (70% yield) | 1.045 kg ≈1.05 kg |

$$k\left(\frac{\partial^2 T}{\partial x^2}+\frac{\partial^2 T}{\partial y^2}+\frac{\partial^2 T}{\partial z^2}\right)=\frac{\partial h}{\partial t} \tag{1}$$

Heat energy is given by

$$h=\int \rho c dT \tag{2}$$

Therefore, by applying the boundary conditions to the above equation, Eq. (3) is obtained.

$$q=k\frac{\partial T}{\partial x}|_{x=0}=k\frac{\partial T}{\partial y}|_{y=0}=k\frac{\partial T}{\partial z}|_{z=0}=h(T-T_{atm}) \tag{3}$$

From these Eqs. (1) and (3), the heat transfer during the solidification process can be calculated [12].

## 3.1 *Calculation to Estimate the Energy Recovered*

The list of properties needed for the calculation of the energy required to melt the metal is given in Table 1.

The energy required to melt the metal is given by the Eq. (4)

$$\mathrm{E}=\Sigma \mathrm{mcTi} \tag{4}$$

Therefore,

$$\mathrm{E}=\mathrm{mcT1}+\mathrm{mcT2}+\mathrm{mL} \tag{5}$$

The required energy to melt 1.1 kg of aluminum Eb = 1081.5 kJ.

## 3.2 *Retrieval of Energy Per Kilogram*

From the experiment, it is known that the heat gets transferred from the shots to the scrap aluminum metal and rises the temperature of the scrap metal up to 90 °C.

Energy recovered from 1.1 kg aluminum

$$= \mathrm{Er} = \mathrm{mc}\Delta\mathrm{T} = 1.05 * 0.91 * (363 - 298) = 61.82\mathrm{kJ}. \quad (6)$$

Therefore, energy recovered per kg of aluminum is 58.88 kJ. The percentage of energy recovery is given by the Eq. (7) [10]

$$= (\mathrm{Er}/\mathrm{Eb}) \times 100 = (61.82/1081.5) * 100 = 5.71\% \quad (7)$$

Hence, an energy recovery of 5.71% of the input energy in the form of heat is possible to obtain from the above-conducted experiment.

### 3.3 *Cooling Curve Comparison*

The cooling curve comparison study for this method is highly significant because the shots are introduced to acquire the heat given out from the solidifying casting and retain it. Temperature is an intensive property [13]. When an external agent such as a metal shot is introduced to acquire heat and retain it, there are possibilities for changes in the cooling curve [14]. It is essential to have control over the cooling rates of the solidifying casting because this will lead to [15].

- Improved mechanical properties.
- Improved response to the heat treatment.
- Improved casting feeding characteristics, thus reducing shrinkage porosity.
- Reducing hot tearing.

Figure 6 depicts the cooling curve of the cast aluminum LM25 is given by Fig. 6. This is taken under normal conditions without the aluminum shots being used to harvest the heat given out by the solidification process [16]. In Fig. 5, the solidification occurs between 200 and 800 s when the latent heat is lost from the liquid metal as it solidifies has been plotted.

Figure 7 gives the experimental cooling curve obtained after the use of aluminum shots along with the casting sand to harvest the heat given out by the solidifying metal casting in the interval between 200 and 800 s in which solidification occurs. From the above figures, it can be noted that the cooling curves are identical as the solidification occurs in the same intervals between similar temperature ranges.

Thus, this minor deviation in solidification rate which is a subset of the cooling curve, will not affect the microstructure of the solidifying metal casting [17] causing major changes in the metallurgic and mechanical properties of the cast product. In the future research, a comparative study in cooling rates can be further done to investigate the effects of changes to cooling rates in the mechanical properties of the cast product.

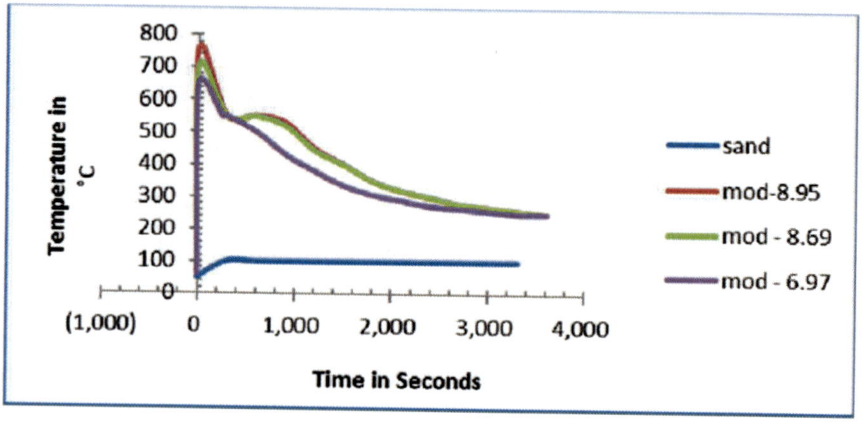


**Fig. 6** Cooling curve of aluminum LM25

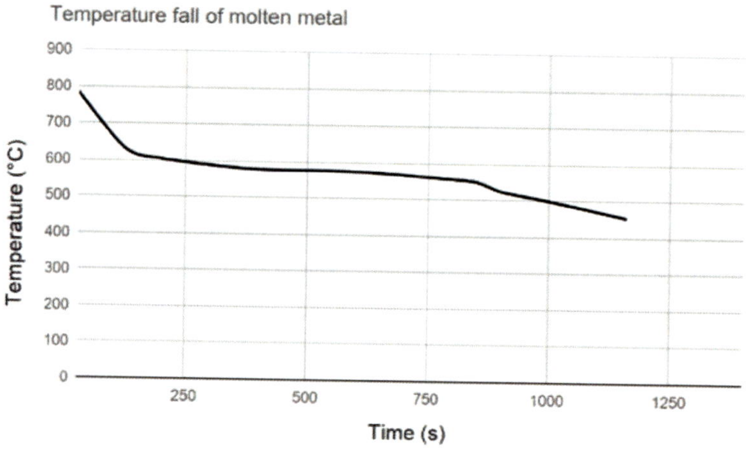


**Fig. 7** Cooling curve with aluminum shots

## 3.4 *Total Energy Savings Per year*

Increasing demands for cast products in the market keep the foundry industry growing significantly every year (Fig. 8) [18]. In the year 2020, the total cast products that were processed globally by a foundry recorded a total of 11.49 million metric tons [19] of both ferrous and non-ferrous metals. So, when production keeps increasing, the energy required for production will also increase. There is a vital need to reduce the energy consumed by the foundry industry to have a positive impact on the environment and the economy. The total number of aluminum castings produced per year (in metric tons) is given by $N$. The symbol $P$ denotes the average power required

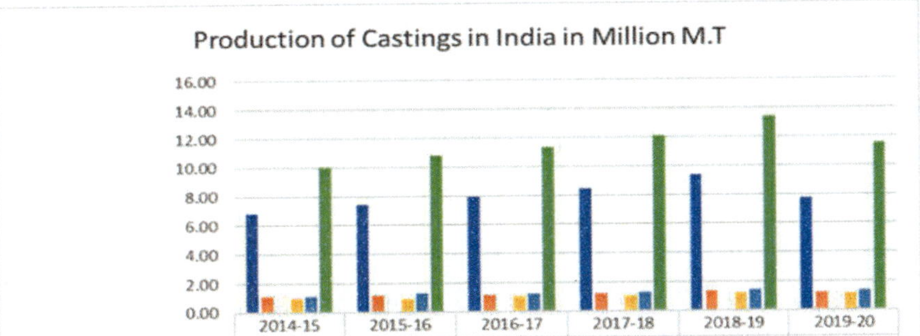


| | 2014-15 | 2015-16 | 2016-17 | 2017-18 | 2018-19 | 2019-20 |
|---|---|---|---|---|---|---|
| Grey Cast Iron | 6.83 | 7.41 | 7.89 | 8.44 | 9.41 | 7.72 |
| Spheroial Graphite Iron | 1.07 | 1.18 | 1.18 | 1.23 | 1.31 | 1.22 |
| Malleable | 0.06 | 0.05 | 0.05 | 0.05 | 0.05 | 0.05 |
| Steel | 0.97 | 0.88 | 1.01 | 1.03 | 1.21 | 1.14 |
| Non-Ferrous | 1.09 | 1.25 | 1.22 | 1.30 | 1.40 | 1.36 |
| Total | 10.02 | 10.77 | 11.35 | 12.05 | 13.38 | 11.49 |

**Fig. 8** Increase in production of casting in India from 2014 to 2020

in melting one ton of aluminum. The symbol $E$ denotes the percentage of energy recovered by the process. By the application of this method, total energy savings per year is given by the Eq. (8)

$$= N * P * E = (1{,}300{,}000 * 1276) * 5.7\% = 94{,}551{,}600\text{kWh} \tag{8}$$

A total annual savings of 94.55 GWh of electric power [20] can be achieved in the aluminum casting industries in India alone. Extension of this method to the casting processes of other metals all over the world will save a humongous amount of electric power. This method can still be optimized by the use of insulating materials like fiber glass to reduce further heat loss, thereby increasing the percentage energy recovery of about 9% being achievable proves the underlying importance of this method.

### *3.5 Economic Benefits of the Technique*

A new method is profitable only when it turns out to be economically beneficial to the manufacturer. Since the harvesting and reuse of heat energy are directly related to the overhead charges of the company, any changes to the energy consumption will have a direct impact on the expenditure of a company. The recovery of 5.7% of the input energy brings down the energy consumption of a commercial casting industry to a considerable extent [18, 20]. Considering a company producing aluminum castings which consume 1276 kWh of electric power per ton [21], the application of 5.7% heat recovery saves 72.7 kWh of electric energy required in the next batch. In the year 2020, India produced 1,300,000 metric tons of aluminum castings. The cost of 1 kWh electric power for medium and large scale industries in India ranges from INR 6 to INR 10. By applying this cost to the energy saved in the aluminum casting industries in India, expenses of INR 91 crores per year can be saved by this energy conservation method. This will turn out to be enormously economically beneficial to the manufacturing companies.

## 3.6 *Ecological Benefits of This Technique*

Production companies have an important standard to be maintained concerning the environmental impact of the company. This includes the carbon footprint that is left behind during production processes. This corresponds to minimizing the emission of carbon dioxide and other greenhouse gasses. The fact is that still, 65.3% of the global electric power produced is primarily from coal, oil and natural gas [20]. These fossil fuels when burnt emits large amounts of greenhouse gases. It is a well-known fact that the foundry industry is a rapidly growing industry and is one of the major energy-consuming industries in the manufacturing sector. By the implementation of this method to reuse heat energy, the consumption of electric power in foundries is reduced. This reduces the power demand in the energy market, ultimately reducing power production. The carbon dioxide emission by the source of power production globally is given in Table 2 [22].

From the calculations, it is known that the production of 1 kWh of electric power, on average emits 0.49 kg of carbon dioxide. On application of this process, the carbon dioxide emission associated with the power production of the conserved energy is cut down. Hence, the $CO_2$ emissions cut down is calculated from Eq. (9).

**Table 2** Carbon dioxide emission by the source of power production

| Generation source | Power production by source (%) | Average $CO_2$ emissions (g/kWh) | Weighted contribution of $CO_2$ by sources (g/kWh) |
|---|---|---|---|
| Coal | 38.3 | 909 | 348.147 |
| Natural gas | 23.1 | 465 | 107.415 |
| Oil/Diesel | 3.7 | 821 | 30.377 |
| Nuclear | 10.4 | 6 | 0.624 |
| Hydro-electric | 16.6 | 4 | 0.664 |
| Geo-thermal | 0.9 | – | 1 |
| Solar/PV | 2 | 105 | 2.1 |
| Wind | 5 | 13 | 0.65 |
| The weighted average contribution of $CO_2$ during power production by source | | | 490.977 |

$$=N * P * E * CO_2 \text{ emission per kWh } //$$
$$=(1,300,000 * 1276) * 5.7\% * 0.490977 = 46.42 \text{ kilotons} \quad (9)$$

As high as 46.42 kilotons of $CO_2$ emissions are cut down when the calculations are applied to the aluminum cast production industries in India. When the method is applied to casting other metals in most of the countries, this method is sure to bring a change in the carbon footprint left by the foundry industries in the world. Considering all these factors and the increasing global warming effect, this method turns out to be very highly beneficial for the welfare of the environment [23].

## 4 Conclusion

An innovative method to harvest and reuse the waste heat emitted by the solidifying casting to preheat the raw material ingots is proposed in this paper. The experimental data confirms that 5.7% of the input power is recovered as heat during the cooling process. Since the most predominant manufacturing process is casting, the application of this method saves about 90.91 GWh of electric power used by the aluminum casting industries in India. This has a major impact on the reduction of production costs ultimately benefiting the economy and reducing emissions. It is to be noted that the experimental conditions were not industrial, if this is applied in an industrial environment with proper insulation of the components and optimized workflow, this method can recover more of the energy input. This innovative method not only applies to casting, similar methods can be developed for other manufacturing process in which there is potential to recover the heat dissipating into the environment. The method discussed in this paper cannot be extended to die casting and is specific to sand casting. There is also no intermediate medium of heat transfer from the molten metal to the external surroundings where such shots can be placed. In the case of the ~~s~~and mold, the shots can be mixed along with the molding sand for heat absorption. Further, research is required in extending this technique to the die casting process.

## References


1. Schifo JF, Radia JT (2004) Theoretical/best practice energy use in metal casting operations
2. Thenarasu M, Ramnadh L, Leeladhar PVVVK, Arun kumar M (2015) Investigation of mechanical and wear behaviour of LM 24 aluminium alloy for different types of casting. International Journal of Applied Engineering Research 10(19):14838–14842
3. Mathew MM, Thenarasu M, Aravind G, Selvaraj J (2016) Performance comparison for aluminium, copper and steel Shots in waste heat recovery and scrap preheating from solidifying molten metal. ARPN Journal of Engineering and Applied Sciences 11(9):6094–6099
4. Larrinaga P, Campos-Celador Á, Legarreta J, Diarce G (2021) Evaluation of the theoretical, technical and economic potential of industrial waste heat recovery in the Basque Country. Journal of Cleaner Production 127494

5. Zheng J, Chen A, Zheng W, Zhou X, Bai B, Wu J, Ling W, Ma H, Wang W (2020) Effectiveness analysis of resources consumption, environmental impact and production efficiency in traditional manufacturing using new technologies: case from sand casting. Energy Convers Manage 209:112671
6. Arink T, Hassan MI (2017) Metal scrap preheating using flue gas waste heat. Energy Procedia 105:4788–4795
7. Selvaraj J, Jawahar CC, Bhatija KA, Thenagan S (2015) Preheating metal scrap in foundries using solar thermal energy. Advances in mechanical engineering. Trans Tech Publications, pp 760–767
8. Prabhuram K, Subrammaniyan V, Thenarasu M (2020) Design and implementation of product embodied riser for energy conservation in aluminium casting process. Advances in materials and manufacturing engineering. Lecture notes in mechanical engineering, pp 449–456
9. Website [cited 24 Nov 2018]. Available: https://doi.org/10.4262/denkiseiko.72.21
10. Srinivasan K, Siddharth CSK, Arun Kaarthic LV, Thenarasu M (2018) Evaluation of mechanical properties, economic and environmental benefits of partially replacing silica sand with biomass ash for aluminium casting. In: Proceedings of materials today, pp 12984–12992.
11. Pariona MM, Mossi AC (2005) Numerical simulation of heat transfer during the solidification of pure iron in sand and mullite molds. J Brazil Soc Mech Sci Eng 27
12. Cengel YA, Boles MA (2006) Thermodynamics: an engineering approach. McGraw-Hill Science, Engineering and Mathematics
13. Arnberg L, American Foundrymen's Society, Bäckerud L, Chai G (1996) Solidification characteristics of aluminium alloys: dendrite coherency
14. Procter RJ (2017) Cutting carbon emissions from electricity generation. Electr J 30:41–46
15. Selvaraj J, Thenarasu M, Aravind S, Ashok P (2015) Waste heat recovery from castings. Appl Mech Mater 813–814:776–781
16. Gopalakrishna VS, Marimuthu P (2008) Optimization of parameters effecting the heat recovery from a sand casting process. Periodicals of Engineering and Natural Sciences 6(2):100–108
17. Zhang LY, Jiang YH, Ma Z, Shan SF, Jia YZ, Fan CZ et al (2008) Effect of cooling rate on solidified microstructure and mechanical properties of aluminium-A356 alloy. J Mater Process Technol 207:107–111
18. Foundry Informatics Centre, Production of Casting. Available: http://foundryinfo-india.org/Production_of_castings.aspx
19. Foundry-info Census-2016 [cited 13 Nov 2019]. Available: http://www.foundryinfo-india.org/statistics/Census-2016-Modern-Casting.pdf
20. None, none, BCS, Inc, Laurel, MD (United States) (2005) Advanced melting technologies: energy saving concepts and opportunities for the metal casting industry. https://doi.org/10.2172/1218650
21. International Energy Agency (2018) Key world energy statistics 2018
22. Clayton J (2019) 1 kilowatt-hour BlueSkyModel. [cited 12 Nov 2019]. Available: https://blueskymodel.org/kilowatt-hour
23. Jiliang Z, Junting LI (2011) Study on the ecologic network system of energy-intensive industries. Energy Procedia 5:1987–1992

# Electric Vehicles

# Design and Comprehensive Analysis of Synchronous Reluctance Motor for Automotive Trike Applications

**V. S. Nagarajan, V. Rajini, M. Harish Babu, P. Akash, S. Sivaramakrishnan, and S. Babu Venkatesh**

## 1 Introduction

### *1.1 A Subsection Sample*

The traditional three phase induction motors [1, 2] have been the choice of electric motor for three-wheeler automotive applications owing to their ruggedness, economical price, self-starting capability and ease of service [3, 4]. However, the challenges include maintaining international efficiency (IE) standards through appropriate sizing which in turn affects torque density and necessitates devising proper cooling mechanisms. These problems are overcome by a transverse laminated synchronous reluctance motor [6–8], which has higher torque density. Also, the added advantages include that of utilization of similar stator and power electronic circuit as employed for a three-phase induction motor drive [9, 10].

The existence of anisotropy in the rotor of transverse synchronous reluctance motor [5, 8] underlines the need to evolve a novel design procedure which involves evaluation of designs suited to automotive application in terms of performance measures such as average torque [8], torque ripple [8, 11, 12] and efficiency in a specified range.

The process initially involves iterative coupled statistical sensitivity analysis and electromagnetic analysis of the design variables by level sensitivity analysis [11] and electromagnetic FEA to appropriately choose geometrical design variables in achieving the required performance criteria. Finite element analysis (FEA) [11] tool, MagNet and MotorSolve are used for electromagnetic-thermal performance analysis involved in geometric parameter variations. The designs obtained through the design procedure are evaluated for vibrational [13] performance by the use of Ansys. Finally,

---

V. S. Nagarajan (✉) · V. Rajini · M. H. Babu · P. Akash · S. Sivaramakrishnan · S. B. Venkatesh
Department of Electrical and Electronics Engineering, Sri Sivasubramaniya Nadar College of Engineering, Kalavakkam, Chennai, India
e-mail: nagarajanvs@ssn.edu.in



N. L. Narasimhan et al. (eds.), *Recent Advances in Energy Technologies*, Lecture Notes in Mechanical Engineering, https://doi.org/10.1007/978-981-19-3467-4_21

a design is fabricated and tested indicating the achievement of required performance through the design procedure.

The organization of the paper is as follows: Sect. 2 deals with the overall design procedure. Electromagnetic analysis is presented in Sect. 3. Sensitivity analysis based on level sensitivity approach and FEA is discussed in Sect. 4. It is followed by Sect. 5, which deals with coupled electromagnetic-thermal-vibrational performance analysis. Experimental test facility for the fabricated prototype is presented in Sect. 6. Results of experimental analysis are presented in Sect. 7. The performance of TL-SyncRel motor is compared to performance of induction motors in Sect. 8. Section 9 presents the conclusion.

## 2 Design Procedure for Automotive Application

The overall design procedure involving electromagnetic analysis, sensitivity analysis, thermal analysis and vibration analysis for TL-SyncRel [14] motor with respect to automotive application is represented in Fig. 1.

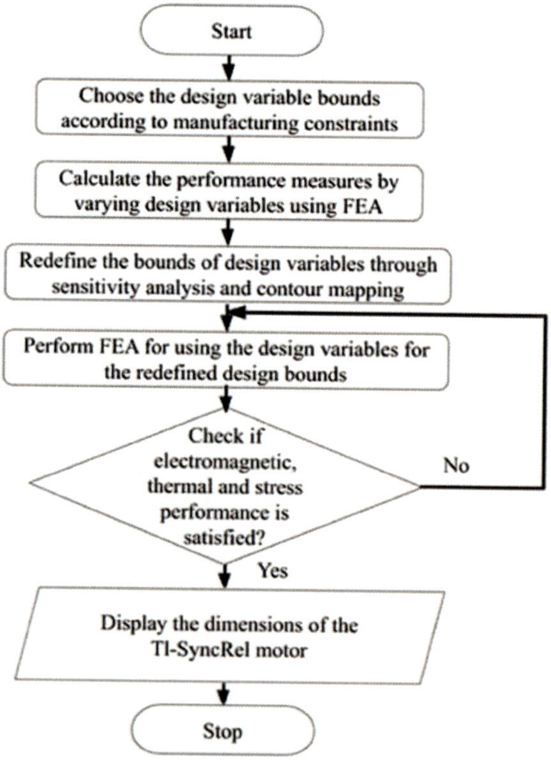


**Fig. 1** Flowchart for overall design procedure of TL-SyncRel motor

**Table 1** TL-SyncRel motor performance requirement

| Performance measure | Requirement |
|---|---|
| Torque output | ≥7 Nm |
| Efficiency | ≥89% |
| Torque ripple | ≤0.15 Nm |

For the automotive application chosen in this work, the performance requirements are specified in Table 1.

## 3 Electromagnetic Analysis

The TL-SyncRel motor [12] for electromagnetic analysis is chosen based on number of stator slots and number of flux barriers. The choice of slot/barrier combination made is 36/4 in order to obtain designs with improved torque density. The 1/4th cross section of TL-SyncRel motor featuring the design variables which affects the performance of the TL-SyncRel motor used in electromagnetic analysis is represented in Fig. 2.

The average torque is obtained from FEA. The torque ripple is estimated using

$$T_{\text{ripple}} = \frac{T_{\max} - T_{\min}}{T_{\text{av}}} \tag{1}$$

The scripting existing in MagNet FEA software is used to calculate the total loss, in order to determine the efficiency of the motor. The design variables chosen for the TL-SyncRel motor with the initial bounds are listed in Table 2. The range of the parameters is limited to manufacturing process.

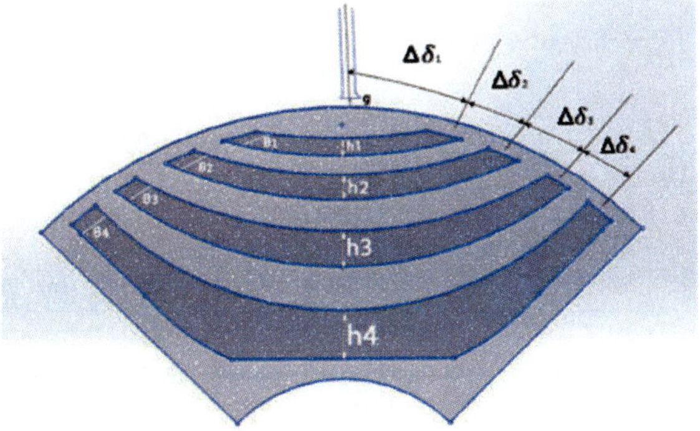


**Fig. 2** TL-SyncRel motor design variables for analysis

**Table 2** Initial bounds of design variables

| Design parameter | Symbol | Design range |
|---|---|---|
| Barrier height (s) | $h_1, h_2, h_3, h_4$ | 2–6 mm |
| Flux barrier angle (s) | $\theta_1, \theta_2, \theta_3, \theta_4$ | 0°–7° |
| Barrier Angular position (s) at the air gap | $\Delta\delta_1, \Delta\delta_2, \Delta\delta_3, \Delta\delta_4$ | 7°–16° |
| Airgap length | $g$ | 0.3–0.6 mm |

The scripting existing in MagNet FEA software is used to calculate the total loss, in order to determine the efficiency of the motor. The design variables chosen for the TL-SyncRel motor with the initial bounds are listed in Table 2. The range of the parameters is limited to manufacturing process.

## 4 Sensitivity Analysis and Optimal Design Space Establishment

The sensitivity perturbation [11] across the target and design variables can be studied as a measure of the relative changes in these variables. In this technique, the relative change in the optimization objective with and without any disturbance in the design variable is measured, and the ratio of these two quantities is calculated as a sensitivity coefficient for that particular pair of the target variable and design variable. This coefficient takes into account the relative difference rather than the absolute difference to measure a design variable's impact in our optimization objective's outcome. The procedure to be followed in estimating the sensitivity indices by this method along with grouping of design variables is shown in Fig. 3.

Sensitivity index is obtained using the formula,

$$\text{Sensitivity index} = \frac{\text{percentage changes in the objective}}{\text{percentage change in the value of design variable}} \tag{2}$$

Also, using sensitivity indices, cumulative sensitivity function can be determined using

$$S_{\text{com}}(x_{\text{i}}) = \lambda_1|S_{\text{y1}}(x_{\text{i}})| + \lambda_2|S_{\text{y2}}(x_{\text{i}})| + \ldots \lambda_3|S_{\text{y3}}(x_{\text{i}})| \tag{3}$$

The sensitivity analysis results are presented for level sensitivity analysis method discussed in the previous section in Table 3, considering the case of four-barrier Tl-SyncRel structure. Also, in Table 3, the strongly sensitive design variables are highlighted in comparison with the mildly sensitive design variables. The sensitivity indices using level sensitivity analysis are represented as $S_{\text{av}}(x_{\text{i}})$, $S_{\text{ripple}}(x_{\text{i}})$ and $S_{\text{eff}}(x_{\text{i}})$, respectively.

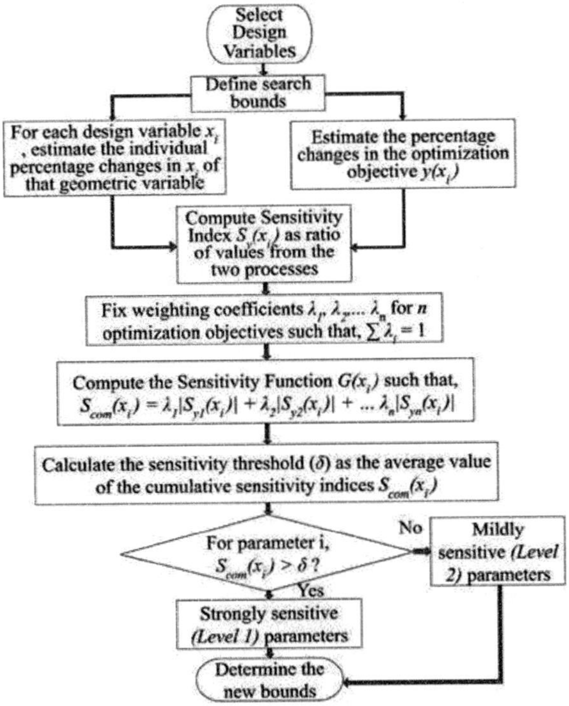


**Fig. 3** Flowchart of sensitivity analysis by level sensitivity method

**Table 3** Sensitivity indices of design variables from Level sensitivity method

| Design variables | $S_{av}(x_i)$ | $S_{ripple}(x_i)$ | $S_{eff}(x_i)$ | $S_{com}(x_i)$ |
|---|---|---|---|---|
| $g$ (mm) | 0.4427 | 0.0937 | 0.0272 | 0.1878 |
| $h_1$ (mm) | −0.3820 | −0.3518 | 0.1446 | 0.2928 |
| $h_2$ (mm) | −0.5002 | −0.2955 | 0.1284 | 0.3080 |
| $h_3$ (mm) | −0.4416 | −0.2758 | 0.0629 | 0.2601 |
| $h_4$ (mm) | −0.4561 | −0.3058 | 0.0729 | 0.1501 |
| $\theta_1$ (degree) | −0.6937 | 0.3846 | 0.1568 | **0.4117** |
| $\theta_2$ (degree) | −0.7619 | 0.4317 | 0.1703 | **0.4546** |
| $\theta_3$ (degree) | −0.4761 | 0.2209 | 0.1635 | 0.2868 |
| $\theta_4$ (degree) | −0.4891 | 0.2390 | 0.1785 | 0.2958 |
| $\Delta\delta_1$ (degree) | 0.6713 | −0.3561 | −0.1322 | **0.3865** |
| $\Delta\delta_2$ (degree) | 0.6227 | −0.2064 | −0.1319 | 0.3203 |
| $\Delta\delta_3$ (degree) | 0.6419 | −0.3291 | −0.1725 | **0.3811** |
| $\Delta\delta_4$ (degree) | 0.6591 | −0.3355 | −0.1865 | 0.2921 |

**Table 4** Refined bounds of design variables based on sensitivity analysis

| Parameter | Symbol | Design bounds |
|---|---|---|
| Barrier height | $h_1$ | 2–4 mm |
| | $h_2$ | 2.5–4.5 mm |
| | $h_3$ | 3.5–5 mm |
| | $h_4$ | 4.5–6 mm |
| Flux barrier angles | $\theta_1$ | 3°–5° |
| | $\theta_2$ | 3°–5° |
| | $\theta_3$ | 2°–4° |
| | $\theta_4$ | 3.5°–4.5° |
| Angular positions at the air gap | $\Delta\delta_1$ | 10°–15° |
| | $\Delta\delta_2$ | 8°–13.5° |
| | $\Delta\delta_3$ | 7°–12° |
| | $\Delta\delta_4$ | 7°–10° |
| Airgap | $g$ | 0.45–0.55 mm |

Based on the sensitivity analysis, the design range is further refined establishing a search space for the combined electromagnetic analysis—thermal analysis—vibration analysis to be adopted to meet application specific requirements as shown in Table 4.

## 5 Electromagnetic-Thermal-Vibration Analysis

Finite element modelling (FEM) is done with computer aided design (CAD) based software packages, MagNet 2021 and MotorSolve 5.1 [11]. The steps involved in the design are

- Preprocessing
- Problem definition
- Solver details specification
- Post processing

Preprocessing involves geometric modelling, choice of material for the components of the motor including stator, rotor, winding, magnet and shaft, boundary condition specification and meshing of the motor to be analysed.

Problem definition involves provision of electrical source for winding excitation defining the operating condition and specification of motion component, if time stepping (or motion) analysis is used. The solver method, consideration of non-linearities in material used, tolerances, maximum number of iterations, frequency of the source and time steps for solving can also be specified.

After the completion of problem definition and solver details specification, the software carries out the solving. Once the simulation is run, the results in terms of torque, magnetic flux density, current, etc., can be obtained from the output tab window.

Postprocessing in terms of charts, animations of magnetic field distribution, excel reports of instantaneous torque, speed, etc., can also be done using the software package.

The finite element model (FEM) and magnetic flux density distribution at the operating conditions is represented in Figs. 4 and 5 respectively.

The temperature plot corresponding to optimal design solution obtained using coupled electromagnetic—thermal analysis [11] is represented in Fig. 6. From Fig. 6, it is justified that thermal loading is within the limit of Class F insulation as the maximum temperature attained is 64 °C in the stator winding.

The vibration analysis [11] of design solution is determined using ANSYS FEA software at 1500 rpm for the rated torque. The construction of flux barriers causes

**Fig. 4** Finite element model of SyncRel motor

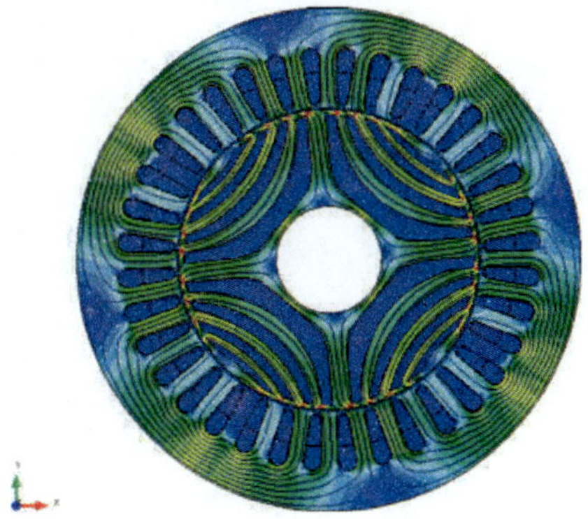

**Fig. 5** Magnetic flux density distribution in SyncRel motor

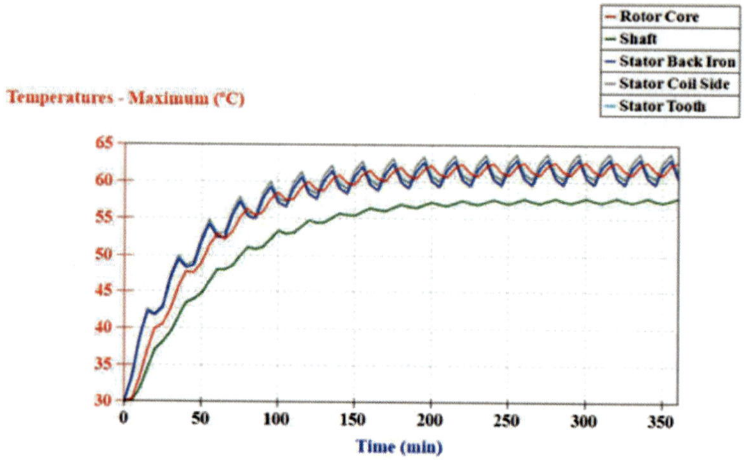


**Fig. 6** Temperature response of TL-SyncRel motor

stresses on the rotor of TL-SyncRel motor. The vibrational performance metric, i.e. von Mises stresses on design solution of TL-SyncRel rotor is represented in Fig. 7.

It is observed from Fig. 7 that the stress induced in rotor of TL-SyncRel motor is within the limit of 295 MPa (yield strength of CRNO M1000 steel used in rotor),

**Fig. 7** Von Mises stress distribution of design solution TL-SyncRel motor

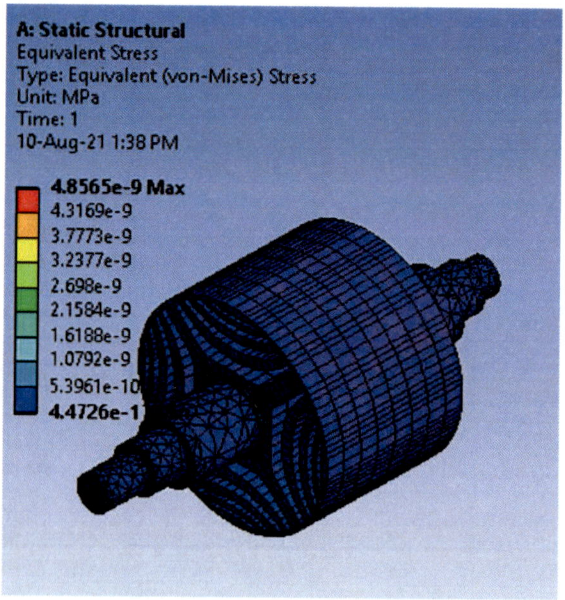

justifying the design procedure resulting in decreased stress as a result of reduced torque ripple.

## 6 Prototyping and Experimental Test Facility

The prototyping of design solution is carried out and a lamination used for rotor is represented in Fig. 8.

The prototype motor is tested using the experimental setup shown in Fig. 9.

**Fig. 8** Rotor lamination of SyncRel motor prototype

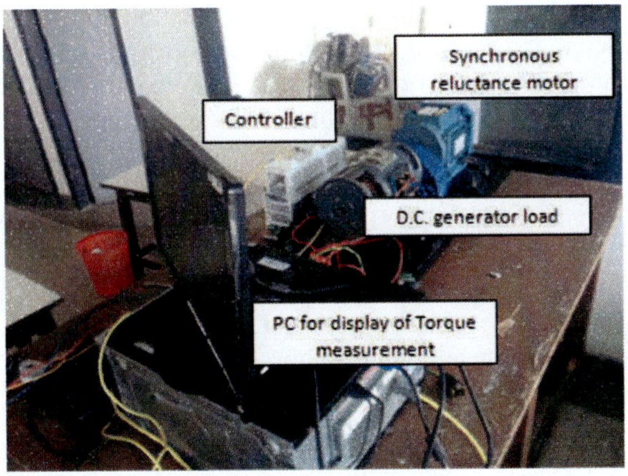


**Fig. 9** Test bench for performance analysis

**Table 5** Optimal design solution

| Parameter | Symbol | Value |
|---|---|---|
| Barrier height | $h_1$ | 2 mm |
| | $h_2$ | 2.7 mm |
| | $h_3$ | 3.7 mm |
| | $h_4$ | 5.2 mm |
| Flux barrier angles | $\theta_1$ | 4.974° |
| | $\theta_2$ | 4.465° |
| | $\theta_3$ | 3.440° |
| | $\theta_4$ | 4.299° |
| Angular positions at the air gap | $\Delta\delta_1$ | 14.537° |
| | $\Delta\delta_2$ | 8.790° |
| | $\Delta\delta_3$ | 8.518° |
| | $\Delta\delta_4$ | 7.199° |
| Airgap | $g$ | 0.5 mm |

**Table 6** Comparison between experimental and FEA results

| Performance parameter | FEA results | Experimental results |
|---|---|---|
| Average torque | 7.58 Nm | 7.44 Nm |
| Torque ripple | 0.13 Nm | 0.15 Nm |
| Efficiency | 91.43% | 89.4% |

## 7 Results and Discussion

Based on the requirements specified in Table 1, particle swarm algorithm is used to determine the optimal solution. The corresponding optimal design solution is chosen for experimental performance evaluation with the parameters as listed in Table 5.

The comparison between FEA and experimental results for rated operating conditions is listed in Table 6.

Results of Table 6 indicate the closeness of the FEA analysis results with the experimental values representing the validity of the design procedure to obtain designs of TL-SyncRel motor as per performance requirement.

The efficiency map corresponding to the design solution is shown in Fig. 10.

From the efficiency map, it is observed that TL-SyncRel motor has a wide operating range suited to automotive application.

## 8 Comparative Analysis with Induction Motor

The TL-SyncRel motor is compared with three-phase induction motor of peak efficiency 82% with same rating in terms of volume and is tabulated in Table 7.

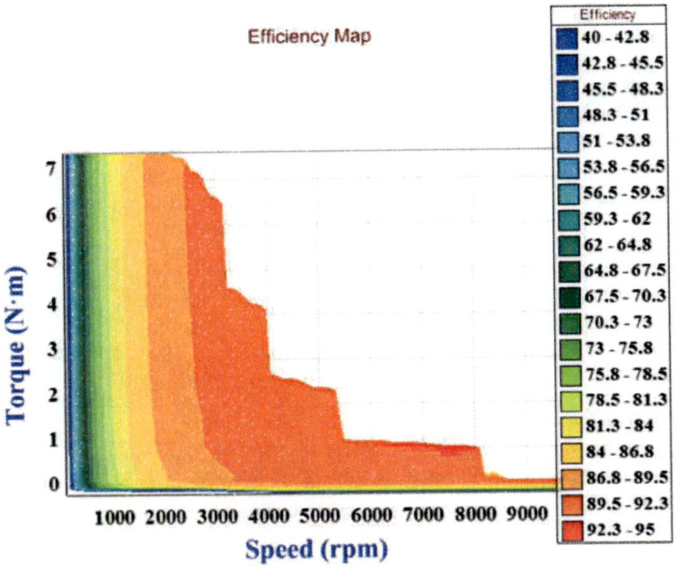


**Fig. 10** Efficiency map

**Table 7** Comparison of volume between Tl-SyncRel motor and induction motor of same power rating

| Performance parameter | Tl-SyncRel motor | Induction motor |
|---|---|---|
| Volume | 1270.117 $cm^3$ | 1627.337 $cm^3$ |

**Table 8** Comparison of volume between Tl-SyncRel motor and Induction motor of same size

| Performance parameter | Tl-SyncRel motor | Induction motor |
|---|---|---|
| Average torque (Nm) | 7.66 Nm | 5.084 Nm |

The performance of TL-SyncRel motor is compared with induction motor of same size and the torque performance result is represented in Table 8.

Results of Tables 7 and 8 indicates the improved torque density of the TL-SyncRel motor in comparison with three phase induction motor making it a suitable alternative for automotive applications.

## 9 Conclusion

A detailed design procedure for TL-SyncRel motor is formulated and evaluated with respect to automotive applications. The bounds of design variables are established by performing a detailed sensitivity study considering torque and efficiency as performance measures. The experimental result on the designed prototype justifies the adopted methodology. The torque density comparison with conventional induction motor signifies the choice of TL-SyncRel motor, which occupies 78% of volume as that of an induction motor and has an efficiency improvement of 7% with respect to induction motor. The generalized design procedure discussed in this work would form the basis of designing TL-SyncRel motor for industrial applications.

**Acknowledgements** This research was supported and funded by SSN Trust.

## References


1. Dmitrievskii V, Prakht V, Kazakbaev V (2019) IE5 energy-efficiency class synchronous reluctance motor with fractional slot winding. IEEE Trans Ind Appl 55(5):4676–4684
2. https://www.mahindraelectric.com/vehicles/treo-electric-auto/
3. Palmieri M, Cascella G, Cupertino F (2019) Design methodologies for the output power maximisation of synchronous reluctance machines. IET Electr Power Appl 13(8):1131–1140
4. Bao Y, Degano M, Wang S, Chuan L, Zhang H, Xu Z, Gerada C (2020) A novel concept of ribless synchronous reluctance motor for enhanced torque capability. IEEE Trans Industr Electron 67(4):2553–2563
5. Chen K, Yu W, Wen C (2019) Rotor optimization for synchronous reluctance motors. China Electrotechnical Society Transactions on Electrical Machines and Systems 3(3):279–284
6. Castagnaro E, Bacco G, Bianchi N (2019) Impact of geometry on the rotor iron losses in synchronous reluctance motors. IEEE Trans Ind Appl 55(6):5865–5872
7. Gamba M, Pellegrino G, Cupertino F (2014) Optimal number of rotor parameters for the automatic design of Synchronous reluctance machines. In: Proceedings of 2014 international conference on electrical machines, ICEM 2014, pp 1334–1340
8. Bianchi N, Degano M, Fornasiero E (2015) Sensitivity analysis of torque ripple reduction of synchronous reluctance and interior PM motors. IEEE Trans Ind Appl 51(1):187–195
9. Bacco G, Bianchi N (2019) Design criteria of flux-barriers in synchronous reluctance machines. IEEE Trans Ind Appl 55(3):2490–2498
10. Oliveira F, Ukil A (2019) Comparative performance analysis of induction and synchronous reluctance motors in chiller systems for energy efficient buildings. IEEE Trans Industr Inf 15(8):4384–4393
11. Nagarajan VS, Kamaraj V, Sivaramakrishnan S (2019) Geometrical sensitivity analysis based on design optimization and multiphysics analysis of PM assisted synchronous reluctance motor. Bull Acad Pol Sci 67(1):155–163
12. Nagarajan VS, Kamaraj V, Balaji M, Arumugam R, Ganesh N, Rashmi R, Sivaramakrishnan S, Rahul R (2018) Synchronous reluctance motor development, modeling, optimization, and analysis—a review. In: 4th international conference on electrical energy systems (ICEES), pp 416–424. https://doi.org/10.1109/ICEES.2018.8442390
13. Nagarajan VS, Balaji M, Kamaraj V, Subash K, BabuVenkatesh S, Babu A, Anirhudh S, Arunkumar R (2019) Design, co-simulation and analysis of reluctance motor drive aided by ferrite for electric vehicle application. In: National Power Electronics Conference (NPEC), pp 1–6. https://doi.org/10.1109/NPEC47332.2019.9034815
14. https://new.abb.com/motors-generators/iec-low-voltage-motors/process-performance-motors/synchronous-reluctance-motors

# A Comparative Life Cycle Assessment of an Electric, a Hybrid, and an Internal Combustion Engine Vehicle Using Monte Carlo Simulation

**Sricharan Dwijesh Kurada, Mirza Imtiaz Ali, and J. Gokulachandran**

## 1 Introduction

### *1.1 General*

Environmental requirements set by governments continue to be a challenge for the automotive industry. Consequently, manufacturers are spending an increasing amount of time and effort on designing cars such that they meet environmental standards throughout their lifecycle (material selection, production, use, and disposal or recycling). Life cycle assessment (LCA) can provide a more objective understanding of resource consumption and environmental impacts during the life cycle of a car. LCA provides the opportunity for improving future cars not only from an ecological standpoint, but also from a technological and economic perspective.

### *1.2 Life Cycle Assessment*

Life cycle assessment (LCA) as defined by the encyclopedia of Ecology in 2019 is "a systematic, standardized approach to quantifying the potential environmental impacts of a product or process that occur from raw materials extraction to end of life". LCA can be done in design stage as well, we can assess the impact that a product may cause in the entirety of its life cycle. This would help to make any necessary changes and ensure sustainability.

LCA's importance lies in products or services that cost less, consume less energy, and produce fewer emissions in the usage phase yet pose a threat to the environment

S. D. Kurada (✉) · M. I. Ali · J. Gokulachandran
Department of Mechanical Engineering, Amrita School of Engineering, Amrita Vishwa Vidyapeetham, Coimbatore, India
e-mail: cb.en.u4mee17135@cb.students.amrita.edu



N. L. Narasimhan et al. (eds.), *Recent Advances in Energy Technologies*, Lecture Notes in Mechanical Engineering, https://doi.org/10.1007/978-981-19-3467-4_22

during other phases of their lives. This is because LCA is a comprehensive approach; it takes into account the impacts throughout a product's life cycle. Imagine that someone develops a process for minimizing the costs, energy consumption, and emissions involved in manufacturing a hair dryer. However, when used, the hair dryer consumes more electricity than it did previously. Depending on how much more electricity is being used, it could countervail the costs, energy units consumed, or emissions savings in the manufacturing of the hair dryer, in turn compromising the objective. LCA can either be performed using software or entirely with a formulative approach, which involves a lot of work on the part of the user.

LCA relies on data that is fed into it, which can either be experimental data or data from reliable sources. Nevertheless, there may be some deviations from the actual values. Monte Carlo Simulation can be used to evaluate such differences. This will help quantify the uncertainties associated with the results.

### *1.3 Monte Carlo Simulation*

In most studies finding emissions of a product, an absolute value is assigned as the input variable, and thus, an absolute value is returned as the output. While this value may be relevant in some scenarios, it might not be relevant to all the possible scenarios. There are a lot of factors that need to be taken into consideration while calculating vehicular emissions. It is not always feasible to account for all the different factors. Monte Carlo Simulation (MCS) may prove useful here. In MCS, variable distributions are used instead of absolute values to take into account any real-time deviations.

Monte Carlo simulation (MCS) is a problem-solving technique that provides the approximate probability of certain outcomes using random variables. With each iteration, it separates out few uncertain parameters and produces the distribution function of all the sampled values.

## 2 Literature

More studies have been done on a specific component of a vehicle than on the vehicle as a whole. Shanbag and Manjare [1] discusses environmental impacts of the manufacturing process for tires. The configuration and all other relevant information were extracted from MRF manufacturers in India. An LCA is performed using SimaPro software. Nitrogen dioxide and sulfur dioxide accounted for the majority of the emissions. The categories with the highest impacts are respiratory inorganics, aquatic acidification, aquatic eutrophication, and terrestrial acidification. Saur et al. [2] gives an overview of the fender of the car and how four versions made of different materials: steel, aluminum, polypropylene, and sheet mold compound, are evaluated for sustainability and recyclability throughout their life cycle. Compounds made from sheet metal have the lowest emissions. Among all materials, aluminum emitted most during the production phase but considerably less during the use phase. In [3], a wind shield wiper made by the famous manufacturer, Bosch, is examined for its life cycle impact. Based on the results of the analysis, improvement options for the wiper are proposed.

In other studies, the focus is on new products with lower carbon footprints that may replace conventional products. Schau et al. [4] analyzes the sustainability of remanufactured alternator from a life cycle perspective. The authors assess remanufactured alternators from a sustainability point of view, evaluating whether they are more efficient than new alternators. For a perspective on the environment, life cycle assessment (LCA) is used. Economic and social dimensions are addressed by Social-LCA and life cycle costing (LCA). The main thing to consider before jumping into the remanufactured alternatives is to assess whether the same benefits follow in the usage phase. It uses a life cycle sustainability assessment approach to remanufacture alternators as a decision support tool for managers and product developers. During the use phase, remanufactured parts are economically and environmentally preferable to new parts. According to this study, some significant measures are needed to improve the sustainability of remanufacturing.

The role of carbon fiber reinforced polymer (CFRP) in reducing environmental emissions throughout a vehicle's life cycle is discussed in [5]. Due to its low weight, CFRP is used for lightweighting automobiles. Considering the difference in emissions between gasoline and electric vehicles, CFRP employment would be more significant for gasoline vehicles.

Luglietti et al. [6] compares the environmental impacts and levels of implementation of different end-of-life alternatives including reuse, recycling, and remanufacturing. In terms of the actual process itself, remanufacturing has the most impact due to the high resource consumption. Overall, the results indicate that remanufacturing is the most environmentally friendly method, while reuse came in second, followed by recycling.

LCA has also been used by authors to compare several products or services. Authors in [7] compare four types of fuels used in automobiles, namely, petrol, diesel, natural gas, and electricity based on their environmental impacts during each of the life stages. It is suggested that in comparison to coal-based electricity produced and used in China, when photovoltaic energy is used to make electric vehicle batteries, the environmental impact is reduced by 69%. [8] compares heavy-duty trucks and hybrid trucks based on their life cycle emissions. Simulation of the drive trains is done using MATLAB Simulink. A break-even analysis is performed to evaluate high $CO_2$ equivalent emissions. The study finds that compared to diesel trucks, hybrid trucks released 4.34 g of $CO_2$ per kilometer fewer emissions.

In a study worth mentioning [9], Monte Carlo simulation was used in interpreting LCA results. Life cycle assessment is performed on an IC engine vehicle and three electric vehicles with different battery types (Ni–Cd, lead acid, and NiMH). Microsoft Crystal Ball is the software used. Moreover, Monte Carlo simulations are performed to identify data uncertainty. Interestingly, emissions were similar between EVs and

ICEVs. EVs, however, greatly reduce annual volatile organic compounds and carbon dioxide emissions.

## 3 Methodology

LCA is an established and widely used tool for evaluating the environmental impact of products and services. A life cycle assessment typically consists of four phases: goal identification, inventory analysis, impact assessment, and result interpretation [10, 11].

The first phase involves defining the objective, scope, and boundary of the study. A life cycle inventory process includes a detailed analysis of the product's materials and energy inputs and outputs. A system component's environmental impact is classified into different impact categories during the assessment phase. Final interpretation involves assessing the impacts to make conclusions or recommendations. An approach of this type can help identify potential reduction opportunities.

### *3.1 Goal Identification*

Life cycle assessment points out the environmental hotspots in a good or a service. It establishes a unique standard against which improvements or new technologies can be measured. Often, vehicle emission estimation is conducted considering only its use phase. However, this neglects to factor in the energy use and emissions emitted during the other phases of its life cycle (e.g., for raw material extraction or end-of-life phases). LCA gives us the opportunity to assess the impacts caused by a vehicle throughout its life cycle [12]. This is motive behind the choice to conduct our study in this field. The objective is to compare three vehicles: an electric vehicle (EV), a hybrid vehicle (HV), and an internal combustion engine vehicle (ICEV). The intent is also to develop a model that can substantiate the use of Monte Carlo simulation in finding out the disparities in life cycle emissions and energy consumption between the electric vehicle (EV), the hybrid vehicle (HV), and the internal combustion engine vehicle (ICEV) per kilometer (km) traveled basis.

The software used for this study is OpenLCA 1.10.3 by GreenDelta, an independent sustainability consulting and software company. OpenLCA provides a range of support solutions and databases.

As a primary data source, this study used version 3.4.1 of the ecoinvent database, while data from the literature was also utilized. Ecoinvent is an established LCI database that has been compliant with ISO 14040 since 2000. The database is regarded as the world's most consistent and transparent life cycle inventory database. As this study is carried out in India and is restricted to the Indian subcontinent, the available data from the data sets will need to be modeled based on real world. Accordingly, the vehicles chosen are Indian cars popular among the general population.

The study focuses on the differences between an EV, a HV, and an ICEV. The comparisons were conducted among mid-size hatchbacks. Studies have shown that consumers prefer petrol-powered cars to diesel-powered ones for internal combustion engines. The main reasons for its popularity were its ease of maintenance and smooth driving experience. Thus, a design that is affordable to most budget shoppers was chosen. Maruti Suzuki's Swift petrol version has been chosen for these reasons. Since there are currently very few electric cars available in India, there were very few choices when choosing an electric car. The Nexon EV by Tata was a top choice since it has proven to be a low cost and reliable compact vehicle. Therefore, this model of electric vehicle has been considered. Drive trains with hybrid techniques combine a traditional internal combustion engine with an electric motor for improved power and fuel efficiency. Hybrid technologies are relatively new in India.

Maruti Suzuki offers mild hybrid versions of many of its vehicle models. The vehicles are commonly referred to as smart hybrid vehicles by suzuki (SHVSs). Compared with fully electric vehicles, the battery on these vehicles is much smaller. Based on the few choices available, Maruti Suzuki's Baleno Dual Jet mild hybrid car was selected for the study.

In order to gather reliable data, the companies that manufacture these cars were contacted, and several official data sources were referred by means of the Internet. For reasons of information security, the companies decided not to provide information.

To obtain data, we rely on literature and websites of automobile manufacturers. Detailed data of vehicles that have been chosen for the study are presented in Table 1. The system boundaries for the vehicles selected are represented as flowcharts in Figs. 1 and 2.

In hybrid and internal combustion engine vehicles, system boundaries are identical. Figure 1 displays the system boundaries for both hybrids and internal combustion engine vehicles. On the other hand (Fig. 2), electric vehicles have many more problems because, unlike hybrid vehicles, which use a very small battery, the batteries used in electric vehicles are larger and are not made in India.

**Table 1** Details of the three vehicles selected

| Model | Internal combustion (petrol) | Electric | Hybrid (Mild hybrid) |
|---|---|---|---|
| Source | Maruti Suzuki | Tata | Maruti Suzuki |
| Capacity | 1197 cc | 30.18 kWh | 1197 cc |
| Model of Ref | Swift | Nexon EV | Baleno Dualjet |
| Weight of model | 1000 kg | 1300 kg | 900 kg |
| Place of Prod | Ahmedabad | Pune | Ahmedabad |
| Life expectancy | 200,000 km | 160,000 km (Range) | 200,000 km |

**Table 2** Results for Internal combustion engine vehicle (Suzuki Swift)

| Impact category | Reference unit | Production (%) | Transport (%) | Use (%) | End of life (%) | Total |
|---|---|---|---|---|---|---|
| Fine particulate matter formation | kg PM2.5 eq | 25.13 | 0.10 | 74.71 | 0.06 | 70.13418 |
| Freshwater Ecotoxicity | kg 1,4-DCB | 41.22 | 0.01 | 58.20 | 0.57 | 8839.918 |
| Freshwater eutrophication | kg P eq | 37.27 | 0.10 | 62.61 | 0.02 | 11.59161 |
| Global warming | kg $CO_2$ eq | 13.48 | 0.07 | 86.18 | 0.27 | 63,149.54 |
| Ionizing Radiation | kBq Co-60 eq | 30.59 | 0.21 | 69.17 | 0.02 | 1968.605 |
| Marine ecotoxicity | kg 1,4-DCB | 40.69 | 0.02 | 58.66 | 0.63 | 11,052.44 |
| Marine eutrophication | kg N eq | 37.91 | 0.07 | 61.98 | 0.04 | 1.341434 |
| Ozone formation, Terrestrial ecosystems | kg NOx eq | 21.91 | 0.22 | 77.81 | 0.05 | 112.6447 |
| Stratospheric ozone depletion | kg CFC11 eq | 11.55 | 0.07 | 88.36 | 0.02 | 0.035764 |
| Terrestrial acidification | kg SO2 eq | 23.04 | 0.11 | 76.81 | 0.05 | 155.6405 |
| Terrestrial ecotoxicity | kg 1,4-DCB | 11.32 | 0.12 | 88.50 | 0.06 | 221,792 |
| Ozone formation, Human health | kg NOx eq | 22.25 | 0.23 | 77.46 | 0.06 | 104.7964 |
| Human carcinogenic toxicity | kg 1,4-DCB | 40.77 | 0.07 | 59.14 | 0.03 | 12,842.79 |
| Human non-carcinogenic toxicity | kg 1,4-DCB | 33.30 | 0.04 | 63.14 | 3.52 | 54,919.01 |
| Fossil resource scarcity | kg oil eq | 11.66 | 0.07 | 88.26 | 0.01 | 18,616.56 |
| Mineral resource scarcity | kg Cu eq | 40.86 | 0.01 | 59.07 | 0.04 | 399.1062 |
| Land use | m2a crop eq | 17.35 | 0.13 | 82.38 | 0.14 | 1596.942 |
| Water consumption | m3 | 39.87 | 0.09 | 62.91 | 0.12 | 249.3335 |

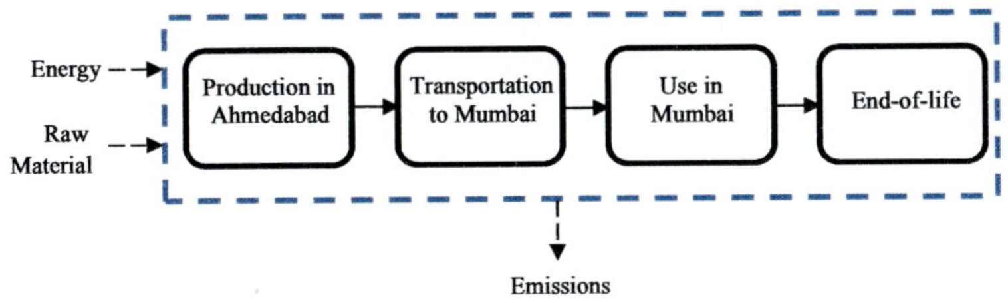


**Fig. 1** System boundaries (hybrid and IC engine vehicle)

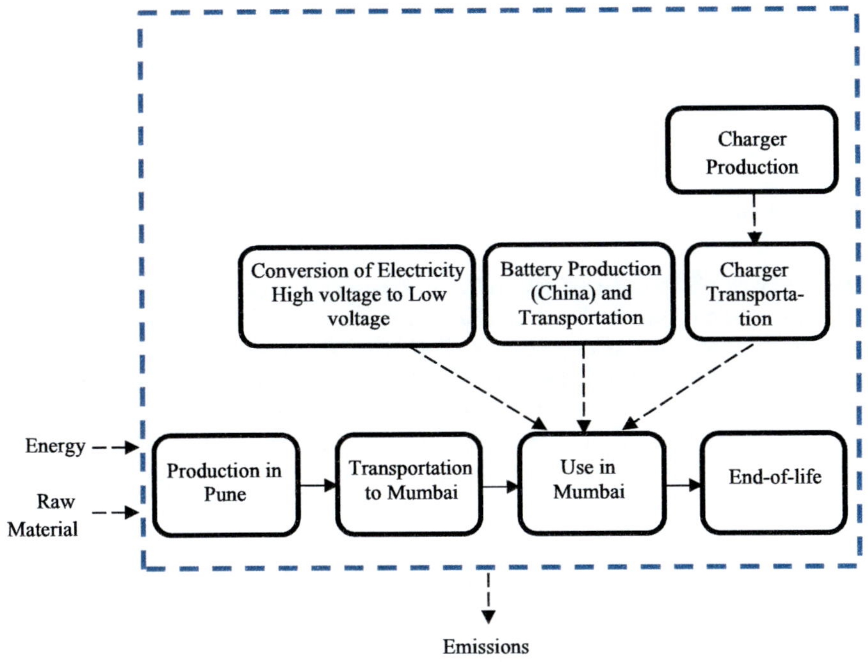


**Fig. 2** System boundaries (electric vehicle)

## *3.2 Assumptions*

Some assumptions have been established in an attempt to simplify the modelling process. First, the ecoinvent database does not have information on the production of hybrid vehicles. It was, therefore, assumed that assembling individual components separately would make the car complete. Secondly, it was presumed that the mass of each raw material in the HV in relation to the ICEV must be calculated. This is necessary for computing the energy and emissions required for manufacturing the HV. Additionally, based on the company's claim that the hybrid vehicle (Maruti Suzuki Baleno dual jet) has a lifespan of 200,000 km, it is assumed that, in its entirety, it uses a petrol engine 85% of the time (170,000 km) and defers use of batteries the remaining 15% (30,000 km). This is based on the fact that the car being used in the study is a mild hybrid. This is necessary for computing the energy and emissions required for manufacturing the HV. In part, this is due to the numerous similarities between hybrid vehicles and the conventional internal combustion engine vehicles.

The energy needed to generate a certain amount of electricity in India, as well as the transmission losses and battery charging efficiencies, are all taken into consideration under the electric vehicle model [13]. The ICEV model takes into account the emissions and energy required to make petrol available, including crude oil extraction, production, refining, shipping, etc. (taken from the database ecoinvent). The assumption is that both electricity generation and petrol production are taken into account in the use phase of the models. The reason being that these fuels are consumed by cars during their use. Furthermore, it is assumed that a vehicle will last its entire lifespan without being damaged or involved in any accidents.

## 4 Inventory Analysis

The life cycle of the three vehicles were modeled and evaluated in OpenLCA software, using the ecoinvent 3.7.1 database. Datasets within ecoinvent data base are categorized into flows and processes. Following the system boundaries (Figs. 1 and 2) and the assumptions previously outlined, various new processes and flows were created, complemented with many already existing processes and flows. As a result, three product systems were designed for the three types of vehicles. The three products systems were then compared under one project [14].

### *4.1 Production of the Vehicles*

The ecoinvent processes "passenger car production, petrol/natural gas|passenger car, petrol/natural gas|APOS, U" and "passenger car production, electric, without battery|passenger car, electric, without battery|APOS, U" were used to produce both internal combustion engine vehicles and electric vehicles, respectively. Using the ecoinvent process, a battery was added to the electric vehicle, "batteries, Li-ion, rechargeable, prismatic, APOS, U". Since ecoinvent provided a weight of reference for the cars considered by the study, each model's weight was considered in this step. The output was a car of 1 unit in quantity.

Ecoinvent did not have a hybrid vehicle production process. The car was presumed to be a whole when each of the requisite components were put together separately. This was accomplished through the use of four ecoinvent processes; "glider production, passenger car|glider, passenger car|APOS, and U" represents the production of a passenger car glider (the chassis, the steering, braking and suspension system, tires, cockpit equipment and electronic components for non-propulsion systems), "internal combustion engine production, passenger car|internal combustion engine, for passenger car|APOS, U" is for the production of an internal combustion engine for a passenger car, "battery production, Li-ion, rechargeable, prismatic|battery, Li-ion, rechargeable, prismatic|APOS, U" for the production of lithium-ion batteries, and "powertrain production, for electric passenger car|powertrain, for electric passenger car|APOS, U" for the electric drive train (electric motor, a converter, an inverter, a charger, a power distribution unit and cables) [15].

## *4.2 Transportation of the Vehicles*

As the internal combustion engine cars are manufactured in India, the transportation to Mumbai was divided into three steps. The first part considered the transportation from Maruti Suzuki's Hansalpur plant until Detroj railway station by lorry, after that the transport from Detroj railway station to Mumbai Central station by train was considered. Finally, from the railway station until Maruti Suzuki Arena, the transport with lorry was considered. For the electric car, the car was produced in Pune city, and then it was transferred by lorry to Mumbai's Maruti Suzuki Arena. The hybrid car was manufactured in Ahmedabad. From Ahmedabad's Hansalpur plant to Detroj railway station, the car was transported by lorry. Then from Detroj railway station until Mumbai Central station, a train was used to transport the car. Lastly, it was transported from Mumbai Central station to Maruti Suzuki Arena by a lorry. Figure 3 summarizes the distance and the weight considered in each transportation step for each vehicle.

## *4.3 Use of the Vehicles*

In the usage phase, life expectancy (as provided on company official websites) is accounted for. Both ICEV and HV can travel for up to 200,000 under ideal circumstances. While the average life expectancy of an EV is 160,000 km. Therefore, all inputs and outputs were calculated in accordance with this number. The inventory of each model's use phase is divided into two parts:

**Fuel and Electricity Consumption.** During use, an ICEV requires fuel. ICEV uses the fuel flow data from ecoinvent 3.7.1 database, "petrol, and four-stroke blend". For the ICEV considered, the fuel consumption ("km/liter of fuel") is 23.20 [Maruti Suzuki, 2020].

Electric vehicles, on the other hand, require electricity. Tata Nexon model's battery voltage ("batteryvoltage") of 320 V and battery capacity ("batterycapacity") of 30.18 kWh was considered to calculate electricity consumption [Tata Motors, 2020]. The battery can be recharged 2000 times ("numberof charges") over the life cycle of the EV (160,000 km), given that each charge lasts for ~300 km [Tata Motors, 2020]. In this way, the number of charges required during the battery's lifetime multiplied by

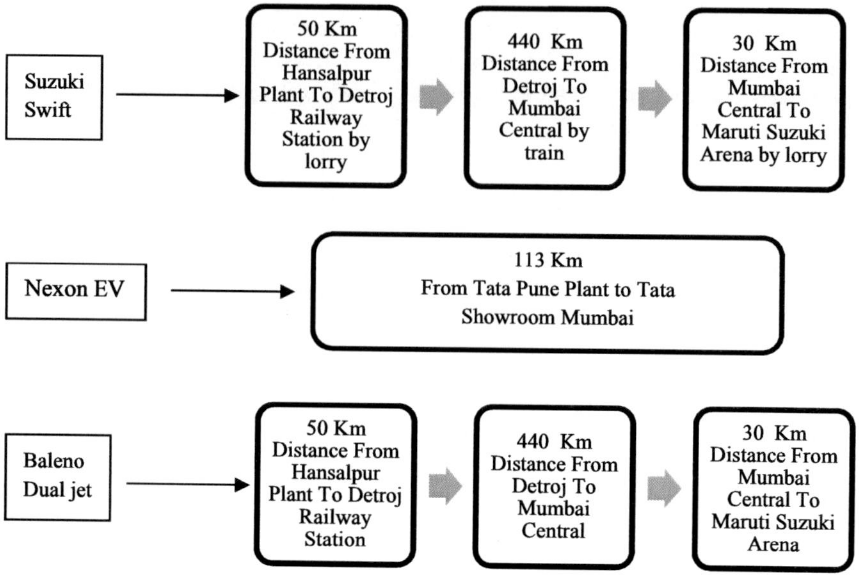


**Fig. 3** Distances covered in the transportation phase

the amount of electricity consumed by the car during use results in the electricity consumed by the car.

**Emissions of vehicles**. The emissions for the electric vehicle were based on the ecoinvent process "transport, passenger car, electric|transport, passenger car, electric|APOS, U". In the same way, for the internal combustion engine vehicle the emissions were based on ecoinvent process "transport, passenger car, small size, petrol|APOS, U".

## 4.4 End of Life

All three vehicles were disassembled manually based on the ecoinvent process "manual dismantling of used passenger car|manual dismantling of passenger car|APOS, U-GLO". Additionally, the batteries of the electric and hybrid vehicles were presumed to have been dismantled but not recycled.

# 5 Results and Discussion

## 5.1 Life Cycle Impact Assessment (LCIA)

The life cycle impact assessment is split into two parts. First, the results of each of the individual cars are discussed. Next, a chart comparing all three is presented. The impact assessment method chosen for this study is ReCiPe 2016 Midpoint (H). An impact method consolidates all the results of an inventory analysis into a limited number of indicators called impact categories.

Emissions from internal combustion engine vehicles were at their highest during the use phase, as shown in Table 2. Table 3 reveals a similar pattern with the use phase being largely responsible for impacts in all categories. This is primarily because India relies on nonrenewable resources for most of its electricity generation. A large part of global warming is also attributed to the production of electricity and its conversion from high to low voltages. It is noteworthy that 15% (35.20 kg Cu eq.) of the total emissions (234.54 kg Cu eq.) from the production phase in the mineral resource scarcity category are driven by the battery manufacturing process.

The hybrid variant has produced similar results to those of EV's and ICEV's. Interestingly, in the human carcinogenic toxicity category (Table 4), the emissions attributed to use phase (8685.72 kg 1,4-DCB) are nearly twice the emissions attributed to production phase (4197.20 kg 1,4-DCB).

**Comparison of vehicles.** The life cycle impacts of the three vehicles are compared in Table 5. There is a vehicle that performs better than others in each impact category, but no vehicle consistently outperforms the other two [16]. Its interesting to note, however, that the HV produced lower emissions than any one of the other two in almost every impact category.

The results from Table 5 are presented as a bar chart in Fig. 4. This chart displays the emissions of the vehicles under all impact categories. On the X-axis are the percent emissions for each of the three vehicles; on the Y-axis are the impact categories. Vehicle contributing the most to an impact category is displayed with the maximum value (100% emission), while the rest are grouped relatively. In the case of stratospheric ozone depletion, for example, the bar that takes 100% value on the Y-axis represents ICEV. Within the same impact category, the bar with the lowest emission value (approximately 55%) corresponds to the HV. It is therefore evident that the HV contributes roughly half as much to ozone depletion as the ICEV.

Figure 4 illustrates, at a glance, how each vehicle fares against other vehicles in various impact categories.

**Monte Carlo Simulation.** Results are displayed in box and whisker plots illustrating the probability distribution function of the output (Fig. 5). In total, 10,000 iterations were completed. Crosses within boxes indicate 50th percentile values. Dashes in the boxes are medians. The 25th and 75th percentiles are edges of the box.

The 2.5th and 97.5th percentile values are endpoints of the plots. Figure 5, for example, shows the difference in carbon emissions between the three vehicles. As for ICEV, the mean value or 50th percentile is 58,265.92 kg $CO_2$ Eq. Twenty fifth and

**Table 3** Results for electric vehicle (TATA Nexon EV)

| Impact category | Reference unit | Production (%) | Transport (%) | Use (%) | End of life (%) | Total |
|---|---|---|---|---|---|---|
| Fine particulate matter formation | kg PM2.5 eq | 18.71 | 0.02 | 81.24 | 0.04 | 98.02575 |
| Freshwater ecotoxicity | kg 1,4-DCB | 34.86 | 0 | 64.54 | 0.59 | 11,300.63 |
| Freshwater eutrophication | kg P eq | 23.12 | 0.01 | 76.87 | 0.01 | 27.94527 |
| Global warming | kg $CO_2$ eq | 20.18 | 0.05 | 79.30 | 0.47 | 45,586.04 |
| Ionizing radiation | kBq Co-60 eq | 14.42 | 0.01 | 85.55 | 0.02 | 4467.025 |
| Marine ecotoxicity | kg 1,4-DCB | 34.92 | 0.01 | 64.42 | 0.65 | 14,301.59 |
| Marine eutrophication | kg N eq | 34.49 | 0.01 | 65.48 | 0.02 | 3.229286 |
| Ozone formation, Terrestrial ecosystems | kg NOx eq | 18.44 | 0.03 | 81.48 | 0.05 | 143.3239 |
| Stratospheric ozone depletion | kg CFC11 eq | 21.70 | 0.08 | 78.18 | 0.04 | 0.020663 |
| Terrestrial acidification | kg SO2 eq | 20.58 | 0.03 | 79.34 | 0.05 | 166.1377 |
| Terrestrial ecotoxicity | kg 1,4-DCB | 30.59 | 0.21 | 69.12 | 0.08 | 136,398.7 |
| Ozone formation, Human health | kg NOx eq | 19.19 | 0.03 | 80.74 | 0.05 | 130.7717 |
| Human carcinogenic toxicity | kg 1,4-DCB | 45.46 | 0.01 | 54.50 | 0.03 | 12,022.7 |
| Human non-carcinogenic toxicity | kg 1,4-DCB | 31.57 | 0.01 | 66.01 | 2.40 | 104,583.9 |
| Fossil resource scarcity | kg oil eq | 19.48 | 0.07 | 80.44 | 0.02 | 11,850.91 |
| Mineral resource scarcity | kg Cu eq | 41.67 | 0.01 | 58.29 | 0.03 | 562.8446 |
| Land use | m2a crop eq | 19.26 | 0.06 | 80.54 | 0.14 | 1615.391 |
| Water consumption | m3 | 22.56 | 0.01 | 77.35 | 0.08 | 424.052 |

**Table 4** Results for hybrid vehicle (Baleno dual jet)

| Impact category | Reference unit | Production (%) | Transport (%) | Use (%) | End-of-life (%) | Total |
|---|---|---|---|---|---|---|
| Fine particulate matter formation | kg PM2.5 eq | 19.50 | 0.07 | 80.39 | 0.04 | 94.50052 |
| Freshwater ecotoxicity | kg 1,4-DCB | 36.32 | 0.03 | 63.06 | 0.59 | 11,308.7 |
| Freshwater eutrophication | kg P eq | 23.82 | 0.05 | 76.12 | 0.01 | 22.40563 |
| Global warming | kg $CO_2$ eq | 14.37 | 0.07 | 85.09 | 0.47 | 58,536.35 |
| Ionizing radiation | kBq Co-60 eq | 16.98 | 0.11 | 82.89 | 0.02 | 3703.146 |
| Marine ecotoxicity | kg 1,4-DCB | 35.95 | 0.03 | 63.37 | 0.65 | 14,201.97 |
| Marine eutrophication | kg N eq | 38.68 | 0.06 | 61.23 | 0.03 | 2.61607 |
| Ozone formation, Terrestrial ecosystems | kg NOx eq | 17.99 | 0.14 | 81.82 | 0.05 | 142.7767 |
| Stratospheric ozone depletion | kg CFC11 eq | 14.16 | 0.08 | 85.72 | 0.04 | 0.030209 |
| Terrestrial acidification | kg SO2 eq | 20.93 | 0.08 | 78.94 | 0.05 | 180.068 |
| Terrestrial ecotoxicity | kg 1,4-DCB | 17.56 | 0.12 | 82.24 | 0.08 | 190,990.8 |
| Ozone formation, Human health | kg NOx eq | 18.61 | 0.15 | 81.19 | 0.05 | 131.2596 |
| Human carcinogenic toxicity | kg 1,4-DCB | 32.56 | 0.04 | 67.38 | 0.02 | 12,890.65 |
| Human non-carcinogenic toxicity | kg 1,4-DCB | 27.75 | 0.03 | 69.82 | 2.40 | 86,632.92 |
| Fossil resource scarcity | kg oil eq | 13.38 | 0.07 | 86.53 | 0.02 | 16,380.08 |
| Mineral resource scarcity | kg Cu eq | 38.18 | 0.04 | 61.75 | 0.03 | 523.0054 |
| Land use | m2a crop eq | 16.89 | 0.11 | 82.86 | 0.14 | 1775.209 |
| Water consumption | m3 | 24.77 | 0.06 | 75.09 | 0.08 | 377.519 |

**Table 5** Comparison of all three vehicles

| Impact category | Reference unit | ICEV | HV | EV |
|---|---|---|---|---|
| Fine particulate matter formation | kg PM2.5 eq | 70.13418 | 94.50052 | 98.02575 |
| Freshwater ecotoxicity | kg 1,4-DCB | 8839.918 | 11,308.7 | 11,300.63 |
| Freshwater eutrophication | kg P eq | 11.59161 | 22.40563 | 27.94527 |
| Global warming | kg $CO_2$ eq | 63,149.54 | 58,536.35 | 45,586.04 |
| Ionizing radiation | kBq Co-60 eq | 1968.605 | 3703.146 | 4467.025 |
| Marine ecotoxicity | kg 1,4-DCB | 11,052.44 | 14,201.97 | 14,301.59 |
| Marine eutrophication | kg N eq | 1.341434 | 2.61607 | 3.229286 |
| Ozone formation, Terrestrial ecosystems | kg NOx eq | 112.6447 | 142.7767 | 143.3239 |
| Stratospheric ozone depletion | kg CFC11 eq | 0.035764 | 0.030209 | 0.020663 |
| Terrestrial acidification | kg SO2 eq | 155.6405 | 180.068 | 166.1377 |
| Terrestrial ecotoxicity | kg 1,4-DCB | 221,792 | 190,990.8 | 136,398.7 |
| Ozone formation, Human health | kg NOx eq | 104.7964 | 131.2596 | 130.7717 |
| Human carcinogenic toxicity | kg 1,4-DCB | 12,842.79 | 12,890.65 | 12,022.7 |
| Human non-carcinogenic toxicity | kg 1,4-DCB | 54,919.01 | 86,632.92 | 104,583.9 |
| Fossil resource scarcity | kg oil eq | 18,616.56 | 16,380.08 | 11,850.91 |
| Mineral resource scarcity | kg Cu eq | 399.1062 | 523.0054 | 562.8446 |
| Land use | m2a crop eq | 1596.942 | 1775.209 | 1615.391 |
| Water consumption | m3 | 249.3335 | 377.519 | 424.052 |

75th percentiles are 39,054.66 kg $CO_2$ eq. and 64,801.55 kg $CO_2$ eq., respectively. The minimum (32,650.91 kg $CO_2$ eq.) and maximum (66,212.06 kg $CO_2$ eq.) values represent the 2.5th percentile and 97.5th percentile, respectively.

Figure 6 talks about fine particulate matter formation (PM2.5). Here, it can be said just by looking at the chart that fine particulate matter formation is generally lower for the ICEV than for the EV. This is evident from the fact that the maximum value of the particulate matter formation emissions generated by the ICEV is less than that of the minimum emitted by the EV. The median value for the hybrid vehicle (HV) is just over 97.5 kg PM2.5 eq., while the 25th percentile for EV is just over 95.21 kg PM2.5 eq.. Note, however, that the upper edge of the box, or 75th percentile, represents the ICEV close to its mean value. It is evident that a little less than 25% of the 10,000 random simulations resulted in a higher fine particulate matter formation for HV than for the EV.

Considering the broad range of LCI variables, one can see their inherent uncertainty. The Monte Carlo Simulation proves to be a powerful tool in such case. Based on the model, if one claims that HVs produce more fine particulate matter (PM2.5) than EVs, there is a 25% likelihood that the claim is correct.

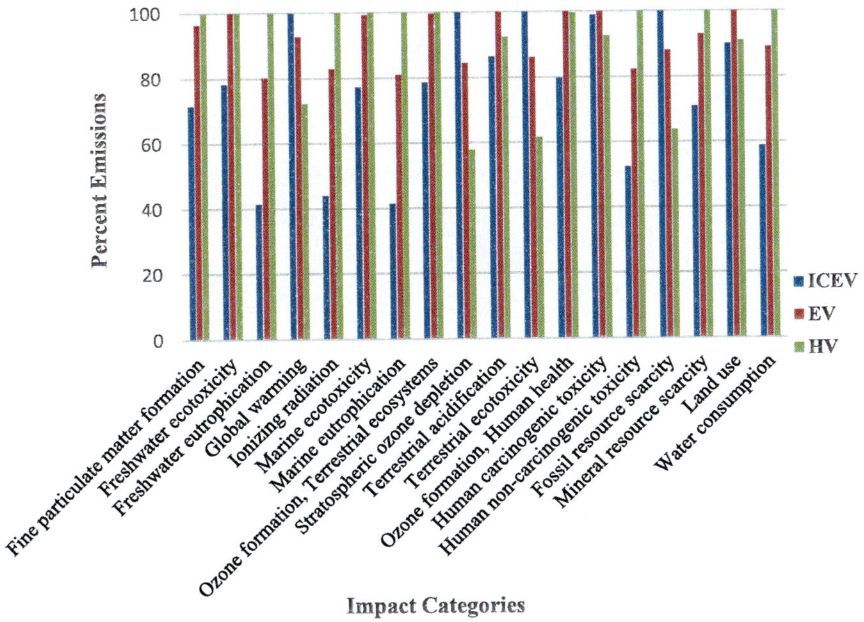


**Fig. 4** Comparison of all three vehicles

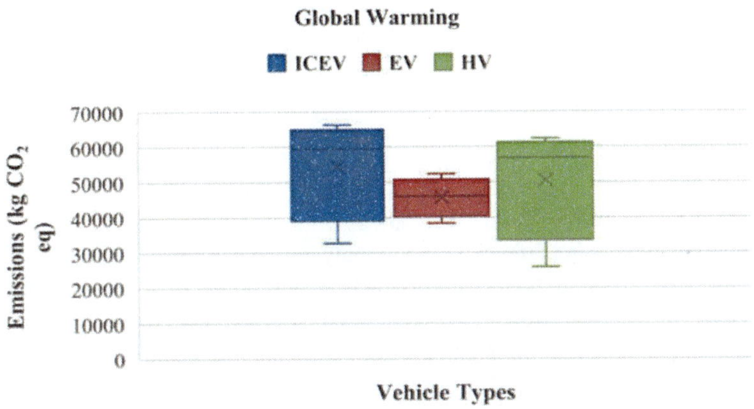


**Fig. 5** Global warming impact category—all three vehicles

# 6 Conclusion

This study contributes to the problem of rethinking urban mobility in order to understand the cause of societal problems. Identifying the emission hotspots was vital to illustrating the necessity of technologies that would reduce the emissions of internal

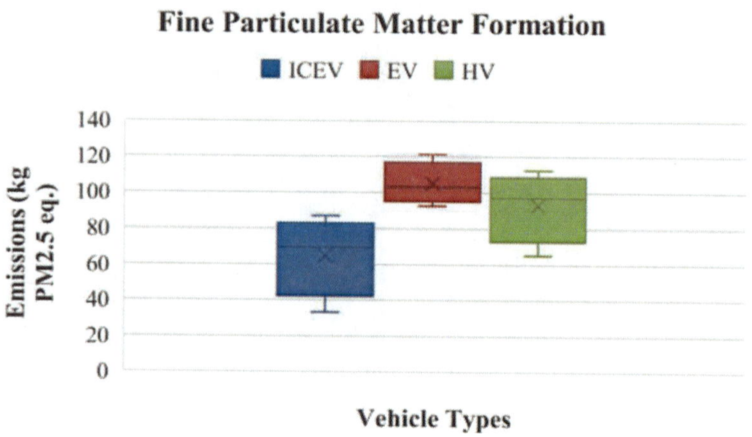


**Fig. 6** Fine particulate matter formation impact category—all three vehicles

combustion engines and hybrid vehicles, as well as move toward greener electricity sources [17], which could help reduce the impact of electric vehicles. According to the findings of this study, Monte Carlo simulation can provide a proper understanding of complex comparisons like those between EVs and ICEVs and HVs. The issue of selecting an "appropriate" value for an input variable such as the life expectancy of a vehicle or emission factor is contested. Variability and uncertainty are associated with each of these variables. Monte Carlo Simulation provides insight into uncertainty and variability that is difficult to observe using deterministic methods.

## 7 Future Research

The extended possibilities of this work are as follows: Further analysis of this study can be carried out using life cycle cost analysis (LCCA) [18]. This would help determine all the costs associated with each vehicle in its lifecycle. It is possible to also analyze the end-of-life processes such as remanufacturing (Schau et al.), reusing [3], and recycling, if reliable data can be found. Life cycle analysis opens a large pool of tools and methods to assess every product and process based on the impacts they have on the environment. Mandating LCA for production processes would contribute to sustainable development and greener technologies. Currently, life cycle approaches are being developed for the social and economic dimensions (Schau et al.), which will only yield more avenues for future research.

## References


1. Shanbag A, Manjare S (2019) Life cycle assessment of tyre manufacturing process. J Sustain Dev Energy Water Environ Syst 8(1)
2. Saur K, Gediga J, Hesselbach J, Schuckert M (1996) Life cycle assessment as an engineering tool in the automotive industry. Int J LCA 15–21
3. Dewulf W, Sas P, Van Brussel H (1999) Life cycle assessment of a windshield wiper—a case study. In: Proceedings first international symposium on environmentally conscious design and inverse manufacturing, pp 120–123
4. Schau EM, Traverso M, Finkbeiner M (2012) Life cycle approach to sustainability assessment: a case study of remanufactured alternators. J Remanufacturing 2:5
5. Zhang X, Yamauchi M, Takahashi J (2011) Life cycle assessment of CFRP in application of automobile. In: Conference: 18th international conference on composite materials
6. LugliettiR, Taisch M, Magalini F (2014) Environmental and economic evaluation of end-of-life alternatives for automotive engine. In: International conference on engineering, technology and innovation (ICE)
7. Helmers E, Dietz J, Weiss M (2020) Sustainability analysis in the LCA of Electric v ICE cars under approximate real-world conditions. Sustainability 12(3):1241
8. Rupp M, Kuperjans I, Schulze S (2018) Comparative life cycle analysis of conventional and hybrid heavy-duty trucks, FH Aachen, NOWUM-Energy
9. McCleese DL, LaPuma PT (2002) Using Monte Carlo simulation in life cycle assessment for electric and internal combustion vehicles. Int J LCA 7:230–236
10. Wenzel H, Hauschield M, Alting L (2001) Environmental assessment of products, volume 1 methodology, tools and case studies in product development
11. European Commission (2010) Institute for environment and sustainability ILCD handbook. Analysis of existing environmental impact assessment methodology for use in life cycle assessment, 1st edn
12. Tagliaferri C, Ekins P, Evangelisti S, Acconcia F, Barletta D, Domenech T, Lettieri P (2006) Life cycle assessment of future electric and hybrid vehicles: a cradle-to-grave systems engineering approach. Chem Eng Res Des 112:298–309c
13. Anandan M, Ramaswamy S (2013) Energy uses in India: a case of electricity, 27–33
14. Manmek S, Kara S (2009) Application of the simplified life cycle inventory for a product life cycle. In: Proceedings of the 7th Australian life cycle assessment conference 22–24 November, Melbourne, Australia
15. Majeau-Bettez G, Hawkins TR, StrØmman AH (2011) Life cycle environmental assessment of lithium-ion and nickel metal hydride batteries for plug-in hybrid and battery electric vehicles. Environ Sci Technol 45(10):4548–4554
16. Poornesh K, Nivya, Pannickott K, Sireesha K (2020) A comparative study on electric vehicle and internal combustion engine vehicles. In: International conference on smart electronics and communication, Trichy, India
17. Eldho RP, Deepa K (2020) A comprehensive overview on the current trends and technological challenges in energy storages and charging mechanism in electric vehicle. J Green Eng 10:4679–6713
18. Klöpffer W (2008) Outlook—role of environmental life cycle costing in sustainability assessment: In: Environmental life cycle costing. CRC Press

# Optimization of Solar and Hybrid Electric Tricycle Design Features Based on Stakeholders' Requirements

**K. Anane-Fenin, W. E. K. Agbesi, N. Y. S. Sarpong, R. N. Ossei-Bremang, C. E. O. Oppon, F. K. Appiah, I. N. Amanor, S. Garriba, J. Boakye, and D. R. Akwada**

## Nomenclature

| | |
|---|---|
| PV | Photovoltaic |
| TAM | Technology Acceptance Model |
| QFD | Quality Function Deployment |
| HOQ | House of Quality |
| GHG | Greenhouse Gas |
| CIIO | Cost of Initial Investment and Operation |
| FSPR | Frequency of Spare Parts Replacement |
| CR | Customer Requirement |
| CIR | Customer Importance Rating |
| SPPH EVs | Solar Powered Plug-in Hybrid Electric vehicles |
| SPEVs | Solar Powered Electric Vehicles |
| $TR_i$ | Ith requirement by the end-user, $i = 1, 2, 3 \ldots m$ |
| $ETC_b$ | Bth technical characteristic, $b = 1, 2, \ldots, n$ |
| $tc_i$ | Degree of satisfaction based on customers perception $KR_i$ |
| $v_i$ | Relative importance of $KR_i$, with a scale of $0 < r_i \leq 1$ and $\sum_{i=1}^{m} r_i = 1$ |
| $OTC_a$ | Degree of the overall customer satisfaction of the ath competitor |
| $TPDC$ | Total product development cost |
| $KR_i$ | Customer perception |

K. Anane-Fenin (✉) · W. E. K. Agbesi · N. Y. S. Sarpong · R. N. Ossei-Bremang · C. E. O. Oppon · I. N. Amanor · S. Garriba · J. Boakye · D. R. Akwada
Department of Mechanical Engineering, Cape Coast Technical University, Cape Coast, Ghana
e-mail: kwame.anane-fenin@cctu.edu.gh

K. Anane-Fenin
Engineering Design and Innovation Centre, Cape Coast Technical University, Cape Coast, Ghana

F. K. Appiah
Energy Commission, Accra, Ghana



N. L. Narasimhan et al. (eds.), *Recent Advances in Energy Technologies*, Lecture Notes in Mechanical Engineering, https://doi.org/10.1007/978-981-19-3467-4_23

| | |
|---|---|
| EVs | Electric Vehicles |
| HEVs | Hybrid Electric Vehicle |
| ICE | Internal Combustion Engine |
| FP | Fuel Prices |
| 3Ws | 3 Wheel |
| RA | Road Accessibility |
| A/SL | Ability/Space for Luggage |
| EO | Ease of Operation |
| ASP | Availability of Spare Parts |
| SPHEVs | Solar Powered Hybrid Electric vehicles |
| EC | Engineering Characteristics |
| $CT_a$ | Ath competitor, $a = 1, 2, \ldots l$ |
| $Ta_j$ | Level of attainment of $ETC_b$, |
| $v_j$ | Target value of $ETC_b$ |
| $ck_j$ | Cost coefficient |
| $CF$ | The cost function of achieving $Ta_j$ |
| $BDP$ | Budget for the development of the product |
| EVs | Electric Vehicles |

## 1 Introduction

As the world continues to make a conscious effort toward sustainability and a safe environment, there has been an urgent need to confront a multitude of challenges such as poor air pollution, high emissions, congestion, traffic volume and fossil fuel dependency most of which stems from the transport sector. Electric mobility addresses the above-mentioned concerns as a sustainable and greener option for transport [1]. The infiltration of the technology of Electric Vehicles (EV) has seen significant growth across the globe [2] and envisions to claim a huge percentage of all consumer vehicles by 2030 [3]. Currently, 1% of light consumer vehicles and 2.5% of all sales of vehicles from 2019 are EVs and by 2020, more than 2-million-unit sales of electric vehicles were made worldwide [4, 5].

As the price of electric vehicle components dwindles each year, researchers predict the dominance of electric vehicles soon depending on governmental regulations [6] and the relentless decision by manufacturers of cars to completely dedicate a huge portion of their manufactured fleet to electric and hybrid vehicles [2]. This shift toward EVs and progressive technological advancements have made many developed countries like the UK and Paris set targets to stop selling conventional cars by 2040 and 2050, respectively, to achieve zero-emission road environments while Scotland aims to stop the sale of internal combustion engine vehicles by 2032 [7]. Some developing countries such as Ghana have also attached great importance to the development of the EV industry. Currently, the national power grid is insufficient and incapable of meeting the increased energy demand associated with electric vehicles

[8] and motivates the provision of renewable energy-based charging options for the transport sector to meet the high population of EV's [9]. An electric charging system (PV module) integrated with power supply from renewable energy (solar) to an energy storage system (battery) in EVs decreases the dependence on high electrical energy demand load on the grid [10]. A more adaptable option is the Hybrid Electric Vehicle (HEV) which is defined by [11], as the combination of the Internal Combustion Engine (ICE) and the green nature of EVs. The former solely depends on PV modules as a substitute for internal combustion engines while the latter integrated the two. The two technologies possess great options when it comes to powering the next century of inexpensive small, lightweight and aerodynamic vehicles in the transport industry making it fully sustainable [12].

Electric two-wheel and three-wheel vehicles are more feasible and suitable alternatives for developing countries because they offer low purchase price, low operational cost and the ability to maneuver on congested roads in comparison to electric four-wheel vehicles [13]. These commercial vehicles popularly known as trikes though not permitted for use in urban cities in Ghana have become attractive to passengers [14] and could be a starting point for electric vehicle accessibility and potential dominance in Ghana. These trikes are powered by fossil fuels, which are detrimental to the climate, ecosystems and the environment through the harmful emissions generated.

EVs and HEVs are sustainable solutions that can promote sustainable cities and consumption patterns. The decision to introduce these trikes into the transport system is a multi-criteria decision problem that requires diverse techniques and approaches. As shown in the methodology, the relative importance of customer requirement/perceptions was extracted from the TAM structured survey to arrive at specific engineering requirements. The consumer preference and acceptability of this shift in the automotive industry would influence current and future studies. This makes it a vital topic for the stakeholders, governments and vehicle manufacturing industries. The novel hybrid framework presented in the current study uses the Technology Acceptance Model (TAM) together with Quality Function Deployment (QFD) in a model to arrive at an ideal design for the end-users without compromising the standard engineering requirements.

The model is well computed to assist countries in sub-Saharan regions in the formulation of appropriate policies geared toward the promotion of electrification in transportation [15] and encourage the development of vehicles that enhance a broader market acceptance of the technology [16, 17].

Previous researchers including [18] present different approaches and methods to access the mindset that influences the acceptance and dissemination of new technologies and also optimize the technology to fit into the transportation ecosystems in countries without deviation from the engineering standards and requirements. Notwithstanding, literature utilizing hybrid TAM and QFD approaches are scarce particularly in relation to the transport sector.

Given the background presented on the state of tricycles in Ghana coupled with the scarce literature for the proposed methodologies, the study did not focus on developing new tricycles but rather provides a framework for the optimization of the already existing systems to integrate renewable energy into it. This optimization was achieved via TAM and QFD to obtain an ideal EV and HEV for efficient operation and transportation. The systematic methodological framework introduced in this research can be adopted globally to optimize the use and acceptance of any new technology, specifically for the transportation systems considering the desires of the end-users.

This paper, therefore, investigates the level of electric tricycles adoption and provides a new framework for the optimization of the EVs and HEV's acceptance in Ghana. To realize the above objective, a TAM and QFD analysis was conducted to assess the perception of key stakeholders and optimize the design and operability of EVs and HEVs, respectively. Data-driven policies were consequently formulated based on recommendations. To the best of our knowledge, this paper demonstrates the first hybrid mathematical model based on TAM and QFD strategic decision-making tools designed for transportation systems optimization and technology acceptance in Ghana. The model prioritizes the overall customer satisfaction under some technical/ engineering constraints.

## 2 Methodology

### 2.1 *Study Area*

The study was conducted in Cape Coast, which is widely recognized as a major tourism and education hub in the country [18]. Cape Coast is a coastal city in the Central Region of Ghana and is located within longitudes latitudes 5° 70–5° 200 north and 1° 110–1° 410 west. The four communities which were selected for the survey were *Abura*, *Kotokuraba*, *Kakumdo* and *Ensuekyir*. These communities were selected because they have the largest commercial and informal transportation hubs within the Municipality. These communities have had several challenges in the sustainable management of traffic and fleet congestion.

### 2.2 *Data Collection Tool*

This study employed a self-administered questionnaire as a research instrument. The questionnaire was structured based on the Technology Acceptance Model (TAM). Results obtained from TAM was used in the HOQ. Access link: (https://docs.google.com/forms/d/1ONmQqfA9e6qcyFf8GDfEicJfqF-_I55FMKMZUrdXZZ0/prefill).

### 2.3 *Translating Customers Voice into Engineering Requirements*

Quality function deployment (QFD) was first introduced by [19]. It is an approach that prioritizes the need of the customer for service and product development, design or improvement as shown graphically in Fig. 1. QFD is defined as "a method to transform customer requirement/perceptions into design quality, to deploy the functions forming quality, and to deploy methods for achieving the design quality into subsystems and component parts and ultimately to specific elements of the manufacturing process" [19].

QFD is a strategic decision-making tool for quality management, quality product development and a versatile planning tool. Over the past few decades, QFD has been widely recognized and successfully applied to many fields such as transportation [20].

QFD is, therefore, adopted in the current study to configure the overall framework for optimizing the design of EVs and HEVs. The QFD methodology has as its core function the house of quality (HOQ), which comprises elements such as end-user or customer requirement/perceptions, importance rating, competition analysis of engineering characteristics, a relationship matrix and a correlation matrix among other engineering requirements.

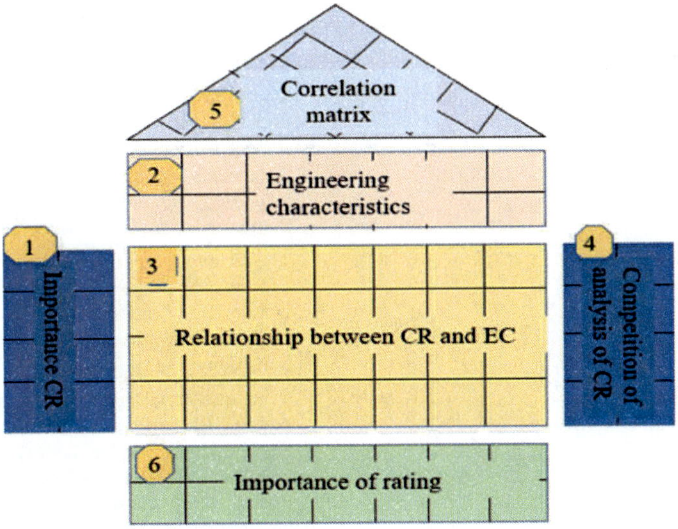


**Fig. 1** House of quality

## 2.4 Modeling the Assessment of Tricycles

The principal indicators considered in this model were the customer's desire, competitive assessment and the engineering characteristics of the various tricycles. This was done with the help of the TAM structured survey to assess the customers' perspectives on the existing tricycles and their readiness to integrate new technology into what they already know. The procedure for the final model configuration has been described in Sects. 3.3.1–3.3.5.

**Customer Requirement/Perceptions.** Table 1 presents the customer requirement /perceptions (CR), their objectives and their relevance in the current study. The benefit criteria are the ones to be improved on while the cost criteria are to be mitigated or gotten rid of completely. Their relative importance for evaluating the transportation system was identified through a digital survey. The frequency of occurrence was used to measure each requirement. The scale is explained as follows; (Above 60%, very important, 50–59% = important, 40–49% = fairly important, 30–39% = less important, below 30% = not important).

**Table 1** Customer requirements/perceptions used in TAM

| SN | Requirements | Objectives | Meaning |
|---|---|---|---|
| 1 | Fuel consumption | Cost | Even though their fuel consumption is not as high as that of cars, they are affected by an increase in fuel prices [21] The instability in fuel prices threatens the economic climate of the tricycle business |
| 2 | Affordability | Benefit | This criterion measures the initial capital cost, the cost required to maintain and replace worn out parts<br>The money required to afford this technology should be moderate, especially in the case of first-time users [22] |
| 3 | Ease of repair and transition | Cost | The technical expertise required by the end-user should be basic and simple, irrespective of the robustness in the engineering requirements so as to aid swift transitioning [23] |
| 4 | Safety | Benefit | Vehicle safety design requirements should be considered to protect the driver from harsh environmental and climatic conditions. These requirements should be integrated into the final design of the EVs and the HEV's [24] |
| 5 | Durability of spare parts | Benefit | The spare parts should be robust and reliable [25] |
| 6 | Stability | Benefit | This criterion measures how the vehicles can balance appropriately when passengers are on board |

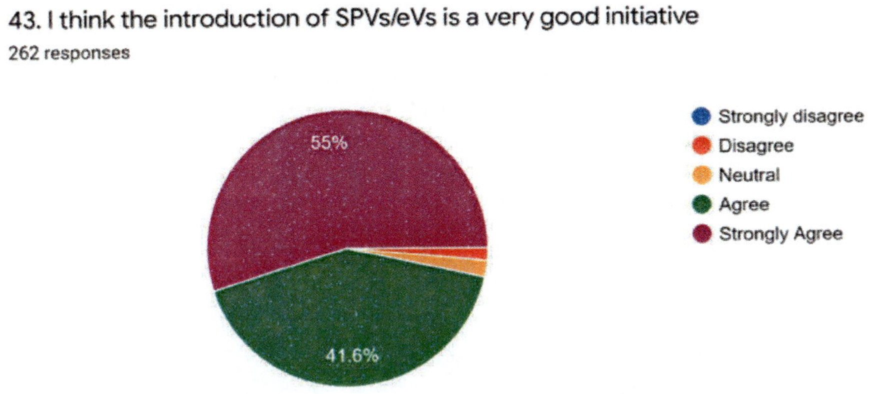


**Fig. 2** Response to question 43

**Engineering Characteristics.** The difference in the ability of a tricycle to meet the customer requirement/perceptions above varies, depending on the engineering characteristics associated with each technology, be it solar, hybrid or complete fossil fuel-powered system. Figure 2 shows the shortlisted engineering characteristics from literature, and consultations with experts such as mechanical, renewable energy, environmental and automobile engineers for the tricycle operation optimization process.

*Clean and Affordable Energy.* Energy that is renewable, sustainable and always available for use. It can be used directly through the photovoltaic cells or can be stored in batteries or capacitors for future use after the peak sunshine hours [26].

*Longevity of Spare Parts .* This indicator brings to bare the robustness and life cycle of the photovoltaics, batteries, charge controllers, inverters and other critical components that would be used in the new technology [27].

*Removal of Clutch.* This criterion measures an improvement in the design features via the removal of the clutch system from the existing tricycles.

*Economic Viability.* This indicator measures comparative analysis of the net present value, internal rate of return, the breakeven points and the gross annual income among other economic indicators of the tricycle business [28].

*Driver Safety Compartment.* A windshield would be provided for the driver and seat belt belts for the passengers on board [23].

*Vehicle Ergonomics.* Focuses on the role of human factors in the design and use of automobiles. This includes analysis of accommodation of driver and/or passengers; their comfort; vision inside and outside the vehicle; control and display design; pedal behavior, information processing and cognitive load during driving etc.

*Environmental Sustainability.* The new intervention in the transportation system should contribute to the global goal of mitigating climate change. Different spectrums of air, environmental and climate pollutants are emitted from vehicles during the combustion of fossil fuels. One of the most dangerous threats to climate change has been linked to GHG gas emissions [29].

*Storage Compartment*. The carrier serves as a space for commuters with luggage [25].

**Mathematical Model Formulation**. The QFD approach proposed by [30] was used to formulate the mathematical model. Optimization of the tricycle energy consumption process was achieved via QFD to determine a set of variables ($X_1$, … $X_2$, … $X_n$) for the technical characteristics of the newly improved EV and HEVs that align with overall end-user satisfaction to ensure a sustainable environment, promote clean and affordable energy and mitigate climate change. Several multi-criteria attributes with conflicting variables, which requires trade-off were configured within the House of Quality.

*Defining the Parameters in the EV's and HEV's*. The mathematical rendition of the problem description can be found summary below.

| | |
|---|---|
| $TR_t$ | ith requirement by the end-user, $i = 1, 2, 3\ldots m$ |
| $ETC_b$ | bth technical characteristic, $b = 1, 2, \ldots, n$ |
| $CT_a$ | ath competitor, $a = 1, 2, \ldots l$ |
| $Ta_j$ | level of attainment of $ETC_b ETC_b$, |
| $KR_i$ | customer perception |
| $tc_i$ | degree of satisfaction based on customers perception $KR_i KR_i$ |
| $v_i$ | relative importance of $KR_i KR_i$, with a scale of $0 < r_i \leq 1$ and $\sum_{i-1}^{m} r_i = 1$ |
| $OTC_a$ | degree of the overall customer satisfaction of the ath competitor |
| $v_j$ | target value of $ETC_b ETC_b$ |
| $ck_j$ | cost coefficient |
| $CF$ | The cost function of achieving $Ta_j$ |
| $TPDC$ | total product development cost |
| $BDP$ | budget for the development of the product |

The basic stage in the QFD is to translate end-users' desires into meaningful $ETC_b$ : Let $T_i$ $(i = 1, 2\ldots m)$ be the same as a functional relationship between $tc_i$ and the attainment levels of $TR_i$, then $tc_i = T_i\ (x_1, x_2, x_n)$ and $u_j = (j = 1, 2, \ldots n)$, represents the functional relationship between $Ta_j$ and other levels of attainment of $ETC_b$. Such that $Ta_j = u_j(x_1, \ldots x_{j-1}, \ldots, x_n)$. Assuming $Y$ is the degree of overall end-user satisfaction for $(y_1, y_2, \ldots\ y_n)$. Therefore, the computation below was used to achieve the attainment levels of target values of ETCb for the EV and HEVs in QFD:

$$\begin{cases} \max Y((y_1, y_2, \ldots y_n) - \max O_a), r = 1, \ldots l \\ \text{s.t} : y_i = u_i(x_1, x_2, \ldots x_n), i, \ldots, m \\ x_j = g_j(x_1, \ldots x_{j-1}, \ldots, x_n) j = 1, \ldots n \\ N_e(x) \leq 0, e = 1, \ldots, p \end{cases} \tag{1}$$

where $N_e$ is the eth EV and HEVs resource needed constraint and $N$ is the number of EV and HEVs constraints.

*Objective Function.* The aggregation of all satisfaction on end-users, thus $\mathrm{OTC}(y_1, y_2, \ldots y_m)$ can be analyzed by computing the levels of end-user satisfaction in Eq. 2 with the respective individual $\mathbf{TR_s}$ that is:

$$\mathrm{OTC}(y_1, y_2, \ldots y_m) = \sum_{i=1}^{m} r_i \mathrm{OTC}_i(cp_i) \tag{2}$$

where $OTC_i(tc_i)$ is an individual value function on $TR_i$. For each $TR$, a numerical value of $tc_i = (i = 1, 2 \ldots, m)$ is allocated to indicate the level of satisfaction of $KR_i$ in comparison with a sustainable EV and HEVs. This numerical code is mostly chosen from a scale of positive numbers. Therefore, $O_i(tc_i)$ can be scaled such that $O_i\left(tc_i^{\min}\right) = 0$ and $O_i\left(tc_i^{\max}\right) = 1$ and can be formulated in Eq. 3 as:

$$\mathrm{OTC}_i(tc_i) = \left(tc_i - tc_i^{\min}\right)/\left(tc_i^{\max} - tc_i^{\min}\right) i = 1, 2, 3, \ldots m \tag{3}$$

and Eq. 4 shows the overall end-user satisfaction $O(y_1, y_2, \ldots\ldots. y_m)$ in a configuration, as follows:

$$\mathrm{OTC}(y_1, y_2, \ldots y_m) = \sum_{i=1}^{m} v_i(tc_i - tc_i^{\min})/(tc_i^{\max} - tc_i^{\min}) \tag{4}$$

This explains, the value of $OTC(y_1, y_2, \ldots\ldots. y_m)$ ranges between 0 and 1, with 0 being the worst and 1 the best. Hence, the objective function can be modeled in Eq. 5 as:

$$\max O' = \sum_{i=1}^{m} r_i(tc_i - tc_i^{\min})/(tc_i^{\max} - tc_i^{\min}) - \max \mathrm{OTC}_a, r = 1, \ldots, t \tag{5}$$

*Normalizing the Target Values of $KR_i$.* $\mathbf{TR_i}$ is further normalized such that the value of $\mathbf{v_j} = 1, 2, 3, \ldots .\mathbf{n}$ the which present target value of$\mathbf{TR_i}$, into the attainment level, $\mathbf{ta_j} = 1, 2, 3, \ldots .\mathbf{n}$ such that $0 \leq \mathbf{ta_j} \leq 1, (1, 2, 3, \ldots .\mathbf{n})$. Target values of $\mathbf{KR_i}$ for the EV and HEVs can be classified into two that is negative and positive. For positive $\mathbf{TR_i}$, the performance of the ETCs is positively proportional to the target value of the ETCs, and for negative $\mathbf{TR_i}$, the performance of ETCs is the reverse of the ETCs target values. The two categories of target values of $\mathbf{TR_i}$ can be normalized using Eqs. 6 and 7 as seen below:

$$ta_j = (s_j^{\max} - s_j)/(s_j^{\max} - s_j^{\min}) \tag{6}$$

$$ta_j = (s_j - s_j^{\min})/(s_j^{\max} - s_j^{\min}) \tag{7}$$

where $s_j^{\max}$ and $s_j^{\min}$ is estimated by considering the requirement and feasibility of the new technology. For the first category of ETCs, $s_j^{\max}$ is the maximum target value of the $TR_i$ to EVs and HEVs performance, and $s_j^{\min}$ is the minimum obtainable value. On the other hand, the second category of ECs, $s_j^{\min}$ is a minimum target value of the $TR_i$ which matches the EVs and HEVs performance, and $s_j^{\max}$ is the maximum that can be obtained.

*Costs Constraint.* The economic aspect, inclusive of all resources involved in the strategic decision-making of the new EVs and HEVs, can be summed up in financial rendition using Eq. 8. Assume that the cost function **CF** for achieving the attainment level EVs and HEVs is scaled linearly to the level of attainment $\mathbf{ta_j}$ which results in:

$$CF_j(ta_j) = \sum_{j=1}^{n} CFta_j \tag{8}$$

where the cost coefficient $CF_j$ is defined as the cost needed to improve $TR_i$, thus when one unit of attainment level of ETCs the cost is c $k_j$ and is determined according to the level of experience of the engineering team or by testing. Assuming that $TPDC$, which the total cost of the budget EVs and HEVs is constrained to a budget BDP, it is formulated in Eq. 9 as:

$$\text{TPDC} = \sum_{j=1}^{n} CF_j(ta_j) = \sum_{j=1}^{n} CF_j(ta_j) \leq \text{BDP} \tag{9}$$

*Final Model Configuration for HOQ.* Considering cost constraints, the QFD decision-making problem can be computed using the configuration shown in Eq. 10 as follows:

$$\begin{cases} \max O' = \sum_{\mathrm{i}=1}^{\mathrm{m}} \mathrm{r_i}\left(\mathrm{tc_i} - \mathrm{tc_i^{min}}\right) / \left(\mathrm{tc_i^{max}} - \mathrm{tc_i^{min}}\right) - \max \mathrm{O_a} \\ s.t.: y_i = Z_i(x_1, x_2, \ldots x_n), i = 1, \ldots m \\ x_j = g_i\left(x_1, \ldots, x_{j-1}, \ldots, x_n\right) j = 1, \ldots n \\ \text{TPDC} = \sum_{j=1}^{n} CF_i(Ta_j) = \sum_{j=1}^{n} CF_i(Ta_j) - \text{BDP} \leq 0 \\ tc_i^{\min} \leq tc_j \leq tc_i^{\max} i \quad i = 1, 2, 3, \ldots, m \\ 0 \leq Ta_j \leq 1, \quad j = 1, \ldots, m \end{cases} \tag{10}$$

**Table 2** Demography of respondents

| Demographic | Range | Modal | Mean Max |
|---|---|---|---|
| Age | 13 to 41 | 25 | |
| Others | | Frequency (n) | Percentage (%) |
| *Gender*<br>Male<br>Female | | 262 | 100 |
| *Marital Status*<br>Single<br>Married<br>Divorced<br>Widowed<br>Have dependents | | 190<br>48<br>13<br>1<br>10 | 72.5<br>18.3<br>5<br>3.8<br>0.4 |
| *Nativity*<br>Natives<br>None natives<br>Do not know | | 207<br>52<br>3 | 79<br>19.8<br>1.1 |
| *Educational Level*<br>No formal education<br>Basic<br>Secondary<br>Tertiary<br>TVET<br>Currently in school | | 17<br>144<br>73<br>14<br>11<br>1 | 6.5<br>55<br>27.9<br>5.3<br>4.2<br>0.4 |

# 3 Results and Discussion

## *3.1 Demography of Respondents*

The questionnaires were distributed within the study area by the staff of the Department of Mechanical Engineering of Cape Coast Technical University. Respondents were randomly selected in terms of education level, age, the status of respondents and others. The demography of respondents can be seen in Table 2.

## *3.2 Technical and Economic Data*

From the TAM model, the following information was collected and grouped into technical and economic data as shown in Table 3.

Table 3 shows the questions posed to respondents, frequent response and the percentage of respondents that agreed to the response. It is evident that for ICEs, spare parts are locally available, petrol is the most used form of energy and the part that is mostly replaced is the clutch for the technical aspect of the survey. However, since EVs and HEVs do not depend on fuel and is completely clutchless, the major

**Table 3** Technical and economic data from the TAM survey

| Data type | Responses | Highest response (%) |
|---|---|---|
| *Technical data* | | |
| Most Common brand | Apsonic | 94.3 |
| Availability of parts | Yes | 90.1 |
| Fuel type | Petrol | 96 |
| Part most frequently replaced | Clutch | 47.9 |
| *Economic indicators* | | |
| Average cost of part replacement | $10 | 43 |
| Cost of servicing | $10 or more | 27.5 |
| Daily fuel purchase | ($6–$10) | 37 |
| Average life span | 5 years | 47 |

target for the technical data to be considered will be making parts available for EVs and HEVs.

It is also evident in the above table that, the average cost of part replacement, cost of servicing, daily fuel purchase and average life span of a vehicle are of major concern to end-users of these vehicles. However, since electric vehicles do not require fuel and a major component such as the clutch which has the highest replacement frequency is absent in an electric vehicle, these two expenditures could be directed toward settling the initial cost of investment for the electric vehicles.

## 3.3 Economic Assessment of ICEs Versus EVs/HEVs

Table 4 breaks down the uncommon expenditure in EVs/HEVs compared to ICEs and how this expenditure could be channeled into the payment of the initial cost of investment in an electric vehicle. This was an attempt to prove the cost-effectiveness of EVs/HEVs to customers that found the initial cost of investment of EVs/HEVs outrageous. It is shown in Table 4 that the average annual cost of investment and the average cost of clutch replacement is absent in EVs/HEVs which makes the payback period for an EV/HEV a year lesser as compared to an ICE. Furthermore, it is also evident that owning EVs/HEVs is more economical compared to ICE if the payback method of payment is applied. This is observed in Fig. 2 below that showed over 90% of respondents agreeing and strongly agreeing with question 43. This asks the question "I think the introduction of EVs and HEVs is a very good initiative".

**Table 4** Economic assessment of ICEs and EVs/HEVs in a payback approach

| Parameter | ICEs | EVs/HEVs |
|---|---|---|
| Initial cost of investment | $1689.66 | $3448.28 |
| Average annual cost of fuel | $2484.00 | – |
| Average annual cost of servicing | $103.45 | $103.45 |
| Average cost of clutch replacement | $25.00 | – |
| Average income/year | $3475.80 | $3475.80 |
| Actual yearly profit (ICI exclusive) | $862.55 | $3372.35 |
| Payback period | **3 Years 10 Month** | **2 Years 1 Month** |

## *3.4 Degree of the Importance of the Customer Requirement/Perceptions*

The customer ratings influence the decision of manufacturers and key stakeholders to make their product more attractive to end-users and for easy dissemination. After the results were obtained from the TAM, data analysis further continued into the QFD model. To ensure robustness and accuracy in decision-making, customer requirement/perceptions with benefit criteria were further considered as the WHATS for the QFD analysis whiles avoiding the cost criteria as shown in the house of quality. The reason for the selection of such variables is the global urgency attached to mitigating climate change, ensuring sustainable consumption patterns and clean cities, via renewable energy integration and adaptation. To the best of our knowledge, green transportation provides a sustainable metric for a cleaner and cooler continent in terms of global warming. Very important variables such as durability of spare parts, affordability, safety, fuel consumption, ease of repair and transition, and stability which form the core indicators to achieve these goals, may present several bottlenecks to the acceptance and integration of EVs and HEVs in the transport sector if considerable attention is not given to them.

Unlike the traditional QFD ratings which use a scale of 1 to 5 to access customers ratings, this study used the frequency of occurrence of a response to determine its importance. Customer requirements/perception for the new technology was rated on a scale of 100 to 40. With 100 being the most important and 40 or less being the least. The variable which achieved a maximum response via the TAM model was considered and used in calculating the relative importance. The six prioritized indicators of much interest to the customer were selected for further consideration in the engineering requirements. These indicators are as shown in the radar diagram below.

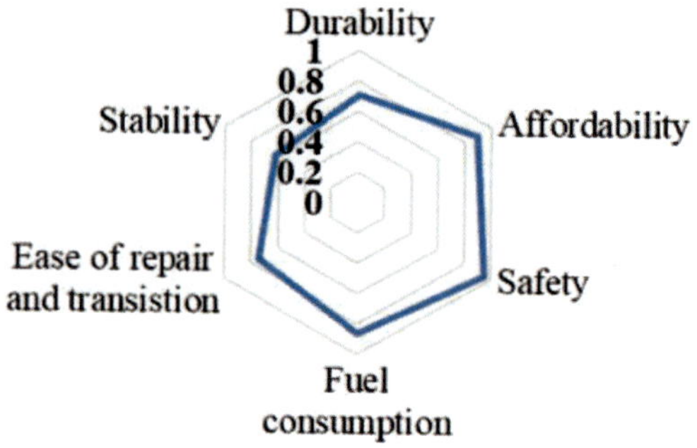


**Fig. 3** Summary of results from the TAM survey

**Table 5** Summary of engineering approaches

| Clean and affordable energy | Economic viability | Absence of clutch (clutchless) | Driver safety compartment | Storage compartment | Vehicle ergonomics | Environmental sustainability |
|---|---|---|---|---|---|---|

Figure 3 gives a graphical summary of selected customer requirements/perceptions and their rate of importance. These six criteria formed the foundation blocks for stakeholders in the transportation industry to reject or accept EVs and HEVs technology into the existing systems. In other words, to have a good market acceptance of electric vehicles, the following attributes should be contained in them considering 3 W EVs and HEVs. 71.0, 90.0, 95.1, 86.0, 75.0 and 62.0% of respondents answered positively by agreeing and strongly agreeing to questions that relate to the following WHATS (Safety, Affordability, Fuel consumption Ease of repair and transition, Durability and Stability) in HOQ.

For ease of calculation, they were expressed as a fraction of 1 as shown in Fig. 3. Stakeholders in the 3 W vehicle market chain attach great importance to the safety of the vehicle, followed by how affordable these types of vehicles are. The least important for which a customer might reject the use of EVs and HEVs is the stability of the vehicles. Different engineering approaches were obtained as shown in Table 5 which led to the construction of the house of quality in Fig. 4.

## 3.5 *Building House of Quality*

As comprehensively discussed in the methodology, the body of the house of quality, focused on the relationship matrix which highlights the relationships between

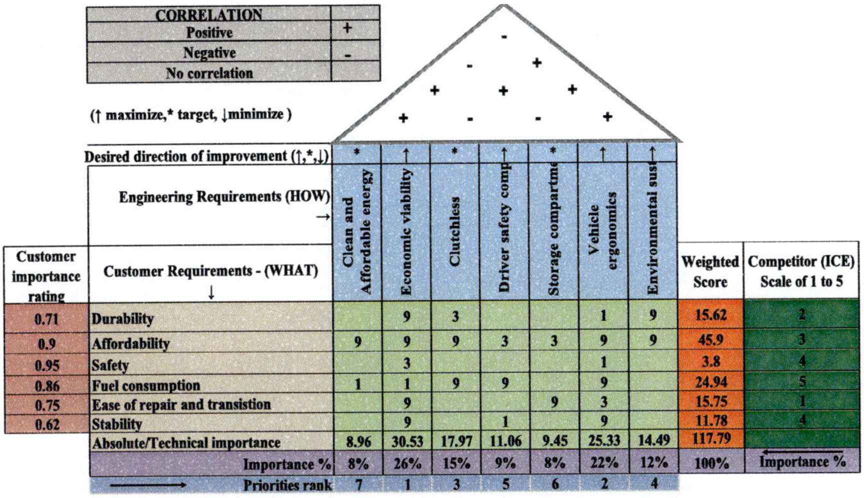


**Fig. 4** Completed House of quality

the customer requirement/perceptions without compromise on the engineering characteristics. Expert's results are important at this stage [31].

Expert results for the relationship matrix between CR and Engineering Characteristics (EC) as well as the correlation matrix is as shown in Fig. 4. The following degrees of strength of the relationship is shown with the symbols below with (9)—strong relationship, (3)—medium relationship, (1)—eak relationship.

**Engineering Characteristics Importance Rating (CIR)**. This study showed that economic viability ranks the highest for the 7 HOWs hence it takes the top solution for the fulfillment of most customer requirements/perceptions of EVs and HEVs and should be a major consideration in the design and integration of electric vehicles. Vehicle ergonomics was the second most prioritized engineering requirement. This may be due to the vulnerability of the driver to accidents as there are no protection shields. Obviously, the absence of the clutch cannot be neglected. The order of prioritization can be seen in the HOQ below.

## 3.6 *Technical Competitive Comparison*

This section discussed the performance of ICEs against EVs and HEVs of 3 W electric vehicles as an inhibiting factor for its penetration into the market. The need to establish the advantages of EVs over ICEs in comparison to the customer requirement/perception is necessary. Certain applicable measurable parameters were assigned to the individual HOWs. The advantages of current ICEs that might threaten

**Table 6** Summary of technical important rating between EVs/HEVs over ICEs

| CIR | Measured parameter | Value on scale |
|---|---|---|
| Durability of spare part | Frequency of replacement | 4.0 |
| Affordability | Capital and operation cost | 3.0 |
| Safety | Seat belt, shield | 4.0 |
| Fuel consumption | Fuel used per day | 5.0 |
| Ease of repair and transition | Technical know-how | 1.0 |
| Stability | Robustness | 4.0 |

the acceptance of EVs and HEVs were valued on a scale of 5. Expert advice and literature influenced the scales and is depicted in Table 6.

The table above highlights the several benefits of 3Ws. Fuel consumption has the highest value on the scale and is closely followed by safety, stability and durability of spare parts. Fuel consumption is the major advantage EVs and HEVs have over ICEs. Technical know-how has the lowest value since there is little to no knowledge on EVs and HEVs in Ghana and is a major disadvantage compared to ICEs.

## 3.7 *Proposed Design for EV and HEV's*

A high market share of electric vehicle integration can be achieved when stakeholders are assured of the economic viability, vehicle ergonomics and absence of clutch as shown in Fig. 5. Vehicle operators are widely concerned with how quickly they can recover their initial cost of investment through their profit and so no fuel consumption will mean that amounts of money used for the purchase of fuel in ICEs could be saved in EV and used for the payment of the initial cost of investment. Also, vehicle ergonomics which looks at certain safety measures like the provision of seat belt shield for driver compartment is ranked second most important attribute to be incorporated into new designs. This solely addresses customer requirements for safety. Furthermore, EVs do not require clutch systems and implicitly addresses the customer requirement for clutchless vehicles.

## 3.8 *Policy Recommendations*

Ghana like most developing countries lags in the manufacture of vehicles in general and has a non-existent EV manufacturing Industry in particular. However, the importation of second-hand vehicles from Europe and the USA is on the ascendency. These used vehicles are mostly repaired accident vehicles all of which are ICE-based. With

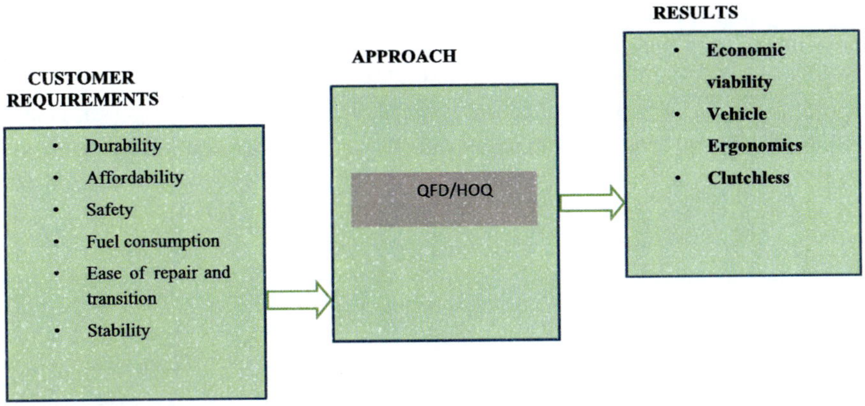


**Fig. 5** Summary of the proposed design for EV's and HEV's

the focus now on renewable energy, government must focus on motivating local and international EV manufacturing companies to bridge the huge gap in vehicle ownership deficit within the country. Having such companies would be important in introducing EVs to stakeholders and thus make its penetration in the local market easy [32]. To address these concerns, significant policy and regulatory instruments focused on incentivizing the emerging EV companies through tax waivers and incentives should be strategically introduced. A study by [33], showed that incentives have played a vital role in the introduction of EVs in many countries. Furthermore, such incentives do not only limit the importation of ICEs but also ensure the production of quality vehicles that suit the environmental, social, economic and human requirements pertaining to the country in question. The initial cost of investment can be tackled at this point. This will also provide jobs for the general masses of unemployed persons in developing countries. It is further recommended to policymakers to put in place measures that encourage and enforce the use of electric vehicles through the provision of technical know-how to local artisans on proper maintenance and services of these vehicles to increase their durability and life cycle so their economic viability can be achieved. To reduce the initial cost of purchase, car-sharing options could also be implemented between the buyer and the manufacturer [34]. The buyer could make a partial payment if he or she is committed to buying an electric car. These recommendations if implemented would ensure a smooth shift from ICE to EVs and HEVs.

EVs, in general, are more environmentally prudent and sustainable. However, for third world countries such as Ghana, where electricity energy demand is far greater than supply, the introduction of purely plug-in EVs may rather be detrimental to its power needs. Some of the negative effects of plug-in HEVs and EVs on power grids include induction of higher circuit currents, Voltage decrease blow regulated limits and higher power demand [35]. Plug-n EVs will adversely introduce significant stress on the national grid during peak charging periods. Therefore, a solution that integrates solar power to fully recharge batteries as in Solar Powered Electric Vehicles

(SPEVs), Solar Powered Hybrid Electric Vehicles (SPHEVs) or partially recharge the batteries as in Solar Powered Plug-in Hybrid Electric vehicles (SPPHEVs) will be more suitable. Furthermore, the advantage of SPPHEVs is that the batteries can be charged utilizing a variety of energy sources such as the IC engine power, Photovoltaic systems and Grid power, hence making them a much more flexible system [35].

## 4 Conclusion

The shift to green mobility has come to stay as the future transportation mode which is clean and safe. Most developed countries are ahead of this technology. Some developing countries look forward to joining this train but are threatened with the acceptance and adoption of this technology by key stakeholders in this sector (drivers, repairers and suppliers).

This paper aimed to first identify the problems that would discourage the use of electric vehicles in a TAM survey by exploring customer requirements/perception on 3 W vehicles (Pragyia) which is fast becoming a preferred mode of transportation in developing countries. The most important of these attributes were selected and converted into customer requirements/perceptions for better acceptance of electric vehicles through various engineering or design requirements in a QFD approach. This hybrid approach resulted in the selection of locally acquired spare parts of EVs and HEVs, economic viability, vehicle ergonomics and absence of clutch as factors to consider increasing the level of acceptance and adoption of electric vehicles in the developing countries market. From our point of view, the analysis revealed that customers are influenced greatly by the value they could get from a product.

Based on the results from this novel hybrid approach, it is recommended to policymakers of developing countries to provide incentives and tax waivers in other to boost the purchase of these vehicles. Future studies could focus on the estimations of the cost of the proposed engineering approaches if adopted by manufacturing companies. In conclusion, this hybrid approach can be applied to other sectors to achieve similar conclusions but in a different context through simple modifications.

**Acknowledgements** The authors wish to acknowledge Energy Commission, Ghana, for providing the funding for this study.

## References

1. Schlüter J, Weyer J (2019) Car sharing as a means to raise acceptance of electric vehicles: an empirical study on regime change in automobility. Transport Res F: Traffic Psychol Behav 60:185–201

2. Vidhi R, Shrivastava P, Parikh A (2021) Social and technological impact of businesses surrounding electric vehicles. Clean Technol 3(1):81–97
3. Kah M (2019) Electric vehicle penetration and its impact on global oil demand: a survey of 2019 forecast trends. Columbia University Center on Global Energy Policy
4. IEA (2020, June) Global EV Outlook 2020. https://www.iea.org/reports/global-ev-outlook-2020
5. Gersdorf T, Hertzke P, Schaufuss P, Schenk S, McKinsey S (2020) Electric vehicle index: Europe cushions a global plunge in EV sales
6. Nurhadi HQA, Nurcahyo R, Gabriel DS (2021) Strategic development for a filter automotive component company in facing the electric vehicles era in Indonesia
7. Bose P, Mandal DK (2021) The future has arrived, are we ready for EV? IOP Conf Ser Mater Sci Eng 1080
8. Moon H, Park SY, Jeong C, Lee J (2018) Forecasting electricity demand of electric vehicles by analyzing consumers' charging patterns. Transp Res Part D: Transp Environ 62:64–79
9. Prajapati MSG, Vyas SR (2021) Energy management strategies of grid connected renewable source for EV charging station
10. Ma CT (2019) System planning of grid-connected electric vehicle charging stations and key technologies: a review. Energies 12(21):4201
11. Rajini V, Anand M (2021) Investigations on interleaved and coupled split-Pi DC-DC converter for hybrid electric vehicle applications. Int J Renewable Energy Res (IJRER) 11(2):808–817
12. Abdelhamid M, Singh R, Qattawi A, Omar M, Haque I (2014) Evaluation of on-board photovoltaic modules options for electric vehicles. IEEE J Photovoltaics 4(6):1576–1584
13. Rajper SZ, Albrecht J (2020) Prospects of electric vehicles in the developing countries: a literature review. Sustainability 12(5):1906
14. Akaateba MA, Yakubu I, Akanbang BAA (2015) Correlates and barriers associated with motorcycle helmet use in Wa, Ghana. Traffic Inj Prev 16(8):809–817
15. Ma SC, Fan Y, Guo JF, Xu JH, Zhu J (2019) Analysing online behaviour to determine Chinese consumers' preferences for electric vehicles. J Clean Prod 229:244–255
16. Dijk M, Yarime M (2010) The emergence of hybrid-electric cars: Innovation path creation through co-evolution of supply and demand. Technol Forecast Soc Chang 77(8):1371–1390
17. Choi YK, Totten JW (2012) Self-construal's role in mobile TV acceptance: extension of TAM across cultures. J Bus Res 65(11):1525–1533
18. Danso SY, Ma Y, Adjakloe YDA, Addo IY (2020) Application of an index-based approach in geospatial techniques for the mapping of flood hazard areas: a case of cape coast metropolis in Ghana. Water 12:3483
19. Yoji A, Mazur GH (1998) Using QFD to assure QS9000 compliance. In: 4th international symposium on quality function deployment, Sydney, 2005–2013
20. Wu SM, Liu HC, Wang LE (2017) Hesitant fuzzy integrated MCDM approach for quality function deployment: a case study in electric vehicle. Int J Prod Res 55(15):4436–4449
21. Afukaar F, Damsere-Derry J, Peters K, Starkey P (2019) Rural Transport Services Indicators: using a new mixed-methods methodology to inform policy in Ghana. Transp Res Interdisc Perspect 3:100074
22. Afful-Arthur P, Filson CK (2019) Information needs and Information-seeking behaviour of auto mechanics in Siwdu, Cape Coast, Ghana. Int J Knowl Content Dev Technol 9(1):7–18
23. Jack JKA, Amoah EK, Hope E, Okyere F (2021) Should Ghana legalize the commercial use of motor bikes and tricycles as means of public transport? A case study of five selected regions in Ghana. J Econ Bus 4(1)
24. Uda K (2017) Proposed personal green mobility (PGM) tricycle. Mech Eng J 4(5):16–00677
25. Roger C, El Omari H (2019) Solar-E-cycles, empowering people project 2014–2019. In: 2019 7th international renewable and sustainable energy conference (IRSEC), 1–8
26. Mallick P (2019) Design of solar assisted tricycle. Doctoral dissertation, Indian Institute Of Engineering Science And Technology, (SHIBPUR)
27. Yadav SK, Bajpai U (2020) Energy, economic and environmental performance of a solar rooftop photovoltaic system in India. Int J Sustain Energ 39(1):51–66

28. Ossei-Bremang RN, Kemausuor F (2021) A decision support system for the selection of sustainable biomass resources for bioenergy production. Environ Syst Decisions 1–18
29. Sulemana A, Donkor EA, Forkuo EK, Asantewaa J, Ankrah IN, Musah AMO (2020) Optimized routing of trucks for institutional solid waste collection in Kumasi, Ghana. Detritus (9):50
30. Ping YJ, Liu R, Lin W, Liu HC (2020) A new integrated approach for engineering characteristic prioritization in quality function deployment. Adv Eng Inform 45:101099
31. Rezvani Z, Jansson J, Bengtsson M (2018) Consumer motivations for sustainable consumption: The interaction of gain, normative and hedonic motivations on electric vehicle adoption. Bus Strateg Environ 27(8):1272–1283
32. Biresselioglu ME, Nilsen M, Demir MH, Røyrvik J, Koksvik G (2018) Examining the barriers and motivators affecting European decision-makers in the development of smart and green energy technologies. J Clean Prod 198:417–429
33. Secinaro S, Brescia V, Calandra D, Biancone P (2020) Employing bibliometric analysis to identify suitable business models for electric *cars*. J Clean Prod 264:121503
34. Dulău LL, Bică D (2020) Effects of electric vehicles on power networks. Procedia Manuf 46:370–377
35. Eragamreddya G, GopiyaNaikb S (2021) Design requirements of solar powered plug in hybrid electric vehicles. Turk J Comput Math Educ 12(3):4635–4641

# Simulation and Prototype Development of Solar Assisted Electric Trolley

**Seyezhai Ramalingam, S. Harika, A. Sowmya, N. Ramakrishnan, and S. Purushothaman**

## 1 Introduction

In India, the majority of the country's economic growth is shared by road transportation. However, in recent days the percentage of load share by transportation sector has been increased to 87% [1]. This transportation sector depends mainly on the petroleum products which increases the greenhouse effect and also, it need to be imported. Also, it has been observed that the mobility share of the buses and railways have been deduced, whereas the share of three wheelers has been grown up [2, 3]. Since, the share of auto rickshaws has increased, the production and demand of the same is also been increased. This vehicle has both motorized and non-motorized types. The vehicle consists of internal combustion engines comes under the motorized type vehicle and the electric vehicle comes under the non-motorized transport mode [4].

In India, the areas like urban, suburbs, and townships, the three-wheeled vehicle has major demand which has been used as private, and public transportation. In West Bengal, recently an e-tricycle has been introduced as public mode of transportation [5]. Though, these vehicle has so many advantages, still, it has some social and technical challenges for proper implementation of electric vehicle. The research survey has been conducted in the urban and suburban areas of West Bengal state in India, where the implementation of e-trike as a public mode of transportation is going on in the past two years.

For such electric trolley, an efficient power electronics converter is required in order to make sure that the proper charging and discharging of battery [6]. The li-ion battery is used in this work to ensure the proper operation of motor drive employed in the trolley [7, 8]. In industries, one of the widely used motors are BLDC motors. The

S. Ramalingam (✉) · S. Harika · A. Sowmya · N. Ramakrishnan · S. Purushothaman
Renewable Energy Conversion Laboratory, Department of EEE, SSN College of Engineering, Kalavakkam, Chennai 603110, India
e-mail: seyezhair@ssn.edu.in



N. L. Narasimhan et al. (eds.), *Recent Advances in Energy Technologies*, Lecture Notes in Mechanical Engineering, https://doi.org/10.1007/978-981-19-3467-4_24

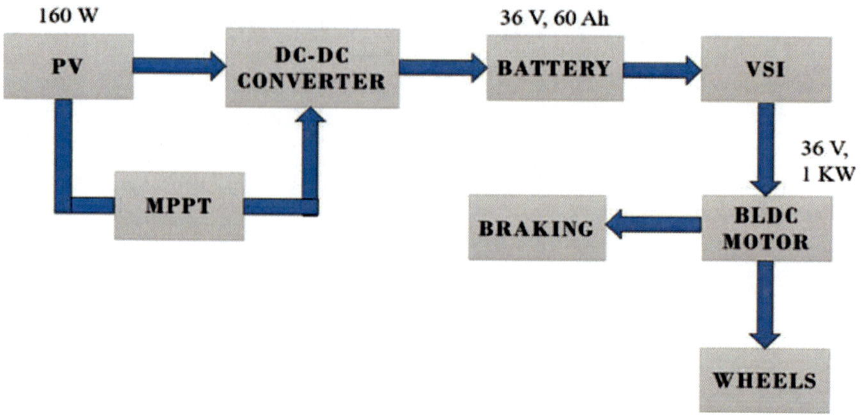


**Fig. 1** Overall block diagram

need for electric motor drives are brought by the development of electric vehicles [9, 10].

The rating of solar array is 160 W. The suggested converter is used to stabilize the variable output from the solar PV. In this way, the output from the PV is regulated and charges the 36 V, 60Ah lithium ion battery. To charge the lithium ion battery of 36 V, 60Ah, the minimum charging current of 3 A is required, hence, the solar rating of 160 W has been utilized. The battery is connected to the controller which drives the BLDC motor. The power rating of the motor is 1 KW, 36 V. Overall block diagram is depicted in Fig. 1. Section 2 deals with boost-buck converter and Sect. 3 discuss about the li-ion battery characteristics. Section 4 describes the BLDC motor drive design and the simulation and hardware results are depicted in Sect. 5.

## 2 Boost-Buck Converter

The vital advantage of the projected converter is that it has reduced output ripple against to the conventional converters. The output voltage can be lesser or higher than the input voltage. Here, the design is done to obtain the voltage at a higher level and for achieving the required current. The suggested converter is coupled with 160 W PV panel (two 80 W Panel in series with $V_{oc} = 21.5$ V and $I_{sc} = 4.7$ A) and circuit diagram is depicted in Fig. 2. The results are depicted in Figs. 3, 4 and 5.

From the waveforms, it is clear that 35 V has been boosted to 40 V, and the input current of 4.28 A has been stepped down to 3.68 A.

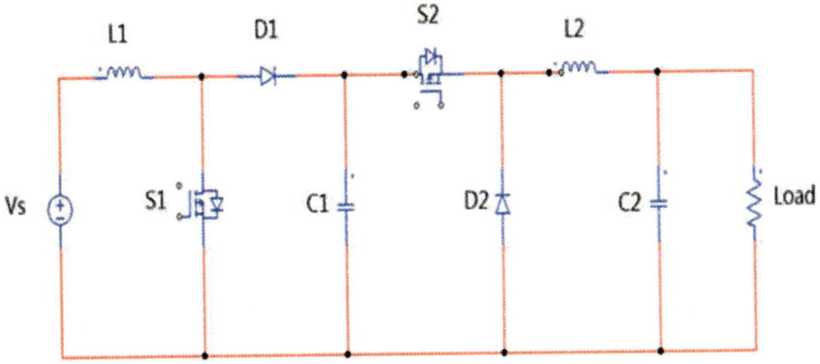


**Fig. 2** Circuit diagram of the projected converter

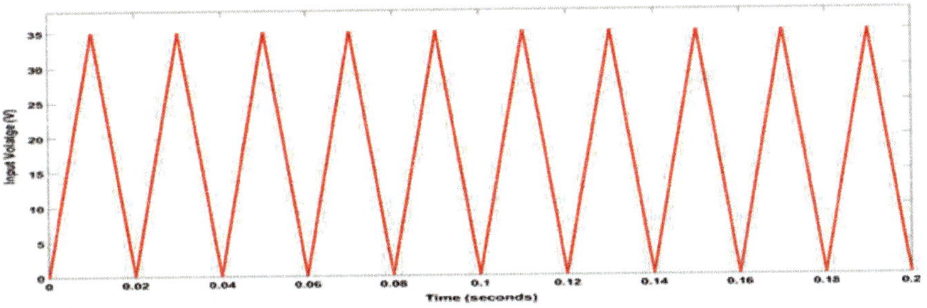


**Fig. 3** Simulated results of applied source voltage waveform of the projected converter coupled with PV

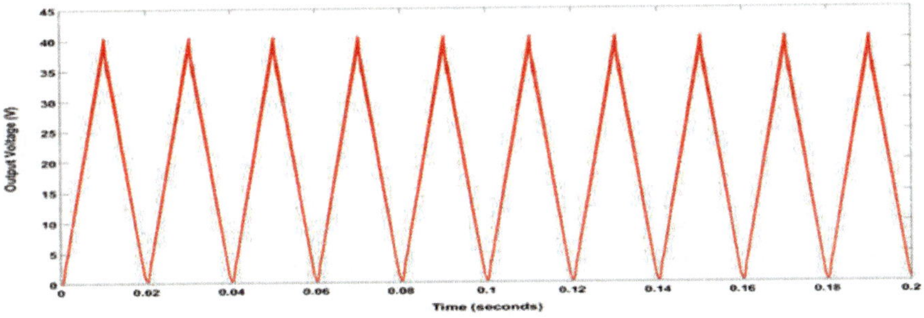


**Fig. 4** Simulated results of the end voltage waveform of the projected converter interfaced with PV

## 3 Li–Ion Battery (LiFePO4)

In electric vehicle, one of the major power source is li-ion batteries as it provides an optimum solution. They have comparatively less charging time and increased

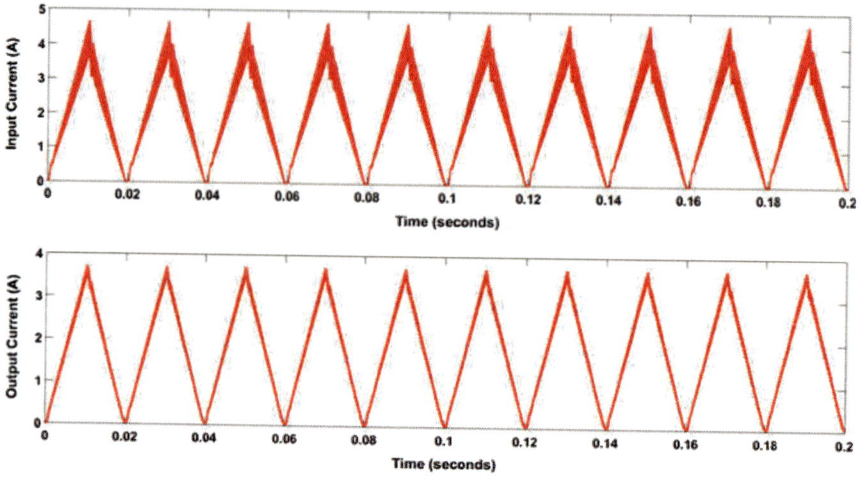


**Fig. 5** Simulated current waveform of projected converter interfaced with PV

tolerance to heavier loads. The measurement of state of charge (SoC) and transient response time ensures the cell performance efficiency. Practical analysis gives an accurate battery SoC. From the results, it is easy to select the proper cell battery for the vehicle. The SOC varies from the range of 0–100%. Normally, the battery is said to be recharged when the SOC reaches 50%. Battery cell voltage, cell current, and cell temperature are the key parameters for the battery SOC estimation. SOC can be defined as the ratio of available capacity and maximum charged that can be stored in the battery. The battery specification employed is 36 V, 60Ah lithium ion battery. The CC-CV method has been used to charge the battery, and the SOC characteristics of LiFePO4 battery is shown in Fig. 6.

From Fig. 6, it is noted that when the voltage greater than the rated voltage of 36 V, the battery starts charging. When the voltage is lesser than the rated voltage, the battery is discharging. The discharge characteristics of li-ion battery is depicted in Fig. 7.

From Fig. 7, it is observed that the battery is discharging at the rate of 60Ah. The discharge characteristics describes amount of charge removed from the battery related to maximum charge that can be stored in the battery.

## 4 BLDC Motor

The BLDC motor power rating taken is 1 KW, 36 V motor. Trapezoidal wave shape of back emf of the BLDC motor is depicted in Fig. 8.

Figure 9 indicates angular position of the rotor when the switching pattern is given to the motor controller. In BLDC motor, the motor is controlled by three-phase

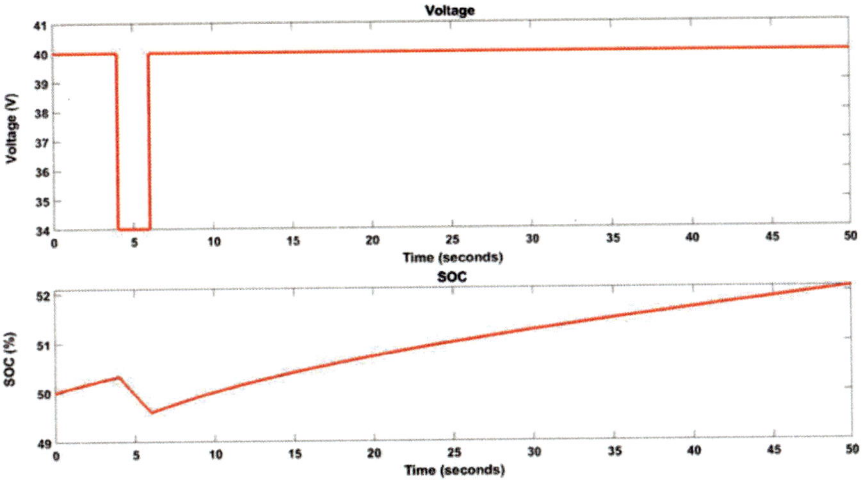


**Fig. 6** Input voltage and SOC characteristics of li-ion battery

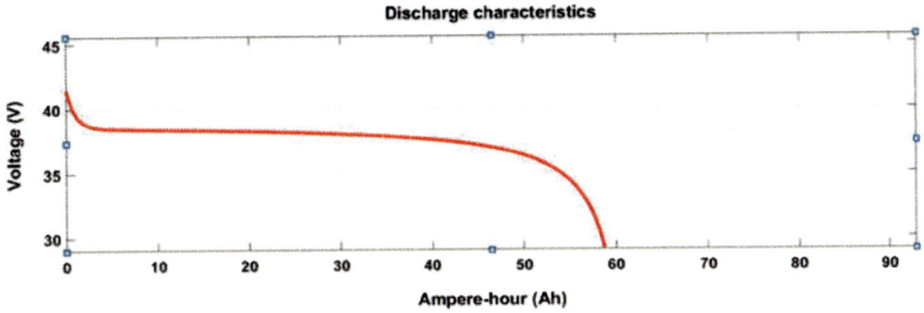


**Fig. 7** Discharge characteristics of li-ion battery

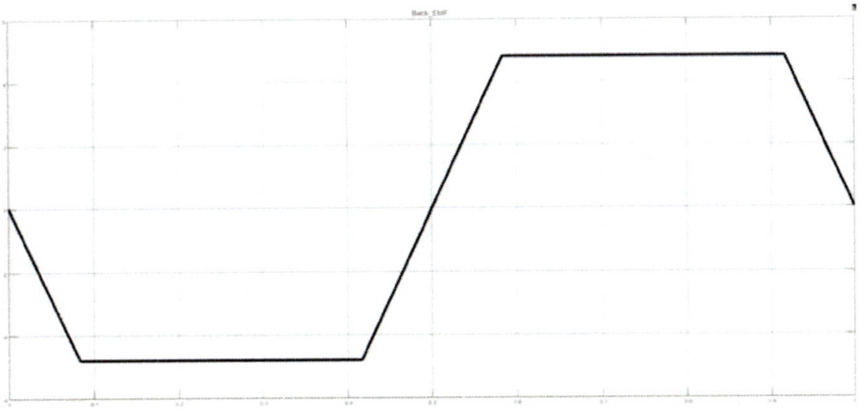

**Fig. 8** Trapezoidal back emf output of BLDC motor

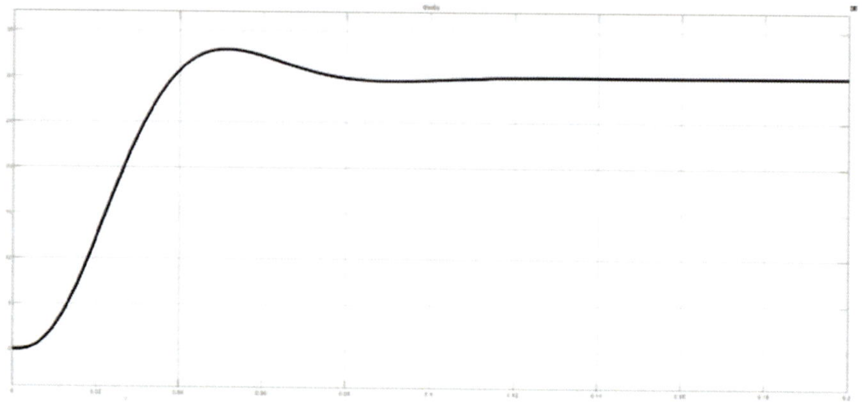

**Fig. 9** Rotor angular position of BLDC motor

inverter which gives pulses to excite the motor phases based on the rotor position. When the rotor position is sensed by the hall sensor, the controller feeds the logic to the appropriate switches in the inverter which in turn powers the appropriate phases. The commutation logic of BLDC motor is presented in Fig. 10.

Figure 10 shows the pulse pattern to excite the motor phases based on the hall sensor output which is based on the rotor position of a BLDC motor. The first graph indicates the sector value based on the pulse pattern. For example, when the sector is activated the phases 'a' high and 'c' low will be energized. In this way, the BLDC motor is controlled.

# 5 Results and Discussion

## *5.1 Simulation Results*

As specified in the block diagram, the overall simulation model has been developed for the electric trolley is shown in Fig. 11. A simulation model for PV is done, and it is coupled to the boost-buck Dc-Dc converter which regulates the PV output. The battery is charged by the solar energy. The battery provides stable output which energizes the motor through controller. The controller excites the motor phases based on the rotor position. The controller feeds the commutation logic to the inverter. The inverter supplies the BLDC motor. With enough torque produced, the vehicle is driven on both no load and on road conditions.

Figure 11 shows the overall MATLAB Simulink model of the electric trolley. With the help of the simulation model, the output waveforms can be determined.

Figure 12 shows the back emf output of the BLDC motor for all the three phases, and the stator current waveform for the BLDC motor is portrayed in Fig. 13.

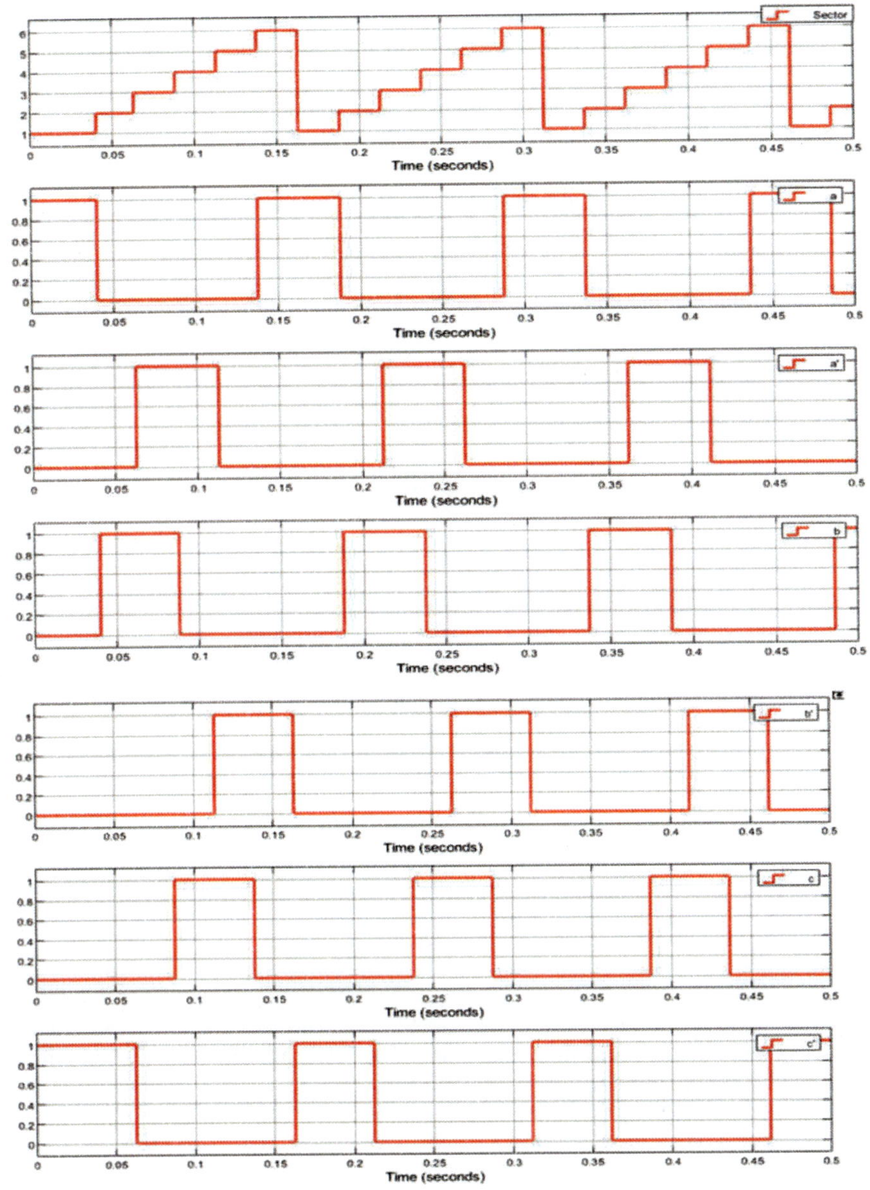


**Fig. 10** Trapezoidal commutation logic of BLDC motor

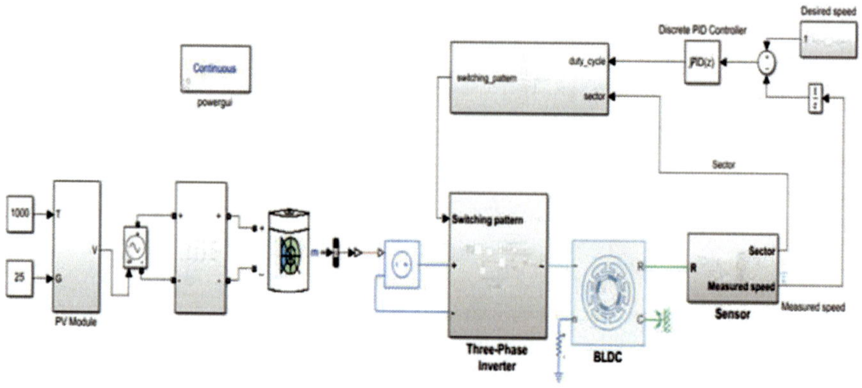


Fig. 11 Overall simulation model for electric trolley

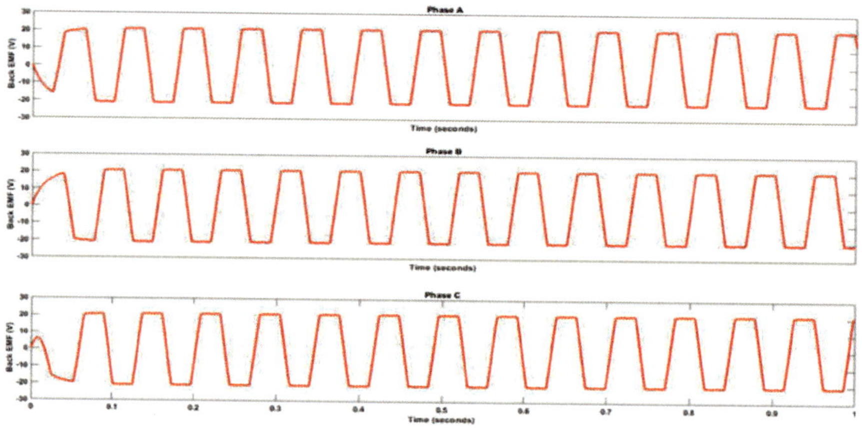


Fig. 12 Three-phase back emf output of BLDC motor

The rated speed for the BLDC motor is 2400 rpm. The graph in Fig. 14 shows the settling of the speed waveform at rated value. The controller makes sure the motor is operated at rated speed.

## 5.2 *Hardware Results*

To verify the proposed optimization strategy, the hardware setup of boost-buck converter was built. Figure 15 depicts the experimental setup of proposed converter.

Figure 16 shows the measurement of battery voltage and ampere hour with the help of battery meter. The battery is charged about 13.4 V, and it is observed from

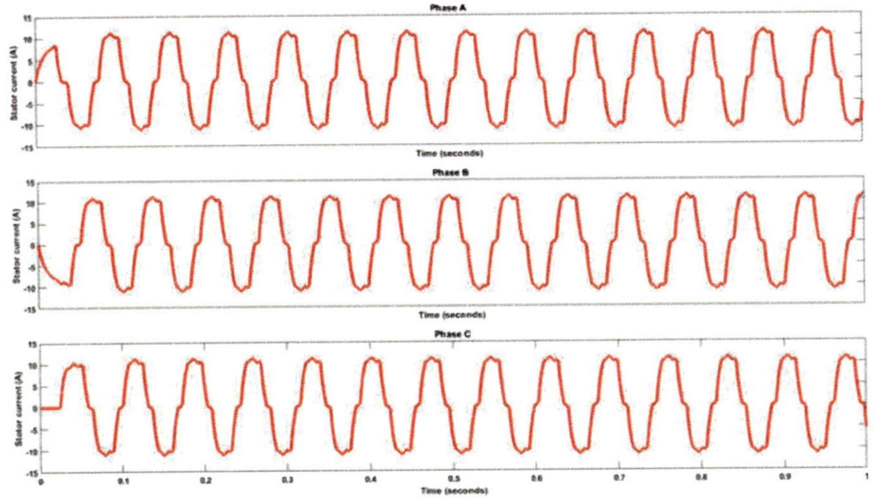


**Fig. 13** Stator current of a BLDC motor

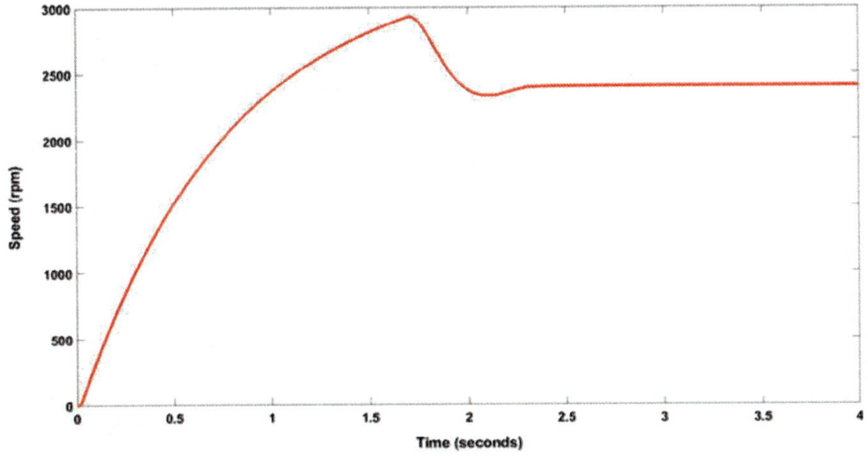


**Fig. 14** Speed waveform of a BLDC motor

the battery meter. The battery meter indicates state of the battery whether the battery needs to be immediately charged and fully charged. And the Li-ion battery and charger controller setup is depicted in Fig. 17. In this work, LiFePO4 battery has been used as it has higher energy density and life cycles than the li-ion battery. The energy density of the battery used in this work is 216 Wh.

Figure 18 shows the BLDC motor setup. The torque produced from the motor is transmitted to the wheels through chain transmission. The comparison between the simulation result and hardware result has been presented in Table 1. List of

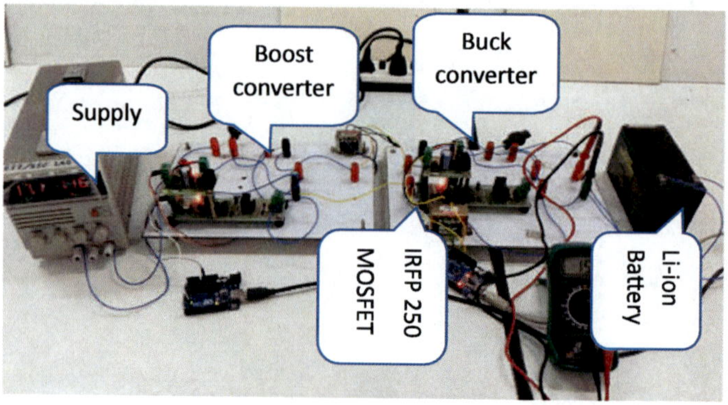


**Fig. 15** Prototype of boost-buck converter

**Fig. 16** Li-ion battery with battery meter

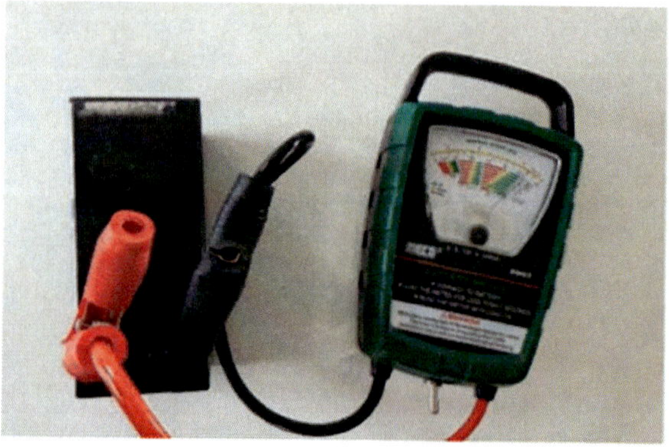

**Fig. 17** Li-ion battery and controller

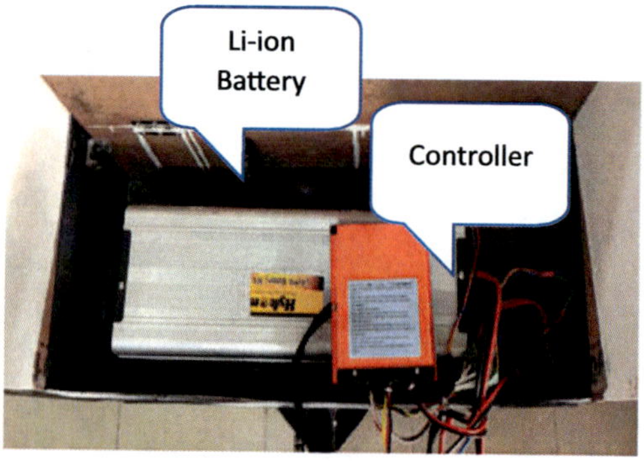

**Fig. 18** BLDC motor setup and wheel transmission of electric trolley

**Table 1** Comparison between the simulation and hardware results

| Boost-buck converter | Simulation result (V) | Hardware result (V) |
|---|---|---|
| Input voltage | 35 | 17 |
| Boost voltage | 41 | 20 |
| Buck voltage | 40 | 15 |

**Table 2** List of components used in the electric trolley setup

| Motor type | BLDC motor |
|---|---|
| Power (W) | 1 kW |
| System voltage | 36 V |
| Charging time | 7–8 h |
| Battery type | Li-ion |

components used in the electric trolley is illustrated in Table 2. Figure 19 shows the overall structure of the electric trolley.

From Table 1, it has been observed that the results of prototype model of the suggested converter validates the simulation results. Thus, the recommended converter has been used to charge the LiFePO4 battery which in turn runs the BLDC motor which has been shown in Fig. 19.

## 6 Conclusion

In this work, simulation and hardware setup of electric trolley system is developed. The performance of various electrical components of electric trolley is closely observed and analyzed. The simulation is done with MATLAB Simulink. Application

**Fig. 19** Electric trolley

of BLDC motor enhances the efficiency of the electric trolley system. The simulation results of boost-buck converter and li-ion battery characteristics are presented. The prototype of the recommended converter has been developed and confirmed the simulation results.

## References


1. Matwankar CS, Alam A (2019) Solar powered closed-loop current controlled DC-DC buck converter for battery charging application. In: 2019 international conference on vision towards emerging trends in communication and networking (ViTECoN), Vellore, India, 2019, pp 1–5
2. Macahig NA (2020) A 6-wire 3-phase inverter topology for improved BLDC performance and harmonics. In: 2020 IEEE applied power electronics conference and exposition (APEC), New Orleans, LA, USA, pp 741–744. https://doi.org/10.1109/APEC39645.2020.9124358
3. Wang Y, Li Y, Jiang L, Huang Y, Cao Y (2019) PSO-based optimization for constant-current charging pattern for li-ion battery. Chin J Electr Eng 5(2):72–78. https://doi.org/10.23919/CJEE.2019.000013
4. Fadli MR, Mushthofa Musyasy M, Furqani J, Purwadi A (2019) Modelling of field orientation control (FOC) method in 120 kW brushless DC motor (BLDC). In: 2019 6th international

conference on electric vehicular technology (ICEVT), Bali, Indonesia, pp 383–389. https://doi.org/10.1109/ICEVT48285.2019.8993973
5. Yasko MA (2018) Analysis, design and simulation of buck converter for photovoltaic system. In: 2018 22nd international conference electronics, Palanga, Lithuania, pp 1–6
6. Boujelben N, Masmoudi F, Djemel M, Derbel N (2017) Design and comparison of quadratic boost and double cascade boost converters with boost converter. In: 2017 14th international multi-conference on systems, signals & devices (SSD), Marrakech, pp 245–252
7. Mikhail T, Tatyana S, Petr S (2018) Usage efficiency of renewable energy sources for charging passenger electric transport. In: 2018 Renewable energies, power systems & green inclusive economy (REPS-GIE), Casablanca, pp 1–5. https://doi.org/10.1109/REPSGIE.2018.8488844
8. Karmaker AK, Roy S, Ahmed MR (2019) Analysis of the impact of electric vehicle charging station on power quality issues. In: 2019 international conference on electrical, computer and communication engineering (ECCE), Cox'sBazar, Bangladesh, pp 1–6. https://doi.org/10.1109/ECACE.2019.8679164
9. Ranawat D, Prasad MPR (2018) A review on electric vehicles with perspective of battery management system. 2018 international conference on electrical, electronics, communication, computer, and optimization techniques (ICEECCOT), Msyuru, India, pp 1539–1544. https://doi.org/10.1109/ICEECCOT43722.2018.9001321
10. Bhuiyan MF, Rejwan Uddin M, Tasneem Z, Hasan M, Salim KM (2018) Design, code generation and simulation of a BLDC motor controller using PIC microcontroller. In: 2018 international conference on recent innovations in electrical, electronics & communication engineering (ICRIEECE), Bhubaneswar, India, pp 1427–1431. https://doi.org/10.1109/ICRIEECE44171.2018.9008910

# Design and Development of Power Electronic Booster to Extend the Range of Supercapacitor-Powered Electric Bicycles

**A. Bharathi Sankar Ammaiyappan and Seyezhai Ramalingam**

## 1 Introduction

Electric bicycles have been a transportation mainstay in developed countries due to the ease of the vehicle maintenance, well-developed infrastructure, systematic driving conditions, and most importantly its eco-friendly nature. The electric bicycle, a self-explanatory term meaning the power, either partially or fully comes from an electric motor. The electric bicycles are currently used for short distances. Advanced research on both battery and drive technology benefits the market regarding the practicality of electric bicycles [1–4]. Environmental issues related to increased emanations and conventional fuel demand have resuscitated the manufacture of electric vehicle industrial and research sector and research community to work on new vehicular systems to improve the efficiency of city driving. Electrically-assisted bicycles have a more advantage of additional power, say during acceleration and cranking yet retaining characteristics of conventional geared bicycles [5–12].

Initially, Maxwell SC cells with the module specifications of 2.85 V, 3400 F, and 3.85 Wh (energy stored), and the individual supercapacitor cell was welded with aluminum bus bar using laser welding. Later, 18 supercapacitor cells were connected in series pack to obtain an SC module with 51.4 V and 69 Wh which is required to run the brushless DC motor (48 V, 250 W, and 10 A) in the bicycle. The design and developed of supercapacitor management system (CMS) to regulate and monitor the supercapacitor voltage and current in each cell during charge and discharge cycles. The developed SC-powered bicycle has been successfully tested in real conditions

A. Bharathi Sankar Ammaiyappan (✉)
School of Electronics Engineering, Vellore Institute of Technology, Chennai Campus, Chennai, India
e-mail: bharathisankar.1987@gmail.com; bharathisankar.a@vit.ac.in

S. Ramalingam
Department of EEE, Renewable Energy Conversion Lab, Sri Sivasubramaniya Nadar College of Engineering, Chennai, India



N. L. Narasimhan et al. (eds.), *Recent Advances in Energy Technologies*, Lecture Notes in Mechanical Engineering, https://doi.org/10.1007/978-981-19-3467-4_25

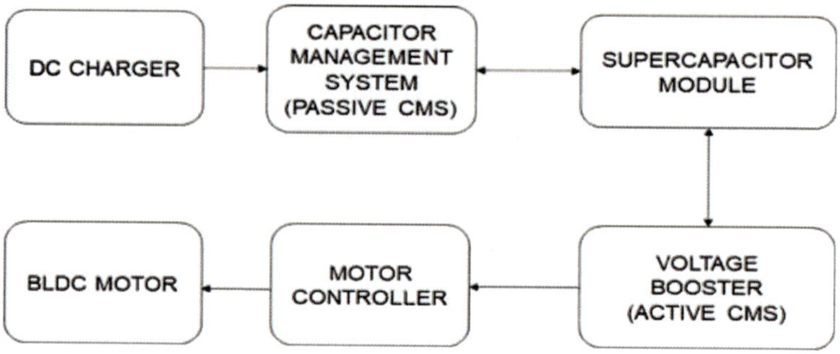


**Fig. 1** Block diagram of supercapacitor-powered electric bicycle

with the driving range of 2 km as shown in Fig. 1. The self-discharge of SC module has been monitored periodically, and the data generation of E-bicycle with parameters such as motor current, driving range, and energy consumption under different load conditions and speed was systematically carried out. In order to extend the driving range of E-bicycle, boost converter in addition to CMS, has been designed to boost up the voltage by recovering the remaining stored energy in SC module. Finally, designed and developed supercapacitor-powered E-bicycle with voltage booster has successfully demonstrated on-road with an extended driving range of 2.5 km.

## 2 Methodology

### *2.1 Design of Power Electronic Voltage Booster*

This research work is based on three approaches to the development of voltage booster simulations: circuit construction, physical construction of PCB, and virtual booster program development. MATLAB software simulations are a valuable asset for the design of electronic circuits, and we apply the same for charging circuit and booster circuit as shown in Fig. 2 [13–17]. The converter is used for operating the booster to develop MATLAB software simulation model and further tests external program coding component for the FPGA program [18–22]. In order to set the size of the DC to DC boost converter appropriately, the electric bicycle was powered with the 51 V, 3400 F, and 71 Wh supercapacitor module pack. This is mentioned in the specification of power DC-DC converter voltage 51.0 V, current 11.2 A, power 250 W.

Inputs from the voltage and current sensor are needed to turn ON the MOSFET of the power converter which is used further to estimate the switching duty cycle of the pulse width modulation. This supercapacitor module is on cut off voltage and ready for voltage booster operation and recovers stored energy of the supercapacitor module. The input (supercapacitor module) and output (power electronic

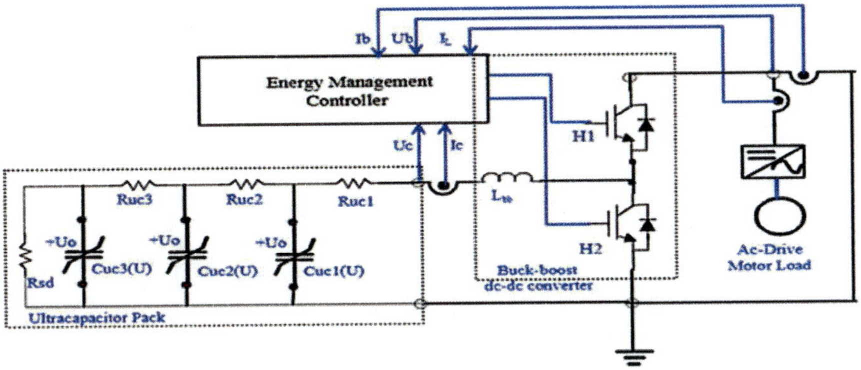


**Fig. 2** Circuit diagram of power electronic voltage booster with supercapacitor pack

booster) sides contain voltage sensor circuits to feed signals to the controller [23–27]. The duty cycle can be favorably adjusted based on the comparative studies of the abovementioned signals.

## 3 Results and Discussion

### *3.1 Simulation Results*

The MATLAB Simulink model of the supercapacitor-powered electric bicycle with power electronic booster is shown in Fig. 3. The smart power electronic booster to extending the range of supercapacitor-powered electric bicycle. The proposed work is to design and development of a pulse width modulation-based power converter with the specifications mentioned below: Fig. 4a–e,

- Input voltage (ultracapacitor): 43.2 V
- Output voltage (booster): 48.5 V
- Output power: 250 W
- Switching frequency: 50 kHz

Figure 4a and c shows that initial and final voltage of supercapacitor module which are about 42.5 V and 49.5 V, respectively. Figure 4e shows that the driving current of BLDC is about 7–9.5 A. Figure 4b and d depicts the values of SC charging booster current (16 A) and on the road, driving boosting current (23.5 A).

State of charge (SOC) level of supercapacitor is about 95%, and the present energy of the SC is 69 Wh. 44 Wh is the remaining amount of energy after discharge from a top-of-charge condition. Another vital parameter to be considered is depth of charge (DOD) which is defined as the percentage of energy to which a supercapacitor is discharged. Maximum temperature the Supercapacitor can withstand is around 36 °C.

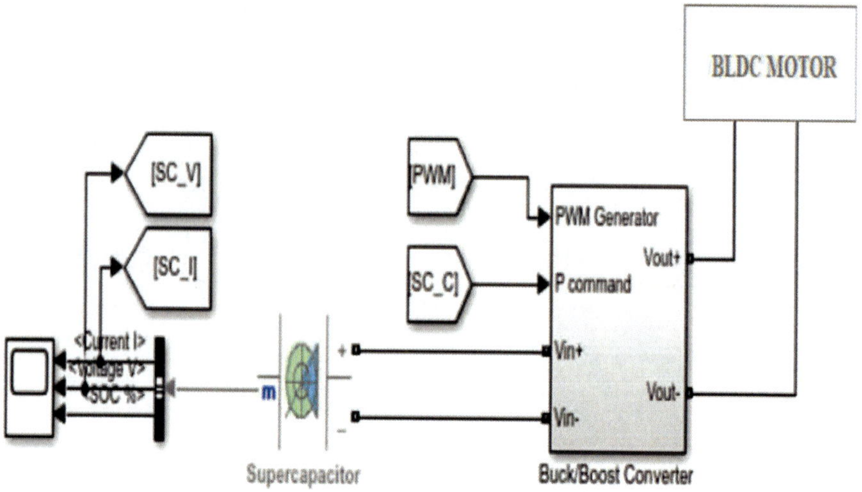


**Fig. 3** MATLAB Simulink model of supercapacitor-powered electric bicycle with power electronic booster

The simulated results of BLDC motor speed, BLDC motor electromagnetic torque, BLDC motor stator current, and back EMF are shown in Fig. 5a–d, respectively.

Figure 5a and b shows that the BLDC motor speed is settled at 300 rpm and starting torque is about 15 Nm. Figure 5c and d shows that the BLDC motor stator current and back EMF voltage.

### 3.2 *Hardware Implementation of Supercapacitor-Powered E-bicycle*

Design, development, and demonstration of supercapacitor-powered electric bicycle using commercial Maxwell SC cells is done. The supercapacitor cell specifications, C = 2.85 V, 3400 F, stored energy each cell, 3.85 WH, capacitor nodule nominal voltage, V = 51.4 V, total stored energy in capacitor module, E total: 69 Wh (18 S cells), and drive range Supercapacitor module voltage: 2 km. As the number of capacitor units increases, the requirement of capacitor management system along with voltage balancing is necessary to prevent individual cells from going into overvoltage. Development of capacitor management system to regulate each cell voltage management, current management, thermal management, and equalization using passive balancing system. Maximum charging voltage (V): 2.85 V (single cell), maximum charging current (A): 4.5 A, maximum discharge current (A): 13 A, overload protection: 20 A (Fuse) and voltage equalization startup voltage (V): 2.50 V. Supercapacitor cells are tightly encapsulated into plastic fixtures at three layers (top,

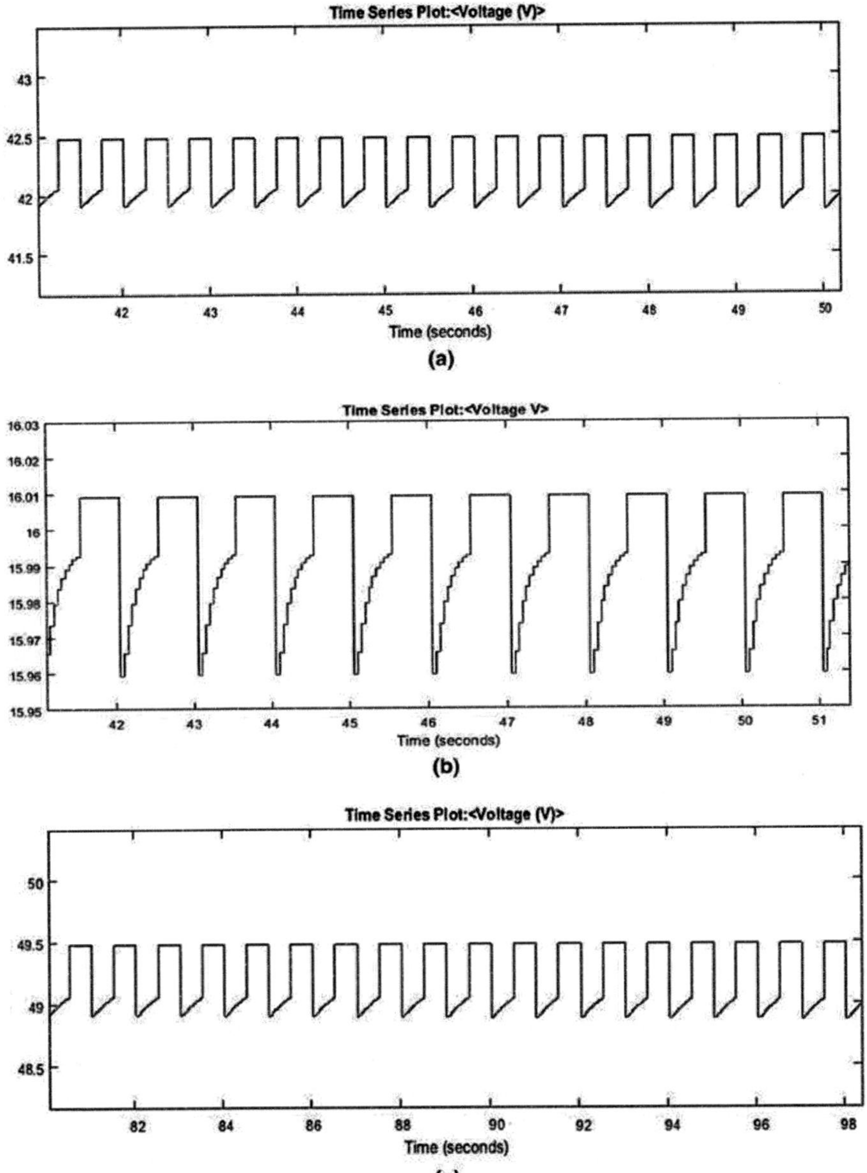


**Fig. 4** **a** Initial voltage characteristics of supercapacitor module, **b** Charging booster voltage characteristics of supercapacitor module, **c** final voltage characteristics of supercapacitor module, **d** discharging booster current characteristics of supercapacitor module, and **e** driving current characteristics of supercapacitor module

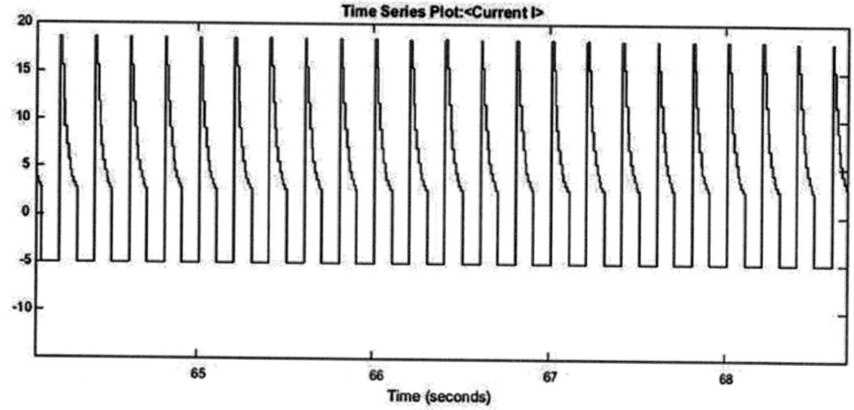

(d)

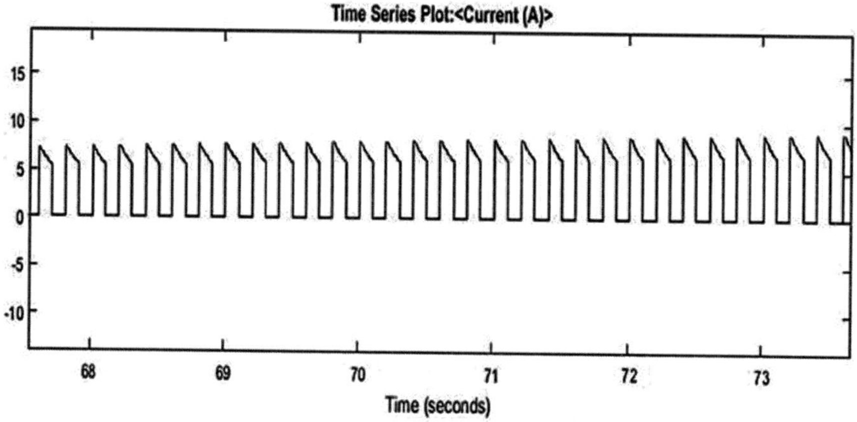

(e)

**Fig. 4** (continued)

middle, and bottom) to arrest any vibration. Inter supercapacitor cell welding is done under the encapsulated condition to provide more sturdiness. Any gap between plastic fixtures and cells on top and bottom is sealed with silicone adhesive. The whole module is protected again from vibration from the outer cover, using rubber beadings. The minimal amount of heat is generated in the supercapacitor during high-duty which tends to extend the service life of all electronic components associated with the module. Heat sinks and exhaust are accommodated internally to provide proper cooling at the rated current. A simple forced air cooling arrangement is also provided. Besides, a fan is also provided for effecting cooling of the capacitor management system. The Supercapacitor bike should have a minimum of 2 km range at 120 kg gross loading capacity (40 kg vehicle weight + 80 kg rider weight) and maximum

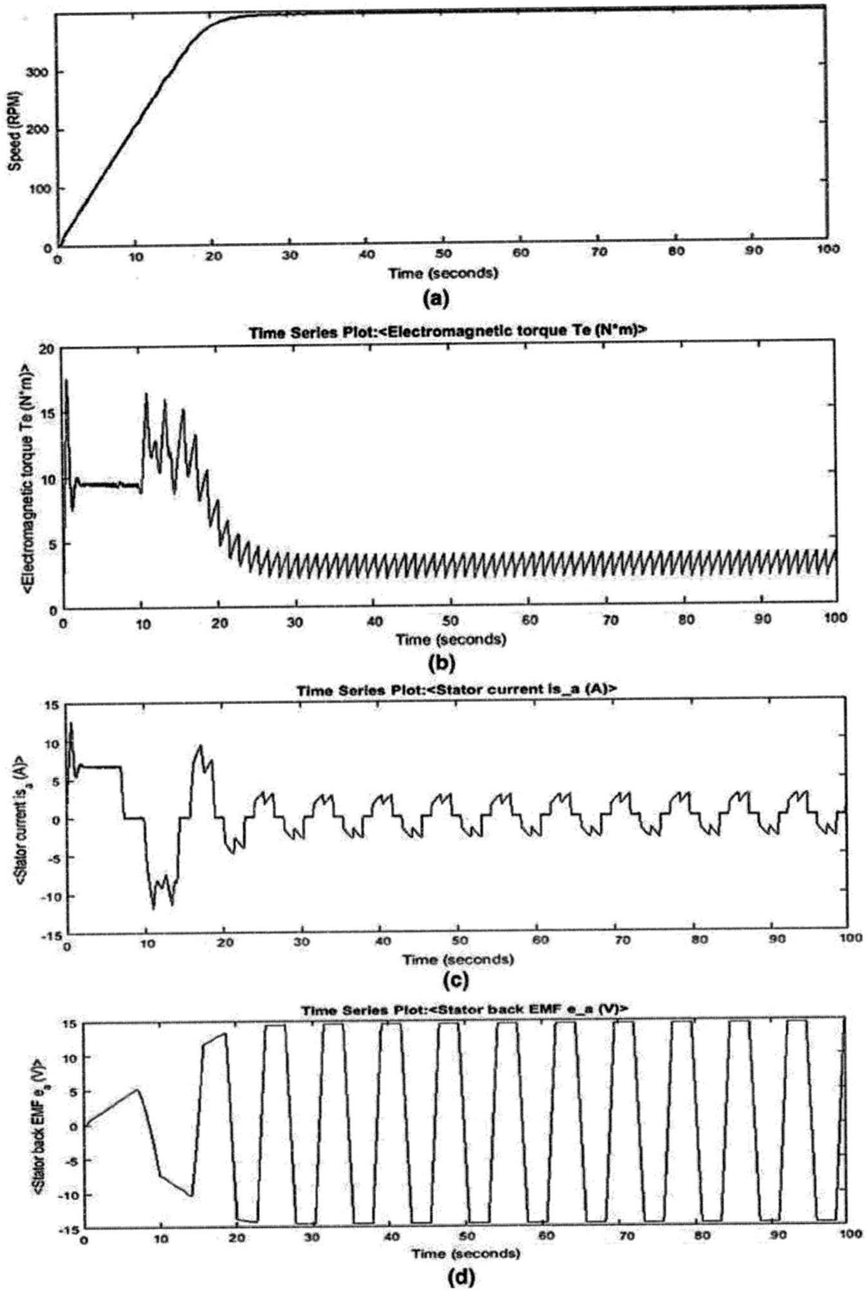


**Fig. 5** **a** Speed characteristics of BLDC motor, **b** torque characteristics of BLDC motor, **c** stator current characteristics of BLDC motor, and **d** back EMF characteristics of BLDC motor

speed is 25 km/h. BLDC motor—48 V–250 W (motor minimum working voltage: 41 V), sine wave motor controller—48 V–12A, DC charger—54 V–4.5 A.

Designing of electronic power booster with a capacitive management technique being the proposed work is to optimize the flow of power from the supercapacitor module to BLDC motor. Further, E-bicycle with the driving range of 2 km at 120 kg gross loading with the maximum speed of 25 km/h has been successfully demonstrated supercapacitor module (51.4 V, 3400 F, and 71.1 Wh). To further extend the driving range of supercapacitor E-bike from 2 to 2.5 km, an effort has been made to design a power electronic converter to boost up the voltage of the supercapacitor module. After consumption of 29 WH out of 69 WH energy for 2 km riding, the remaining stored energy (42 WH) in supercapacitor module can be partly (11 WH) recovered using active capacitor management system (CMS) with a boost converter. Before assembling of boost converter for supercapacitor module, simulation and verification of voltage boost up concept have been checked in a single supercapacitor cell using PCB boards. Development of field programmable gate array and integration of digital voltage booster into the supercapacitor module has been done. Finally, the driving range of supercapacitor E-bike has been successfully extended from 2.0 to 2.5 km using a developed voltage booster. During the discharging time, supercapacitor voltage dropout the energy from the supercapacitor is used to boost converter that can supply 15A of output current at 43.2–49 V. This output can be used to hold up the DC bus voltage for extending the E-bicycle range 2–2.5 km range as shown in Figs. 6 and 7.

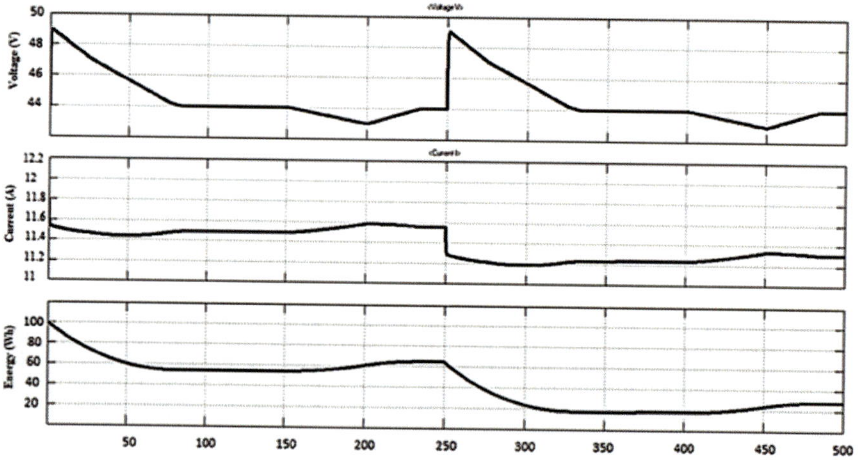


**Fig. 6** Hardware prototype verification of power electronics booster with SC-powered E-bicycle

**Fig. 7** Supercapacitor-powered electric bicycle

## 4 Conclusion

A 48 V, 250 W rear BLDC hub motor-based electric bicycle powered by a supercapacitor pack. Serial connections of supercapacitor cell were made between supercapacitor module (48 V and 71 WH) and power electronic booster with controlled field programmable gate array controller-based DC-DC power converter which will interface power between the supercapacitor and BLDC motor. The proposed work method for the power DC-DC converter was developed using various inputs of BLDC motor and supercapacitor pack (current and voltage sensor). Supercapacitor-powered electric bicycle has been successfully demonstrated. Further, self-discharge of supercapacitor module and data generation of supercapacitor E-bicycle parameters such as motor current, driving range, and energy consumption during various load and speed have been collected. The charge and discharge cycles with voltage booster have been monitored on a regular basis.

## References


1. Ates Y, Erdinc O, Uzunoglu M, Vural B (2010) Energy management of an FC/UC hybrid vehicular power system using a combined neural network-wavelet transform based strategy. Int J Hydrogen Energy 35(2):774–783
2. Marzougui H, Amari M, Kadri A, Bacha F, Ghouili J (2016) Energy management of fuel cell/battery/ultra capacitor in electrical hybrid vehicle. Int J Hydrogen Energy 42(13):8857–8869
3. Amari M, Bacha F, Ghouili J (2015) Average model of dual active bridge interfacing ultracapacitor used in electrical vehicle. Int J Energy Optim Eng 4:36–54
4. Payman A, Pierfederici S, Meibody-Tabar F (2008) Energy control of supercapacitor/fuel cell hybrid power source. Energy Convers Manage 49(6):1637–1644
5. Zhang L, Wang Z, Hu X, Sun F, Dorrell DG (2015) Acomparative study of equivalent circuit models of ultracapacitors for electric vehicles. J Power Sources 274:899–906

6. Stefan S (2009) Power sources for hybrid electric vehicles. M.thesis, Lund University, Faculty of Engineering, USA
7. Cultura AB, Salameh ZM (2015) Modeling, evaluation and simulation of a supercapacitor module for energy storage application. In: Proceedings of the international conference on computer information systems and industrial application, pp 876–882
8. Zhang L, Zhenpo W, Sun F, Dorrell D (2014) Ultra capacitor modelling and parameter identification using the extended kalmanfilter. ITEC Asia Pacific 7:3204–3217
9. Shi L, Crow ML (2008) Comparison of ultra capacitor electric circuit models. In: Proceedings of the IEEE power and energy society 2008 general meeting: conversion and delivery of electrical energy in the 21 st Century, PES, Pittsburgh, PA,USA, July 2008
10. Ban S, Zhang J, Zhang L, Tsay K, Song D, Zou X (2013) Charging and discharging electrochemical supercapacitors in the presence of both parallel leakage process and electro chemical decomposition of solvent. ElectrochimicaActa 90:542–549
11. Jin L-Q, Zheng Y, Li J-H, Liu Y-L (2015) A study of novel regenerative braking system based on supercapacitor for electric vehicle driven by in-wheel motors. Adv Mech Eng 7(3):1–12
12. Hala S, Khalil Faris N, Khaled Faris N (2015) Ultra capacitor: modelling and characterization for electric vehicle application. Int J Eng Res 5:681–686
13. Fletcher S, Kirkpatrick L, Drink R, Puttock R, Thring R, Howroyd S (2017) The modelling of carbon-based supercapacitors: distributions of time constants and pascal equivalent circuits. J Power Syst 345:247–253
14. Fuyuan Y, Languang L, Yuping Y, He Y (2010) Characterization, analysis and modelling of an ultracapacitor. World Electr Veh J 4:358–369
15. Zhang L, Wang Z, Sun F, Dorrell DG (2014) Online parameter identification of ultra capacitor models using the extended Kalmanfilter. Energies 7(5):3204–3217
16. Plett GL (2004) Extended Kalman filtering for battery management systems of Li PB-based HEV battery packs—part 1. Background. J Power Sources 134(2):252–261
17. Michalczuk M, GrzesiakLech M, Ufnalski B (2012) Alithium battery and ultra capacitor hybrid energy source for an urban electric vehicle. Politechnika Warszawska 4:158–162
18. Ashar M (2016) Integration of ultra capacitor with battery using DC-DC bidirectional buck boost converter in an electric vehicle. Int Res J Eng Technol 3:1513–1517
19. Bottu M, Crow ML, Elmore AC, Atcitty S (2013) A Power electronic conditioner using ultracapacitors to improve wind turbine power quality. Smart Grid Renewable Energy 4:69–75
20. Parker AA (2006) Electric power-assisted bicycles reduce oil dependence and enhance the mobility of the elderly electric power assisted bicycle. Presented at 29th Australian Transport Research Forum
21. Tuite D (2007) Get the lowdown on ultracapacitors. Technology report electronic design online, Nov 2007
22. Schneuwly A, Maher B, Auer J (2004) Ultracapacitors, the new thinking in the automotive world. Maxwell Technologies Inc
23. Pay S, Baghzouz Y (2003) Effectiveness of battery-supercapacitor combination in electric vehicles. In: Power tech conference proceedings, 2003 IEEE Bologna, vol 3, pp 6, 23–26 June 2003
24. Dixon JW, Ortuzar ME (2002) Ultracapacitors + DC-DC converters in regenerative braking system. Aerosp Electron Syst Mag IEEE 17(8):16–21
25. Paymen A, Oierfederici S (2008) Energy control of supercapacitor/fuel cell hybrid power source. Energy Convers Manage 49:1637–1644
26. Farzanehfard H, Beyragh DS (2008) A Bidirectional soft switched ultracapacitor interface circuit for hybrid electric vehicles. Energy Convers Manage 49:3578–3584
27. George M, Choi S (2003) Contrasting hybrid electric bicycles and electric bicycle. Inter-Professional Project 315 Spring, pp 20–25

# Alternate Fuels

# Analysis of Performance and Emission Properties of Biodiesel Using Corn Oil Blended with Ethanol

K. S. Karthi Vinith, P. Sathiamurthi, C. Gowrishankar, S. Shaachin, and D. K. Karthi

## 1 Introduction

The transportation medium that moves on its own is invented first with the steam boilers that use the water as fuel. Later, fossil fuels popularly known as gasoline and diesel are invented. But over hundreds of years of usage, fossil fuels stock has started draining. So, for further survival, a major research is going on to find an alternative for complete or partial replacement of fossil fuels. So, people are doing various research for the best biodiesel as an alternative fuel. Additives are used in today's scenario to satisfy the International Standard and to minimize the problems of biodiesel. According to Karthi Vinith et al. [1], there is a decrease in calorific value for the biodiesel that is made up of two different oil and also shows the lower brake thermal efficiency (BTE) for those biodiesel blends. Biodiesel blends give lower HC and NOx emissions. Madiwale et al. [2] stated that ethanol can be obtained easily from biomass using the fermentation process using different feedstocks like sugarcane, sugar beet, etc. Balamurugan et al. [3] produced corn oil-based biodiesel and experimented with the same biodiesel, and there are similar performance, combustion, and emission characteristics with pure diesel. According to Madiwale et al. [4], ethanol concentration should be 99% and the total amount of ethanol used in the biodiesel blend should not exceed 5% by volume. Sathiamurthi et al. [5] have used Al2O3 nanoparticles along with diesel and found that brake thermal efficiency has increased considerably, and there is a substantial reduction in NOx and unburnt hydrocarbons. Ravishankar Sathyamurthy et al. [6] says that corn oil-based biodiesel exhibits the BTE similar to that of diesel but HC and NOx emissions were high. There are many of these kinds of literature on the performance

K. S. Karthi Vinith (✉) · C. Gowrishankar · S. Shaachin · D. K. Karthi
Department of Automobile Engineering, Kongu Engineering College, Erode 638316, India
e-mail: vinithmech10@gmail.com; karthivinith.auto@kongu.edu

P. Sathiamurthi
Department of Mechanical Engineering, Kongu Engineering College, Erode 638316, India



N. L. Narasimhan et al. (eds.), *Recent Advances in Energy Technologies*, Lecture Notes in Mechanical Engineering, https://doi.org/10.1007/978-981-19-3467-4_26

of biodiesel with ethanol available. But there is no significant research for the usage of corn oil which is available in abundance with ethanol as an additive has been made.

## 2 Materials and Methods

In this research, pure diesel, corn oil, and ethanol are made available from the market. Table 1 gives the various blends of diesel with corn oil and ethanol. These blends are properly mixed with the stirrer for 15–20 min before using in the engine for proper blending.

Here B[XX] means B for biodiesel and xx for % of biofuel blended with pure diesel. So, the term B15, B20, and B25 indicate the 15% of biofuel blended with diesel, 20% biofuel blended with diesel, and 25% of biofuel blended with diesel, respectively. Here biofuel stands for the blend of corn oil and ethanol additives.

Figure 1 shows a multi-fuel single cylinder 4—stroke engine fitted with an eddy current dynamometer has been used for the performance and emission measurement, and Table 2 replicates the specifications and make of the engine. The engine is calibrated by measuring the properties of the standard diesel fuel. Later, the abovementioned blends are tested for their performance and emission properties.

Biodiesel has been tested for certain specifications and quality to meet regulatory standards for use. The fuel properties for the blends B15, B20, and B25 are given in Table 3.

**Table 1** Blends of biodiesel with diesel

| Blend | Diesel (%) | Corn Oil (%) | Ethanol (%) |
|---|---|---|---|
| B15 | 85 | 11.25 | 3.75 |
| B20 | 80 | 15 | 5 |
| B25 | 75 | 18.75 | 6.25 |

**Fig. 1** Engine used for testing purpose

**Table 2** Specification of engine

| Product | 4 stroke single cylinder diesel engine (computerized) |
|---|---|
| Engine | **Make:** Kirloskar, **Model:** TV1, **Type:** Single cylinder 4 stroke diesel, water cooled, **Power:** 5.2 kW at 1500 rpm, **Stroke:** 110 mm, **Bore:** 87.5 mm–661 cc, and CR 17.5 |
| Dynamometer | Eddy current and water-cooled setup |
| Propeller shaft | Universal joint |
| Fuel Tank | Capacity 15 L fitted with glass fuel metering column |
| Load Indicator | Digital type with range 0–50 kg and supply 230 V AC |
| Load Sensor | Load cell of strain gauge type with range 0–50 kg |
| Fuel Flow Transmitter | DP transmitter and range 0–500 mm WC |
| Air Flow Transmitter | Pressure transmitter and range (–) 250 mm WC |

**Table 3** Fuel properties of the biodiesel blends

| | Biodiesel blends | | | |
|---|---|---|---|---|
| Properties | Diesel | B15 | B20 | B25 |
| Corrosion Tendency | 1–4 | 1 a (Slight Tarnish) | 1 a (Slight Tarnish) | 1 a (Slight Tarnish) |
| Cloud Point °C | –11 | –10.7 | –12.5 | –10.2 |
| Pour Point °C | –2 | –16.8 | –18.2 | –17.4 |
| Flash Point °C | 59 | 29.8 | 32.4 | 40.5 |
| Fire Point °C | 65 | 32.2 | 33.6 | 44.5 |
| Calorific value (MJ/kg) | 43 | 42.313 | 41.307 | 40.61 |
| Density kg/m$^3$ | 835 | 816 | 816 | 824 |

Based on the above table, the calorific value of B15 > B20 > B25, and approximately equal to diesel, and the density is also nearer. The compression ratio (CR) is fixed as 17.5.

# 3 Results and Discussion

## *3.1 Performance Characteristics of Biodiesel Blends*

The CI engine is made to operate with three different blends of biodiesels B15, B20, B25, and B0. The performance for each blend is determined and discussed in this section.

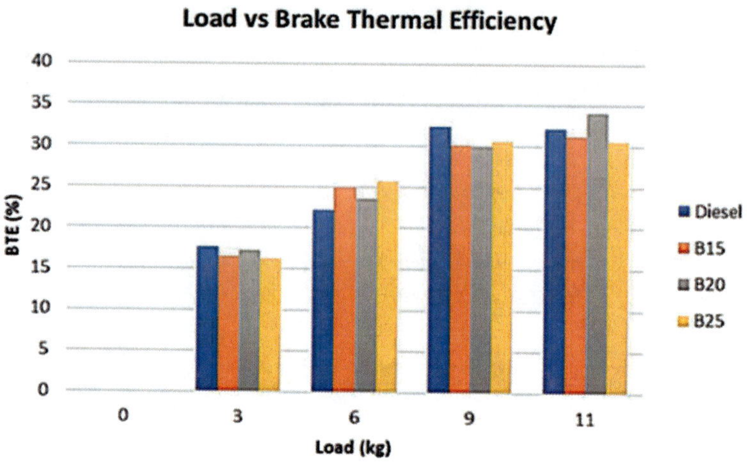


**Fig. 2** Brake thermal efficiency (BTE)

**Brake Thermal Efficiency (BTE)**

The Brake Thermal Efficiency (BTE) for different blends for the complete operation of the engine is shown in Fig. 2. It is noted that the BTE for B15, B20, and B25 is performing well at all loads among that B20 is comparatively good in this aspect.

**Brake Specific Fuel Consumption (BSFC)**

BSFC or SFC is the measure of the fuel efficiency of any prime mover that produces rotational or shaft power. It is known that SFC decreases as the load increases. Now, by looking into Fig. 3, the SFC for all blends of fuel gets reduced as the load gets increased. Now, by looking particularly, B20 is good at loads 3, 9, and 11 kg, whereas B15 is good at 6 kg. So, on average it can be concluded that B20 is good.

## 3.2 *Combustion Characteristics*

**Net Heat Release Rate**

The combustion characteristics for the biodiesel blended with ethanol can be determined by the neat heat release that is used during the combustion between the crank angles –40° to 40° as shown in Fig. 4. It is observed that the maximum heat release is obtained by the B25 blend. The next high heat release is observed for the B20 blend and next to that B15 blend has the high heat release is observed.

**Pressure Angle versus Crank Angle**

The combustion character for the biodiesel with ethanol can be obtained by measuring cylinder pressure which is obtained during the combustion process between the crank

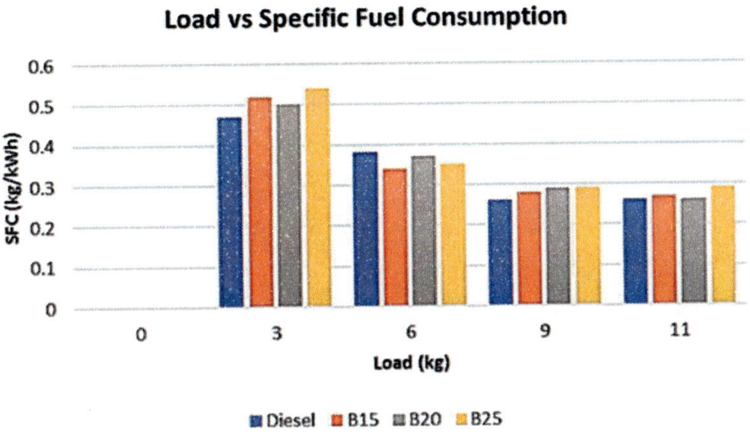


**Fig. 3** Specific fuel consumption (SFC)

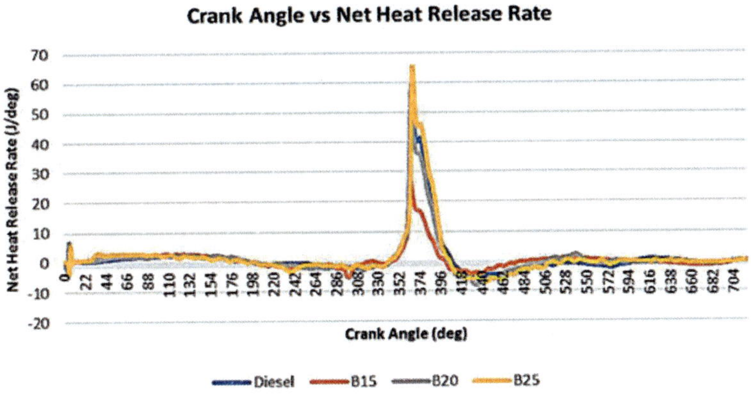


**Fig. 4** Net heat release rate

angles –360° to 360°. From Fig. 5, it is clear that the pressure is exerted by the biodiesel blends is of B25 > B20 > B15.

## 3.3 *Emission of Diesel with Corn Oil*

**Emission of NOx**

From Fig. 6, it is seen that B25 has lesser emission of oxides of nitrogen at all loads. Hence, it is obvious that adding more ethanol with diesel reduces the emission of oxides of nitrogen at all load.

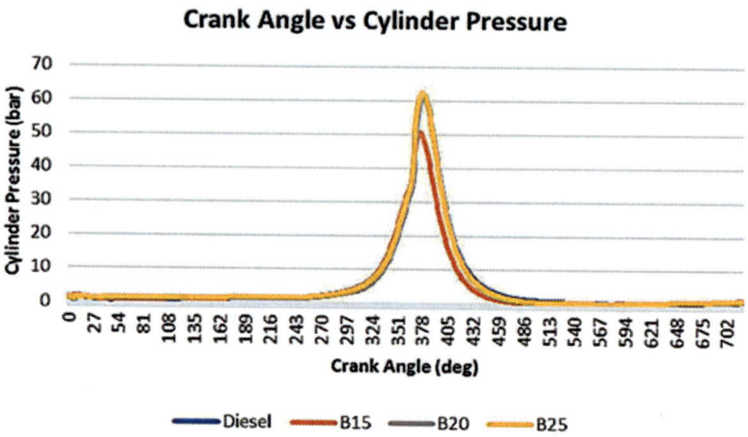


**Fig. 5** Cylinder pressure angle with various crank angle

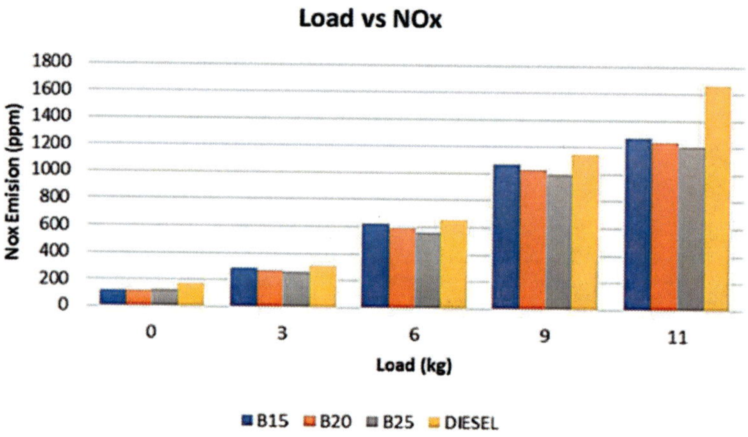


**Fig. 6** Emission of NOx

## Emission of HC

From Fig. 7, it is seen that biodiesel blends have lower emissions than diesel. Particularly, noted that B15 blends have lower emissions at lower loads and as the load increases, and the higher blends of biodiesel with ethanol additive have lower emissions of hydrocarbon.

## Emission of CO

From Fig. 8, it is seen that CO emissions of biodiesel blends with ethanol as an additive are lesser than the CO emissions of pure diesel. This is mainly due to the presence of ethanol, as it oxidizes properly, the CO emission gets reduced. CO

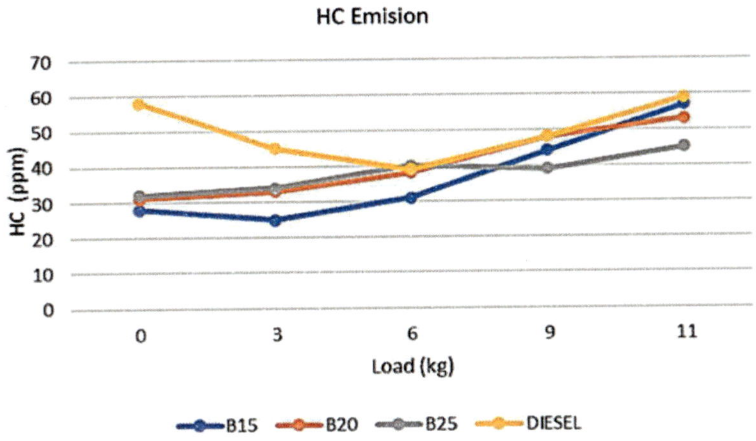


**Fig. 7** Emission of HC

emission at lower loads is comparatively higher and at higher loads, the emissions are lower. This is due to a higher load; the temperature is higher and complete combustion is possible.

Thus, corn oil with ethanol as an additive can be used as biodiesel as its performance and combustion characteristics are almost nearer or better than diesel. While looking into the emission characteristics, it has minimum emissions at higher loads while compared to the emission at lower loads.

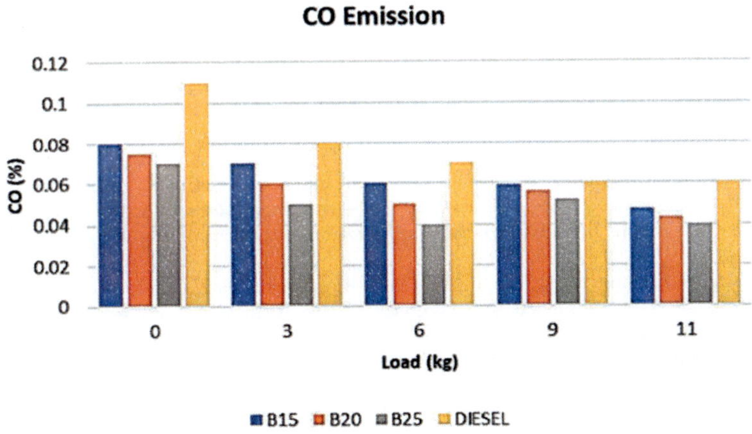


**Fig. 8** Emission of CO

# 4 Conclusion and Future Work

The analysis which is aimed to study the performance of the CI engine by preparing biofuel with corn oil + ethanol + diesel and is assessed for biodiesel with blends of 15, 2, and 25% by volume shows that,

- The Break Thermal Efficiency, while compared to diesel, is low for the lower blends of the biodiesel, whereas it shows the higher value compared to diesel at higher blends at higher loads.
- While considering the CO emission, blended fuels have lower emissions than diesel.
- The value of hydrocarbon tested in the exhaust gas was comparatively less for all blends than with diesel. Moreover, as the percentage of corn oil in the biodiesel increases, there is an increase in HC emission in the exhaust, but due to addition of ethanol as an additive, the HC emission gets reduced.
- While comparing to diesel, the NOx present in the blended fuels is lesser only. And also, it is noted that the blend ratio of corn oil increases, there is a decrease in NOx level.
- With the analysis of overall blends performance, it can be concluded that B25 blend performs well in the diesel engine.

**Future Scope**

Since the automobile industry is becoming electrical and fully automated, our future generation vehicles will be mostly electrical and hybrid. Hence to meet the present demand for diesel fuels, it is better to prefer alternate fuels as best fuels. It can be further improved by using the fuel as fumigated in diesel engines.

# References

1. Karthi Vinith KS, Soundar A, Mahalingam S, Sujai S, Guru Prasad PK (2021) Experimental investigation for the usage of diesel—Jatropha—Rice bran biodiesel mixture blends in four stroke diesel engine. In: IOP conference series: materials science and engineering, vol 1055(1), pp 1–12
2. Madiwale S, Karthikeyan A, Bhojwani V (2018) Properties investigation and performance analysis of a diesel engine fuelled with Jatropha, Soybean, Palm and Cottonseed biodiesel using Ethanol as an additive. Mater Today: Proc 5:657–664
3. Balamurugan T, Arun A, Sathishkumar GB (2018) Biodiesel derived from corn oil—a fuel substitute for diesel. J Renewable Sustain Energy Rev 94:772–778
4. Madiwale S, Karthikeyan A, Bhojwani V (2017) A comprehensive review of effect of biodiesel additives on properties, performance, and emission. In: IOP conf. series: materials science and engineering, vol 197, pp 01–07
5. Sathiamurthi P, Karthi Vinith KS, Sivakumar A (2019) Performance and emission test in CI engine using magnetic fuel conditioning with nano additives. Int J Recent Technol Eng (IJRTE) 8(3):7823–7826

6. Sathyamurthy R, Balaji D, Gorjian S, Jenoris Muthiya S, Bharathwaaj R, Vasanthaseelan S, Essa FA (2021) Performance, combustion and emission characteristics of a DI-CI diesel engine fueled with corn oil methyl ester biodiesel blends. Sustain Energy Technol Assess 43:100981

# An Investigation of Emissions of Jojoba Oil as a Supplemental Fuel for Direct Injection Diesel Engines

M. Nagappan and J. M. Babu

## 1 Introduction

Biodiesel fuels have a number of disadvantages, including lengthy chemical processes and high manufacturing costs [1]. Researchers expect that after suitable modifications to diesel engines, biofuel will be incorporated into the engine once they insist on their idea [2]. In order for neat oil to be suitable for combusting heavier molecular structures without adverse effects on performance or emissions, the engine must have the amount of combustion capacity needed [3]. The researchers suggested many modifications to the clean oil in the CI engine when it was applied in higher proportions. Air pollution, abrupt climate change, and oil spills will directly impact the environment by using fossil fuel products on a regional and global scale via the use of fossil fuel products [4].

Chemical reactions and manufacturing costs are the two most significant drawbacks of biodiesel fuel [5]. Researchers expect that after suitable modifications to diesel engines, biofuel will be incorporated into the engine once they insist on their idea. Engines that use neat oil should be able to combust heavier molecules without performance loss or emission problems [6]. As a result, the researchers have suggested many modifications to the compression ignition (CI) engine while utilizing neat oil of higher proportions. It is possible to make biodiesel from a variety of sources, including edible and non-edible oils, animal fats, and cooked oil sludge [7]. As a result of using biodiesel in CI engines, it is associated with few disadvantages, including higher density, low cloud and pour points, heavy density, and

M. Nagappan (✉) · J. M. Babu
Department of Mechanical Engineering, Vel Tech Rangarajan Dr. Sagunthala R&D Institute of Science and Technology, Chennai, India
e-mail: mnagappan@veltech.edu.in

J. M. Babu
e-mail: jmbabu@veltech.edu.in



N. L. Narasimhan et al. (eds.), *Recent Advances in Energy Technologies*, Lecture Notes in Mechanical Engineering, https://doi.org/10.1007/978-981-19-3467-4_27

problems with piston rings, fuel efficiency deduction of nearly 10%, a hard time starting in cold conditions, and high $NO_X$ emissions [8].

The cause of the increased value of fossil fuels and the massive rise in demand for global energy is due to developing countries relying on traditional fuels like coal, natural gas, and petroleum-based products [9]. Asian countries that import fuel face economic difficulties and a lack of equal distribution of fuel across all of their countries [10]. Fuel in the combustion chamber is incompletely burned due to the presence of volatile organic compounds in exhaust gases. Its chemical compositions, such as formaldehyde, benzene, and acetaldehyde, are harmful to human health because of its chemical properties. Vehicles running on fuel or diesel, particularly those in urban areas, produce the majority of volatile organic compounds [11]. These emissions will cause more severe health problems such as cancer if they are composed of polycyclic aromatic hydrocarbons [12].

Diesel engines are utilized to generate electricity for irrigation pumps in addition to large vehicles such as ships, trains, and airplanes [13]. Ecosystems and human beings are affected directly by engine emissions. Adding alternative fuels would be a smart way to deal with these problems because they are abundant in nature and easy to use in engines [14]. Many analysts have suggested that non-conventional sources could be used as a replacement for existing fossil fuels in the face of increasing energy demand and environmental problems caused by petroleum-based fuels [15]. Nature has a tremendous amount of sustainable energy resources, and they have a larger capacity, which produces the amount of energy people need [16]. Considering that biodiesel is made from renewable resources, it is a suitable substitute for diesel fuel [17].

The world's crude oil needs typically supply 84% of its energy needs. The cold war between countries along with its inadequacy led to the development of alternative energy sources [18]. Fossil fuels have become more popular in recent years, as have their demands. Research is being conducted very hard to find suitable oil for alternative fuels, which are currently a promising field for automobile sectors [19]. Researcher is working tirelessly to find a solution to the problem in this field. The development of a fuel that will completely replace diesel and gasoline still remains a challenge [20]. Several undesirable characteristics of fuel and a lack of supply play a part in this. Researchers believe that plant oils still offer the greatest hope. Using plant oils instead of diesel fuel is the best way to eliminate it fully [21].

In addition to growing from 1 to 5 m tall, jojobas are dioecious, perennial shrubs that can live between 100 and 200 years. In addition to goat nuts, pignuts, wild hazel, quinine nuts, coffeeberries, and gray box shrubs, this plant is also known as Simmondsia Chinensis—a species of plant in the Simmondsian family. When engine loads are low, there is a little bit more fuel consumption for blends and less energy consumption for blends. Emissions are lower at maximum engine loads compared to minimum engine loads. There is a small decrease in NOx emissions at low engine loads, and there is no difference at high engine loads. In all ranges of engine loads, the emissions of carbon monoxide and hydrocarbons are higher for the blends.

The present study is an investigation of the performance and emissions of jojoba oil as a supplemental fuel for direct injection diesel engines. The performance of a single cylinder, direct injection, naturally aspirated diesel engine was evaluated using both diesel and diesel with jojoba oil blends. At different engine loads and as a function of engine speed, 100% diesel was compared with other fuel blends in terms of engine performance parameters.

## 2 Setup for Experiments

Kirloskar TV1 is a single cylinder diesel with direct injection, four stroke, water cooled, 3.8 kW, and 2000 rpm. An engine test rig system with an eddy-current dynamometer was utilized to perform experimental research. The setup also includes a variety of measurement equipment for determining the input and end outputs. An anti-pulsating drum, an air temperature monitoring device, an air heater, and an energy meter make up the engine's intake section. Similarly, the engine's output section has an exhaust gas temperature (EGT) indicator, an exhaust gas analyzer, and a smoke meter.

Burette and stopwatch were used to measure the diesel input. An anti-pulsating drum was also used to assess air usage. The eddy-current dynamometer attached to the engine aids in measuring and varying the engine's power output. The setup also had a pressure sensor and crank angle encoder device to calculate the combustion pressure at each degree of crank angle. EGT indicator and gas analyzer were used to measure the exhaust gas temperature and pollutant content, respectively. In order to measure the opacity of the smoke, a smoke opacity meter was connected to the setup. The setup also had a computer system with a data acquisition (DAQ) facility to acquire cylinder pressure data. The cylinder pressure and crank angle data were collected using a LabVIEW-based program named "engine soft." This program was also utilized to calculate the performance of the temperature and pressure release rates, as well as other metrics.

This work has used jojoba oil as a test fuel. Studies were conducted on the feasibility and performance of jojoba oil in its natural and modified forms (various blends). Various properties of such fuels are given in Table 1.

In this research, 10, 20, 30, and 40% jojoba blends were admitted directly into the compression ignition engine through the regular fuel admission device of the CI engine. The engine used for this method has no special modifications. At each load stage, all mixes were evaluated for operational performance, combustion, and emission characteristics (Table 2).

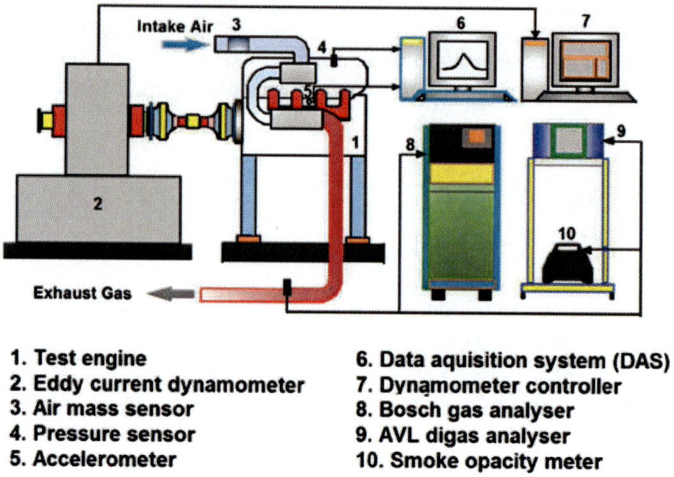


**Fig. 1** Diagram of the experimental setup

**Table 1** Specifications of the engine

| S.No | Details | Descriptions |
|---|---|---|
| 1 | Make and model | Kirloskar and Direct injection, four stroke and CI Engine |
| 2 | Compression ratio | 16.6 |
| 3 | Rated output | 4.4 kW |
| 4 | Rated speed | 1500 rpm |
| 5 | Injector opening pressure | 170 bar |
| 6 | Type of cooling | Water cooled |

**Table 2** Jojoba's properties versus diesel versus mustard versus neem oil

| S.No | Properties | Diesel | Mustard oil | Neem oil | Jojoba oil |
|---|---|---|---|---|---|
| 1 | Density in kg/m$^3$ | 820 | 844 | 870 | 905 |
| 2 | Specific gravity | 0.82 | 0.85 | 0.926 | 091 |
| 3 | Boiling point in °C | 180–340 | 410 | 350 | 398 |
| 4 | Viscosity in c St | 4–5 | 15 | 18 | 22 |
| 5 | Flash point in °C | 74 | 141 | 245 | 225 |
| 6 | Flammability limit in % volume | 1 | 2.5 | 3.6 | 1–5.2 |
| 7 | Cetane number | 41–45 | 40–45 | 51 | 45–48 |

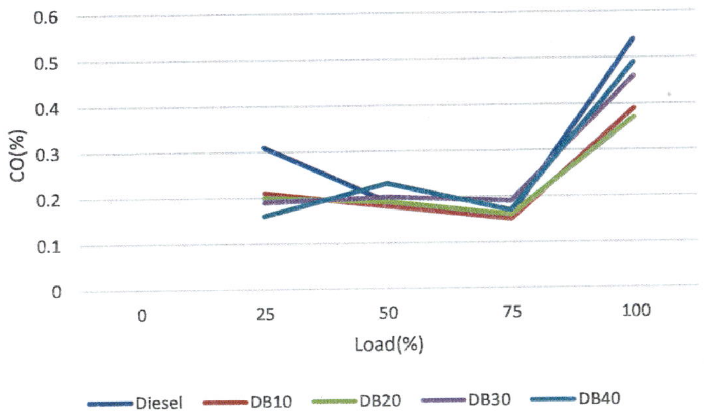


**Fig. 2** Carbon monoxide variation

# 3 Results and Discussion

## 3.1 CO Emission

Carbon monoxide (CO) emissions are most common when the air-to-fuel equivalency ratio for a rich fuel combination is less than one and carbon dioxide cannot be produced if there is not enough oxygen. Figure 2 depicts DB10 (diesel 90% and jojoba 10%), DB20 (diesel 80% and jojoba 20%), DB30 (diesel 70% and jojoba 30%), DB40 (diesel 60% and jojoba 40%), and diesel CO emissions vs load. As jojoba blends are increased, CO emission reductions are significant compared to diesel.

## 3.2 HC Emission

Longer ignition delay periods and a lean mixture over the reaction zone tend to create hydrocarbon (HC) emissions. DB10, DB20, DB30, DB40, and diesel engines' HC emissions are illustrated in Fig. 3. At the apex load, diesel emissions reached 49 ppm, DB10 reached 32 ppm, DB20 reached 35 ppm, DB30 reached 40 ppm, and DB40 reached 44 ppm.

## 3.3 Emission of Smoke

By increasing the engine load, smoke emissions are formed more firmly. Fuel injection occurs more frequently as engine load increases, causing smoke to form. During

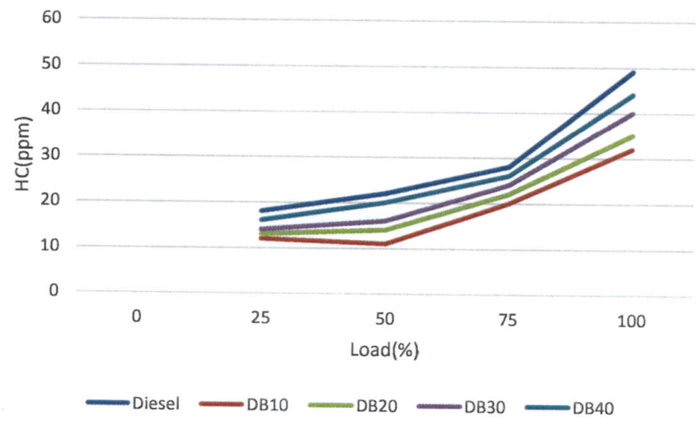


**Fig. 3** HC variability

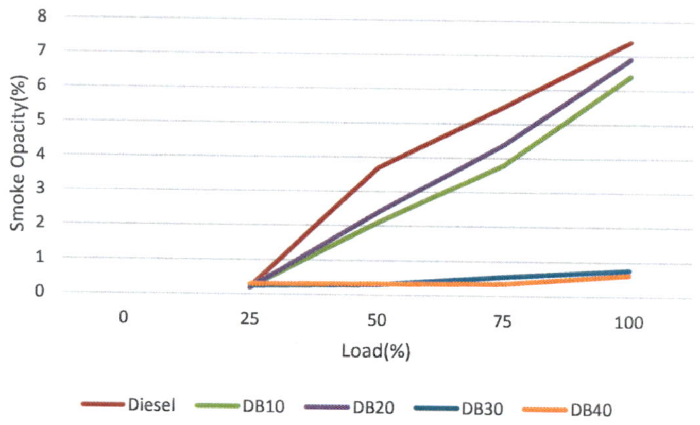


**Fig. 4** Smoke variations

diffusion combustion, rich fuel–air occurs for longer periods of time, which may be the reason for smoke formation. In full load conditions, biodiesel has a high oxygen content which improves combustion and reduces smoke emissions.

## 3.4 Oxides of Nitrogen Emissions

For various jojoba-diesel mix fractions, the braking power and nitrogen oxides (NOx) emissions are depicted in Fig. 5. The graph depicts a decrease in NOx emissions as the percentage of jojoba grows. Oxides of nitrogen are usually formed after nitrogen reacts at 1000 °C. Directly proportional to temperature is the amount of nitrogen

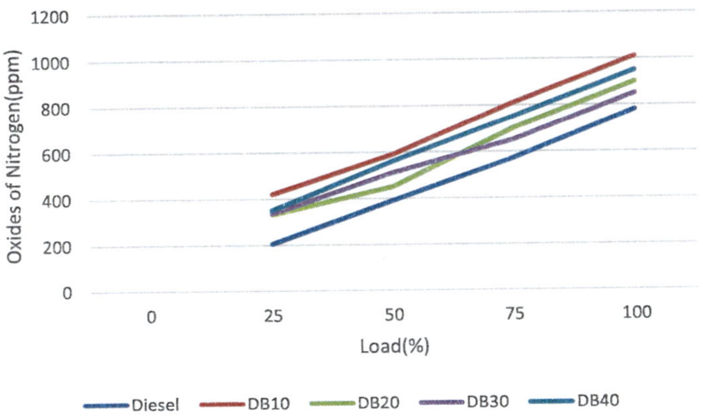


**Fig. 5** NOx emission variation with load for jojoba-diesel blends

oxides generated. Consequently, the combustion temperature of the mix that emits more NOx is higher. Previous findings have shown that the mix with less jojoba fuels combustion at a higher temperature. Because jojoba releases less NOx than a blend with a lower jojoba concentration, blends with a large percentage of jojoba produce less NOx.

## 4 Conclusions

As per the procedure, the investigations were carried out and the following major conclusions were drawn. Based on the experimental results, it was observed that the engine has capability to operate upto DB30 jojoba oil through diesel. The results shows 3% higher BTE, 4% lower CO, 8% lower HC, 4.6% higher NOx, and 15% reduced smoke than diesel. It is 3.5% higher BTE, 17% lower CO, 30% lower HC, and 12% higher NOx smoke than neat jojoba oil (100% jojoba oil) cast off the normal engine. According to the findings of the aforementioned research, the blend DB30 performed admirably in typical CI engines in terms of performance, combustion, and exhaust emission characteristics. As a result, the mix DB30 was established as the maximum blend and 30% adroit jojoba oil may be used in a standard CI engine without causing performance or emission problems. The blend of more than DB30 was unable to perform well and emitted an unacceptable level of smoke and other emissions.

## References

1. Babu JM et al (2018) Analysis on the effect of pilot injection strategies on combustion and

emission characteristics of palm-munja biodiesel/diesel blend on CRDI diesel engine. Energy Convers Manag 0(0):1026–1038. https://doi.org/10.1016/j.enconman.2018.11.047
2. Ravichandra Ganesh P, Hemachandra Reddy K, Babu JM, Sarath Chandra M (2021) Experimental investigation of performance, emission and combustion characteristics of a di-diesel engine fuelled with aqueous cerium oxide and aqueous aluminium oxide nanoparticle additives. Lect Notes Mech Eng 26(X):85–96. https://doi.org/10.1007/978-981-15-7557-0_8
3. Babu Jonnalagadda M, Sai Deepak Raj NV, Bharmal P, Balaji M (2020) Experimental investigation on the production of linseed biodiesel yielding and properties evaluation. In: AIP Conf. Proc., vol 2311, Dec 2020. https://doi.org/10.1063/5.0034924
4. Akashkumar P, Nagesh C, Babu JM, Aatmesh J, Vora KC (2019) Production and engine performance and emission evaluation of karanja and jatropha-based biodiesel. Lect Notes Electr Eng 486:1119–1133. https://doi.org/10.1007/978-981-10-8506-2_76
5. Devaraj A, Devarajan Y, Vinoth Kanna I (2021) Investigation on emission pattern of biodiesel and Nano-particles. Int J Ambient Energy 42(10):1103–1107. https://doi.org/10.1080/01430750.2019.1586765
6. Tamil Selvam N, Devaraj A (2019) Emission study on MgO 2 nano-additive doped biodiesel on immobile diesel engine. Int J Ambient Energy 0(0):1–5. https://doi.org/10.1080/01430750.2019.1611654
7. Devaraj A, Devarajan Y, Vinoth KI (2021) Effect of di-ethyl-ether on biodiesel fuelled diesel engine. Int J Ambient Energy 42(5):495–499. https://doi.org/10.1080/01430750.2018.1557546
8. Devaraj A, Yuvarajan D, Vinoth Kanna I (2020) Study on the outcome of a cetane improver on the emission characteristics of a diesel engine. Int J Ambient Energy 41(7):798–801. https://doi.org/10.1080/01430750.2018.1492452
9. Devaraj A, Vinoth Kanna I, Tamil Selvam N, Prabhu A (2020) Emission analysis of cashew nut biodiesel-pentanol blends in a diesel engine. Int J Ambient Energy 0(0):1–5. https://doi.org/10.1080/01430750.2020.1725633
10. Ganesan S, Devaraj A, Devarajan Y (2020) Emission characteristics on single cylinder diesel engine using biofuels. Int J Ambient Energy 41(14):1613–1616. https://doi.org/10.1080/01430750.2018.1517694
11. Vinoth Kanna I, Devaraj A, Subramani K (2020) Bio diesel production by using Jatropha: the fuel for future. Int J Ambient Energy 41(3):289–295. https://doi.org/10.1080/01430750.2018.1456962
12. Prabhu A, Reddy BK, Nagappan M, Bharath N (2020) Effect of Al2O3 nano-additives on the performance and emission characteristics of jatropha and pongamia methyl esters in compression ignition engine. Int J Ambient Energy 41(5):528–532. https://doi.org/10.1080/01430750.2018.1477688
13. Nagappan M, Ravichandran T (2019) A comprehensive study on the effect of thermal barrier on diesel engine efficiency. Int J Ambient Energy 0(0):1–6. https://doi.org/10.1080/01430750.2019.1694989
14. Devaraj A, Nagappan M, Yogaraj D (2020) Emission pattern of higher alcohol-biodiesel blends in compression ignition engine. Mater Today Proc 37(Part 2):1876–1879. https://doi.org/10.1016/j.matpr.2020.07.455
15. Nagappan M, Ravichandran T (2017) Performance and emission characteristics of palm oil as an alternate fuel in diesel engine. Int J Mech Prod Eng Res Dev 7(6):417–424. https://doi.org/10.24247/ijmperddec201747
16. Nagappan M, Vinoth Kanna I (2020) A novel technique and detailed analysis of cars in Indian roads to adopt low ground clearance. Int J Ambient Energy 41(10):1089–1095. https://doi.org/10.1080/01430750.2018.1501753
17. Ravichandran T, Nagappan M, Arunraj R (2018) Experimental study of performance and emission characteristics of a single cylinder diesel engine using karanja biodiesel and diesel blends. Int J Mech Prod Eng Res Dev 8(4):677–684. https://doi.org/10.24247/ijmperdaug201870
18. Ravichandran T, Nagappan M (2021) A novel technique and detailed study of concepts of low-cost hybrid vehicle design. Int J Ambient Energy 42(11):1252–1257. https://doi.org/10.1080/01430750.2019.1592777

19. Murugan N, Venu H, Jayaraman J, Appavu P (2019) Emission and performance characteristics study on nanographene oxide additives doped palm oil methyl ester blend in a diesel engine. Int J Ambient Energy 0(0):1–7. https://doi.org/10.1080/01430750.2019.1697361
20. Nagappan M et al (2020) Emission characteristics study of biodiesel-alcohol blends in a compression ignition engine. In: AIP Conf. Proc., vol 2311, Dec 2020. https://doi.org/10.1063/5.0034430
21. Murugan N, Appavu P (2020) Investigation on low temperature biogas generation. Int J Ambient Energy 41(1):5–7. https://doi.org/10.1080/01430750.2018.1443283

# Studies on Hydrogen Production for Enhancing Performance of Spark Ignition Engine

**F. Adritowin and V. Christus Jeya Singh**

## 1 Introduction

We all know that India is a leading automobile sector in the world. This impact on the growth in the automobile sector is towards the carbon footprint. At present, fuel has become a necessary thing for human beings. The majority of the fuel that we use nowadays depends mainly on fossil fuel. The decrease in the supply of hydrocarbon leads to the proportional rise in the fuel price which leads to the search for an increased alternative fuel. This is because; people won't be willing to buy the fuel at an extremely higher price. The renewable energy that supplies as previously uneconomic sources become sufficiently economical to exploit. Using present technology, artificial gasoline and other biofuels require more expensive production. So, there is a need to overcome this problem with an alternative fuel.

Based on various studies, hydrogen seems to be a better alternative fuel. Sustainable energy has become a major concern world-wide. Fossil fuel consumption leads to the rapid change in the global climate along with the fear of energy supply shortage. It demands to explore of the potential benefits of a hydrogen economy. Sustainable development focuses on the use of various resources of sustainable sources, sustainable usage, etc. The usage of hydrogen as an alternate fuel is focussed as a view for sustainable development [1, 2].

There were various methods of hydrogen production such as steam reforming, electrolysis, thermolysis, photo-catalytic water splitting and alkaline hydrolysis. Out of these, alkaline hydrolysis of aluminium seems to be the most suitable one. The hydrogen production was first observed whilst cutting aluminium and its alloys in contact with water. This occurs as a result of the chemical reaction between the water and the fresh surface of the aluminium. But, this process is not continuous due to

F. Adritowin (✉) · V. Christus Jeya Singh
Department of Mechanical Engineering, St. Xavier's Catholic College of Engineering, Nagercoil, Kanyakumari, Tamil Nadu, India
e-mail: adritowin10@gmail.com

 
N. L. Narasimhan et al. (eds.), *Recent Advances in Energy Technologies*, Lecture Notes in Mechanical Engineering, https://doi.org/10.1007/978-981-19-3467-4_28

the formation of a passive layer [3]. So, there is a need to break this passive layer in order to make the reaction continuously for hydrogen production [4–8]. This can be achieved by using alkaline materials like NaOH and KOH. In this process, NaOH is used because it makes the reaction faster, compared to KOH. Also, the NaOH acts as a catalyst since it is regenerated in the final process [8–12]. The $Al(OH)_3$ produced as a by-product of this process can be used as a raw material for aluminium production by Haber's process. $Al(OH)_3$ also has various other applications in the manufacturing of abrasives, waterproofing fabrics, cosmetics, antacids, etc. The reactions are as follows:

$$2Al + 2NaOH + 6H_2O \rightarrow 2NaAl(OH)_4 + 3H_2 \quad (1)$$

$$NaAl(OH)_4 \rightarrow NaOH + Al(OH)_3 \quad (2)$$

$$2Al + 6H_2O \rightarrow 2Al(OH)_3 + 3H_2 \quad (3)$$

In order to control the reaction rate, the reaction should be carried out in a reactor [13]. So, a reactor should be selected that can withstand the extreme high corrosion. The hydrogen thus produced could be used to power the SI engine [10, 14–17]. Hydrogen is a cleaner and eco-friendly fuel. It burns to produce only heat and water. It reacts with air as per the following equation:

$$2H_2 + O_2 + 3.76N_2 \rightarrow 2H_2O + 3.76N_2 \quad (4)$$

The thermal efficiency of hydrogen is higher since its calorific value is higher than any other fuel. Several attempts were already made which justify the increase in thermal energy on using hydrogen [18, 19]. The temperature of the cylinder becomes higher which leads to a higher heat transfer rate [20, 21]. Since hydrogen is a carbon-free fuel, there is no chance of carbon emission [22–25]. However, there may be slight traces of carbon emission due to the combustion of lubrication oil. Hydrogen, on combustion, produces only water vapours which form condensation inside the engine. This results in a reduction in the temperature of the engine due to the water's cooling effect [26].

Based on the above descriptions, this research paper explains the production of hydrogen using alkaline hydrolysis of aluminium. Also, the improvements in performance and emission characteristics of an SI engine using hydrogen were discussed.

## 2 Experimentation

Initially, various hydrogen producing methods were studied. On analysing, alkaline hydrolysis of hydrogen method is selected because of producing hydrogen from waste aluminium scraps, waste to energy concept. The methodology is shown in Fig. 1.

The hydrogen reactor is designed and fabricated using suitable materials to withstand high alkaline corrosion. The experimentation layout is shown in Fig. 2.

It consists of the reactor in which the waste aluminium scraps are fed into it. The water mixed with 0.1% NaOH is also fed into the reactor using the inlet pipe. The solution starts to react with aluminium and started producing hydrogen. Since this reaction is an exothermic reaction, heat is liberated. This system is maintained at 70 °C to maintain efficient hydrogen production. The reactor is immersed in the water bath during operation. Air is passed through this water bath using an air blower to remove the excess heat. Thus, the system is maintained at a constant temperature. The hydrogen thus produced contains traces of water vapour. This is passed through the bubbler filter to remove the water vapour. Hydrogen flow rate is measured by an

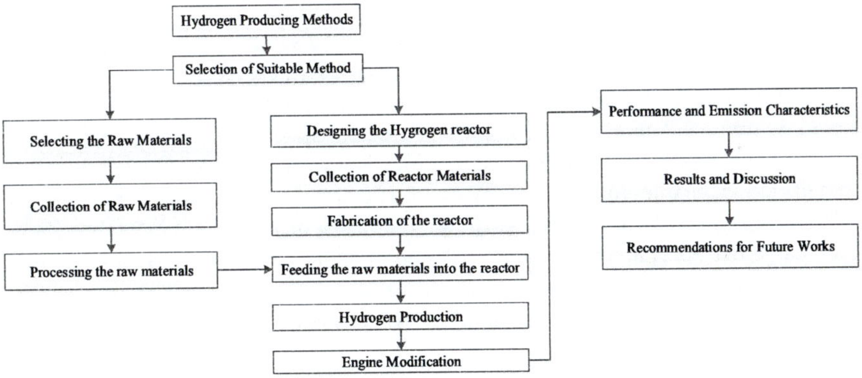


**Fig. 1** Methodology

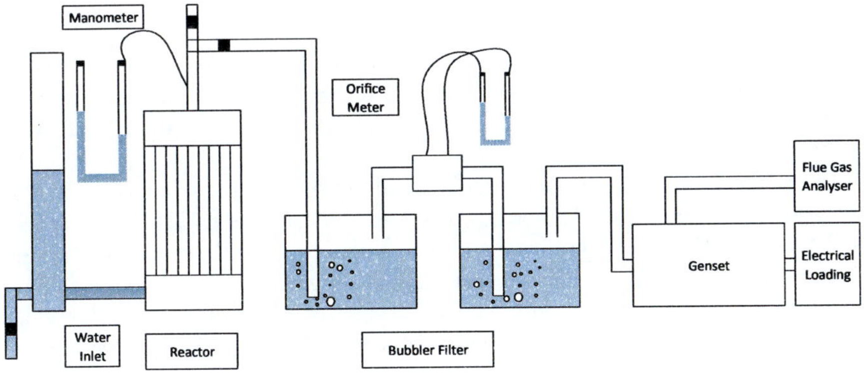


**Fig. 2** Experimental layout

**Fig. 3** Experimental setup

orifice metre. An additional bubbler setup is used as a safety measure to prevent the backfire. The setup for this experimentation is shown in Fig. 3.

The engine selected for this experimentation is EBK 650, a single-cylinder air-cooled SI engine. The engine here used is a petrol start kerosene run engine. The hydrogen fuel has to be delivered just near the fuel inlet valve. So, a small hole has been drilled in the inlet fold and a small metallic calliper through which the hydrogen is delivered is inserted and sealed. As the engine runs with hydrogen perfectly without any change in the default parameters of the engine, there is no need for any complex modification. The engine is then made to run using the produced hydrogen. The performance and emission characteristics are compared for various electric loads with hydrogen and petrol.

## 3 Results and Discussion

The usage of hydrogen has a great improvisation in both performance and emission characteristics on compared with petrol. The characteristics of the engine using both hydrogen and petrol were presented and discussed. The experimentation was carried out for various load conditions.

From Fig. 4, the brake thermal efficiency of petrol is higher when compared to hydrogen. The brake thermal efficiency increases about 16% at 0.5 KW electric load. Brake thermal efficiency signifies the effective conversion of heat energy from the combustion oerf fuel to mechanical energy by the engine.

From Fig. 5, the brake mean effective pressure of the hydrogen is 5% greater than petrol at 0.5 KW electric load. The brake mean effective pressure is the average (mean) pressure imposed on the piston uniformly from top to bottom of each power stroke. As the brake power increases, the brake mean effective pressure increases. As the speed increases, the brake mean effective pressure decreases.

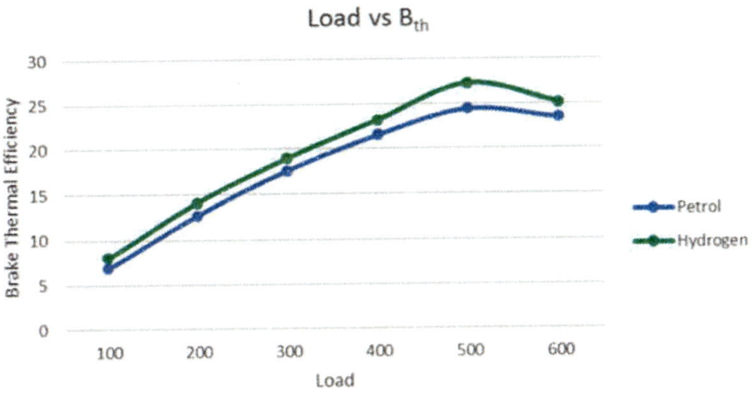


**Fig. 4** Load versus brake thermal efficiency

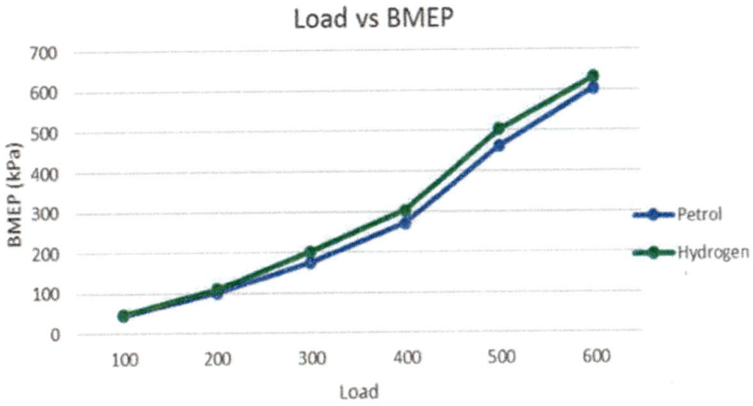


**Fig. 5** Load versus brake mean effective pressure

Fuel consumption is the rate at which an engine consumes the fuel. From Fig. 6, we can see that there is a great reduction in the mass of fuel consumed by using hydrogen fuel. The mass of fuel consumption has also decreased to about 15–23% on using hydrogen compared to petrol.

The specific fuel consumption signifies the amount of energy produced per unit mass of fuel consumed. Figure 7 shows that the specific fuel consumption of hydrogen was 23–33% lower than petrol. This illustrates that hydrogen is more efficient compared to petrol.

Figure 8 shows that the volumetric efficiency of hydrogen is about 25–30% lower than petrol. This is because hydrogen is the lightest gas, and it needs a large volume to store per unit mass of hydrogen.

In a petrol engine, the $CO_2$ is emitted because it is the product of complete combustion. But, in the case of hydrogen, there is no emission of $CO_2$ in the combustion

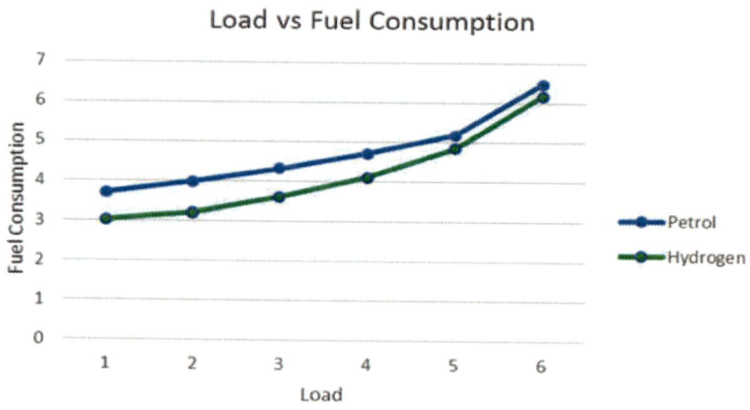


**Fig. 6** Load versus fuel consumption

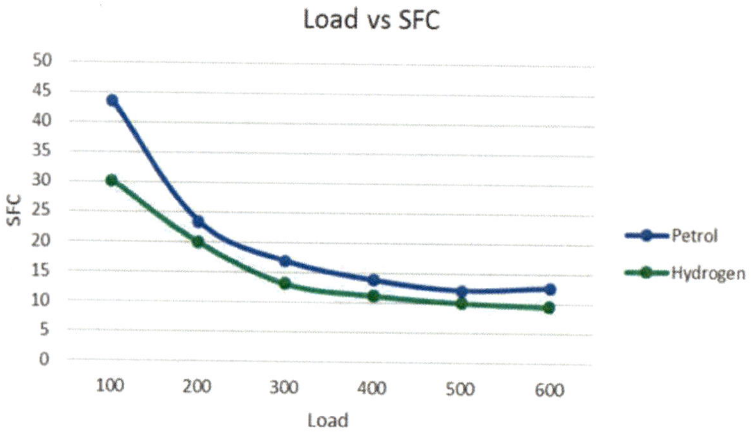


**Fig. 7** Load versus SFC

product. But, the carbon in the air mixture may be oxidised, and the emission of $CO_2$ is possible during lean fuel mixture. Also, due to the lubrication oil in the walls of the combustion chamber, it gets combusted and liberates a trace of $CO_2$.

Figure 9 shows that, on using hydrogen as a fuel, the $CO_2$ emission has reduced to a greater level. As mentioned earlier, there is no carbon component in the fuel supplied. Therefore, hydrogen can be termed as a zero-emission fuel.

Figure 10 shows that the emission of carbon monoxide is less or negligible in hydrogen when compared to petrol. This is because the combustion in hydrogen engines leaves only water or water vapour as an exhaust waste. There is no possibility of emission of any carbon monoxide in a hydrogen-powered engine. As mentioned above, there is no carbon component in the fuel supplied. Therefore, hydrogen can be termed as a zero-emission fuel.

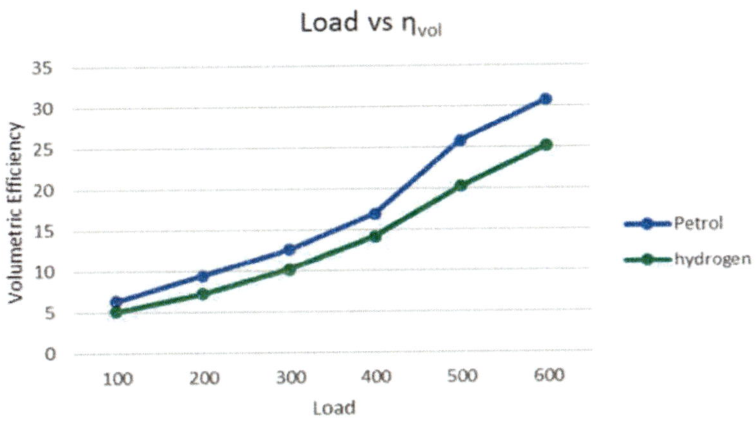


**Fig. 8** Load versus volumetric efficiency

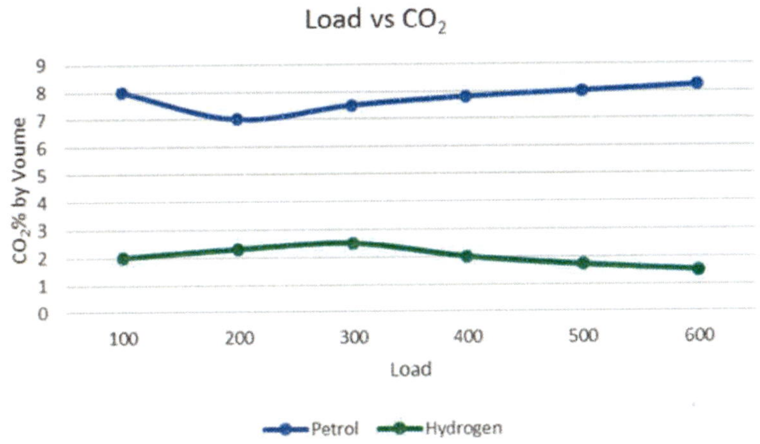


**Fig. 9** Load versus $CO_2$

The oxygen consumption levels on using hydrogen and petrol are compared in Fig. 11. In a hydrogen engine, the emission of $O_2$ is greater than the petrol engine because in a hydrogen engine the combustion by-product is water. The flue gas analyser shows a value of around 21%. In the case of petrol, the air–fuel is supplied at the stoichiometric ratio. Therefore, the oxygen in the air–fuel mixture gets fully combusted, and there is a negligible amount of oxygen in the exhaust gas.

From Fig. 12, the emission of $NO_x$ is less in hydrogen when compared to petrol. This may occur because the combustion of hydrogen with air may produce oxides of nitrogen. The emission of $NO_x$ is because of the combustion of hydrogen with air at a higher temperature. There is a significant $NO_x$ emission in both cases; but in hydrogen, it is 43–50% lower.

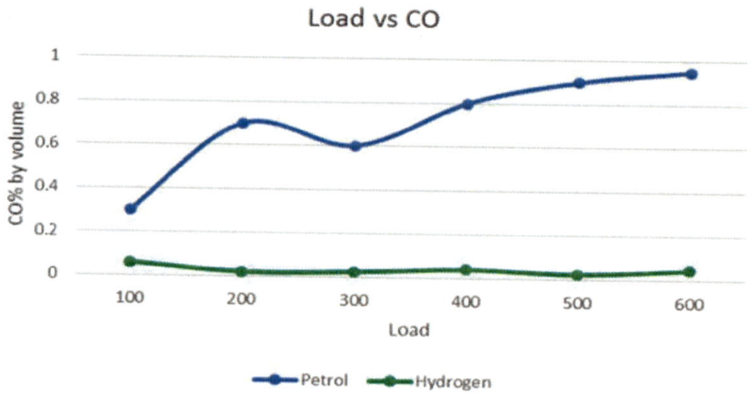


**Fig. 10** Load versus CO

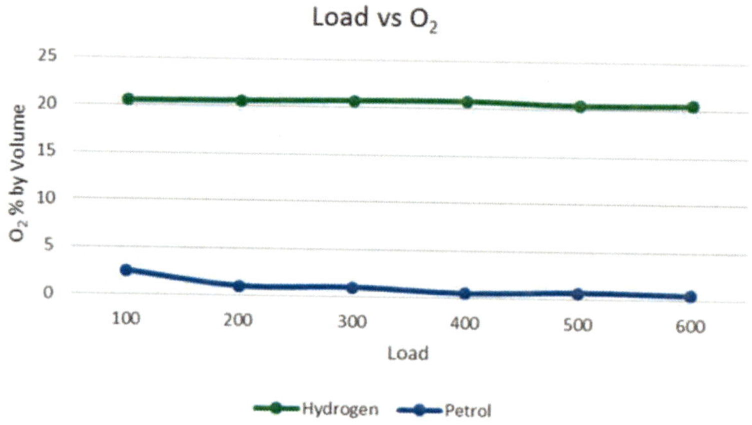


**Fig. 11** Load versus $O_2$

## 4 Conclusion

The increasing awareness about the depletion of fossil fuels and the effects on the environment creates a look towards alternate energy resources and how to utilise them effectively. The various methods of hydrogen production were studied, and the alkaline hydrolysis method is selected. Brake thermal efficiency and brake mean effective pressure were 16 and 5% higher than petrol in hydrogen mode at 0.5 KW electric load. The mass of fuel consumption has also decreased to about 15–23% on using hydrogen compared to petrol. Also, the specific fuel consumption of hydrogen is about 23–33% lesser than petrol. The volumetric efficiency of hydrogen is 25–30% lower since hydrogen is the lightest gas. Moreover, the CO, $CO_2$ and HC levels were negligible on using hydrogen as a fuel. There is a significant $NO_x$ emission in both

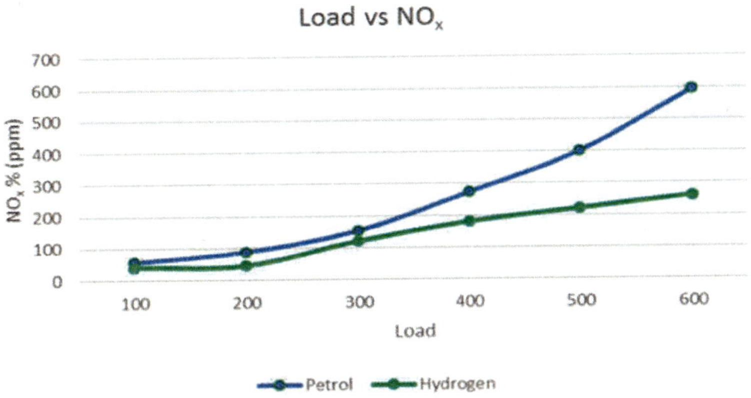


**Fig. 12** Load versus $NO_x$

cases; but in hydrogen, it is 43–50% lower. This shows that hydrogen can be a better alternative to fossil fuels in near future.

## References

1. Khan BH (2017) Non-conventional energy resources, 3rd edn. McGraw Hill Education
2. Singh SN (2017) Non-conventional energy resources, 1st edn. Pearson Education India
3. Uehara K, Takeshita H, Kotaka H (2002) Hydrogen gas generation in the wet cutting of aluminium and its alloys. J Mater Process Technol 127:174–177
4. Ivanov VG, Safronov MN, Gavrilyuk OV (2001) Macrokinetics of oxidation of ultradisperse aluminum by water in the liquid phase. Combust Explosion Shock Waves 37:173–177
5. Tang Y, Lu L, Roesky HW, Wang L, Huang B (2004) The effect of zinc on the aluminum anode of the aluminum-air battery. J Power Sources 138:313–318
6. Abedin SZ, Enders F (2004) Electrochemical behaviour of Al, Al–In and Al–Ga–In alloys in chloride solutions containing zinc ions. J Appl Electrochem 34:1071–1080
7. Kravchenko OV, Semenenko KN, Bulychev BM, Kalmykov KB (2005) Activation of aluminum metal and its reaction with water. J Alloy Compd 397:58–62
8. Appels L, Baeyens J (2008) Jan Degre've and Raf Dewil, Principles and potential of the anaerobic digestion of waste-activated sludge. Prog Energy Combust Sci 34:755–778
9. Fan MQ, Xu F, Sun LX (2007) Studies on hydrogen generation characteristics of hydrolysis of the ball milling Al-based materials in pure water. Int J Hydrogen Energy 32:2809–2815
10. Verhelst S, Sierens R (2001) Hydrogen engine-specific properties. Int J Hydrogen Energy 26:987–990
11. Soler L, Candela AM, Macanás J, Munoz M, Casado J (2009) In situ generation of hydrogen from water by aluminum corrosion in solutions of sodium aluminate. J Power Sources 192:21–26
12. Liu K, Song C, Subramani V (2010) Hydrogen and syngas production and purification technologies. Wiley
13. Andersen ER, Andersen EJ (2003) Method for producing hydrogen. US Patent 6506360
14. Escalante Soberanis MA, Fernandez AM (2010) A review on the technical adaptations for internal combustion engines to operate with gas/hydrogen mixtures. Int J Hydrogen Energy 35:12134–12140

15. Nieminen J, D'Souza N, Dincer I (2010) Comparative combustion characteristics of gasoline and hydrogen fuelled ICEs. Int J Hydrogen Energy 35:5114–5123
16. Hari GR, Subramanian V, Balasubramanian V, Mallikarjun JM, Ramesh A, Sharma RP (2008) Hydrogen fueled spark ignition engine with electronically controlled manifold injection: an experimental study. Renewable Energy 33:1324–1333
17. Szwaja S, Bhandary KR, Naber JD (2007) Comparisons of hydrogen and gasoline combustion knock in a spark ignition engine. Int J Hydrogen Energy 32:5076–5087
18. Kim YY, Lee JT, Choi GH (2005) An investigation on the causes of cycle variation in direct injection hydrogen fueled engines. Int J Hydrogen Energy 30:69–76
19. Trijselaar A (2012) Knock prediction in gas-fired reciprocating engine: development of a zero-dimensional two zone model including detailed chemical kinetics. University of Twente, Faculty of Engineering Technology, Department of Thermal Engineering
20. Verhelst S, Wallner T (2009) Hydrogen-fueled internal combustion engines. Prog Energy Combust Sci 35:490–527
21. Sáinz D, Diéguez PM, Sopena C, Urroz JC, Gandía LM (2012) Conversion of a commercial gasoline vehicle to run bi-fuel (hydrogen–gasoline). Int J Hydrogen Energy 37:1781–1789
22. Heywood JB (1988) Internal combustion engine fundamentals. McGraw-Hill, Maidenherd
23. Mathur HB, Khajuria PR (1986) A computer simulation of hydrogen fuelled spark ignition engine. Int J Hydrogen Energy 11:409–417
24. Das LM (2016) Hydrogen-fueled internal combustion engines. Indian Institute of Technology, New Delhi, India
25. Das LM, Rohit Gulati R, Gupta PK (2000) Performance evaluation of a hydrogen-fuelled spark ignition engine using electronically controlled solenoid actuated injection system. Int J Hydrogen Energy 25:569–579
26. White CM, Steeper RR, Lutz AE (2006) The hydrogen-fuel internal combustion engine. Int J Hydrogen Energy 31:1292–1305

# Effect of Intake Air Preheat on Diesel Engine Performance Fueled with Coconut Fatty Acid Distillate Biodiesel

**K. Rajesh, M. P. Natarajan, P. K. Devan, S. Ponnuvel, V. Dillibabu, and S. Arun Kumar**

## 1 Introduction

The developing nations greatly depend on the energy sector especially fossil fuels like petroleum, natural gas, coal, etc. Transportation sector greatly rely on the petroleum-based fuel which is non-renewable. Owing to the rapid increase in the petroleum products cost such as gasoline and diesel, also the disappearance of fossil fuels, research and development of alternative fuels is on the rise. Government policy differs from country to country, and biodiesel research is underway to meet the requirements of less polluting fuels [1]. Petroleum fuels are a powerful key aspect that influences the global economy, agriculture sectors, shipping and road transportation. Biodiesel produced from agricultural resources reduces global oil imports, supports local agriculture, raises farmer incomes and reduces emissions [2]. To accomplish this, CFAD is used to make biodiesel. The use of internal combustion engines which require petroleum-based fuels is the most widespread mode of transportation in India. As compared to fossil fuels, biodiesel releases less harmful contaminants such as UHC,

K. Rajesh (✉) · P. K. Devan · S. Arun Kumar
Department of Mechanical Engineering, RMK College of Engineering and Technology, Chennai, Tamil Nadu, India
e-mail: rajeshmechh@gmail.com

M. P. Natarajan
Department of Mechanical Engineering, Bharath Institute of Higher Education and Research, Chennai, Tamil Nadu, India

S. Ponnuvel
Department of Mechanical Engineering, Sri Venkateswara College of Engineering, Sriperambadur, Tamil Nadu, India

V. Dillibabu
Department of Mechanical Engineering, KCG College of Engineering and Technology, Karapakkam, Tamil Nadu, India

 
N. L. Narasimhan et al. (eds.), *Recent Advances in Energy Technologies*, Lecture Notes in Mechanical Engineering, https://doi.org/10.1007/978-981-19-3467-4_29

smoke and carbon monoxide when used in internal combustion engines [3]. In addition to the above emissions, ash residue keeps adding to the environment's high concentration of small particles suspended in the air, making it even more hazardous to human health. These microscopic particles are extremely carcinogenic, and they have the potential to impair cognitive functioning, which can result in neurological diseases [4]. To minimize the drawbacks associated with conventional fuels, one of the alternatives is the usage of biodiesel produced using vegetable oils. But vegetable oils have high viscosity and methods such as chemical blending, heating, transesterification and emulsion are used to reduce the viscosity. The viscosity and latent heat of vaporization of biodiesel remain greater than diesel even after conversion from vegetable oils. Various methods are used to eliminate the drawback of poor latent heat of vaporization and high viscosity of biodiesel, one of the methods employed is preheating the intake air [5].

Yao et.al investigated intake air temperature and its effect in a diesel engine operating in dual fuel mode with diesel and methanol. It was reported that the engine showed improved performance characteristics caused by methanol injection and intake air preheating. The indicated thermal efficiency, higher ignition delay and EGT temperature dropped for lower intake air temperatures along with increasing methanol percentage. Due to methanol induction, the NOx and smoke decreased whereas the hydrocarbon and carbon monoxide emission increased. Instead, increasing the intake air temperature lowered HC and CO emissions while simultaneously increasing NOx emissions and smoke emissions. With a higher methanol concentration and higher intake air temperatures, the emissions are more and hence optimum condition was selected for better performance [6]. Mekonan et al. in their paper investigated the influence of preheated intake air and fuel in diesel engine for its performance and emission. Palm oil biodiesel is preheated with the exhaust which enhances the performance and emission characteristics. It resulted in improved fuel properties, BTE and BSFC. HC, CO and smoke emissions were observed to be lower, while NOx emissions were higher than diesel with increasing the intake air temperature [7].

An investigation on the influence of preheating of the intake air into the diesel engine on the performance and emissions was conducted by Nadir Yilmaz. The test fuels used were biodiesel methanol blends and tested at two different temperatures that are 85 and 30 °C. The test fuels were tested at room temperature and compared with the methanol blended fuel at an increased intake air temperature. Findings of the work implied that higher methanol content increased HC and CO emissions, whereas higher intake air temperature reduced all emissions except NOx [8]. Kumar and Raj et al. experimented with the effect of injection timing and intake air preheat. The biodiesel and ethanol blended fuel were used to run the engine. The blend selected for the study was B5E15 (5% biodiesel, 15% Ethanol and 80% diesel). The intake air was preheated to 60, 40 and 30 °C. It was found that advancing the injection timing improved engine combustion and performance. The intake air preheating had a great effect in improving the premixed combustion thereby resulting in higher HRR and higher cylinder pressures [9].

Senthur Prabu et al., in their work, Palm Oil (PO) and diesel blends were heated to 60 °C utilizing waste heat from exhaust gases utilizing a heat exchanger, ensuring that the mixture was homogeneous. On comparison, PO20 was shown to be more effective than other produced biodiesel blends. PO20 was mixed with n-butanol and butylated hydroxytoluene to reduce NOx emission [10]. Lal Jain et al., tested the diesel engine for performance and emission with varying percentages of thumba in heated and unheated condition. The B20 blend which is preheated has been found to provide better output and lower emissions than all other tested biodiesel fuels and diesel. The BTE of the optimized biodiesel blend showed an increase of 1.27% when it is preheated [5]. Pradhan et al. in their paper, the engine exhaust gas was circulated through a helical tube heat exchanger to heat the Jatropha oil to improve the fuel property. As engine load rises, a decrease in BSFC and an increase in BTE were noted for all of the fuels studied. The unheated and preheated Jatropha oil brake thermal efficiency was just 2.44 and 5.18% lower than diesel fuel. On average, CO, HC and NOx emissions were reduced by 5.28, 2.67 and 37.2% when the engine was driven by Preheated Jatropha oil [11]. Based on the literature review preheating the engine's intake air improves efficiency and reduces emissions. A comparison of diesel and CFAD biodiesel blends at room temperature and 60 °C with air preheating were performed in this study.

## 2 Material and Methods

### 2.1 *Materials*

The CFAD oil was sourced from Raasi energy limited, Tamil Nadu, India. The required chemicals such as methanol, ethanol, sulfuric acid, potassium hydroxide and phenolphthalein indicator for biodiesel extraction were purchased from Lab chemicals limited, Tamil Nadu, India. The biodiesel was extracted by two-step process, i.e., first processed by acid esterification due to higher FFA (24.5%) and then followed by transesterification process. The raw biodiesel has been cleaned with distilled water, it is heated to 110 °C so that the water content can be drained out of the biodiesel. [12]. Table 1 gives the list of properties of CFAD biodiesel (B100).

### 2.2 *Experimental Procedure*

The experimental setup used in this work is given in the Fig. 1. It is a DI single cylinder 5.2 kW diesel engine; the engine's features are listed in Table 2. The load variation was achieved by the use of an eddy current dynamometer. Fuel flow meters to measure fuel consumption and K-type thermocouples were utilized to detect the various temperatures in the system, and an air mass sensor was employed to monitor

Table 1 Properties of B100 CFAD biodiesel

| Properties | Units | B100 values |
|---|---|---|
| Acid value | mg KOH/g | 0.8 |
| Flash point | °C | 90 |
| Density | kg/m$^3$ | 879 |
| Kinematic viscosity @40 °C | cSt | 4.30 |
| Saponification value | – | 221.1 |
| Cloud point | °C | –1 |
| Pour point | °C | –3 |
| Gross calorific value | kJ/kg | 36,069 |
| Acid value | mg KOH/g | 0.8 |

the amount of air flowing into the system. An electric heater was fitted at the air inlet of the diesel engine and was used to preheat the intake air entering into the engine. At the time of test, the temperature of the intake air is maintained at 60 °C plus or minus 2 °C throughout the duration of the test. The engine performance characteristics were studied under various loads. Each test is persisted three times to obtain average results and reduce uncertainty and finally, the data was recorded with a data acquisition system.

Measurement of exhaust emissions was done using the AVL DI GAS 444 analyzers depicted in Fig. 2. The gas analyzer is a system that analyses the amount of emissions in an engine's exhaust gas. Carbon monoxide, unburned hydrocarbon and NOx content were measured contained in exhaust gas. Figure 3 depicts the AVL smoke meter used which measures the soot content in diesel engine exhaust gas using the filter paper process. The apparatus collects samples from the engine's exhaust stream and filters them through clean filter paper. The filtered soot causes the filter paper to blacken, which is measured using a photoelectric meter.

**Fig. 1** Experimental diesel engine setup

**Table 2** Features of the test engine

| Properties | B100 values |
|---|---|
| Model | TAF 1 |
| Engine type | Single cylinder, water-cooled |
| Dynamometer | Eddy current |
| Maximum power | 5.2 kW @ 1500 rpm |
| Stroke length | 110 mm |
| Bore diameter | 87.5 mm |
| Connecting rod length | 234 mm |
| Compression ratio | 17.5:1 |
| Injection pressure | 210 bar |
| Injection timing | 23° bTDC |

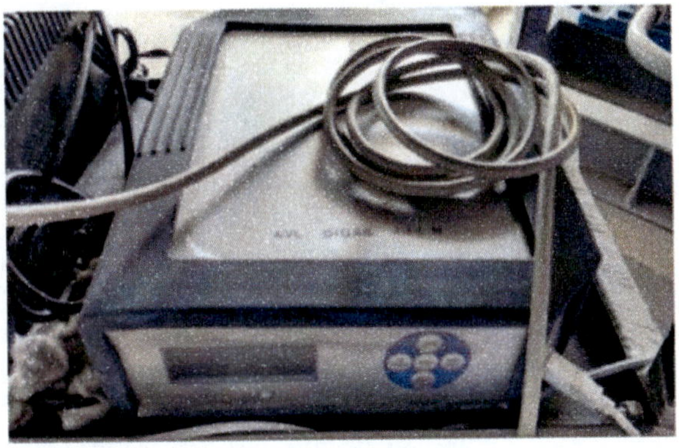

**Fig. 2** AVL DI 444 analyzers

**Fig. 3** AVL smoke meter

Fig. 4 BSFC versus Load

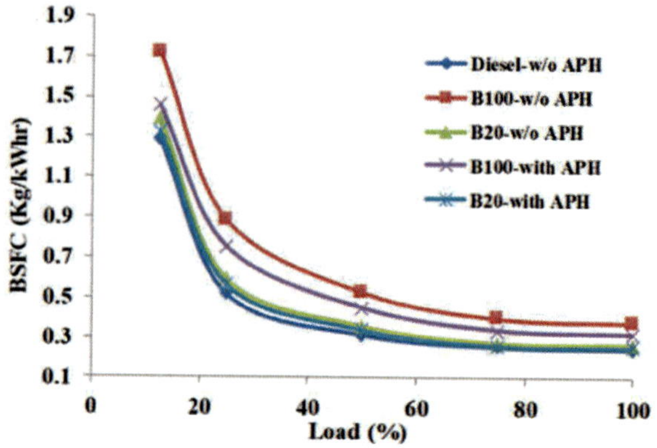


# 3 Results and Discussion

## 3.1 *Performance*

### 3.1.1 Brake Specific Fuel Consumption (BSFC)

The BSFC is the proportion of the tested fuel's mass flow to its effective capacity. For all of the fuels that were tested, the BSFC went down as the engine's load is increased as shown in Fig. 4. The brake power increased with respect to engine load due to reduced heat loss at higher loads leading to lower fuel consumption [13]. This was owing to the increased power production associated with increased engine load, as increased engine load results in relatively lower heat losses, resulting in decreased BSFC [8]. Without preheating the intake air for each load, the BSFC is higher for biodiesel fuels and this is due to the creation of a poor air–fuel combination as a result of the high viscosity of biodiesel causing it to burn poorly. When the air is preheated the BSFC is decreased for B20 and B100 in relation to load due to better air–fuel mixing. From Fig. 4, it is clear that the BSFC of B20 at intake air preheated condition is comparable to diesel. At 100% load condition, the BSFC of B20 at air preheated condition is decreased by 9.24% than unheated condition resulting from the elevated temperature that exists within the combustion chamber [14].

### 3.1.2 Brake Thermal Efficiency (BTE):

The BTE is a unit of measurement for the amount of power generated in proportion to the amount of energy supplied by fuel injection [15]. The variation between the load and the BTE when CFAD biodiesel and diesel tested with unheated and preheated intake air condition is shown in Fig. 5. As the load increases due to increased supply of fuel which leads rich fuel–air mixture resulting in higher BTE [16]. When the load varied from low load to full load, the BTE produced by CFAD biodiesel and

**Fig. 5** BTE versus Load

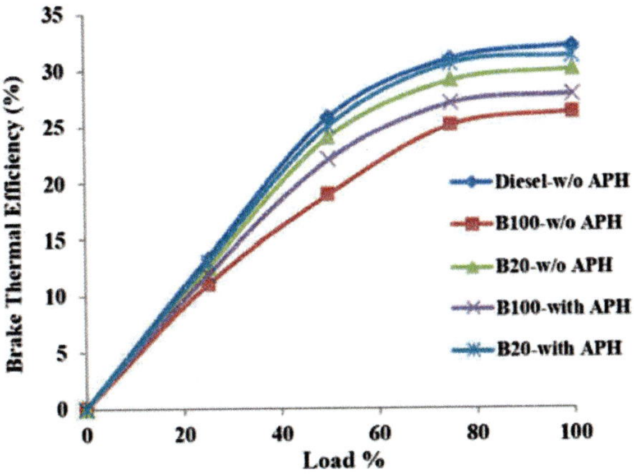


its blend is lower than the diesel for without preheating and preheated air. Due to the reduced calorific content of biodiesel compared to diesel, it has a lower BTE. But the biodiesel fuels with preheated intake air showed higher BTE than unheated conditions. A greater combustion zone temperature allows for improved evaporation and mixing of the biodiesel fuels, resulting in increased BTE [14]. When compared to unheated conditions, the BTE increased by 5.9% and 7.2% for B20 and B100 with intake air preheat, respectively.

### 3.1.3 Exhaust Gas Temperature (EGT)

Figure 6 depicts the relationship between the load and the EGT of the test fuels under consideration. Exhaust gas temperatures rise when the load increases because more fuel is being injected to match the need for power, resulting in higher exhaust temperatures. Hence fuel particles continue to burn during the expansion stroke resulting in higher EGT [8]. The EGT of biodiesel fuels is higher for both preheated intake air condition and unheated condition. The CFAD biodiesel produces higher exhaust gas temperature than diesel, reduced ignition lag results in fewer particles taking part in premixed combustion and the remaining fuel is burnt during diffusive combustion. As a result, the fuel particles proceed to burn during the expansion stroke resulting in higher EGT. But in the case of intake air preheat condition, the fuel particles taking part in premixed combustion is increased due to improved air and fuel mixture formation which reduces the fuel quantity burning during the expansion stroke [16, 17]. At maximum load, the EGT of B20 and B100 at preheated air intake conditions is 5.81 and 11.15% higher, whereas at the normal operating condition it is 9.09 and 21.32% higher than diesel.

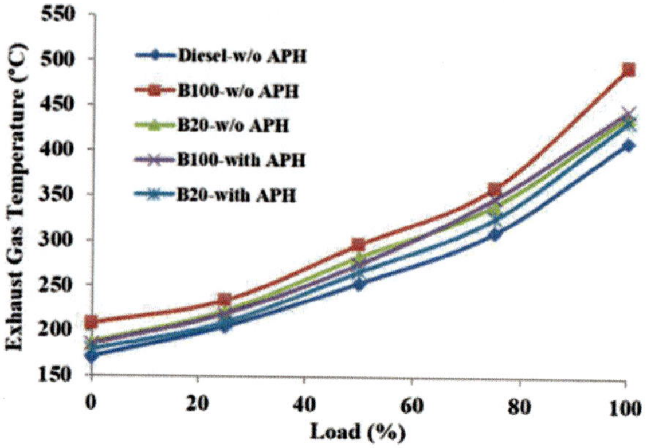


**Fig. 6** EGT versus Load

## 3.2 *Exhaust Emissions*

### 3.2.1 Carbon Monoxide Emission

The incomplete combustion in the engine produces carbon monoxide [13]. As Fig. 7 shows that when the load keeps on increasing, the CO emission is decreased. It was observed that when CFAD biodiesel when compared to diesel, it produced much reduced CO emissions. This is related to the fact that biodiesel has a higher combustion efficiency than conventional diesel because of the presence of oxygen. When biodiesel blended fuels are used, it encourages the carbon to react with oxygen in the fuel, which reduces the amount of CO emitted at different loads in comparison with diesel [16]. When compared to the temperature of the ambient air, preheating the intake air can significantly reduce CO emissions from biodiesel blends. The intake air preheated B100 blend showed lower emission in comparison with all the tested fuels. It showed an average reduction of CO by 40.5% than diesel.

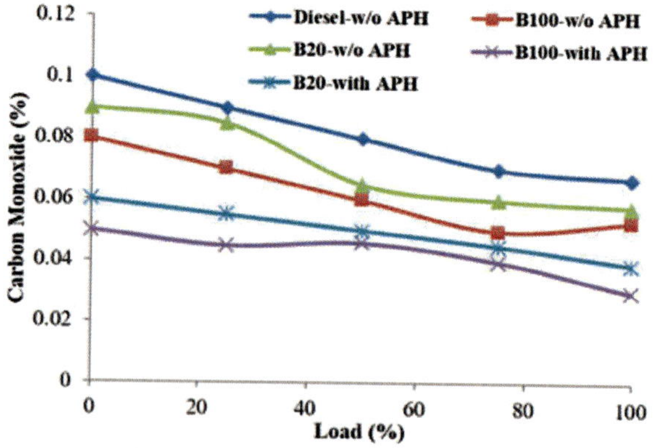


**Fig. 7** Carbon monoxide versus Load

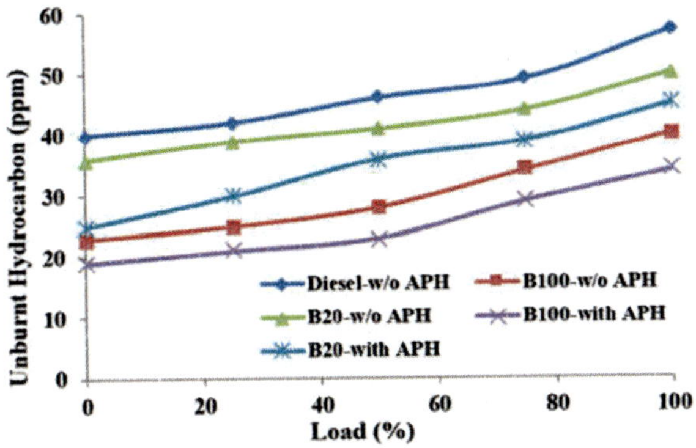


**Fig. 8** Unburnt Hydrocarbon versus Load

### 3.2.2 Unburned Hydrocarbon (UHC)

Figure 8 demonstrates the relationship between UHC and load for the tested fuels. The combustion chamber's inhomogeneity of the fuel–air combination is directly responsible for engine hydrocarbon emissions [16]. With biodiesel fuel, the quantity of UHC reduction is attributed due to enriched oxygen, reduced ignition delay and higher combustion chamber temperature. It has been shown that increasing the temperature of the input air for CFAD biodiesel fuels greatly lowers UHC emissions, which may be attributable to better mixing of the fuel introduced inside the cylinder. In addition to the increased cylinder temperature caused by biodiesel, the enhanced oxygen in the fuel results in decreased hydrocarbon emissions under both heated and unheated intake air circumstances. It showed an average reduction of UHC by 32.1% than diesel.

### 3.2.3 Oxides of Nitrogen (NOx)

Figure 9 depicts the relationship between load and oxides of nitrogen emission of diesel, CFAD fuels tested under preheated and unheated intake air condition. At various engine loads, NOx emissions from inlet air heated and unheated CFAD biodiesel fuels were found to be greater than diesel fuel. Oxygen concentration, reduced ignition delay and longer combustion duration are the key factors influencing NOx emission [18]. Preheating the inlet air further enhances the combustion chamber temperature leading to the formation of higher thermal NOx. The NOx emission of B20 and B100 at intake air preheated condition showed an average increase of 15.25% and 23.56% than diesel, respectively.

Fig. 9 Oxides of Nitrogen versus Load

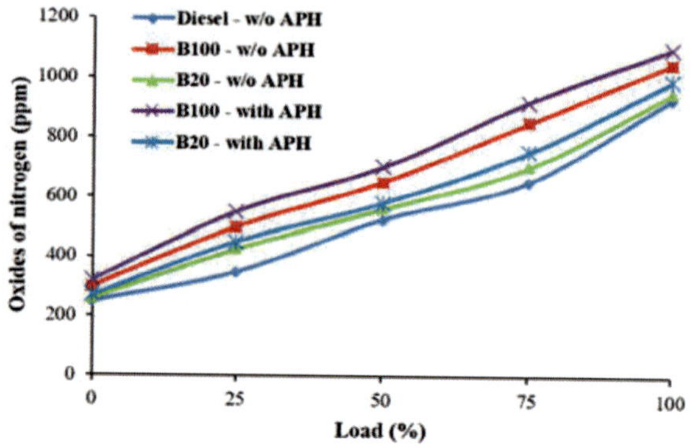


### 3.2.4 Smoke (%)

The relationship between the load and the smoke produced by diesel, B20 and B100 under various test settings is depicted in Fig. 10. Incomplete combustion is the primary cause of smoke emission [14] since more fuel is used to achieve higher power output at higher engine loads, the intensity of smoke increases as the load on the engine is increased [19]. Oxygen-rich nature of the biodiesel fuel support oxidation of soot particles thereby reducing smoke emission. It is clear that for intake air preheat condition the smoke emission was greatly reduced compared to the unheated condition for biodiesel fuels. Inlet air preheating raises the temperature of the cylinder, allowing the fuel particles to vaporize, better fuel mixing and participate in the process of combustion. It also increases the oxidation of soot particles, which reduces the amount of smoke produced. When compared to diesel, it demonstrated an average reduction in the smoke of 27.4%.

Fig. 10 Smoke versus Load

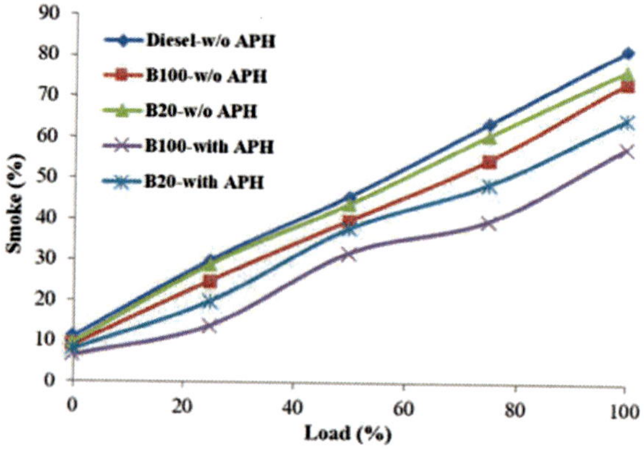

## 4 Conclusion

When preheating the input air, it has been demonstrated that it influences the performance and emission behavior of CFAD biodiesel, according to the current investigation. Following are some of the conclusions reached as an outcome of the study:

- Increasing the air intake temperature decreases the brake specific fuel consumption of B20 and B100. The BSFC of B20 at preheated condition showed a 9.24% reduction compared to unheated condition
- Preheating enhances cold-start performance, reducing ignition delay and improving BTE and combustion. The BTE rose 5.9% for the B20 and 7.2% for the B100 when compared to without preheat condition.
- Preheating improves cold-start performance, which in turn reduces the ignition delay, resulting in a higher BTE and better combustion. The BTE increased by 5.9 and 7.2% for the B20 and B100 compared to without preheating conditions.
- The EGT of B20 and B100 at preheated air intake conditions is 5.81 and 11.15% higher, whereas at the normal operating condition it is 9.09 and 21.32% higher than diesel at full load conditions.
- The exhaust emissions such as CO, UHC and smoke were significantly reduced with air preheating.
- The NOx emission of B20 and B100 at intake air preheated condition showed an average increase of 15.25% and 23.56% than diesel, respectively.

## References

1. Campos-Fernández J, Arnal JM, Gómez J, Dorado MP (2012) A comparison of performance of higher alcohols/diesel fuel blends in a diesel engine. Appl Energy 95:267–275
2. Rajesh K, Devan PK, Lokesh B, Rajesh Kumar K, Vignesh R, Yogesh Kumar R (2020) Investigation on performance and emission of pongamia biodiesel using diethyl ether and zinc oxide as additive in diesel engine. In: IOP Conf Ser.: Mater. Sci. Eng. Vol 923, p 012069
3. Rajesh Kumar B, Saravanan S, Rana D, Anish V, Nagendran A (2016) Effect of a sustainable biofuel—N-octanol—on the combustion, performance and emissions of a di diesel engine under naturally aspirated and exhaust gas recirculation (EGR) modes. Energy Convers Manag 118:275–286
4. Aravindhan N, Maclin John Vasanth K, Vignesh Kumar R, Jayasurya M, Suriya Prakash S, Sabareeshwaran V (2020) A novel approach for improving the performance of air engine to achieve zero-emission for a pollution-free environment. Mater Today: Proc **33**(Part 1):39–43
5. Jain NL, Soni SL, Poonia MP, Sharma D, Srivastava AK, Jain H (2017) Performance and emission characteristics of preheated and blended thumba vegetable oil in a compression ignition engine. Appl Therm Eng 113:970–979
6. Pan W, Yao C, Han G, Wei H, Wang Q (2015) The impact of intake air temperature on performance and exhaust emissions of a diesel methanol dual fuel engine. Fuel 162:101–110
7. Mekonena MW, Sahoo N (2018) Combined effects of fuel and intake air preheating for improving diesel engine operating parameters running with biodiesel blends. J Renewable Sustain Energy **10**:043103

8. Yilmaz N (2012) Effects of intake air preheat and fuel blend ratio on a diesel engine operating on biodiesel–methanol blends. Fuel 94:444–447. ISSN 0016-2361
9. Senthil Kumar K, Thundil Karuppa Raj R (2013) Effect of fuel injection timing and elevated intake air temperature on the combustion and emission characteristics of dual fuel operated diesel engine. Procedia Eng 64:1191–1198
10. Zhu R, Wang X, Miao H, Yang X, Huang Z (2011) Effect of additives on performance, combustion and emission behavior of preheated palm oil/diesel blends in DI diesel engine. Fuel 90(5):1731–1737
11. Pradhan P, Raheman H, Padhee D (2014) Combustion and performance of a diesel engine with preheated Jatropha curcas oil using waste heat from exhaust gas. Fuel 115:527–533
12. Rajesh K, Devan PK, Bharth Sai Kumar GK (2021) Parametric optimization and biodiesel production from coconut fatty acid distillate. Iran J Chem Chem Eng (IJCCE) 40(1):343–355
13. Senthil Kumar S, Purushothaman K (2012) High FFA rubber seed oil as an alternative fuel for diesel engine—an overview. Res Inventy: Int J Eng Sci 1(10):16–24. ISBN: 2319-6483, ISSN: 2278-4721
14. Bibin C, Seeni Kannan P, Devan PK (2020) Performance, emission and combustion characteristics of a Di diesel engine using blends of punnai oil biodiesel and diesel as fuel. Therm Sci 24:1–18
15. Gopinath S, Devan PK, Pitchandi K (2020) Production of pyrolytic oil from ULDP plastics using silica-alumina catalyst and used as fuel for DI diesel engine. RSC Adv 10:37266–37279
16. Rajesh K, Natarajan MP, Devan PK, Ponnuvel S (2021) Coconut fatty acid distillate as novel feedstock for biodiesel production and its characterization as a fuel for diesel engine. Renewable Energy 164:1424–1435
17. Devan PK, Mahalakshmi NV (2009) Utilization of unattended methyl ester of paradise oil as fuel in diesel engine. Fuel 88:1828–1833
18. Rajesh K, Devan PK, Saran R, Vasanthakumar S, Vignesh K (2020) Performance and emission evaluation of diesel engine fueled with Karanja Methyl Ester. IOP Conf Ser Mater Sci Eng 923:012068
19. Rajesh K, Aravindhan N, Manimaran B, Elavarasan D, Dinesh Kumar L, Mohan Kumar T (2021) Study on performance and emission characteristics of VCR diesel engine fuelled with palm fatty acid distillate biodiesel. J Phys Conf Series 2054:012034. https://doi.org/10.1088/1742-6596/2054/1/012034

# An Experimental Analysis in a DICI Engine Powered with MWCNT Blended Emulsions

J. Sadhik Basha, Abdul Rahman Al Musalami, Basmah Al Noufali, Sara Al Balushi, Baraah Al Basti, Zahra Al Ajmi, Ranim Al Balushi, and Marwa Al Maqbali

## Abbreviations

| | |
|---|---|
| CO | Carbon monoxides |
| D | Diesel |
| 98D2WS | Surfactants + 98% diesel + 2% water |
| 96D4WS | Surfactants + 96% diesel + 4% water |
| 98D2WS50MWCNT | Surfactants + 50 ppm MWCNT + 98% diesel + 2% water |
| 96D4WS100MWCNT | Surfactants + 100 ppm MWCNT + 96% diesel + 4% water |
| HC | Hydrocarbons |
| HRTEM | High-resolution transmission electron microscopes |
| MWCNT | Multi-walled carbon nanotubes |
| $NO_x$ | Nitrogen oxides |
| TEM | Transmitting electron microscopy |

## 1 Introduction

There is no doubt that air pollution and environmental degradation are two major problems that the world is facing critically in the recent times, due to the activities practiced by humans in the industrial and technological fields. In this regard, many researchers and professional experts have made their efforts to protect the environment, and to ensure the continuity of the Earth's ecological cycle from pollution

J. Sadhik Basha (✉) · A. R. Al Musalami · B. Al Noufali · S. Al Balushi · B. Al Basti · Z. Al Ajmi · R. Al Balushi · M. Al Maqbali
Department of Process Engineering, National University of Science and Technology (International Maritime College Oman), Sohar, Oman
e-mail: mailsadhik@gmail.com



N. L. Narasimhan et al. (eds.), *Recent Advances in Energy Technologies*, Lecture Notes in Mechanical Engineering, https://doi.org/10.1007/978-981-19-3467-4_30

that was particularly caused by different sectors (such as industry, transportation, and power generation). In all those sectors, mostly, diesel engines are considered as the main prime mover to generate electrical energy (due to its superiority in thermal efficiency while that of petrol engines). Despite with some diesel engine advantages, it also emits harmful pollutants (such as nitrogen oxides, smoke, unburned hydrocarbons) because of the post-combustion process [1–3]. To get rid of these harmful pollutants (which are basically classified main precursors of global warming, acid rain, depletion of the ozone layer, melting of icebergs in Antarctic and Greenland, etc.), many efforts have been devised by many researchers to overcome those global hazards. Several research attempts have been made to reduce unsafe pollutants from CI engines in order to comply with EuroVI emission regulations. One of the most well-established studies research attempts is the fuel modification technique which were tried out by many researchers [4–6]. The technical communities [7–15] have made significant contributions to the adoption of emulsion fuels to reduce hazardous emissions produced by diesel engines. The addition of an emulsifier to emulsion fuel causes evaporation (due to the elevated temperature in the cylinder during combustion, which transforms water to steam) and so reduces unsafe emissions, and simultaneously lessening corrosion on the engine cylinder surfaces [16–20].

In a diesel engine, micro-explosion effect occurs when $H_2O$ droplets in the fuel break-down into smaller ones, resulting in quicker vaporization and better fuel–air mixing during combustion [21–23]. Moreover, a number of studies [24–35] have found significant limitations in the usage of emulsion fuels in diesel engines (such as substandard calorific value, lengthy ignition delay, and bumpy engine operation problems). Scientists have explored a variety of ways to enhance the quality of emulsion fuels in order to overcome these constraints. Recent studies [36–40] have explored the use of emulsion fuels in conjunction with nano-additives (such as CNT, alumina, and so on). Ghojel et al. [20] found that employing water-diesel emulsions stemmed lower pollutants of NO and HC as well as poorer performance than using plain diesel fuel. They also stated that adding a cetane improver to water-diesel emulsion fuels (in the form of an additive) may reduce ignition time and improve performance. In general, the lower value of cetane number of water-diesel emulsion results in a longer igniting delay. Several scientists and researchers [41–46] have recently added distinct additives into emulsion fuels to improve their fuel attributes. Nanoparticles are now being developed for use in emulsion fuels as an additive/fuel carried catalyst [16–19, 39]. According to Moy et al. [46], CNT blended diesel fuel can boost blazing rate, cetane number, and reduce smoke emission during combustion. Mehrdad et al. [47] used a hybrid nano-catalyst with diesel–biodiesel blends, mixing ceria, and amide-functionalized multi-wall carbon nanotubes (MWCNT). When compared to normal diesel, they detected a considerable reduction in the soot, NO, CO, and HC pollutants as well as enhanced performance. According to researchers [39, 40, 43, 46, 48, 49], nano-additives improved fuel quality (such as enhanced cetane number, quick ignition characteristics, and a bigger surface area/volume ratio). According to a recent study [39, 40, 43, 46, 48, 49], employing fuel-borne catalytic nano-additives with biodiesel fuels, emulsion fuels, and diesel resulted in a shorter ignition delay, improved performance characteristics, and less hazardous emissions. Furthermore, combining water and additives with diesel (in the company of emulsifiers) considerably decrease hazardous emissions while preserving the diesel engine's braking thermal efficiency, according to current research results [16–19, 22, 23, 39, 40]. Furthermore, there has been very few research have been carried out on combining multi-walled carbon nanotubes (MWCNT) as additives with water-diesel emulsions in CI engines. Henceforth, MWCNT will be used as a nano-additive with water-diesel emulsion fuels in the current research investigation to examine the operational performance attributes of a single cylinder diesel engine.

## 2 Experimental Methodology and Fuel Preparation Technique

The working traits (viz., emission and performance) of a 4 stroke naturally aspirated constant speed single cylinder diesel engine are evaluated in this study (Table 1). In a research-oriented experimental configuration, an eddy current dynamometer is utilized in conjunction with a compression ignition engine to impart loads from zero to full load (as shown in the Fig. 1). The gravimetric technique was used to determine the volumetric flow rate of the fuels. The emissions of the 4 stroke naturally aspirated constant speed single cylinder diesel engine were measured using a calibrated KANE AutoPlus 5–2 gas analyzer cum calibrated KANE Auto600 Smoke meter. ASTM standards were used to examine the fuel properties (Table 2). To start and warm up the diesel engine, pure diesel was used, and then, the study was conducted on the test fuels (pure diesel emulsions and MWCNT additive mixed diesel emulsions). An average of three trials were executed under steady-state circumstances to calculate all the recorded data during the experiment.

**Table 1** Diesel engine test rig. specs

| | |
|---|---|
| Brand name | M/s. Kirloskar |
| Diesel engine type | Four stroke constant speed diesel engine |
| Cylinder type | Vertical position |
| Cooling type | Air-cooled |
| Engine's brake power | 4.41 kW @ 1500 rpm |
| Speed | 1500 rpm (rated) |
| Fuel | Diesel |
| Cylinder dimensions | 87.5 mm × 110 mm (bore × stroke) |

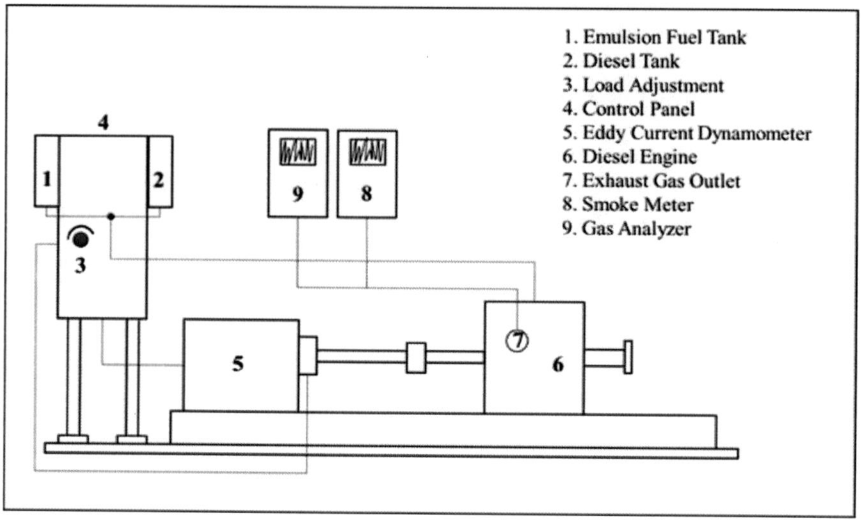


**Fig. 1** Layout of diesel engine-eddy current dynamometer setup

**Table 2** Prepared emulsion fuels details

| Fuel codes | Pure diesel (%) by vol | Water (%) by vol | MWCNT (in grams) | Emulsifiers (%) by vol | |
|---|---|---|---|---|---|
| | | | | Emulsifier Span 80 | Emulsifier Tween 80 |
| 98D2WS | 98 | 2 | – | 1 | 1 |
| 96D4WS | 96 | 4 | – | 2 | 2 |
| 98D2WS50MWCNT | 98 | 2 | 0.2 | 1 | 1 |
| 96D4WS100MWCNT | 96 | 4 | 0.4 | 2 | 2 |

## 2.1 *Experimental Error Analysis*

Observation, instrumentation, and test planning all contribute to systematic and random mistakes during experimentation. Based on the specific uncertainties of instruments accustomed during the experiment, percentages of uncertainties for distinct physical variables were determined. An error study [50] was used to ensure that the experiments were accurate. Taylor's formula was used to guide the error analysis procedure. $x_1, x_2, x_3, \ldots, x_n$ are independent variables. Following that, the error in the value of '$E$' can be determined as follows:

$$\frac{\partial E}{E} = \left\{ \left(\frac{\partial x_1}{x_2}\right)^2 + \left(\frac{\partial x_2}{x_2}\right)^2 + \cdots + \left(\frac{\partial x_n}{x_n}\right)^2 \right\}^{1/2} \tag{1}$$

Uncertainty error (%)

$$= \sqrt{\begin{array}{l}(NO_x)^2 + (CO)^2 + (CO_2)^2 + (HC)^2 \ + (\text{Burette Reading})^2 \\ +(\text{Digital Overhead Stirrer})^2 + (\text{Stop Watch})^2 + (\text{Smoke})^2 + \ (\text{Digital load })^2 \\ +(\text{Thermocouples })^2 + (\text{Speed Sensor})^2 + (\text{Ammeter AReading})^2\end{array}}$$

$$= \sqrt{\begin{array}{l}(1)^2 + (1)^2 + (0.01)^2 + (0.1)^2 + (1)^2 + (1)^2 + (1)^2 + (0.05)^2 \\ +(0.1)^2 + (1)^2 + (1)^2 + (0.1)^2\end{array}}$$

$$= 2.791\%$$

The experiment's total uncertainty was determined to be 2.791%.

## 2.2 *Preparation of Water-In-Diesel Emulsions*

To combine two immiscible fluids for a specific need, some chemicals in terms of surfactants are accustomed. Surfactants were added to the water-diesel mixture in a systematic manner to lower surface tension and enhance minor contact regions, producing in stable emulsions [51]. The physical preparation notion of the water-in-diesel emulsion fuel made with a mixture of two non-ionic surfactants, Span80 and Tween80, are shown in Table 2. In the presence of emulsifiers, a set of two water-diesel emulsion fuels was formed by varying the water ratio (say, 2, and 4% by volume basis) with the diesel fuel (in line with the report stated by Selim Ghanim [52]).

## 2.3 *Synthesis of MWCNT Blended Water-Diesel Emulsions*

MWCNT blended water-diesel emulsions were prepared in three stages using a high-speed mechanical stirrer. The MWCNT was weighed individually using a weighing device to a preset dose of say 50 ppm and dispersed in normal water (say 2%) for thirty minutes using a digital stirrer. Figures 2 and 3 illustrate the XRD and TEM morphology [53] of MWCNT, respectively. In the second stage, a combined Span80 and Tween80 solution was created by combining the two surfactants with an HLB (value of 8). In the third stage, a high-speed stirrer was used to mix the plain diesel (98% by volume) with the surfactant combination, while an injection nozzle injected the MWCNT dispersed in normal water solution that had been produced previously for 30 min at 35 degrees Celsius. As a result, the MWCNT blended water-diesel emulsions fuel (98DS2W50MWCNT) was created, and the 96DS4W100MWCNT was prepared using the same technique. Figures 4, 5, 6, 7, 8 and 9 show the photographic images of the prepared plain water-diesel emulsion fuels and MWCNT mixed water-diesel emulsions. The emulsion fuel samples were a rich white color and were finally

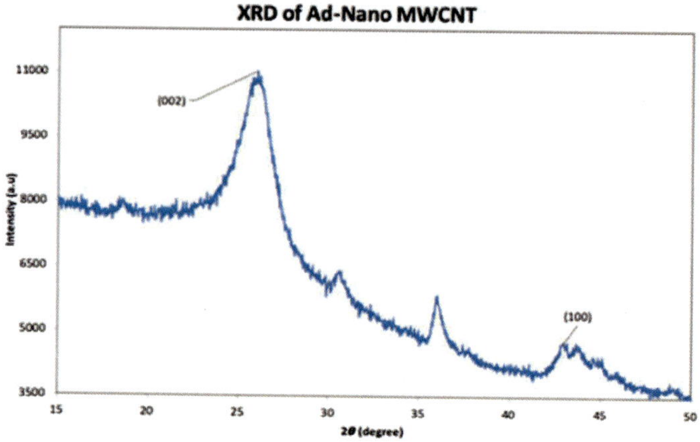


**Fig. 2** XRD image of MWCNT [53]

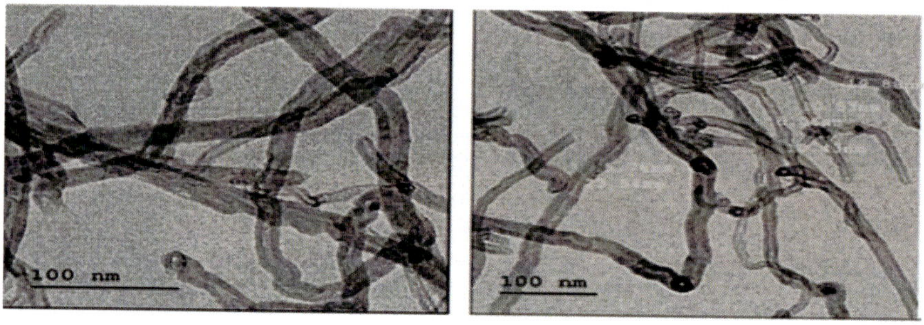


**Fig. 3** HRTEM images of MWCNT [53]

stored in transparent test tubes for stability tests (in line with Selim and Ghannam 2010 [52] stability studies).

# 3 Results and Discussion

The subsequent portion reflects the experimental work data (i.e., performance and emission traits) collected from the air-cooled diesel engine experimental test setup.

**Fig. 4** Diesel and Span 80

**Fig. 5** Diesel and 98D2WS

## 3.1 *Characteristics of Performance Features of Diesel Engine*

Figures 10 and 11 illustrate the correlation between the performance parameters (viz., brake-specific fuel consumption and brake thermal efficiency) for each tested fuel and brake power. All emulsion cum additive fuels exhibited lower brake-specific fuel consumption than pure diesel, according to the findings of the testing carried out in the research investigation. This is most likely due to the MWCNT's enhanced combustion properties and the water present in the additive mixed fuels. The MWCNT additive water molecules in diesel fuel may have absorbed energy instantaneously to generate water vapors and so burst through the surrounding oil layer (Fig. 12) due to speedy

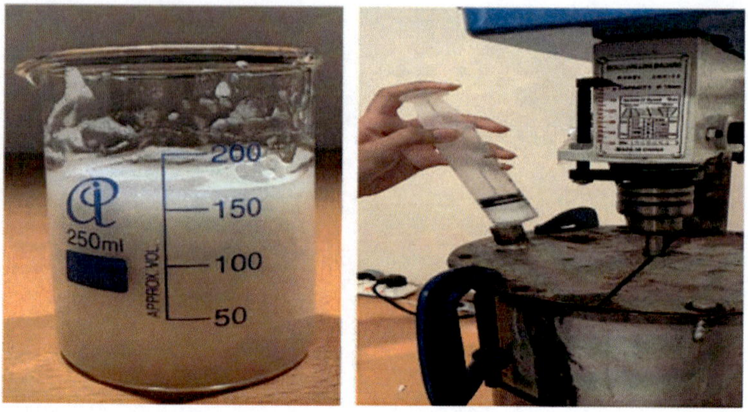


**Fig. 6** Water mixed with Tween 80 surfactant

**Fig. 7** Tween 80

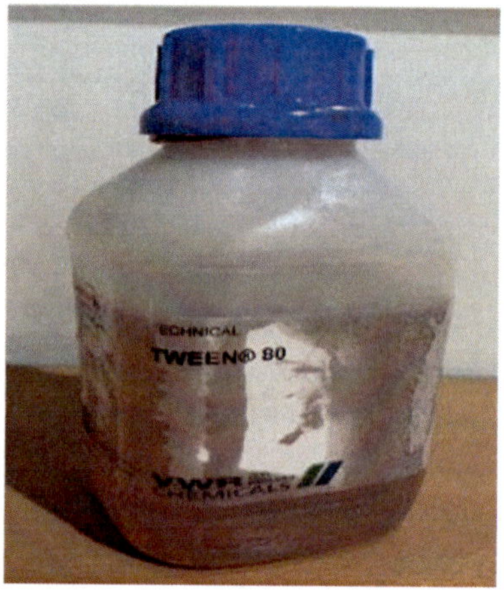


evaporation of H2O droplets encapsulated within the fuel droplet [22, 23, 31, 54–58]. 0.345 kg/kWh of brake-specific fuel consumption (at full load) was noted for the diesel fuel, but for 98D2WS50MWCNT, it was 0.289 kg/kWh. Because the qualities of brake thermal specific fuel consumption and brake thermal efficiency are inversely related, 98D2WS50MWCNT fuel had a better brake thermal efficiency than plain diesel fuel. In the presence of MWCNT, these processes (both secondary atomization and micro-explosions) may have resulted in a higher degree of fuel–air mixing, resulting in improved performance features of than that of diesel fuel.

**Fig. 8** MWCNT

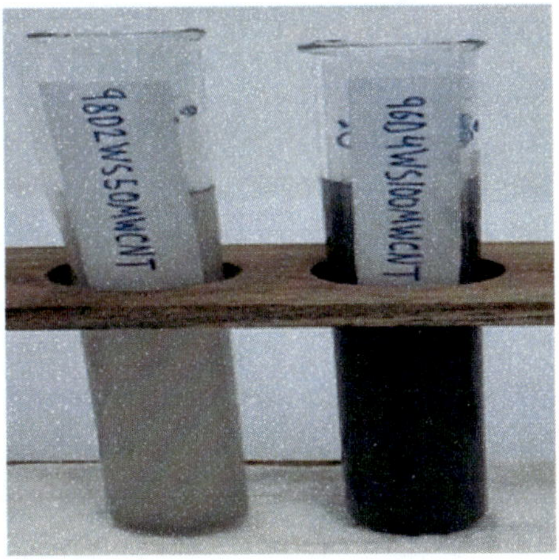

**Fig. 9** 98DS2W50MWCNT and 96DS4W100MWCNT fuels

## 3.2 *Variation of Nitrogen Oxides ($NO_x$) and Exhaust Gas Temperature*

Nitrogen oxides are a series of highly reactive and poisonous gases that are seen as a threat to the global environment. Several hazardous gases are created when the diesel fuel is burned at high temperatures [59]. In case to reduce these exhaust gases, the combustion process should be modified. The graph Fig. 13 shows the change in the amount of nitrogen oxides emissions for all prepared fuels in the experiment compared to the brake power. In reality, when the applied load on the diesel engine is raised, the $NO_x$ level of the fuels will grow owing to the fuel burning temperature inside the combustion chamber of the engine (as per Zeldovich thermal $NO_x$ concept [18, 54]). However, as compared to regular diesel fuel, modified fuels had lower $NO_x$

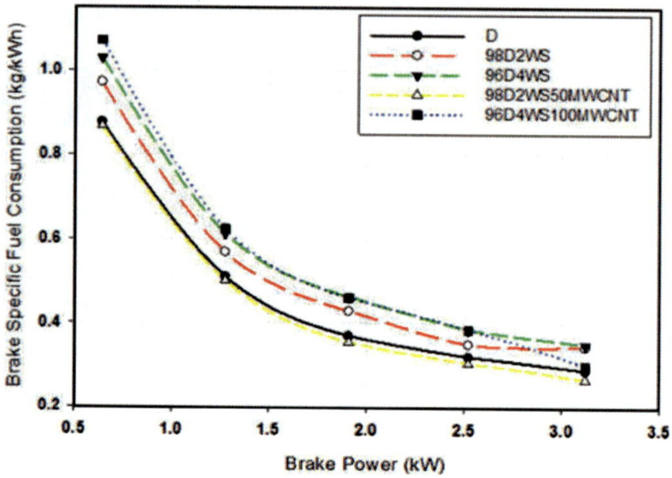


**Fig. 10** Brake power with brake-specific fuel consumption

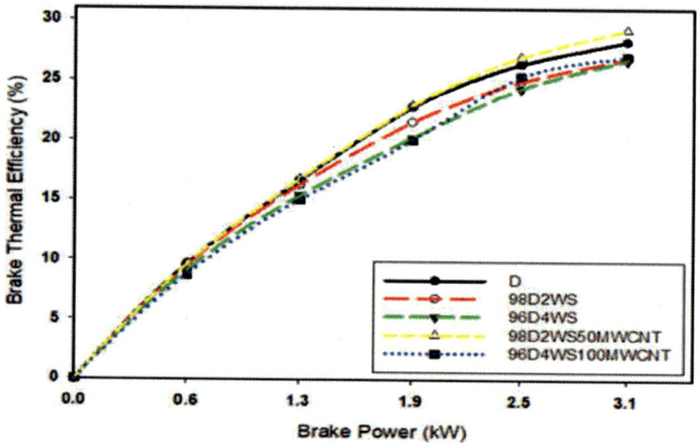


**Fig. 11** Brake power with brake thermal efficiency

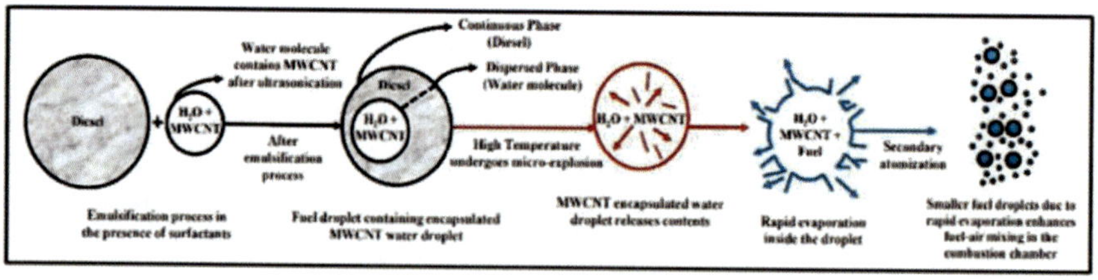


**Fig. 12** Emulsion fuel's micro-explosion phenomenon [23]

emissions (due to the substantial surface area/volume ratio of MWCNT existence in the emulsions). The graph clearly shows that when MWCNT modified fuels are used instead of plain diesel, the ($NO_x$) percentage is significantly lower, especially as the load increases. The same trend of increased exhaust gas temperature was reflected (as illustrated in Fig. 14). At the full load, the $NO_x$ emission value for the 98D2WS50MWCNT was 998 ppm, which was the lowest value among the tested fuels.

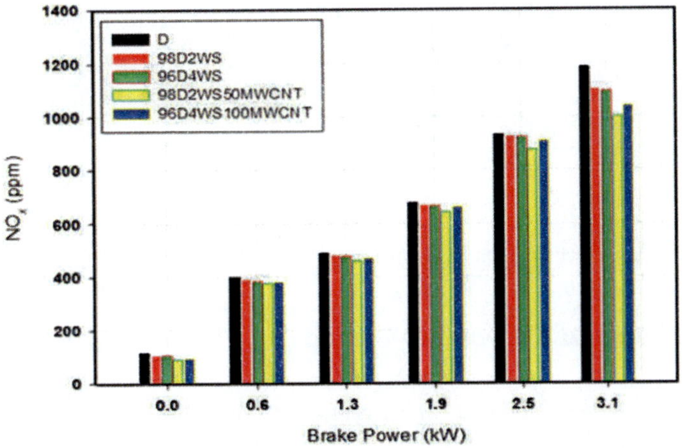


**Fig. 13** Brake power with nitrogen oxides ($NO_x$)

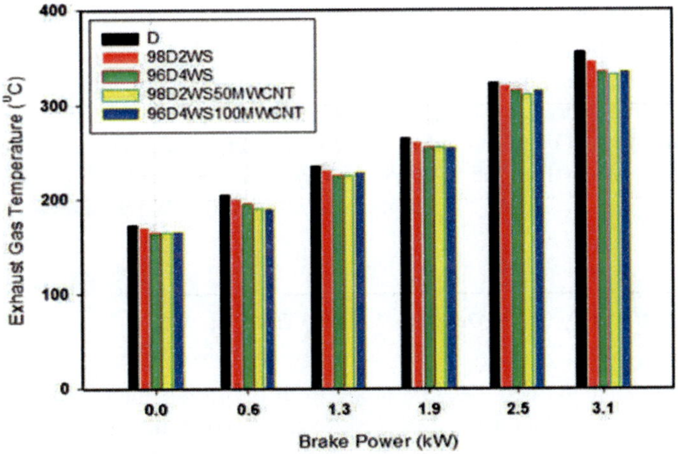


**Fig. 14** Brake power with exhaust gas temperature

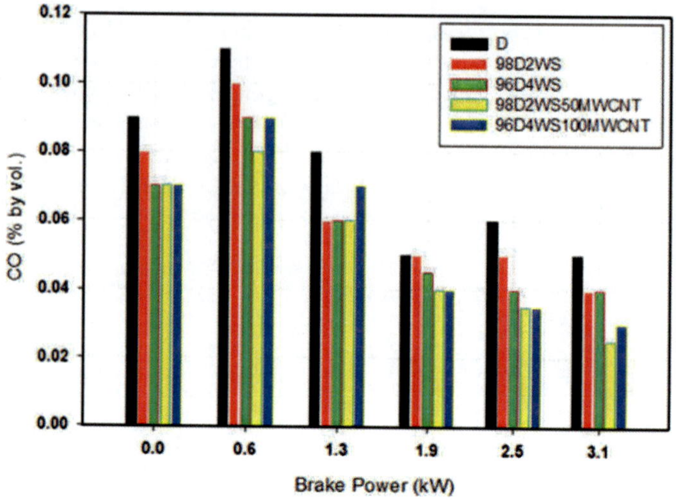


**Fig. 15** Brake power with carbon monoxides (CO)

## 3.3 *Variation of Carbon Monoxides*

In general, CO are one of the lethal gases that can impair the ecological environment and affect people's health substantially. Thus, one of the objectives needs to be achieved in this experimental work is to reduce these harmful emissions. The change in the percentage of CO level of the tested fuels as a function of brake power is depicted in Fig. 15. CO had unstable values at each brake power for all of the tested fuels, as depicted in the Fig. 15. In addition, it is clearly understood from the above figure that additive fuels emitted less CO than diesel fuel (particularly at the higher loads of the diesel engine). This is most likely due to a mechanism that combines secondary atomization and micro-explosion events [16, 18, 54], which may have resulted in lower CO emissions for additive fuels than for pure diesel fuel. At full load, CO emissions from the 98D2WS50MWCNT fuel (0.025% by volume) were marginally lower than those from normal diesel fuel (0.05% by volume).

## 3.4 *Variation of Hydrocarbon (HC) Emissions*

The quantity of unburned HC emissions from the tested fuels versus the diesel engine's brake power is depicted in Fig. 16. At all loads, the amount of HC emissions from all of the studied fuels was gradually increased [31, 41]. As a result, the neat diesel fuel emits somewhat more unburnt HC than the other tested fuels, as shown in the figure above. It is also ascertained that among the tested fuels, the

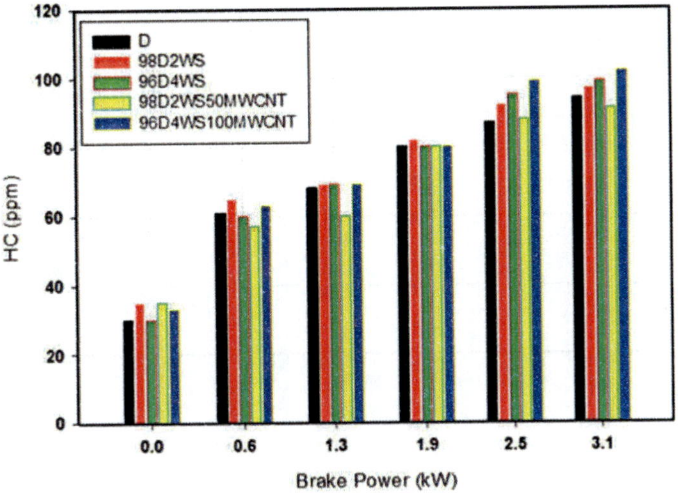


**Fig. 16** Brake power with hydrocarbon (HC)

98D2WS50MWCNT fuel had the lowest value of unburned HC emissions (particularly between the intermediate to higher loads). This could be owing to the high-fuel–air mixing owing to the existence of MWCNT in the emulsion fuel. The additive mixed fuels had decreased unburnt HC emissions as a result of this impact. The magnitude of HC emissions at full load for diesel fuel was 94 ppm, whereas it was 89 ppm for 98D2WS50MWCNT.

## 3.5 *Variation of Smoke Opacity*

Black smoke is basically made up of elemental carbon resulted from the incomplete combustion of diesel fuel with some traces of engine oil. In addition, when there is an excess of fuel and inadequate residence time in the combustion zone, and/or inadequate oxidants, incomplete combustion occurs, and soot forms [10, 11]. During the testing at various applied loads, it was found that the aforesaid effect was reflected for plain diesel fuel, and therefore, higher smoke contents were reflected than for emulsion fuels (Fig. 17). The percentage of smoke opacity in emulsion fuels was reflected in lower in magnitude at all the loads of the engine while that of plain diesel fuel. The proportion of smoke opacity for 98D2WS50MWCNT fuel was 3.4% at full load, compared to 4.5% for plain diesel fuel, as seen in the Fig. 17. Both emulsion fuels and MWCNT emulsions have amended combustion properties (owing to the combined effects micro-explosions followed by secondary atomization effects [16, 18, 54]).

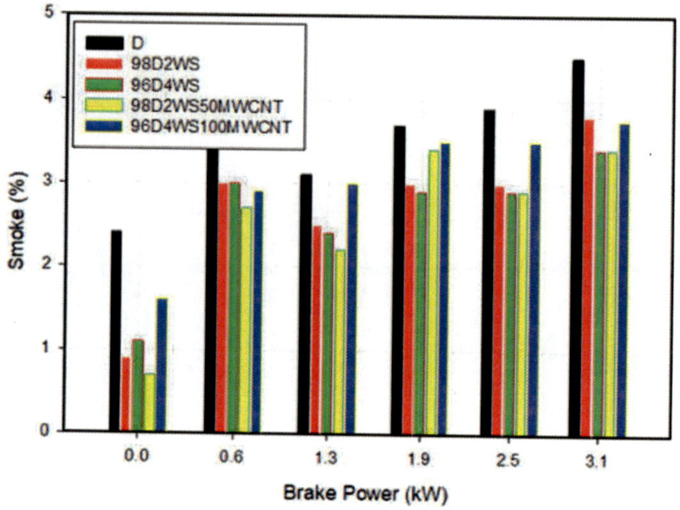


**Fig. 17** Brake power with smoke opacity

## 4 Conclusions

The following prominent conclusions were drawn on using water and MWCNT with diesel fuel in an air-cooled diesel engine:

- The effective integration of MWCNT into diesel fuel was discovered, and the fuels were proven to be stable for more than three weeks.
- Adding MWCNT to emulsion fuels raised the flashpoint, fire point, and calorific values. Owing the enhancement MWCNT emulsion fuel's properties and exuberant effects of secondary atomization, the performance and emission levels were improved.
- The fuel consumption of the DICI engine was 0.345 kg/kWh for diesel fuel, against 0.289 kg/kWh for 98D2WS50MWCNT at the full load. Because both performance parameters are inversely related. The brake thermal efficiency of 98D2WS50MWCNT fuel was higher than that of regular diesel fuel.
- The $NO_x$ emission value for 98D2WS50MWCNT was 998 ppm, which was the lowest among the studied fuels due to improved combustion feature in the diesel engine. A similar impact was seen with regards to smoke emissions. The percentage of smoke opacity for 98D2WS50MWCNT fuel was 3.4 at full load, compared to 4.5 for plain diesel fuel.

**Acknowledgements** The authors are grateful to MOHERI, Oman, Dr. Amer Al Habsi (DDAA), and the management of the International Maritime College of Oman (IMCO) for providing us with the resources that we needed to accomplish the research work in the stipulated period of time.

## References


1. Nubia MR, Angelo CP, Cristina MQ, Gisele OR, Leonardo SG, Teixeira et al (2007) The role of additives for diesel and diesel blended (Ethanol or Biodiesel) fuels: a review. Energy Fuels 21:2433–2445
2. Ullman TL, Spreen KB, Mason RL (1994) Effects of cetane number, cetane improver, aromatics and oxygenates on 1994 heavy-duty diesel emissions. SAE Technical Paper no. 941020
3. Sadhik Basha J, Anand RB (2012) Effects of nanoparticle additive in the water-diesel emulsion fuel on the performance, emission & combustion characteristics of a CI engine. Int J Veh Design 59:164–181
4. Prakash R, Singh RK, Murugan S (2013) Experimental investigation on a diesel engine fueled with bio-oil derived from waste wood–biodiesel emulsions. Energy 55:610–618
5. Lapuerta M, Fernández JR, Contreras RG (2015) Effect of a glycerol-derived advanced biofuel–FAGE (fatty acid formal glycerol ester)–on the emissions of a diesel engine tested under the New European Driving Cycle. Energy 93:68–79
6. Rakopoulos DC, Rakopoulos CD, Giakoumis EG, Papagiannakis RG, Kyritsis DC (2014) Influence of properties of various common bio-fuels on the combustion and emission characteristics of high-speed DI (direct injection) diesel engine: Vegetable oil, bio-diesel, ethanol, n-butanol, diethyl ether. Energy 73:354–366
7. Tsukahara M, Yoshimoto Y (1989) W/O emulsion realizes low smoke and efficient operation of DI Engines without high pressure injection. SAE Technical Paper no 890449
8. Afify HM, Korah NS, Dicky DW (1987) The effect of air charge temperature on performance, ignition delay and exhaust emissions of diesel engines using W/O emulsions as fuel. SAE Technical Paper no 870555
9. Hountalas DT, Mavropoulos G, Zannis T (2007) Comparative evaluation of EGR, intake water injection and fuel-water emulsion as NO reduction techniques for heavy duty diesel engines. SAE Technical Paper no 2007-01-0120
10. Tzirakis E, Karavalakis G, Schinas P, Korres D, Karonis D, Stournas S et al (2006) Diesel-water emulsion emissions and performance evaluation in public buses in Attica Basin. SAE Technical Paper no 2006-01-3398
11. Yoshimoto Y (2006) Performance of DI diesel engines fuelled by water emulsions with equal proportions of gas oil-rapeseed oil blends and the characteristics of the combustion of single droplets. SAE Technical Paper no 2006-01-3364
12. Hall D, Thorne C, Goodier D (2003) An investigation into the effect of a diesel/water emulsion on the size and number distribution of the particulate emissions from a heavy-duty diesel engine. SAE Technical Paper no 2003-01-3168
13. Ithnin AM, Ahmad MA, Abu Bakar MA, Rajoo S, YahyaIthnin WJ (2015) Combustion performance and emission analysis of diesel engine fuelled with water-in-diesel emulsion fuel made from low-grade diesel fuel. Ener Conv Mgt 90:375–382
14. Koc AB, Abdullah M (2013) Performance and NO emissions of a diesel engine fueled with biodiesel-diesel-water nanoemulsions. Fuel Proc Tech 109:70–77
15. Rao MS, Anand RB (2016) Performance and emission characteristics improvement studies on a biodiesel fuelled DICI engine using water and AlO(OH) nanoparticles. App Ther Eng 98:636–645
16. Sadhik Basha J, Anand RB (2011) Role of nano-additive blended biodiesel emulsion fuel on the working characteristics of a diesel engine. J Renew Sustain Energy 1–17
17. Sadhik Basha J (2015) Preparation of water-biodiesel emulsion fuels with CNT & Alumina nano-additives and their impact on the diesel engine operation. SAE Technical Paper no 2015-01-0904
18. Sadhik Basha J, Anand RB (2014) Performance, emission & combustion characteristics of a diesel engine using carbon nanotubes blended jatropha methyl esters emulsions. Alex Eng J 53:259–273
19. Sadhik Basha J, Anand RB (2010) Effects of nanoparticle blended water-biodiesel emulsion fuel on working characteristics of a diesel engine. J Global Warming 2:330–346

20. Ghojel J, Damon H, Al-Khaleefi K (2006) Performance, emissions and heat release characteristics of direct injection diesel engine operating on diesel oil emulsion. App Ther Eng 26:2132–2141
21. Kweonha P, Inseok K, Seungrnook O (2004) The effect of water emulsified fuel on a motorway-bus diesel engine. KSME Int J 18:2049–2057
22. Sadhik Basha J, Anand RB (2011) An experimental study in a CI engine using nano-additive blended water-diesel emulsion fuel. Int J Green Energy 8:332–348
23. Sadhik Basha J, Anand RB (2011) An experimental investigation in a diesel engine using CNT blended water-diesel emulsion fuel. Proc IMechE J Power Energy 225:279–288
24. Kadota T, Yamasaki H (2002) Recent advances in the combustion of water fuel emulsion. Prog Ener Comb Sci 28:385–404
25. Dong HQ, Hao C, Chia FL, Li MG, Yao ZB (2010) Experimental studies of a naturally aspirated DI diesel engine fuelled with ethanol-biodiesel-water micro-emulsions. Energy Fuels 24:652–663
26. Wagner U, Eckert P, Spicher U (2008) Possibilities of simultaneous in-cylinder reduction of soot and NO emissions for diesel engines with direct injection. Int J Rotat Mach Article ID 175956
27. Selim MYE, Ghannam MT (2007) Performance and engine roughness of a diesel engine running on stabilized water diesel emulsion. SAE Technical Paper no 07 NAPLES–63
28. Nadeem M, Rangkuti C, Anuar K, Haq MRU, Tan IB, Shah SS (2006) Diesel engine performance and emission evaluation using emulsified fuels stabilized by conventional and gemini surfactants. Fuel 85:2111–2119
29. Armas O, Ballesteros R, Martos FJ, Agudelo JR (2005) Characterization of light duty diesel engine pollutant emissions using water-emulsified fuel. Fuel 84:1011–1018
30. Ling YH, Wei FB (2005) Study on the mechanism of saving fuel consumption for emulsified fuel with wall catalytic reforming reaction. Comb Sci Tech 177:671–689
31. Subramanian KA, Ramesh A (2002) Use of diethyl ether along with water-diesel emulsion in a DI diesel engine. SAE Technical Paper no 2002-01-2720
32. Samec N, Kegl B, Dibble RW (2002) Numerical and experimental study of water/oil emulsified fuel combustion in a diesel engine. Fuel 81:2035–2044
33. Park JW, Kang Y, Huh Lee JH (2001) Reduction of NO, smoke and brake specific fuel consumption with optimal injection timing and emulsion ratio of water-emulsified diesel. Proc Inst Mech Eng Part D J Auto Eng 215:83–93
34. Sheng HZ, Chen L, Wu CK (1995) The droplet group micro-explosions in W/O diesel fuel emulsion sprays. SAE Technical Paper no 950855
35. Harbach JA, Agosta V (1991) Effect of emulsified fuel on combustion in a four-stroke diesel engine. J Ship Res 35:356–363
36. Sadhik Basha, J (2014) An experimental analysis of a diesel engine using alumina nanoparticles blended diesel fuel. SAE Technical Paper no 2014-01-1391
37. Sadhik Basha J, Anand RB (2010) Performance and emission characteristics of a DI compression ignition engine using carbon nanotubes blended diesel. J Adv Ther Sci Eng 1:67–76
38. Lin CY, Wang KH (2004) Effects of a combustion improver on diesel engine performance and emission characteristics when using three-phase emulsions as an alternative fuel. Energy Fuels 18:477–484
39. Kao MJ, Chen-Ching T, Bai-Fu L, Tsing-Tshih T (2008) Aqueous aluminium nanofluid combustion in diesel fuel. J Test Eval 36:PA 19428-2959
40. Farfaletti A, Astorga C, Martini G, Manfredi U, Mueller A, Rey M et al (2005) Effect of water/fuel emulsions and a cerium-based combustion improver additive on HD and LD diesel exhaust emissions. Environ Sci Tech 39:6792–6799
41. Sadhik Basha J (2015) Applications of functionalized carbon based nanomaterials. In: Vijay KT, Manju KT (eds) Chemical functionalization of carbon nanomaterials: chemistry and applications. CRC-Taylor and Francis Inc, Boca Raton, pp 573–588

42. Sajith V, Sobhan CB, Peterson GP (2010) Experimental investigations on the effects of cerium oxide nanoparticle fuel additives on biodiesel. Adv Mech Eng ID 581407:1–6
43. Tyagi H, Patrick EP, Ravi P, Robert P, Taewoo L, Jose RP et al (2008) Increased hot plate ignition probability for nanoparticle-laden diesel fuel. Nano Lett 8:1410–1416
44. Sabourin JL, Daniel MD, Yetter RA, Frederick LD, Ilhan AA (2009) Functionalized graphene sheet colloids for enhanced fuel/propellant combustion. ACS Nano 3, 3945–3954
45. Roos WJ, Duncan R, David JC (2008) Diesel fuel additives containing cerium or manganese and detergents. US Patent No US2008/0066375 A1
46. Moy D, Niu C, Tennent H et al (2002) Carbon nanotubes in fuels. US Patent 6419717
47. Mirzajanzadeh M, Tabatabaei M, Ardjmand M, Rashidi A, Ghobadian B, Barkhi M et al (2015) A novel soluble nano-catalysts in diesel–biodiesel fuel blends to improve diesel engines performance and reduce exhaust emissions. Fuel 139:374–382
48. Shaafi T, Velraj R (2015) Influence of alumina nanoparticles, ethanol and isopropanol blend as additive with diesel–soybean biodiesel blend fuel:combustion, engine performance and emissions. Renew Energy 80:655–663
49. Karthikeyan S, Prathima A (2016) Characteristics analysis of carbon nanowires in diesel: Neochloris oleoabundans algae oil biodiesel–ethanol blends in a CI engine. Energy Sour Part A Recovery Utili Environ Effects 38:3089–3094
50. Moffat RJ (1988) Describing the uncertainties in experimental results. Exp Ther Fluid Sci 1:3–17
51. Chiaramonti D, Bonini M, Fratini E, Tondi G, Gartner K, Bridgwater AV, Grimm HP, Soldaini I, Webster A, Baglioni P (2003) Development of emulsions from biomass pyrolysis liquid and diesel and their use in engines—Part 1: emulsion production. Biomass Bioenergy 25:85–99
52. Selim MYE, Ghannam MT (2010) Combustion study of stabilized water-in-diesel fuel emulsion. Energy Sour Part A Recovery Util Environ Eff 32:256–274
53. Ad-Nano Technologies (n.d) Multi-Walled Carbon Nanotubes. Retrieved on 17 June 2021 from https://ad-nanotech.com/
54. Sadhik Basha J (2018) Impact of Carbon Nanotubes & Di-Ethyl Ether as additives with biodiesel emulsion fuels—an experimental investigation. J Energy Inst 91:289–303
55. Abu-Zaid M (2004) An experimental study of the evaporation characteristics of emulsified liquid droplets. Heat Mass Trans 40:737–741
56. Sadhik Basha J (2017) Impact of Nano-additive blended biodiesel fuels in Diesel Engines. Springer Pub, ISBN 9783319454597, Chap 14, pp 325–339
57. Sadhik Basha J, Al Balushi M (2020) The influence of water and DEE as additives with diesel fuel in a light duty diesel engine generator-an experimental investigation. J Eng Sci Technol 1:54–61
58. Sadhik Basha J, Anand RB (2020) Recent technologies for enhancing performance and reducing emissions in diesel engines, IGI–Global Publishers, USA, ISBN13: 9781799825395
59. EPA (2019) Nitrogen Oxides ($NO_x$) control regulations. Retrieved on 16 June 2021 from Nitrogen Oxides Control Regulations|Ozone Control Strategies|Ground-level Ozone|New England|US EPA

# Automotive Applications

# A CFD Cold Flow Analysis of Different Piston Configurations for Internal Combustion Engine

S. C. Amith, R. Prakash, D. Arun, and S. Cyril Joseph Daniel

## 1 Introduction

Nowadays, the emission standards are strict, thus the engines are becoming more futuristic in the IC engine and SI engine to suggest the perception like lean-burn, stratification, and direct-injected to increase the fuel recession and decrease the exhaust emissions. The performance of an engine is improved by proper stratification at varying load. Air–fuel interaction and inside the flow of the cylinder is regulating the stratification in the engine by compelling criterion. By changing the pattern of the compression ratio, the speed of the engine, combustion chamber, etc.; the air–fuel interaction and inside the flow of the cylinder is modified. Due to propagation rates and lower flame initiation with higher combustion, deviation is correlated with lean-burn. Lean-burn approach is used in IC engines to enhance means flow and overwhelm internal cylinder flow turbulence. Performance of the IC engine, combustion, and emission has an extensive impact in inside the flow of the cylinder. Swirl and tumble flow have a two type of flow in inside the cylinder, one is a cylinder axis is parallel to rotational motion of fluid and another axis of cylinder is perpendicular to rotational motion of fluid. SI engines have a higher swirl and tumble flow to attain faster burning rates. The speed of the engine, the position of the piston cavity, the pattern of the piston top, direction of the intake manifold, etc., are an important criterion of inside the flow of the cylinder [1].

CFD technique can be well-known to learn and analyze the shape of flow which involves many complex calculations and large variables. IC engine concerns combustion, moving geometrics, and air flow interaction between complex fluid dynamics. Inside cylinder combustion simulation, full-cycle simulation, cold flow analyses, and port flow analysis executed in CFD. [2]. In this research, cold flow simulations

S. C. Amith (✉) · R. Prakash · D. Arun · S. Cyril Joseph Daniel
Department of Mechanical Engineering, Sri Sivasubramaniya Nadar (SSN) College of Engineering, Kalavakkam, Chennai, Tamilnadu 603110, India
e-mail: amithsc@ssn.edu.in

 
N. L. Narasimhan et al. (eds.), *Recent Advances in Energy Technologies*, Lecture Notes in Mechanical Engineering, https://doi.org/10.1007/978-981-19-3467-4_31

are used, which contain the flow of air design and conceivably the injection of fuel in the transient engine cycle without reactions. The objective was to catch the air–fuel mixture ratio by precisely computing for the induction process. The modifying attribute of the flow of air in the jet to determine whether exhaust valves are open and close through the exhaust jet and the swirl intake valves with the tumbles and the cylinder, and by cause of compression and squish turbulence generation in swirl and tumble [3].

The turbulent kinetic energy (TKE) as high as 12.56% in the flat piston compared to flat piston with a center bowl, pent-roof offset bowl piston, dome piston with a center bowl, when TKE is assigned to all over the combustion chamber [4]. Inside the cylinder, tumble flow relies on crank position in attentive of engine speed on direct injection spark ignition (DISI) engine using simulation; this can be done by varying a piston crown pattern under motoring condition using particle image velocimetry (PIV) technique [5]. In an inquiry by PIV technique, turbulence intensity compared to concave-shaped piston crown, it urges higher tumble ratio (TR) in flat piston crown [6]. This numerical simulation performed homogeneous gasoline direct injection (GDI) engines for air–fuel mixture ratio at both high speed and low engine speed in full load conditions. It shows that the combination between inside the cylinder of air flow and spray stream has pretentious the space trajectory and flow filed structure [7]. Miloket et al. [8] investigated numerically gasoline direct injection engine (GDI) using three different piston patterns varying engine speeds. In that air–fuel interaction was heavily influenced by piston patterns, GDI ratios and direct injection timing for fuel injection during compression stroke. Adbullah et al. [9] noted that delay the injection and ignition timings reduce the CO and NOx emission levels with better performance. CFD results predicted that the emissions are more during higher engine speed when compared to values measured by gas analyzer. The large turbulence is created in the intake stroke by initiating a vortex flow with combination of swirl and tumble, which helps in faster combustion in the end of compression stroke [10]. The experimental study by using 2D laser Mie-scattering technique on DISI engine for two different pressures indicates dense fuel film was deposited over the hollow-cone pressure-swirl piston crown during compression stroke [11]. Recent studies were carried out in a three-dimensional CFD model which emphasized on dual phase flow characteristics with crank angle degree movements [12]. The role of piston progression for analyzing the parameters like air–fuel interactions and their thermodynamics within the combustion chamber has been an important area of research [13]. There were several more studies [14] on the CFD analysis of internal combustion engine with double-lobed piston head [15], free piston internal combustion engine linear generator [16], pent-roof piston profile SI engine [17], diesel engine [18], heat exchangers [19], and different profile configurations in diesel engine [20]. In this work, four piston head designs were investigated, and CFD simulations were used to assess the performance of each piston profile. The in-cylinder flow was simulated at various crank angles, and variables such as turbulent kinetic energy, tumble ratio, and turbulence inside the engine combustion chamber were compared.

## 2 Design and Mesh Creation

The cold flow simulation is performed for four different profiles of piston crown, that is, flat, convex, concave, and recessed type in a DISI four-stroke engine with 1500 rev/min speed of engine. Cold flow analysis modeling involves moving geometry and air flow in transient engine cycle without reactions and fuel injection. The motion of air inside the cylinder is analyzed through considering the dynamics of the fluid and the geometry of the chamber. The following three steps were used in the study for IC engine simulation:

- The geometry of piston head and the combustion chamber was decomposed into different zones, and proper mesh was applied.
- The mesh motion strategies were applied to different decomposed geometry regions.
- In ANSYS FLUENT, the engine case was setup and transient IC simulation was performed.

The specifications considered for the engine are shown in Table 1. The studies of fluid dynamics inside the combustion chamber were started at 360°, from the intake stroke initiation and carried out till 720°, which is the end of compression stroke.

### 2.1 Geometry Decomposition

The initial step for simulation is importing the design geometry of the engine. The design is split into various lesser volume parts prior to the meshing process. This helps in meshing the individual parts. Further, sub-divisions of the volume were done for the ease of meshing each and every sub-volume. The mesh was either hexagonal or tetragonal-shaped elements based on the volume available for meshing.

**Table 1** Engine parameters used for simulation

| Type of the engine | Four-stroke SI engine, two-valve configurations |
|---|---|
| Compression ratio | 10:1 |
| Bore dia | 87.5 mm |
| Speed of the engine | 1500 rev/min |
| Stroke length | 110 mm |
| Number of cylinders | 1 |
| Minimum valve lift | 0.2 mm |

## 2.2 Mesh Creation

The mesh arrangements vary for the piston geometry and the valves. Before breaking down the geometry, the valves were kept at top dead center (TDC) and the piston was in closed position. Simulations with piston and valves at TDC are comparatively difficult for mesh creation, but upon successful creating the mesh at TDC position, the change in piston movement from TDC to bottom dead center (BDC) and vice versa have proper mesh throughout the simulation. Figure 1 shows the types of SI engine combustion chamber that are analyzed in this work.

The valves were initially kept at TDC for all these designs. The cells having layers were kept in the gap of valve seat and the valve, as some gap has to be maintained between them for simulation purposes. An inconsistent interface was introduced to close the valve. Simulations were carried out in ANSYS FLUENT software, in which valve lift values were 0.05 mm to 0.5 mm.

Dynamic meshing was used specifically for defining the moving parts like piston and the valves. Since there is a use of dynamic mesh, any improper meshing in other volumes may alter the simulation results and eventually provoke analysis failure. The grid for the hexahedral cells was generated with the help of a grid generating tool. The mesh size varied from 0.495 mm to 4.5 mm. The fluid zone names and the mesh requirements for straight single valve SI engine are shown in Fig. 2. The meshing and solver setup were done for all the parts shown in Table 2, and the simulations were carried out henceforth.

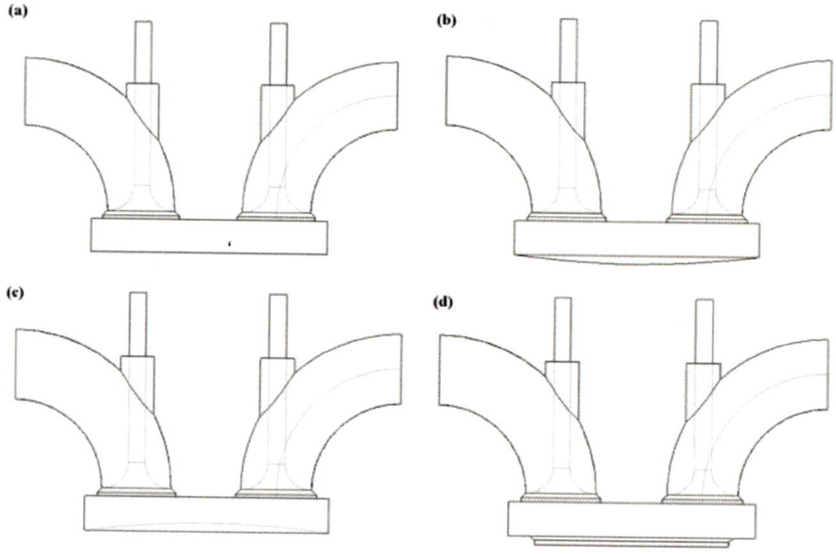


**Fig. 1** Piston head geometries are **a** flat; **b** dome; **c** concave; and **d** recessed piston

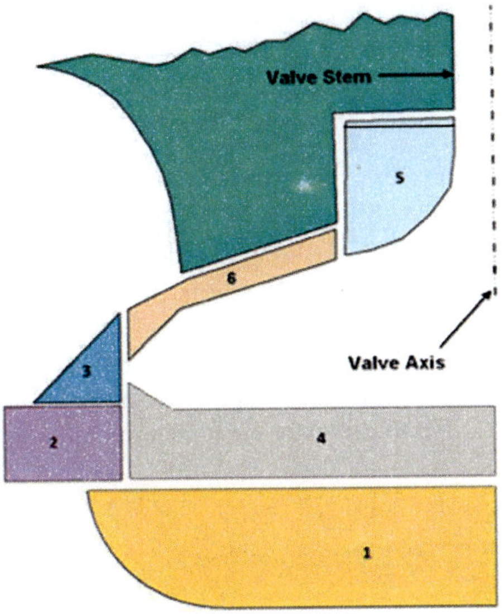


**Fig. 2** Sub-volume zone names and mesh requirements for straight valve

**Table 2** Fluid zone names and mesh requirements for straight valve

| S. No | Fluid zone name | Mesh requirement |
|---|---|---|
| 1 | Piston | Any mesh |
| 2 | ch-lower | Layered mesh |
| 3 | ch-upper | Any mesh |
| 4 | ch-valveID | Layered mesh |
| 5 | valveID-ib | Mesh with at least one layer at the top |
| 6 | valveID-vlayer | Layered mesh |

## 3 Results and Discussion

The atmospheric pressure and the ambient temperature were used as the input parameters. The simulation studies on the suction and the compression stroke were done without considering the thermal behavior and the fuel interactions, thus it was a cold flow analysis. The layers used in this simulation varied, as the piston moves from TDC to BDC there were increase in layers and decreased for the piston movement from BDC to TDC. The motions of layers were important factor for identifying the air dynamics within the combustion chamber. The validation of the model has been done with the experimental trial in the form of in-cylinder pressure during intake and compression stroke at 1500RPM, and comparing with the CFD values.

The simulations were carried out in transient state. The parameters measured for analysis of this study are turbulent kinetic energy, contour of velocity, cross tumble ratio, and swirl ratio. The basic conditions were defined before simulation, such as the fluid properties, material details, and the boundary definition, in ANSYS

FLUENT software. The model included an unstable (transient) calculation, moving meshes and boundaries, high compressible Reynolds number, high fluid dynamics features (turbulence intensity), momentum, heat and mass transfer, and complicated geometries.

The dynamics effects were ignored by using constant pressure boundary conditions at both the intake and exhaust ports. The adiabatic state was defined as the walls of the intake and exhaust ports, as well as the lateral walls of the valves. The constant temperature boundary conditions for the cylinder head, cylinder wall, and piston crown that delineate the combustion chamber's walls were assigned separately. In the dynamic mesh settings, the engine specifications were mentioned. The valve displacement and piston movement values were given as default in the valve lift program option. The crank angle degrees of 0.25° were given as the time step number to carry out the transient simulation process.

## 3.1 *Velocity Contours*

From Fig. 3, the maximum tumble vortex was formed in all the piston profile at around 540 crank angle degrees (CAD). The velocity magnitude was the highest in flat piston head profile and the least in recessed shaped piston head profile.

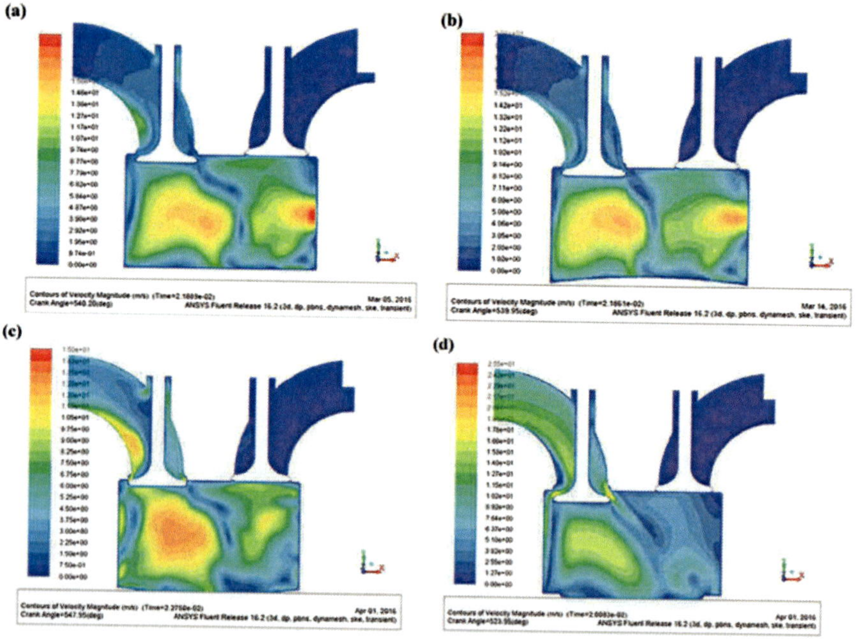


**Fig. 3** Contours of velocity magnitude in m/s for **a** flat; **b** concave; **c** dome; and **d** recessed piston

The contours of the velocity magnitude were slightly shifted away from the longitudinal axis. The dome-shaped piston configuration showed maximum tumble vortex around 520 CAD, which was before the piston reached the bottom dead center (BDC). This phenomenon was caused due to the slight volume decrease in the combustion chamber with dome-shaped piston configuration. Similarly, in the concave-shaped piston head profile, the maximum velocity magnitude was obtained around 590 CAD. This delay in maximum velocity magnitude was due to the slight increase in the volume for concave-shaped piston head.

## 3.2 Swirl Ratio

The ratio of the in-cylinder flow's angular momentum about each of the three orthogonal axes is known as the swirl ratio. From Fig. 3, the maximum tumble vortex was formed in all the piston profile at around 540 crank angle degrees (CAD). The swirl ratio was higher for concave and recessed piston configurations as shown in Fig. 4.

In the flat and dome-shaped piston head configurations, the swirl ratio is fluctuating within 380 CAD to 650 CAD but with very less value of swirl ratio. The swirl motion inside the combustion chamber was less for flat and dome-shaped profile due

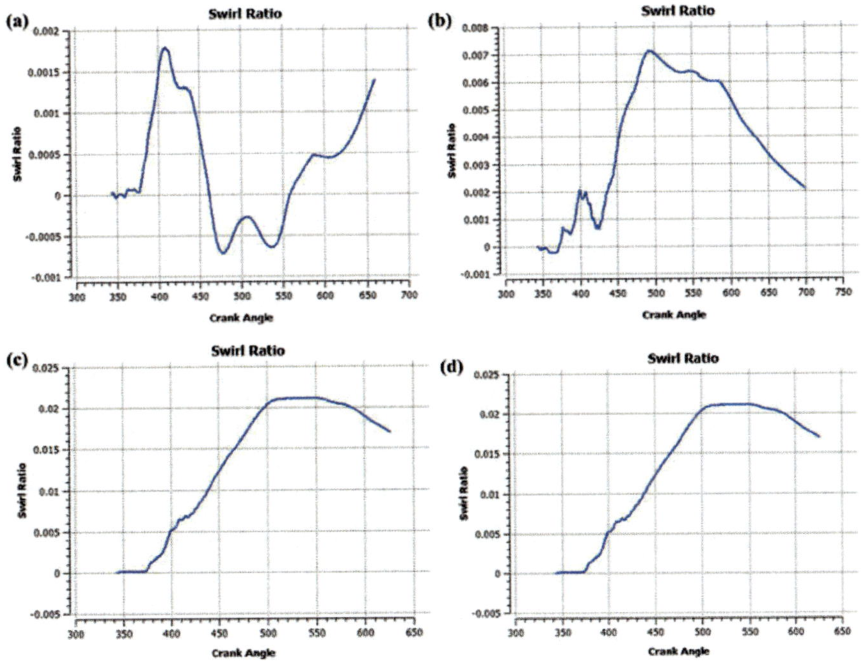


**Fig. 4** Swirl ratio for **a** flat; **b** dome; **c** concave; and **d** recessed piston

to the air in-flow characteristics, and pressure variations just over the piston heads were almost uniform throughout the crank movement. The fluctuations of the pressure at various locations just above the piston head of concave and recessed piston configurations were the significant factor determining the higher value of swirl ratio in these two piston head profiles.

### 3.3 Cross Tumble Ratio

The cross tumble ratio is the rotational ratio of airflow on the cross tumble axis to the other axes. The highest cross tumble ratio was achieved by the flat piston geometry as shown in Fig. 5. It can be clearly seen from the graphs that flat piston head configuration has better curve which depicts the motion for air is high.

The flat piston moves from TDC to BDC, the gradual increases in the volume of the combustion chamber causes the tumble flow to accelerate in a controlled manner. Thus, the rotation motion of air in-cylinder flow perpendicular to the cylinder axis was higher in flat type piston head configuration.

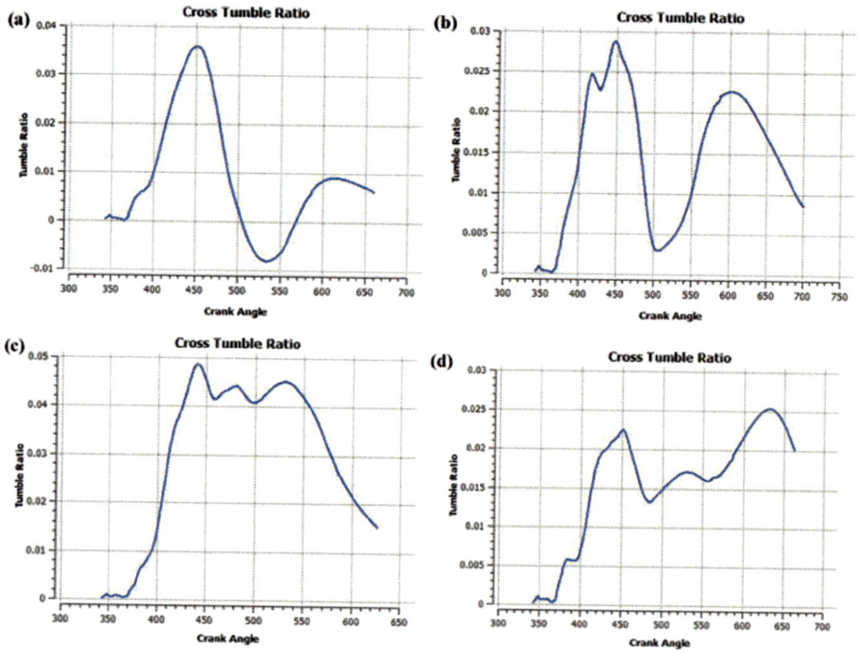


**Fig. 5** Cross tumble ratio for **a** flat; **b** dome; **c** concave; and **d** recessed piston

## 3.4 Turbulent Kinetic Energy

The mean kinetic energy per unit mass associated with eddies in turbulent flow is known as turbulence kinetic energy (TKE). The mass average turbulent kinetic energy was the key factor which depicts the air motion within the combustion chamber and thus more the TKE, better combustion of the charges and higher efficiency with cleaner exhaust gases are obtained in the flat piston head profile. From Fig. 6, the mass average turbulent kinetic energy was marginally better for piston geometry with flat profile. The flat piston configuration had the maximum velocity magnitude at 540 CAD, which was exactly at the BDC and the cross tumble ratio was also high for flat piston head configuration. Overall, the turbulence was maximum in flat piston configuration, which translates that the in-cylinder air flow and the proper stratification was achieved in flat piston head profile. The slight change in volume and pressure fluctuations inside the combustion chamber of other three piston head configurations were the major factors that did not contribute in increasing the overall turbulent kinetic energy and enhancing the performance of the engine.

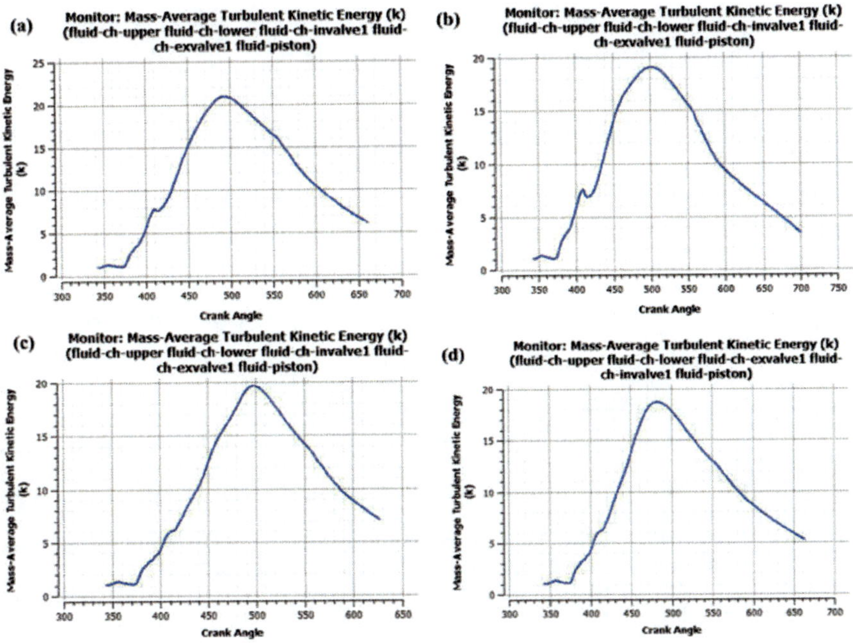


**Fig. 6** Turbulent kinetic energy for **a** flat; **b** dome; **c** concave; and **d** recessed piston

## 4 Conclusions

The cold flow simulation analysis for flat, convex, concave, and recessed type piston configurations in SI combustion engine was carried out. The vortex and turbulence formation of the fluid were studied, and the conclusions are as follows:

- At 540 crank angle degrees, the cross tumble vortex was formed for all geometry of piston; it was situated slightly away from center axis.
- The contours of velocity magnitude were the highest in flat piston head profile and the least in recessed shaped piston head profile.
- Concave and recessed piston configurations have high swirl ratio compared to other two piston configurations.
- The highest tumble ratio was achieved by the piston geometry with flat profile.
- The turbulent kinetic energies are distributed all over the combustion chamber, which includes the swirl and the cross tumble motions. The TKE of flat piston achieved higher value compared to remaining three piston configurations.

The flat piston head configuration was more efficient with better fluid dynamics and turbulence inside the combustion chamber of SI engine. Thus, it is the most suitable piston head profile for SI engine in terms of better TKE, high tumble ratio, better performance, and ease of piston manufacturing.

## References

1. Hill PG, Zhang D (1994) The effects of swirl and tumble on combustion in spark-ignition engines. Prog Energy Combust Sci 20:373–429
2. Harshavardhan B, Mallikarjuna JM (2015) Effect of combustion chamber shape on in-cylinder flow and air-fuel interaction in a direct injection spark ignition engine—a CFD analysis. SAE Technical Paper 26-0179
3. Alger T, Hall M, Matthews RD (2000) Effects of swirl and tumble on in-cylinder fuel distribution in a central injected DISI engine. SAE Paper 01-0533
4. Harshavardhan B, Mallikarjuna JM (2013) CFD analysis of in-cylinder flow and air-fuel interaction on different combustion chamber geometry in DISI engine. Int J Theor Appl Res Mech Eng 3:104–108
5. Krishna BM, Mallikarjuna JM (2009) Tumble flow analysis in an unfired engine using particle image velocimetry. World Acad Sci Eng Technol 54:430–435
6. Huang RF, Yang HS, Yeh C-N (2008) In-cylinder flows of a motored four-stroke engine with flat-crown and slightly concave-crown pistons. Exp Therm Fluid Sci 32:1156–1167
7. Liu Y, Shen Y, You Y, Zhao F (2012) Numerical simulation on spray atomization and fuel-air mixing process in a gasoline direct injection engine. SAE Technical Paper
8. Joh M, Huh KY, Noh SH, Choi KH (1999) Numerical prediction of stratified charge distribution in a gasoline direct-injection engine-parametric studies. SAE Technical Paper
9. Abdullah SH, Kurniawan WH, Shamsudeen A (2008) Numerical analysis of the combustion process in a compressed natural gas direct injection engine
10. Heywood JB (1988) Combustion engine fundamentals. 1ª Edição. Estados Unidos. 25:1117–1128
11. Stanglmaier RH, Hall MJ, Matthews DR (1998) Fuel spray/charge-motion interaction of a direct-injected 4-valve SI engine. SAE Paper 980155

12. Zhao HY, Ming PJ, Zhang WP, Liu Q, Qi WL (2020) Three-dimensional simulation of two-phase flow in a complex gallery and telescopic pipe coupled system. Appl Therm Eng 169:114918
13. Bavandla KC, Tripathi A, Sun Z, Yang S (2019) Numerical simulation of a controlled trajectory rapid compression and expansion machine. In: 11th US national combustion meeting, pp 1–10
14. Jamil A, Baharom MB, Aziz ARA (2021) IC engine in-cylinder cold-flow analysis—a critical review. Alexandria Eng J 60:2921–2945
15. Bibu B, Rajan V (2019) Numerical simulation of cold flow analysis of internal combustion engine with double-lobed piston head. In: Advances in fluid and thermal engineering. Springer, pp 657–668
16. Guo C, Zuo Z, Feng H, Jia B, Roskilly T (2020) Review of recent advances of free-piston internal combustion engine linear generator. Appl Energy 269:115084
17. Shafie NM, Said MFM (2017) Cold flow analysis on internal combustion engine with different piston bowl configurations. J Eng Sci Technol 12:1048–1066
18. Azad AK, Halder P, Nanthagopal K, Ashok B (2019) Investigation of diesel engine in cylinder flow phenomena using CFD cold flow simulation. In: Advanced Biofuels. Elsevier, pp 329–336
19. Plotnikov LV, Zhilkin BP, Brodov YM (2017) The influence of piston internal combustion engines intake and exhaust systems configuration on local heat transfer. Procedia Eng 206:80–85
20. Subramanian S, Rathinam B, Lalvani JIJ, Annamalai K (2016) iston bowl optimization for single cylinder diesel engine using CFD. SAE Technical Paper

# Combustion and Emissions

# Numerical Analysis of Heterogeneous Methanol Flames Over Porous Sphere Surfaces Using Short Kinetics Mechanism

**Sharanya Nair and Vasudevan Raghavan**

## 1 Introduction

The porous sphere technique is a simplified representation of fuel particle or droplet burning in a forced convective environment. It consists of an experimental setup with a porous sphere placed in an air flow, with fuel being supplied at the center of the sphere at the rate of its consumption in the flame. It is ensured that the entire sphere surface is uniformly wet with the fuel, and dripping of the liquid is avoided. This technique has been often used to examine the mass burning characteristics and chemical structure of flames from various liquid fuels. Two types of flames are observed to be formed based on the convective air velocity; one is an envelope flame, which surrounds the sphere, and the second is a wake flame formed in the wake region of the sphere. In spray combustion, the flames could sustain in the wake of the droplet, when droplet moves with higher velocities from the fuel injector. The flame further transitions into the envelope regime as the droplet decelerates due to drag. Spalding [1] reported a critical velocity for the air flow, at which a flame does not sustain in the upstream face of the sphere. This transitional (critical) velocity, at which there is a rapid reduction in mass loss rate, was seen to be proportional to the sphere diameter [1]. Sami and Ogasawara [2] conducted experiments on porous spheres in a heated (300–900 K) and pressurized (0.4–16 bar) air stream, up to the transitional velocity to wake regime. The mass burning rate was seen to increase significantly with pressure and temperature. The high temperature was found to increase the transitional velocity. Gollahalli and Brzustowski [3] further verified the envelope and wake regimes of flames around porous spheres. König et al. [4] presented the structure of the flame

S. Nair (✉)
Worcester Polytechnic Institute, Worcester, MA 01609, USA
e-mail: snair2@wpi.edu

V. Raghavan
Indian Institute of Technology Madras, Chennai, Tamilnadu 600036, India

 
N. L. Narasimhan et al. (eds.), *Recent Advances in Energy Technologies*, Lecture Notes in Mechanical Engineering, https://doi.org/10.1007/978-981-19-3467-4_32

over fuel fed porous sphere by measuring OH radical using Laser Induced Predissociation Fluorescence (LIPF) under micro-gravity conditions. Experimental and numerical studies on mass burning rates and flame shapes around fuel fed porous spheres were reported by Balakrishnan et al. [5]. The effects of air velocity on burning porous sphere in upward and downward configurations were analyzed, and the flame shapes and mass loss rates were reported. A numerical model with constant properties (excluding density) and infinite rate kinetics was used to simulate the experimental cases. A flattened flame front was seen for higher downward velocity, which opposed the natural convection. Raghavan et al. [6] conducted experimental and numerical studies on porous spheres in convective air to identify these two flame regimes. For the numerical model, a global single-step reaction with finite rate kinetics was used along with varying thermo-physical properties, to analyze the flame shape and mass burning rates in envelope and wake flame regimes. The predicted results were reasonably close to the experimental data in the envelope flame regime [6]. Transition velocity and wake flame features were predicted by finite rate chemistry, but not with infinite rate chemistry. An experimental study using porous sphere technique was recently conducted by Das et al. [7] to study the mass loss rate variation and transition from envelope flame to wake regimes under atmospheric conditions. A theoretical model considering mixed convective transport for mass burning rate was validated for methanol. A power law for dimensionless mass burning rate was proposed with respect to the effective Reynolds number.

Literature survey portrays that experimental results are available for burning of liquid methanol fuel over porous sphere surfaces in convective environment, in terms of flame shapes and mass burning rates in both envelope and wake regimes. The numerical models in literature have incorporated infinite rate chemistry or a global single-step reaction kinetics to model oxidation of methanol. This is sufficient to bring out the characteristics of envelope flames under different conditions, which are non-premixed flames, controlled by transport processes. However, a single-step reaction kinetics are not accurate enough to predict the characteristics of a wake flame, which is a partially premixed flame. Intermediate species and radicals, and elementary reactions involving them, are required for this purpose. Therefore, at least a short kinetics mechanism with sufficient number of species and elementary reactions is required for analysis of the flame shape and mass burning rates. Such a mechanism is reported in Tarrazo et al. [8], and this is used to validate various types of homogeneous and heterogeneous flames by Nair and Raghavan [9]. In this study, heterogeneous methanol flames over single and two porous spheres are simulated using this numerical model [9]. Detailed analysis of the characteristics of envelope and wake flames is reported. Interference effects between two equal and unequal spheres burning one over the other are also studied.

## 2 Numerical Methodology

Numerical simulations are performed using a pressure-based axisymmetric ($r$-$x$ coordinates) solver in Ansys Fluent. A short chemical mechanism [8] with 18 species ($CH_3OH$, $O_2$, O, H, OH, $H_2$, $H_2O$, $HO_2$, $H_2O_2$, CO, $CO_2$, HCO, $CH_3$, $CH_4$, $CH_2O$, $CH_3O$, $CH_2OH$, and $N_2$) and 38 elementary reactions is implemented to model the methanol oxidation. To consider radiation losses due to species such as $CH_4$, $CO_2$, CO, and $H_2O$, an optically thin radiation model [10] is incorporated in the energy equation as a negative source term using a user-defined function (UDF). Variable thermo-physical properties and full multi-component diffusion including Soret effect are included. In experiments, since methanol is fed at the rate at which it is consumed, the flames are quasi-steady in nature. For flames over a condensed fuel surface, interface coupling conditions are prescribed at the sphere surface and only the gas-phase is solved.

A non-reactive cold flow simulation is executed for a low air velocity of 0.4 m/s. The convergence criteria are set as normalized values for the residuals less than $10^{-3}$ for continuity and $10^{-4}$ for momentum and species equations. After convergence, a small region near the front stagnation point of the sphere is initialized with a temperature of 1500 K and volumetric reactions are enabled. Energy equation and volumetric reactions are solved with convergence criterion of normalized energy residual to be less than $10^{-6}$. Moreover, normalized mass imbalance ratio of net outgoing flow to the lowest incoming mass flow rate is ensured to be less than 1%. Similarly, it is ensured that the normalized heat imbalance is less than 5%. In the case where the flow is transient, unsteady cases are executed with the time step of $10^{-5}$ s, providing adequate iterations within the time step to converge these equations. Air inlet velocity is changed in small increments of 0.2 m/s, and converged reactive solutions are obtained for each velocity.

### 2.1 Computational Domain and Grid Independence

The computational domains are shown in Fig. 1a, b, for single and two porous spheres, respectively. Here $D$ is diameter of the porous sphere and $S$ is the spacing between the centers of the two spheres. Ambient air enters through the air inlet at a uniform velocity.

At the surface of a porous sphere, coupled interface conditions are used based on the energy balance and mass and momentum conservation. In Ansys Fluent, these are defined using a user-defined function (UDF) with latent heat of vaporization ($h_{fg}$) and the temperature-dependent properties of mixture density ($\rho$), mixture thermal conductivity ($k$), and binary mass diffusion coefficient, being the input parameters at the boundary. The detailed explanation of the coupled interface boundary condition including the equations has been listed below.

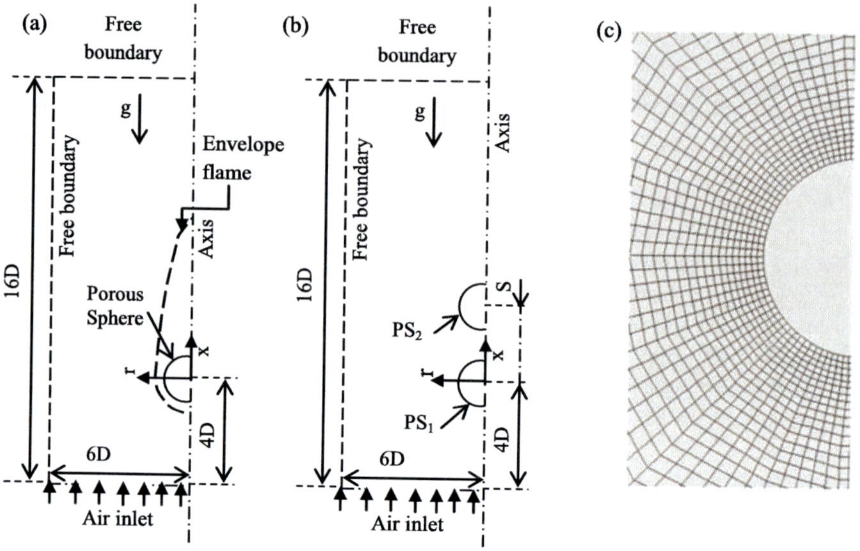


**Fig. 1** Computational domain for **a** single porous sphere, **b** two porous spheres, and **c** structured, multi-block mesh near the porous sphere surface

(i) Heat transfer by conduction from the ambient to the sphere surface is considered for calculating the liquid surface temperature, $T_s$ (surface denoted by subscript $s$). The heat addition to the sphere surface due to radiation has not been considered following literature [6, 11]. The energy conservation at the interface can be written by equating the product of mixture density, $\rho$, normal component of velocity, $v$, and latent heat of vaporization of the liquid fuel, $h_{fg}$ to the product of the normal temperature gradient at the surface and mixture thermal conductivity, $\lambda$ as shown in Eq. (1). Thus, for steady equilibrium evaporation,

$$\rho_s v_s h_{fg} = \lambda_s \left.\frac{\partial T}{\partial n}\right|_s \quad (1)$$

Here, subscript $s$ denotes the interface quantity. Using a 3-point interpolation, the temperature gradient is evaluated using surface temperature of the fuel and temperature values in two adjacent interior nodes.

(ii) The partial pressure, $p$ in mm Hg, of the methanol vapor above the fuel surface is calculated as the saturation pressure at the surface temperature using the Antoine equation. Including methanol specific constants [12], $A = 8.08097$, $B = 1582.27$, and $C = 239.7$, the Antoine equation is written as,

$$\log_{10} p = A - B/(C + T_s - 273.15) \quad (2)$$

From the partial pressure (or the mole fraction) of the species, its molecular mass, and the mixture molecular mass, the mass fraction of fuel denoted as $Y_F$ can be calculated.

(iii) The mass flux of the fuel due to evaporation is evaluated from Fick's law of ordinary diffusion as shown in Eq. (3). It considers the molecular diffusion between the fuel vapor and nitrogen, $D_{Fs}$, and the gradient of fuel mass fraction at the interface normal to the sphere surface. A 3-point interpolation is used to evaluate this derivative of mass fraction of the fuel at the surface.

$$\rho_s v_s = \rho_s v_s Y_{Fs} - \rho D_{Fs} \left.\frac{\partial Y_F}{\partial n}\right|_s \tag{3}$$

(iv) Using Fick's law, the mass fractions of all other species, denoted as $Y_i$, are calculated such that each of their net mass flux at the interface is zero. This is written as,

$$\rho_s v_s Y_i = \rho D_{is} \left.\frac{\partial Y_i}{\partial n}\right|_s \tag{4}$$

Again, 3-point interpolation is used to calculate the derivative of mass fraction of species $i$ at the interface normal to the surface. For species $i$, the binary diffusivity with respect to its diffusion with nitrogen, $D_{is}$, is calculated using Lennard–Jones parameters at the surface temperature, $T_s$. These equations are solved iteratively within a main iteration so that the output values of $T_s$, $v_s$, $Y_{Fs}$, and $Y_{is}$ at all the nodes of the interface converge to required accuracy (normalized difference between two inner iterations for each variable is within 0.1%).

The free boundaries in the domain are pressure-specified. When the flow goes out of the computational domain, the variables are extrapolated from the interior. If due to adverse pressure gradient, the flow comes into the domain, then ambient air (21% $O_2$ and 79% $N_2$) at 300 K is assumed to enter the domain.

Structured, multi-block, and quadrilateral cells have been used to divide the domain in a nonuniform manner, as shown in Fig. 1c, maintaining orthogonality of the grids at the sphere surface and boundaries. Grid independence study has been done. For the single porous sphere case, coarse (7774 cells), intermediate (10,558 cells) and fine (15,857 cells) meshes are used for free stream velocity of 0.4 m/s.

The profiles of temperature at $r = 0$ and mass fraction of oxygen and carbon-dioxide at $x = 0$ are used as the criteria and have been compared for the three meshes as shown in Fig. 2. It is seen in Fig. 2 that the profiles almost coincide as grids are refined, with the maximum normalized difference in temperature being about 2% and in mass fraction being about 2.2% and 4.2%, for $O_2$ and $CO_2$, respectively between the intermediate and fine grids. The differences between prediction by the three meshes at other axial and radial locations are seen to be smaller. Therefore, the intermediate grid has been used for the analysis.

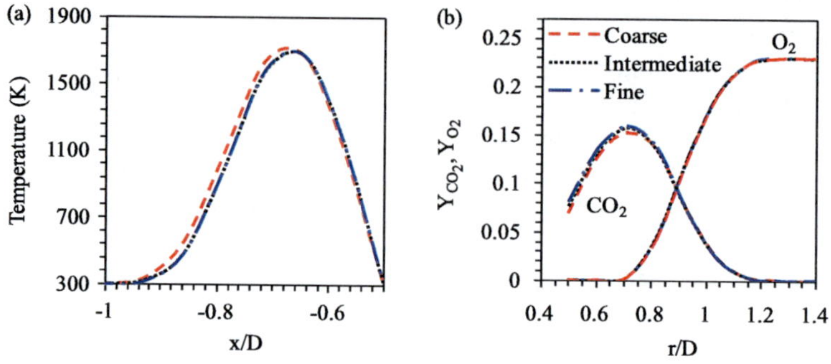


**Fig. 2** Grid independence check in terms of profiles of **a** temperature at $r = 0$ and **b** mass fraction at $x = 0$

## 2.2 Numerical Model Validation

The numerical model is validated for single porous sphere of diameter 8 mm, for which experimental data is available [6, 7]. The effects of varying the air velocity from 0.4 m/s to 1.6 m/s on the mass burning rate and the flame stand-off are shown in Fig. 3a and b, respectively. The mass burning rate is evaluated as an area integral of the mass flux along the entire sphere surface. The flame stand-off distance is evaluated as the distance from the location of the stoichiometric oxygen to methanol ratio at the axis measured from the front stagnation point of the sphere. The stoichiometric oxygen to methanol ratio is calculated from mass fractions ($Y$) as $Y_{\text{methanol}} - Y_{\text{oxygen}}/\nu = 0$, where $\nu$ kg represents the mass of oxygen required to burn 1 kg of methanol.

For the air velocity less than 1.25 m/s, a flame envelopes the sphere. In this range, with increase in air velocity, the flame stand-off is seen to move closer to the sphere surface. As the free stream velocity is further increased to a critical value (predicted as 1.25 m/s for 8 mm sphere diameter), the reaction zone moves downstream of the

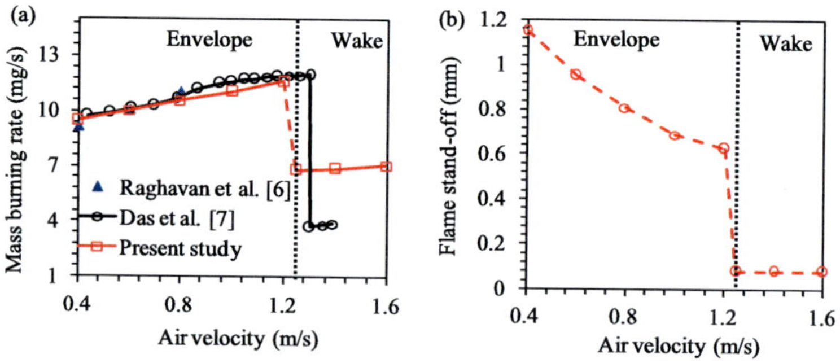


**Fig. 3** Variation of **a** predicted and measured mass burning rate and **b** predicted flame stand-off with air velocity for a sphere of 8 mm diameter; vertical line indicates transition from envelope to wake flame regime

sphere to sustain as a wake flame. The value for critical velocity is fine-tuned by reducing the air velocity increment from 0.2 m/s to 0.05 m/s.

The flame in the front portion of the sphere extinguishes due to insufficient flow residence time (Damköhler number (Da) < 1). Damköhler number is defined as the ratio of flow residence time to the chemical reaction time. The flow residence time is obtained by dividing the porous sphere diameter with the free stream air velocity. The chemical reaction time is evaluated as the ratio of concentration of the fuel to its kinetic rate. Even though fuel and oxygen are available near the front stagnation region, reactions occur only in the wake region, where sufficient time is available for them to complete. The case of envelope flame represents diffusion mode of combustion. The rate at which fuel and air mixes determines formation of the flame zone. However, the case of wake flame is like a partially premixed flame, where methanol vapor and air premix around the front stagnation point. The critical air velocity, measured by the experimental study of Das et al. [7] is 1.3 m/s and the numerical model is able to predict this as 1.25 m/s. A lower critical velocity of around 0.9 m/s has been reported for 8 mm sphere in Raghavan et al. [6] even though the mass burning rates in the envelope regime (air velocity < 0.9 m/s) reported by Raghavan et al. [6] are quite closer to those reported by Das et al. [7] (Fig. 3a).

At sub-critical air velocities, where an envelope flame is seen, the mass burning rate (MBR) increases with an increase in the free stream velocity (Fig. 3a). There is a gradual decrease in the flame stand-off distance (Fig. 3b) in the envelope regime due to increased strain caused by the increasing momentum of incoming air near the front stagnation region. At the critical velocity of 1.25 m/s, a wake flame is formed in the rear half of the sphere and a rapid reduction in the mass burning rate is observed when this transition occurs. The location of stoichiometric oxygen to methanol ratio coincides almost with the surface at the front portion of the sphere, indicating no reaction in its front part. The flame stand-off rapidly decreases to a value close to zero (Fig. 3b). As a result of sudden reduction in the flame surface area, a drastic reduction in heat transfer to the sphere surface occurs. Thus, the mass burning rate rapidly decreases as shown in Fig. 3a. The trend and the values of MBR closely follow the experimental values of Raghavan et al. [6] and Das et al. [7]. The wake flame burning rate is over-predicted since it is much difficult to measure the burning rate of wake flames due to the oscillatory nature of the flame and the difficulty in maintaining uniformly wet surface precisely. However, as the transition is predicted well, the results are considered as useful qualitative data. Dynamics of wake flame in the gas-phase, predicted by the model are also quite useful in understanding the nature of partial premixed flames. It was noticed that the rapid decrease in the value of MBR during transition to wake flame and the MBR variation with air velocity in the wake flame regime has not been predicted well by the single-step chemistry [6].

## 3 Results and Discussion

### 3.1 Flames Over Single Porous Sphere

An analysis of flames over single 8 mm porous sphere is presented in this section. The temperature contours along with mixture velocity vectors is illustrated in Fig. 4 for four air velocities. Two envelope flames, at air velocities of 0.4 m/s and 1.2 m/s, and two wake flames at air velocities of 1.25 m/s and 1.6 m/s, have been analyzed. The diffusion fluxes of $CH_3OH$ and $O_2$, as evaluated from ordinary and thermal diffusion, are shown in Fig. 5, along with the contour line of stoichiometric oxygen to methanol ratio. For an 8 mm sphere, the critical velocity is 1.25 m/s. At a velocity slightly lesser than this, say at 1.2 m/s, a clear envelope flame is formed. For 0.4 m/s and 1.2 m/s, as shown in Fig. 4a and b, respectively, an envelope flame is present over the sphere, because sufficient residence time is available for the reactions to complete in the front part of the sphere. The flow is found to curve around the sphere, and it accelerates across the flame zone. As the air velocity is increased within the critical limit, high temperature zone moves closer to the rear portion of the sphere surface, as observed by the isotherm of 1400 K. Around the front stagnation region, the temperature gradient increases due to increased momentum of the incoming air. The maximum temperature in the front stagnation region decreases as a result of higher air velocity that causes convective cooling of the flame. However, the isotherm of 1400 K has come much closer to the front stagnation point, indicating more heat transfer to the sphere surface.

In Fig. 5a, b, for envelope flame cases, fuel vapor and oxygen diffuse into the combustion zone across the stoichiometric contour line, slightly upstream of $x = -0.5D$, to mix and react. At the front stagnation point, the scenario is equivalent to that of an opposed flow flame, where the fuel vapor and air flow in direction opposite to each other. A nonlinear variation of mass flux is observed along the surface of the sphere depending on the flame stand-off distance from the fuel surface, as indicated by the diffusion fluxes. For an envelope flame, the mass flux is the highest at the front stagnation point, due to the availability of fresh air and the presence of highest gradients in that location. The fuel diffusion flux reduces along the sphere surface. The diffusion fluxes are seen to be larger for 0.8 m/s due to higher gradients caused by lower flame stand-off. The stoichiometric contour line shifts toward the front stagnation point as the free stream velocity is increased in the envelope flame regime. These result in an increasing trend of mass burning rate as the free stream velocity is increased (Fig. 3a) in the envelope flame regime.

As the fuel and air velocities are increased, the flame zone becomes thinner due to increased strain rate at the front stagnation point. At the critical air velocity, the strain rate attains a critical value. As a result, the flame extinguishes in the front portion, forming a wake flame (Fig. 4c,d). At 1.25 m/s, when a wake flame sustains in the rear half of the sphere, there is a rapid reduction in the mass loss rate, as shown in Fig. 3a. The stoichiometric contour line almost coincides with the surface at the front portion of the sphere (Fig. 5c,d). In a wake flame, fuel vapors produced in small quantities

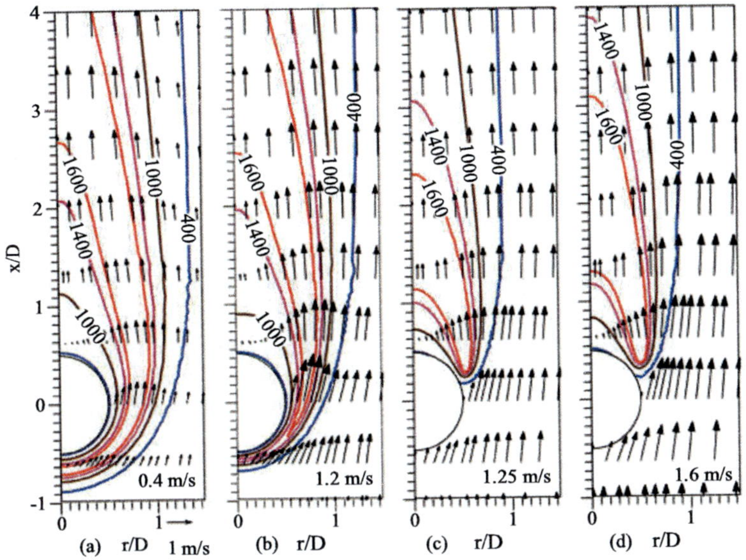


**Fig. 4** Temperature contours with mixture velocity vectors for air velocity of **a** 0.4 m/s, **b** 1.2 m/s, **c** 1.25 m/s, and **d** 1.6 m/s; maximum temperature in the domain are **a** 1816 K, **b** 1779 K, **c** 1846 K, and **d** 1934 K

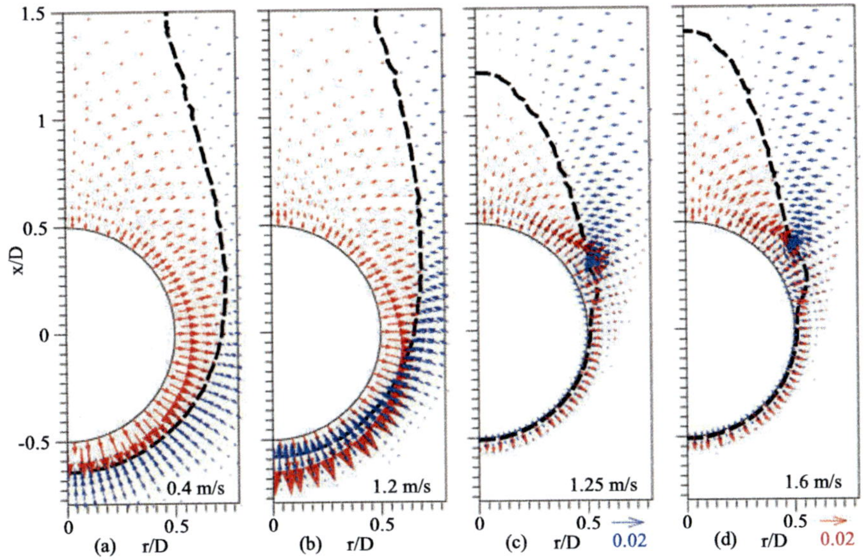


**Fig. 5** Stoichiometric contour line (dash line) with vectors of diffusion fluxes (in kg/m$^2$-s) of $CH_3OH$ and $O_2$ for air velocity of **a** 0.4 m/s, **b** 1.2 m/s, **c** 1.25 m/s, and **d** 1.6 m/s

in the front part of the sphere are transported to the wake region by the convective air. This results in partial premixing of the reactants. As the free stream velocity is increased in the wake flame regime, the extents of maximum temperature increase, and the flame radius also slightly increases, as shown by the isotherms of 1600 K and 1000 K, in Figs. 4c,d. In Fig. 5c,d, oxygen penetration and partial premixing are observed around $x = 0.25D$, where the flame base sustains. The highest mass flux is located somewhere in the rear half of the sphere, around the flame base. It is also clear that the fuel vapor diffusion occurs at a lower rate in wake flames and the fuel evolved in the front portion of the sphere is carried toward the rear portion as indicated by the mixture velocity vectors, where it burns.

In Fig. 6, contours of the net consumption rate of $CH_3OH$ (kg/m$^3$-s) are shown in grayscale along with line contours of $H_2O$ and OH mass fractions, for the four air velocities. In Fig. 6a,b, the $H_2O$ mass fraction lines of 0.13 envelope the sphere, and the major fuel consumption zone is visible near the front stagnation of the sphere. The maximum rate of consumption of $CH_3OH$ increases from 31 kg/m$^3$-s to 55 kg/m$^3$-s, when air velocity is increased from 0.4 m/s to of 1.2 m/s.

Reaction rate being dependent on the concentrations, due to increased concentrations of methanol vapor (higher burning rate) and oxygen (higher air velocity), a higher consumption rate of $CH_3OH$ is observed. Two pathways for $H_2O$ formation from $CH_3OH$ and OH are

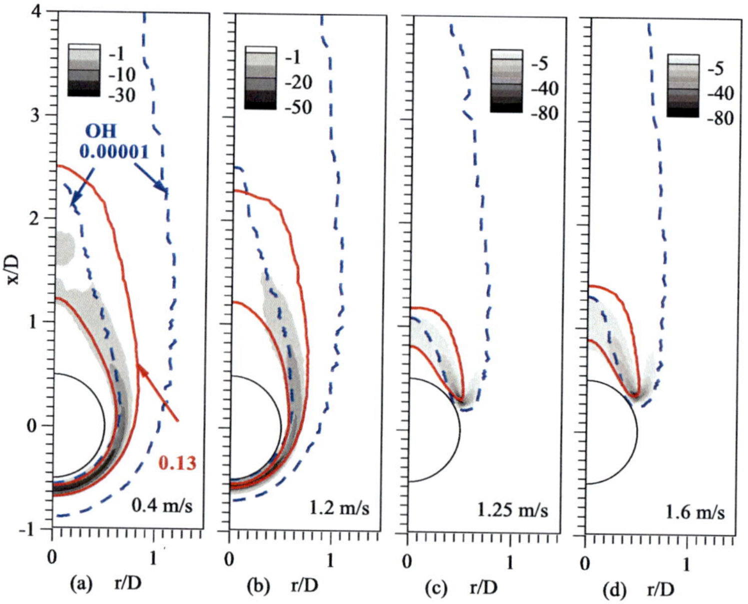


**Fig. 6** Contours of net kinetic rate of $CH_3OH$ (negative indicates consumption) with mass fraction lines of OH (dash line) and $H_2O$ (solid line) at **a** 0.4 m/s, **b** 1.2 m/s, **c** 1.25 m/s, and **d** 1.6 m/s

$$CH_3OH + OH \Leftrightarrow CH_2OH + H_2O$$
$$CH_3OH + OH \Leftrightarrow CH_3O + H_2O.$$

These equations are included in the short mechanism. The inner OH mass fraction line with a much low value of 0.00001, which shows its consumption, delineates the consumption rate of $CH_3OH$ to form $H_2O$ in the region. The reaction zone is bound by $H_2O$ mass fraction contour line of 0.13, showing its formation in this region. The OH and $H_2O$ mass fraction lines are seen to shift closer to the sphere surface, at higher air velocity.

At 1.25 m/s, in the wake flame regime, the net consumption rates of $CH_3OH$ along with line contours of mass fractions of $H_2O$ and OH are as shown in Fig. 6c,d. Clearly, the consumption of methanol occurs only in the wake region, although some amount of methanol vapor forms the front portion of the sphere to mix with air. In Fig. 6c,d, two clear branches of methanol consumption are observed. One, branching radially inwards, corresponds to the non-premixed combustion, where the additional methanol vapor from the rear sphere surface mixes with oxygen and burns. The other corresponds to the combustion of partially premixed methanol vapor and air transported from the front part of the sphere. Due to the higher air velocity, the partially premixed mixture is able to burn only at the wake region, where Damköhler number is close to unity. The consumption of OH, as shown by the inner mass fraction line with a value of 0.00001, coincides with the $CH_3OH$ consumption zones. The maximum consumption rate of methanol has increased to 84 kg/m$^3$-s. This is due to effect of partial premixing in the front portion of the sphere.

### 3.2 *Flames Over Two Porous Spheres in Tandem Arrangement*

Interaction of flames from multiple porous sphere surfaces presents interesting features and can reveal the capability of the numerical model with short reaction mechanism to predict such flame interactions. For this, three case studies have been carried out for flames over two porous spheres of: (case 1) equal diameter ($D_1 =$ 8 mm, $D_2 = 8$ mm, and $S = 2D_1 = 16$ mm), (case 2) unequal diameter ($D_1 = 4$ mm, $D_2 = 8$ mm, and $S = 2D_2 = 16$ mm), and (case 3) unequal diameter ($D_1 = 8$ mm, $D_2$ $= 4$ mm, and $S = 2D_1 = 16$ mm). Here, $D_1$ and $D_2$ represent the diameters of porous spheres, $PS_1$ and $PS_2$, respectively [Fig. 1b]. Considering integer multiples of the sphere diameter, separation distance of 2D creates maximum interference, without making the spheres touch each other.

For free stream velocity of 0.4 m/s, the contours of temperature (lines) and OH mass fraction (grayscale) along with mixture velocity vectors are illustrated in Fig. 7 for two spheres burning in the envelope flame regime. For the same cases, line contours of oxygen mass fraction and stoichiometric contour line (dashed line) are shown in Fig. 8. When a larger or equal-sized sphere burns in the bottom, a single

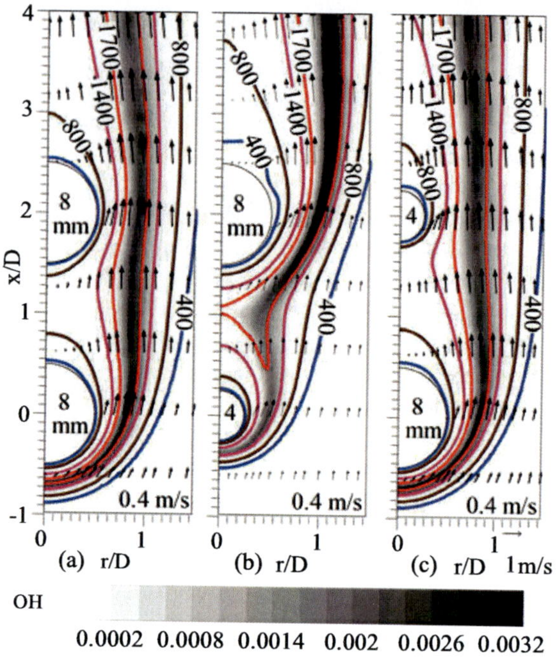


**Fig. 7** Temperature and OH mass fraction (grayscale) contours with mixture velocity vectors for air velocity of 0.4 m/s for the case of **a** 8 mm–8 mm, **b** 4 mm–8 mm, and **c** 8 mm–4 mm

envelope flame is observed around the two spheres as indicated by the 1400 K isotherms in Fig. 7a,c and the stoichiometric contour lines in Fig. 8a,c. The mixture velocity vectors are seen to accelerate across the flame (high temperature) zone due to the buoyancy. For the top sphere, it is clear that oxygen is unable to reach its front portion, as indicated by oxygen contour line of 0.02. This is the reason for the absence of individual envelope flame in the top sphere.

In the case of smaller sphere (4 mm diameter) burning at the bottom (Fig. 7b), distinct 1400 K isotherms are observed slightly upstream of the front face for both top and bottom spheres. Deeper oxygen penetration in the wake of the bottom 4 mm sphere is shown by the oxygen contour line of value 0.02 (Fig. 8b). Also, a high temperature zone of 1700 K is clearly visible (Fig. 7b) in between the two spheres, around $x = 1D$, indicating the presence of individual envelope flames around the two spheres. This is ascertained by the stoichiometric contour line in Fig. 8b, which shows that stoichiometric mixing happens in the front face of the 8 mm sphere, in the wake of the 4 mm sphere. The stand-off distances of stoichiometric lines at the front stagnation point for unequal spheres are lesser than that in the equal spheres.

In the envelope flame regime, the mass burning rate (MBR) of the bottom sphere ($PS_1$) increases with the air velocity, as shown in Fig. 9a–c, because of the movement of the high temperature zone closer to the front stagnation point of the bottom sphere, enhancing the heat and mass transfer processes, like in the isolated sphere case. The trend for flame stand-off location, as evaluated from the location of the stoichiometric line with respect to the front stagnation point of the bottom sphere, shows a gradual

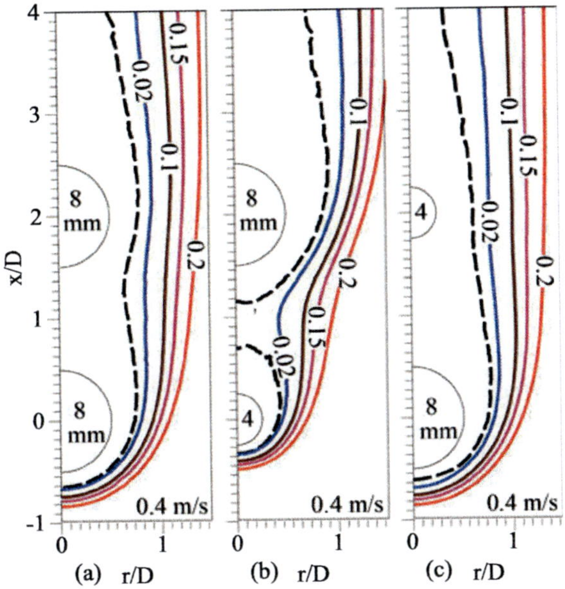


**Fig. 8** Oxygen mass fraction contours with stoichiometric line (dash line) for air velocity of 0.4 m/s for the case of **a** 8 mm–8 mm, **b** 4 mm–8 mm, and **c** 8 mm–4 mm

decrease with an increase in the air velocity (Fig. 9d). MBR (Fig. 9a) and flame stand-off (Fig. 9d), in the bottom sphere for 8 mm–8 mm (case 1), are seen to almost coincide with those of the single sphere case, until the transition occurs for the single sphere case. The presence of equal diameter sphere in its wake does not interfere with the flow field in the front part of its surface. For the bottom sphere in 4 mm–8 mm (case 2), shown in Fig. 9b, MBR of the bottom 4 mm sphere ($PS_1$) is lower than that of the isolated 8 mm sphere, because of its smaller surface area. MBR is seen to increase slightly with an increase in the air velocity until the transition occurs.

For the 8 mm–4 mm (case 3), the variation trend of MBR of bottom 8 mm sphere is similar to that of the single sphere. However, the presence of the smaller sphere in the wake of the larger sphere leads to a flow field that facilitates higher rate of burning (Fig. 9c) for the 8 mm sphere burning at the bottom, when compared to that of the isolated 8 mm sphere. The flame stand-off at each air velocity is seen to be lower for this case as compared to the single sphere case (Fig. 9d).

In the envelope flame regime of the bottom sphere, the mass burning rates of the top spheres are discussed subsequently. The 8 mm sphere at the top in case 1 burns almost like a single sphere having only a wake flame over its rear surface (Fig. 9a). This is because of non-availability of oxygen to its front portion. For 4 mm–8 mm (case 2), MBR is comparable to that of an isolated 8 mm sphere. It has a higher value at 0.4 m/s as compared to that at 0.6 m/s (Fig. 9b). This is due to the presence of individual envelope flames for the two spheres ensuring high temperature gradient near the front face of the top sphere at the velocity of 0.4 m/s. MBR slightly decreases at 0.6 m/s because of the formation of single flame front around both spheres. MBR beyond free stream velocity of 0.6 m/s is seen to almost overlap with that of the single sphere

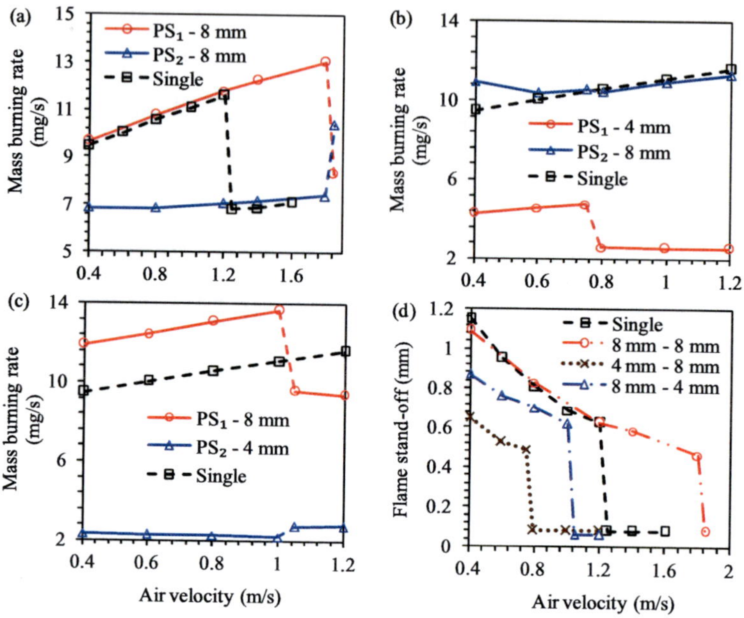


**Fig. 9** Mass burning rate variation with air velocity for the case of **a** 8 mm–8 mm, **b** 4 mm–8 mm, **c** 8 mm–4 mm, and **d** flame stand-off variation with air velocity for the bottom sphere

case, indicating lesser interference from the small flame over the smaller sphere at the bottom. For the top sphere in 8 mm–4 mm (case 3), the mass burning rate is much lower (Fig. 9c), since it has a smaller diameter (smaller surface area), and it burns in the wake region of the 8 mm sphere that shadows the incoming flow field with fresh oxygen.

Due to the presence of either an equal or an unequal sphere at the wake region, transition to wake flame occurs for the bottom sphere at a different air velocity when compared to that of isolated sphere case. In the wake regime, contours of temperature (lines) along with mixture velocity vectors and OH mass fraction (grayscale) at the critical velocity for each of the two sphere cases are shown in Fig. 10.

The oxygen mass fraction contour lines and the stoichiometric line (dashed line) for the same cases are portrayed in Fig. 11. For 8 mm–8 mm (case 1), transition to wake flame occurs at a critical velocity of 1.85 m/s (Fig. 10a), which is notably higher than the isolated sphere (1.25 m/s). Isotherm of 1700 K, bounding the OH mass fraction contours, is observed close to the top sphere. As opposed to the wake flame in the single sphere (Fig. 4c), the wake flame in this case sustains at a location farther from the rear point of the bottom sphere, due to higher air velocity that pushes the flame zone further downstream. Due to the occurrence of flame transition for the bottom sphere, oxygen penetration is observed near the front half of the top sphere indicating the formation of the flame zone. Thus, its MBR increases rapidly as shown in Fig. 9a.

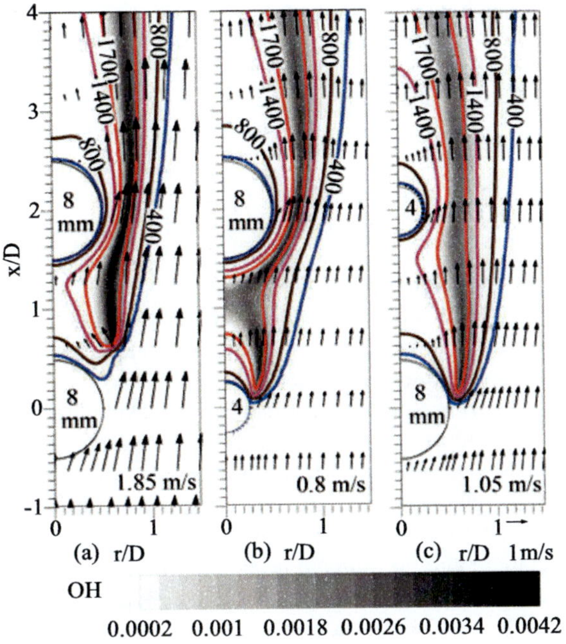


**Fig. 10** Temperature and OH mass fraction (grayscale) contours with mixture velocity vectors at critical velocity for the case of **a** 8 mm–8 mm, **b** 4 mm–8 mm, and **c** 8 mm–4 mm

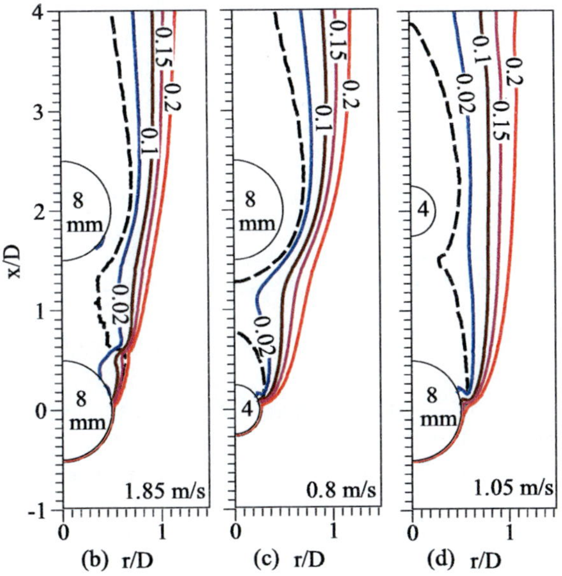


**Fig. 11** Oxygen mass fraction contours with stoichiometric line (dash line) at critical velocity for the case of **a** single sphere, **b** 8 mm–8 mm, **c** 4 mm–8 mm, and **d** 8 mm–4 mm

At 0.8 m/s, for 4 mm–8 mm (case 2), transition occurs for 4 mm sphere burning at the bottom. Critical velocity of transition is dependent on sphere diameter and interference effects. An individual envelope flame is seen around the top 8 mm sphere. The wake flame of the bottom sphere just merges with the base of the envelope flame over the top sphere in the spacing between the two spheres, as denoted by the OH mass fraction contours and 1700 K isotherm (Fig. 10b). This is also indicated by distinct stoichiometric lines for bottom and top spheres in Fig. 11b. In the wake flame, oxygen penetration is seen near $x = 0.2D$, as shown by oxygen mass fraction line of 0.02. The stoichiometric line is seen to move closer toward the top sphere with an increase in air velocity from 0.4 m/s (Fig. 8b) to 0.8 m/s (Fig. 11b). As a result, MBR increases with air velocity for the top sphere (Fig. 9b), comparable to that of the isolated sphere, with a least interference from the smaller sphere burning in wake regime at the bottom.

The transition to the formation of a wake flame for the bigger sphere at the bottom is delayed in 8 mm–4 mm (case 3) and occurs at 1.05 m/s (Figs. 10c and 11c), as opposed to the case 2. However, this is a lower velocity when compared to that of the single sphere case. At 1.05 m/s, a single flame front is observed in the wake regime of the bottom sphere, enveloping the top sphere, as portrayed by the OH mass fraction contours and the stoichiometric line in Fig. 10(c). Oxygen penetration is observed near $x = 0.25D$ for the bottom sphere following the $O_2$ mass fraction line of 0.02 in Fig. 11c. Individual envelope flame does not form around the 4 mm sphere in the wake regime due to oxygen starvation, unlike the wake flame shown in Fig. 11b.

At the critical velocity, there is a rapid decrease in the flame stand-off distance (Fig. 9d) due to absence of the flame in the front half of the bottom sphere. MBR of the bottom sphere reduces rapidly (Fig. 9a–c). For the top sphere, there is an increase in MBR with the air velocity, after the occurrence of wake flame transition in the bottom sphere, due to an increased oxygen penetration to the front portion of the top sphere and existence of high temperature zone near that region as shown in Figs. 10 and 11. Reaction zones are analyzed subsequently (Fig. 12).

The net consumption rate of $O_2$ in kg/m$^3$-s is shown as grayscale contours in Fig. 12, for envelope flame cases, at an air velocity of 0.4 m/s. $CO_2$ primarily forms from OH through the chain propagation reaction,

$$CO + OH \Leftrightarrow CO_2 + H$$

The mass fraction contour line of $CO_2$ that branch out from the maximum consumption zone of $O_2$ has a value of 0.14. This shows the bounds where $CO_2$ is formed. Similarly, OH mass fraction lines of 0.00001 (low value) bound the $O_2$ consumption zone. Major oxygen consumption is observed near the front stagnation of the bottom sphere for all the cases. A single continuous consumption zone for $O_2$ around both the spheres is seen Fig. 12a and c, due to shadowing of the incoming air flow by the bottom 8 mm sphere, reducing the oxygen availability near the front portion of the top sphere. Merging of the $O_2$ consumption zones is observed near the rear of the bottom 4 mm sphere in Fig. 12b.

**Fig. 12** Contours of net kinetic rate of $O_2$ (negative indicates consumption) with mass fraction lines of $CO_2$ (solid line) and OH (dash line) for air velocity of 0.4 m/s for the case of **a** 8 mm–8 mm, **b** 4 mm–8 mm, and **c** 8 mm–4 mm

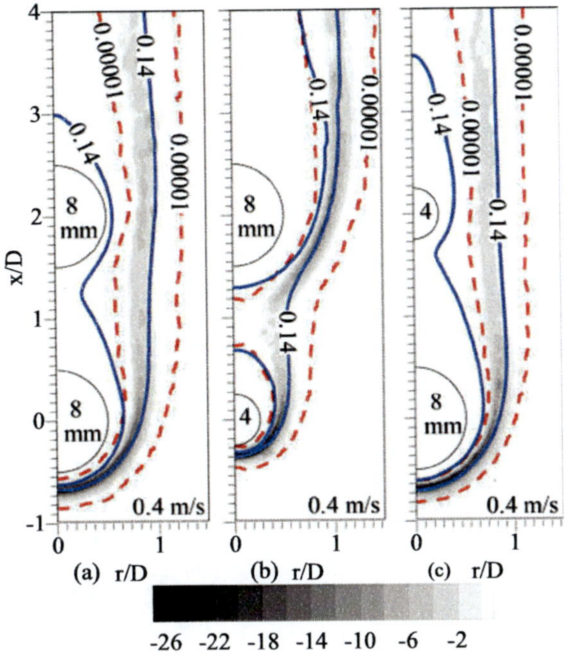


In the wake regime, the net consumption rates of $CH_3OH$ in kg/m$^3$-s (grayscale) for the three cases are portrayed in Fig. 13, at their respective critical velocities. Two-branched consumption zone of methanol, as discussed in the single sphere case, is observed, due to partial premixing occurring in the rear part of the bottom sphere in all the cases. The mass fraction contour lines of 0.13 for $H_2O$ and 0.00001 for OH are seen to branch out from the maximum fuel consumption zone. Individual reaction zones are seen in the wake region of smaller bottom sphere and enveloping the 8 mm top sphere as in Fig. 13b. Merging of reaction zones of bottom and top spheres is seen in Fig. 13c for the 8 mm–4 mm configuration. OH mass fraction line of 0.00001 clearly bounds the maximum consumption and the outer branch of methanol consumption.

## 4 Conclusions

Heterogeneous methanol flames over porous sphere surfaces are simulated in Ansys Fluent in the laminar mixed convection regime. Fuel vaporization using interface boundary conditions is predicted using a user-defined function. A short reaction mechanism, (18 species and 38 elementary reactions) is incorporated to model the

**Fig. 13** Contours of net consumption rate of $CH_3OH$ with mass fraction lines of $H_2O$ (solid line) and OH (dash line) at critical velocity for the case of **a** 8 mm–8 mm, **b** 4 mm–8 mm, and **c** 8 mm–4 mm

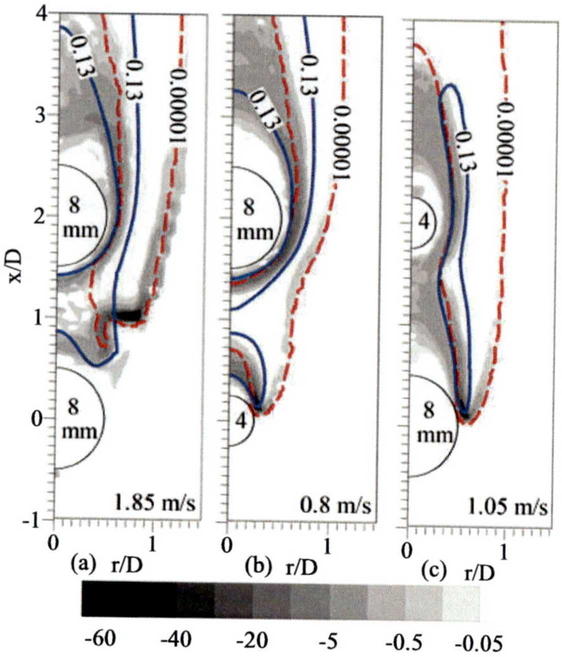


kinetics. An optically thin radiation model accounts for the radiation losses of the participating species.

Two regimes of burning, dependent on sphere diameter and free stream velocity, are predicted by the model close to experimental data from literature at appropriate air velocities. Values of mass burning rates and its increase with an increase in the air velocity has been predicted quite reasonably by the model. Causation for this trend in terms of movement of stoichiometric contour line has been discussed. Critical velocity of flame transition from front part of the sphere to its rear part has been predicted reasonably closer to the experimental value for a single sphere case. In this case, the reactant mixture moves toward the rear portion of the sphere to form a partially premixed reaction zone in addition to the non-premixed flame zone. Branched reaction zones of methanol have been illustrated in the wake flame regime. There is a rapid reduction in the mass loss rate due to the absence of the flame in the front half of the porous sphere.

In the case of burning of two porous spheres in tandem arrangement, the resultant flow field as prescribed by the interaction between the two spheres, plays an important role in the determining the oxygen availability, especially for the sphere in the wake region, apart from dictating the overall flow field around both the spheres. The interference effects of the convective burning of two spheres and the resultant flame characteristics have been discussed with evidence of flow, temperature, species, and reaction fields, and compared with those of a single sphere. In the case where the bottom porous sphere shadows the top sphere from the incoming flow, especially when the bottom sphere has a same or bigger diameter, oxygen deficiency is observed near the front half of the top porous sphere, resulting in lower mass burning rate. The values for the mass burning rates of the top sphere in this case are seen to be of the same order as that for a single porous sphere burning in the wake regime. Individual envelope flames for each of the spheres are not seen in this case and a single flame front forms.

The critical velocity of transition from envelope to wake flame is dependent on the arrangement of the two porous spheres and their interaction. When a wake flame is formed for the bottom sphere, its mass burning rate reduces drastically. Partial premixing in the rear half of the bottom sphere is observed. However, the mass burning rate for the top sphere increases due to the availability of oxygen and the presence of high temperature zone near its front part. The consumption zones of fuel and oxygen and production of $CO_2$ and $H_2O$ through intermediates such as OH have been depicted for the different configurations. With the help of temperature, flow, species, and reaction fields, interpretations of the physical phenomena, which affect the flame structure and mass burning characteristics, have been explained.

**Acknowledgements** The authors thank P.G. Senapathy Center, IIT Madras for providing computing resources.

## References

1. Spalding DB (1953) The combustion of liquid fuels. Symp (Int) Combust 4:847–864
2. Sami H, Ogasawara M (1970) Study on the burning of a fuel drop in heated and pressurized air stream: 1st report, experiment. Bull JSME 13(57):395–404
3. Gollahalli SR, Brzustowski TA (1973) "Experimental studies on the flame structure in the wake of a burning droplet. Symp (Int) Combust Combust Inst 14(1):1333–1344
4. König J, Eigenbrod C, Tanabe M, Renken H, Rath HJ (1996) Characterization of spherical hydrocarbon fuel flames: laser diagnosis of the chemical structure through the OH radical. Symp (Int) Combust 26(1):1235–1242
5. Balakrishnan P, Sundararajan T, Natarajan R (2001) Combustion of a fuel droplet in a mixed convective environment. Combust Sci Technol 163(1):77–106
6. Raghavan V, Babu V, Sundararajan T, Natarajan R (2005) Flame shapes and burning rates of spherical fuel particles in a mixed convective environment. Int J Heat Mass Transf 48(25–26):5354–5370
7. Das M, Chakraborty A, Datta A, Santra AK (2017) Experimental studies on burning characteristics of methanol, diesel, and sunflower biodiesel fuels. Combust Sci Technol 189(2):213–230
8. Tarrazo EF, Sanz MS, Sánchez AL, Williams FA (2016) A multipurpose reduced chemical-kinetic mechanism for methanol combustion. Combust Theor Model 20(4):613–631
9. Nair S, Raghavan V (2019) Numerical study of methanol flames in laminar forced convective environment using short chemical kinetics mechanism. Combust Theory Model 24(2):279–306

10. Barlow RS, Karpetis AN, Frank JH, Chen JY (2001) Scalar profiles and NO formation in laminar opposed-flow partially premixed methane/air flames. Combust Flame 127(3):2102–2118
11. Marchese AJ, Dryer FL (1997) The effect of non-luminous thermal radiation in microgravity droplet combustion. Combust Sci Technol 124(1–6):371–402
12. Dortmund Data Bank (DDBST GmbH) http://ddbonline.ddbst.com/AntoineCalculation/AntoineCalculationCGI.exe?component=Methanol

# A Study of Structure and Entropy Generation in Confined Biogas Coflow Diffusion Flames

**R. Nivethana Kumar, S. Muthu Kumaran, and V. Raghavan**

## 1 Introduction

Biogas is synthesized from the anaerobic digestion of organic, sewage and animal waste. Due to its ease of availability, low cost and emission characteristics, biogas is considered as a good alternative fuel. Biogas is chiefly composed of $CH_4$ and $CO_2$, besides small amount of nitrogen, oxygen, water vapour and other trace gases. The presence of carbon dioxide varies from 30 to 45% by volume. The amount of $CO_2$ strongly affects the flame characteristics, combustion efficiency and stability of biogas flames [1]. There are several methodologies reported in the literature to improve the combustion efficiency and stability of biogas, such as preheating the reactants, partial premixing of fuel-oxidizer mixture, addition of hydrogen and exhaust gas recirculation. Mahallawy et al. [2] observed the partial premixed LPG-air flame to be more stable than diffusion and premixed flames. Also, there was an optimum level of primary air in the mixture, beyond which, the flame was less stable. Sahu et al. [3] studied partially premixed methane flame by varying the equivalence ratio of mixture. An increase in flame height and soot formation was seen with an increase in equivalence ratio. Cha and Chung [4] investigated the characteristics of lifted turbulent flames from confined jets. The lift off height varied proportionally with the nozzle diameter and flow velocity in confined jets, whereas it was independent of nozzle diameter for unconfined jets. Karbasi and Wierzba [5] analysed the stability limits of hydrogen added to methane flames in confined coflow burner. It was reported that the flame stability in confined burner depends upon the velocities of coflow stream and fuel jet, and its diameter. Hutchins et al. [6] studied the features of confined turbulent lifted methane flames and reported that fully confined burner provides stable flame than an unconfined burner at high jet velocities. Baudoin et al. [7] studied the effect of primary air on the stabilization of methane flames. It was found that the optimum

R. Nivethana Kumar · S. Muthu Kumaran (✉) · V. Raghavan
Department of Mechanical Engineering, Indian Institute of Technology Madras, Chennai, India
e-mail: kumaranms92@gmail.com



N. L. Narasimhan et al. (eds.), *Recent Advances in Energy Technologies*, Lecture Notes in Mechanical Engineering, https://doi.org/10.1007/978-981-19-3467-4_33

level of partial mixing can improve combustion stability. The combustion efficiency of the system can be quantified by evaluating entropy generated rate in the system. Ismail et al. [8] observed that the entropy generation of premixed fuel–air mixture increases proportionally with an increase in equivalence ratio in micro porous media burner.

It is clear from literature that partial premixing and presence of confinement have considerable effects on the flame characteristics. Analysing the stability and characteristics of confined biogas flames is important for utilizing biogas in large industrial burners and furnaces where it is burnt in a confined manner. Investigating the entropy generation due to various processes in these flames is essential for identifying the optimum conditions for burning biogas. Such data are not available in the literature and forms the motivation of the present study.

This work mainly focusses on studying the effect of partial premixing in a confined coflow burner with varying rate of coflow air fed around core biogas jet. In the first set of cases, the primary air in the fuel stream has been increased from 0 to 40% of the required stoichiometric air. Here, the coflow air stream has been kept at the stoichiometric value corresponding to the fuel flow rate. Thus, in this set of cases, the total air flow into the burner is more than the required stoichiometric value when primary air is fed. In the next set, the primary air is varied as before, however, the coflow rate is kept such that the total air fed to the burner is equal to the stoichiometric value corresponding to the fuel flow rate. The entropy generation during chemical reactions and during transport processes such as heat transfer, mass transfer and coupled heat and mass transfer is analysed for all the cases presented above. The flame stability and characteristics have been systematically analysed through flow, temperature, species and entropy generation fields.

## 2 Numerical Methodology

Numerical simulations are carried out using *OpenFOAM*. The governing equations for laminar reactive flow are given below

*Mass conservation equation*

$$\frac{\partial \rho}{\partial t} + \nabla.(\rho \vec{V}) = 0 \tag{1}$$

*Momentum conservation equation*

$$\frac{\partial (\rho \vec{V})}{\partial t} + \nabla.(\rho \vec{V} \vec{V}) = -\nabla p + \nabla.\vec{\tau} + \rho \vec{g} \tag{2}$$

*Species equation*

$$\frac{\partial(\rho Y_i)}{\partial t} + \nabla.(\rho \vec{V} Y_i) + \nabla.\vec{J_i} = \dot{\omega}_i''' \quad (3)$$

*Energy equation*

$$\frac{\partial(\rho h)}{\partial t} + \nabla.(\rho \vec{V} h) = \nabla.(k\nabla T) - \nabla.\left[\sum_{i=1}^{n} h_i \vec{J_i}\right] - \sum_{i=1}^{n} \dot{\omega}_i''' \Delta h^o_{f,i} + \dot{q'''}_{rad}. \quad (4)$$

Here, $\rho$ is the mixture density (kg/m$^3$), $\vec{V}$ is the mixture velocity (m/s), $\vec{\tau}$ is the fluid stress tensor (N/m$^2$), $p$ is the pressure (N/m$^2$), $\vec{g}$ is the gravitational acceleration (m/s$^2$), $Y_i$ is the mass fraction of species $i$, $\vec{J_i}$ is the species diffusion flux (kg/m$^2$-s), $\dot{\omega}_i'''$ is the net reaction rate of species $i$ (kg/m$^3$-s), $h$ is the enthalpy (J/kg), $T$ is the temperature (K), $\Delta h^o_{f,i}$ is the enthalpy of formation of species $i$ (J/kg), $q_{rad}$ is the radiative heat loss term per unit volume (W/m$^3$) and k is the thermal conductivity (W/m–K).

An optically thin approximation-based model is used to calculate the thermal radiation absorbed by species such as $CH_4$, CO, $CO_2$ and $H_2O$ [9]. Heat energy radiated per unit volume (*W/m³*) is included as a sink term in energy equation and expressed as

$$\dot{q'''}_{rad} = 4\sigma(T^4 - T_b^4)\sum(p_i \times a_{pi}) \quad (5)$$

The species diffusion flux, $\vec{J_i}$, is calculated as

$$\vec{J_i} = -\rho D_{i,m} \nabla Y_i - D_{T,i}\frac{\nabla T}{T}. \quad (6)$$

Here, $D_{i,m}$ is the multi-component mass diffusion coefficient (m$^2$/s) for species $i$ in the mixture and $D_{T,i}$ is the thermal diffusion coefficient (kg/ms). The mass diffusion coefficient, $D_{i,m}$, is determined from the binary mass diffusivity (m$^2$/s), $D_{ij}$, which governs the diffusion between any two species '*i*' and '*j*'. The binary mass diffusivity, $D_{ij}$, calculated using Chapman–Enskog theory [10] as

$$D_{i,m} = \frac{1 - X_i}{\sum_{j \neq i}\left(\frac{X_j}{D_{ij}}\right)} \quad (7)$$

The thermal diffusion coefficient is determined as [11]

$$D_{T,i} = -2.59 \times 10^{-7} T^{0.659}\left[\frac{M_{w,i}^{0.511} X_i}{\sum_{i=1}^{n} M_{w,i}^{0.511} X_i} - Y_i\right]\left[\frac{\sum_{i=1}^{n} M_{w,i}^{0.511} X_i}{\sum_{i=1}^{n} M_{w,i}^{0.489} X_i}\right]. \quad (8)$$

Here, $M_{w,i}$ is the molecular weight of species $i$ (kg/kmol) and $X_i$ is mole fraction of species $i$.

The reacting flow problems in *OpenFOAM* are solved using a module named *reactingFoam*. The chemical kinetics scheme is obtained after eliminating all higher order hydrocarbons apart from $C1$ and $C2$ species and their associated reactions from GRI MECH 2.11. The final short mechanism with 25 species distributed among 121 elementary reactions is used in this work. The fuel and air are allowed to mix until required convergence is achieved, and then, the reactant mixture is ignited just above the burner rim by patching a high temperature zone of 1500 K to initiate a flame. Steady-state simulations are run for all the cases. Convergence is achieved when the normalized residuals fall below $10^{-6}$ for all solution variables.

The total entropy generated in a reactive flow is obtained as a sum of entropy generation during processes such as heat transfer ($S_h$), mass transfer ($S_m$), coupling between heat and mass transfer ($S_c$), chemical reactions ($S_r$) and viscous dissipation ($S_v$). The total rate of entropy generation ($S_t$) of the system is given as (in W/m$^3$-K) [12]

$$S_t = S_h + S_m + S_c + S_r + S_v \tag{9}$$

The entropy generation rate during viscous dissipation ($S_v$) is usually negligible [12] and hence neglected.

Entropy generated during transport of heat, per unit volume, is calculated as

$$S_h = \frac{k(\nabla T)^2}{T^2} \tag{10}$$

Entropy generated during transport of mass, per unit volume, is expressed as

$$S_m = \frac{-\rho R_u}{\overline{W}} \sum_i v_i.\nabla X_i. \tag{11}$$

Here, the ordinary mass diffusion velocity ($v_i$) is given as

$$v_i = \frac{-D_{i,m}}{Y_i} \nabla Y_i, \tag{12}$$

which satisfies the constraint

$$\sum_i Y_i v_i = 0. \tag{13}$$

Here, $R_u$ is the universal gas constant (kJ/kmol-K) and $\overline{W}$ is the average molecular weight (kg/kmol). Entropy generated during coupled heat and mass transfer processes, per unit volume, is evaluated as

$$S_m = \frac{-\rho R_u}{\overline{W}T} \sum_i w_i.\nabla X_i \tag{14}$$

Here, the thermal diffusion velocity ($w_i$) is given by

$$w_i = \frac{-D_{T,i}}{\rho Y_i} \frac{\nabla T}{T}, \tag{15}$$

subjected to the constraint

$$\sum_i Y_i w_i = 0 \tag{16}$$

Entropy generated during chemical reactions, per unit volume, is given as

$$S_r = \frac{-1}{T} \sum_i \mu_{c,i} \dot{\omega}_i'''. \tag{17}$$

Here, $\mu_{c,i}$ is the chemical potential calculated as $\mu_{c,i} = h_i - Ts_i$ ($h_i$ and $s_i$ are the enthalpy (J/kg) and entropy (J/kg-K) of $i$th species, respectively).

## 2.1 Computational Domain

The coflow burner is made of two concentric tubes in which the fuel and air flows at prescribed velocities, temperature and composition of species. The dimensions for coflow burner are obtained from a similar burner used in Mokhov et al. [13]. The fuel tube is made of stainless steel and has an inner diameter of 5 mm and thickness 1.5 mm. The annular coflow tube has an inner diameter of 95 mm. The computational domain is shown in Fig. 1.

## 2.2 Boundary Conditions

(1) Fuel inlet: A uniform value for fuel velocity corresponding to the power rating of the burner, and temperature of 300 K, is specified. Appropriate mass fraction of fuel components ($CH_4$ and $CO_2$) and air (for partial premixing cases) is also specified at this boundary.
(2) Coflow air inlet: Air velocity corresponding to the required air flow rate (stoichiometric value required for the mass of fuel fed, as in the first set of cases, or stoichiometric air minus the primary air, as in the second set of cases) is

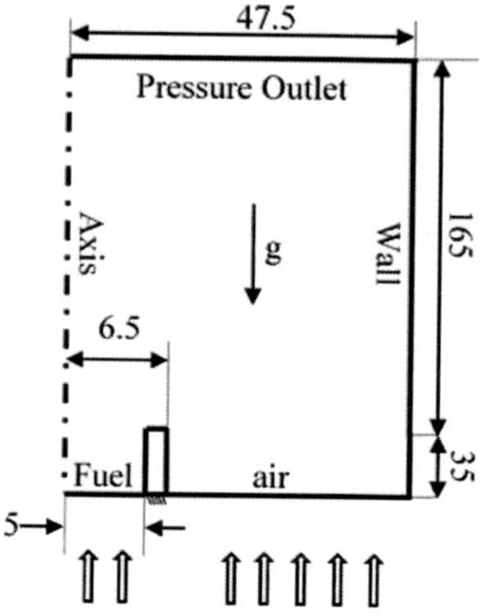


**Fig. 1** Schematic of computational domain (all dimensions in mm)

supplied at this boundary. Temperature (300 K) and composition of $O_2$ and $N_2$ (0.23 and 0.77) are also specified at this boundary.

(3) Pressure outlet: Atmospheric pressure is imposed at this boundary. The first derivatives of all the flow variables normal to this boundary are set to zero. In case of reverse flow at outlet, atmospheric air (23% $O_2$ and 77% $N_2$ by mass) enters the domain through this boundary.
(4) Walls: No slip condition for velocity, zero heat flux for temperature and a zero gradient for species mass fraction are specified at this boundary.
(5) Fuel tube: This boundary is coupled with the fluid region, and conjugate heat transfer is solved between the gas and solid phases. No slip condition is specified for velocity.
(6) Axis: The radial velocity, $v$, and the flow, species and temperature gradients in the radial direction are set to zero.

## 2.3 *Grid Independence Study and Validation*

Grid independence study is performed using three meshes. Coarse mesh having 16,245 cells with a minimum cell size of 0.5 mm in the radial and axial directions, intermediate mesh with 32,775 cells and a minimum cell size of 0.25 mm in the radial direction and 0.5 mm in the axial direction and a fine mesh consisting of 56,350 cells with a minimum cell size of 0.25 mm in both directions have been used.

Since experimental data on biogas confined flames have not been reported in the literature, experimental results of unconfined diffusion flame fuelled by methane, as reported in Mokhov et al. [13], has been used for validation. The radial temperature profile at two axial locations given by Mokhov et al. [13] are used for validating the numerical model. Numerical results along with experimental data are shown in

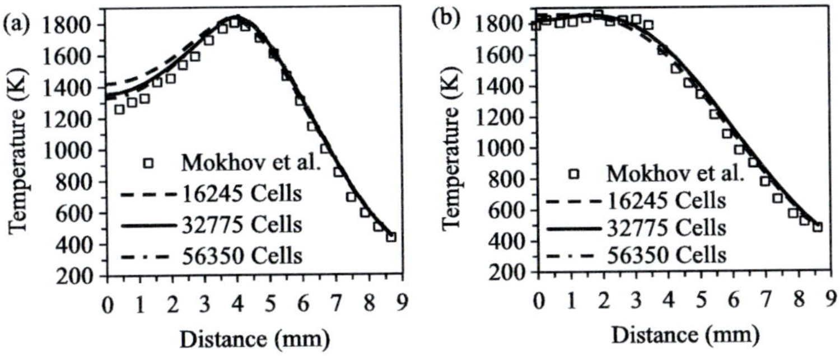


**Fig. 2** Predicted radial temperature profile validated against Mokhov et al. [13] at two axial locations **a** 14 mm and **b** 24 mm for three mesh sizes

Fig. 2. It is observed that the results do no change notably on increasing the mesh size from 32,775 cells to 56,350 cells. Hence, the domain with 32,775 cells is considered for further parametric study. Also, the numerical results are seen to be much closer to the experimental data, proposing a good validation of the numerical model.

# 3 Result and Discussions

## *3.1 Parametric Study*

In general, the amount of $CO_2$ in biogas varies from 30 to 45% by volume. Biogas with maximum $CO_2$ content is reported to have the least flame stability. Hence, biogas with 45% $CO_2$ and 55% $CH_4$ by volume is considered for the present study to understand the flame structure and entropy generation in these flames. The burner is operated at 0.25 kW for all the cases. Primary air is mixed with fuel and fed through the fuel inlet, and the coflow air is supplied through the coflow inlet (Fig. 1). A parametric study is carried out by varying the amount of primary air in the fuel stream in the range of 0% to 40% of stoichiometric value required for the fuel. For one set cases, 100% stoichiometric air is supplied in the coflow stream, making these cases to be rich in air when primary air is supplied. For second set of cases, the coflow air is reduced appropriately such that the total air supplied to the burner is equal to the stoichiometric value required for the corresponding fuel flow rate. These cases are listed in Table 1.

**Table 1** List of cases with varying primary and coflow air

| Case Number | Primary air (% stoichiometric) | Coflow air (% stoichiometric) | Mass flow rate of fuel (kg/s) | Total mass flow rate of air (kg/s) |
|---|---|---|---|---|
| Effect of primary air (Set 1) | | | | |
| 1, 2, 3, 4, 5 | 0, 5, 10, 20, 40 | 100 | $1.598 \times 10^{-5}$ | $8.49 \times 10^{-5}$ to $11.84 \times 10^{-5}$ |
| Effect of coflow air (Set 2) | | | | |
| 6, 7, 8, 9 | 5, 10, 20, 40 | 95, 90, 80, 60 | $1.598 \times 10^{-5}$ | $8.49 \times 10^{-5}$ |

### 3.2 *Effect of Primary Air—Set 1 Cases*

The effects of primary air (varied from 0 to 40% of stoichiometric value) on biogas flames are discussed in this section. It should be noted that partial premixing in this range will not result in the formation of a reactant mixture that is ignitable (premixed reactants within the flammable limits) and additional air from coflow stream is required for flame formation. The coflow air is fixed at 100% of stoichiometric value in these cases.

Figure 3 shows the temperature contours as well as the vectors of mixture velocity and stoichiometric contour line ($\phi = 1$), focussing a region closer to the burner exit. The stoichiometric contour line is defined by the equation, $Y_f - Y_{\text{ox}}/\nu = 0$. Here, $Y_f$ and $Y_{\text{ox}}$ are the fuel and oxidizer mass fractions, respectively, and $\nu$ is the stoichiometric oxidizer to fuel ratio. It is observed that the axial extent of the isotherm of 500 K increases with an increase in primary air flow rate. The maximum flame temperature is almost the same for all the cases as indicated below the plots because of the same amount of fuel injection. The contour line of $\phi = 1$, shown as a dashed line, almost intersects with the high temperature zone for all the cases.

The radial extent of $\phi = 1$ line slightly decreases with an increase in primary air flow rate due to oxygen availability within the fuel stream. The magnitude of velocity vectors shows a slight increase in mixture velocity with a gradual increase in the amount of the primary air, which is attributed to higher velocity of the core jet. The vectors also indicate the flow acceleration in the flame zone.

Figure 4 presents the axial temperature profiles and methane mass fraction for cases 1–5. Temperature remains constant inside the core tube, increases sharply to attain a peak value in the flame zone and then gradually decreases in the plume zone after major reactions are complete. The axial extent of the peak temperature shifts towards the exit of the burner with an increase in primary air flow rate, indicating a reduction in combustion/flame zone height. The steepness in the temperature rise increases with an increase in the flow rate of primary air. Similarly, the axial location where methane mass fraction tends towards zero, also decreases with an increase in primary air indicating faster fuel consumption rate at higher level of premixing.

Figure 5 shows the predicted radial profiles $CH_4$, $O_2$, CO, OH, $CO_2$ and $H_2O$ mass fraction at the burner exit and at the axial location of 30 mm from the burner exit for cases 1–5. The intensification of mixing with an increase in the mass flow rate

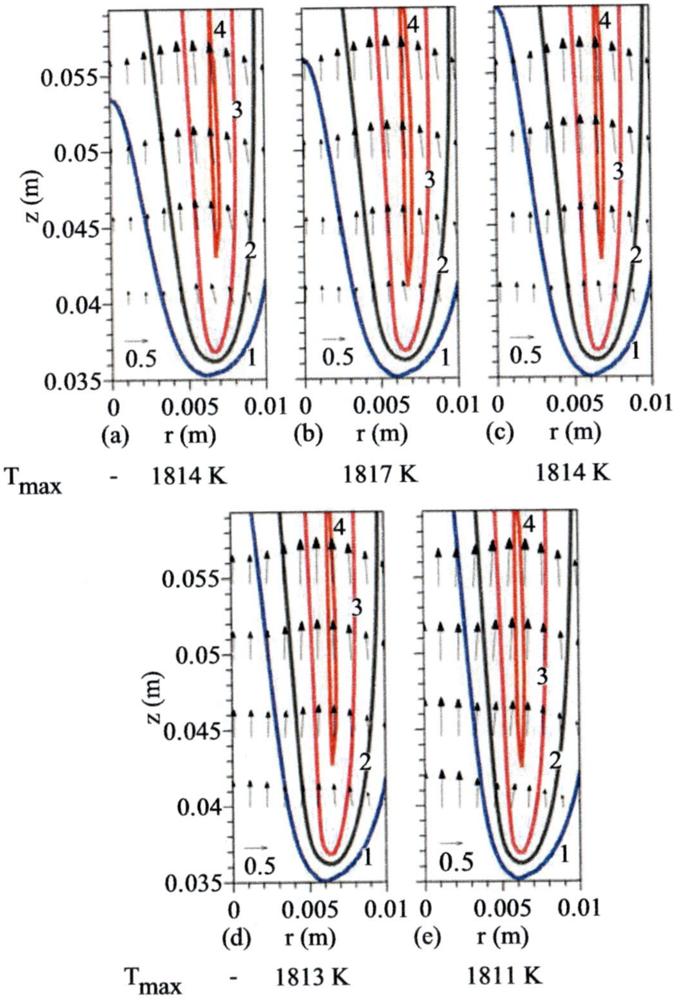


**Fig. 3** Line contours of temperature with velocity vectors for primary air of **a** 0% (Case 1), **b** 5% (Case 2), **c** 10% (Case 3), **d** 20% (Case 4) and **e** 40% (Case 5) of stoichiometric value, with coflow air of 100% stoichiometric value. Dashed line indicates stoichiometric contour line [Label: Temperature (K): 1—500, 2—1000, 3—1500, 4—1800]

of primary air in the core region is shown through the profiles of $CH_4$ and $O_2$ mass fractions in Fig. 5a. The methane mass fraction tends to zero value just outside the burner rim at the burner exit. Thus, there is a decrease in the gradients of the profiles with an increase in primary air. Oxygen from the ambient also reaches the value in the core stream asymptotically. Even at the burner exit, CO and radical formation are present, as shown in Fig. 5b. The mass fractions of CO and OH peak just to the left of fuel consumption zone. The peak mass fraction of CO increases up to 10%

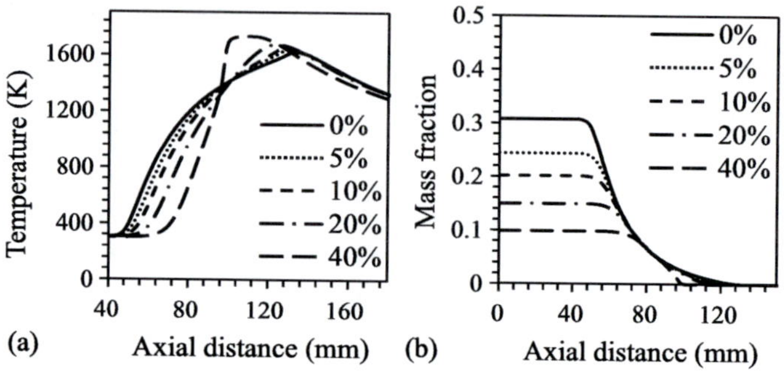


**Fig. 4** Predicted axial profiles of **a** flame temperature and **b** methane mass fraction for cases with primary air 0, 5, 10, 20 and 40% of stoichiometric value and 100% coflow air

primary air and then decreases with further increase in primary air up to 40% of stoichiometric value.

On the other hand, the OH mass fraction continuously increases with the amount of primary air, which helps in formation of radicals even at the burner exit, where the temperature is much lower (Fig. 5b). An increase in the OH peak value is observed for 20% and 40% primary air cases, which showed reduction in peak CO mass fraction. The mass fraction of $H_2O$ peaks just left to the location where $CO_2$ mass fraction tends towards zero (Fig. 5c). A marginal increase in the peak value of $H_2O$ mass fraction for primary air 20% and 40% is due to an increase in the peak OH mass fractions for these cases (Fig. 5c).

At the axial location of 30 mm from burner exit, the consumption zones of methane and $O_2$ shift radially inwards with a gradual increase of primary air. It is evident that both coflow and core air are consumed around the flame zone along with methane (Fig. 5d). The peak value of CO mass fraction increases with increase in primary air flow rate, however, the peak value of OH mass fraction remains almost the same with an increase in the flow rate of primary air (Fig. 5e). The mass fraction of $CO_2$ is maximum at the axis and continuously decreases in radial direction, except for the case with 40% primary air, where it remains almost constant up to a distance of 6 mm and then decreases radially away from the axis (Fig. 5f). This is due to inherent $CO_2$ present in the core flow as well as due to its formation in the flame zone and its further diffusion towards the ambient. The $H_2O$ mass fraction increases, attains a maximum around the flame zone, where the profile of $CO_2$ shows a changing slope, and then decreases with an increase in radial distance irrespective of the amount of primary air in the fuel mixture.

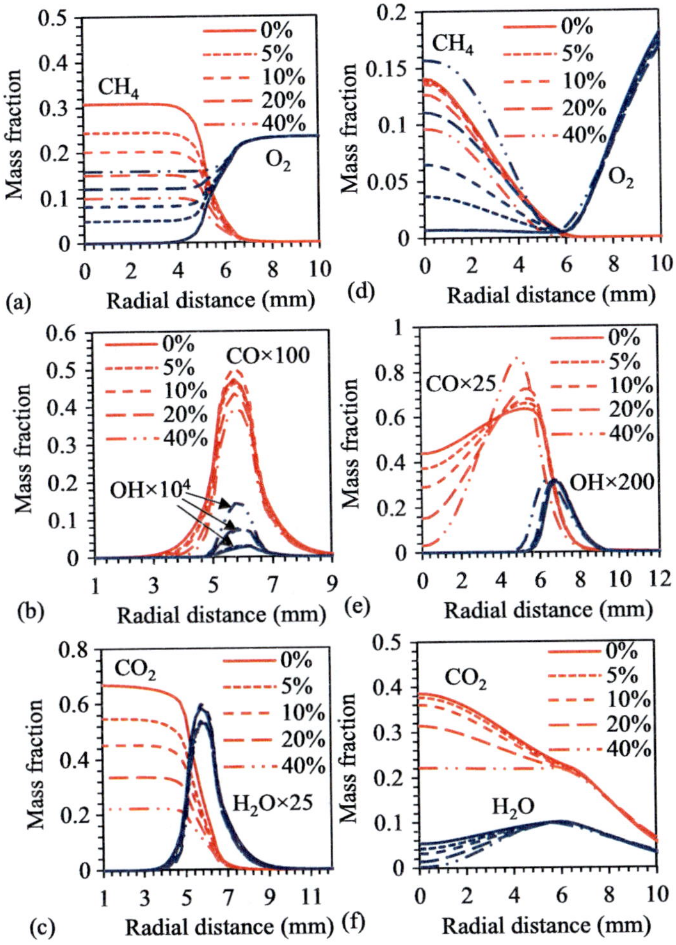


**Fig. 5** Predicted radial profiles of species mass fraction at (left) burner exit and (right) 30 mm from the burner exit **a, d** $CH_4$, $O_2$ **b, e** CO, OH and **c, f** $CO_2$, $H_2O$ for cases with primary air 0, 5, 10, 20 and 40% of stoichiometric value and 100% coflow air

## *3.3 Effect of Primary Air—Set 2 Cases*

In this section, the primary air content in the fuel mixture is increased from 0 to 40% of stoichiometric value and rate of coflow air is adjusted such that the total air supplied is equal to the stoichiometric air requirement. Figure 6 shows temperature contours and velocity vectors along with stoichiometric contour line of $\Phi = 1$ focussing near the burner exit. Even though the plots are similar to those presented in set 1 cases in Fig. 3, there are intrinsic differences between set 1 and set 2 cases. For example, the isotherm of 500 K intersects within the burner rim when primary air is increased up to 10% of stoichiometric value (Fig. 6b, c) indicating the movement of flame base towards the burner rim. Flame anchors around the burner rim and has higher stability in these cases. The flame penetration towards the burner rim can be attributed to a decrease in the coflow air to 95% (for 5% primary air) and 90% (for primary air 10%) of stoichiometric value.

At higher amounts of primary air (20% and 40%), core jet velocity increases and this pushes the flame away from the burner rim. However, the flame anchoring

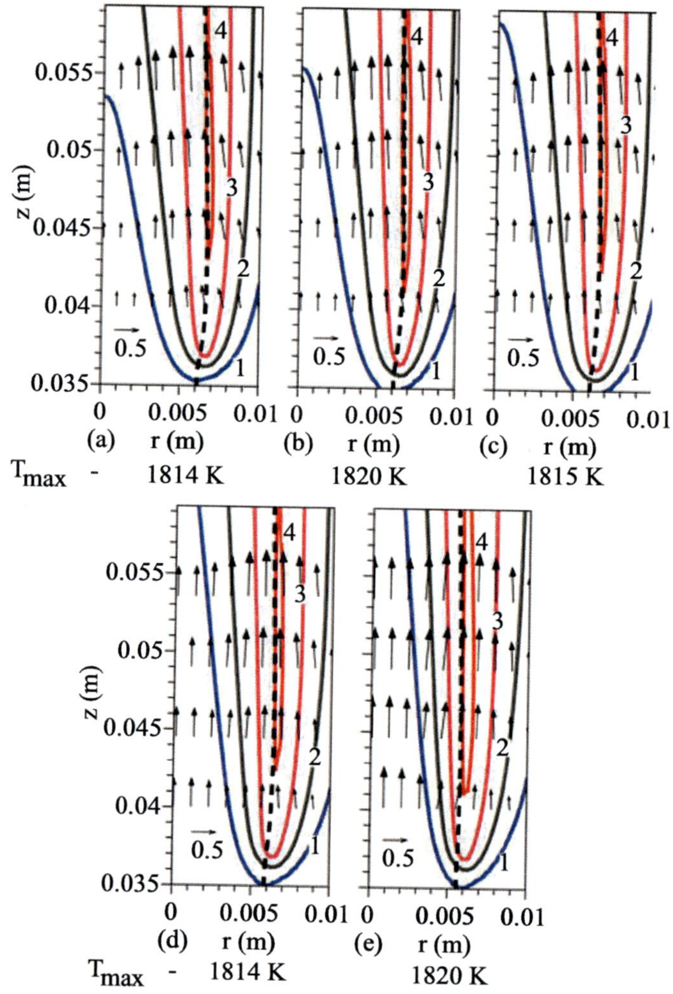


**Fig. 6** Contours of temperature (lines) with velocity vectors for coflow air of **a** 100% (case 1), **b** 95% (case 6), **c** 90% (case 7), **d** 80% (case 8) and **e** 60% (case 9) of stoichiometric value. Dashed line indicates stoichiometric contour line ($\phi = 1$) [Label: Temperature (K): 1—500, 2—1000, 3—1500, 4—1800]

in these cases (cases 8 and 9), indicated by the downward movement of isotherm of 500 K, is better as compared to cases 4 and 5. The stoichiometric contour line intersects with the peak temperature zone as observed earlier. Similarly, the increase in mixture velocity and acceleration of the mixture in the plume region are clearly indicated by the velocity vectors in all the cases.

Figure 7 presents the radial profiles $CH_4$, $O_2$, CO, OH, $CO_2$ and $H_2O$ mass fraction at the burner exit and at 30 mm from the burner exit for cases with varying amount of coflow air (cases 1, 6, 7, 8 and 9). With a decrease in coflow air, the gradients in the profiles of $CH_4$ and $O_2$ mass fractions decrease (Fig. 7a). This is due to increase in the amount of air in the core stream that reduces the mass fraction of methane as well as the air required for combustion. The peak value of CO mass fraction increases with an increase in the flow rate of primary air and attains a maximum for 10% primary air (90% coflow air), as shown in Fig. 7b. However, it decreases with a gradual increase in primary air to 20% and further to 40% of stoichiometric value, and a corresponding increase in OH mass fraction is also observed (Fig. 7b). The $CO_2$ mass fraction approaches the zero value at almost the same locations, where OH reaches the zero value (Fig. 7b, c). At the burner exit plane, the mass fraction of $H_2O$ peaks around the same location where OH attains its maximum value.

At the axial location of 30 mm from the burner exit, the fuel and oxidizer are almost consumed at the flame zone with a small amount of leakage across the flame zone.

With a gradual increase in primary air, the flame zone shifts towards the axis, as seen in set 1 cases (Fig. 7d). The peak value of CO mass fraction also increases, and the peak value of OH mass fraction remains almost the same with an increase in primary air (Fig. 7e). However, peak locations vary with a variation in primary and coflow air. Mass fractions of $CO_2$ and $H_2O$ (Fig. 7f) show similar trends as observed in the cases of set 1, with distinctly varying slope in $CO_2$ profile and peaking of $H_2O$ around those locations. These results show that based on the confinement geometry, the transport of coflow air towards the flame zone is almost unaffected with the rate of coflow air.

### 3.4 *Entropy Generation*

Figure 8 shows the regions of volumetric entropy generation during heat transfer, chemical reaction, mass transfer and coupled heat and mass transfer processes for 0% primary air and 100% coflow air (Case 1). Integrated value of entropy generation over the entire domain is also indicated within the plots. Entropy generation during heat transfer is higher near the burner rim on the preheat zones, and the maximum value is found on air side of the flame. The chemical entropy generation is found exactly at the base of the reaction zone. The entropy generation during mass transfer process also occurs near the burner rim adjacent to the reaction zone where oxygen diffuses towards the fuel to sustain a flame. The generation of entropy during chemical reaction forms the major part followed by that during heat transfer. The entropy

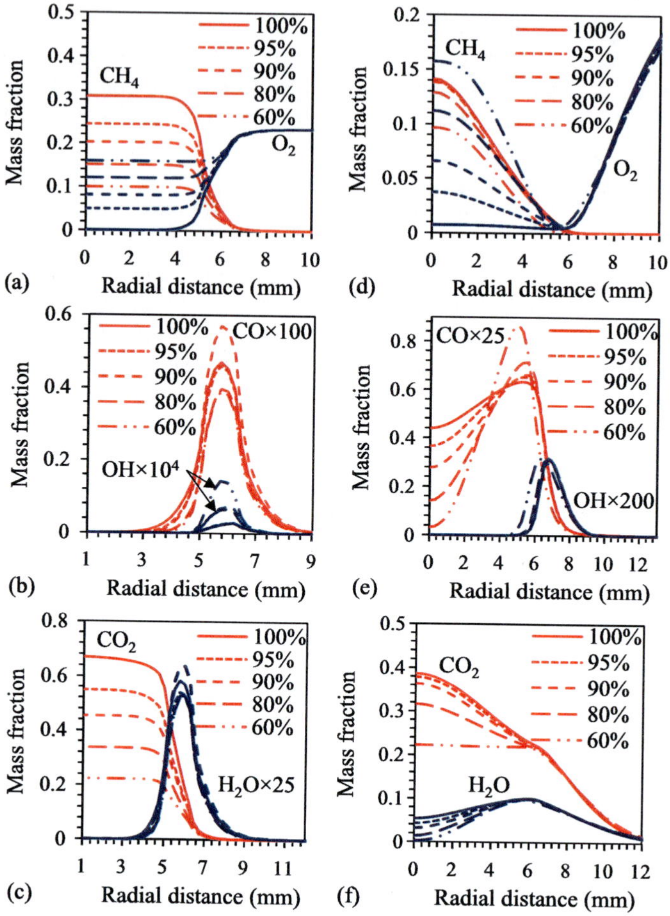


**Fig. 7** Predicted radial profiles of species mass fraction at (left) burner exit and (right) 30 mm from the burner exit **a**, **d** $CH_4$, $O_2$ **b**, **e** CO, OH and **c**, **f** $CO_2$, $H_2O$ for cases with coflow air 100% (case 1), 95% (case 6), 90% (case 7), 80% (case 8) and 60% (case 9) of stoichiometric value

generation during coupled heat and mass transfer processes is the least among all modes of entropy generation.

Figure 9 shows the volumetric entropy generation rate for Case 2 and Case 5 (primary air–5% and 40%). Maximum entropy generation during heat transfer increases slightly with an increase in primary air. The region of maximum entropy generation is found on the air side for Case 2 (5% primary air), and it extends to the fuel side with increase in primary air to 40%.

On the other hand, entropy generation during mass transfer and coupled heat and mass transfer decreases with an increase in primary air. The maximum entropy generation found on the fuel side for primary air 5% is shifted to the air side, as

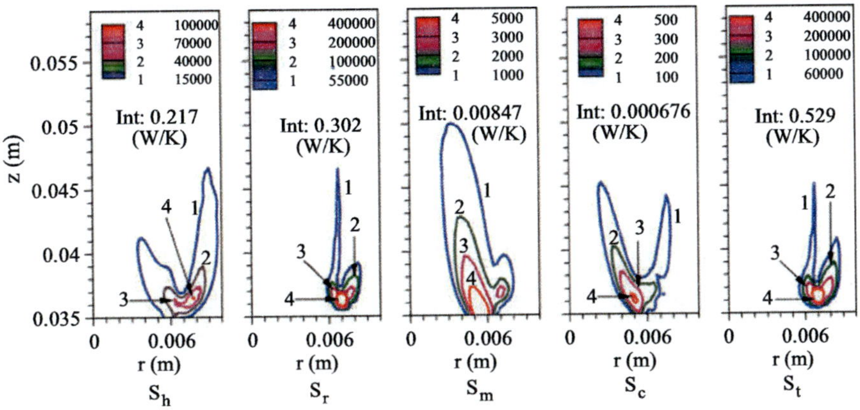


**Fig. 8** Rate of entropy generation (W/m$^3$-K) during heat transfer ($S_h$), reaction ($S_r$), mass transfer ($S_m$), coupled heat and mass transfer ($S_c$) and total entropy ($S_t$), for 0% primary air and coflow air of 100% of stoichiometric value (Case 1); integrated values (W/K) are shown within plots

the primary air flow rate is increased to 40%. The axial extent of entropy generated during mass transfer and coupled heat and mass transfer processes also reduces with an increase in primary air content. The entropy generation during chemical reactions and the integrated values increase with an increase in primary air, due to increased rates of chemical reactions due to partial premixing. As a result, the total entropy generation per unit volume increases with an increase in the flow rate of primary air.

Figure 10 shows the volumetric entropy generated for Case 6 and Case 9 with reduced rate of coflow air (95% and 60% of stoichiometric air). The entropy generation during heat transfer and chemical reactions increase with an increase in the primary air flow rate as discussed earlier. The axial extents of entropy generation during heat transfer and chemical reactions increase with an increase in primary air. Similarly, the entropy generations during mass transfer and coupled heat and mass transfer, as well as their axial extents decrease with an increase in primary air. It is clear that the primary air affects the entropy generation rate irrespective of the rate of coflow air.

Figure 11 shows the volume integrated entropy generation rate (in W/K) for all the cases. The entropy generation during mass transfer (Fig. 11a) decreases continuously with an increase in primary air irrespective of whether stoichiometric amount or reduced amount of coflow air is fed. This is due to decrease in the diffusion as a result of premixing. The entropy generation due to coupled heat and mass transfer processes (Fig. 11a) increases slightly for 5% primary air and then decreases continuously with further increase in primary air; however, its value is much lower. The entropy generation during heat transfer slightly increases with increase in primary air irrespective of the amount of coflow air (Fig. 11b). This is due to higher heat release at the flame base as a result of partial premixing. The entropy generation due to chemical reactions (Fig. 11b) presents a two-slope behaviour. It increases with a higher slope when primary air flow rate is increased from 0 to 10%, and the slope

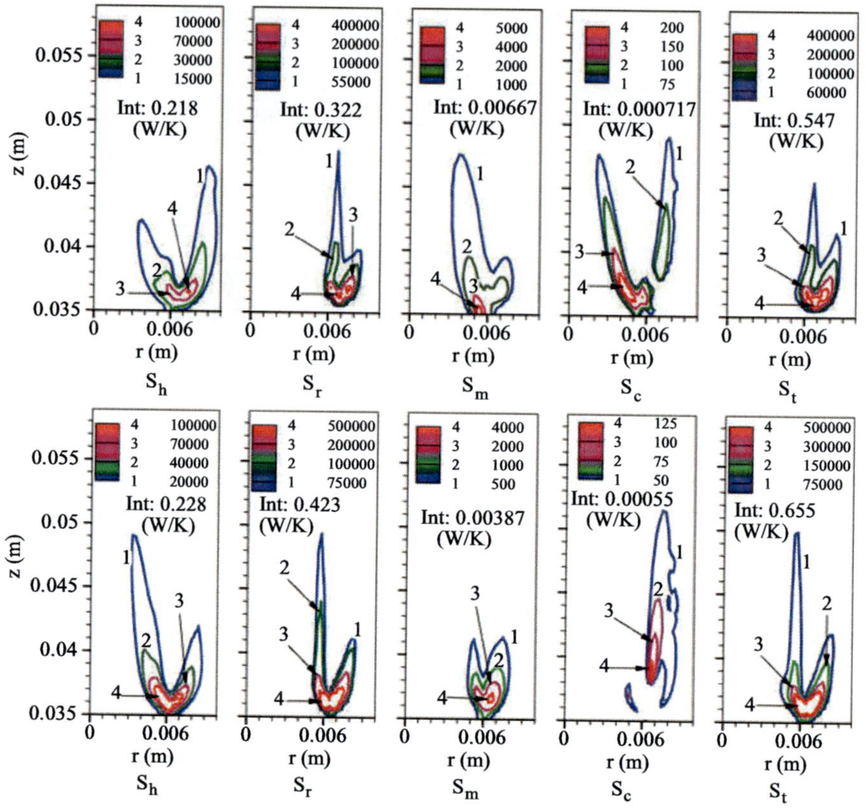


**Fig. 9** Rate of entropy generation (W/m$^3$-K) during heat transfer ($S_h$), reaction ($S_r$), mass transfer ($S_m$), coupled heat and mass transfer ($S_c$) and total entropy ($S_t$) for Case 2 (top) and Case 5 (bottom) with coflow air of 100% of stoichiometric value; integrated values (W/K) are shown in plots

decreases with further increase in the primary air from 10 to 40%. This suggests that primary air of around 20% can be an optimum value for the present configuration for biogas with 45% $CO_2$, for both stoichiometric and reduced coflow air. Total entropy generation rate follows the trend of entropy generation from reactions.

## 4 Conclusions

In this work, confined biogas coflow flames with varying amount of primary air in the fuel stream are analysed using OpenFOAM. The coflow air is kept equal to the stoichiometric value in one set of cases and varied appropriately depending on the level of premixing in another set of cases. The flame stability marginally increases for cases with primary air 5% and 10% of stoichiometric value and in

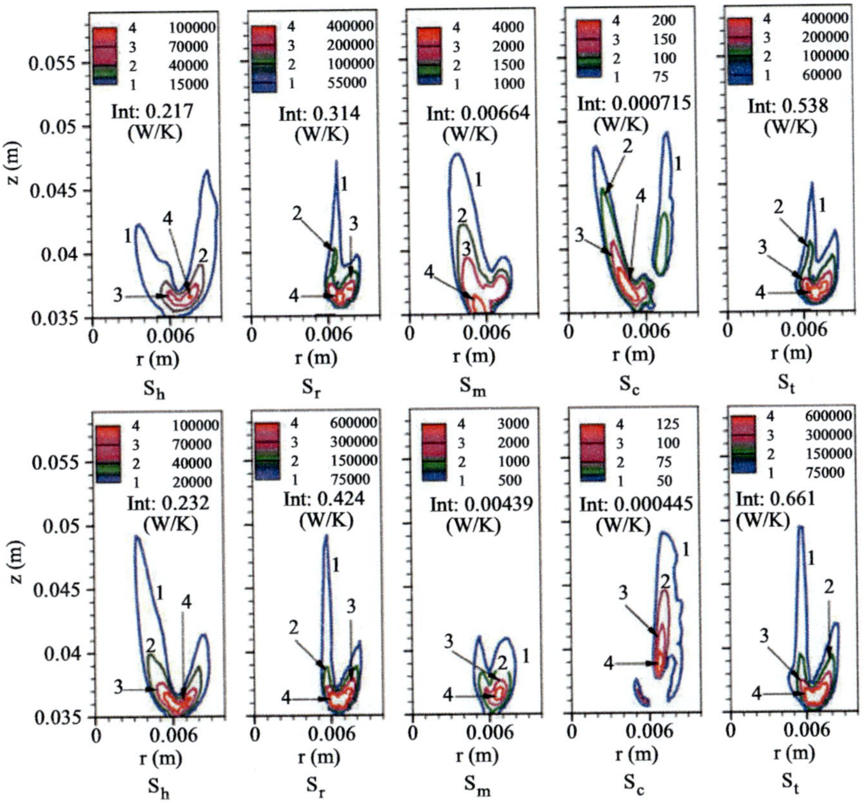


**Fig. 10** Rate of entropy generation (W/m$^3$-K) during heat transfer ($S_h$), reaction ($S_r$), mass transfer ($S_m$), coupled heat and mass transfer ($S_c$) and total entropy ($S_t$) for cases with primary air 5% (case 6) and 40% (case 9) of stoichiometric value with reduced coflow air; integrated values (W/K) are shown in plots

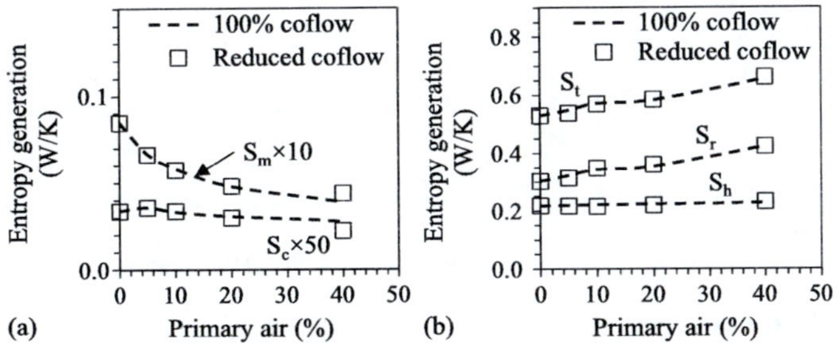


**Fig. 11** Variation of integrated entropy generation rate (W/K) with primary air 0, 5, 10, 20 and 40% of stoichiometric value, entropy generation due to **a** mass transfer ($S_m$), coupled heat and mass transfer ($S_c$) and **b** heat transfer ($S_h$), chemical reactions ($S_r$) and total entropy generation rate ($S_t$)

the presence of reduced coflow air. The flame temperature remains almost constant irrespective of partial premixing. The profiles of species mass fraction indicate higher CO consumption as the primary air flow rate is increased above 20% of stoichiometric value. The entropy generation through chemical reactions is higher followed by heat transfer, mass transfer and coupled heat and mass transfer processes. The region of maximum entropy generation during heat transfer and chemical reactions extends to the fuel side with increase in partial premixing. Subsequently, the region of maximum entropy generation during mass transfer and coupled heat and mass transfer processes shifts to the air side of the flame. The total entropy generation in each mode is almost the same irrespective of the amount of coflow air used. Entropy generation study suggests that primary air of around 20% can be an optimum value for the present configuration for biogas flames with 45% $CO_2$ in the fuel mixture, for cases with stoichiometric and reduced coflow air.

## References


1. Giurcan V, Movileanu C, Musuc AM, Mitu M (2021) Laminar burning velocity of Biogas-containing mixtures—a literature review. Processes 9:996
2. Mahallawy FE, Abdelhafez A, Mansour MS (2007) Mixing and nozzle geometry effects on flame structure and stability. Combust Sci Technol 179:249–263
3. Sahu KB, Ghose P, Mohapatra N, Puthal MK (2015) Numerical investigation on structure of a partially premixed methane flame at various equivalence ratios. Int J Adv Res Sci Eng 4(1):50
4. Cha MS, Chung SH (1996) Characteristics of lifted flames in nonpremixed turbulent confined jets. In: 26th international proceedings of the combustion institute, pp 121–128
5. Karbasi M, Wierzba I (1998) The effects of hydrogen addition on the stability limits of methane jet diffusion flames. Int J Hydrogen Energy 23(2):123–129
6. Hutchins AR, Kribs JD, Muncey RD, Lyons KM (2013) Assessment of stabilization mechanisms of confined, turbulent, lifted jet flames: Effects of ambient coflow. In: Proceedings of the ASME 2013 Power conference, Boston (USA)
7. Baudoin E, Bai XS, Yan B, Liu C, Lantz A, Hosseini SM, Li B, Elbaz A, Sami M, Li ZS, Collin R, Chen G, Fuchs L, Alden M, Mansour MS (2011) Effect of partial premixing on stabilization and local extinction of turbulent methane/air flames. Flow Turbul Combust 90(2):269–284
8. Ismail NC, Abdullah MZ, Mazlan NM, Mustafa KF (2020) Entropy generation and exergy analysis of premixed fuel-air combustion in micro porous media burner. Entropy 22(10):1104
9. Barlow RS, Smith NSA, Chen JY, Bilger RW (1999) Nitric oxide formation in dilute hydrogen jet flames: Isolation of the effects of radiation and turbulence–chemistry sub models. Combust Flame 117(1–2):4–31
10. Hirschfelder JO, Curtiss CF, Bird RB (1954) Molecular theory of gases and liquids, 1st edn. Wiley, New York
11. Kuo KKY (1986) Principles of combustion, 1st edn. Wiley, New York
12. Pope DN, Raghavan V, Gogos G (2010) Gas-phase entropy generation during transient methanol droplet combustion. Int J Thermal Sci 49:1288–1302
13. Mokhov AV, Bennett BAV, Levinsky HB, Smooke MD (2007) Experimental and computational study of $C_2H_2$ and CO in a laminar axisymmetric methane-air diffusion flame. Proc Combust Inst 31(1):997–1004

# A Study of Entropy Generation in Coflow Diffusion Flames Fueled by Liquefied Petroleum Gas

S. Muthu Kumaran, H. Surya Shashank Sekhar, and V. Raghavan

## 1 Introduction

LPG flames are generally long and sooty. LPG coflow diffusion flames and partially premixed flames are commonly employed in large industrial burners. Coflow improves the mixing of reactants and flame stability, and partial premixing reduces the flame length and soot. Combustion of gaseous fuels involves several processes such as mixing of reactant species, heat generation during chemical reactions, and transport of heat and mass to and from the flame zone. The addition of coflow or primary air to the fuel stream increases the entropy generation in these flames. Analysing the entropy generation associated with these processes is important for optimising combustion in gas burners.

Datta [1] developed a theoretical model for calculating entropy generation in axisymmetric laminar diffusion flame. The variation in air–fuel ratio and inlet air temperature on entropy generation were investigated. Later, a numerical study of entropy generation for the same configuration was reported [2]. The second law efficiency of the process was analysed through exergy destruction between the inlet and outlet streams. It was found that the rate of entropy generation during chemical reactions and heat transfer remained almost constant, irrespective of variation in gravity. Further, the entropy generation rate was lower at reduced gravity when compared to normal gravity flames. Chen et al. [3] studied the entropy generation in hydrogen-air opposed jet flames. Reynolds number of flow and volumetric percentage of $H_2$ in the fuel mixture were varied systematically to study the entropy generation. Briones et al. [4] reported the entropy generation in $H_2$–$CH_4$-air triple flames. The entropy generation due to various processes was calculated on a transient flow field.

S. Muthu Kumaran (✉) · V. Raghavan
Department of Mechanical Engineering, Indian Institute of Technology Madras, Chennai, India
e-mail: Kumaranms92@gmail.com

H. Surya Shashank Sekhar
University College of Engineering, Osmania University, Hyderabad, India

 
N. L. Narasimhan et al. (eds.), *Recent Advances in Energy Technologies*, Lecture Notes in Mechanical Engineering, https://doi.org/10.1007/978-981-19-3467-4_34

The entropy generation increased with $H_2$ addition as a result of enhanced heat conduction and reactivity. They observed that the second law efficiency remained almost constant with the addition of $H_2$, as the increase in entropy was balanced by an increase in available energy of the fuel mixture. Tehrani et al. [5] predicted the entropy generation in sooty diffusion flames fueled by kerosene. The Moss-Brookes-Hall model was employed to estimate the soot formation in these flames. Results revealed that chemical entropy generation played a major contribution in entropy generation as compared to heat conduction, mass transfer and viscous dissipation processes. It was found that the major soot production region in flames could be predicted with entropy generation fields within acceptable deviation. Morsli et al. [6] analysed the entropy generation in propane-air coflow flames. The effects of equivalence ratio, temperature and Bejan number were studied. It was also showed that mixing $H_2$ with propane reduced the flame temperature and CO emissions.

Arjmandi and Amani [7] numerically studied the effects of swirl number, inlet reactant flow rate and distance between air and fuel nozzles on entropy generation in gas turbine combustor. The entropy generation calculated based on overall entropy balance was compared with the local entropy generation rate. They observed that entropy generation during chemical reactions and heat transfer contribute significantly to the total entropy generation in the system. Hosseini et al. [8] investigated the entropy generation in biogas flames mixed with $H_2$. The entropy generation increased with $H_2$ addition due to increased irreversibilities associated with higher flame temperature. Jiang et al. [9] numerically investigated the entropy generation in premixed $CO/H_2$/air flames. The effects of $CO/H_2$/air flow rates and CO concentration on entropy generation were analysed systematically. The entropy generation rate due to chemical reactions were suppressed by the addition of CO to the fuel mixture. However, entropy generation in the process of heat conduction increased.

In this work, entropy generation in LPG-Air flames when burnt in non-premixed and partially premixed modes, with and without the presence of coflow air is studied systematically. This can be useful in optimising the operating conditions for achieving maximum efficiency with minimum emissions when LPG is burnt in large industrial burners.

## 2 Numerical Model

Characteristics of LPG coflow diffusion flames and flames with varying amount of primary air are reported in Kumaran et al. [10]. Entropy generation in these flames have been reported in the present study. ANSYS-FLUENT v16.1 is used for numerical simulations. Figure 1 shows the schematic of the computational domain. It has a core tube of diameter 10 mm and coflow tube of 42 mm inner diameter. A kinetic mechanism consisting of 43 species and 396 elementary reactions is used for modelling the flame [11]. Full multicomponent diffusion along with thermal diffusion and diffusion energy source is used. Soot formation and its oxidation are modelled using the Moss-Brookes model and Fennimore-Jones model, respectively.

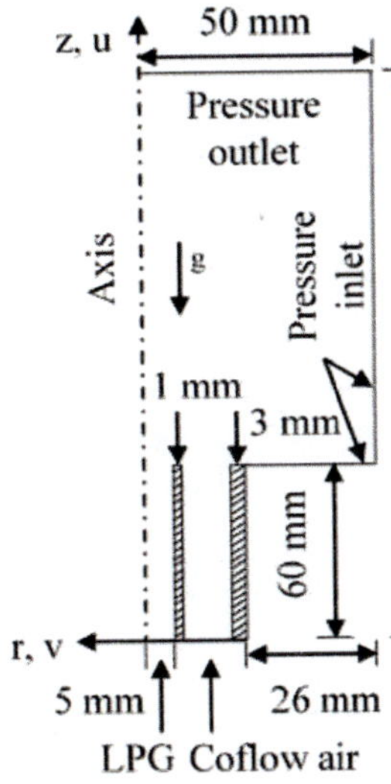


**Fig. 1** Schematic of computational domain

An optically thin radiation model is used to model the radiation absorbed by participating species and soot. The details of the numerical model, soot model, radiation model and flow rates of fuel, coflow air and primary air are presented in Ref. [10].

The processes contributing to entropy generation in reactive flows are [12]: entropy generation due to transport of heat ($S_h$), mass ($S_m$), coupling between heat and mass transport processes ($S_c$), chemical reactions ($S_r$) and viscous dissipation ($S_v$). The entropy generated due to viscous dissipation and coupled heat and mass transport processes is negligible as compared to other processes, hence not considered here. The total volumetric rate of entropy generation ($S_t$) in reactive flows (in W/m$^3$-K) is given as:

$$S_t = S_h + S_m + S_r \tag{1}$$

Entropy generated during transport of heat, per unit volume, is calculated as,

$$S_h = \frac{k(\nabla T)^2}{T^2} \tag{2}$$

Entropy generated during transport of mass, per unit volume, is expressed as,

$$S_m = \frac{-\rho R_u}{\overline{W}} \sum_i v_i.\nabla X_i \tag{3}$$

Here, the ordinary mass diffusion velocity ($v_i$) is written as,

$$v_i = \frac{-D_{i,m}}{Y_i}\nabla Y_i, \tag{4}$$

which satisfies the constraint,

$$\sum_i Y_i v_i = 0. \tag{5}$$

Here, $R_u$ is the universal gas constant (kJ/kmol-K), and $\overline{W}$ is the average molecular weight (kg/kmol). Entropy generated due to coupling between heat and mass transfer processes, per unit volume, is evaluated as,

$$S_m = \frac{-\rho R_u}{\overline{W}T}\sum_i w_i.\nabla X_i \tag{6}$$

Here, the thermal diffusion velocity ($w_i$) is given by,

$$w_i = \frac{-D_{T,i}}{\rho Y_i}\frac{\nabla T}{T} \tag{7}$$

subjected to the constraint,

$$\sum_i Y_i w_i = 0 \tag{8}$$

Entropy generated during chemical reactions, per unit volume, is given as,

$$S_r = \frac{-1}{T}\sum_i \mu_{c,i}\dot{\omega}_i'''. \tag{9}$$

Here, $\mu_{c,i}$ is the chemical potential calculated as $\mu_{c,i} = h_i - Ts_i$ ($h_i$ and $s_i$ are the enthalpy (J/kg) and entropy (J/kg-K) of $i$th species, respectively). Equations (2)–(9) are solved using an in-house MATLAB code. Table 1 shows the list of cases considered in the present study.

## 3 Results and Discussions

The cases listed in Table 1 are transient in nature. Therefore, the entropy generation during heat transfer is evaluated from the time-averaged fields of temperature and its gradients. Similarly, the entropy generation during mass transfer is calculated from the time-averaged species mass fraction and species gradient fields. The entropy generation during chemical reactions should be calculated from the enthalpy, entropy

**Table 1** List of cases

| Fuel flow rate (kg/s) | Coflow air × $10^{-5}$ (kg/s) (% stoichiometry) | Primary air × $10^{-5}$ (kg/s) (% stoichiometry) |
|---|---|---|
| $6.25 \times 10^{-6}$ | Coflow diffusion flames | |
| | 0 (0%) | 0 (0%) |
| | 9.65 (100%) | |
| | 14.475 (150%) | |
| | 19.3 (200%) | |
| | 38.6 (400%) | |
| | Partially premixed flames | |
| | 0 (0%) | 0.965 (10%) |
| | | 1.93 (20%) |
| | | 3.86 (40%) |
| | | 4.825 (50%) |
| | Partially premixed flames in coflow | |
| | 9.65 (100%) | 0.965 (10%) |
| | | 1.93 (20%) |
| | | 3.86 (40%) |
| | | 4.825 (50%) |

and net reaction rate data of all species evaluated at each time step. This process is time consuming and costly. Hence, the entropy generation during chemical reactions is calculated from the instantaneous fields of enthalpy, entropy and net reaction rate of species at every 0.1 s for a time period of 1 s (each case is run for at least 1 s) and this average value of entropy generation field is reported.

## 3.1 Entropy Generation in Coflow Diffusion Flames

Figure 2 depicts different modes of entropy generation in pure jet diffusion flame (0% coflow). Volume integrated values over the entire domain are shown within the plots. The entropy generation in the process of heat transfer is higher on the fuel side of the flame. The entropy generation during mass transfer also occurs in the same region, which corresponds to the preheat zone on the fuel side. However, its integrated value is quite low as compared to entropy generation during heat transfer. Maximum entropy is generated near the burner rim, where fuel and air diffuse to sustain a flame. The entropy generated during chemical reactions is observed in the core reaction zone and is the highest among all modes of entropy generation.

Figure 3 presents the entropy generation during the process of heat transfer in coflow diffusion flames with the rate of coflow air varied as 0, 100, 150, 200 and 400% of stoichiometric value (Table 1). The entropy generation during heat transfer

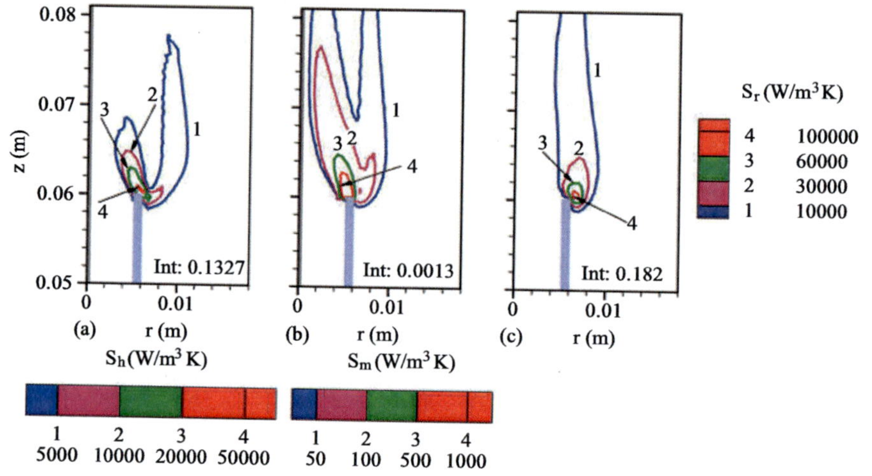


**Fig. 2** Rate of entropy generation (W/m$^3$-K) during **a** heat transfer ($S_h$) **b** mass transfer ($S_m$) and **c** chemical reactions ($S_r$) for pure jet diffusion flame. Integrated values (W/K) are shown within the plots

increases with a systematic increase in coflow air. The enhanced heat transfer to the preheat zone is the reason for higher entropy generation in these flames. The contours of entropy generation are distributed on both air and fuel sides, indicating entropy generation in the preheat zone on either side of the flame. However, the contour of maximum entropy generation shifts to the air side with an increase in coflow rate. Further, the axial extent of the contours notably increases with a gradual increase in the rate of coflow air. The radial extent of the contours remains almost a constant, irrespective of the rate of coflow air.

Figure 4 shows the entropy generation in the process of mass transfer in coflow diffusion flames with the rate of coflow varied as 0, 100, 150, 200 and 400% of stoichiometric value. Maximum entropy generation occurs near the burner rim for all cases.

The maximum entropy generation contours are inclined towards the axis. This is due to steep gradients in reactant mass fractions adjacent to the burner rim. The axial extent of the entropy generation contours, as shown by the contour line of 100 W/m$^3$K, increases with an increase in the flow rate of coflow air. The integrated value of entropy generation increases with an increase in coflow rate up to 100% of stoichiometric air. Then, it remains almost constant up to 150% coflow air and increases with a further increase in coflow air up to 400% of stoichiometric value.

Figure 5 shows the entropy generation during chemical reactions in coflow diffusion flames with 0, 100, 150, 200 and 400% of stoichiometric air in coflow stream. The contours of entropy generation exist in a narrow zone close to the air side of the flame. The integrated value of entropy generation increases when the coflow rate is increased to 100% of stoichiometric value. It has a local minimum at 150% coflow air and gradually increases with a further increase in coflow air to 400%.

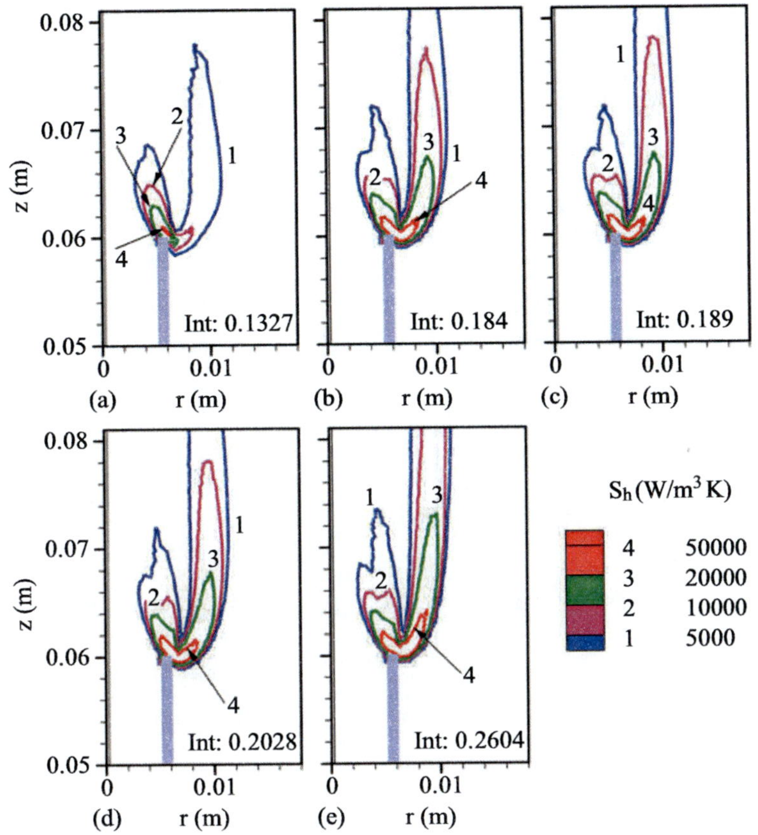


**Fig. 3** Rate of entropy generation (W/m$^3$-K) during heat transfer ($S_h$) in flames with **a** 0%, **b** 100%, **c** 150%, **d** 200% and **e** 400% stoichiometric air in coflow. Integrated values (W/K) are shown within the plots

Maximum entropy is generated near the burner rim on the air side of the flame. The entropy generation during chemical reactions is higher as compared to entropy generation during heat transfer up to 100% coflow air. With further increase is coflow air, entropy generation in the process of heat transfer dominates over other modes of entropy generation.

## 3.2 Entropy Generation in Partially Premixed Flames

Figure 6 presents the different modes of entropy generation in flame with 10% primary air. The entropy generation during heat transfer is higher on the fuel side of the flame. The entropy generation during mass transfer occurs in preheat zone

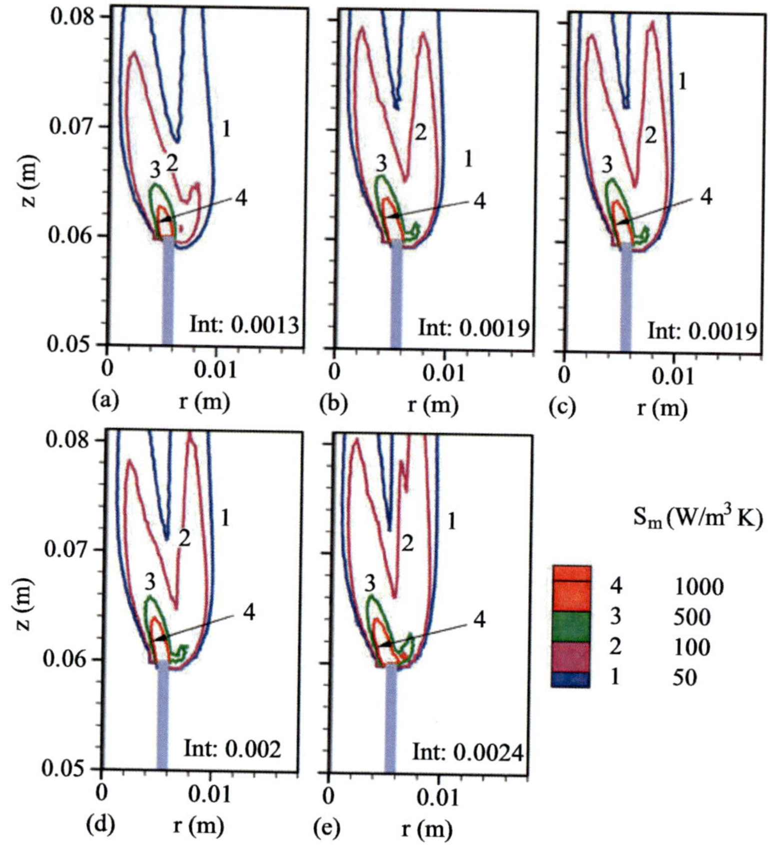


**Fig. 4** Rate of entropy generation (W/m$^3$-K) during mass transfer ($S_m$) in flames with **a** 0%, **b** 100%, **c** 150%, **d** 200% and **e** 400% stoichiometric air in coflow. Integrated values (W/K) are shown within the plots

on either side of the flame. As observed earlier, its integrated value is quite low as compared to entropy generated through heat transfer and reactions. Entropy generation during chemical reactions is the highest among all modes of entropy generation. Further, the entropy generated during heat transfer and chemical reactions are higher as compared to the pure jet diffusion flame.

Figure 7 presents the entropy generation in flames with primary air in the fuel mixture gradually increased from 0 to 50% of the stoichiometric value. The entropy generation in the process of heat transfer increases initially as the primary air is increased to 10% of stoichiometric value. Then it attains a local minimum with around 20% primary air in the reactant mixture. Subsequently, it increases with further increase in primary air up to 50% of stoichiometric value. At high degree of partial premixing (40% and 50%), the contour lines of maximum entropy generation move towards the burner axis, where significant temperature gradients are observed.

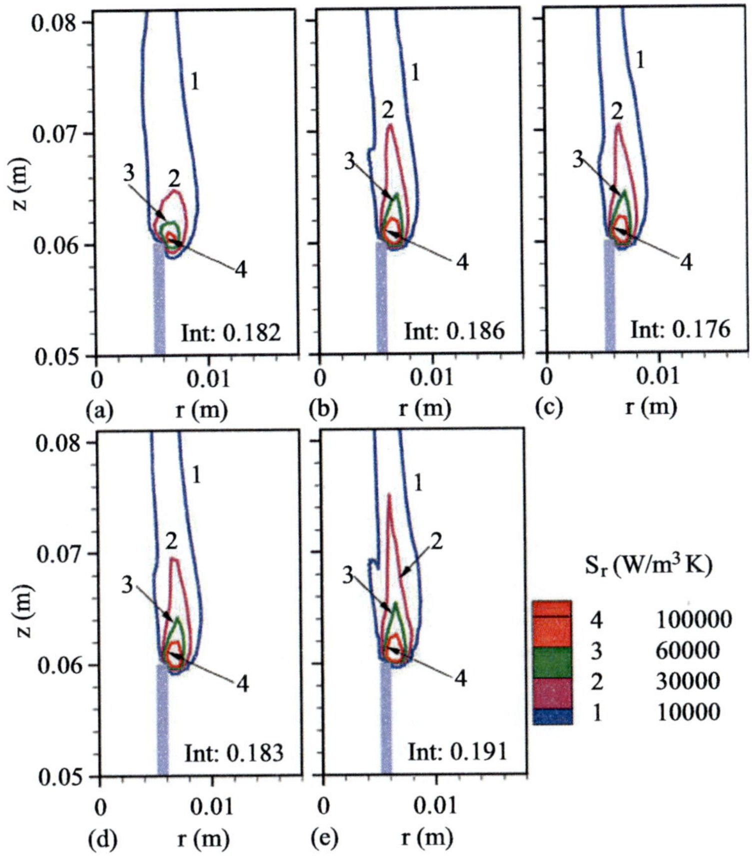


**Fig. 5** Rate of entropy generation (W/m$^3$-K) during chemical reactions ($S_r$) in flames with **a** 0%, **b** 100%, **c** 150%, **d** 200% and **e** 400% stoichiometric air in coflow. Integrated values (W/K) are shown within the plots

The axial extent of the contours also increases significantly when primary air flow rate is increased above 40% of stoichiometric value. The entropy generation contours in the preheat zone on the air side also displays an increasing trend with primary air.

Figure 8 shows the entropy generation during mass transfer in flames with primary air flow rate varied as 0, 10, 20, 40 and 50% of the stoichiometric value. The entropy generation decreases with an increase in the rate of primary air up to 20% of stoichiometric value. Then, it increases gradually with a further increase in primary air up to 50%. Maximum entropy generation is found on the fuel side of the flame as observed earlier. For 10% primary air, although the entropy generated in the process of heat transfer increases from baseline case (0% primary air) the entropy generated during mass transfer remains almost a constant. The entropy generation is the least

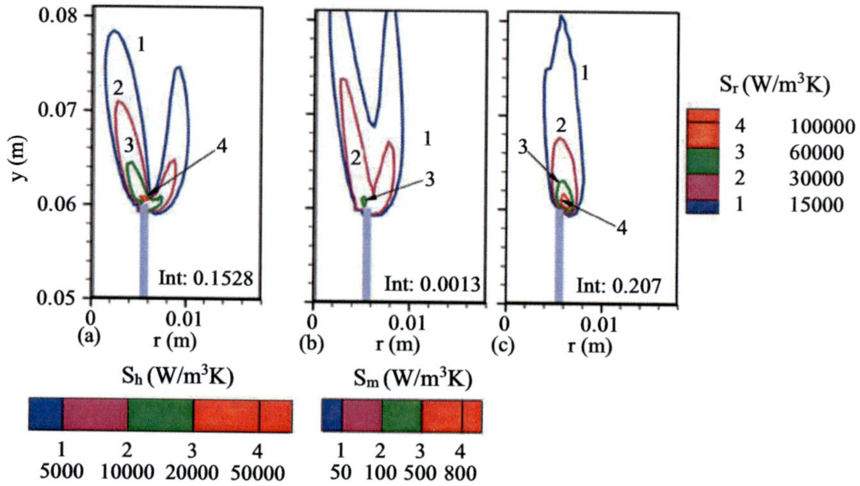


**Fig. 6** Rate of entropy generation (W/m$^3$-K) during **a** heat transfer ($S_h$) **b** mass transfer ($S_m$) and **c** chemical reactions ($S_r$) in flame with 10% primary air. Integrated values (W/K) are shown within the plots

for the case with 20% primary air due to low flame temperature observed for this case [10].

Entropy generation during chemical reactions in flames with varying amount of primary air is shown in Fig. 9. It increases when the primary air is initially increased to 10% of stoichiometric value and then gradually decreases with a further increase in primary air. The axial extent of the contour increases with a gradual increase in primary air in the mixture. Even though the entropy generated per unit volume increases with primary air, its distribution in the flame zone decreases, as indicated by the low values of integrated entropy generation. The entropy generation field clearly shows the movement of the reaction zone close to the burner axis with an increase in primary air flow rate.

### 3.3 *Entropy Generation in Partially Premixed Flames with Stoichiometric Coflow*

Figure 10 presents the entropy generation during heat transfer in flames with varying primary air and stoichiometric air in coflow. The entropy generation increases notably when primary air flow rate is increased to 10% of stoichiometric value. It then increases gradually with a further increase in primary air and produces a local minimum around 40% of stoichiometric value. The axial extent of the contours increases on the fuel side but decreases on the air side. The region of maximum entropy generation moves from the burner rim to the fuel side, as observed earlier

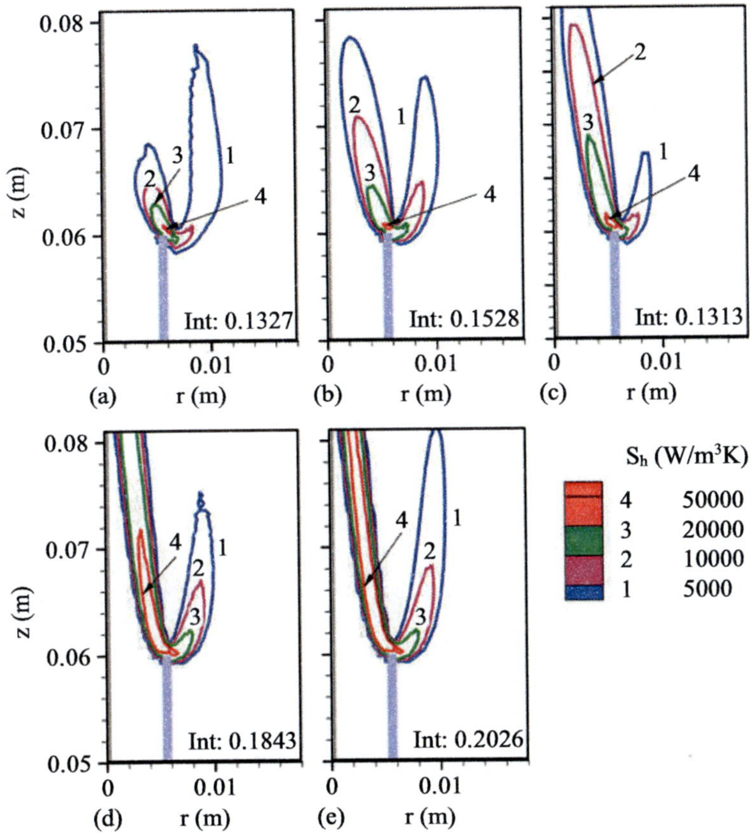


**Fig. 7** Rate of entropy generation (W/m$^3$-K) during heat transfer ($S_h$) in flames with **a** 0%, **b** 10%, **c** 20%, **d** 40% and **e** 50% primary air. Integrated values (W/K) are shown within the plots

(Fig. 7). The radial extent of the flame on the air side remains almost a constant for all cases.

The entropy generation during mass transfer is negligible and shows a similar trend as in partially premixed flames in the absence of coflow (not shown here).

Figure 11 shows the entropy generation during chemical reactions in flames with varying primary air and stoichiometric air in coflow stream. The integrated values of $S_r$ do not show a significant variation with primary air. However, the axial extent of the entropy generation contours and its distribution depends on the amount of primary air in the mixture. The radial extent of the contours also decreases with an increase in the flow rate of primary air.

Figure 12 presents the integrated values of entropy generation in coflow diffusion and partially premixed flames. In coflow flames (Fig. 12a), entropy generation during heat transfer and mass transfer increases almost continuously with an increase in the rate of coflow air. On the other hand, the entropy generation during chemical reactions

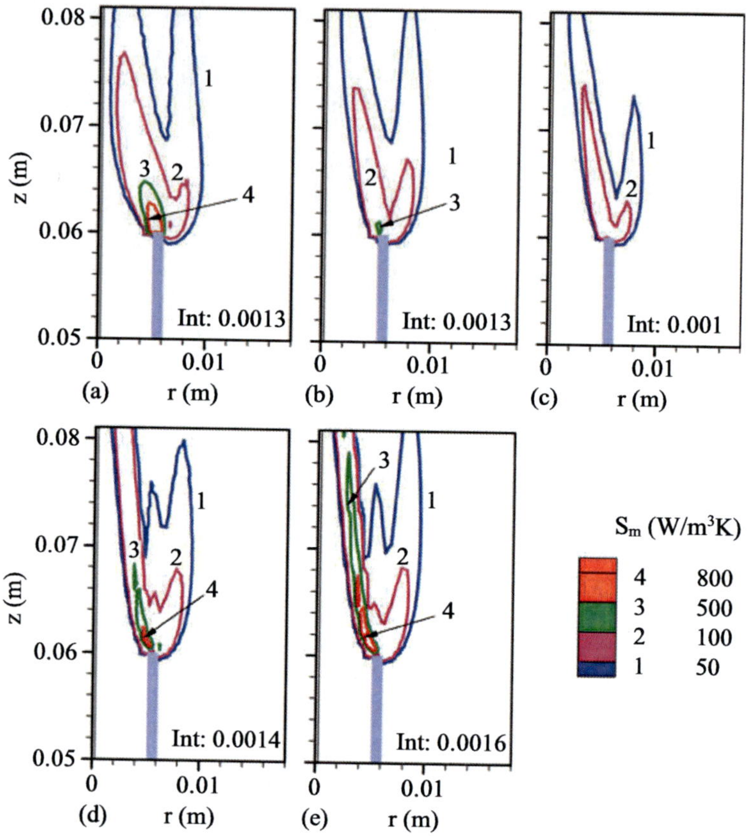


**Fig. 8** Rate of entropy generation (W/m$^3$-K) during mass transfer ($S_m$) in flames with **a** 0%, **b** 10%, **c** 20%, **d** 40% and **e** 50% primary air. Integrated values (W/K) are shown within the plots

shows a non-monotonous trend and displays a local minimum around 150% coflow air. The total entropy generation notably increases with 100% coflow. It presents a local minimum around 150% coflow air, as that of $S_r$, and then increases gradually with further increase in the rate of coflow air. It should be noted that soot production was reported to be the maximum for coflow air of around 150% and it decreased significantly for coflow air of 200% [10]. Therefore, the burner can be operated with 150–200% coflow air, keeping in mind the entropy generation data.

Comparison of entropy generation in partially premixed flames (Fig. 12b) shows that entropy generation through all modes shows a local maximum and minimum around 10% and 20% primary air in the fuel mixture, respectively. As the primary air flow rate is increased above 20% of stoichiometric value, the entropy generation during chemical reactions decreases, but the entropy generations during heat transfer and mass transfer processes increase. For the case with 40% primary air, the entropy

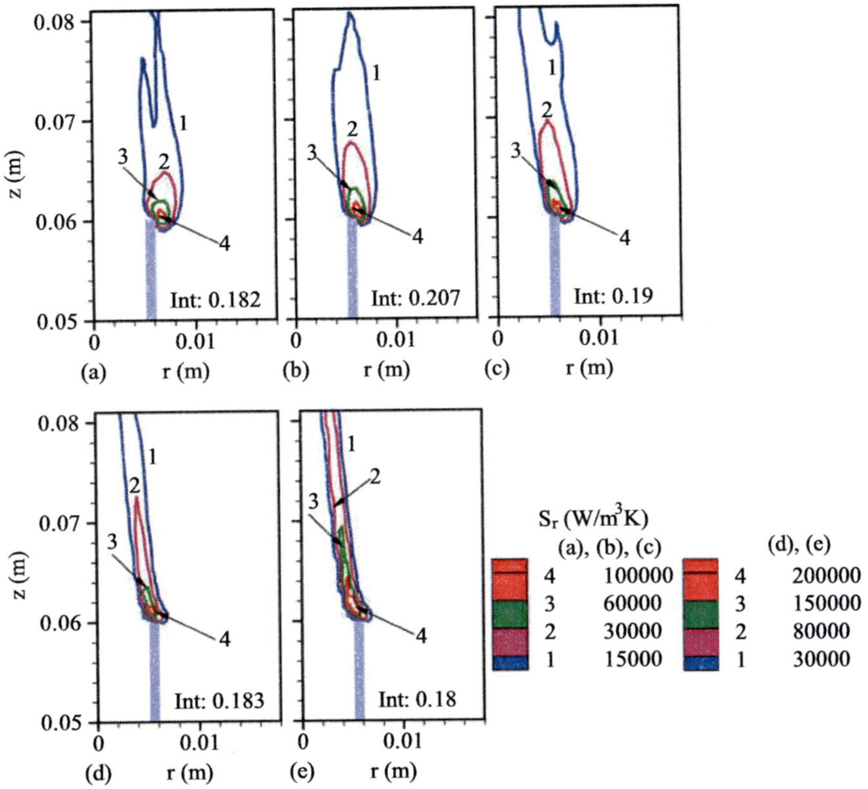


**Fig. 9** Rate of entropy generation (W/m$^3$-K) during chemical reactions ($S_r$) in flames with **a** 0%, **b** 10%, **c** 20%, **d** 40% and **e** 50% primary air. Integrated values (W/K) are shown within the plots

generation during heat transfer and chemical reactions is almost comparable to each other.

In partially premixed flames with stoichiometric coflow rate (Fig. 12c), the entropy generation in the process of mass transfer decreases with an increase in primary air flow rate up to 40% of stoichiometric value. However, entropy generations during heat transfer and chemical reactions do not change notably till primary air is 40% of the stoichiometric value. Both $S_h$ and $S_m$ increase notably when primary air further increased to 50%. However, the entropy generation during chemical reactions shows negligible variations with further increase in primary air. The total entropy generation follows the same trend as entropy generation in the process of heat transfer and has a maximum value at 50% stoichiometric air.

For partially premixed flames, with and without coflow air, the soot production was reported to be negligible when primary air was increased to 40% of stoichiometric value [10]. Keeping the note of total entropy generation, the LPG burner with 100%

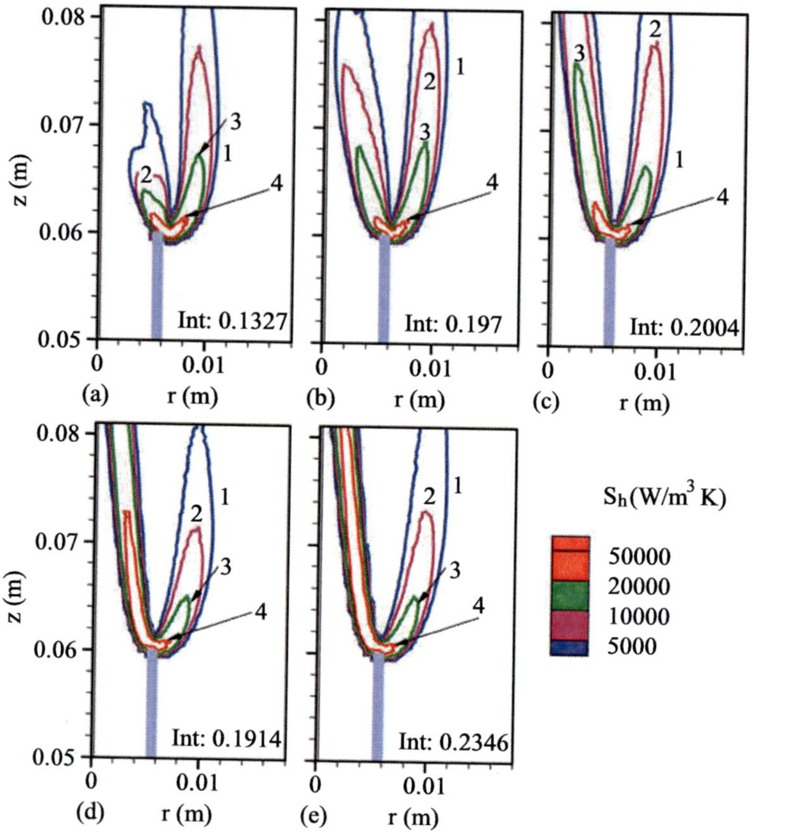


**Fig. 10** Rate of entropy generation (W/m$^3$-K) during heat transfer ($S_h$) in flames with **a** 0%, **b** 10%, **c** 20%, **d** 40% and **e** 50% primary air and 100% coflow air. Integrated values (W/K) are shown within the plots

coflow air can be operated at around 40% primary air. In the absence of coflow air, partial premixing in the range of 10% to 20% of primary air seems to be beneficial.

## 4 Conclusions

A systematic study of entropy generation in LPG-air coflow flames is carried out to calculate the total entropy generation in various configurations. The entropy generations during heat transfer, mass transfer and chemical reactions have been thoroughly investigated for each case. The important conclusions from the present study are:

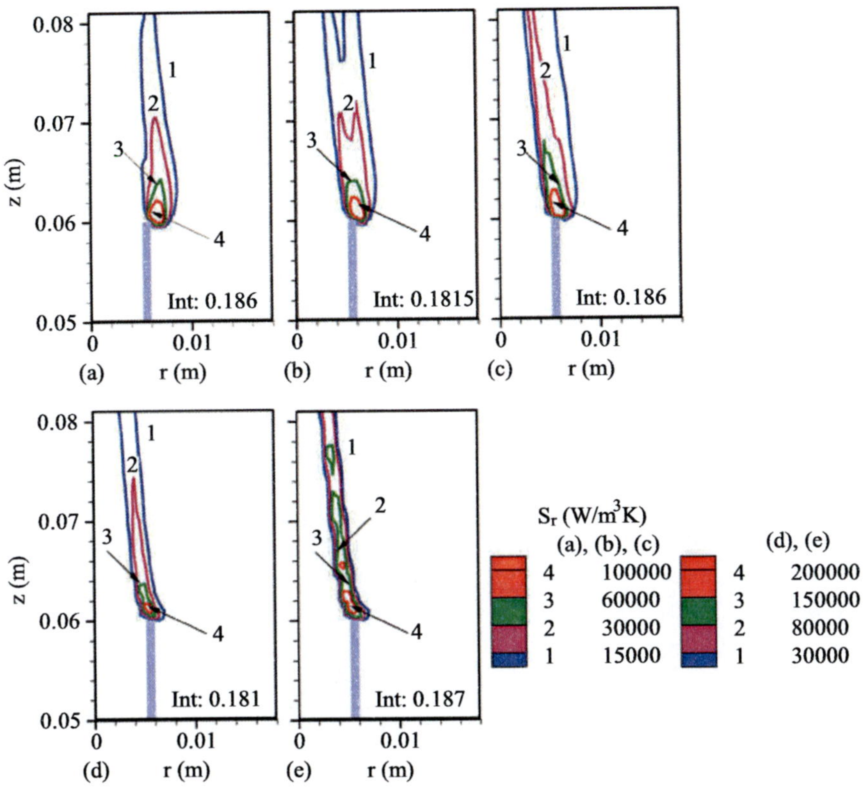


**Fig. 11** Rate of entropy generation (W/m$^3$-K) during chemical reactions ($S_r$) in flames with **a** 0%, **b** 10%, **c** 20%, **d** 40% and **e** 50% primary air and 100% coflow air. Integrated values (W/K) are shown within the plots

1. Entropy generation during chemical reactions is dominant in pure jet diffusion flame, but entropy generation in the process of heat transfer equally dominates in coflow diffusion flames.
2. Entropy generation during chemical reactions has a major contribution in partially premixed flame up to 40% primary air.
3. In coflow partially premixed flames, the total entropy generation remains almost a constant up to 40% primary air and increases notably for 50% primary aeration.
4. Entropy generation during mass transfer is the lowest in all flames.
5. Total entropy generation is minimum for cases with around 150% coflow air (no partial premixing), with around 20% primary air (without air coflow) and with around 40% primary air (with stoichiometric air coflow).

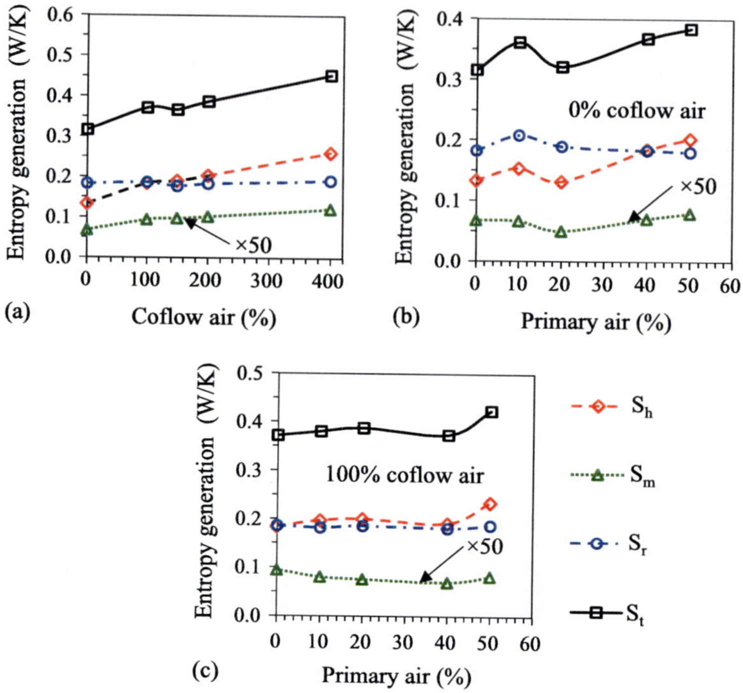


**Fig. 12** Entropy generation (W/K) in LPG-air **a** coflow air flames, **b** partially premixed flames and **c** partially premixed flames in stoichiometric coflow

## References

1. Datta A (2000) Entropy generation in a confined laminar diffusion flame. Combust Sci Technol 159:39–56
2. Datta A (2005) Effects of gravity on structure and entropy generation of confined laminar diffusion flames. Int J Therm Sci 44(5):429–440
3. Chen S, Han H, Liu Z, Li J, Zheng C (2010) Analysis of entropy generation in non-premixed hydrogen versus heated air counter-flow combustion. Int J Hydrogen Energy 35:4736–4746
4. Briones AM, Mukhopadhyay A, Aggarwal SK (2009) Analysis of entropy generation in hydrogen-enriched methane-air propagating triple flames. Int J Hydrogen Energy 34(2):1074–1083
5. Bazdidi-Tehrani F, Abedinejad MS, Mohammadi M (2019) Analysis of relationship between entropy generation and soot formation in turbulent kerosene/air jet diffusion flames. Energy Fuels 33(9):9184–9195
6. Morsli S, Sabeur A, El Ganaoui M, Ramenah H (2018) Computational simulation of entropy generation in a combustion chamber using a single burner. Entropy 20(12):922
7. Arjmandi H, Amani E (2015) A numerical investigation of the entropy generation in and thermodynamic optimization of a combustion chamber. Energy 81:706–718
8. Hosseini SE, Bagheri G, Wahid MA (2014) Effect of hydrogen addition on the entropy generation of biogas conventional combustion. Jurnal Teknologi 66(2)
9. Jiang D, Yang W, Teng J (2015) Entropy generation analysis of fuel lean premixed $CO/H_2$/air flames. Int J Hydrogen Energy 40(15):5210–5220

10. Kumaran SM, Raghavan V (2020) Numerical study of characteristics of liquefied petroleum gas flames in coflow air. Thermal Sci Eng Prog 20:100759
11. Kumaran SM, Shanmugasundaram D, Narayanaswamy K, Raghavan V (2021) Reduced mechanism for flames of propane, n-butane, and their mixtures for application to burners: development and validation. Int J Chem Kinet 53(6):731–750
12. Pope DN, Raghavan V, Gogos G (2010) Gas-phase entropy generation during transient methanol droplet combustion. Int J Therm Sci 49:1288–1302

# Effect of Addition of Carbon Monoxide and Hydrogen in Syngas Inverse Diffusion Flames

**Mohd. Ibrahim, S. Muthu Kumaran, and V. Raghavan**

## 1 Introduction

Concerns about depleting fossil fuels and environmental effects are addressed by the usage of alternative fuels. Syngas composition is affected by feedstock, operating parameters, reactor type and oxidizer used [1]. Different compositions of syngas have effects on burning and emission characteristics of syngas flames [2]. Inverse coflow diffusion burner is reported to have several advantages [3] such as high combustion efficiency and better emission characteristics than the normal coflow diffusion burner.

Wu and Essenhigh [4] identified six types of flames after observing the appearance and stability of different flames. They also reported the axial and radial distribution of temperature and species concentrations measured in inverse diffusion flame fueled by methane. Park et al. [5] studied the effect of $CO_2$ dilution in syngas ($H_2$–CO mixture) flames. With the increase of CO and $CO_2$, loss of heat by flame radiation increases. It was also reported that the addition of $CO_2$ affects the oxidation pathway of $H_2$ and CO. Stelzner et al. [6] carried out experimental and numerical study of confined IDF by employing pure oxygen. Radiation and diffusion effects strongly influenced the structure of these flames. Soret effect was found important for such flames and must be included in the numerical simulations. Huanhuan et al. [7] reported the effects of $H_2O$ and $CO_2$ diluted oxidizer on the flame shape and structure of syngas ($H_2$–CO mixture) coflow diffusion flames. Replacement of $N_2$ with $CO_2$ and $H_2O$ reduced the peak temperature. Radiation heat loss also increased with an addition of $CO_2$ and $H_2O$. Addition of $H_2O$ enhanced the production of OH radicals and the addition of $CO_2$ reduced the OH concentration. Flame height and radius decreased with addition of $H_2O$, and it increased with the addition of $CO_2$. Patel and Shah [8] investigated the consequences of hydrogen addition on the characteristics of methane IDF (swirling and non-swirling). Flame height decreased with hydrogen

Mohd. Ibrahim · S. M. Kumaran (✉) · V. Raghavan
Department of Mechanical Engineering, Indian Institute of Technology Madras, Chennai, India
e-mail: kumaranms92@gmail.com

 
N. L. Narasimhan et al. (eds.), *Recent Advances in Energy Technologies*, Lecture Notes in Mechanical Engineering, https://doi.org/10.1007/978-981-19-3467-4_35

enrichment and CO emission decreased significantly. Effects were more predominant in non-swirling IDF. With increase of hydrogen content, combustion noise increases. Mahesh and Mishra [9] experimentally studied turbulent IDF of liquefied petroleum gas (LPG) in a burner having backward-facing step. As fuel velocity increases, blow off velocity was reported to rise monotonically. As air to fuel velocity ratio increases, visible flame length decreased, but length of visible blue flame increases. This was due to improved fuel–air mixing.

It is clear that several studies on syngas inverse diffusion flames are reported in literature. Effects of diluents such as $CO_2$, $H_2O$ and $N_2$ on flame characteristics have also been reported. However, flame characteristics of varying $H_2$ and CO in fuel are scarce. Variation of air flow rate on the burning characteristics of syngas mixture composed of different $H_2$ and CO content are also scanty. These form the motivation of the present study. Here, a comprehensive numerical model has been employed to study effect of $H_2$, CO content and air flow rate in inverse coflow diffusion flames of typical syngas fuels produced during gasification of solid fuels using different reactors. Fields of temperature, flow and species are investigated and reaction rate profiles are plotted to understand the characteristics of these flames.

## 2 Numerical Model

Commercial CFD software, ANSYS-FLUENT 17.2 [10] is used to solve steady state governing equations for conservations of mass, momentum, energy and species in a segregated manner. The model includes full multi-component diffusion, Soret diffusion and diffusion energy source to model species diffusion. Thermo-physical parameters that are dependent on temperature and species concentration, a detailed kinetics mechanism and an optically thin approximation-based radiation model are also included. For convection terms, a second-order upwind method is used. Coupling of pressure and velocity is solved using the SIMPLE algorithm. Arrhenius rate parameters are used to calculate net rate of generation or annihilation of species during chemical processes, which is then solved concurrently using a stiff-chemistry solver. Chemical kinetic mechanism based on GRI 2.11, with 49 species and 279 elementary reactions [11] have been employed in the simulations. An optically thin radiation model reported by Barlow et al. [12] has been used to model the thermal radiation absorbed by species such as $CH_4$, CO, $CO_2$ and $H_2O$. The radiation heat loss per unit volume ($W/m^3$) is written as,

$$q_r = 4 \times \sigma \times \left(T^4 - T_b^4\right) \times \sum \left(p_i \times a_{pi}\right). \tag{1}$$

Here, $\sigma$ is Stefan-Boltzmann constant ($5.67 \times 10^{-8}$ $W/m^2K^4$), $T$ represents local flame temperature in $K$, $T_b$ shows ambient temperature in $K$, $p_i$ is the partial pressure of species '$i$' in $atm$ and $a_{pi}$ is the Plank mean absorption coefficient ($atm^{-1}.m^{-1}$) for every species '$i$', which further depends on temperature and species mole fractions

[12]. To account this radiation loss, an extra sink term is added in the energy equation, through a user defined function (UDF) written in C (Eq. (1)).

## 2.1 Computational Domain

The schematic of axisymmetric computational domain with boundary conditions is shown in Fig. 1. The inverse diffusion flame burner used in this study is similar to that reported by Wu and Essenhigh [4]. The burner consists of central tube of diameter 11 mm and a coflow tube of diameter 27 mm. Fuel and air inlets are specified with a given velocity, temperature and corresponding mass fractions of species involved. Free boundary is open to atmosphere. Pressure is specified here and based on the pressure difference between domain and the ambient, either an inflow of ambient air or outflow of product gases occur at this boundary.

In case of outflow, all the flow variables are extrapolated from the interior cells adjacent to this boundary. For inflow, the oxygen mass fraction is set as 0.232 (rest nitrogen) and a temperature of 300 K have been specified. At the axis, first derivatives of all the variables in normal direction and normal component of velocity are set to zero. Walls are considered adiabatic, and a zero value is specified for the velocity components and first derivatives of species mass fractions. For energy, convergence criteria is given as normalized residual value of $10^{-6}$, for continuity $10^{-3}$ and $10^{-3}$ for all other variables. In addition, mass imbalance is kept within 1%.

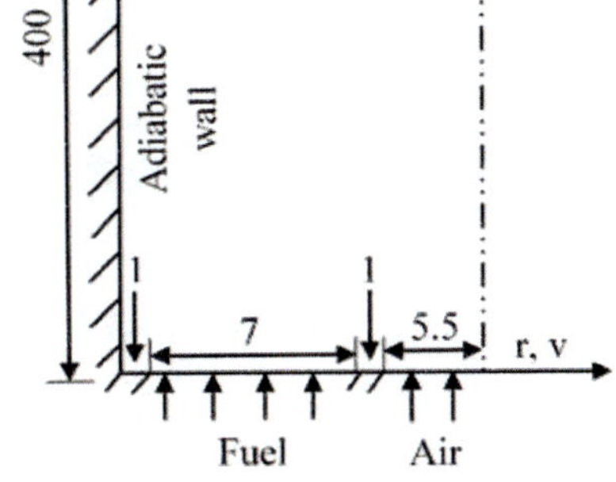


**Fig. 1** Schematic of the computational domain with boundary conditions (Dimensions in mm)

## 2.2 *Grid Independent Study and Validation*

Computational domain is divided into uniform cells in both directions (i.e. radial and axial). Three grids with minimum sizes of 1 mm, 0.5 mm and 0.25 mm with 6000, 23,200 and 92,800 cells, respectively, have been used. The boundary conditions, as reported in the experiments in Wu and Essenhigh [4], have been used.

Figure 2a, b shows the comparison of numerical predictions of temperature and species profiles with the experimental data reported in Wu and Essenhigh [4]. It is clear that numerical model is capable of predicting trends quite reasonably. The usage of simplified radiation model and considering pyrex glass confinement used in the experiments as an adiabatic wall are the reasons for the over prediction of experimental data. Moreover, uncertainties in the temperature and species measurements have not been reported in Wu and Essenhigh [4]. Furthermore, results reveal that changing the size from 0.5 mm to 0.25 mm has little effect on temperature and species profiles. As a result, a mesh with a minimum cell size of 0.5 mm is selected for further investigation, keeping in mind computational time and economy.

# 3 Results and Discussion

Four synthetic gases, produced from different gasification processes, as reported by Couto et al. [13], have been considered. The compositions of these syngas fuels and their corresponding lower heating value (LHV) are presented in Table 1. All these reactors employ air as the gasification agent. Table 1 also reports average values of the components in these fuels.

Characteristics of inverse coflow diffusion flames of these syngas fuels have been reported elsewhere [14]. Even though the LHV is of similar order, due to variations in $H_2$ and CO contents, these fuels can display different characteristics. Therefore, the effects of $H_2$ and CO in syngas have been analyzed. For this purpose, $H_2$ and

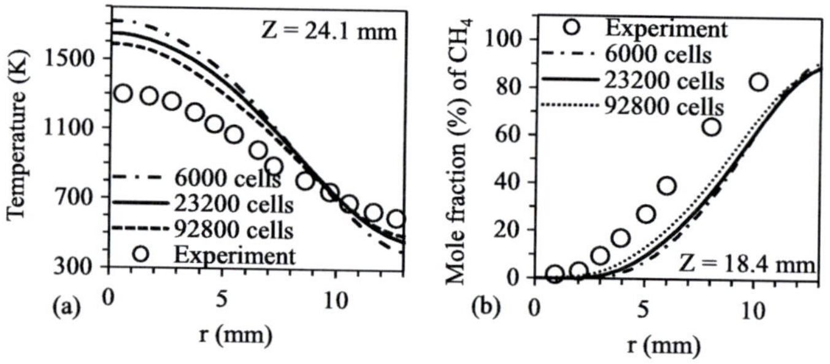


**Fig. 2** Comparison of predicted radial profiles of **a** temperature and **b** mole fractions of $CH_4$ in inverse diffusion flame (IDF) at different height from the burner exit with experimental data [4] using different grids

**Table 1** Composition of typical syngas fuels [13]

| Syngas | Reactor | Syngas composition (% volume) | | | | | LHV ($MJ/m^3$) |
|---|---|---|---|---|---|---|---|
| | | $H_2$ | CO | $CO_2$ | $CH_4$ | $N_2$ | |
| A | BFB | 9 | 14 | 20 | 7 | 50 | 5.24 |
| B | Updraft fixed bed | 11 | 24 | 9 | 3 | 53 | 5.29 |
| C | CFB | 14.1 | 18.7 | 14.7 | 3.5 | 49 | 5.13 |
| D | Downdraft fixed bed | 17 | 21 | 13 | 1 | 48 | 4.84 |
| | Average values | 12.8 | 19.4 | 14.2 | 3.6 | 50 | |

CO are varied individually and keeping average values reported in Table 1 for others components. Table 2 reports the cases (1–4) used to study the effect of varying $H_2$ on flame characteristics. Here, hydrogen is varied exactly as listed in Table 1, from 9 to 17%, and average values, as shown in Table 1, are used for CO, $CO_2$ and $CH_4$.

Similarly, Table 3 presents the cases (5–8) used to study the effect of varying CO on flame characteristics. Here, CO is varied exactly as listed in Table 1, from 14 to 24%, and average values are used for $H_2$, $CO_2$ and $CH_4$ as in Table 1. The power rating is kept as 0.25 kW for all the cases. Air flow rate has been varied as 100%, 125%, 167% and 200% of stoichiometric air required for each case. Based on the LHV of the fuel mixture, fuel flow rate and its velocity are calculated. Stoichiometric air

**Table 2** List of cases with varying $H_2$

| Cases | Syngas composition (%) | | | | | Fuel velocity (m/s) | % stoichiometric air | Air velocity (m/s) |
|---|---|---|---|---|---|---|---|---|
| | $H_2$ | CO | $CO_2$ | $CH_4$ | $N_2$ | | | |
| Case 1 | 9 | 19.4 | 14.2 | 3.6 | 53.8 | 0.121 | 100 | 0.573 |
| | | | | | | | 125 | 0.717 |
| | | | | | | | 167 | 0.956 |
| | | | | | | | 200 | 1.146 |
| Case 2 | 11 | 19.4 | 14.2 | 3.6 | 51.8 | 0.115 | 100 | 0.570 |
| | | | | | | | 125 | 0.713 |
| | | | | | | | 167 | 0.951 |
| | | | | | | | 200 | 1.140 |
| Case 3 | 14.1 | 19.4 | 14.2 | 3.6 | 48.7 | 0.108 | 100 | 0.571 |
| | | | | | | | 125 | 0.713 |
| | | | | | | | 167 | 0.951 |
| | | | | | | | 200 | 1.142 |
| Case 4 | 17 | 19.4 | 14.2 | 3.6 | 45.8 | 0.102 | 100 | 0.570 |
| | | | | | | | 125 | 0.712 |
| | | | | | | | 167 | 0.950 |
| | | | | | | | 200 | 1.140 |

**Table 3** List of cases with varying CO

| Cases | Syngas composition (%) | | | | | Fuel velocity (m/s) | % stoichiometric air | Air velocity (m/s) |
|---|---|---|---|---|---|---|---|---|
| | CO | $H_2$ | $CO_2$ | $CH_4$ | $N_2$ | | | |
| Case 5 | 14 | 12.8 | 14.2 | 3.6 | 55.4 | 0.128 | 100 | 0.583 |
| | | | | | | | 125 | 0.730 |
| | | | | | | | 167 | 0.973 |
| | | | | | | | 200 | 1.166 |
| Case 6 | 18.7 | 12.8 | 14.2 | 3.6 | 50.7 | 0.113 | 100 | 0.573 |
| | | | | | | | 125 | 0.717 |
| | | | | | | | 167 | 0.956 |
| | | | | | | | 200 | 1.146 |
| Case 7 | 21 | 12.8 | 14.2 | 3.6 | 48.4 | 0.107 | 100 | 0.569 |
| | | | | | | | 125 | 0.712 |
| | | | | | | | 167 | 0.949 |
| | | | | | | | 200 | 1.138 |
| Case 8 | 24 | 12.8 | 14.2 | 3.6 | 45.4 | 0.100 | 100 | 0.565 |
| | | | | | | | 125 | 0.706 |
| | | | | | | | 167 | 0.941 |
| | | | | | | | 200 | 1.130 |

required for the estimated fuel flow rate is then calculated. Air velocity is calculated based on this.

## 3.1 *Effects of Varying $H_2$ and CO*

In this section, $H_2$ and CO in syngas are varied systematically, keeping the other species at their average values, and the resultant flame characteristics are studied. First, hydrogen is varied as per the composition and flow rates listed in Table 2. The power rating of the flame is chosen as 250 W, and the air flow rate is kept at 125% of stoichiometric air.

Figure 3 (top) shows the contours of temperature (lines) along with mass fraction of OH (grayscale) for cases listed in Table 2 with varying $H_2$. At this air flow rate, all flames are attached to the burner rim. The amount of $H_2$ in syngas has a direct impact in the temperature field and the maximum temperature attained by the flame; the maximum temperature and the extent of 1600 K isotherm increase with $H_2$ content. As the amount of $H_2$ in the fuel mixture increases, the axial extents of the flame (as indicated by isotherm of 800 K in Fig. 3) decreases. This is because of low density and high diffusivity of hydrogen as compared to other components in the fuel.

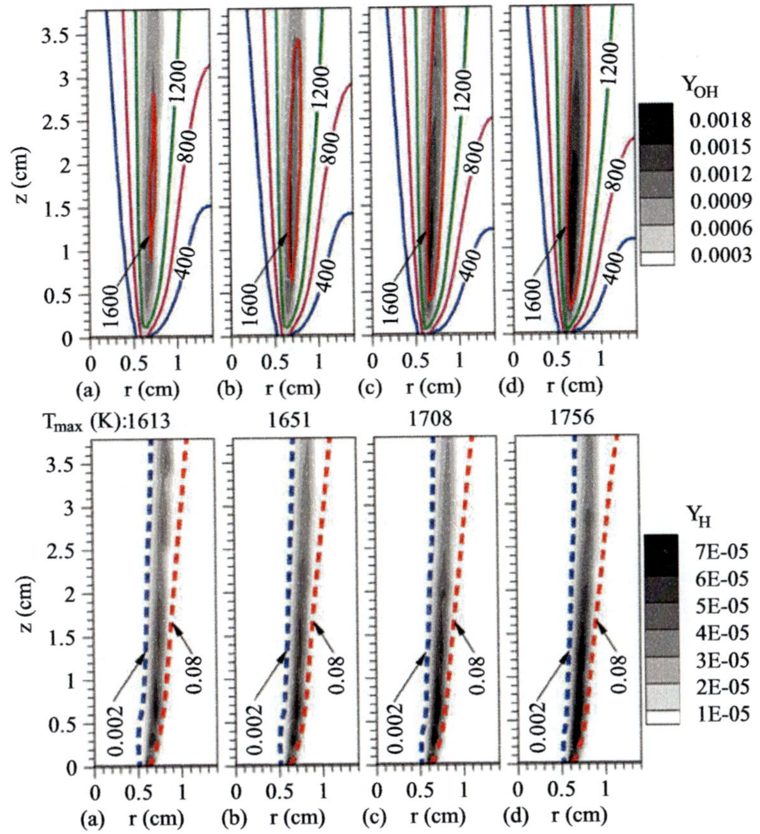


**Fig. 3** (Top) Temperature contours (lines) and mass fraction of OH (greyscale) and (bottom) contours of mass fractions of CO (lines) and H (greyscale) for **a** Case 1 (9% $H_2$), **b** Case 2 (11% $H_2$), **c** Case 3 (14.1% $H_2$) and **d** Case 4 (17% $H_2$) with 125% stoichiometric air

The intensity and extent of OH mass fraction contours increase notably, as $H_2$ in the fuel mixture increases. Figure 3 (bottom) shows distribution of H mass fractions (gray scale) and line contours of CO. It is clear that CO contour line of 0.002 binds the H field on the air side and OH field extends to left of CO contour line of 0.002, indicating the reaction of CO and OH on the air side to produce $CO_2$. The reaction between $H_2$ and OH takes place on the fuel side to produce $H_2O$.

The effect of varying hydrogen content in IDF flames is further investigated by studying the reaction rates profiles of important reactions given as:

$$O + H_2 \leftrightarrow H + OH \tag{R1}$$

$$H + O_2 \leftrightarrow O + OH \tag{R2}$$

$$CO + OH \leftrightarrow CO_2 + H \tag{R3}$$

$$H_2 + OH \leftrightarrow H_2O + H \tag{R4}$$

The chain propagation reactions, R1 and R2, are important for sustaining the flame. Reactions R3 and R4 are important for the formation of stable products, $CO_2$ and $H_2O$, and also account for major heat release in the flame. A significant amount of the OH radicals produced through R1 and R2 is consumed in the reactions R3 and R4.

Figure 4 shows the net reaction rate profiles of major species such as $H_2$, $O_2$, OH, CO, $CO_2$ and $H_2O$, as well as the rates of reactions R1, R2, R3 and R4 at 5 mm from burner exit for Case 1 (9% $H_2$) and Case 4 (17% $H_2$) with 125% stoichiometric air.

The net consumption rate of oxygen and net production rate of $H_2O$ increases with an increase in $H_2$ content (Fig. 4a, b). On the other hand, the net consumption rate of CO and the net production rate of $CO_2$ decreases with an increase in $H_2$ concentration. Even though the production of OH through reactions R1 and R2 increases with an increase in hydrogen content (Fig. 4c, d), it reacts with $H_2$ to form $H_2O$ through reaction R4 as indicated by the high rate of reaction R4.

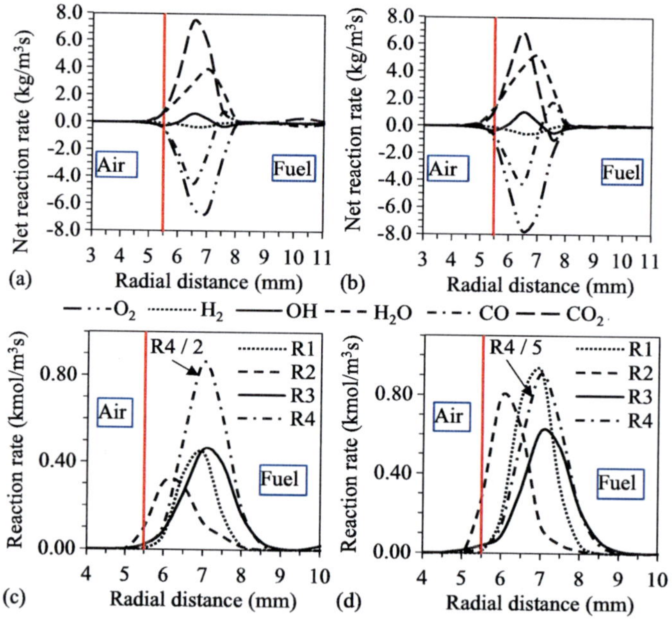


**Fig. 4** Predicted radial profiles of **a**, **b** net reaction rate and **c**, **d** rate of reactions R1, R2, R3 and R4 at an axial distance of 5 mm for **a** Case 1 (9% $H_2$) and **b** Case 4 (17% $H_2$) with 125% stoichiometric air [Vertical line indicates boundary between the air and fuel streams]

Next, CO is varied as per the composition and flow rates listed in Table 3. Figure 5 (top) shows the temperature contours for cases 5, 6, 7 and 8, as listed in Table 3, with a supply of 125% of stoichiometric air. Power rating is kept at 250 W.

Also, in these cases, the flames are attached to the burner rim. Even though the maximum temperatures are of similar order as observed in varying $H_2$ cases, the extent of 1600 K isotherm is higher in these cases. Higher stability is obtained with increasing CO as the reaction zone move towards the burner rim. As CO is converted to $CO_2$, due to the exothermic reaction, the maximum flame temperature increases with increasing CO content. The intensity of OH mass fraction increases relatively gradually with increasing CO content. The maximum OH mass fraction is slightly lower than that observed in cases with higher hydrogen content (Fig. 3). Figure 5 (bottom) portrays H field along with CO contour lines. Intensity of H is relatively lesser for cases 5–8, when compared to cases 1–4.

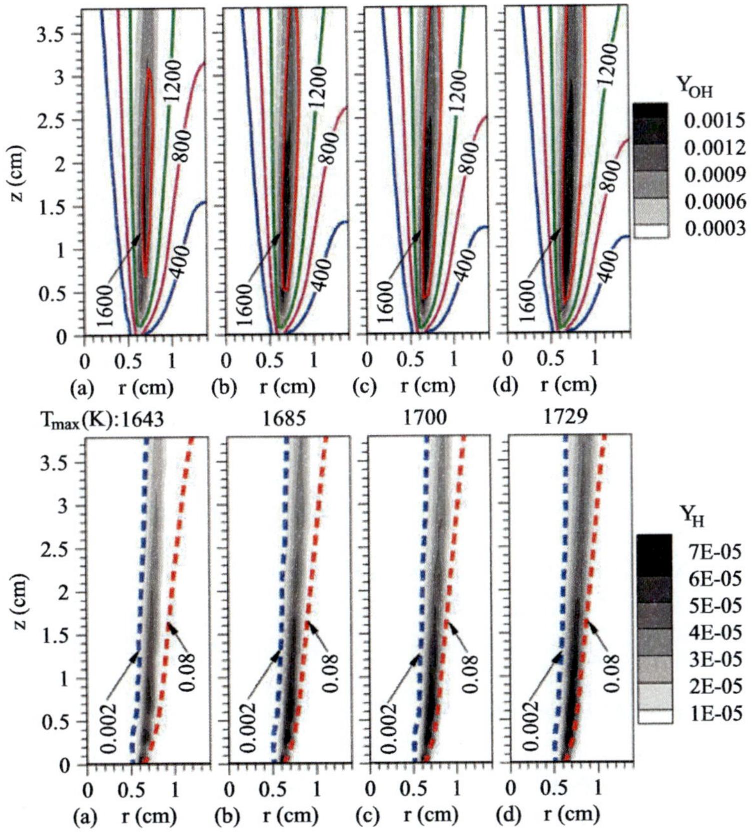


**Fig. 5** (top) Temperature contours (lines) and mass fraction of OH (greyscale) and (bottom) contours of mass fractions of CO (lines) and H (greyscale) for **a** Case 5 (14% CO), **b** Case 6 (18.7% CO), **c** Case 7 (21% CO) and **d** Case 8 (24% CO) with 125% stoichiometric air

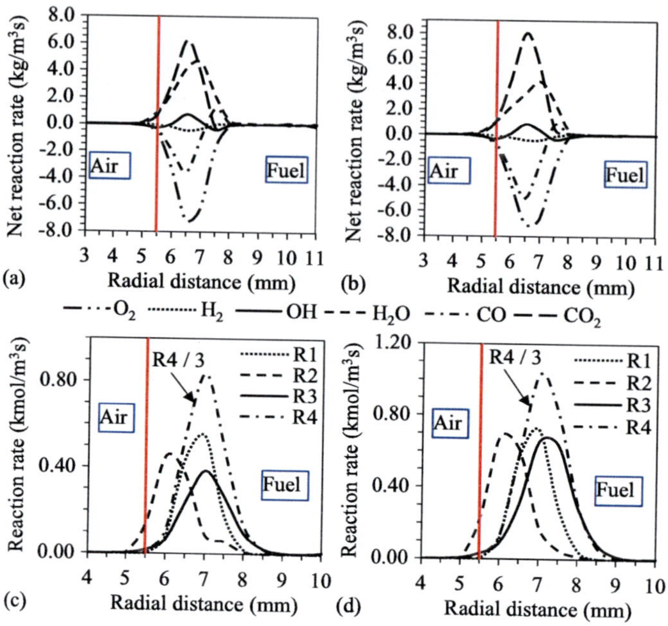


**Fig. 6** Predicted radial profiles of **a**, **b** net reaction rate and **c**, **d** rate of reactions R1, R2, R3 and R4 at an axial distance of 5 mm for **a** Case 5 (14% CO) and **b** Case 8 (24% CO) with 125% stoichiometric air [Vertical line indicates boundary between the air and fuel streams]

Figure 6 shows the net reaction rate profiles of major species and rates of reactions R1, R2, R3 and R4 at 5 mm from burner exit for Case 5 (14% CO) and Case 8 (24% CO) with 125% stoichiometric air. The net consumption rate of oxygen is almost the same in both the cases (Fig. 6a, b). However, the net consumption rate of CO and the net production rate of $CO_2$ increases with an increase in CO concentration. The rate of production of OH through reaction R1 is almost the same for both the cases, whereas, its production through reaction R2 is higher for the case with high CO content (Fig. 6c, d). Therefore, the net production rate of OH is marginally higher for the case with 24% CO. Similarly, the rate of production of $CO_2$ through reaction R4 is higher for case with high CO content.

## 3.2 *Effect of Air Flow Rate*

In this section, the effect of air flow rate on the characteristics of IDF is studied. Case 1 (Table 2) with lesser hydrogen is chosen, and air flow rate is varied as 100%, 125%, 167% and 200% of stoichiometric air requirement. Power rating is chosen as 250 W.

Figure 7 (top) shows the temperature contours for these cases. It is clear that the maximum temperatures do not vary much in these cases, due to fixed composition of the fuel. The extent of maximum temperature decreases slightly, as indicated by the contour line of 1600 K, when air flow rate is increased from 100 to 125% of stoichiometric value. As air flow rate is increased to 167% of the stoichiometric value, the extent of the maximum flame temperature decreases due to convective cooling by increased air flow rate. For the given $H_2$ content in Case 1, flame is not able to stabilize itself close to the burner rim, when the air flow rate is increased to 200% of stoichiometric air; the increased air velocity decreases the residence time for reactants to mix at stoichiometric proportions close to the burner exit. The flame lifts-off to about 7 mm from the burner rim. The flame temperature also increases marginally due to partial premixing happening at the flame base. The intensity of H shown along with isotherms, remains almost the same for all cases.

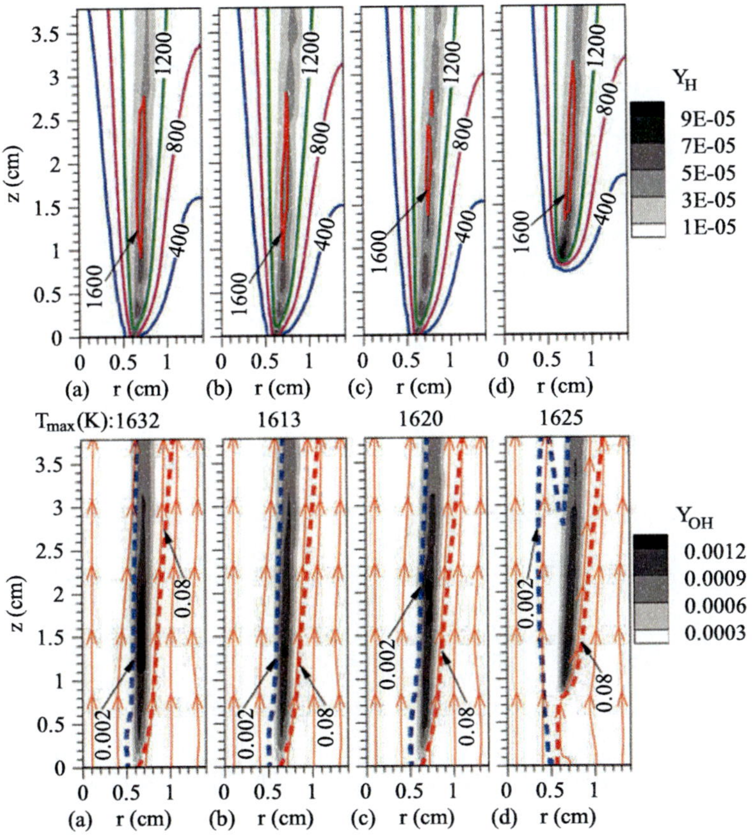


**Fig. 7** (Top) Temperature contours (lines) and mass fraction of H (greyscale) and (bottom) contours of mass fractions of OH (greyscale) and CO (lines) for Case 1 with **a** 100%, **b** 125%, **c** 167% and **d** 200% of stoichiometric air

Figure 7 (bottom) shows the distribution of CO mass fractions (greyscale) and OH (lines). In Fig. 7 (bottom) it is shown that the CO contour line of 0.002 lies almost near the OH field (close to the middle region). This clearly indicates that OH reacts with H in the fuel side and it reacts with OH in the CO side to produce $H_2O$ and $CO_2$, respectively, within a narrow region. The curvature in the CO contour line of 0.08 indicates the flame anchoring position. The intensity of OH field is almost unaffected as the fuel composition is unaltered and it follows the extent of maximum temperature region shown in Fig. 7 (top).

Figure 8 shows the net reaction rate profiles of major species as well as rates of reactions R1, R2, R3 and R4 at 10 mm from burner exit for Case 1 (9% CO) with 100% and 200% stoichiometric air. The net consumption rate of oxygen is significantly higher for the case with 200% stoichiometric air (Fig. 8b). This is due to the presence of a lifted flame in this case. Subsequently, the net consumption rate of CO and the net production rate of $CO_2$ increases for this case. The rate of production of OH through reaction R1 is higher for the case with 200% stoichiometric air (Fig. 8d). The production of OH through reaction R2 is higher for case with 100% stoichiometric air (Fig. 8c). Similarly, the high rate of reaction R4 is the reason for the high net production reaction rate of $H_2O$ in case with 200% stoichiometric air (Fig. 8b, d).

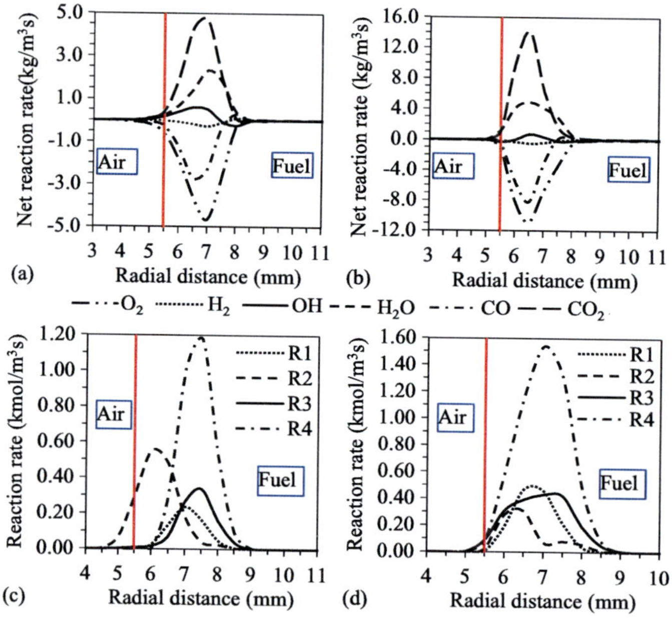


**Fig. 8** Predicted radial profiles of **a**, **b** net reaction rate and **c**, **d** rate of reactions R1, R2, R3 and R4 at an axial distance of 10 mm for Case 1 (9% $H_2$) with (left) 100% and (right) 200% stoichiometric air [Vertical line indicates boundary between the air and fuel streams]

Figure 9 shows the temperature, H, CO and OH contours for Case 4 with 17% hydrogen, with various air flow rates. Due to the higher percent of hydrogen in the fuel, maximum temperature is higher for this case by around 100 K, as compared to that of Case 1. The higher percent of hydrogen also helps in improving the stability of the flame, as seen in the case of 200% stoichiometric air (Fig. 9d), where the flame anchors close to the burner rim, as opposed to the lifted flame observed for Case 1 in Fig. 7d. The intensities of OH and H have increased for this case. In summary, higher hydrogen content in the fuel is able to offer increased stability for the primary reaction zone at air flow rates as high as 200% of stoichiometric value.

A case with varying CO, case 5 (Table 3), is selected to analyze the effects of air flow rate on flame characteristics. As compared to case 1 (Table 2), case 5 has higher $H_2$ and lower CO content, and percentages of $CO_2$ and $CH_4$ are the same in both. Figure 10 (top) shows the contours of temperature. Slight variations in the

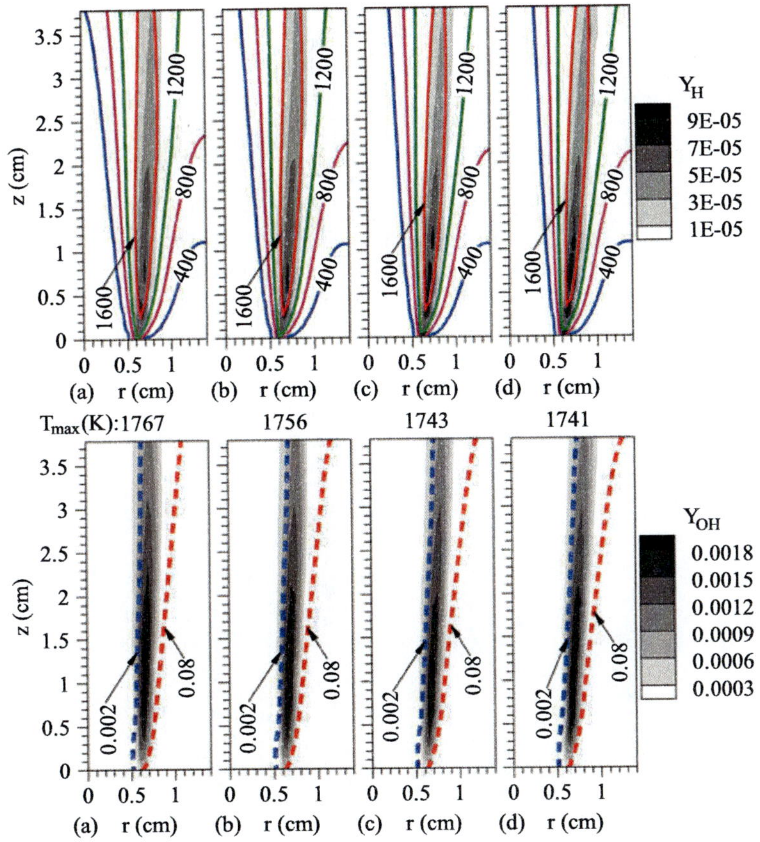


**Fig. 9** (Top) Temperature contours (lines) and mass fraction of H (greyscale) and (bottom) contours of mass fractions of OH (greyscale) and CO (lines) for Case 4 with **a** 100%, **b** 125%, **c** 167% and **d** 200% of stoichiometric air

maximum temperature values are observed. In the flame of case 1 (Fig. 7), the extent of maximum temperature is seen to decrease with an increase in air flow rate and then it increases when the flame lifted-off. On the other hand, in case 5 (Fig. 10), the extent of maximum temperature decreases initially and remains almost a constant with increment in air flow rate, as shown by isotherm of 1600 K.

In case 5, the flame moves away from the burner tip when air flow rate is increased to 167% of stoichiometric air itself, whereas the flame in case 1 is seen to be stable at 167% of stoichiometric air. The lift-off height at 167% air is about 1.5 mm. For 200% stoichiometric air, it is around 3.5 mm, only about half of the value observed for case 1. Even though the flame lifts-off due to lower CO in case 5, its lift-off height is lower than that in case 1, due to higher $H_2$ in case 5. The intensity of H field does not change notably with air flow—however, it has a slightly higher intensity than case 1, due to higher $H_2$ for this case. The line contours of CO and greyscale contours

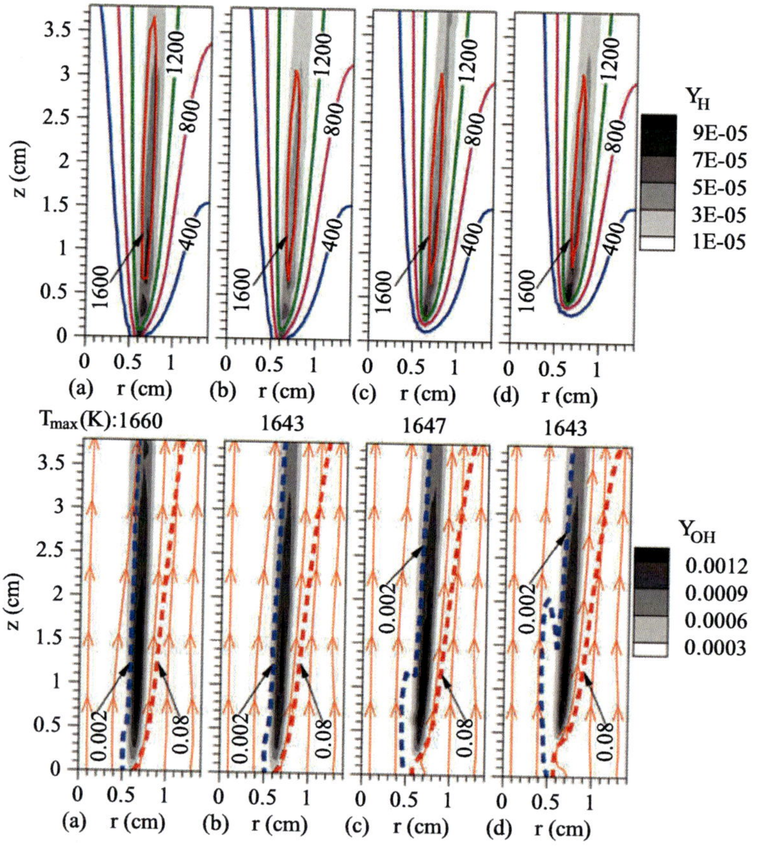


**Fig. 10** (Top) Temperature contours (lines) and mass fraction of H (greyscale) and (bottom) contours of mass fractions of OH (greyscale) and CO (lines) for case 5 with **a** 100%, **b** 125%, **c** 167% and **d** 200% of stoichiometric air

of OH for case 5 are shown in Fig. 10 (bottom). As observed in the contours of H (Fig. 10, top), the intensity of OH field for case 5 is also relatively higher than that of the case 1. The curvature in the contour line of 0.08 indicates the flame anchoring position in lifted flames. The CO contour line of 0.08 is curved more towards the fuel stream side, when compared to that seen for case 1, because of lower CO in the fuel stream.

Figure 11 presents the net reaction rate profiles of important species and rates of reactions R1, R2, R3 and R4 at 10 mm from burner exit for case 5 (14% CO) with 100% and 200% stoichiometric air. The net consumption rate of oxygen is slightly higher for the case with 200% stoichiometric air as a result of partial premixing at the flame base due to the lifted flame (Fig. 11b).

The net consumption rate of CO attains the peak value at the same location where the net production rate of OH peaks (Fig. 11a, b). The peak value of the net production rate of $CO_2$ is positioned slightly towards the left of the peak net reaction rates of CO and OH. The rate of reactions R1, R2 and R3 are almost the same in both the cases (Figs. 11c, d). A slightly higher reaction rate of R4 may be the reason for high net reaction rate of $H_2O$ for the case with 200% stoichiometric air.

Figure 12 presents the fields of temperature, H, CO and OH for case 8, which has 24% CO at various air flow rates. Higher stability due to higher CO content is observed. The maximum temperature has increased due to higher amount of heat

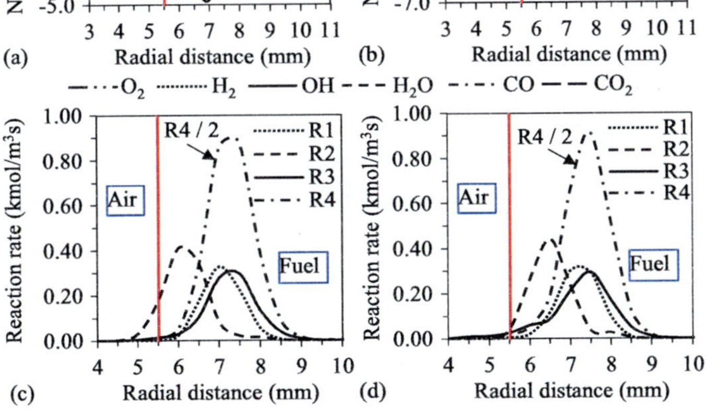


**Fig. 11** Predicted radial profiles of **a**, **b** net reaction rate and **c**, **d** rate of reactions R1, R2, R3 and R4 at an axial distance of 10 mm for case 5 (14% CO) with (left) 100% and (right) 200% stoichiometric air [Vertical line indicates boundary between the air and fuel streams]

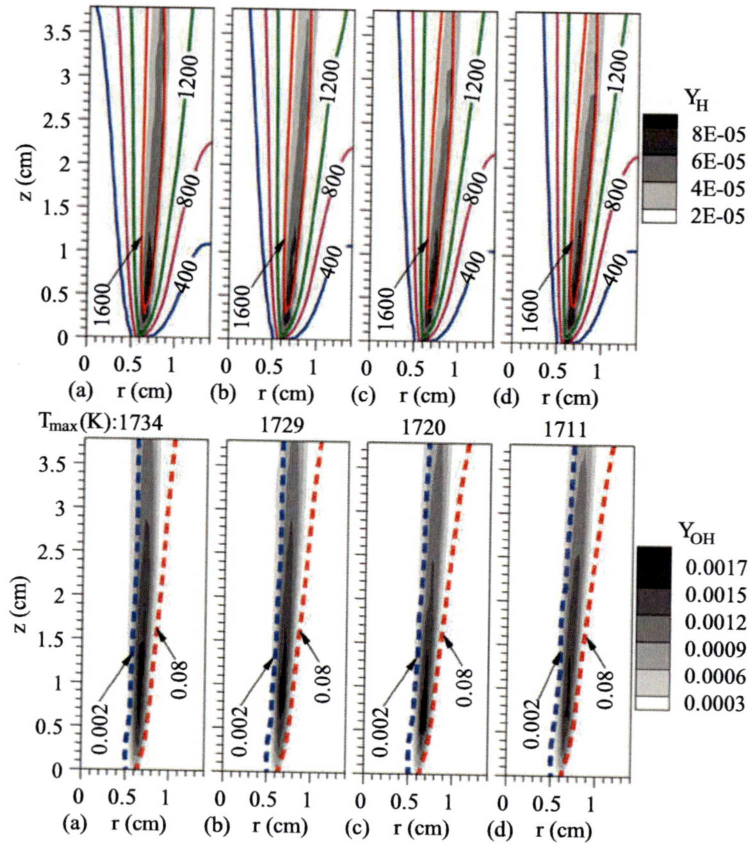


**Fig. 12** (Top) Temperature contours (lines) and mass fraction of H (greyscale) and (bottom) contours of mass fractions of OH (greyscale) and CO (lines) for case 8 with **a** 100%, **b** 125%, **c** 167% and **d** 200% of stoichiometric air

released from the oxidation of additional CO. Increase in air flow rate does not significantly affect the maximum temperature or its extent. This indicates that CO also helps in flame stabilization as well as enhancing heat release like hydrogen, for air flow rate up to 200% stoichiometric value.

## 4 Conclusions

Comprehensive numerical simulations of inverse coflow flames, where air is supplied from the core pipe and fuel is supplied through the annular region, are presented. Typical synthetic gases from air-based gasification of solid fuels have been used. Variations of temperature, flow and species fields with variations in hydrogen and

CO in the fuel mixture are presented systematically. Profiles of net reaction rates of several species and rates of important reactions have been presented to understand the effect of $H_2$ and CO addition on flame stability. Increase in hydrogen content in fuel mixture is seen to increase the flame stability, extent of reaction zone and the flame temperature. Flame is anchored close to the burner rim for the case with fuel containing 17% hydrogen, even with 100% excess air. Increase of CO also results in an increase in flame temperature and stability. At a given fuel composition, increasing air flow rate does not contribute to notable change in the maximum temperature.

However, the extent of temperature field is affected. Moreover, the flames from fuels with lower hydrogen and CO contents lift-off from the burner rim at higher air flow rate.

## References


1. Heidenreich S, Foscolo PU (2015) New concepts in biomass gasification. Prog Energy Combust Sci 46:72–95
2. Safer K, Tabet F, Ouadha A, Safer M, Gökalp I (2013) Combustion characteristics of hydrogen–rich alternative fuels in counter–flow diffusion flame configuration. Energy Convers Manage 74:269–278
3. Elbaz AM, Roberts WL (2014) Flame structure of methane inverse diffusion flame. Exp Thermal Fluid Sci 56:23–32
4. Wu KT, Essenhigh RH (1985) Mapping and structure of inverse diffusion flames of methane. Symp (Int) Combust 20(1):1925–1932
5. Park J, Bae DS, Cha MS, Yun JH, Keel SI, Cho HC, Kim TK, Ha JS (2008) Flame characteristics in $H_2$/CO synthetic gas diffusion flames diluted with $CO_2$: effects of radiative heat loss and mixture composition. Int J Hydrogen Energy 33(23):7256–7264
6. Stelzner B, Hunger F, Voss S, Keller J, Hasse C, Trimis D (2013) Experimental and numerical study of rich inverse diffusion flame structure. Proc Combust Inst 34:1045–1055
7. Huanhuan X, Fengshan L, Shaozeng S, Yijun Z, Shun M, Wenbo T (2017) Effects of $H_2O$ and $CO_2$ diluted oxidizer on the structure and shape of laminar coflow syngas diffusion flame. Combust Flame 177:67–78
8. Patel V, Shah R (2019) Effect of hydrogen enrichment on combustion characteristics of methane swirling and non-swirling inverse diffusion flame. Int J Hydrogen Energy 44:28316–28329
9. Mahesh S, Mishra DP (2008) Flame stability and emission characteristics of turbulent LPG IDF in a backstep burner. Fuel 87(12):2614–2619
10. Fluent user guide, Fluent 17.2.0 (2016)
11. Bowman CT, Hanson RK, Davidson DF, Gardiner JWC, Lissianski V, Smith GP, Golden DM, Frenklach M, Goldenberg M, http://combustion.berkeley.edu/gri-mech/new21/version21/text21.html
12. Barlow RS, Smith NSA, Chen JY, Bilger RW (1999) Nitric oxide formation in dilute hydrogen jet flames: isolation of the effects of radiation and turbulence–chemistry sub models. Combust Flame 117(1–2):4–31
13. Couto N, Rouboa A, Silva V, Monteiro E, Bouziane K (2013) Influence of the biomass gasification processes on the final composition of syngas. Energy Procedia 36:596–606
14. Ibrahim M, Kumaran SM, Raghavan V (2021) Numerical study of characteristics of confined diffusion flames of synthetic gases in coflow and inverse coflow configurations. Prog Comput Fluid Dyn

# Three-Way Catalyst System Design and Emission Characteristics Study with Precious Group Metal Loadings for CNG Vehicles

S. Karthikeyan, A. Raj, A. L. Suresh, and S. Krishnan

## 1 Introduction

Gas fueled heavy-duty vehicles (HDV) are already available within the market since several years based on the studies carried out by Fontaras et al. [1]. Among all alternative fuels, one of the simplest solutions for fuel substitution is CNG, primarily due to fuel availability in all over the globe, naturally uncontaminated combustion, inexpensive fuel, and gasoline and diesel engines adaptableness. Furthermore, due to their environmental benefits especially due their reduced emissions levels of particulate matter (PM), particulate number (PN), and nitrogen oxides ($NOx$), this compressed gas fueled engines are actively used for heavy-duty vehicles as substitute for diesel fuel. As mentioned by Yoon et al. [2, 3] within the United States in the last decade, the gas urban bus population got doubled. Furthermore, in the state of California the usage of gas urban bus population is also expanded from 24 to 45% from 2002 to 2011. Other studies also reported, utilization of latest buses powered by CNG engines worldwide is progressively increased according to the studies made by Guo et al., Wang et al., and Svenskt et al. [4–6]. Furthermore, CNG trucks are also used extensively for trash collection purposes in US as well as other countries worldwide more than a decade according to Bradley et al., Clark et al., Weaver et al., and Lopez et al. [7–11]. Besides the above- mentioned points, lower $CO_2$ emissions in CNG fuel are due to its high H/C ratio compared to diesel fuel. However, due to performance as well as cost benefit reasons, the CNG engines are preferred considerably. Nevertheless, the automobile manufacturers are challenged to come up with innovative ideas, for further reduction of the gaseous emissions from a conventional CNG engine with the introduction of the Bharat Stage-VI (BS-VI) emission norms. Hence,

S. Karthikeyan (✉) · A. Raj · A. L. Suresh · S. Krishnan
Department of Engine and Exhaust After-Treatment System, Ashok Leyland Technical Centre, Chennai, India
e-mail: dr.karthikeyan@ashokleyland.com



N. L. Narasimhan et al. (eds.), *Recent Advances in Energy Technologies*, Lecture Notes in Mechanical Engineering, https://doi.org/10.1007/978-981-19-3467-4_36

to accomplish this, enhancing the after-treatment system design along with superior catalyst technologies becomes significant.

Usually, the exhaust stream from a gas combustion primarily comprises emissions. Mostly with the use of commonly used three-way catalyst system, along with combination of platinum (Pt), rhodium (Rh), and palladium (Pd), the emissions are effectively controlled. However, the studies show that natural gas vehicles operated over stoichiometric combustion provide better performance over the older technology vehicles equipped with oxygen catalyst operated with lean-burn combustion to meet the NO*x* emission standards according to Karavalakis et al. [12]. Therefore, the three-way catalyst technology with stoichiometric combustion usage is the best preference for oxidization of exhaust gas. The pollutants like hydrocarbons (HC) and carbon monoxide (CO) are oxidized with Pt/Pd combination catalyst and with Rh catalyst the reduction of nitrogen oxides (NO*x*) accomplished. Hence, with the adequately designed three-way formulation, the catalyst promotes two kinds of reactions: oxidation reaction in the presence of oxygen to reduce HC and CO (given by Eqs. (1) and (2)) and reduction reaction to reduce NO*x* with presence of CO (given by Eq. (3))

$$\mathrm{CO} + 1/2\,\mathrm{O_2} \rightarrow \mathrm{CO_2} \tag{1}$$

$$\mathrm{CH_4} + 2\mathrm{O_2} \rightarrow \mathrm{CO_2} + \mathrm{H_2O} \tag{2}$$

$$\mathrm{NO} + \mathrm{CO} \rightarrow 2\mathrm{N_2} + \mathrm{CO_2} \tag{3}$$

Furthermore, another important emission is methane ($CH_4$) gas, emitted by CNG vehicles. This $CH_4$ plays a major role with regards to the relative $CO_2$ equivalent GHG emissions. In actual driving duty cycle, the combustion of $CH_4$ completely is not possible. Thus, a major amount of unburned $CH_4$ is present within the exhaust gases of the engines. Under ideal conditions, the combustion of $CH_4$ produces $CO_2$ and $H_2O$, hence for this tetrahedron structured molecule, precise heat energy required to oxidize the $CH_4$ species is mentioned in Fig. 1. Therefore, based on the suitable catalyst formulation, the catalyst induces the specified outputs during specific high temperature operations are recognized. Nevertheless, only based on the engine operating conditions as well as the exhaust gas composition, emission conversion efficiency can increase to 98%. However, at close to stoichiometric (lambda = 1) conditions, only CO and HC enhanced oxidation can be accomplished. The typical lambda conditions for emission conversion are shown in Fig. 2.

Though, the lean excursions (during excess amount of oxygen, lambda > 1) reduce NO*x* conversion considerably, and it is absolutely indispensable to ensure high emission conversion efficiency over the lifetime of the vehicle. For this, optimized lambda control between catalysts is must. However, simultaneous reduction of aforementioned three significant gases within the CNG fuel exhaust is accomplished by frequently varying the air–fuel ratio (AFR). This strategy is usually known as fuel-dithering. During rich operation, for achieving greater efficiency in reducing NO*x* and CO and during considerable lean operation for HC reduction, this process

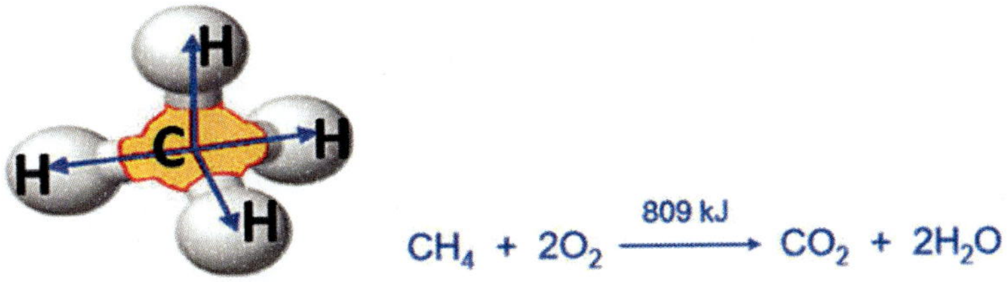


Fig. 1 $CH_4$ structure and heat energy

Fig. 2 Lambda (air to fuel ratio) condition

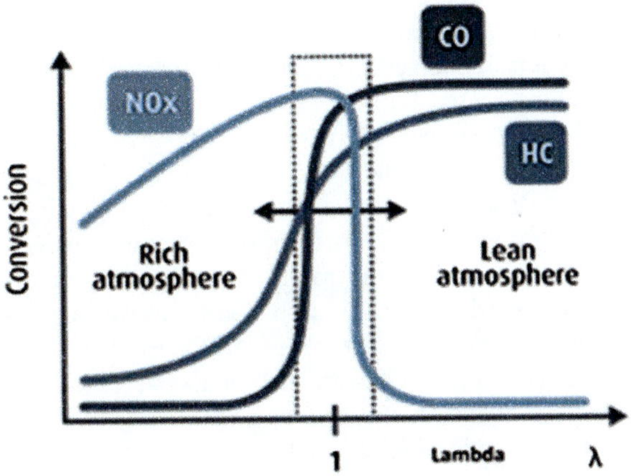


has shown to be a good approach according to Defoort et al. [13]. The fuel-dithering optimization study for natural gas engine showed that the balanced dithering lambda-midpoint resides in slightly rich combustion, and precise control of frequency and amplitude of the fuel-dithering tends to reduce the overall emission performance characteristics of the TWC system based on studies carried out by Xian et al. [14]. Additionally, to the key chemical change mechanism, two more reactions like water–gas shift reaction and steam reforming reactions have also known to occur over the TWC system. Under suitable rich operating conditions of the engine when there is oxygen deficiency within the exhaust stream favors un-oxidized CO and HC to form $CO_2$ and hydrogen ($H_2$) according to Barbier et al. and Ohtsuka et al. [15, 16], and expressed by Eqs. (4) and (5). Precursor $NH_3$ is then generated over the catalyst in the presence of molecular $H_2$ with available NO and CO [17], expressed mainly by the Eqs. (6) and (7).

$$CO + H_2O \rightarrow CO_2 + H_2 \quad (4)$$

$$CH_4 + 2H_2O \rightarrow CO_2 + 4H_2 \quad (5)$$

$$2NO + 5H_2 \rightarrow 2NH_3 + 2H_2O \tag{6}$$

$$2NO + 2CO + 3H_2 \rightarrow 2NH_3 + 2CO_2 \tag{7}$$

Moreover, the above-mentioned reaction mechanisms, $NH_3$ formation over the TWC could be a chain effect of species and their kinetic behavior, due to AFR changes, either fuel rich combustion or oxygen rich combustion. Hence, there always exist a challenge in optimizing the air–fuel ratio on one hand. On the other hand, for the faster heat-up of the TWC, proper flow optimization is critical. This flow optimization influences to attain the required reaction conditions at less time after engine cranking, specifically imperative for urban driving conditions especially during frequent start–stop events. Hence, the most significant design factors for design optimization focused for the pressure drop across the substrate, flow uniformity index, and light-off temperature performance of catalyst as per studies carried out by Martin et al, Zhang et al., and Yedddi et al. [18–20]. According to Julia Windmann et al. [21], studies described the impact of the catalyst light-off temperature performance due to flow uniformity. Furthermore, other influencing system that affects the exhaust flow distribution and uniformity inside the after-treatment system is exhaust manifold design, cone design both inlet and outlet, as well as substrate dimensions, and configuration based on studies carried out by Salasc et al. [22]. Usually, the muffler-integrated box exhaust system design encompasses ceramic/metallic substrate structure. This substrate surface is incorporated with wash coat material and PGM. Furthermore, the PGM loading, wash coat thickness on the substrate material (ceramic or metal), substrate size, and layout arrangement for box exhaust system on vehicle also play a foremost role in the design and emission characteristics. Therefore, due to aforementioned ATS influencing system's behavior, the exhaust after-treatment system optimization by experimentation is time-consuming and expensive due to the prototype development with different kind of geometries. To reduce the time and cost, usage of catalyst simulation software to optimize the catalyst and flow dynamics has considerably supported to enhance the emission performance. According to Benjamin et al. [23] computational fluid dynamics (CFD) technique is utilized to analyze the uniform conversion efficiency across the monolith of catalytic converter. Furthermore, Tsinoglou et al. [24] by using CFD code techniques, the exhaust flow distribution of the catalytic converter and radial velocity profiles are studied. As per, Soo-Jin Jeong et al. [25], studies also reported that an enhanced design optimization among the pressure drop across the substrate and flow distribution uniformity is indispensable for improving the design efficiency of an auto-catalyst. Based on the studies by Lun et al. [26], AVL CFD software is used to perform the CFD analysis. According to Kumar et al. [27], studies are reported with complex heterogeneous chemistry using full-scale monolithic catalytic converters simulation. Based on the studies reported by Hayes et al. [28] with full size catalytic converter, computational study results are developed as well as validated against experimental data for design enhancements. Based on the aforementioned literature, the flow distribution uniformity and pressure drop analysis for the after-treatment system are focused for this research work. Furthermore, for this exhaust flow analysis, the box-type exhaust system model is developed using advanced CAD software CATIA V5. By using the suitable boundary input conditions, as well as importing the CAD data inside the CFD software, CFD analysis is carried out. The substrate modeled as porous medium and defined with appropriate porosity % for analysis purpose.

Section succinctly presents overview of CNG emission norms and design requirements. Section 3 explains experimental results for CFD methodology and three-way catalyst PGM loading necssities for the proposed after-treatment system. Section 4 provides conclusion for CFD analysis and the experimental study of emission. Section 5 provides summary of overall work related to this paper.

## 2 CNG Emission Norms and Design Requirments

The upcoming Bharat stage emissions legislation call for an exhaust system with a very high NO*x* reduction efficiency. The suggested BS-VI level is given in Table 1, and NO*x* should be reduced by 95% (for emission during world harmonized transient cycle (WHTC) test) compared to the BS-IV level. Consequently, during this TWC, new product development for meeting the emission legislations, the balanced design requirements considerations are must. Figure 3 shows balanced requirement for TWC after-treatment system.

**Table 1** WHTC emission norms for BS-VI emissions

| | Cycle | NO*x* (mg/kW-hr) | NMHC (mg/kW-hr) | $NH_3$ (ppm) | $CH_4$ (mg/kW-hr) |
|---|---|---|---|---|---|
| Norms as per BS-VI | WHTC | 460 | 160 | 10 | 500 |
| DF factors | WHTC | 1.15 | 1.4 | – | 1.4 |

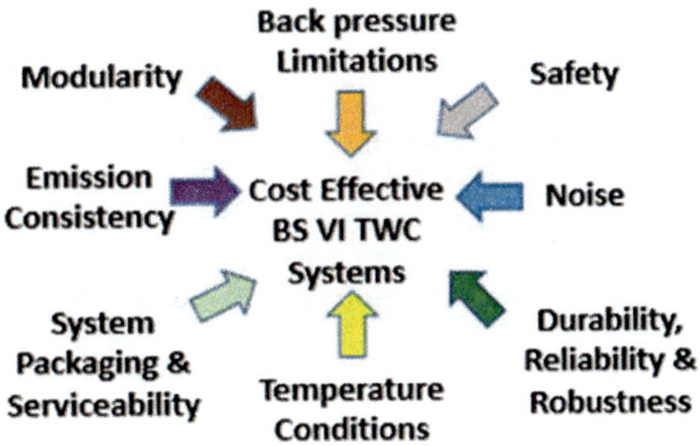


**Fig. 3** Balanced ATS system design requirements for BS-VI

This part provides a brief overview of methodology followed for design of after-treatment system and PGM selection for TWC. By using CATIA V5, the design of exhaust system is completed. However, the time delay for model updation is reduced during ATS design changes by use of this 3D software. This box-type after-treatment exhaust system contains three-way catalytic convertor with integrated muffler. For attenuating, the exhaust noise mufflers are normally used. In this box system, reflective type of muffler arrangement is developed and adequate for low frequencies conditions. The three-way catalytic convertor (after-treatment box system) contains inlet perforated tube at the entry of the substrate inlet, (metallic substrate) as well as perforated tube at outlet.

Usually, in the metallic substrate, there are large number of small channels (cell densities) which are encompassed, and PGM is coated inside these channels. Furthermore, this box-type after-treatment system is designed using the parametric modeling approach using CATIA V5. This 3D model of ATS systems using CFD pre-processing tool is imported and meshed subsequently. However, to get the accurate results and to reduce computation time, in this present work, polyhedral cell mesh element used. The total numbers of polyhedral cell mesh elements were around 1.5 million. For this flow analysis, following assumptions are made for the model. This analysis is performed to study the maximum flow rate at steady state condition and TWC substrate with porous domain for this study is assumed. Though, the focus is only ATS design enhancements for exhaust gas flow, the chemical reactions, and the heat transfer studies are not considered. However, the inertial and viscous coefficients are considered for this porous media structure. For Solver setting, analysis type is steady state, compressible, and ideal gas law. For turbulence model, K-Epsilon model is used, and wall type is set as adiabatic wall condition. Other boundary conditions like fixed mass flow used with turbulence intensity of 10%, as well as the reference pressure is assumed as 101,325 Pa for this analysis.

Regarding three-way catalyst selection, the use of P-diagram development process is significant. From the P-diagram (process diagram), it is perceived that while selecting the catalyst formulation, the emission performance durability is considered for performance. The P-Diagram for PGM selection is shown in Fig. 4. Hence, based on the process control factors for TWC, it is must that essential requirements should be controlled precisely to meet the emission performance durability. The three-way catalyst PGM used for this study is shown in Fig. 5a, b which shows advanced TWC coatings (Pt: Pd: Rh with <180 g/Cuft) on metallic substrate with cell density of 600 Cpsi and metallic substrate volume <4.0 L. Furthermore, even from DFMEA top RPN should be checked and implementing necessary action is significant for meeting emission durability. For the TWC, highest RPN is given for sintering effect of catalyst, during high temperature operations. To avoid the sintering effects and to improve the emission performance durability, TWC is designed with suitable catalyst promoters for achieving stabilized active sites for emission conversion.

Figure 6 shows process improvement for catalyst durability. The high PGM loadings can adversely affect the entire catalyst formulations, as a result it is anticipated there could be drop in performance. Hence, to enhance the PGM performance for

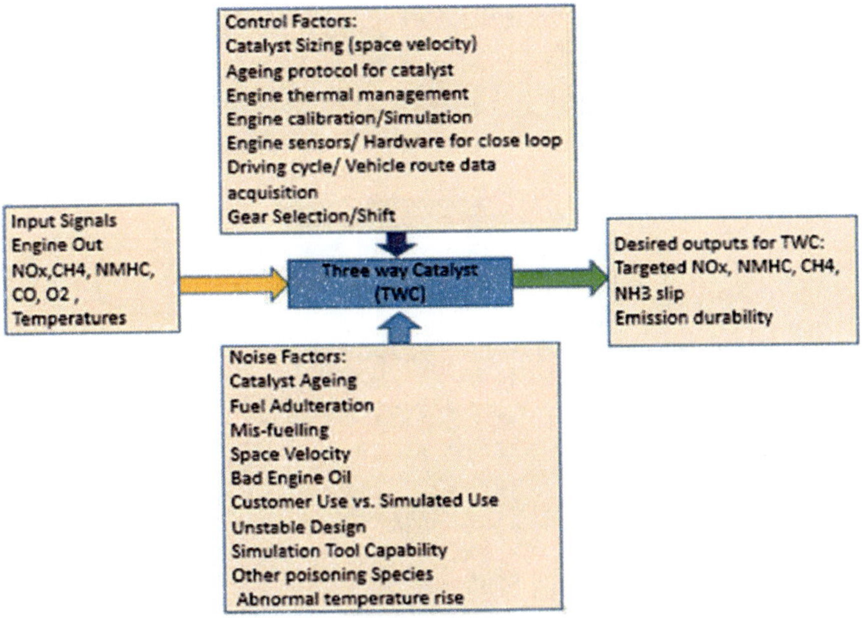


**Fig. 4** P-Diagram for TWC system

emission conversion, promoters are included. This advanced catalyst formulation for TWC is for extending durability by reducing the sintering effect.

Figure 7 shows catalyst surface comparison with and without promoters. The catalyst without catalyst promoter shows more sintering effect.

Figure 8 shows aged TWC catalyst with catalyst promoter is having less catalyst sintering when compared to aged catalyst without catalyst promoter.

Figure 9 shows aged TWC catalyst without catalyst promoter with more sintering effect in aged catalyst. Catalyst aging condition is decided based on promoters in catalyst during high loading requirements as well as vehicle operating duty cycle conditions.

## 3 Experimental Results

This experiment investigates the performance of TWC after-treatment system design based on CFD as well as emission performance verification of after-treatment with selected catalyst on emission test bench.

Figure 10 shows back pressure across box-type exhaust systems with TWC. In the ATS system, the inlet side of the after-treatment system is highlighted in red color. Based on optimized design concept at maximum exhaust gas mass flow condition, the total back pressure across the system is 83 mbar. This pressure drop is within the

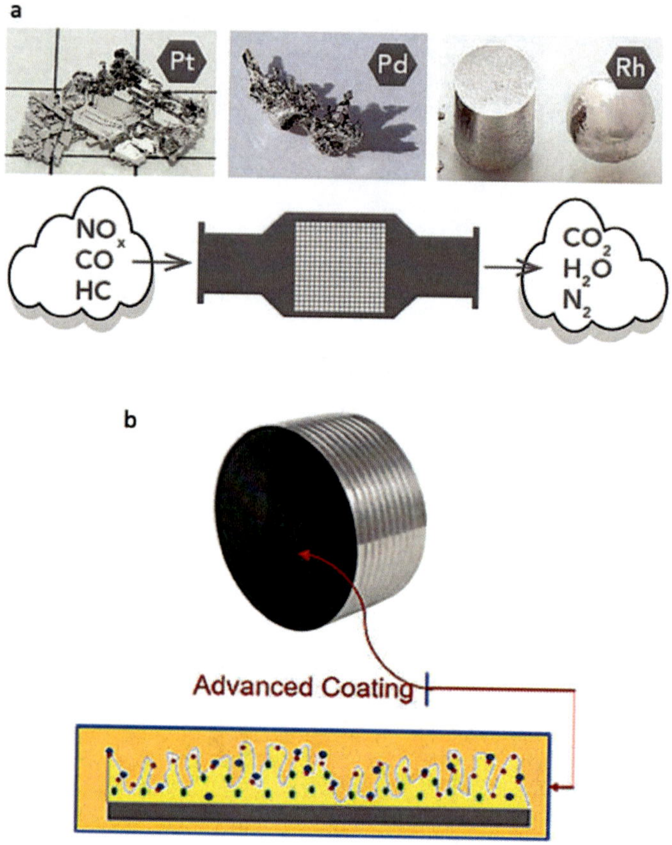


Fig. 5 a PGM for TWC system and b advanced TWC coating on metallic substrate

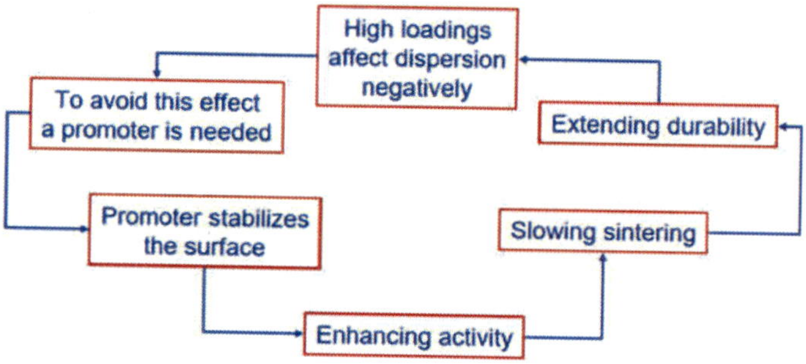


Fig. 6 Catalyst durability improvement process

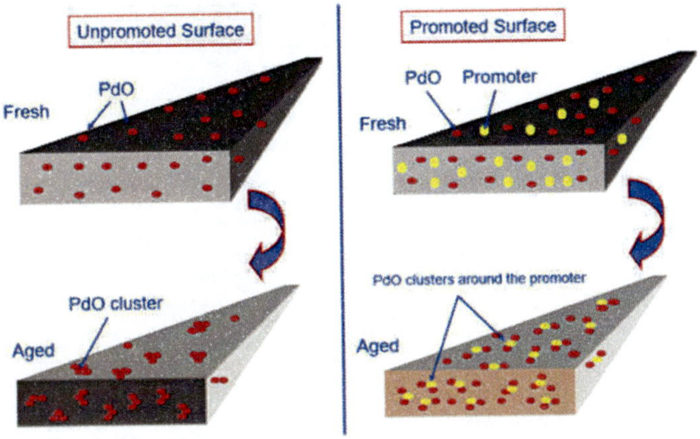


**Fig. 7** Catalyst surface with and without promoters

**Fig. 8** TWC catalyst with catalyst promoter

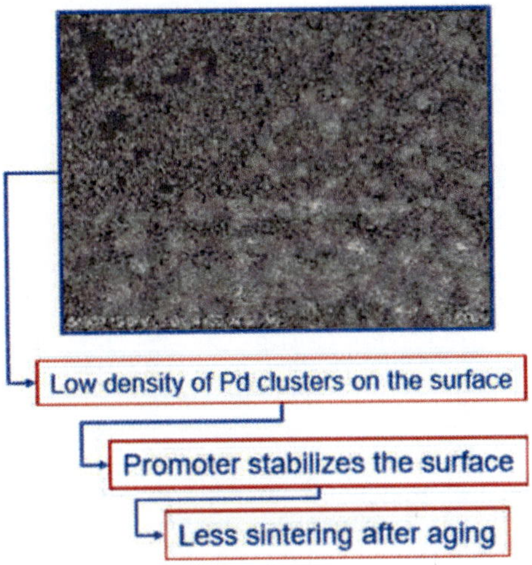


target design specification of 120 mbar, and this ensures the performance of system. For the CFD analysis, following input conditions are used, and one of the most common turbulence models K-epsilon model is used. The ATS flow gas (density is 0.5508 kg/m$^3$, viscosity is 3.814e-5 Pa. s). The flow rate of exhaust gas is 358 kg/hr and exhaust gas temperature is 717° Celsius as inlet conditions.

Figure 11 shows exhaust flow uniformity index at inlet of TWC. For the optimized backpressure design, the maximum uniformity index of 0.98 is achieved. It shows catalyst utilization effectively at maximum flow rate condition to meet the emission performances durability.

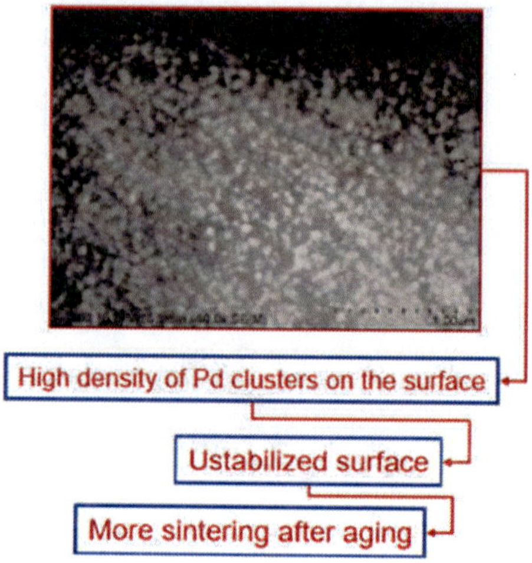


**Fig. 9** TWC catalyst without catalyst promoter

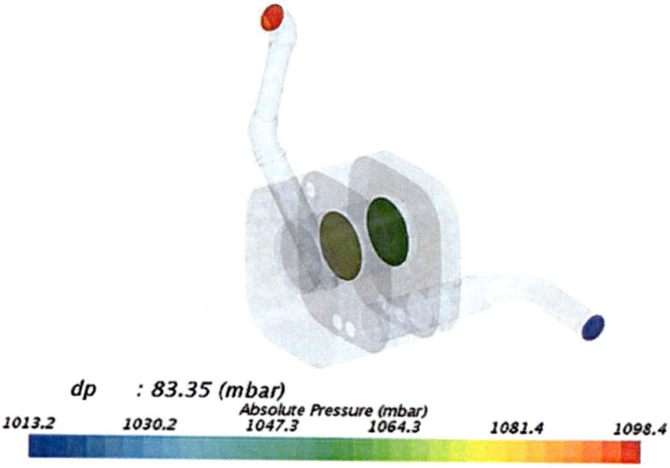


**Fig. 10** Backpressure across box-type system with TWC

Figure 12 shows velocity contour plots of box-type exhaust after-treatment system with TWC. The exhaust flow distribution inside the system is analyzed using streamline contour plots. From the results, exhaust gas reverse flow nearby substrate inlet is not observed as well as no formation of eddies in the TWC system. Therefore, in this design, flow circulation is well optimized for effective utilization of catalyst.

Furthermore, Fig. 13 shows velocity contour around inlet lambda sensor, the sensor location is well optimized for required flow around the sensor. Hence, this will support in adequate lambda control for the CNG engine.

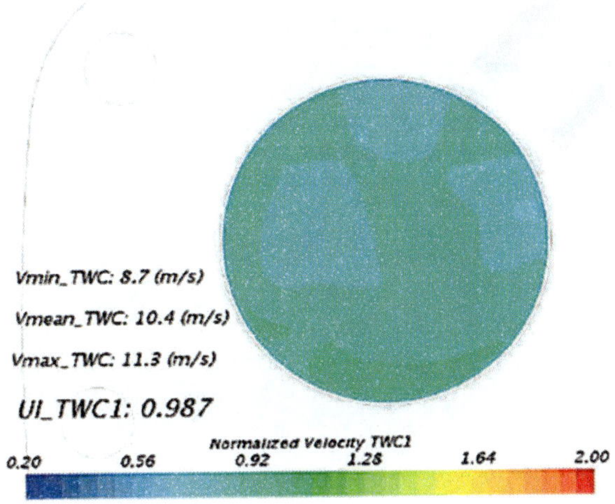


**Fig. 11** Exhaust flow uniformity index at inlet of TWC

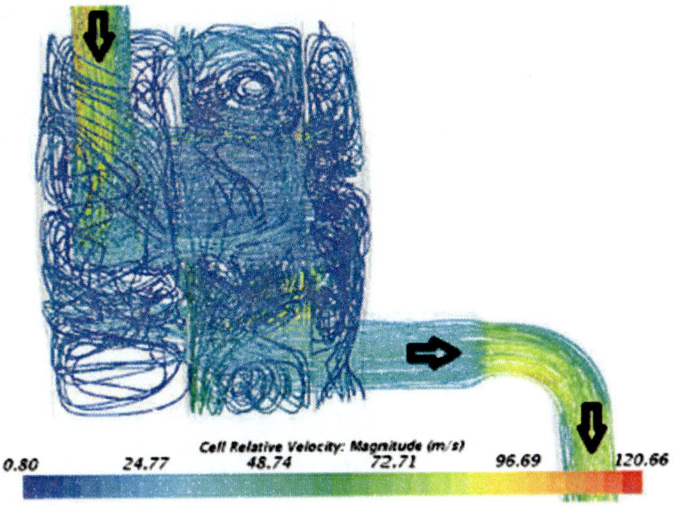


**Fig. 12** Velocity contour of box-type exhaust system with TWC

Gas bench emission conversion efficiency is shown below for the selected catalyst with enhanced PGM loading formulation at steady state condition. Figures 15 and 16 show cold and hot WHTC test $CH_4$ traces, and emissions are marginally higher in cold WHTC cycle compared to hot WHTC. However, based on the optimized lambda control, the selected TWC is able to meet the $CH_4$ emission conversion and meet the WHTC limits. Due to better temperature in hot WHTC, $CH_4$ is oxidized considerably more in hot compared to cold WHTC, this could be due to catalyst formulation. Overall, the TWC is adequately designed to oxidize $CH_4$.

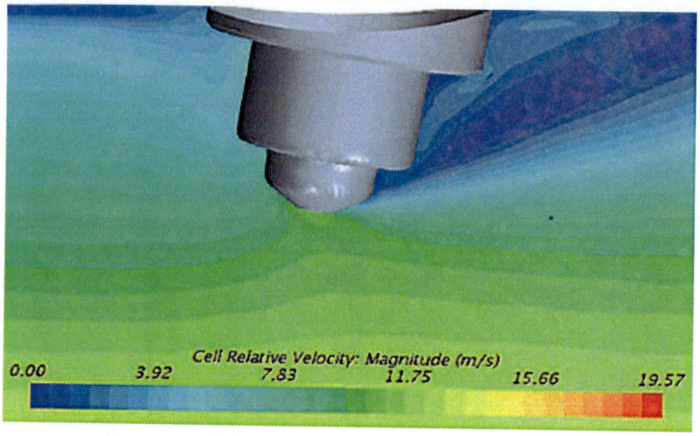


**Fig. 13** Velocity contour for lambda sensor in box-type system

**Fig. 14** Gas bench emission conversion efficiency

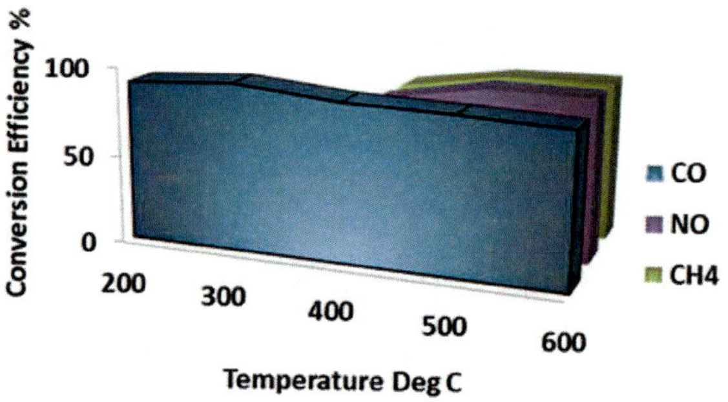


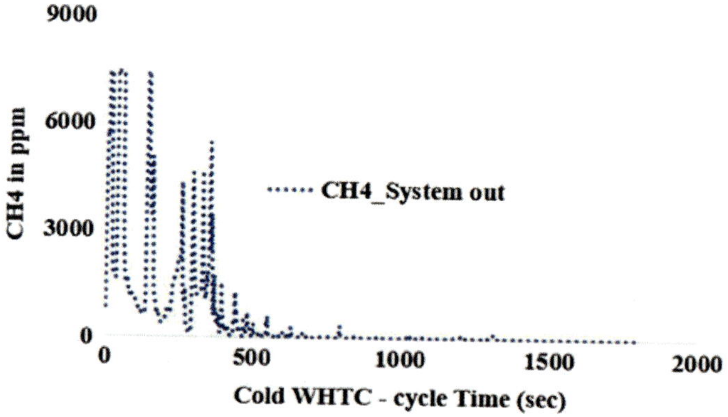


**Fig. 15** Cold WHTC emission traces for $CH_4$

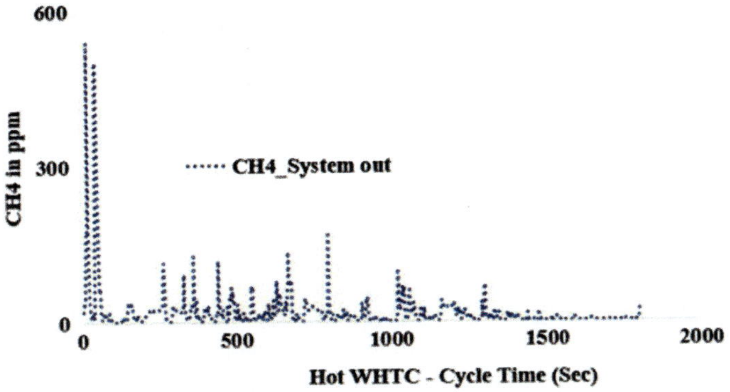


**Fig. 16** Hot WHTC emission traces for $CH_4$

Figures 17 and 18 show cold and hot WHTC NO*x* and $NH_3$ traces. Emissions are marginally higher in cold WHTC. Based on the optimized lambda control for the selected TWC, the NO*x* emission conversion with less $NH_3$ slip is achieved to meet the regulation limit.

In cold-start WHTC cycle, $NH_3$ slip is marginally higher compared to hot WHTC, and this could be due to steam reforming reactions over the TWC system. However, the TWC is adequately designed as well as capable for reducing NO*x* to $N_2$ in presence of CO.

Figure 19 shows cold and hot WHTC emission conversion efficiency. With the design optimized after-treatment system and selected PGM loading ratios, emission conversion of 98% for NO*x*, 99% for NMHC, and 97% for $CH_4$ is accomplished. Fourier transform infrared spectrometry (FTIR) gas analyzer is used for continuous measurements of NO*x* and $NH_3$ after the TWC systems.

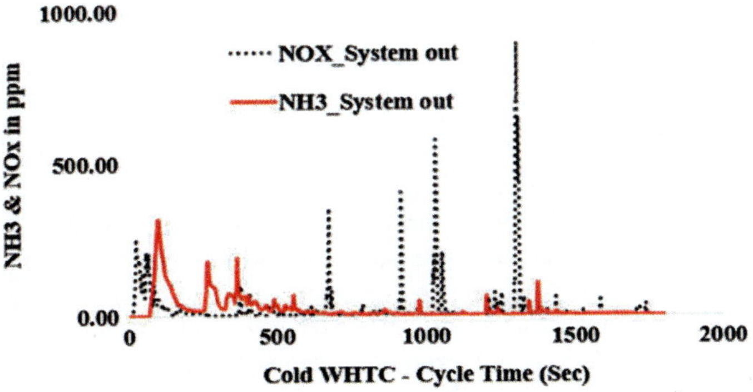


**Fig. 17** Cold WHTC emission traces for $NH_3$ and NO*x*

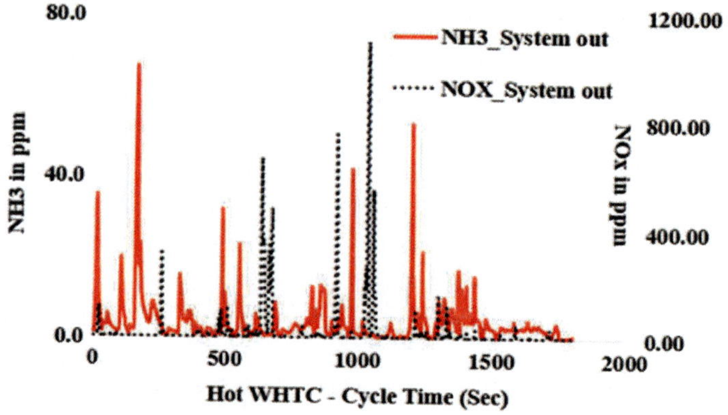


**Fig. 18** Hot WHTC traces for $NH_3$ and NO*x* emissions

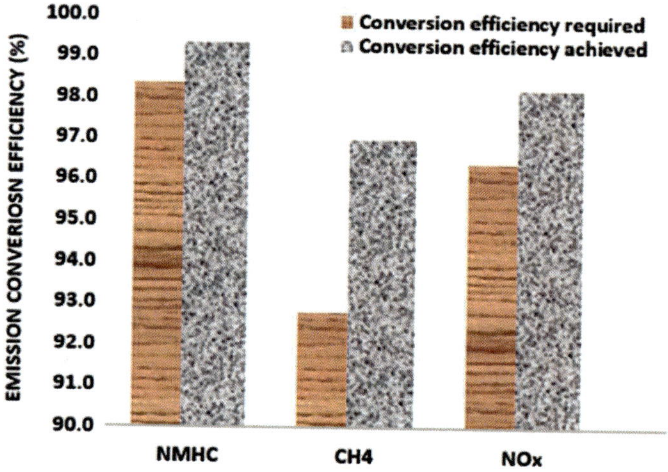


**Fig. 19** Cold and Hot WHTC emission conversion efficiency

## 4 Conclusion

This paper reports the ATS design enhancements with suitable PGM loadings for TWC system suitably to satisfy BS-VI emission norms for CNG vehicle applications. Initially, the 3D model of exhaust after-treatment systems is designed with balanced necessities. Using STAR-CCM + software, ATS design related studies like exhaust gas velocity and flow distribution, exhaust flow uniformity index and flow pressure drop across ATS are analyzed. Furthermore, in this box-type after-treatment design, there are no reverse flow observed near to metallic substrate toward inlet enlarging area. The inlet of ATS is well optimized for exhaust gas flow transfer without any obstruction in the exhaust gas flow path. Hence, there is a low chance for the formation of eddies. As a result, for efficient catalyst utilization, the uniformity index of 0.98 at inlet to TWC is achieved for better emission conversions. Furthermore, the usage of catalyst surface area for exhaust flow is also enhanced. Accordingly, by properly packaging the after-treatment system in the vehicle, the temperature availability for the catalyst is also sustained. Accordingly, for the lifetime of the vehicle, a right balance among catalyst performance and lambda control is accomplished with higher emission conversion levels. From the emission test with and without catalytic converter, the results show that the WHTC conversion efficiency for TWC after-treatment system is >98% for NMHC, $CH_4$ efficiency is >96%, NO$x$ conversion is 98%, and $NH_3$ slip is <8 ppm. Based on the experimental research, the results obtained are favorable in favor of TWC emissions conversion efficiency. Hence, this can be concluded that the after-treatment designed and selected advanced TWC technology is suitable for achieving BS-VI emission with enhanced engineering margin as well as durability requirements.

**Acknowledgements** The author would like to thank Mr. M. Sathyandan, GM—engine testing and Mr. Samir Kale—DGM engine testing for the testing support. The rounds of discussion with them gave fruitful direction to this research work.

In addition, the author would also like to thank Dr. S.S. Tipse Sr. Deputy Director and S.D. Rairikar and team members for their extended support in engine calibrations at powertrain laboratory of ARAI, Pune.

Special thanks to the engine team members of Ashok Leyland, Technical Centre, Chennai for their support in Engine Testing activities.

## References

1. Fontaras G, Manfredi U, Martini G, Dilara P et al (2012) Experimental assessment of a diesel-LPG dual fuel supply system for retrofit application in city busses. SAE Technical Paper 2012-01-1944
2. Yoon S, Collins J, Thiruvengadam A, Gautam M, Herner J, Ayala A (2013) Criteria pollutant and greenhouse gas emissions from CNG transit buses equipped with three-way catalysts compared to lean-burn engines and oxidation catalyst technologies. J Air Waste Manage Assoc 63:923–933
3. Yoon S, Hu S, Kado N, Thiruvengadam A, Collins J, Gautam M, Herner J, Ayala A (2014) Chemical and toxicological properties of emissions from CNG transit buses equipped with three-way catalysts compared to lean-burn engines and oxidation catalyst technologies. Atmos Environ 83:220–228
4. Guo J, Ge Y, Hao L, Tan J, Li J, Feng X (2014) On-Road measurement of regulated pollutants from diesel and CNG buses with urea selective catalytic reduction systems. Atmos Environ 99:1–9
5. Wang H, Fang H, Yu X, Wang K (2015) Development of natural gas vehicles in China: an assessment of enabling factors and barriers. Energy Policy 85:80–93
6. Svenskt Gastekniskt Center AB, SGC (2013) Testing of unregulated emissions from heavy duty natural gas vehicles; SGC Report 2013:289
7. Bradley MJ (2005) A demonstration of diesel particulate filter emission control technologies on refuse collection trucks and deployment of natural gas—powered street sweepers. Associates Inc., Northeast States for Coordinated Air Use Management, Boston, MA

8. Clark NN, Rapp BL, Gautam M, Wang W, Lyons DW (1998) A long-term field emissions study of natural gas fueled refuse haulers in New York City. SAE Technical Paper No. 982456, Society of Automotive Engineers, Warrendale, PA
9. US Department of Energy; Energy Efficiency and Renewable Energy, Case Study-Compressed Natural Gas Refuse Fleets (2014) Available at: http://www.afdc.energy.gov/uploads/publication/casestudy_cng_refuse_feb2014.pdf
10. Weaver CS (2003) Overview of emissions and in-service monitoring; better air quality 2003. Philippines, Manila
11. Lopez JM, Gomez A, Aparicio F, Sanchez FJ (2009) Comparison of GHG emissions from diesel, biodiesel and natural gas refuse trucks of the city of Madrid. Appl Energy 86:610–615
12. Georgios K, Hajbabaei M, Jiang Y, Yang J, Johnson KC, Cocker DR, Durbin TD (2016) Regulated, greenhouse gas, and particulate emissions from lean-burn and stoichiometric natural gas heavy-duty vehicles on different fuel compositions. Fuel 175:146–56. https://doi.org/10.1016/j.fuel.2016.02.034
13. Defoort M, Olsen D, Willson B (2004) The effect of air-fuel ratio control strategies on nitrogen compound formation in three-way catalysts. Int J Engine Res 5(1):115–122. https://doi.org/10.1243/146808704772914291
14. Shi X, Seiser R, Chen J-Y, Dibble R, Cattolica R (2015) Fuel-dithering optimization of efficiency of TWC on natural gas IC engine. SAE Int. https://doi.org/10.4271/2015-01-1043
15. Barbier J, Duprez D (1994) Steam effects in three-way catalysis. Appl Cataly B: Environ 4(2):105–140. https://doi.org/10.1016/0926-3373(94)80046-4
16. Ohtsuka H (2015) Pt-Rh/$CeO_2$-$Al_2O_3$ for controlling emissions from natural gas engines: three-way catalytic activity at low temperatures and effects of $SO_2$ Aging. Emiss Control Sci Technol 1(1):108–116. https://doi.org/10.1007/s40825-014-0009-0
17. Nagashima K, Zhang G, Hirota T, Muraki H (2000) The effect of aging temperature on catalyst performance of Pt/Rh and Pd/Rh TWCs. SAE Int. https://doi.org/10.4271/2000-01-1954
18. Martin AP, Will NS (1998) Effect of flow distribution on emissions performance of catalytic converters. SAE Paper 980936
19. Zhang X, Tennison P (2008) Numerical study of flow uniformity and pressure loss through a catalytic converter with two substrates. SAE Paper 2008-01-0614
20. Ren Y, Lou D, Tan P, Zhang Y, Sun X (2021) Emission reduction characteristics of after-treatment system on natural gas engine: effects of platinum group metal loadings and ratios. J Clean Prod 298:126833
21. Windmann J, Braun J, Zacke P, Tischer S, Deutschmann O, Warnatz J (2003) Impact of the Inlet flow distribution on the light-off behavior of a 3-way catalytic converter. SAE Technical paper 2003-01-0937
22. Salasc S, Barrieu E, Leroy V (2005) Impact of manifold design on flow distribution of a close-coupled catalytic converter. SAE Paper 2005-01-1626
23. Benjamin SF, Liu Z, Roberts CA (2004) Automotive catalyst design for uniform conversion efficiency. Appl Math Model 28:559–572
24. Tsinoglou DN, Koltsakis GC, Missirlis DK, Yakinthos KJ (2004) Transient modelling of flow distribution in automotive catalytic converters. Appl Math Model 28:775–794
25. Jeong S, Kim W (2003) A study on the optimal monolith combination for improving flow uniformity and warm-up performance of an autocatalyst. Chem Eng Process 42:879–895
26. Lun H, Xiaowei N, Liang Z, Yongping L, He Z, Wei H (2010) CFD simulation of the effect of monolith wall thickness on the light off performance of a catalytic converter. Int J Chem React Engg 8:1542–6580
27. Kumar A, Mazumder S (2010) Toward simulation of full-scale monolithic catalytic converters with complex heterogeneous chemistry. Comput Chem Eng 34:135–145
28. Hayes RE, Fadic A, Mmbaga J, Najafi A (2012) CFD modelling of the automotive catalytic converter. Catal Today 188:94–105

# Study on Performance Enhancement Technologies for Two Cylinder Diesel Engine for Off-Road Applications

S. Karthikeyan and K. Annamalai

## 1 Introduction

The off-road engines have been acknowledged to be a substantial cause of pollutant, and this contributes to substandard air quality, negative human impacts and climate change. Therefore, to alleviate the impact of emissions from these off-road engines, emission monitoring control programs play a significant role in implementing air quality management policies worldwide. Accordingly, focus for implementing the emission control legislation for off-road engine is established. Henceforth, the gaseous and particulate emissions for each stage are significant than in the automotive sector. However, several countries round the world have technologically advanced in controlling emissions from on-road vehicles and are in the process of targeting other emission source categories also through the enriched emission control programs [1–3].

Normally, the commercial usage of engines for off-road applications is important for today's economy [4]. Nevertheless, these off-road diesel engines of naturally aspirated types, conservatively suffer from less power output per liter. This could be due to the quantity of fuel that can be burnt efficiently in each combustion chamber, consequently the power output is inhibited [5]. Hence, for increasing the power density, [6] improvement related to injector spray parameters like injector pressure increases with reduced injector orifice diameter [7, 8] minimizes the emissions. However, there is no constraint for the formation of air–fuel mixture [9] based on the injection pressure. Hence, optimizing the fuel injection pressure is supposed to be significant parameter for facilitating enhanced progression in diesel combustion.

S. Karthikeyan (✉)
Department of Engine and Exhaust After-Treatment System, Ashok Leyland Technical Centre, Chennai, India
e-mail: dr.karthikeyan@ashokleyland.com

K. Annamalai
Department of Automobile Engineering, Anna University, M.I.T Campus, Chennai, India

 
N. L. Narasimhan et al. (eds.), *Recent Advances in Energy Technologies*, Lecture Notes in Mechanical Engineering, https://doi.org/10.1007/978-981-19-3467-4_37

Usually, for efficient combustion, not only the quantity of air received is important, but good air and fuel mixing should be accomplished. Subsequently, for designed engine, the capacity of air induction is fixed. If the inducted air can be compressed to an increased density than ambient before introducing it into the engine, this can further improve the power output, simply by adding more fuel in proportion to the additional quantity of air inducted. Therefore, boosting charge air by packing the air molecules tight before inducting into the engine becomes essential [9, 10].

In this study based on accomplishment after exhaust emissions improvement for off-road diesel engine after enhancing the base NA engine with modified fuel injection nozzle parameters along with a waste-gate turbocharger [11] and intercooler is explored. The focus of this study is framed to accomplish acceptable emissions levels without trade-off between engine performance and fuel consumption due to injector design changes [12]. Further, the purpose of this study is for optimizing the injector orifice diameter, nozzle opening pressure, injection timing, fuel delivery to improve fuel consumption and lower the emission levels for TCIC engine [13]. Additionally, knocking is not easy to distinguish from normal combustion. However, in TCIC engine, optimized rate of combustion pressure raise could reduce knock.

Moreover, the off-road emission norms demand considerable decrease in nitrogen oxides (NO*x*) and particulate matter (PM) in exhaust gas of internal combustion engines. Usually, nitric oxide (NO) is more in total NO*x* from engine-out conditions in DI diesel engine exhaust. This NO account for more than 70–90%. Therefore, NO*x* emission reduction by retarded start of injection, fuel nozzle modification, change of compression ratio, water direct injection, water emulsification and exhaust gas recirculation (EGR) is focused on one hand, and on the other hand, selective catalytic reduction (SCR) methods are followed. However, in this study, EGR only method is followed to meet the emissions regulations. Although, it is commonly accepted, the cooled exhaust gas recirculation (EGR) could accomplish the required NO*x* limit [6, 14], the anticipated penalties in terms of particulate emissions and brake specific fuel consumption (BSFC) should be controlled to a great extent [15]. The significances of the EGR cooler requirement in this study are to provide high heat exchange efficiency with compact packaging in the engine assembly [16–21]. Moreover, this study is limited to such an extent that emissions levels with TCIC engine can meet the norms with after-treatment system without any further modifications on the engine performance.

Table 1 gives specification comparison between naturally aspirated and modified engine with turbocharger. Modifications are related to fuel injection pump pressure, injector holes, injector pressure and intercooler and turbocharger installations. Based on the inclusion of the above aforementioned components, the overall weight of the engine could increase. However, the fuel efficiency and back pressure conditions are optimized to meet the required power output for TCIC engine.

**Table 1** The engine specification for naturally aspirated engine and TCIC engine

| Parameter | Naturally aspirated engine | Modified engine specification |
|---|---|---|
| No. of cylinders | 2 cylinder and Inline type | 2 cylinder and Inline type |
| Displacement | 1670 cc | 1670 cc |
| No. of stroke | 4 | 4 |
| Compression ratio | 18.5:1 | 18.5:1 |
| Fuel system Mechanical fuel pump Pressure | 500 bar | 700 bar |
| Injection pressure | 240 bar | 250 bar |
| Aspiration | Naturally aspirated | Turbocharged intercooled |
| Camshaft type | Single overhead cam (SOHC) | SOHC |
| Timing system | Belt drive system | Belt drive system |
| Nozzle holes | 7 | 6 |

## 2 Engine Modifications

### 2.1 *Injector Modification*

From the injector tip, one or more jets of fuel through small orifices or nozzles at higher velocity diesel fuel are typically injected. Mostly, the fuel penetrates into the combustion chamber as smaller droplets due to atomization. Though, the requirement is to improve the engine performance for TCIC engine, the impact of emission is carried out by changing the number of spray holes by keeping common piston bowl design (re-entrant type with lip) and spray cone angle. Accordingly, for this study, the combustion chamber design is kept same for both injectors, as well as nozzles holes are maintained on the same periphery without any offset. Injector location and combustion chamber design for existing and modified injector are shown in Fig. 1. The nozzle with standard 7-hole injector design is shown in Fig. 2 and the nozzle with 6-hole injector design is shown in Fig. 3. Specification comparison of both injectors are provide in Table 2.

With 6-holes injector, the start of the injection is changed to 9 Degree b-TDC (before Top Dead Center) furthermore, the fuel pump delivery also enhanced by 22% along with increased injector pressure from 240 to 250 bar for the better atomization and increased the duration of a combustion. These changes provided marginally better emissions. However, studies show the reduced 6-hole nozzle provides marginally better emission.

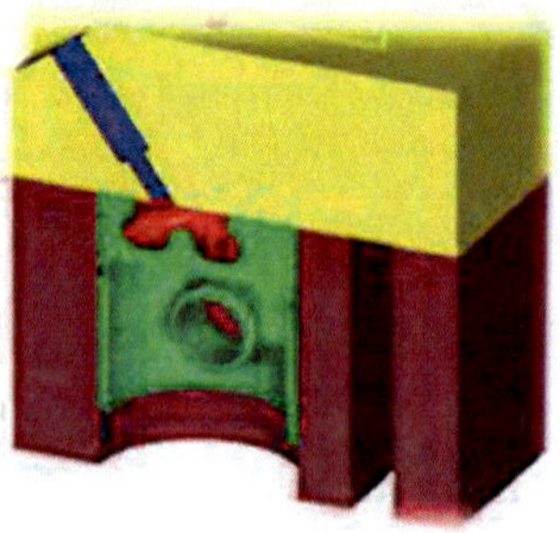

**Fig. 1** Injector location and combustion chamber design

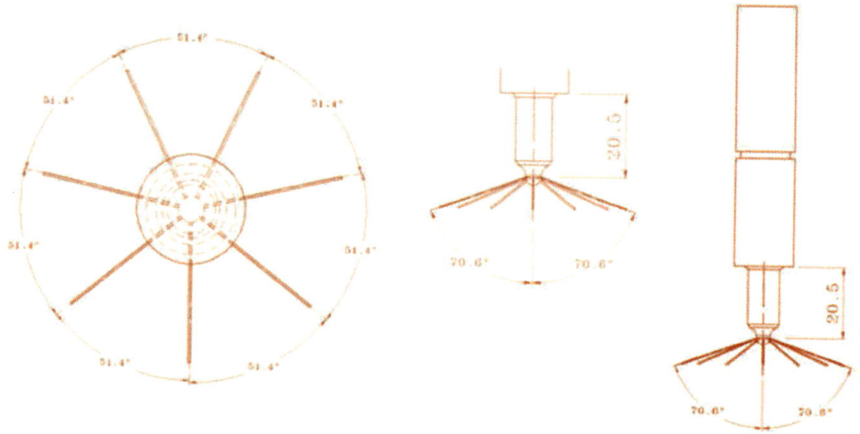


**Fig. 2** Standard injector with 7-holes

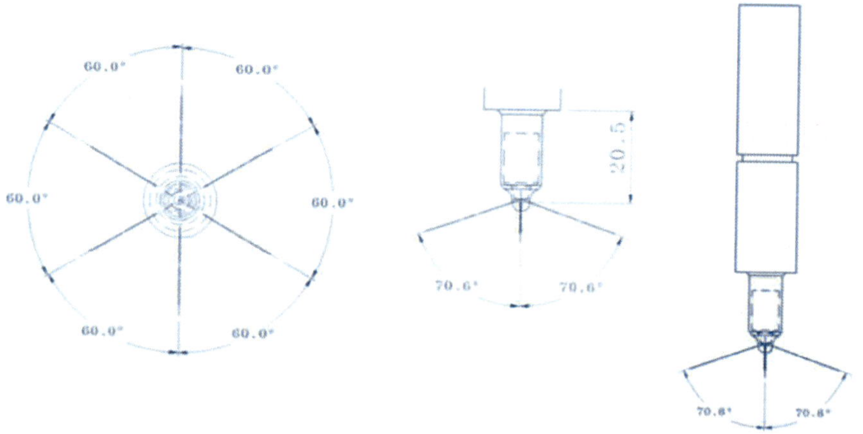


**Fig. 3** Modified injector with 6-holes

**Table 2** Injector specifications

| SI. No. | Injector specifications | | | | | Injection pressure |
|---|---|---|---|---|---|---|
| | Hole size (mm) | Cone angle (degree) | Needle lifting height mm | No. of spray holes | Injection timing | Bar |
| 1 | 0.262 | 146p | 0.30 | 7 | 8.5 Degree before TDC | 240 |
| 2 | 0.262 | 146p | 0.30 | 6 | 9.0 Degree before TDC | 250 |

## 2.2 *Turbocharger Integration*

In a naturally aspirated engine, due to the piston movement downwards toward bottom dead center during the intake stroke, the air for combustion is drawn into the engine's cylinders based on atmospheric pressure condition. However, for increasing the air density, engine modifications with TCIC are required. Naturally aspirated engine setup connected to dynamometer for baseline engine testing is shown in Fig. 4. For converting naturally aspirated engine to turbocharge intercooled engine, few modifications are significant. Considering the modifications, turbocharger and intercooler engine experimental arrangements is shown in Fig. 5. In this study, the exhaust manifold outlet is installed with 30 mm diameter exhaust outlet pipe (refer highlighted red arrow in Fig. 5) to connect turbocharger for transferring the exhaust gases. This exhaust outlet pipe is selected with GG25 material (cast iron alloy with melting temperature of 1120 °C) to withstand 675 °C continuous operating temperature and 815 °C intermittent operating temperature condition. Furthermore, this outlet pipe is adopted with temperature sensor boss to fit K-type thermocouple with maximum measuring capability of 1260 °C. With the exhaust outlet pipe, turbocharger is bolted with 0.8 mm gasket on one side. On the other side, V-band clamp is used to connect with flexible bellow to avoid vibrations transmitting from engine to after-treatment system. For turbocharger system cooling the oil is circulated using separate oil flow pipe made of stainless steel material and connected from the engine block to turbocharger. Oil flow pressure gauge is connected to engine cylinder to incessantly monitor oil pressure during engine running. Oil from turbocharger outlet is also circulated back to crankcase and required provision made in engine. Moreover, the inlet of compressor is connected to air filter, and outlet is connected to intercooler by using the flexible tube.

In this study, two different specifications of waste-gate type turbocharger initially selected for this testing. Turbocharger with, turbine wheel size of 35 mm diameter (specification-1) and 31 mm (specification-2), however, for both the turbocharger common compressor wheel size 33 mm is maintained for this study.

Typically, these maps are created by running the turbocharger at a constant speed and varying the pressure ratio on the compressor side. This can be done by restricting the flow on the compressor outlet by running a turbo on a properly instrumented gas

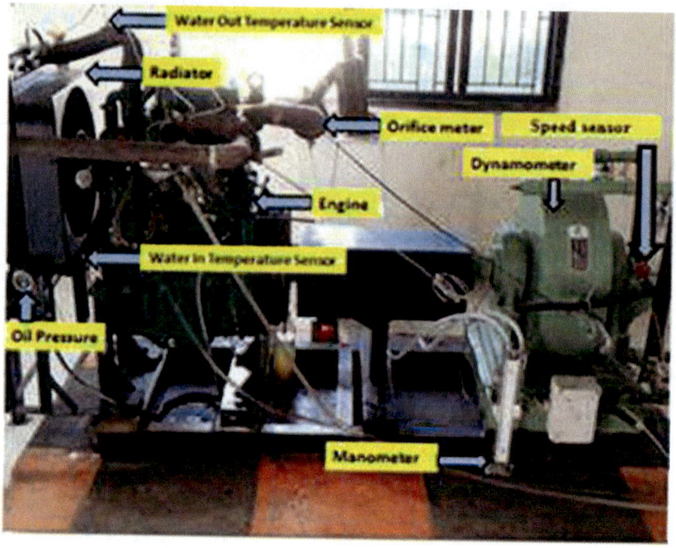


**Fig. 4** Naturally aspirated engine experimental arrangements

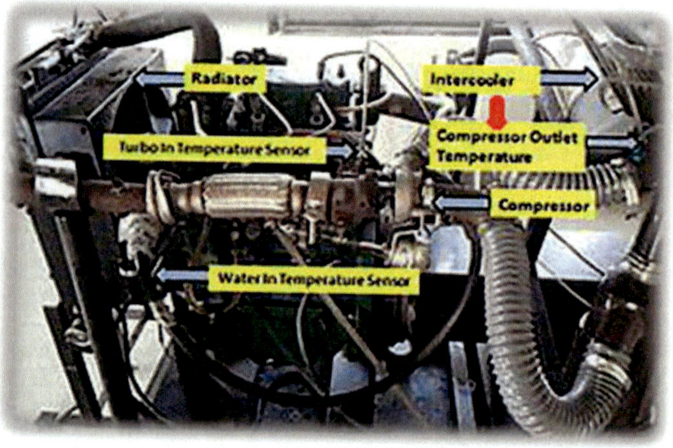


**Fig. 5** Turbocharge intercooled engine experimental arrangements

stand in a test cell. Generally, in gas stand, natural gas burns to maintain the controlled and consistent environment to drive the turbocharger as well to maintain repeatability during testing. Precise measurements of mass flow rates, temperatures, pressures and turbocharger shaft speed are essential requirements. All the speed lines on a compressor map for the selected turbocharger having a matching line on the turbine map. However, the turbine map changes drastically depend on the compressor wheel, bearing system and turbine housing used during testing. However, the compressor

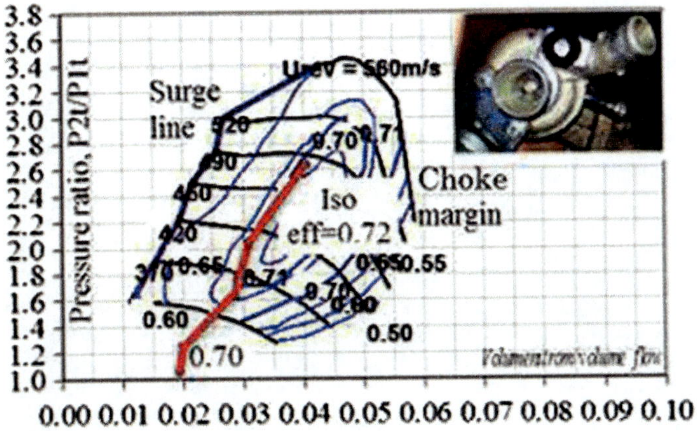


**Fig. 6** Compressor map for turbocharger specification-2

housing itself can also provide a minor effect on the turbine map. Accordingly, based on the adequate match for compressor and turbine, turbocharger's performance is determined. For this reason, it is convenient and logical to test them together in the form of a turbocharger on the engine test rig to understand the efficiency based on the selected test engine.

Based on test results, the surge and choke margin with specification-2 (with reduced turbine housing diameter) is observed to be better compared to specification-1 for this engine application. Compressor map for type specification-2 turbocharger developed for this application is shown in Fig. 6. Specification-2 turbocharger observed with better surge, choke margin and speed conditions.

## 2.3 *Intercooler Integration*

The cross-flow type heat exchanger with louvered fin type design is selected for this study. The water spray cooled intercooler experimental arrangements developed for this investigation is shown in Fig. 7. The intercooler heat dissipation from air is calculated using the formula $Q = m * \mathrm{Cp} * \mathrm{dT}$. The water mass flow rate m [kg/s] is a measurement of the amount of water flowing around the hot water loop for intercooler. The specific heat capacity Cp [kJ/kg/°C] is a thermodynamic property specific to the fluid used to transfer heat. The exhaust gas temperature difference dT [°C] is the difference in temperature before and after heat transfer. For this investigation based on compressor outlet air temperature, intercooler with heat rejection of 5 kW is required. Further, based on the simulation studies to meet the aforementioned heat rejection rate, the necessary design specifications for intercooler is finalized. Table 3 gives critical intercooler design specifications for this finalized intercooler.

**Fig. 7** Water spray cooled intercooler experimental setup

**Table 3** Intercooler specifications

| | |
|---|---|
| Intercooler design type | Cross-flow heat exchanger |
| Quantity of air tube | 9 |
| Inner fin pitch and type | 3.5 mm and triangular type |
| Intercooler heat rejection capacity at 8 m/sec 170 kg/h air flow rate | 5 kW |
| Pressure drop | 3.8 kPa |
| Outside fin pitch | 4.5 mm |
| OFA | 3.71 $m^2$ |
| L * W * H | 115 * 520 * 70 mm |

## 2.4 *Cooled Exhaust Gas Circulation Integration*

Usually, inside the EGR cooler tube, the critical design parameters that affect the exhaust gas flow and heat transfer characteristics are pitch of a spiral tube, the lead angle and the raise of the spiral ridge. Accordingly, the association between the lead angle and the spiral pitch varies depending on the number of spirals. However, it is possible to use multiple spirals tubes instead of the single spiral tubes to reach the desired coolant flow circulation and heat transfer performance. The design parameters of a spiral tube EGR cooler used for this experimental investigation with 6 mm pitch are shown in Fig. 8.

Furthermore, the temperature efficiency ($\varepsilon$) for the designed single spiral tube is further calculated using the below formula [18].

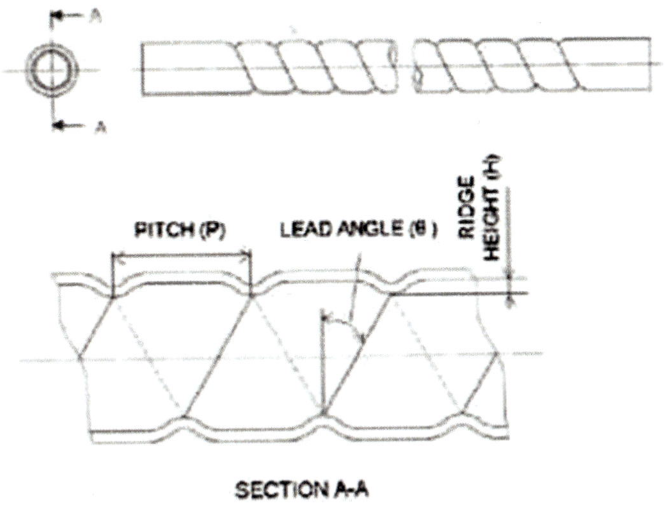


**Fig. 8** Design parameters of a single spiral EGR cooler

$$\varepsilon = (\text{Inlet gas temperature} - \text{Outlet gas temperature}/\\ \text{inlet gas temperature} - \text{Inlet water temperature}) \times 100$$

$\varepsilon$ = Temperature efficiency (%), Inlet EGR gas temperature = Temperature of EGR gas into a tube (°C), Out gas temperature = Temperature of EGR gas out of a tube (°C) and Inlet water temperature = Temperature of coolant water into the shell (°C).

Moreover, to simplify the coolant flow analysis condition, heat flux is calculated due to the EGR flow inside the tubes, and it is applied on the tube's wall as a heat source. Water mass flow rate and pressure outlet are taken as inlet boundary condition. Further, for thermal analysis using ANSYS CFX tool, EGR cooler mesh modeled with tetrahedral mesh type as shown in Fig. 9. Total number of mesh elements used in this model is 0.27 million. The developed mesh quality is 0.3 minimum and skewness 0.38. To decrease the mesh size and solving time, half model with axis symmetric boundary condition is considered. EGR cooler mesh modeled with tetrahedral mesh for thermal analysis. To enhance the hot effectiveness, quantity of tubes increased from 11 to 37 tubes based on this analysis study.

Typically, exhaust vacuum regulating valve (EVRV/modulator) is used to generate vacuum to lift the EGR valve, due to this, the exhaust gas from exhaust manifold outlet pipe flows through the EGR valve to EGR cooler. Cooled EGR schematic layout is shown in Fig. 10.

The EGR valve used for this study is pneumatic type. The operation of the exhaust vacuum regulating valve is shown in Fig. 11. Based on the EGR demand, the ECU generates the PWM signal. Due to this signal, the vacuum regulator or modulator transfers the vacuum generated from vacuum pump to EGR valve.

**Fig. 9** EGR cooler mesh model

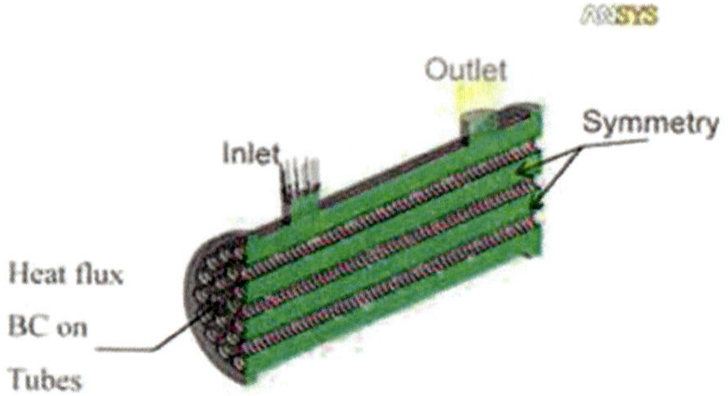


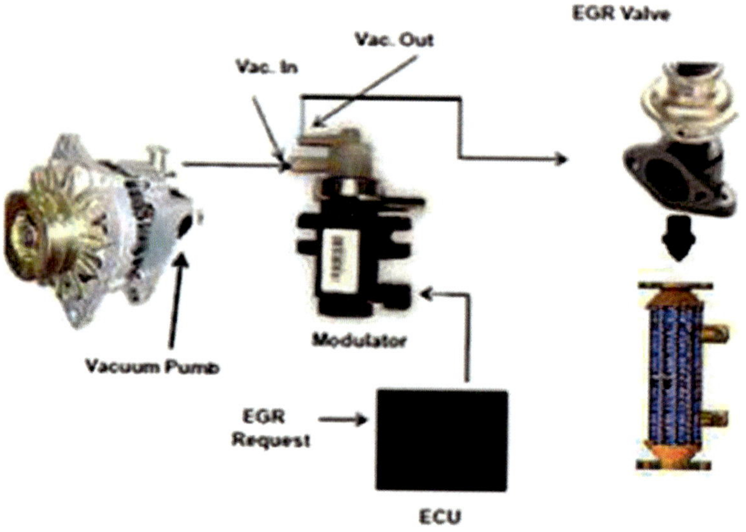


**Fig. 10** Cooled EGR schematic layout

**Fig. 11** Exhaust vacuum regulating valve operation

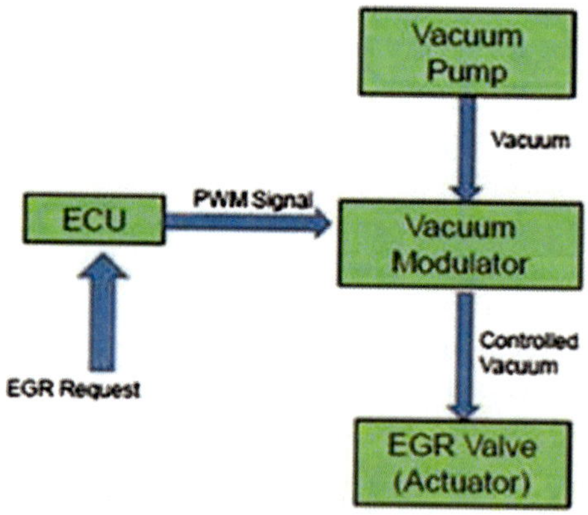

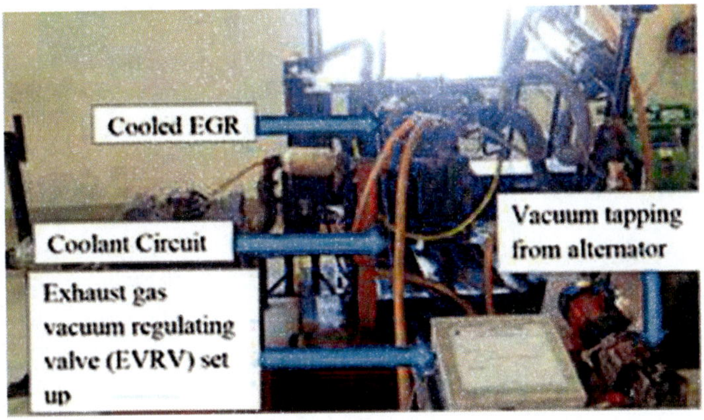


**Fig. 12** Cooled EGR experimental setup on engine

To maintain sufficient coolant flow rate and pressure for flow circulation, separate coolant reservoir tank is used to support flow of coolant inside the EGR cooler for this study. Based on this, the experimental setup of cooled exhaust gas recirculation (EGR) on engine is shown in Fig. 12.

## 3 Engine Optimization Methodologies

Table 4 gives research optimization approach followed for this study. First standard engine is tested. Further, in the second phase, nozzle opening injector pressure (NOP) increased. Then in third phase, spray holes reduced along with NOP increase. In the fourth phase with the above best optimized conditions, turbocharger specification-1 and specification-2 along with the intercooler is tested for performance. Finally, tested with C-EGR tested for emissions.

## 4 Results and Discussion

The brake power comparison for standard and modified engine with TCIC is shown in Fig. 13. For standard engine, the rated power is 15-kW. By increasing the injection pressure to 250 bar, power improves by 1 kW. This might be due to better atomization. Further, decreasing the number of nozzle spray holes from 7 to 6, power improves by another 0.3-kW. Since the hole size is the same by reducing the number of holes by one, fuel jet velocity should increase. As a result, breakup of outer surface of the fuel jet takes place at or before the nozzle exit plane. This might result in smaller droplets whose average diameter may be much smaller than those of the nozzle hole diameter. However, bigger quantity of small size droplets leads to a higher

**Table 4** Research methodologies for engine optimization

| Engine parameters for modification | Standard diesel engine (baseline) | Modified diesel engine (phase 1) | Modified diesel engine (phase 2) | Modified diesel engine (phase 3) | Optimized diesel engine (phase 4) |
|---|---|---|---|---|---|
| Nozzle hole diameter | 0.26 mm | 0.26 mm | 0.26 mm | 0.26 mm | 0.26 mm |
| Number of nozzle hole | 7-holes | 7-holes | 6-holes | 6-holes | 6-holes |
| Injection pressure | 240 bar | 250 bar | 250 bar | 250 bar | 250 bar |
| Fuel pump delivery increase | No | 22% | 22% | 22% | 22% |
| Injection timing b-TDC | 8.5 Degree | 8.5 Degree | 8.5 Degree | 9.0 Degree | 9.0 Degree |
| Engine type | Naturally aspirated | Naturally aspirated | Naturally aspirated | Naturally aspirated | Turbocharged intercooler |
| C-EGR % | 0% | 0% | 0% | 0% | 5% |

surface area that enables heat transfer from the hot compressed air to the small fuel volume contained in those droplets. This smaller droplet evaporates and mixes with the air available in its immediate vicinity which enhances the combustion, and this improves power in naturally aspirated engine. When injection timing advanced to 9.0° b-TDC along with fuel delivery increase, power further improved by 6% from 16.3-kW to 17.5-kW. This condition may be endorsed to slight longer ignition delay. During this period due to the longer ignition delay, larger amount of fuel gets accumulated. Perhaps, this provides better chance for fuel to mix with air. Moreover, with appropriate fuel–air ratio, further increase in fuel quantity due to larger delay may induce knocking and therefore to avoid this proportional increase in mass of air is to be taken care. In this connection, it is found that 2.1 bar boost pressure is more appropriate which increases the power by 40% from baseline along with 7% $NOx$ increase. This improvement is after incorporating turbocharger specification-1. However, by incorporating intercooler, turbocharger specification-2 along with 5% EGR and 37 tubes EGR cooler, power is found to improve by 45%. Further, by using improved version of turbocharger, boost pressure of 2.3 bar and power 27.6-kW could be achieved.

The torque comparison for standard and modified engine with TCIC is shown in Fig. 14. Under standard condition, the torque is found to be 97-Nm at rated load. Further the injection characteristics are improved by changing the pressure to 250 bar, which may provide better atomization. As a result, 4.9% torque is improved. However, duration of fuel injection gets reduced. With the above-mentioned engine condition, along with 6-hole nozzle, 6.7% improvement in torque is found. Moreover, by advancing the injection timing to 9.0° b-TDC, fuel pump delivery increase, torque further improves by 13%. For the longer ignition delay, higher quantity of

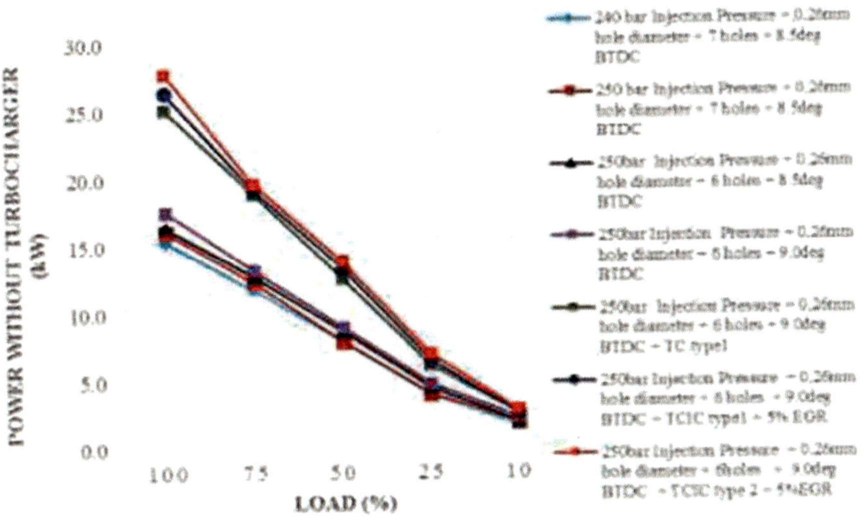


**Fig. 13** Brake power comparison of standard and TCIC engine

fuel is accumulated during this period, and this provides better chance for fuel to mix with air. With extra increase in fuel quantity, proportional enhancement of air mass received into the engine, henceforth higher engine output could be achieved. For this, specification-1 turbocharger torque found to provide 38% with 2.1 bar boost pressure. By intercooling along with 5% C-EGR, torque still improves by 6%. Furthermore, incorporating improved version type-2 turbocharger as well as with boost pressure in 2.3 bar, 45% torque improvement could be achieved.

The intercooler is designed to remove the heat from intake air due to compression into a smaller volume, and the heat the intake air had absorbed from the hot casing. By means of air flow as well as coolant spray over the intercooler air cooling efficiency enhanced. Figure 15 shows intercooler efficiency comparison. Considering the coolant flow rate limitations (500 l/hr.), as well as open frontal area of 3.71 $m^2$, turbocharger speed is limited to avoid further increase in boost air temperature. The boost air temperature at rated load before intercooler is 152 °C for specification-1 or type-1 turbocharger and 145 °C specification-2 or type-2 turbocharger.

The cross-flow heat exchanger type intercooler is used in this investigation, and based on this design, 78% air cooling efficiency with specification-1 turbocharger along with intercooler is achieved. For the specification-2, turbocharger air cooling efficiency at rated load is further improved to 81.6%.

Further, the EGR cooler selection for NO*x* emission reduction is based on heat transfer analysis (1D simulation study). Based on the results generated from 1D simulation studies, the coolant flow rate to improve cooler efficiency is optimized to meet the target requirement of >85% efficiency is shown in Fig. 16. Hence, based on simulation, it is observed that the coolant flow rate greater than 900//hr is required to

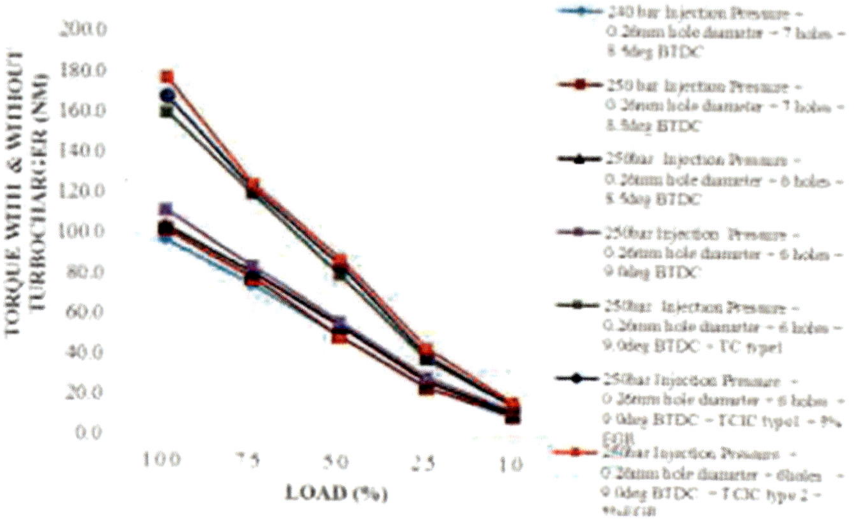


Fig. 14 Torque comparison of standard and TCIC engine

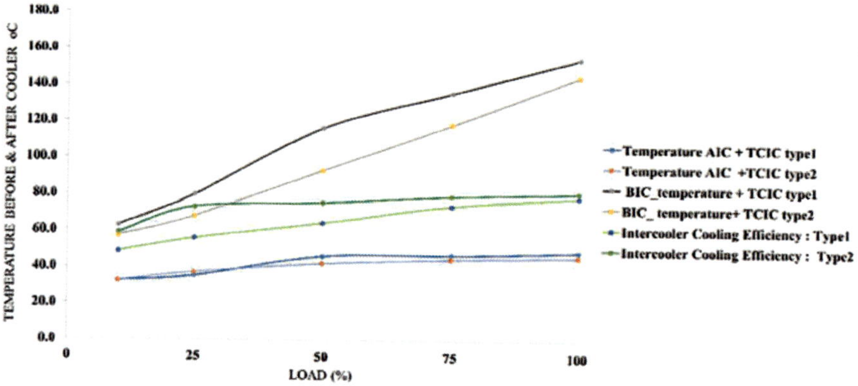


Fig. 15 Intercooler efficiency of specification-1 and 2 turbochargers

avoid risk of water boiling as well as to meet the target exhaust gas cooling efficiency without hotspot formations.

Using Ansys software, temperature contour to identify hot spot formation is generated and shown in Fig. 17. The selected EGR cooler is having 11 tubes in which exhaust gas flow at 680 °C and around the tube coolant enters at 90 °C, to cool the exhaust gas. Furthermore, this coolant removes the heat and then passes through EGR cooler outlet tube. Based on these conditions to simplify the problem for analysis, heat flux is calculated due to the exhaust gas flow in the tubes, and it is applied on the tube's wall as a heat source. Water mass flow rate is taken as inlet boundary

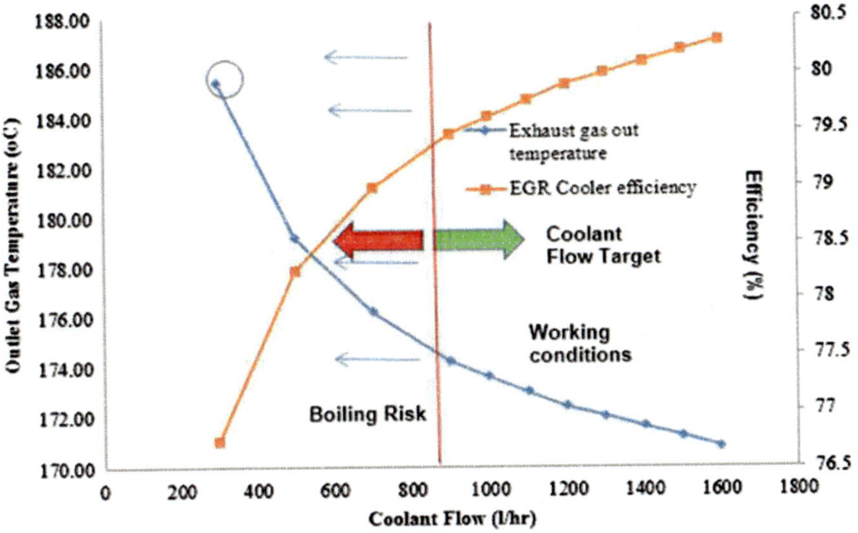


**Fig. 16** Exhaust gas cooling efficiency 1D simulation

condition and pressure outlet condition also considered. Further, based on this hot spot formation with 900 L /hr. coolant flow rate condition, maximum heat dissipation (Q) is 7.18-kW. This provides 79.8% exhaust gas cooling efficiency. But obtained efficiency is lower than target efficiency percentage. Further, hot spot formations are also observed as shown in Fig. 17 with circle mark. Hence, to improve the hot effectiveness, number of tubes could be increased.

Further to enhance the hot effectiveness, number of tubes increased to 37 as shown in above Fig. 18. With 900 L/hour and heat dissipation of 8.3 kW, the hot effectiveness improved to 93%. Further, hot spot formations are also not observed compared to 11 tube design. Hence, this 37 tube EGR cooler, 900 l/hr coolant flow rate could be optimized condition for further NO*x* emissions.

The comparison of NO*x* (oxides of nitrogen) for standard as well as modified TCIC engine is shown in Fig. 19. Weighted NO*x* emission average for all 5 modes with standard engine condition, the NO*x* emission is 6.62 g/kW-h. By increasing the injection pressure from 240 to 250 bar, NO*x* increased slightly to 7.59 g/kW-h. Further, changing only, the injector nozzle hole from 7 to 6 and injector pressure at 250 bar, keeping other engine parameters as standard injection timing, NO*x* emissions reduce to 6.14 g/kW-h. And further advancing the timing to 9.0° b-TDC, NO*x* is increased to 7.05 g/kW-h due to longer ignition delay. This is could be due to advancing the injection timing, pre-mixed combustion gets enhanced and increases NO*x* emission. With this optimized injection parameter, further to enhance the power output, turbocharger is incorporated. In connection with this oxide of nitrogen emission decreased to maximum level of 6.5 g/kW-h for specification-1 turbocharger without intercooler. To enhance NO*x* further, two different modifications are included like

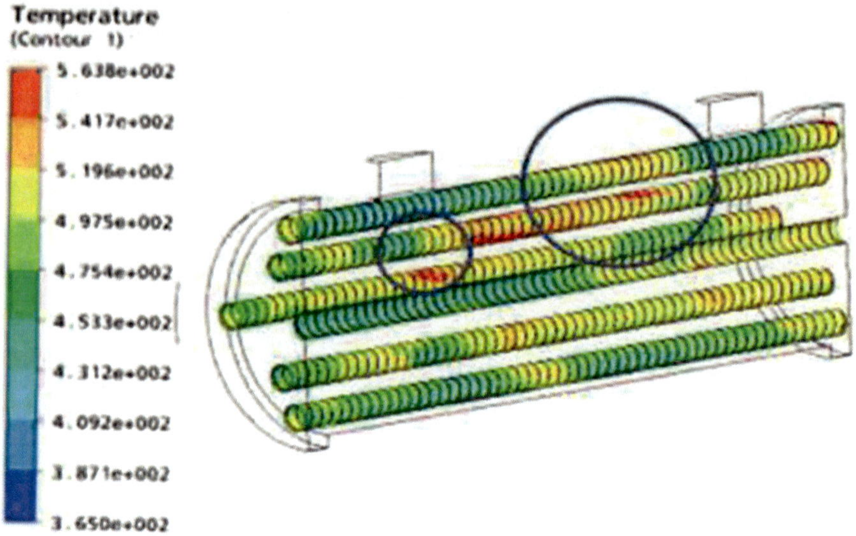


**Fig. 17** Temperature contour for 11 tubes at 900 l/h coolant flow rate

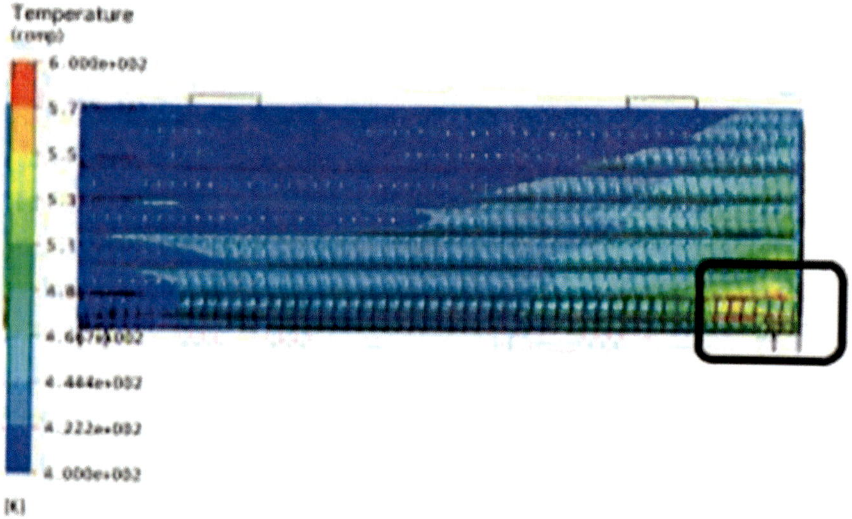


**Fig. 18** Temperature contour for 37 tubes at 900 l/h coolant flow rate

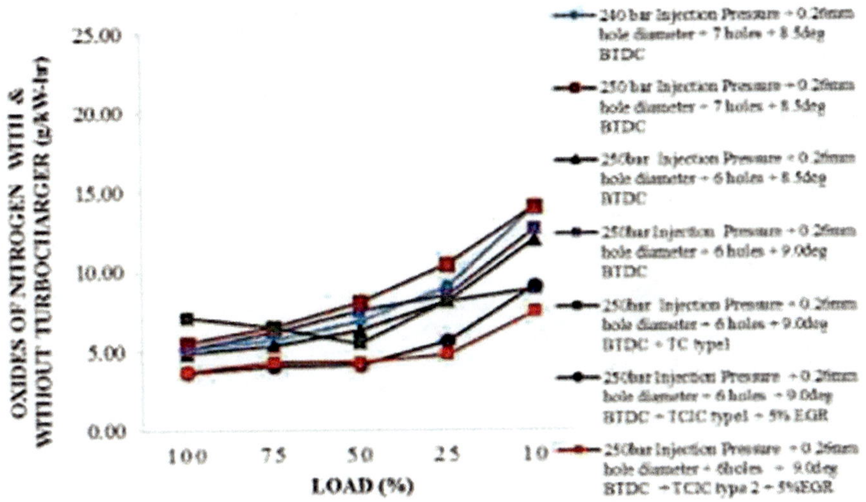


**Fig. 19** Oxides of nitrogen comparison for standard and TCIC engine

specification-1 with intercooler and 5% of EGR, NO*x* reduced to 4.33 g/kW-h. With specification-2 turbocharger and intercooler along with 5% EGR gave 4.13 g/kW-h (37% NO*x* reduction).

Figure 20 shows the comparison of smoke with load for standard as well as modified TCIC engine. Smoke in diesel engine is an indication of poor combustion, resulting from an over-rich A/F ratio or partially evaporated fuel during cold start conditions. The average smoke of all 5 modes with standard engine condition, the smoke emission is 1.18 FSN. By increasing the injection pressure from 240 to 250 bar, smoke decreased slightly to 1.06 FSN. Further, changing only, the injector nozzle hole from 7 to 6 and injector pressure at 250 bar, keeping other engine parameters as standard injection timing, smoke emissions reduce to 0.97 FSN. And further advancing the timing to 9.0° b-TDC, fuel pump delivery increase, smoke is increased to 0.86 FSN due to longer ignition delay. With this optimized injection parameter, further to enhance the power output, turbocharger is incorporated. In connection with this, smoke emission increased to maximum level of 1.13 FSN for specification-1 turbocharger without intercooler. To enhance NO*x* further, two different modifications are included like specification-1 with intercooler and 5% of EGR smoke reduced to 1.11 FSN. With specification-2 turbocharger and intercooler along with 5% EGR gave 1.33 FSN, 11% smoke increased.

The comparison of hydrocarbon for standard and modified TCIC engine is shown in Fig. 21. The average of 5 mode emission with standard engine condition, the hydrocarbon emission is 0.14 g/kW-h. With increasing the injection pressure from 240 to 250 bar, hydrocarbon decreased slightly by 0.12 g/kW-h due to improved atomization and widen lean flame-out region. Due to increasing the injection timing, at 250 bar injection pressure and 6-hole nozzle, this smaller droplet evaporates and

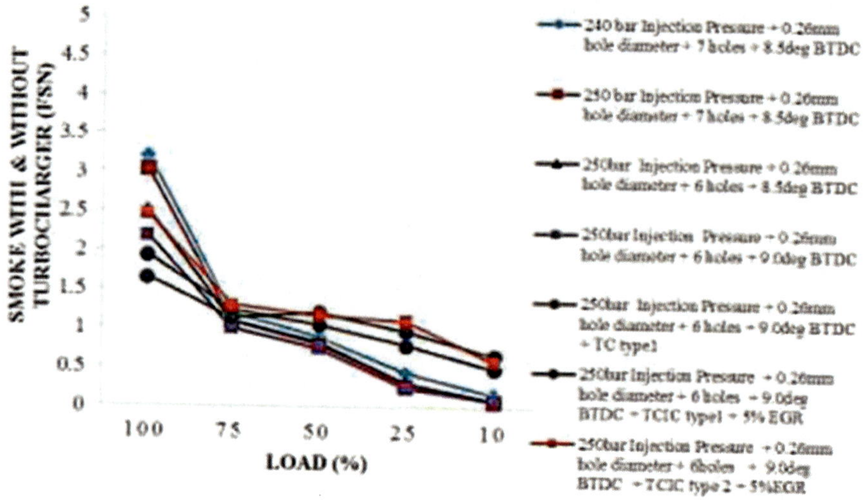


**Fig. 20** Smoke comparison for standard and TCIC engine

mixes with the air available in its immediate vicinity for combustion and this improves hydrocarbon in naturally aspirated engine, hence, decrease of 0.11 g/kW-h could be achieved. And further advancing the timing to 9.0° hydrocarbon increased to 0.13 g/kW-h. With this condition, further to enhance the power output, turbocharger is incorporated. In connection with this, hydrocarbon emission is slightly decreased to 0.10 g/kW-h for specification-1 turbocharger without intercooler due to lean A/F ratio. Furthermore, with same turbocharger (specification-1) with intercooler and 5% of EGR and specification-2 turbocharger with intercooler as well as 5% EGR, hydrocarbon amounts are 0.11 g/kW-h and 0.10 g/kW-h could be achieved.

The comparison of carbon monoxide for standard and modified engine is shown in Fig. 22. The average of 5 mode emission with standard engine condition the carbon monoxide emission is 3.3 g/kW-h load. By increasing the injection pressure from 240 to 250 bar, perceived carbon monoxide is 4.2 g/kW-h. Further, due to increasing the injection timing, at 250 bar injection pressure and 6-hole nozzle, no change is observed. By advancing the timing to 9.0° carbon monoxide not changed. With this optimized injection parameter, further to enhance the power output, turbocharger is incorporated. In connection with this, carbon monoxide emission decreases to 0.5 g/kW-h (>73% improvement) for specification-1 turbocharger without intercooler due to lean A/F ratio. Furthermore, two different modifications are included like specification-1 with intercooler and 5% of EGR and specification-2 turbocharger with intercooler and 5% EGR. Based on optimized turbo waste-gate setting, carbon monoxide is 0.7 g/kW-hr and 0.8 g/kW-hr. Improvement could be >80%.

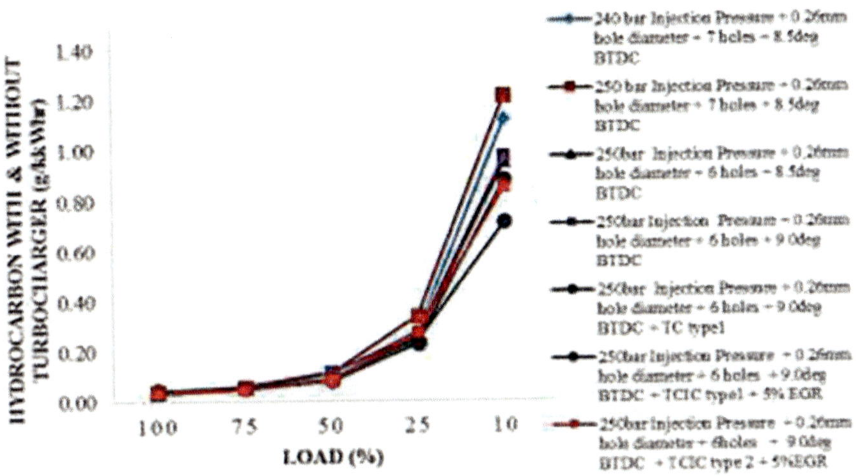


**Fig. 21** Hydrocarbon comparison for standard and TCIC engine

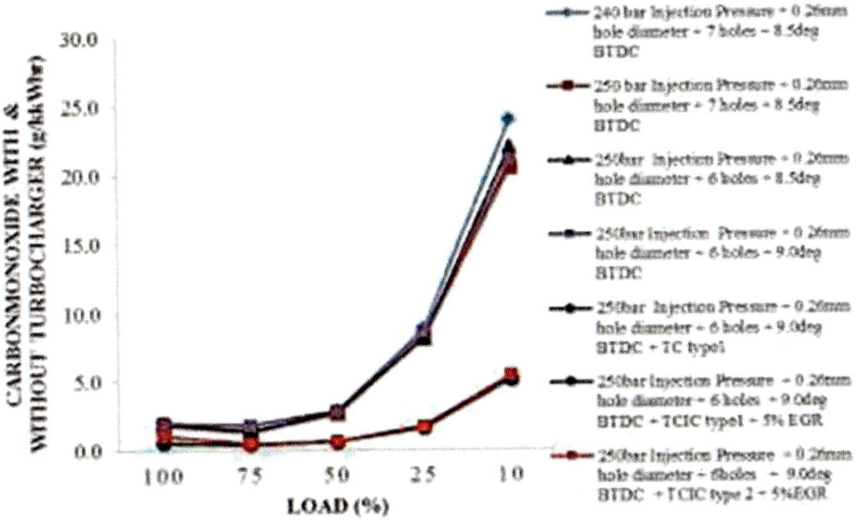


**Fig. 22** Carbon monoxide comparison for standard and TCIC engine

## 5 Conclusion

From the present investigations, the following conclusions are drawn. After converting naturally aspirated to turbocharged intercooler engine, power extensively increased from 15-kW to 27-kW, with specification-2 turbocharger (with turbine wheel size of 31 mm), shows 43% improvement and torque increased from 97-Nm to 176-Nm. With specification-1 turbocharger specification, power is 26-kW and torque 166-Nm only achieved. Normally, higher flow velocity can be achieved by reducing no. of injection holes and modifying k-factor. But here same k-factor maintained for both the hole configurations and therefore, increasing the pressure alone supported to achieve the required minimal power. The nozzle opening pressure increased from 240 to 250 bar. This injection pressure increase is achieved by using the higher shim thickness in injector (to have higher stiffness on spring and load on the needle), and the pressure is measured using nozzle pressure tester. Further, to match the existing turbulence conditions, studies are limited with 6-hole configuration. Hence, the spectrum of pressure is not considered as well as to avoid the NO*x* with turbo charger.

The NO*x* strategy adopted is medium NO*x* where engine-out NO*x* is targeted less than 5 g/kW-h and the remaining NO*x* reduction is targeted with after-treatment strategy. This medium NO*x* is selected to have better specific fuel consumption benefit compared to high NO*x* strategy.

Related to EGR cooler development with 11 tubes computationally shows 78% exhaust gas cooling efficiency. Based on further optimization, 37 tube EGR cooler shows 93% improvement in hot effectiveness. Hence, using 37 tube coolers with 5% EGR (medium NO*x* strategy), rate is selected for reducing NO*x* emission from 6.62 g/kW-h to 4.13 g/kW-h; therefore, 37% is improved in actual testing.

Related to mass (HC: 0.10 g/kW-hr + NO*x*: 4.13 g/kW-hr) emission, limit is well within the Tier 4 mass emission limit <4.23 g/kW-hr norms limit. But to increase in the emission safety margin, engine should be incorporated with after-treatment systems. Hence, in this study, it is demonstrated with proper choice of engine hardware configuration, significant increases in peak power and torque, as well as emission reductions can be obtained for off-road engines ratings between 19 and 37 kW.

## References

1. Dallmann T, Menon A (2016) Technology pathways for diesel engine used in off-road vehicles and equipment. White paper, the international council of clean transportation. Retrieved from www.theicct.org/sites/default/files/publications/Off-road-Tech-Pathways_white-%20paper_vF_ICCT_20160915.pdf
2. Arokiaraj MJ, Anirbandeep S, Murthy P, Parag D, Velusamy R (2013) Intelligent exhaust gas recirculation governing for robust BS-III compliant 2.5L mechanical pump drive diesel vehicle. SAE Paper no 2013-26-0052
3. Ueda M, Takami M, Tanaka T, Miyazaki H (2009) Development of newly concept diesel engine for industrial use. SAE 2009-32-0057/20097057
4. Quazi MA, Dhiman V, Singh S (2013) Development of two-stage turbocharger system for off road application diesel engine in order to achieve 75HP. SAE. Paper no 2013-01-2749. https://doi.org/10.4271/2013-01-2749
5. Heywood JB (1988) Internal combustion engine fundamentals. McGraw-Hill, New York, p 2015

6. Ohashi N, Nakatani K, Asanuma T, Fukuma T, Matsubara H, Sobue Y, Watanabe M (2008) Development of next—generation NO*x* reduction system for diesel exhaust emission. SAE paper 2008-01-0065
7. Tang J, Pischinger S, Grutering U, Keck J (2008) Effects on deposit formation in injection nozzles of direct-injection diesel engines. MTZ 69(9):754–761
8. Woods M, Kamo R, Bryzik W (2001) High pressure fuel injection for high power density diesel engines. SAE Paper no 2001-01-1186
9. Abdullah NR, Mamat R, Rounce P, Tsolakis A, Wyszynski ML, Xu HM (2009) Effect of injection pressure with split injection in a V6 diesel engine. SAE Paper 24-0049
10. Mutta S, Sathiya Narayanan M, Gupta P, Nandhakumar K, Daithankar P (2016) Thermodynamic study of turbocharger matching and combustion optimization for better low-end torque and high-speed power. SAE Paper no: 2016-28-0015
11. Niemi SA, Lauren MJ, Murtonen TT (2002) Effect of waste-gate turbocharging on the exhaust particulate matter of an off-road diesel engine. SAE paper no 2002-01-2159
12. Dreeben T, Millen L, Wells M, Woolworth J (1992) Effect of Sac volume on injector performance. SAE Paper, 920680
13. Iyer H, Shaik R, Vagesh A, Ravisankar M, Srikanth S, Velusamy R (2011) Turbocharging a small two cylinder DI diesel engine-experiences in improving the power, low end torque and specific fuel consumption. SAE paper no-2011-24-0133
14. Eastwood P (2000) Critical topics in exhaust gas after-treatment. Research studies Press. Ltd. ISBN 0863802427
15. Tullis S, Greeves G. Improving NO*x* versus BSFC with EUIZ 200 using EGR and pilot injection for heavy duty diesel engines. SAE paper 960843
16. Banzhaf M, Lutz R (1997) Heat exchanger for cooled exhaust gas recirculation. SAE Paper No. 971822
17. Yao M, Zhang Q, Liu H, Zheng Z, Zhang P (2010) Diesel engine combustion control: medium or heavy EGR? SAE paper no 01-1125
18. Honma J, Murao T, Yamashita Y, Tsujita M, Sugihara H (2004) Development of a highly efficient and reliable multi-tube EGR cooler. SAE Paper no-2004-01-1446
19. Addy MW (2006) Diesel emissions and their control. SAE International, Warren dale. PA, ISBN-10 0-7680-0674*0 and 13 978-0-7680-0674-2
20. Puranen A, Tanska T. A study on off-road diesel engines emission reduction. SAE paper no: 941755.3
21. Onishi T, Sasaki H, Okuda M, Naito K, Yoshida K (2007) Techniques for higher power density and lower exhaust emissions on off-road in direct injection diesel engines. SAE paper 2007-32-0021

# Sustainable Energy Systems

# Sustainable Power Generation with Fluidized Bed Combustion

**M. C. Anand Chakaravarthi, A. Ravinthiran, and S. Vaidyanathan**

## 1 Introduction

The total power produced in India as on September 2020 is 373,029 MWe, and share of Thermal Plant is 61.9%. If the share of biomass and captive power station is added this figure is likely to go up. Contribution of coal to the power sector on this date is nearly 55.2% [1]. A lesser-known fact is that railways pay 30% insurance charges to power supplier for uninterrupted power. This fact indicates that reliability is the prime consideration when power is traded. This reliability comes from the decades of experience spanning over a century with thermal plants (Fig. 1).

Power is needed 24 $\times$ 7 and often it is a political issue. In India, elections have been fought and won on this issue. Globally power is rated as index of the standing of the nation in world community. In the last few decades, technology has gone a vast change. World has also become more conscious about pollution and health hazard. Power should not only be economical but should not create pollution. Solar and wind energy are the prime candidates to cater the world's need of power. Sun God has bestowed us immense power, and wind is available freely. However, these two forms of energy are uncontrollable and solar energy is only partially available and not uniformly distributed across the world. India cannot depend on gas and oil as source of power. Though currently share of gas is nearly 10.0%—thanks to Bombay high—share of oil in power production is less than 1.0%. Besides oil will be always needed for pharmaceutical purposes and should not be used for combustion purpose.

The objective this exercise is to present a model for power generation, which is **economical, reliable** and at the same time reduces greenhouse effect. A heterogeneous model optimally using all the available sources—while reducing use of coal—is suggested here. In the opinion of authors, solar and wind are not the solution in the yonder and coal and combustion devices will have their roles for many

M. C. A. Chakaravarthi · A. Ravinthiran · S. Vaidyanathan (✉)
Department of Mechanical Engineering, Sri Sairam Engineering College, Chennai, India
e-mail: vaidyanathan.mech@sairam.edu.in



N. L. Narasimhan et al. (eds.), *Recent Advances in Energy Technologies*, Lecture Notes in Mechanical Engineering, https://doi.org/10.1007/978-981-19-3467-4_38

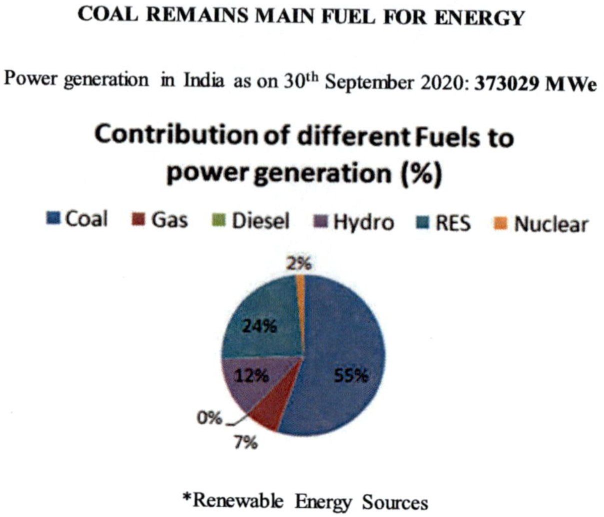


**Fig. 1** Contribution of different fuels to power generation

more decades. Solar energy being a low intensity energy and wind being uncontrollable and unpredictable cannot altogether replace combustion and gasification with reliability. A distributed technology based on fluidized bed combustion (FBC) is suggested which will replace suspension firing. Locally available waste such as bagasse, rice husk, saw dust, pet-coke, local municipal waste, stubble and others are envisaged as source of energy. Power plants need not be of mammoth size, and a few regions can combine and produce power locally. It can be used for local needs, and rest power can be fed into the grid, generating revenue. FBC technology has been envisaged as the solution since it is more flexible, environment friendly and can be downsized unlike pulverized fuel combustion (PFC/PFF) technology. Good grid system available all over the country makes such a distributed system possible. Such an approach will also reduce waste disposal problem.

Article is not a detailed review of CFBC and does not compare BFBC with CFBC. Both have their roles to play depending on the size and fuels to be burned.

## 2 Current Models of Power Plants

Power plants based on Rankine cycle and burning coal in pulverized mode are in vogue all over the world. PFF plants have proved its utility in last hundred years due to its high efficiency, reliability and established technology. PFF/PFC-based power plants having capacity above 1000 MWe are routinely built now [2, 3]. However, PFF/PFC plants have following limitations.

1. Pulverizing mills are fuel specific.
2. System is not ideal for small power plants.
3. High power consumption for pulverizing fuel.

Apart from the above, there are many other limitations of PFF, like high erosion of downstream components, its mammoth size, high skill and specialized knowledge needed for maintenance and operation, etc. Pulverized fuel-fired plants are likely to lose its exalted position in the coming decades due to above limitation. It is likely to be replaced by fluidized bed combustion technology.

## 3 Analysis of Different Sources of Power

The main sources of energy which are in vogue are;

1. Thermal power which includes nuclear plants as well.
2. Solar power.
3. Wind Power.

These three resources together contribute nearly 80% of power produced, remaining coming from hydro plants. Each source has their unique characteristics. Only partial availability of solar energy and uncontrollable nature of wind energy makes these two sources unreliable and one has to fall back on thermal plants. The main disadvantages of solar and wind energy can be stated below.

1. Solar energy being low intensity energy needs lot of space.
2. Solar energy is not available all through the time. Partial availability of solar and its directional nature limits it availability.
3. Needs lot of space. A big engineering industry running umpteen machines cannot depend on solar or wind energy for its operation.
4. High initial cost of energy.
5. Solar energy can either be directly fed into the grid or stored. The former restricts availability and later makes it expensive.
6. Wind energy restricts the location.
7. The initial cost and technological immaturity restrict wind energy.
8. Wind energy is not noise and pollution free as often claimed.
9. All the above restricts the plant size and cannot compete with the large size thermal power plant which is the need.
10. Cost of solar and wind energy enhances with the increase in land cost.

    Failures of solar energy has been reported as it has not been able live up to the expectations [4]. Our own experience to implement solar energy for domestic applications in Coimbatore has not met with success due to engineering short comings.

Nuclear plants and hydro plants are not analysed here, and it is assumed that these plants will continue to occupy their established and important position. This brings one back to thermal power plant, but suitable model of thermal power plants needs to be analysed and established. This is the objective of this article, "**To propose a suitable model of thermal power plant in totality that is commensurate with current scenario of country**".

## 4 Suggested Model

Features of suggested model and assumption made are as under.

1. A distribute model has been envisaged. Instead of having centralized power station, power plants are distributed right up to the village level by combining a few villages. Such plants are called mini power plants in this study
2. Efficiency is not the sole criterion, but cost of power, pollution, reliability with local technical resources and competence will decide the choice of plant.
3. All different forms of energy should be used, thermal, hydro, wind and solar. However, nuclear energy has been excluded due to complexity involved.
4. Power produced will be used locally and extra power will be fed into the grid.
5. **These mini power plants will use the locally available waste material for power generation. No fuel will be outsourced. Use of coal is proscribed.**
6. Plant works on Rankine cycle. Pulverized fuel-fired boiler has been replaced by fluidized bed boiler—circulating or bubbling—and choice of condenser has been left to the local conditions and plant size. It can be air cooled condenser for small size.

These assumptions have been arrived at, based on years of experience in power generation spawning over four decades in this country. A concept based on above said parameters cannot work on pulverized fuel combustion, but needs simpler and acceptable technology. Fluidized combustion, including bubbling and circulating, has been envisaged as an alternative. Circulating fluidized bed boiler can be made in small size from 5 T/h steam generation to 1000 T/h steam depending on need. CFBC has the ability to efficiently burn difficult materials like pet-coke, tyre, washery rejects, stubble and more, while reducing NO*x* emission and capturing sulphur.

## 5 Combustion Systems [5]

Two types of combustion devices were used to burn solid fuel until 1990. These are;

1. Grate combustion.
2. Suspension burning.

**5.1 Grate combustion** works on the same principle as domestic *chullha*, where fuel is spread on a surface, and combustion air is introduced from the bottom.

Combustion takes place in the furnace transferring heat; exhaust gases at low temperature are let out from the chimney. This technology when automated, has served well over centuries, but suitable only in small sizes. Grate combustion has been discarded after nineteenth century in utility sector. Grate combustion cannot burn coal containing ash above 45–50%. It is also uneconomical for power plants due to low power density. Grate combustion is slowly being phased out and replaced by fluidized bed combustion which can give efficiency up to 80% depending on the nature of fuel. FBC can give combustion efficiency 95% and above.

**5.2 Suspension burning** as the name suggests; in this method, fuel is burned in suspension. Developed around 1910 [2], this is one of the most successful methods of burning coal and all coal-fired power plants all over the world work on this technology. In this method, fuel is pulverized to micron size depending on its quality and conveyed pneumatically to furnace. Small sized fuel burn in suspension like oil. Heat generated is absorbed by water surrounding the furnace producing steam. Flue gases leave the system via induced draft fan through chimney. However, suspension burning consumes high power in pulverizing fuel, increases downstream erosion with high ash Indian coal and increases pollution. It also needs standby pulverizer, which calls for high maintenance.

**5.3 Fluidized bed combustion** [6] the third method of combustion invented by Fritz Winkler in 1922, and developed for petroleum cracking in MIT, has been well adopted by combustion engineers. In this method, coal is crushed and not pulverized. This method of combustion has numerous advantages over other two methods of combustion, so much so, that boiler industries even do not get an inquiry for grate-fired boilers. It is also opined by technologist that suspension firing (PFF/PFC) will be replaced by CFBC in the coming years due to its superiority.

Fluidized bed combustion has been extensively studied and described in open literature [7–9]. Beginning from 1980 fluidized bed combustion has almost replaced the grate combustion. Though there are many different types of FBC only two types of fluidized beds viz. bubbling fluidized bed (BFBC) and circulating fluidized bed (CFBC) are applied to combustion. Their ability to burn notoriously famous high ash Indian coal and other solid fuel has made grate combustion disappear from the scene. Figure 2 shows one of the earliest designs of bubbling fluidized bed boiler in India, used for generating 5 T/h steam while burning bio fuel. This unit has been working for nearly 20 years.

Among BFBC and CFBC the former is more suitable in smaller sizes, while latter is preferred when the size of plant is bigger and fuel is difficult to burn such as pet-coke or tyre, etc.

## 6 Circulating Fluidized Bed Boilers

Circulating fluidized bed boiler consists of a tall tower like furnace, followed by a collecting device for arresting elutriated material. Generally, this device is a cyclone,

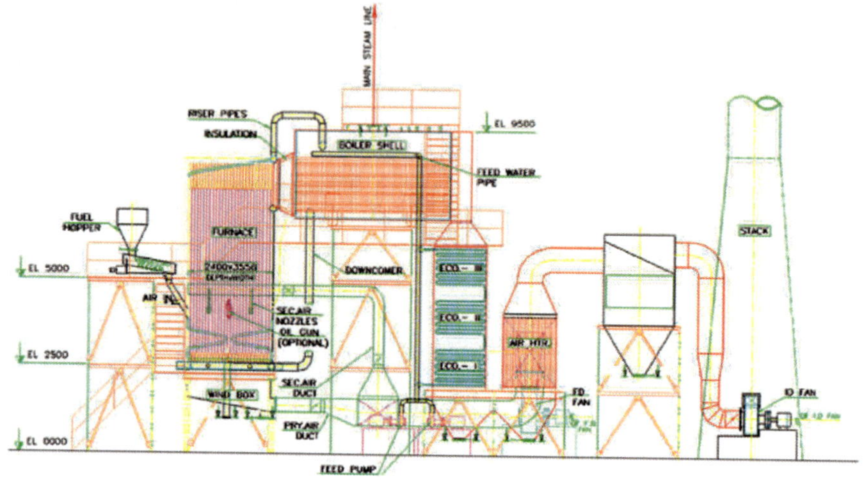


**Fig. 2** Bubbling fluidized bed boiler in small capacity

may be hot or cold. Heat absorbing convection surfaces are located downstream of cyclone. Figure 3 shows the schematic diagram of circulating fluidized bed boiler.

CFBC can tolerate wide variation in fuel characteristics and reduce $NO_x$ emission while capturing sulphur dioxide. This makes fluidized bed combustion systems—CFBC in particular—a better candidate for power plant boiler. However, the key difference in CFBC is its better hydrodynamic resulting in better combustion and its control [7–9]. This is due to high diffusion in CFBC compared to BFBC. Circulating fluidized bed systems have higher slip resulting in higher diffusion which leads to

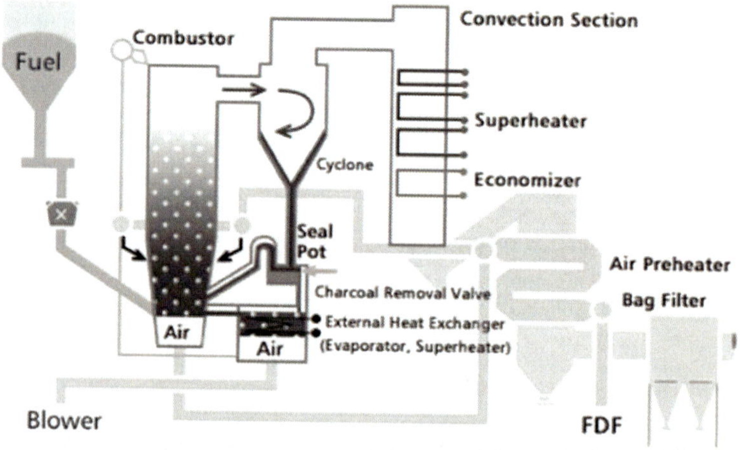


**Fig. 3** Schematic of circulating fluidized bed boiler

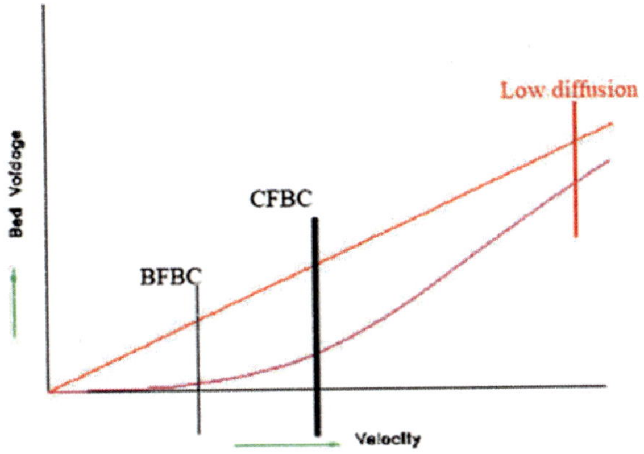


**Fig. 4** Variation of slip in CFBC

better combustion. This is shown in Fig. 4. All three modes of combustion are diffusion controlled. However, high turbulence in circulating fluidized bed boiler increases diffusion bringing its combustion efficiency nearly equal to that of pulverized fuel firing. This unique characteristic of CFBC makes it possible to burn any material in CFBC, which is not possible in PFF.

Circulating fluidized bed combustion is often perceived as circulation of elutriated material vide loop consisting of furnace, cyclone, return leg and material transferring system which is a fluidized bed. Such circulation is only a physical necessity and cannot assure improved combustion efficiency. Improvement in combustion efficiency in CFBC is due to better diffusion which has been explained with the aid of Fig. 4. Figure 4 shows the variation of fluid velocity and particle velocity. Difference between these two velocities is called slip. From the Fig. 4, it can be seen that the slip is initially zero and gradually increases as the fluid velocity is increased. It reaches a maximum and then starts reducing. When the slip is maximum, diffusion is maximum and results in better combustion. While a CFBC operates at the point of maximum slip and hence diffusion, a bubbling fluidized bed operates at lower diffusion. This is the cause of better combustion efficiency in CFBC. However, if the system is operated at far right where though the velocity is high, diffusion will be less due to lower slip. This will result in lower combustion efficiency. It will be further evident when different configuration of CFBC is discussed in the following sections. However, this aspect has not been well understood and often has resulted in poor performance of CFBC.

When the fluidized bed operates at the point of maximum slip it is called turbulent fluidized bed or circulating fluidized bed in industrial and commercial market.

Four different configurations of circulating fluidized bed are in vogue with different degree of success. These are;

1. External cyclone.
2. Compact separator.
3. U-Beam.

4. CYMIC.

CYMIC yet to come in commercial operation but the rest three have their own share of market. A brief description of each configuration is stated below. Though configuration may be different they all consist of the same four components, viz. furnace, particle circulating system, downstream heat absorption system and auxiliaries. The only difference is in particle circulating system.

## *6.1 CFBC with External Cyclone*

This is first generation CFBC pioneered by pyro-power. This design has a thick refractory made cyclone. Figure 5 shows the basic design of this type of CFBC. System consists of furnace, followed by external cyclone made in refractory and convection heat absorbing surfaces. Such a design is no more preferred by users due to heavy spalling of refractory. However, pyro-power is credited for establishing concept of CFBC at commercial level. One such boiler is existing in Aamlai Paper Mills in the state of Chhattisgarh, India.

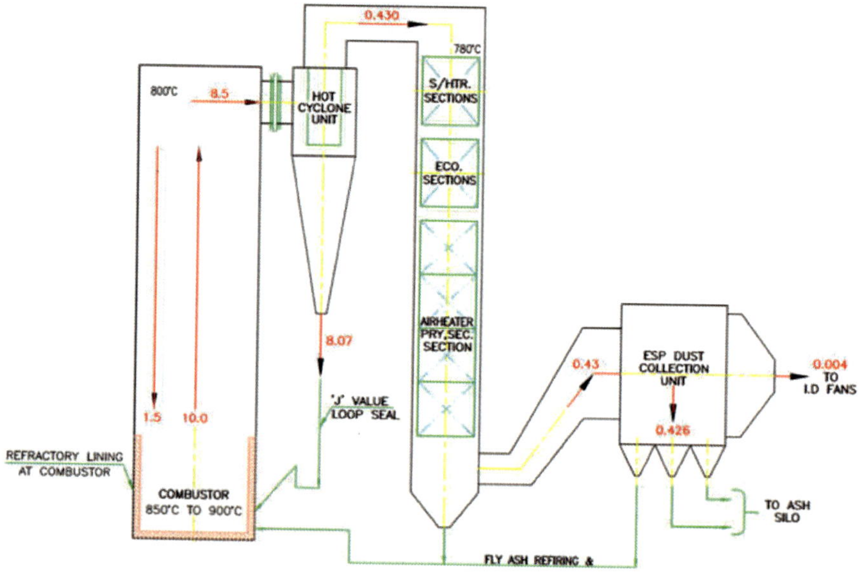


**Fig. 5** CFBC with refractory cyclone

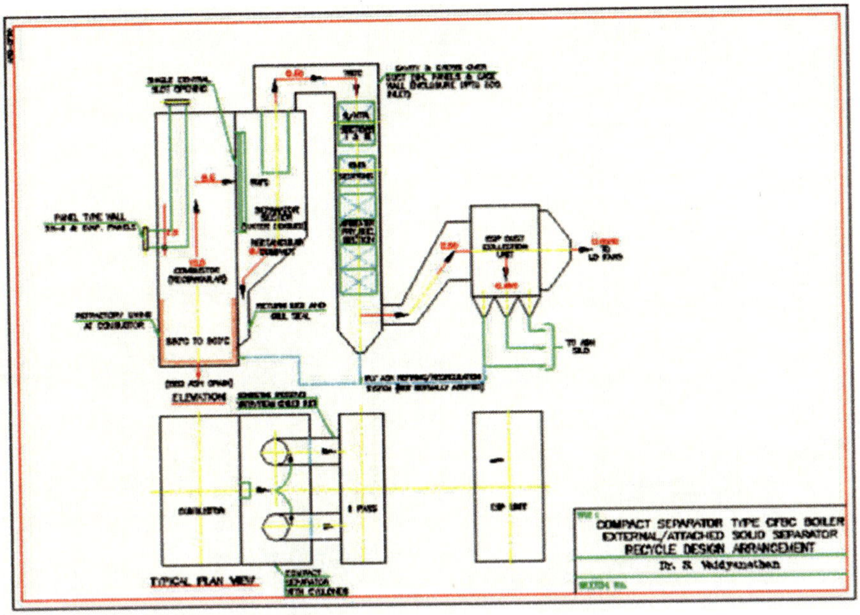


**Fig. 6** CFBC with compact separator

## 6.2 *Compact Separator*

CFBC of this design is offered by Foster Wheeler Inc (USA) and has the largest share of market. In this design, cyclone is integrated with pressure parts and is made in membrane wall construction, thereby reducing quantum of refractory. Design is shown in Fig. 6. Compact separator design has been made in size above 1000 T/h of steam, and it is still growing. Plants of such design exist in India in Mysore Paper Mill, Karnataka, and Rain Calcination, near Hyderabad. Boilers had been operating for more than 20 years burning difficult fuel like pet-coke.

## 6.3 *U-Beam*

U-Beam is the design of Babcock USA and manufactured in India by Thermax Ltd. Pune. Design is shown in Fig. 7. In this design, carry over particle is arrested by impacting it on channel members called U-Beam. Absence of refractory is claimed to be an advantage of this design. Two such boilers are working in Kanoria Chemicals in Uttar Pradesh for more than a decade. Two important deviations in this design are;

1. Absence of critical components like recirculating system
2. Elutriated material falls on the top of the bed.

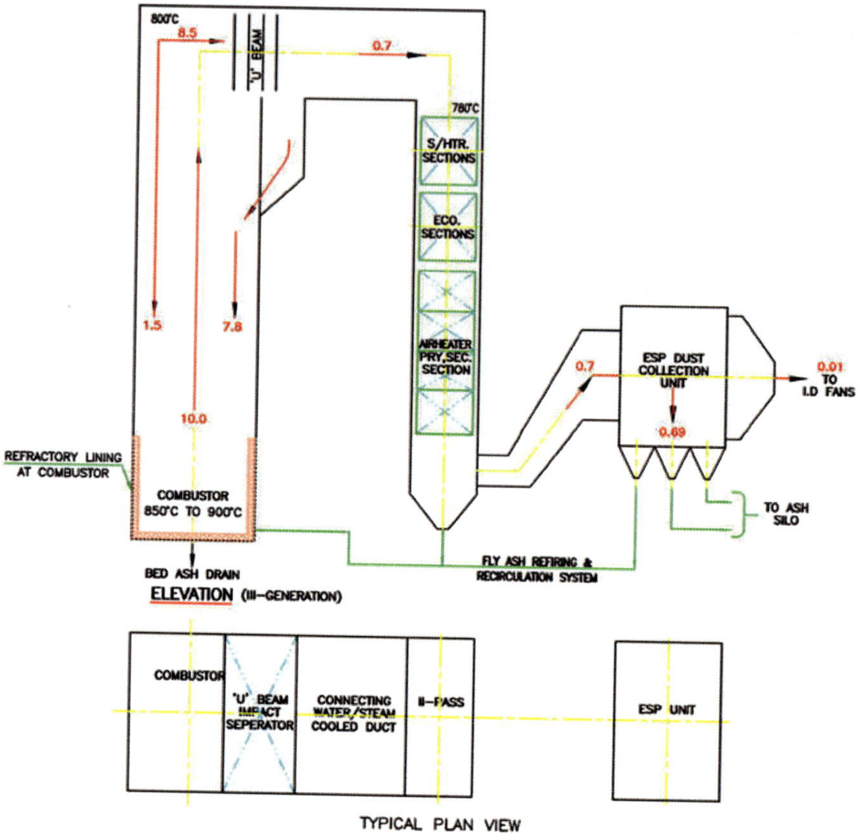


**Fig. 7** CFBC with U-Beam

Elutriated material does not burn in dense bed but in lean phase of the bed. Despite this combustion, efficiency is stated to be high. It confirms that the improved diffusion and not the physical circulation is responsible for better combustion efficiency in CFBC. This agrees with the statement made in Fig. 4.

## 6.4 CYMIC

CYMIC is the acronym for "**cy**lindrical **m**ulti-**i**nlet **c**yclone". This is a novel configuration designed by Kvaerner and is shown in Fig. 8. In this design, there is no external cyclone to arrest the elutriated material. Instead, cyclone is formed within the core of boiler. Annulus works as furnace where fuel is burned. Entire construction is formed in pressure parts. Instead of single entry to cyclone, there are number of inlets to cyclones and hence the name of multi-inlet. Unique feature of this boiler is absence

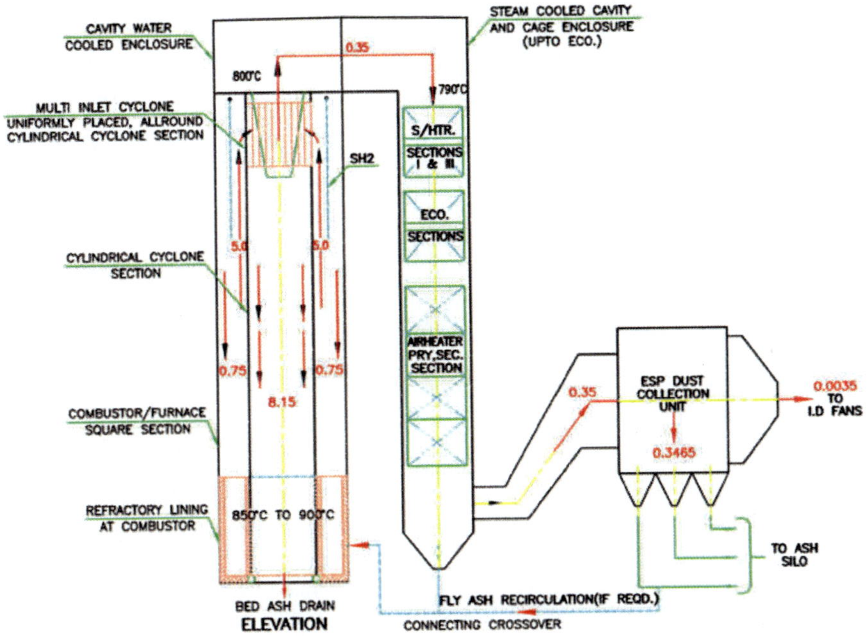


**Fig. 8** CYMIC

of refractory which is always a nuisance. Design is very compact. However, design has not yet been commercialized for some unknown reasons.

## 7 Status of FBC/CFBC in Power Generation

Concept of FBC/CFBC has been put into practice by number of investigators and reported here. It has not yet entered the utility sector, but it has been extensively catering processing sector and cogeneration market. BFBC is considered the system for burning solid fuel, and CFBC is expected to replace PFF/PFC in the decades to come. Some of the major fuels burned in fluidized bed are listed in Table 1.

Only few applications have been mentioned here, and the list is longer and more and more are added. Almost all these units are either BFBC or CFBC boilers producing power between 2 and 50 MWe. Most plants have operating periods more than 10 years, and hence, it can be said that technology has been proved. Some quantitative data are quoted below as this is in line with our subject.

**Washery rejects**: This is first large size BFBC boiler in India operating at TISCO colliery in Jamadhoba, supplied by Bharat Heavy Electricals Ltd, Tiruchirappalli. Fuel contains 65% ash. Despite umpteen problems, boiler was successfully commissioned and operated in the late nineties. Cost of producing power was 0.25 Rs/kWh.

**Table 1** BFBC/CFBC plants using different fuels [10]

| S. No. | Fuel | Capacity [T/h] or MWe | Location (Tentative) | References |
|---|---|---|---|---|
| 1 | Washery reject. 65% ash | 2–2.5 MWe | Jamadhoba, India | [11, 12] |
| 2 | Pet-coke | 55 T/h | Rain calcination, Hyderabad | [13] |
| 3 | Saw dust | 5.0 T/h, steam | Harihar Polyfibres Ltd, Grasim Industries, Karnataka | [14] |
| 4 | Husk | Up to 60 T/h | Rice mills in Ganganagar | [15] |
| 5 | Tyre | – | Research Project | [10] |
| 6 | High moisture lignite | – | NLC Ltd, India | [16] |
| 7 | Anthracite culm | | | [17] |
| 8 | Waste from liquor industries | 30 T/h | Ayodhya, India, Dalkia (UK) | [15] |
| 9 | Agricultural waste | 27 MWe | | [10] |
| 10 | Sewage sludge | 27t/d (dry) | Oneida, New York | [10] |
| 11 | Industrial waste | 32 T/h | E.I.Du Pont, North Carolina | [10] |
| 12 | Paper industry waste | | | [10] |

Success of this unit has given impetus to TISCO to go for higher ash washery rejects plants in West Bokaro [11].

**Pet-coke**: Pet-cokes are the by-products of oil extraction and available in two forms, called delayed coke and fluid coke. Former contains very little volatile matter, and it is difficult to sustain combustion. CFBC boilers have successfully burned this fuel, thereby establishing itself through this technology. Two plants burning pet-cokes built in India are working in Mysore Paper Mills, Karnataka, and the other in Rain Calcination, near Hyderabad. Innumerable plants burning pet-coke are working in western countries. Indian plants were put in operation nearly 20 years back. Pet-coke has also been burnt in bubbling fluidized bed boiler with auxiliary fuel support [13].

**Saw dust**: Saw dust is often a nuisance in ply wood industry. An existing package grate-fired boiler was converted into fluidized bed boiler by this author to burn the freely available saw dust and produces low pressure steam for processing. It is working for nearly 15 years without problem and has resulted in a cleaner plant [14].

**Husk**: Ganganagar and Chhattisgarh are the rice bowl of India. Very large number of boilers are working in these areas burning rice husk. However, further improvement in combustion will make these boilers more lucrative and can generate revenue by selling the bed ash [15].

**Tyre**: Burning tyre in open area is proscribed in India and is a punishable offence. Some researchers think tyre to be perfect fuel to burn in BFBC/CFBC due to its high calorific value which is about 38 MJ/kg, almost equivalent to oil. Tyres cannot be disposed-off by land filling as it is a health hazard. Some bird nest problems have been seen while burning tyre, but CFBC can solve this problem by bed ash drainage [CF 10].

**Liquor waste from liquor industry**: Liquor from industry containing more than 60% water has been burnt in turbulent fluidized bed. One such boiler is operating near Ayodhya [15]. Another liquor waste in semi-solid form is burnt in Dalkia in U.K. This is a bubbling fluidized bed boiler and commissioned recently. Performance figures are awaited.

**Agricultural waste**: India is an agricultural country and will continue to generate agricultural waste which should be our source of renewable energy. Sadly, this has never been our outlook. CFBC boilers can burn almost any agricultural waste when properly designed. Babcock and Wilcox, Pyro-power has built number of boilers burning agricultural waste [16, CF 10]. Burning stubble in open is a common practice in India and had been in the news in the recent past for wrong reasons.

**Industrial and paper industry waste**: Such waste can be both in liquid and solid form. Such boilers must be tailor made for particular waste if the refuse is in liquid form. Paper industry refuse needs special consideration [15, 17].

## 8 Discussion

Versatility of technology has been proven over decades with the heterogeneous fuel. Fuels burnt includes farm waste to liquid waste. Following inference can be drawn.

1. Technology is well established and does not need much research work. It needs attitudinal change and financial support for implementation.
2. FBC is highly versatile and can burn heterogeneous fuel, starting from low calorific value fuels like washery rejects to high calorific value fuels like pet-coke. FBC can thus cater our energy needs.
3. High power density of FBC demands less space for the producing the same power and is reliable. This is the main advantage over solar, which is low intensity power producer and unreliable.
4. FBC boilers can be made in small size like 5 TPH steam to large size like 1000 MWe. This removes the limitations of PFF/PFC which are not suitable in small size.
5. Dynamic response of plant is comparable with PFF/PFC.
6. Above makes decentralization of power production a feasibility.
7. Much less fuel preparation is needed when compared with suspension firing. No preparation is needed if fuel is available in particulate shape and size, like husk, saw dust, pet-coke, etc.

8. CFBC reduces $NO_x$ emission, and thus creates less pollution. Sulphur dioxide can also be retained in bed by feeding limestone as inert material.
9. Fuel flexibility is the biggest advantage of FBC when compared with other modes of firing. It makes FBC suitable for country like India with widely varying occupation, geography and culture.
10. High combustion efficiency, almost in par with PFF/PFC makes the technology ideal.
11. Highly reliable and under full control unlike solar and wind energy. The only method which can cater the emergency needs.
12. Technology is simpler when compared with PFF/PFC. Does not demand too much skill.

Technology has desired features like adaptability, lower emission, versatility, dynamic response and scalability at both ends. Moreover, it is reliable when compared with other forms of energy production. It is said that in coming decades CFBC will occupy entire utility sector all over the world.

## 9 Conclusion

Following conclusions can be drawn from this study.

1. Fluidized bed combustion technology is well developed and has established itself to be an alternative to pulverized coal combustion. In the coming decades, all utility sectors may switch over to CFBC.
2. Heterogeneous waste can be burned in FBC/CFBC, and power can be produced.
3. Coal as a fuel can be largely replaced by biomass and local refuse as a source of power.
4. With technology percolating in India, the country should envisage decentralizing power production using locally available waste and adopt a distributed model.
5. It is advised that educational institutes carry forward basic combustion research by developing combustion set-up that was in vogue in 80 s [18, 19].

## References

1. https://www.google.com/search?q=what+was+the+percnetage+coal+consumption+by+Power+Plant+in+Indai&oq=what++was+the+percnetage+coal+consumption+by+Power+Plant+in+Indai&aqs=chrome..69i57.17009j0j15&sourceid=chrome&ie=UTF-8
2. Field MA, Gill DW, Morgan BB, Hawksley PGW (1967) Combustion of pulverized coal. The British Coal Utilization Research Association
3. Modern Power Station Practices, Central Electricity Generating Board, U.K. (1971)
4. https://www.power-technology.com/features/solar-failed-projects-struggling/
5. Steam/its generation and utilization. Babcock and Wilcox, New York

6. Howard JR (1983) Fluidized beds combustion and application. Applied Science Publishers, London and New York
7. Oka SN (2044) Fluidized bed combustion. Marcel Dekker Inc., New York, Base
8. Kunii D, Levenspiel C (1991) Fluidization engineering. John Wiley and Sons Inc
9. Yaverbaum L (1977) Fluidized bed combustion of coal and waste materials. Noyes Data Corporation, New Jersey, USA
10. Anthony EJ (1995) Fluidized bed combustion of alternative solid fuel. Prog Eng Sci Combustion 21:239–268
11. https://fossil.energy.gov/international/Publications/cwg_april06_wcu_bhel.pdf
12. Personal visit to TISCO colliery and discussion with plant officials (1987)
13. Personal experience in Enmas Process Technology, Chennai (1991–2000)
14. Personal experience with Harihar Polyfibres Ltd, Grasim, Karnataka, India and discussion with their Vice-President, Mr. Shukla (1997)
15. Personal experience and Communication with Rajinder singh, Vice-President Cheema Boilers Limited (1997–2000)
16. file:///E:/08%20Reviews%20&%20Pub/SV%20Tech%20Papers/Sustainable%20Power%20 Model/Anthracite%20coal%20comb.pdf
17. Augevine PA (1982) Fluid bed combustion of sludge. Tappi, Proc of 1982 Engg Conf, pp 506–513, CF 12
18. Vaidyanathan S (1987) Studies on fluidized bed combustion of coal. Ph.D. Thesis, IIT Bombay
19. Suranani S, Goli VR (2011) Combustion characteristics of sawdust in a bubbling fluidized bed. Int Conf Chem Chem Process, IPCBEE 10:167–172

# Miscellaneous Topics

# Computational Modeling of Thermal Phenomenon in Friction Stir Welding of ASTM A710 Steel with AA5083 Alloy

**D. Surya Sinivas Raju, R. Vaira Vignesh, and R. Padmanaban**

## 1 Introduction

The automotive industries focus on the development of lighter and more efficient materials [1] to achieve economic fuel consumption and reduced emissions in line with the environmental norms. For example, a few internal and external panels are replaced by fiberglass in light vehicles. The frame and chassis materials for extreme sports vehicles are replaced by carbon-fiber composites and other rare-earth-based metal alloys. However, a persistent research gap exists in the light-weighting of commercial vehicles. Aluminum alloys are the most utilized material in the automotive industry for their high strength-to-weight ratio [2], relatively good corrosion resistance, and cost-effectiveness.

However, ferrous alloys (steels) are the primary choice for high load-bearing components. Steels are the principal material choice for structural members in heavy-duty vehicles such as trucks, cranes, and other industrial vehicles. The automotive regulations must be getting stringent by the day, which calls for the need for aluminum alloys to be used in heavy machinery. Hence, the use of metals and metal alloys with diverse properties (typically lightweight and high strength) in the automotive industry is inevitable. New classes of alloys and composite are being developed and tested by researchers around the world. The usage of diverse materials for the fabrication of components demands the development of effective joining methodologies [3, 4] (welding, brazing, soldering, mechanical fasting, and adhesion bonding).

Typically, automotive industries have been utilizing aluminum alloy (sheet) for the frame and outer structures. Hence, the joining of aluminum alloy to the steel alloys

D. Surya Sinivas Raju · R. Vaira Vignesh (✉) · R. Padmanaban
Department of Mechanical Engineering, Amrita School of Engineering, Amrita Vishwa Vidyapeetham, Coimbatore, India
e-mail: r_vairavignesh@cb.amrita.edu

R. Padmanaban
e-mail: dr_padmanaban@cb.amrita.edu



N. L. Narasimhan et al. (eds.), *Recent Advances in Energy Technologies*, Lecture Notes in Mechanical Engineering, https://doi.org/10.1007/978-981-19-3467-4_39

(chassis) is crucial. However, only a few kinds of joining processes are capable of joining aluminum alloys and steel. Conventional welding techniques involve fusing two or more parts that are coalesced using heat, pressure, or both with or without filler material. Most of the time, melting and solidification of filler material and workpiece result in microstructural changes [5].

Besides, the development of an erratic temperature profile induces stresses or deformation of the component. Cumulatively, conventional welding compromises the strength of the welded component. In addition, conventional welding of materials that have far apart melting points is a cumbersome process. Because, depending on the temperature, the material with a lower melting point melts first and the one with a higher melting point remains solid or plasticized state. Consequently, the components are poorly fused into each other well [6].

Aluminum not only has a comparatively lesser melting point than steel but also has a more rapid oxidation tendency. Hence, solid-state joining is the primary and reliable choice for conjoining aluminum alloys and steels. One of the solid-state welding processes is friction stir welding (FSW). It is recommended over other solid-state joining processes as it gives rise to lesser residual stresses. However, the weld properties are determined by the peak temperature obtained in FSW. Because the peak temperature influences the degree of recovery and recrystallization. Researchers have been working on the development of effective mathematical and numerical models to simulate the thermal phenomenon in the course of FSW [7]. A summary of the literature related to FSW of aluminum alloys and steel is described in the subsequent section.

The FSW machine converts about half of the total mechanical energy measured into heat energy. As a result, the temperature of the workpiece rises inadvertently during FSW. Depending on the welding conditions, the tool shoulder generates 80–90% of the heat, with the rest going to the other tool surfaces. The joint's tensile strength suffers as a result of a longer friction duration and, as a result, a greater friction temperature. This was linked to the production of FeAl intermetallic phases at the interface, as well as the accumulation of other allowing elements. The production of IMC is inhibited by preventing excessive heat input through the use of a cooling system.

The automotive industry utilizes new alloys to reduce the kerb weight of the vehicle and also possessing the standard strength requirements. Hence, joining of lightweight alloys (aluminum alloys) with high-strength alloys (steels) is the focus of industrial research. In this study, COMSOL Multiphysics software was used to develop a computational model to predict the thermal phenomena that occurred during the FSW of aluminum alloy AA5083 with steel ASTM A710.

## 2 Nomenclature

| Symbol | Description | Symbol | Description |
|---|---|---|---|
| $C_p$ | Heat capacity at constant pressure (kg K) | $\omega$ | Angular velocity (rad/sec) |
| $P$ | Density (kg/m$^3$) | $A_s$ | Surface area (mm$^2$) |
| $K$ | Thermal conductivity (m K) | $q_u$ | Heat flux upside (W/m$^2$) |
| $E$ | Young's modulus | $q_d$ | Heat flux downside (W/m$^2$) |
| $H$ | Poisson's ratio | $h_u$ | Convective heat transfer coefficient upside (W/(m$^2$K)) |
| $q_{\text{tip}}$ | Heat generated at the tip (kJ) | $h_d$ | Convective heat transfer coefficient downside (W/(m$^2$K)) |
| $q_{\text{shoulder}}$ | Heat generated at the shoulder (kJ) | $\epsilon$ | Emissivity (m$^{-1}$) |
| $\mu$ | Friction coefficient | $\sigma$ | Stefan–Boltzmann constant (W·m$^{-2}$·K$^{-4}$) |
| $r_p$ | Pin radius (mm) | $T_0$ | Initial temperature (K) |
| $U$ | Tool transverse speed (m/s) | $T_{\text{amb}}$ | Ambient temperature (K) |

## 3 Computational Methodology

COMSOL Multiphysics software was used to simulate the FSW of the aluminum alloy AA5083 and ASTM A710 steel. MATLAB was used to create a statistical model, which was then utilized to investigate the impact of FSW process parameters on peak temperature.

### *3.1 Numerical Modeling*

The numerical model was built on the assumption that the FSW tool pin was stiff and cylindrical, and that no heat other than frictional heat flowed into the workpiece. During FSW, the FSW tool moves at a constant speed along the workpiece's processing length. A moving coordinate system was fixed on the tool's axis in the generated model.

Hence, the dynamic process of friction stirring near the pin was ignored [8]. The model's coordinate transformation assumes that the workpiece was infinitely long [9]. As a result, the effects at the workpiece's edges were ignored. FSW is applied to the workpiece's upper side. Hence, for the upper side of the workpiece, non-uniform boundary conditions were described that changed in time and the FSW process parameters.

**Table 1** Properties of ASTM A710 steel and AA5083 alloy

| Sl | Property | ASTM A710 Steel | AA5083 alloy |
|---|---|---|---|
| 1 | Density ($\rho$) (kg/m$^3$) | 7850 | 2700 |
| 2 | Young's modulus ($E$) (Pa) | $2.05 \times 10^{11}$ | $69 \times 10^9$ |
| 3 | Poisson's ratio ($\eta$) | 0.29 | 0.33 |
| 4 | Thermal conductivity (K) (W/(m K)) | 36 | 201 |
| 5 | Heat capacity at constant pressure ($C_p$) (J/(kg K)) | 475 | 900 |

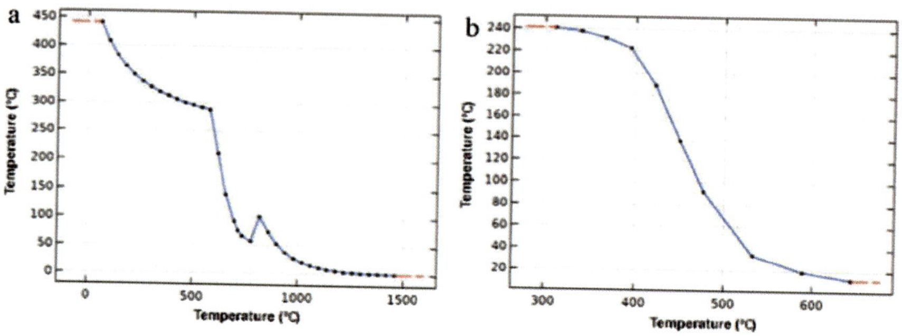


**Fig. 1** Temperature versus tensile strength **a** ASTM A710 steel. **b** AA5083 alloy

### 3.1.1 3D Model

The 3D workpiece was created utilizing the model builder window. The workpieces were designed with an equal thickness of 4 mm. The FSW tool was designed with the following geometrical profile: pin diameter of 4 mm and shoulder diameter of 20 mm.

### 3.1.2 Material Properties

The properties of the workpiece ASTM A710 steel and AA5083 alloy are given in Table 1. The tensile strength of ASTM A710 steel and AA5083 alloy relating to the change in temperature is shown in Fig. 1a, b, respectively.

### 3.1.3 Governing Equations

**Heat Generation.** In the current model, heat generation was taken into consideration at the tool shoulder-workpiece and tool pin-workpiece interfaces. Frictional work was considered to be the source of heat created at the tool shoulder and workpiece interface. Equation (1) describes the heat generation at the interface of tool shoulder and workpiece.

$$q_{\text{shoulder}}(r,T)=\begin{cases}\mu\left(\frac{F_n}{A_s}\right)\omega r, \text{if} T<T_{\text{melt}}\\ 0, \quad \text{if} T>T_{\text{melt}}\end{cases} \tag{1}$$

The friction coefficient ($\mu$) between the FSW tool and the workpiece should differ during FSW. However, a suitable constant friction coefficient was employed in the model. The total heat generated by the tool pin is made up of heat generated by material shearing and heat generated by friction on the pin's surface. It was assumed that the tool pin was cylindrical and non-threaded by profile. Equation (2) defines the heat generation in the pin. The temperature ($T$) at the tool and workpiece contact is set to zero when the workpiece temperature approaches the melting point of the aluminum alloy.

$$q_{\text{tip}}(T)=\begin{cases}\frac{\mu}{\sqrt{3(1+\mu)^2}}r_p\omega Y(T), \text{if} \quad T<T_{\text{melt}}\\ 0, \text{if} T>T_{\text{melt}}\end{cases} \tag{2}$$

**Heat transmission.** The atmosphere encompassed the free surfaces of the workpiece at room temperature, resulting in convective heat transfer between the workpiece surface and the atmosphere. Equation (3) regulates the workpiece's heat transfer. To compensate for the fixation of the coordinate axis, a convection term was added to the heat transfer equation that supplements the conduction term [10].

$$Q=(\rho*Cp*u*\nabla T)+\nabla(-k\nabla T) \tag{3}$$

The computational model did not include the backing plate. The impact of the backing plate, on the other hand, was achieved at the bottom of the workpiece by increasing the heat transfer coefficient. $h_u$ and $h_d$ were considered for the workpiece's upper and lower side. The relevant heat fluxes at the top and bottom sides of the workpiece are given by Eqs. (4) and (5).

$$q_u=h_u(T_0-T)+\epsilon\sigma\left(T_{\text{amb}}^4-T^4\right) \tag{4}$$

$$q_d=h_d(T_0-T)+\epsilon\sigma\left(T_{\text{amb}}^4-T^4\right) \tag{5}$$

### 3.1.4 Meshing Scheme

The free quadrilateral fine mesh was used to mesh the infinite domains and the upper side of the workpieces. The shoulder of the tool and the tool pin meshed to a very fine triangular mesh.

The FSW tool and workpiece's meshing scheme are shown in Fig. 2. To solve the model, a direct stationary solver was used.

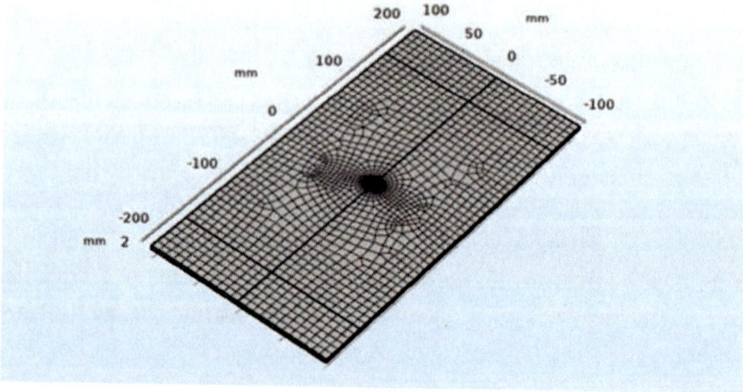


**Fig. 2** Meshed 3D model

### 3.2 *Statistical Model*

Statistical modeling technique was utilized to explore the influence of the FSW process parameters (TRS, WS, and AF) on the peak temperature obtained in the course of FSW of AA5083 and ASTM A710. The process parameters were varied at three levels. The experimental design was based on a face-centered central composite design. AA5083 alloy has a melting temperature of 840 K, and ASTM A710 steel has a melting temperature of 1650 K. Therefore, the highest temperature should be less than 800 K to avoid melting of aluminum alloy in the stir zone [11, 12]. Hence, the process parameters were selected cautiously.

## 4 Results and Discussion

During FSW, the material's microstructure is changed by the spinning tool's plasticizing, stirring, and forging actions. The amount of heat generated has a significant impact on the evolution of the microstructure due to the aforementioned behavior. Therefore, the FSW's basic requirement is to produce frictional heat to plasticize the substance at an acceptable temperature [13].

### 4.1 *Validation of Numerical Model*

The peak temperature of the specimens that were friction stir welded at 800 rpm, WS of 3.5 mm/s, and AF of 1 kN was predicted using a computational model created in MATLAB.

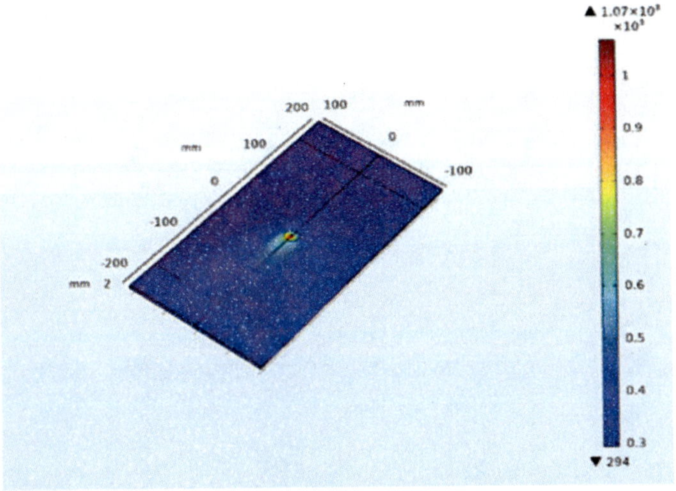


**Fig. 3** Temperature profile of aluminum alloy AA5083 and ASTM A710 steel

Figure 3 depicts the workpiece's surface temperature at the given FSW process settings. In the numerical model, the peak temperature was obtained which was observed to be 792 K. The predicted value of Tp was comparable to the Tp determined experimentally [6, 13–18]. The numerical model developed was therefore positively validated. The validated numerical model was used to forecast Tp at various levels of process parameter combination during the FSW of AA5083 alloy and ASTM A710 steel.

## 4.2 *Numerical Model*

The numerical model was used to simulate the peak temperature in the course of FSW of AA5083 alloy and ASTM A710 steel at the various combination of process parameters. The predicted peak temperature is given in Table 2.

A statistical model was developed using the FSW process parameters (TRS, AF, WS) and the response variable ($T_p$) by linking the process parameters to the response variable, as given by Eq. (6).

$$\begin{aligned}\text{Peak temperature } (T_p) &= 822.3333 + 101.3*\text{TRS} \\ &- 29.9*\text{WS} + 32.1*\text{AF} - 1.166667*\text{TRS}^2 - 9.625*\text{TRS}*\text{WS} \\ &+ 9.875*\text{TRS}*\text{WS} + 9.833333*\text{WS}^2 \\ &- 4.875*\text{WS}*\text{AF} - 5.166667*\text{AF}^2\end{aligned} \quad (6)$$

**Table 2** Experimental design (TRS, TTS, and AF) and predicted peak temperature

| Sl | TRS (rpm) | WS (mm/s) | AF (N) | Peak temperature (K) |
|---|---|---|---|---|
| 1 | 500 | 1.5 | 500 | 716 |
| 2 | 1100 | 1.5 | 500 | 916 |
| 3 | 500 | 3.5 | 500 | 702 |
| 4 | 1100 | 3.5 | 500 | 864 |
| 5 | 500 | 1.5 | 1500 | 756 |
| 6 | 1100 | 1.5 | 1500 | 996 |
| 7 | 500 | 3.5 | 1500 | 723 |
| 8 | 1100 | 3.5 | 1500 | 924 |
| 9 | 500 | 2.5 | 1000 | 721 |
| 10 | 1100 | 2.5 | 1000 | 931 |
| 11 | 800 | 1.5 | 1000 | 901 |
| 12 | 800 | 3.5 | 1000 | 773 |
| 13 | 800 | 2.5 | 500 | 762 |
| 14 | 800 | 2.5 | 1500 | 882 |
| 15 | 800 | 2.5 | 1000 | 803 |

The model efficiency was determined by the statistical parameters $R^2$ and RMSE. RMSE value was 7.506 ($\cong$1%) of the temperatures which were obtained, and the $R^2$ value is 0.933. Hence, the developed model was utilized to observe the effect of the FSW process parameter on peak temperature [13, 19–21].

## 4.3 *Influence of Process Parameters*

Interfacial adhesion between the asperities and tiny plastic deformation during relative motion of the contacting surfaces causes friction between the workpiece and the tool. The frictional energy dissipated at the surfaces during microscopic deformations is completely transformed to heat energy.

In actuality, the frictional force is determined by the properties of the interacting surfaces, as well as their dependency on the load, relative velocities, and temperature. The minute deformations primarily occur on the workpiece surface, as the tool surface more or less is considered non-deformable. As a result, the heat created is unevenly distributed between the two surfaces. With a faster traverse speed, the material ahead of the tool has less time to preheat, which results in less material softening and higher torque. As a result, more energy is required to process the material. In addition, more thermal energy is transmitted to the workpiece when the tool rotates faster. The heat generated to the weld can also be influenced by the localized plastic deformation process that occurs in the bulk of the workpiece.

The influence of FSW process parameters on the peak temperature is plotted as contour plots using Eq. (6). Figure 4a shows the effect of TRS and WS on the $T_p$. As the TRS increased, the peak temperature also increased (TRS $\propto$ T). However, with an increase in WS, the peak temperature decreased (WS $\propto$ 1/T) [21, 22]. FSW of the specimen between 1000 and 1100 rpm and at WS less than 1.5 m/s resulted in peak temperature above 940 K. Inversely at WS above 3 m/s and TRS less than 600 rpm, the peak temperature was below 750 K. Figure 4b shows the effect of TRS and AF on the $T_p$. The peak temperature increased with an increase in AF (AF $\propto$ $T$) [23, 24]. An increase in TRS increased the peak temperature (TRS $\propto$ $T$). At 500 N AF and 500 rpm, the peak temperature was below 750 K. As the AF and TRS increase to 1500 N and 1100 rpm, respectively, the peak temperature reaches above 950 K, which may lead to ineffective weld. Figure 4c shows the effect of WS and AF on the Tp. The peak temperature increases with an increase in AF (AF $\propto$ $T$). But as WS increased, the peak temperature reduced. This is attributed to the less heat input with an increase in WS (WS $\propto$ 1/$T$). Also, the combination of the process parameters causes property changes and defect formation.

## 5 Conclusion

The numerical model for analyzing the thermal phenomenon in the course of FSW of AA5083 alloy with ASTM A710 steel was developed using COMSOL Multiphysics software. A statistical model was developed to correlate the FSW process parameters (TRS, WS, AF) with the peak temperature obtained in the FSW process. The results demonstrate that TRS, WS, and AF have contributed significantly to the generation of heat. TRS is the most influential process parameter influencing the heat generation of aluminum alloy AA5083 and ASTM A710 steel during FSW. FSW at TRS speed between 1000 and 1100 rpm, WS between 1 mm/min and 1.75 mm/min, and AF between results 1300 and 1500 N in peak temperature between 920 and 950 K.

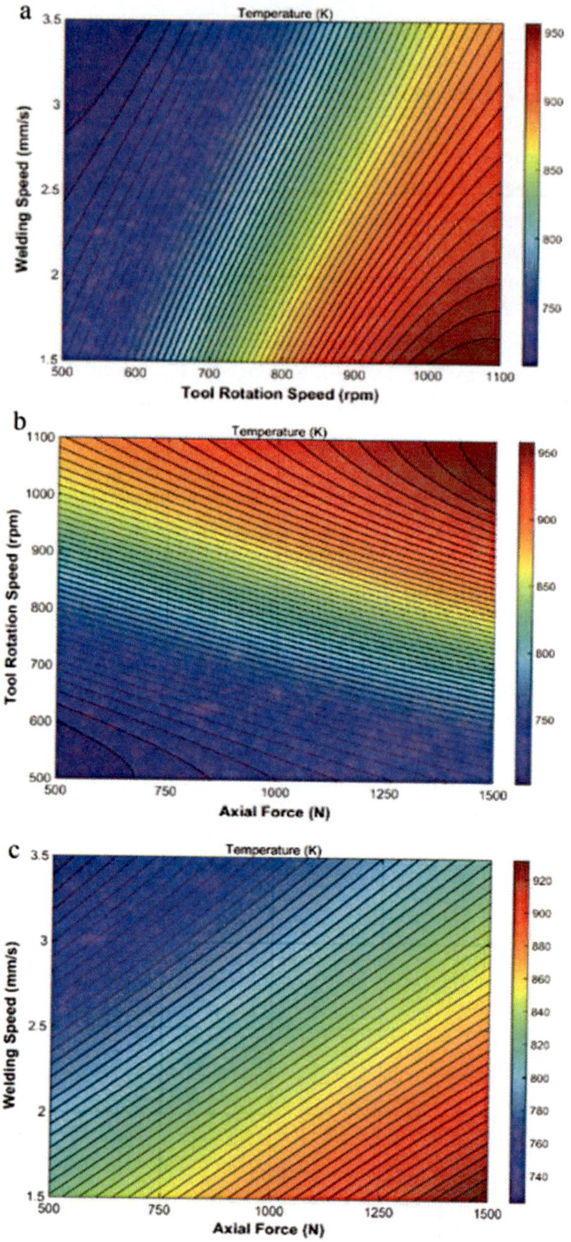


**Fig. 4** Contour plots describing the influence of FSW process parameters on **a** Tool rotation speed (TRS) versus welding speed (WS). **b** Axial force (AF) versus tool rotation speed (TRS). **c** Axial force (AF) versus welding speed (WS)

## References

1. Liedl G, Bielak R, Ivanova J, Enzinger N, Figner G,Bruckner J, Pasic H, Pudar M, Hampel S (2011) Joining of aluminum and steel in car body manufacturing. Phys Procedia 12(Part A):150–156
2. Sahin M (2009) Joining of stainless-steel and aluminum materials by friction welding. Int J Adv Manuf Technol 41:487–497
3. Singha VP, Patel SK, Ranjanb A, Kuriachena B (2020) Recent research progress in solid-state friction-stir welding of aluminum–magnesium alloys: a critical review. Accepted 3 January 2020, Available online 5 February 2020 Science Direct
4. Manjhi SK, Das A, Prasad SB (2020) Review on joining of aluminum alloy by solid-state welding technique. Mater Today: Proc 26(Part 2):1255–1261
5. Gronga Ø, Sandnesa L, Bertoa F (2019) Progress in solid state joining of metals and alloys. Procedia Struct Integrity 17:788–798
6. Zhu XK, Chao YJ (2004) Numerical simulation of transient temperature and residual stresses in friction stir welding of 304L stainless steel. J Mater Process Technol 146(2):263–272
7. Nandan R, Roy GG, Debroy T (2006) Numerical simulation of three-dimensional heat transfer and plastic flow during friction stir welding. Metall Mater Trans A 37:1247–1259
8. Song M, Kovacevic R (2003) Thermal modeling of friction stir welding in a moving coordinate system and its validation. Int J Mach Tools Manuf 43:605–615
9. Schmidt H, Hattel J, Wert J (2003) An analytical model for the heat generation in friction stir welding. Modell Simul Mater Sci Eng 12:143
10. Chen C, Kovacevic R (2003) Finite element modeling of friction stir welding—thermal and thermomechanical analysis. Int J Mach Tools Manuf 43:1319–1326
11. Vignesh RV, Padmanaban R, Arivarasu M, Thirumalini S, Gokulachandran J, Mutyala Sesha Satya Sai R (2016) Numerical modeling of thermal phenomenon in friction stir welding of aluminum plates. IOP Conf Ser: Mater Sci Eng 149:012208
12. Ammouri A, Kheireddine A, Kridli G, Hamade R (2013) FEM optimization of process parameters and in-process cooling in the friction stir processing of magnesium alloy AZ31B. In: ASME 2013 international mechanical engineering congress and exposition: American Society of Mechanical Engineers) pp V02ATA079-V02AT02A
13. Arif A, Pandey KN, Swaroop C (2013 March) Published in a conference temperature validation for friction stir welding (FSW) of dissimilar aluminum alloys
14. Vaira Vignesh R, Padmanaban R (2016) Investigations on the mechanical properties of MWCNT reinforced ASTM A48 by testing & mathematical modelling. In: 10th international conference on intelligent systems and control, pp 761–767. https://doi.org/10.1109/ISCO.2016.7726989 (IEEE Xplore)
15. Hicho GE, Smith LC, Singhal S, Fields RJ (1984) Effects of variations in heat treatment on the mechanical properties and microstructure of ASTM A710 Grade A class 3 steel. J Heat Treating 3:205–212
16. Vaira Vignesh R, Padmanaban R (2018) Influence of friction stir processing parameters on the wear resistance of aluminium alloy AA5083. Mater Today: Proc 5(2):7437–7446. https://doi.org/10.1016/j.matpr.2017.11.415 (Elsevier)
17. Tabana E, Gouldb JE, Lippold JC (2010) Dissimilar friction welding of 6061-T6 aluminum and AISI 1018 steel: properties and microstructural characterization. Mater Des 31:2305–2311
18. Riahi M, Nazari H (2011) Analysis of transient temperature and residual thermal stresses in friction stir welding of aluminum alloy 6061-T6 via numerical simulation. Int J Adv Manuf Technol 55:143–152
19. Silva ACF,De Backer J, Bolmsjö G (2017) Temperature measurements during friction stir welding. Int J Adv Manuf Technol 88:2899–2908
20. Padmanaban R, Balusamy V, Ratna Kishore V (2011) Effect of axial pressure and tool rotation speed on temperature distribution during dissimilar friction stir welding. Adv Mater Res

21. Vaira Vignesh R, Padmanaban R, Povendhan AP, Balakumharen AP (2018) Optimizing the tensile strength of friction stir welded dissimilar aluminium alloy joints using particle swarm optimization. Mater Today: Proc 5(11):24820–24826. https://doi.org/10.1016/j.matpr.2018.10.280 (Elsevier)
22. Alex AJ, Vaira Vignesh R, Padmanaban R, Govindaraju M (2020) Effect of heat treatment on the mechanical and wear behavior of friction stir processed AA5052 alloy. Mater Today: Proc 22(4):3340–3346. https://doi.org/10.1016/j.matpr.2020.03.297 (Elsevier)
23. Zhoua L, Yua M, Liu B, Zhang Z, Liu S, Song X, ZhaoH (2020) Microstructure and mechanical properties of Al/steel dissimilar welds fabricated by friction surfacing assisted friction stir lap welding. J Mater Res Technol 9(1):212–221 (January–February)
24. Padmanaban R, Balusamy V, Vaira VR (2020) Effect of friction stir welding process parameters on the tensile strength of dissimilar aluminum alloy AA2024-T3 and AA7075-T6 joints. Materialwiss Werkstofftech 51(1):17–27

# Theoretical Characterisation of Standing Waves Responsible for Sloshing in Stored and Moving Liquid Containers

S. Karthick, V. Satish, L. Kailash, and S. Peri

## Nomenclature

| | |
|---|---|
| $\eta$ | Interfacial displacement [cm] |
| X | Spatial location for temporal analysis [cm] |
| $\lambda$ | Wavelength [cm] |
| a | Wave amplitude [cm] |
| $\omega$ | Angular frequency [rad/s] |
| c | Phase speed [cm/s] |
| g | Acceleration due to gravity [g/s$^2$] |
| k | Wavenumber [rad/cm] |
| n | Mode number |
| u | Velocity of liquid along $x$-axis [cm/s] |
| t | Instantaneous time [s] |
| Ø | Velocity potential [cm$^2$/s] |
| $\psi$ | Stream function [cm$^2$/s] |
| H | Depth of the liquid in the container [cm] |
| L | Width of container [cm] |
| $\mu$ | Dynamic viscosity of liquid [cm$^2$/s] |
| $\rho$ | Density of liquid [g/cm$^3$] |
| $\sigma$ | Surface tension of fluid–fluid [dyne/cm] |
| $\Omega$ | Temporal growth rate [cm/s |

S. Karthick (✉) · V. Satish · L. Kailash · S. Peri
Department of Mechanical Engineering, Amrita School of Engineering, Amrita Vishwa Vidyapeetham, Coimbatore, India
e-mail: s_karthick@cb.amrita.edu



N. L. Narasimhan et al. (eds.), *Recent Advances in Energy Technologies*, Lecture Notes in Mechanical Engineering, https://doi.org/10.1007/978-981-19-3467-4_40

## 1 Introduction

Liquid sloshing in moving or stationary vessels as a response to external disturbances like inundations in roads remains a great concern in the design of road and ship tankers. This restricts the level of liquid that is to be filled in the container during transportation. Splashing and sloshing are the two possible consequences of instability based on liquid depth in the containers. Intense sloshing creates extremely localised bearing force on tank walls or ceilings, resulting in damage to the vessel. Subsequently, this induces moments affecting the stability of the vehicles, particularly once the external excitation frequency matches the resonance frequency. Of these dual phenomena, sloshing has practical importance. Sloshing denotes to an oscillation of fluid free surface inside a moderately filled container whenever the container is subjected to an external disturbance. It is influenced by internal and external factors. Internal factors include mainly fluid properties like surface tension, viscosity, etc., while external factors include locomotive acceleration, deceleration, turns, and irregularities in road and wave turbulence. Whenever the excitation frequency matches the resonant frequency, container walls break leads to liquid spillage causing accidents and pollution. To avoid such risky situations, anti-slosh devices like baffles are installed in the containers and vessels to dampen sloshing. The sloshing of fluids in sealed ampules has been the focus of numerous investigations over the last epoch. Research works involving experimentation, theoretical studies and numerical simulations have been reported in the last three decades highlighting the dominance of progressive waves over standing modes of instability. To understand the influence of standing modes in instigating the interfacial instability, this analytical model has been formulated and the corresponding temporal growth rate determined.

Experimental investigation of partly filled cylindrical tanks subjected to adjustable acceleration delineated the measure of sloshing rate and hydrodynamic force on the wall for the safe designing of a liquid storage tank. With an increase in longitudinal acceleration, liquid surface displacement escalated resulting in shifting of a moment of inertia and centroid [1]. The effect of a magnetic field in sloshing has been tested by the addition of solid magnetic particles to liquid in a tank. This resulted in a downward shift in peak resonant frequency responsible for sloshing of water, the magnetic particles aligned into parallel lines with back-and-forth movement of water resulting in suppression of liquid motion [2]. Sloshing effects in tanks containing liquid were studied mechanical method and showed that minimum clearance is desired to reduce the negative effects of sloshing it may even cause buckling at the bottom of the tank [3]. An empirical study on pressure distribution due to liquid sloshing in rectangular tanks resulted that without baffles the effect of pressure increases with increase in depth. When the vertical baffles are introduced, greater reduction in pressure is observed while horizontal baffles enhance slamming [4]. Investigation of sloshing in cylindrical ampules with baffles of variable cross-section elucidated the influence of baffles on the fundamental frequencies and open surface. The shape, position and the quantity of baffles inside a reservoir dictate the effectiveness, whereas conical baffles are more efficient (vibrations differ more significantly) [5]. Numerical study of sloshing phenomena in tanks determined the role of ring placed horizontally and blade baffles in vertical orientation in damping the instability [6]. Subsequently, utilizing polar finite elements and involving vibration analysis devoid of external disturbances for horizontal cylindrical shells with and without baffles sloshing rates were compared [7]. The analysis revealed the effect of baffles on the swilling mode, fundamental frequencies and interfacial shape. Numerical investigation on liquid sloshing in road tankers carrying LPG equipped with baffles reveals that containers with baffles have less turbulent kinetic energy, less free surface velocity, and stabilisation time. Containers with "X" shaped baffles are more effective than "+" shaped baffles in constraining sloshing [8]. Kerbing of liquid swilling by foams deliberated that the addition of a few layers of foam on the free surface of water increases the damping coefficient of the foams. It also reduces the displacement amplitude of the liquid-free surface at resonance frequency [9]. Simulations of a moving tank having horizontal and vertical baffles helped in understanding the usage of baffles in suppressing sloshing. The results highlighted the effectiveness of vertical baffles in comparison with horizontal equivalent [10]. The reduction of sloshing loads due to ball baffles in ship tanks has been determined numerically. The results conferred that ball baffles are an effective alternative arrangement available for dipping sloshing of fluid. Using these baffles, tank wall pressure is reduced to half of the initial wall pressure, they predict this can be adopted in tankers and have the potential in emerging as a major sloshing reducing technique in tanker vessels in the future [11].

Sloshing in upright cylindrical containers executing circular orbital motion at steady-state resonant condition describes several free surface phenomenon wave breaking and steady-state liquid rotation, the solution describes insignificance of wave elevation at the wall during liquid rotation [12]. Sloshing control of an automotive fuel tank elucidated generation of sloshing noise during fuel filling. An anti-wave board with a small diameter and a larger number of bleed holes inhibits sloshing [13]. The effect of viscosity of fluid on sloshing in cargos carrying viscous fluids has been experimentally determined. High-viscous liquids exert less pressure compared to liquid with low viscosity. In addition, low-viscous liquids exert more pressure on the container wall [14]. Based on smoothing particle hydrodynamic model, part of crosswise excitation in the X direction in a moderately packed reservoir shown. Beginning from respite, amplitude and frequency amplified over time till they settled [15]. Dynamic analysis on pressure distribution in dissimilarly shaped storage vessels due to liquid sloshing revealed that the impact of pressure distribution and it effect on the walls due to liquid sloshing is maximum for a cylindrical containers. In comparison, rectangular and spherical containers have lesser impact pressure. However, for specially designed LNG containers, this parameter is minimum even at resonance conditions [16].

The above-mentioned works highlighted the need for theoretical understanding in addition to dedicated experiments has been realized. From the seminal works, it is understood that only two types of waves are possible at liquid–air interface, namely: progressive and standing waves. Progressive wave is a basic sinusoidal wave transferring energy from one point to another in the fluid medium. On the other hand, waves with identical amplitude and wavelength moving in opposite direction super impose

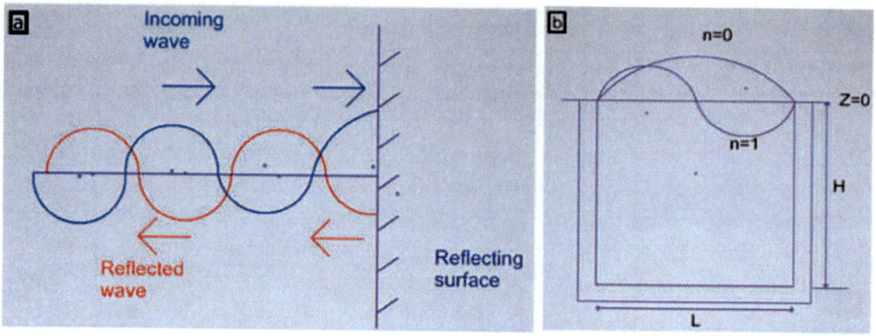


**Fig. 1** **a** Depiction of standing waves; **b** Flow domain involving a tank of width L and depth H

with each other leading to the formation of standing waves. This wave experiences a node at zero displacement and an anti-node at maximum displacement position. It is also known as dilatational mode in stability analysis. Research works based on the influence of progressive waves in liquid sloshing is adequately conferred. An attempt has been made to theoretically characterise sloshing using analogy with gravity waves and water wave mechanics. From a theoretical perspective and to understand the role of standing waves in sloshing phenomena, this analytical model has been attempted and flow domain is illustrated in Fig. 1. Characterising the sloshing phenomena induced by non-propagative standing waves at the liquid–air interface and the results can be useful from academic perspective and for situations where progressive waves are suppressed.

## 2 Formulation

Considering two non-propagating waves of amplitude '**a**' and moving in opposite direction in $x$-axis, their respective surface displacement equations are:

$$\eta(x,t) = a[\cos(kx - \omega t)] \quad \text{in} + x\text{-axis} \tag{1}$$

$$\eta(x,t) = a[\cos(kx + \omega t)] \quad \text{in} - x\text{-axis} \tag{2}$$

For identical right and left travelling waves, applying superposition theorem on Eqs. 1 and 2 results in the expression for interfacial displacement of a standing wave.

$$\eta(x,t) = 2a[\cos(kx)\cos(\omega t)] \tag{3}$$

For mathematical completeness of the interfacial problem, two boundary conditions need to be applied across the interface, namely: kinematic boundary condition where the moving free surface comprises of same fluid particles and dynamic boundary condition which ensures pressure balance across the interface. The flow field is defined using stream function ($\psi$) and the corresponding velocity components $u' = \left(\frac{\partial \psi}{\partial z}\right)$ and $w' = -\left(\frac{\partial \psi}{\partial x}\right)$ can be recognised. The free surface does not proliferate but simply fluctuates up and down with frequency '$\omega$', keeping the nodal points stationary.

Kinematic free surface boundary condition applied at fluid–fluid interface $z = \eta$ approximated as,

$$\left(\frac{\partial \psi}{\partial z}\right)_{z=\eta} \cong \left(\frac{\partial \eta}{\partial t}\right) \tag{4}$$

Utilizing Bernoulli's equation in the linearized form:

$$\left(\frac{\partial \psi}{\partial t}\right) + \left(\frac{P}{\rho}\right) + (g\eta) = 0 \tag{5}$$

Assuming gauge pressure across the interface,

$$P_{z=\eta} = 0 \tag{6}$$

Utilizing Eqs. 5 and 6 simplified form of the dynamic free surface boundary condition at $z = \eta$ is obtained. On linearization, the quantities can be evaluated at z = 0 instead of z = $\eta$.

$$\left(\frac{\partial \psi}{\partial t}\right)_{z=\eta} \cong \left(\frac{\partial \psi}{\partial t}\right)_{z=0} \cong -(g\eta) \tag{7}$$

Solving Eqs. 5 and 6 yields stream function as,

$$\psi = \left(\frac{2a\omega}{k}\right)\left(\frac{\sin h(\mathrm{k}(z + H))}{\sin h(\mathrm{k}H)}\right)\sin(kx)\sin(\omega t) \tag{8}$$

The possible wavelengths of a standing wave are found by setting: $u = 0$ near the walls.

$$u = \left(\frac{\partial \psi}{\partial z}\right) \tag{9}$$

$$u = \left(\frac{2a\omega}{k}\right)\left[\frac{\cos h(k(z + H))}{\sin h(\mathrm{k}H)}\right]\sin(kx)\sin(\omega t) \tag{10}$$

Utilising boundary conditions at walls of the storage vessel to be $x = 0$ and $x = L$ corresponding to condition of no flow through walls requires: $\sin(kL) = 0$ [17]:

$$\mathrm{k}L = (n+1)\pi, \lambda = \left(\frac{2L}{n+1}\right) \tag{11}$$

Effective utilisation of Eqs. 7–10 assist in determining the natural frequency of standing wave as given below:

$$\omega = \sqrt[2]{\frac{\pi g(n+1)}{L}\left(tanh\left(\frac{(n+1)\pi H}{L}\right)\right)} \tag{12}$$

Using the fundamentals of wave theory and definition of wave number, wavelength of the standing wave can be arrived.

$$\lambda = \left(\frac{2\pi}{\mathrm{k}}\right) \tag{13}$$

The phase speed of standing wave(c) is obtained using angular frequency and wave number of the generated disturbance and called as dispersion equation.

$$c = \left(\frac{\omega}{\mathrm{k}}\right) \tag{14}$$

First derivative of interfacial displacement ($\eta$) based on time assists in determination of the temporal growth rate [$\Omega$] of standing wave.

$$\Omega = \left(\frac{\mathrm{d}\eta}{\mathrm{d}t}\right) = -[2a\omega](\cos kx)(\sin \omega t) \tag{15}$$

Equations 11–15 mentioned above enable determining the corresponding dispersion relation, phase speed and temporal growth rate of the introduced wave. These parameters were determined based on Eq. 14 will be carried at a spatial location $x = 0.5$ m, based on chosen tank dimensions: width of 1 m and depth of 1 m. Utilising the obtained growth rate equation (Eq. 15), for each fluid combination, the temporal progression of waves generated at the interface [18–21] can be determined, and subsequently, the wave having maximum growth rate has been delineated.

## 3 Results and Discussions

The temporal analysis for standing wave responsible for sloshing has been carried out at specific times for different fluids of practical importance. A range of fluids of industrial and practical importance, namely: standard fluid [water], industrial chemicals [benzene, butyl acetate, toluene, acetic acid, methanol, chloroform and carbon tetrachloride] and cryogenic fluids [liquid nitrogen and liquid oxygen] have been selected. This selection enables testing the accuracy of the model for a range of thermo-physical properties. The time chosen for the analysis is in the range of 900–3600 s. Based on mathematical understanding, the possible modes for standing waves include fundamental [$n = 0$] and corresponding sub-harmonics [$n = 1$–$10$]. The analysis involves determining the appropriate wave number for selected modes and correspondingly determining the angular frequency and the phase speed. Combining these parameters of interest the growth rate of the standing waves determined. The influencing parameters, namely wavelength, wavenumber, temporal growth rate and phase speed are plotted as a function of mode number for standard fluid water in Fig. 2. The growth rate of the standing wave in the time domain surges with escalation in mode number.

The analysis is extended to fluid combinations of practical relevance. In addition, to improve the significance of this analysis, the influencing parameters are made dimensionless using appropriate physical quantities. Dimensionless growth rate for different fluid combinations is obtained using respective surface tension ($\sigma$) and dynamic viscosity ($\mu$) values. Correspondingly, dimensionless wave number is attained using depth of the liquid in the container ($H$).

Figure 3a, b depicts the dimensionless growth rate of the tested liquids as a function of dimensionless wave number at different time instants covering one hour duration. The independent evolution specifies the consequence of surface tension and viscosity on growth rate measured in temporal domain. It is to be noted that fluids having lesser surface tension, namely butyl acetate and liquid nitrogen, have the highest temporal growth rate in the tested time intervals [$900 < t < 3600$ s]. However, at an instantaneous time of 3600 s, there is no significant difference in the growth rate of tested fluids. The influence of surface tension diminishes over time, and there is no selective amplification of the waves. However, the effect of surface tension is to escalate the temporal growth rate for higher modes as evident in Fig. 2. In the case of industrial liquids tested, similar behaviour is perceived. That is, acetic acid having larger surface tension has a higher growth rate entirely amongst the selected fluids of industrial importance. On the other hand, in the case of milk and liquid oxygen, a higher viscous effect suppresses the growth of disturbances dominating over inherent surface tension.

The mathematical formulation for equivalent progressive wave and its influence in liquid sloshing in container has been attempted [22]. Using the corresponding equations of progressive wave, the parameters of interest are compared in Fig. 4 for water, liquid nitrogen and benzene at 1800s. This is done to confirm the earlier research predictions affirming that progressive wave's dominance over equivalent standing waves. However, for the chosen fluids, in the tested range of $k$ from 0.03 to 0.3 rad/cm, standing waves are dominant over its counterpart progressive wave. This range of wavenumbers has been selected based on the array of mode numbers ($1 < n < 10$) chosen for the study. This rudimentary investigation enables us to comprehend the role of standing waves in sloshing phenomena.

**Fig. 2** Variation of instability parameters as a function of mode number ($n$): (a) wave length, (b) wave number, (c) temporal growth rate and (d) phase speed

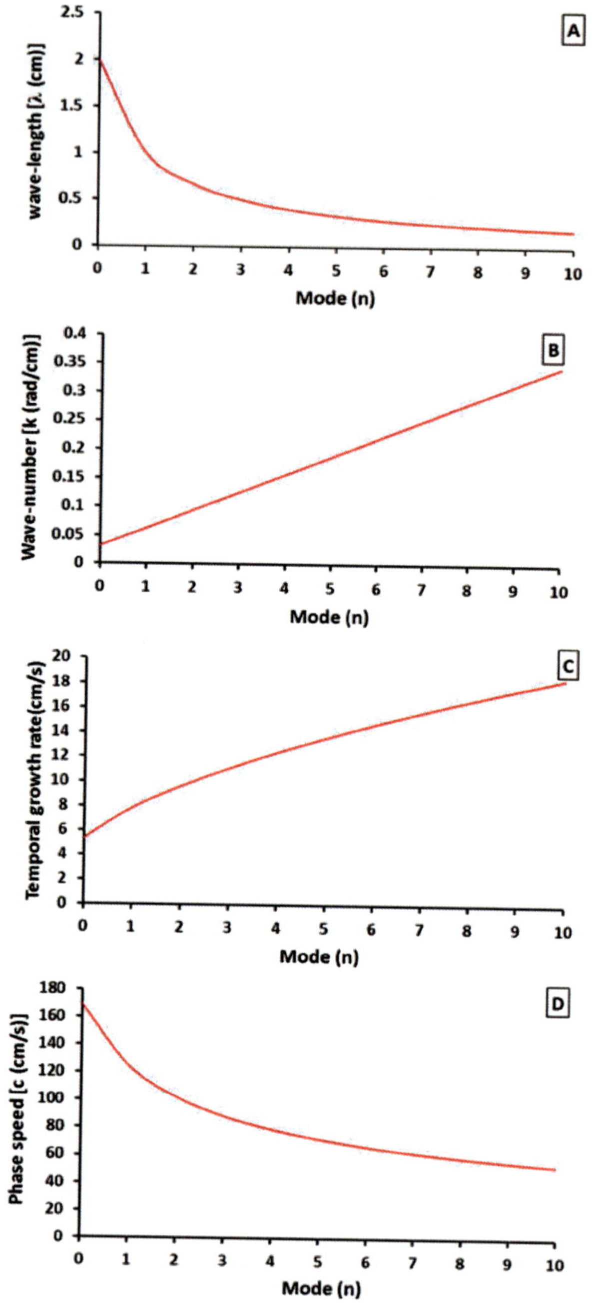

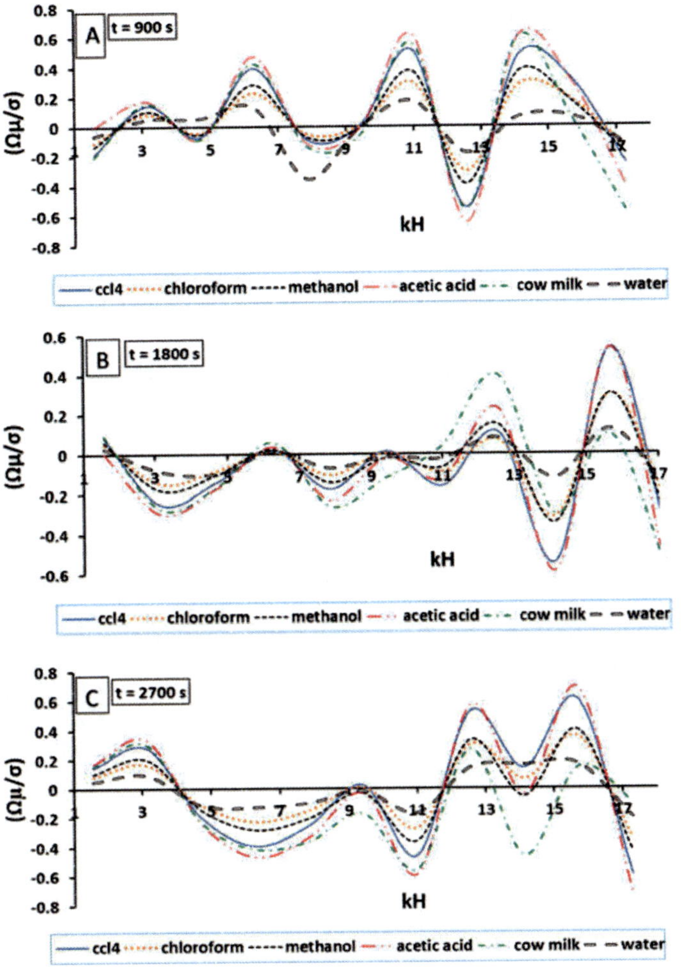


**Fig. 3** (a) Significance of different liquids (carbon-tetrachloride ($\sigma = 26.3$ dyne/cm), chloroform ($\sigma = 26.7$ dyne/cm), methanol ($\sigma = 22.1$ dyne/cm), acetic-acid ($\sigma = 27$ dyne/cm), milk ($\sigma = 51.7$ dyne/cm) and water ($\sigma = 71.87$ dyne/cm)) under the influence of surface tension and viscosity as a function of wave number and temporal growth rate for mode of 0–10 at times (A)900 s, (B)1800s, (C)2700 s and (D)3600 s. (b) Significance of different liquids (benzene ($\sigma = 28.2$ dyne/cm), toluene ($\sigma = 27.9$ dyne/cm), butyl acetate ($\sigma = 24.8$ dyne/cm), liquid oxygen ($\sigma = 13.4$ dyne/cm)) and liquid nitrogen ($\sigma = 8.92$ dyne/cm) under the influence of surface tension and viscosity as a function of wave number and temporal growth rate for mode of 0–10 at times (E)900 s, (F)1800s, (G)2700 s and (H)3600 s

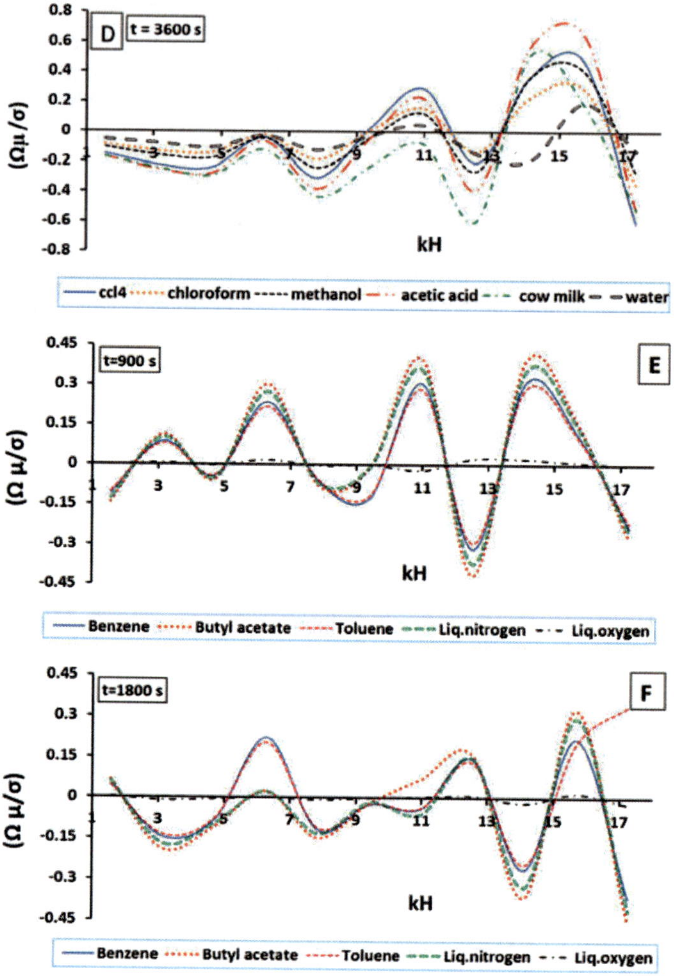


**Fig. 3** (continued)

## 4 Conclusion

The effect of standing waves in enhancing the sloshing phenomena has been theoretically examined using the analogy surface gravity waves. A range of fluids of both theoretical and industrial significance have been tested for sloshing stability. The stability parameters determined include the angular frequency, the phase speed and the corresponding temporal growth rate have been determined. On an average, the impact of surface tension and viscosity has been suppression of the growth rate of perturbations in the tested range. Comparison with an equivalent progressive wave, the dominance of standing wave has been realised for the wave numbers. However,

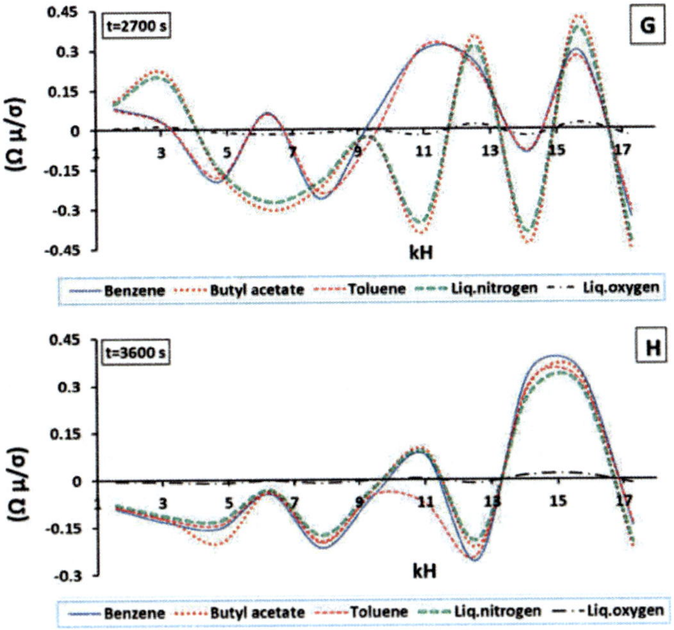


**Fig. 3** (continued)

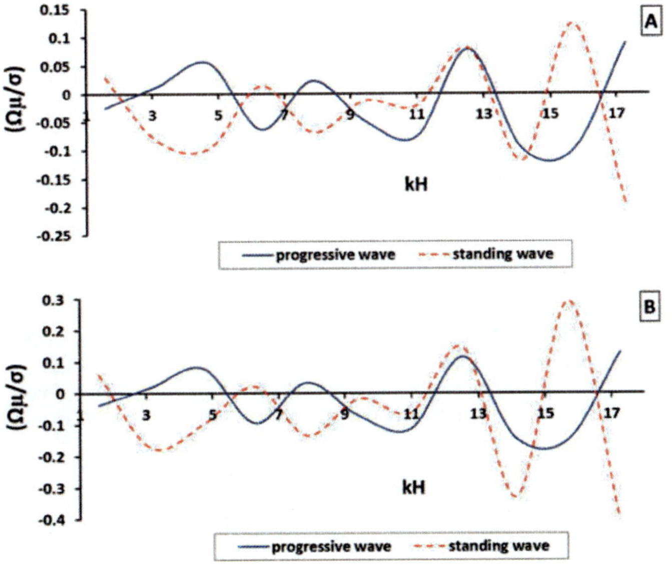


**Fig. 4** Comparison of progressive and standing waves behaviour for different fluids at time 1800s: (a) water, (b) liquid nitrogen, (c) benzene

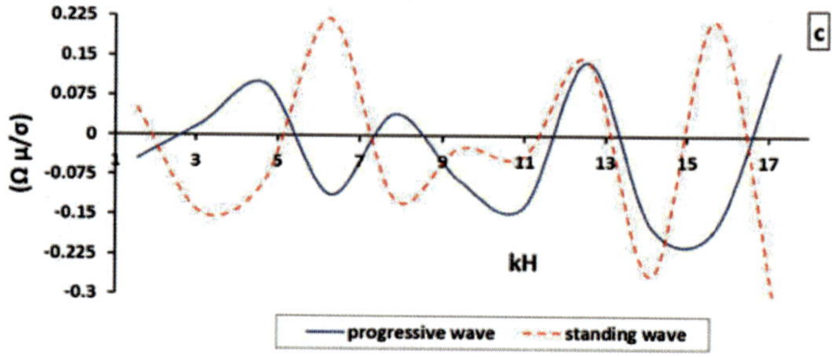


**Fig. 4** (continued)

this dominance of standing wave cannot be expected for all range of wave numbers. The present work is based on linear stability theory while the real phenomenon may involve interaction of different modes illustrating the non-linear nature of sloshing in practical systems. For better insight, a nonlinear analysis involving both progressive and standing modes can provide better insight to the practicality and accuracy of the analytical model.

## References


1. Badran O, Gaith MS, Al-Solihat A (2012) The vibration of partially filled cylindrical tank subjected to variable acceleration. Engineering 540–547
2. Frandsen JB (2008) Sloshing tanks in magnetic field. In: 9th international conference on flow-induced vibration
3. Sivý M, Musil M, Chlebo O, Havelka R (2017) Sloshing effects in tanks containing liquid. In: MATEC web of conferences, vol 107, p 00069
4. Akyildiz H, Unal E (2005) Experimental investigation of pressure distribution on a rectangular tank due to the liquid sloshing. Ocean Eng 1503–1516
5. Gnitko V, Naumemko Y, Strelnikova E (2017) Low frequency sloshing analysis of cylindrical containers with flat and conical baffles. Int J Appl Mech Eng 867–881
6. Liu D, Lin P(2008) A numerical study of three-dimensional liquid sloshing in tanks. J Comput Phys 227:3921–3939
7. Kashani BK, Sani AA (2016) Free vibration analysis of horizontal cylindrical shells including sloshing effect utilizing polar finite elements. Eur J Mech Solids 58:187–201
8. Bautista-Jacobo JL, Rodríguez-Morales E, Montes-Rodríguez JJ, Gámez-Cuatzin H (2015) Effect of baffles on the sloshing in road tankers carrying LPG: a comparative numerical study. Math Problems Eng 1–9
9. Sauret A, Boulogne F, Cappello J, Dressaire E, Stone HA (2015) Damping of liquid sloshing by foams. Phys Fluids 022103
10. Siddapureddy S, Satapathy AK (2016) Liquid sloshing in a moving tank- CFD. In: 6th Intl. and 43rd national conference on fluid mechanics and fluid power
11. Jithu S, Janardhanan S, Asok N (2018) Numerical study on the effect of ball baffles in reducing sloshing loads in ship tanks. In: International conference on computational and experimental marine hydrodynamics
12. Raynovskyy I, Timokha A (2018) Steady state resonant sloshing in an upright cylindrical container performing a circular orbital motion. Math Problems Eng 1–8

13. He Y, Zeng Z, Yu Y, Ruan D (2019) Liquid fuel sloshing control of an automotive fuel tank. Noise & Vibration Worldwide, pp 227–236
14. Chenliang W, Weijun W, Wenfeng W, Changwen Z, Yubin Y, Jialin G, Jiaoyang T, Jiakuo Z, Fan S, Xuxiu W (2019) Experimental analysis of the effect of cargo viscosity on sloshing in tank. In: 2nd world conference on mechanical engineering and intelligent manufacturing
15. Mashy DG, José Y, Dominguez M, Gesteira MG, Peiró J (2021) Smoothing particle hydrodynamic a simulations of 3 dimensional sloshing in tanks. 229:108925
16. Xue M-A, Chen Y, Zheng J, Qian L, Yuan X (2019) Fluid dynamics analysis of sloshing pressure distribution in storage vessels of different shapes. Ocean Eng 192:106582
17. Kundu PK, Cohen IM, Dowling DR (2012) Fluid mechanics, 5th edn. Elsevier Academic Press
18. Meirovitch L (2001) Fundamentals of vibrations, International edition. McGraw hill publications
19. Karthick S, Srivatsan M, Raghav R, Vishnu Supreet K (2015) Theoretical investigation of primary breakup of liquid jet using linear stability analysis in the first wind induced regime based on Kelvin- Helmholtz instability. IJAER, Special issue 10(55):3901–3905
20. Sivadas V, Karthick S, Balaji K (2020) Symmetric and asymmetric disturbances in the Rayleigh zone of an air-assisted liquid sheet: theoretical and experimental analysis. J Fluids Eng ASME 071302
21. Raj N, Karthick S (2020) Characterization of Rayleigh–Taylor Instability at the fluid-fluid interface. Trends in Mechanical and Biomedical design, Lecture notes in Mechanical Engineering, Springer, Singapore, 401
22. Karthick S, Satish V (2021) Theoretical characterization of liquid sloshing in Containers. In: ASME conference, IMECE2021, 70152

# Design and Simulation of Boost Integrated Half-Bridge LLC Resonant Converter for LED Applications

**Sridhar Makkapati, R. Seyezhai, S. Sridhar, S. Srikirthi, B. Sriram, and V. Vikram**

## 1 Introduction

One among the promising source of lighting is LEDs with a high lifetime, efficacy and colour rendering index in lighting applications [1]. LED as a light source is mostly dominated in street lighting, and automotive applications due to its energy-saving and longer life operational capability compared with conventional light sources [2]. In street lighting application, LED needs constant current and constant voltage when driven from the AC source. In two-stage converters [3] as shown in Fig. 1, one as a constant current stage another power factor correction (PFC) stage are employed to regulate the output voltage and to improve the front-end power factor, respectively. However, the cost of the driver circuit with two independent control schemes in two-stage made the driver circuit complex and bulky. Hence, a single stage solution [4] is focussed in the literature and its block diagram is shown in Fig. 2.

In single stage operation, both the PFC and output current control can be done by using a single integrated topology, with reduced device count and control complexity. Nowadays, several DC-DC converter topologies with DCM operation are employed with faster response and low cost. Buck converter [5] is used for low output voltage and electrolytic capacitor-less (EC-less) operation, and different PFC converter with ripple cancellation [6] is employed in single stage LED driver. A fly back circuit [7] with front-end DCM is employed for commercial usage. In all these topologies, hard switching is employed for power switches which increases the losses in the system. Hence, the resonant converters [8] with soft switching capability assist in increase the efficiency of the converter.

In resonant converters, LLC topology with a front PFC circuit increases the efficiency. In traditional two-stage front-end, PFC with HB-LLC converter [9] is

S. Makkapati · R. Seyezhai · S. Sridhar (✉) · S. Srikirthi · B. Sriram · V. Vikram
Sri Sivasubramaniya Nadar College of Engineering, Kalavakkam, Chennai 603110, India
e-mail: sridhar.makkapati2@gmail.com; sridharm@ssn.edu.in



N. L. Narasimhan et al. (eds.), *Recent Advances in Energy Technologies*, Lecture Notes in Mechanical Engineering, https://doi.org/10.1007/978-981-19-3467-4_41

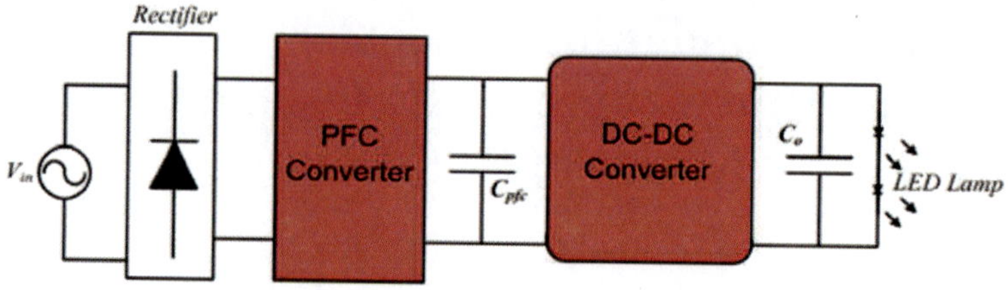


**Fig. 1** Block representation of two-stage LED driver

**Fig. 2** Block representation of single stage LED driver

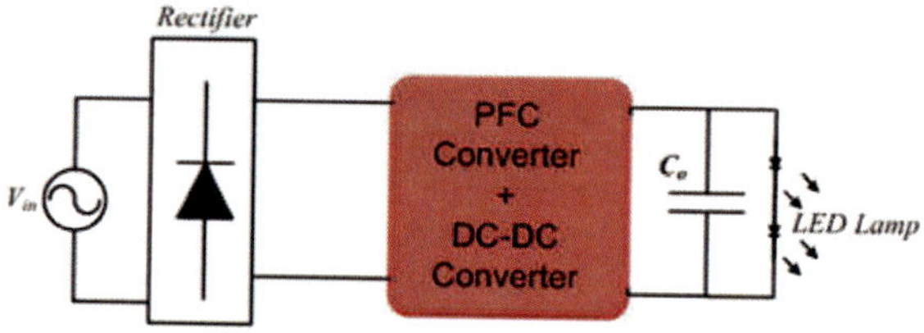


employed which is expensive and complex control. Reference [10] presents the interleaved boost integrated LLC converter with two boost converters which increase the count of the magnetic components. Bridgeless boost integrated HB-LLC is presented in [11], however, the power switches does not undergo full resonant condition. In this paper, boost integrated HB-LLC topology is designed and discussed in detail to achieve ZVS characteristics. LED is driven with 36 V voltage and 2.77 A current from 110 $V_{rms}$.

Section 2 explains the operation of the boost integrated HB-LLC converter. Design equations for DCM boost and HB-LLC for desired output voltage are elaborated in Sect. 3. Simulated results for boost HB-LLC are explained in Sect. 4. Finally, Sect. 5 ends with the conclusion.

## 2 Boost Integrated HB-LLC LED Driver

The single stage boost integrated HB-LLC converter is shown in Fig. 3. Front-end AC input voltage is fed through the diode bridge rectifier to the integrated converter. The boost converter in the figure above comprises $L_{boost}$, switch $S_1$, bus capacitor $C_{bus}$ and diode $D_2$. Half-bridge resonant converter comprises of two switches $S_1$, $S_2$, resonant components such as $L_r$, $C_r$ and $L_m$ are resonant inductance, resonant capacitance and magnetizing inductance, respectively. The output of the isolation transformer is fed to the full-bridge diode rectifier. Here, the switched $S_1$ and $S_2$ are shared to achieve DCM operated boost converter and resonant characteristics in the HB-LLC converter. To analyse the circuit under steady-state analysis, the basic assumption is considered, such as the power switches are loss-less devices, the resonant frequency

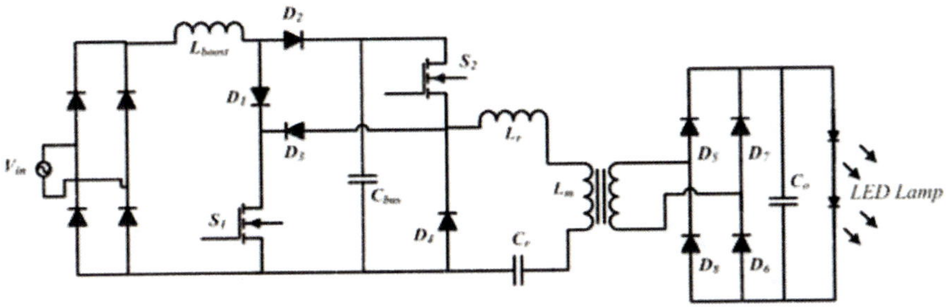


**Fig. 3** Boost integrated HB-LLC single stage LED driver

is greater than the supply frequency. The mode of operation over a switching cycle is explained in detail.

## 2.1 Modes of Operation

During Mode 1 operation both the switches are turned OFF. At this instant, both the magnetizing current ($i_{Lm}$) and leakage current are equal and no power is transferred to the secondary side. Therefore, output power is delivered by the stored energy in the output capacitor Co. The operating circuit is shown in Fig. 4.

In Mode 2 as shown in Fig. 5, switch $S_1$ is turned ON and switch $S_2$ is in an OFF state. The boost inductor current $i_L$ increases linearly, and the output diode $D_7$ and $D_8$ will come into conduction. The difference in current between the $i_{Lm}$ and inductor current $i_{Lr}$ will transfer the power to the secondary side of the high-frequency transformer.

In Mode 3, switch $S_1$ is still turned ON, and the diode $D_1$ and $D_3$ will come into conduction, inductor current $i_L$ increases linearly and the difference in current

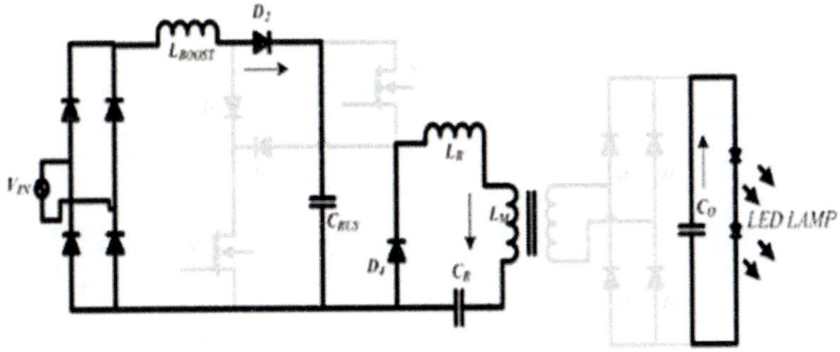


**Fig. 4** Mode 1

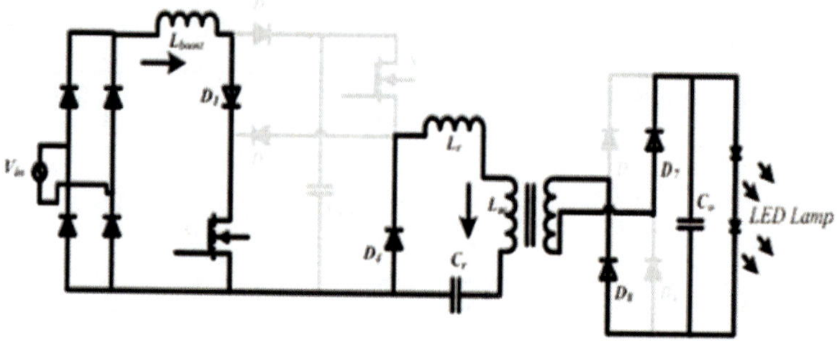


Fig. 5 Mode 2

between the $i_{Lm}$ and inductor current $i_{Lr}$ will transfer the power to the secondary side. The operating circuit is shown in Fig. 6.

During Mode 4 as shown in Fig. 7, switch $S_1$ is still turned ON and the diode $D_1$ and $D_3$ are turned OFF. Output capacitor delivers the power to the load as $L_m$ undergoes resonant condition. Both the switches are turned OFF at the end of this mode.

During Mode 5, the circuit operates during the dead time as shown in Fig. 8. The half-cycle operation ends with Mode 5, and the next half-cycle operation continues from Mode 6 to Mode 8 as shown in Figs. 9, 10 and 11.

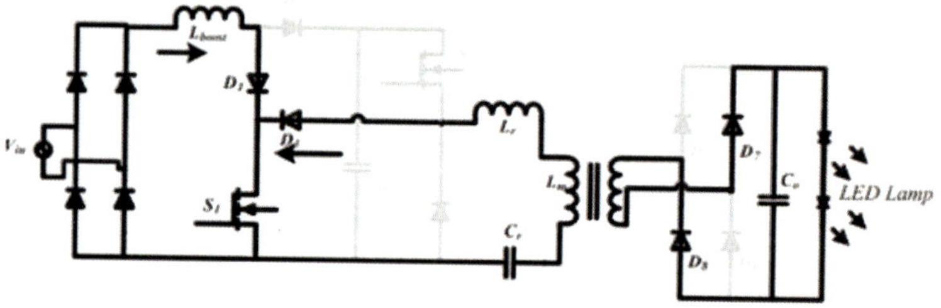


Fig. 6 Mode 3

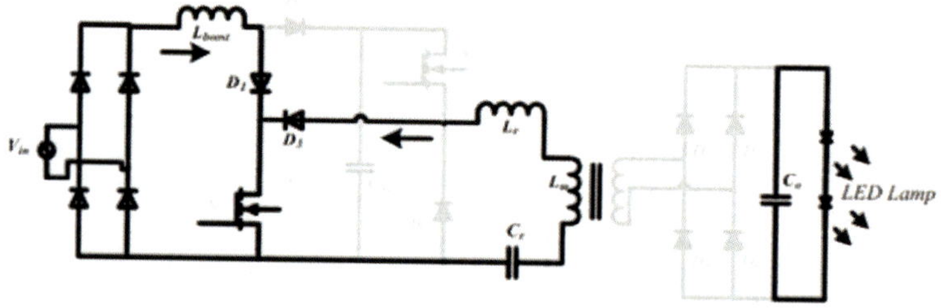


Fig. 7 Mode 4

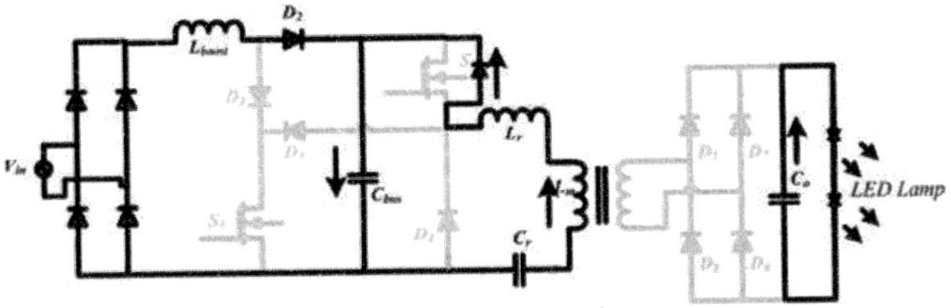


**Fig. 8** Mode 5

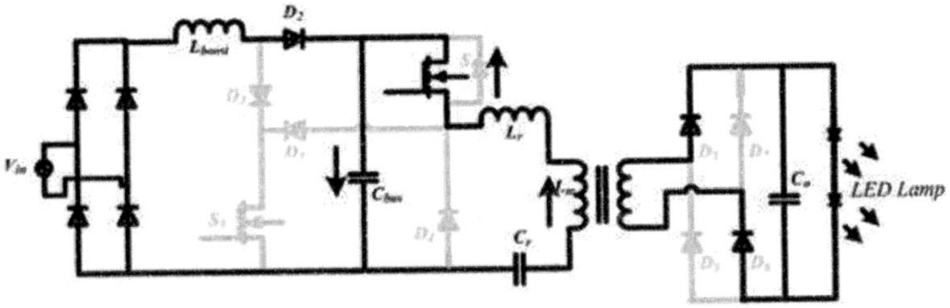


**Fig. 9** Mode 6

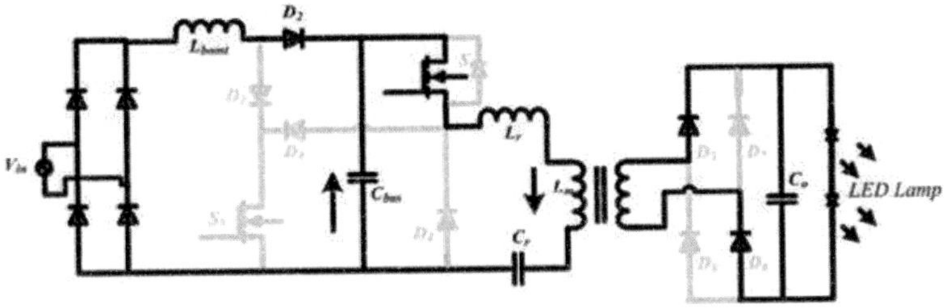


**Fig. 10** Mode 7

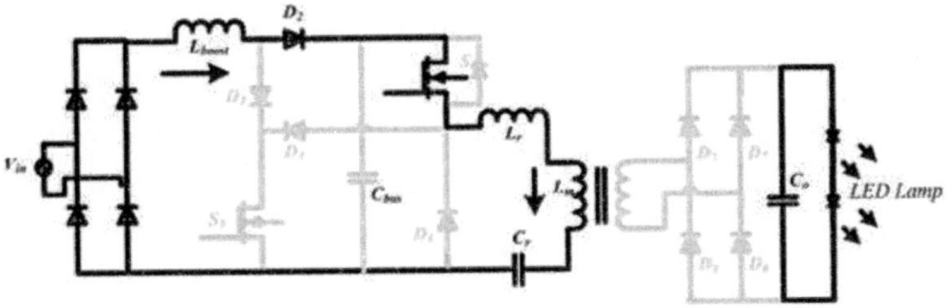


**Fig. 11** Mode 8

# 3 Design of Boost HB-LLC LED Converter

## 3.1 Design of Boost Inductor

To achieve DCM operation of boost converter, constant input line voltage is applied for inherent PFC operation. The obtained DC voltage from the diode bridge is fed to the boost converter. The average input current depends on the peak inductor current ($i_{LP}$) over a switching cycle. Therefore, the peak inductor current is obtained in (1)

$$i_{Lp}(t) = \frac{V_{\text{rect}}}{L_{\text{boost}}} D.T_s \tag{1}$$

Therefore, average input current is obtained as

$$i_{\text{in}}(t) = \frac{D^2 T_s V_m}{2L_{\text{boost}}} \sin \omega t \tag{2}$$

$V_m$ is the peak input voltage and $L_{\text{boost}}$ is the boost inductor. $D$ is operating duty cycle. Equating the input and output power with an assumption that front-end boost cell acts as loss-less converter. Therefore,

$$P_o = P_{\text{in}} = \frac{1}{2} V_m I_{\text{in}} \tag{3}$$

where

$$P_{\text{in}} = \frac{V_{\text{m}}^2 D^2}{4L_{\text{boost}} f_s} \tag{4}$$

Boost inductor value is obtained as

$$L_{\text{boost}} = \frac{V_m^2 D^2}{4P_o f_s} = \frac{\left(110 * \sqrt{2}\right)^2 * 0.5}{4 * 100 * 120 * 10^3} = 162\mu H \tag{5}$$

## 3.2 Design of HB-LLC Resonant Components

Based on [12], the resonant components such as Cr, Lr and Lm are obtained. It involves step by step procedure to estimate the resonant components. LLC equivalent circuit is represented in Fig. 12.

- Maximum and minimum output voltage of boost

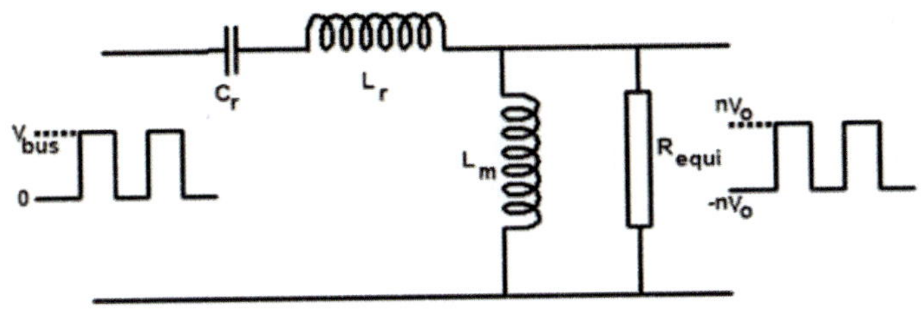


**Fig. 12** LLC resonant converter equivalent circuit

$$V_{\text{DCMAX}} = 2\sqrt{2}V_{\text{acMAX}} = 340\text{ V} \tag{6}$$

$$V_{\text{DCMIN}} = 2\sqrt{2}V_{\text{acMIN}} = 283\text{ V} \tag{7}$$

- Turns ratio (n)
  Turns ratio corresponding to the high-frequency transformer is obtained as

$$n \geq \frac{\text{D}V_{\text{DCMAX}}}{V_{\text{o}} + V_{\text{F}}} = 5 \tag{8}$$

- Equivalent load resistance ($R_{\text{equi}}$)

$$R_{\text{equi}} = \frac{8n^2}{\pi^2}R_L = \frac{8 * 25}{\pi^2} * \frac{36}{2.77} = 263.36\Omega \tag{9}$$

- Maximum and minimum gain ($M_g$)—LLC resonant converter

$$M_{g(\min)} = \frac{5(36(1 + 1\%) + 0.7)}{\frac{340}{2}} = 1.06 \tag{10}$$

$$M_{g(\max)} = \frac{5(36(1 + 1\%) + 0.7 + 1.05)}{\frac{283}{2}} = 1.34 \tag{11}$$

- Determine inductance ratio (m) and quality factor ($Q$)
  From Fig. 13, quality factor is considered as 0.4 to obtain the desired $M_{g\max.}$ Similarly, from Fig. 14, inductance ratio (m) is considered as 5 to obtain the desired gain within the inductive region.

- Determination of resonant parameters

$$C_r = \frac{1}{2\pi Q_e f_o R_e} = \frac{1}{2\pi * 0.4 * 120 * 10^3 * 263.36} = 12.5nF \tag{12}$$

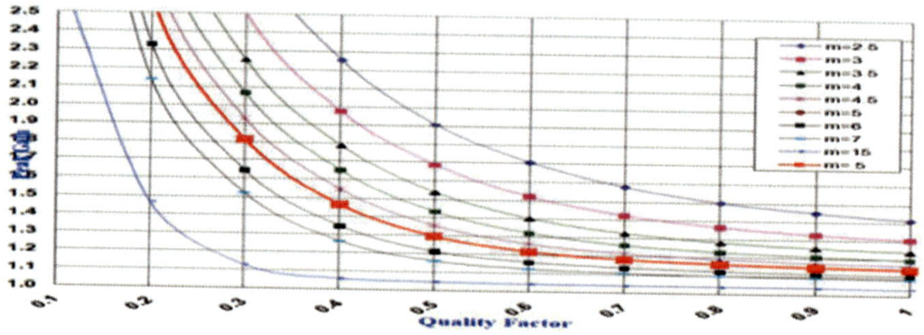


Fig. 13 Peak gain curves for different inductance ratio

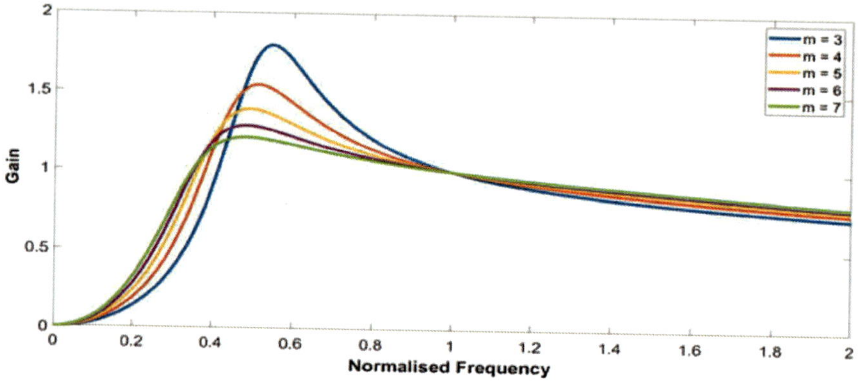


Fig. 14 Gain versus normalized frequency curve

$$L_r = \frac{1}{(2\pi f_o)^2 * C_r} = \frac{1}{4\pi^2 \left(120 * 10^3\right)^2 * 12.5 * 10^{-9}} = 140\mu H \tag{13}$$

$$L_m = m * L_r = 5{*}140\mu H = 700\mu H \tag{14}$$

Based on the equations from (5) to (14), boost integrated HB-LLC converter components values are determined and is shown in Table 1.

## 4 Simulation Results

In this section, based on the design equation discussed in the previous section boost integrated HB-LLC converter is simulated using PSIM.A 110 Vrms is fed to the converter to achieve ripple-free LED operation. The output power required to drive the LED is 100 W with 36 V, 2.77 A constant voltage and current, respectively.

**Table 1** Designed values for boost HB-LLC converter

| Parameter | Notation | Value |
|---|---|---|
| Boost inductor | $L_{boost}$ | 162 μH |
| Isolation transformer turns ratio | N | 5:1 |
| Bus capacitor | $C_{bus}$ | 220 μF |
| Resonant capacitor | $C_r$ | 12.5 nF |
| Resonant inductor | $L_r$ | 140 μH |
| Magnetizing inductance | $L_m$ | 700 μH |

Figure 15 shows the line parameters where the input current waveform aligned with source voltage which indicates the better input power factor. For a constant input voltage under the fully loaded condition, the results are shown. The total harmonic distortion (THD) is obtained as 9.32%, and the corresponding power factor is obtained as 0.995.

Figure 16 represents the front-end boost cell to operate in DCM of operation that can nullify the front-end additional PFC controller. It can act as self-power factor correction circuit. The output from the boost cell is fed to the coupled to HB-LLC through bus capacitance. The average value of bus voltage is 500 V which slightly higher but it does not affect the output voltage and the resonant phenomena (Fig. 17).

Figure 18 represents the pulse pattern for switch $S_1$ and its corresponding voltage and current. Switch $S_1$ undergoes ZVS and is shown clearly during the turn-on instants. The current through switch starts increasing only after the drain to source voltage tends to zero. Therefore, the switching losses are minimized.

Switches in the converter are operated at 0.5 duty cycle, while the frequency of the converter depends on the input and output voltage. Hereunder fully loaded condition with fixed input and an assumption of constant output voltage, the results are shown.

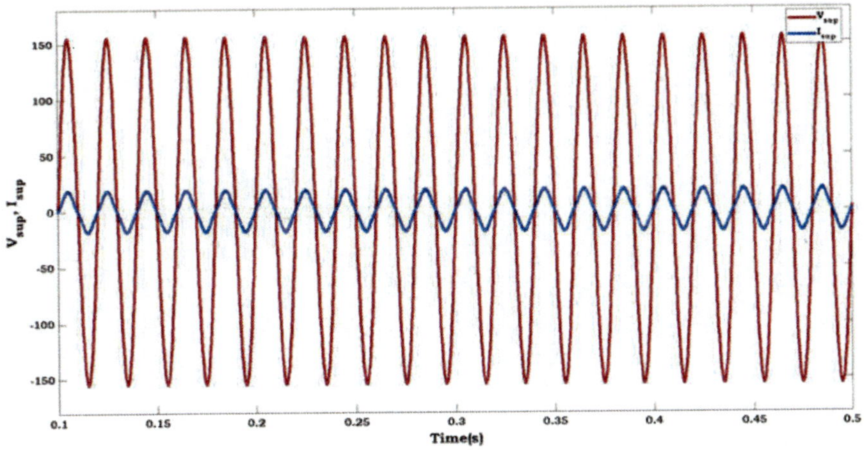


**Fig. 15** Line voltage and current waveform

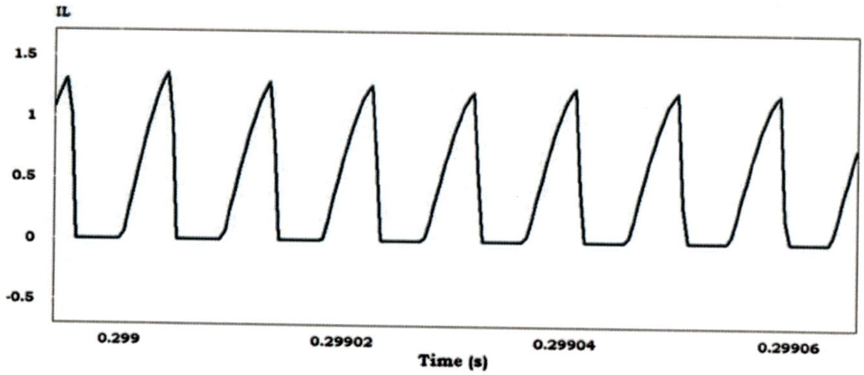


**Fig. 16** Boost inductor current with DCM operation

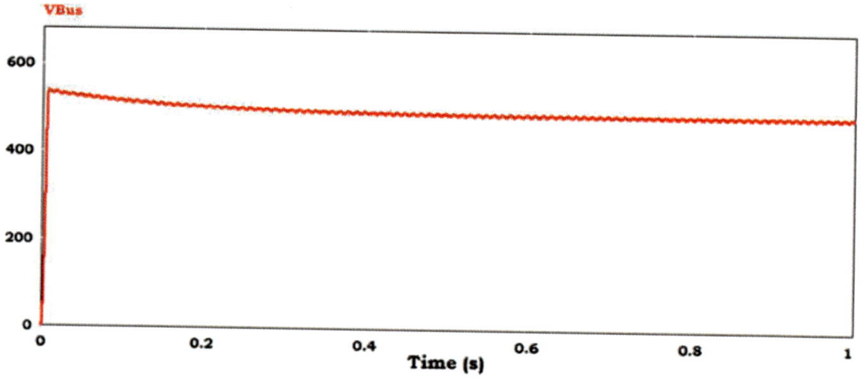


**Fig. 17** Bus voltage waveform

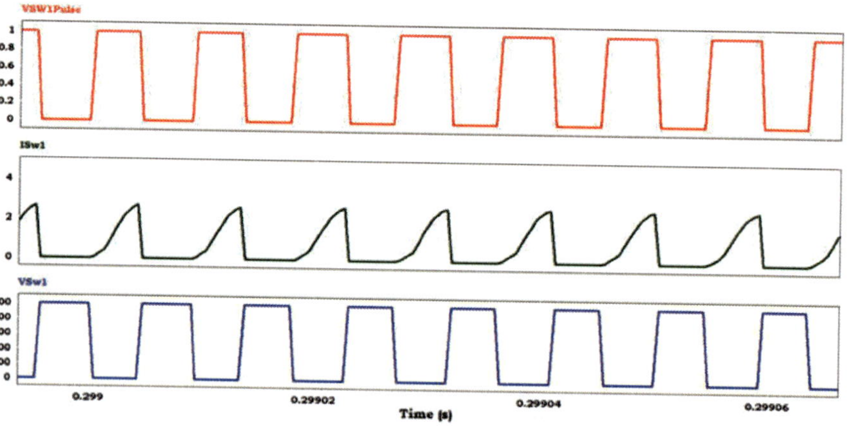


**Fig. 18** **a** Switch $S_1$ pulse pattern, **b** switch $S_1$ current waveform and **c** switch $S_1$ voltage

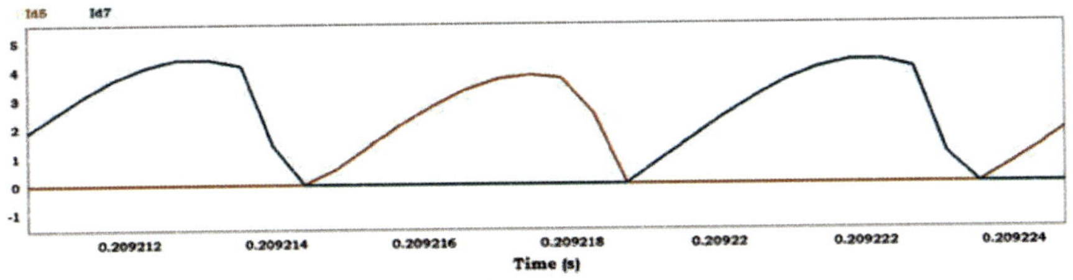


**Fig. 19** ZCS operation of output diodes

Under fully loaded condition, Fig. 19 indicates the current through output diodes $D_5$ and $D_7$ which undergo zero current switching (ZCS) to enhance the efficiency of the driver circuit.

LED output parameters are shown in Figs. 20 and 21, respectively. LED deserves the constant voltage and current with minimal ripple for flicker free operation. The output voltage of 36 V under steady conditions is obtained with an output voltage ripple of 5.7%. The output current ripple in this topology is 1.5% which ensures the ripple-free LED current with a high power factor and low THD.

Figure 22 shows the comparison of harmonic order with the IEC61000-3-2. Boost HB-LLC converter exhibits low input current harmonic content as per the standards required. The harmonic content is calculated under a fully loaded condition with an input voltage of 110 $V_{rms}$. This may vary with the higher input voltage. The corresponding input and output parameter for simulation is tabulated in Table 2.

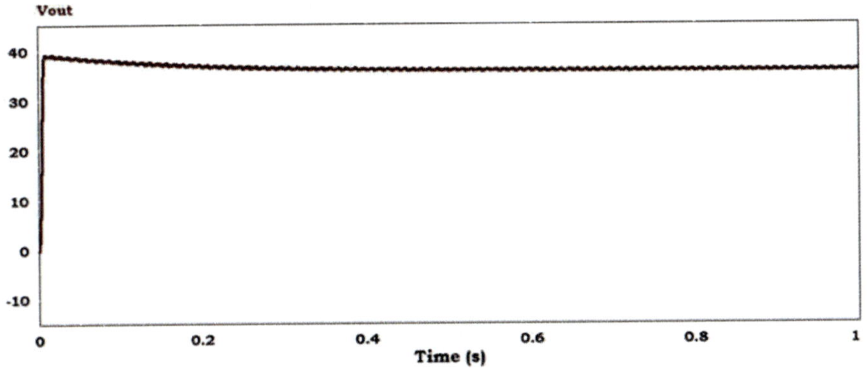


**Fig. 20** Output LED voltage

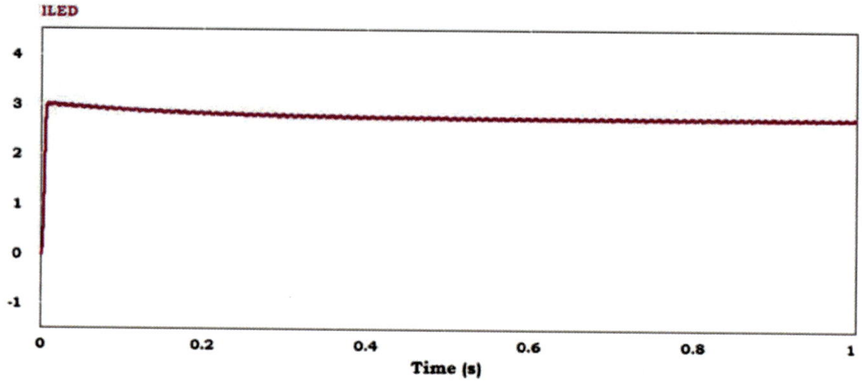


**Fig. 21** Output LED current

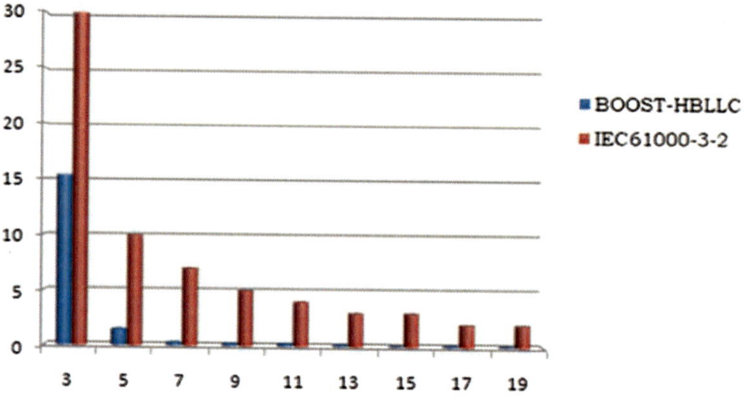


**Fig. 22** Harmonic standards with boost HB-LLC

**Table 2** Parameters for the simulation framework

| S.No | Parameter | Value |
|---|---|---|
| 1 | Source voltage | 110 $V_{rms}$ |
| 2 | Source current | 1.22 $I_{rms}$ |
| 3 | Inductor current | 1.68 $I_{rms}$ |
| 4 | Duty cycle | 0.5 |
| 5 | Input current THD | 9.32% |
| 6 | Input power factor | 0.995 |
| 7 | Bus voltage | 500 V |
| 8 | Average output LED voltage | 36 V |
| 9 | Average output LED current | 2.77A |
| 10 | Output voltage ripple | 5.7% |
| 11 | Output current ripple | 1.5% |

## 5 Conclusion

Single stage boost integrated HB-LLC topology is suggested in this paper for LED lighting of 100 W. Both the boost converter and HB-LLC resonant converter are integrated with the shared switch without losing the resonant characteristics. The front-end boost converter operates in DCM mode which acts as an inherent PFC circuit. The performance of this topology is analysed with input power factor input current THD, output voltage and current ripple. This topology exhibits the lower output current ripple which ensures the flicker free LED operation. The input power factor is nearly unity, and the harmonic content lies below the IEC61000-3-2 standard. Even though the switches are shared among the two converters, the voltage and current stress on the switches are shared equally. ZVS operation is achieved by the power switch S1, and the ZCS is achieved by the output rectifier diodes to enhance the efficiency of the converter.

## References

1. Fang P, Webb S, Liu YF, Sen PC (2019) Single-stage LED driver achieves electrolytic capacitor-less and flicker-free operation with unidirectional current compensator. IEEE Trans Power Electron 34(7):6760–6776
2. Almeida PS, Camponogara D, Dalla Costa M, Braga H, Alonso JM (2015) Matching LED and driver life spans: a review of different techniques. IEEE Ind Electron Mag 9(2):36–47
3. Wu YE (2016) Design and implementation of a novel two-stage LED driver. Int J Energy Power 5:19
4. Mohamadi A, Afjei E (2015) A single-stage high power factor LED driver in continuous conduction mode. In: The 6th power electronics, drive systems & technologies conference (PEDSTC2015)
5. Valchev VC, Mareva DJ, Yudov DD (2017) Analysis and improvements of a buck converter based LED driver. In: 2017 XXVI international scientific conference electronics (ET)
6. Ramanjaneya Reddy U, Narasimharaju BL (2017) Single-stage electrolytic capacitor less non-inverting buck-boost PFC based AC–DC ripple free LED driver. IET Power Electron 10(1):38–46
7. Wang J-M, Wu S-T, Yen S-C, Lin J-Y (2014) A simple control scheme for a single stage flyback LED driver. Int J Circuit Theory Appl 43(12):1879–1898
8. Ma W, Xie X, Jiang S (2017) LLC resonant converter with variable resonant inductor for wide LED dimming range. In: 2017 IEEE applied power electronics conference and exposition (APEC)
9. Spini C (2012) 48 V 130 W high-efficiency converter with PFC for LED street lighting applications. STMicroelectronics, Geneva, Switzerland, Appl.Note AN3106
10. Wang Y, Guan Y, Zhang X, Xu D (2013) Single-stage LED driver with low bus voltage. Eletron Lett 49(7):455–456
11. Wang Y, Guan Y, Liang X, Wang W, Xu D (2014) Two-stage LED street lighting system based on a novel single-stage ac/dc converter. IET Power Electron 7(6):1374–1383
12. Cheng C-A, Cheng H-L, Chung T-Y (2013) A novel single-stage high-power-factor LED street-lighting driver with coupled inductors. 2013 IEEE Industry Applications Society Annual Meeting, Oct

# A Holistic Entropy-Based Design Approach for a Network of Systems

H. Shyam Sundar and A. S. Krishnan

## *Nomemclature*

| | |
|---|---|
| $\Delta S$ | Change in Entropy |
| FE | Fuel Efficiency |
| FC | Fuel Consumption |
| TFC | Total Fuel Consumption |
| Q | Heat Transferred |
| T | Temperature |
| W | Work Done |

## *Subscripts*

| | |
|---|---|
| d | Disposal |
| l | During life |
| m | Maintenance |
| p | Pre-life |
| ro | Regular operation |
| sp | Sub-process |
| eg | Exhaust gases |

H. Shyam Sundar
Department of Mechanical Engineering, Eindhoven University of Technology, Eindhoven, The Netherlands

A. S. Krishnan (✉)
Department of Mechanical Engineering, Coimbatore Institute of Technology, Coimbatore, India
e-mail: krishhttp@alumni.iitm.ac.in



N. L. Narasimhan et al. (eds.), *Recent Advances in Energy Technologies*, Lecture Notes in Mechanical Engineering, https://doi.org/10.1007/978-981-19-3467-4_42

# 1 Introduction

The subject of energy and environment needs no introduction, with the words "go green" being heard or used more often than not. Myriads of top-notch studies have been conducted and being continued on various aspects of energy conversion, transmission, storage, and utilization [1]. Few examples of this topic studied in depth are solar, geothermal, nuclear, MHD power generation, etc. The research progresses in these areas have been well complemented by advances in material science, cutting-edge manufacturing technologies, simulation facilities, and alike [2].

Some of these at times put forth the need for sophistication, which is likely to be achieved at an increased energy budget, driving us away from the objective. A fairly linear solution is available from a philosophical approach to energy crisis, which however, is next to impossible to implement in a differential environment, at least in the foreseeable future. It necessitates, henceforth, a much "broader than present" approach to energy, environment, and sustainability.

The concept of sustainability is relatively a new one, having its inception after concerns of environmental damage through emissions [3]. Sustainable development, which denotes technological development on the lines of sustainability, is now a common phrase in the industrial sector and academia alike. Several studies on the topic of sustainable development have given rise to multifield strategies and algorithms, to combat the dwindling resources alongside curtailing the ever-growing environmental pollution by various means [4, 5]. Engineering optimization does help in sustainability. Sustainability, however, has certain nuances with optimization. While optimization is generally considered a term for efficient use of resources and power, sustainability involves keeping the environmental impact too in check.

While several studies are linked to energy and energy conservation, a relatively lesser attention has been given to the thermodynamic property of entropy. Entropy, unlike energy, can be incorporated to simultaneously define the active and passive amount of energy stored, amount of energy that can be utilized, and subsequently, the amount of energy that cannot be. The fact that negative entropy for a process is theoretically impossible, as dictated by the second law, is more logical and felicitous to compare products across their whole life cycle. This arises as a plausible alternate to the ambiguity of energy analyzes while evaluating a product, considering the natural directionality of any process. Hence, entropy gives a more unconstrained approach in terms of time.

## *1.1 Literature Review*

Sustainability is a subject of intense focus in the current timeline. Energy conservation has become an integral part of all industries with fields of wide diversities vying for efficiency and optimization of its resources. Sustainable development, incipient through the 1960s [6, 7] as a way of thinking to combat the negative human influence on the environment, was made significant in 1972 at the UN Conference on the Human Environment [8] held in Stockholm. Sustainable development rose to prominence following the *Brundtland Commission Report* [9], the definition posited being accepted as a plausible definition of sustainability. Subsequently, exergy [10] and entropy became popular in theoretically defining sustainability.

Quantitatively, entropy is best defined by the second law of thermodynamics, the origins and implications of which are heavily studied upon [11]. Entropy has been described qualitatively, and its implications on biology and several general fields are described by the popular excerpt of *What is Life* [12]. Although introspective and arguing advertently to the philosophers, the theoretical implications of the ideas and perspectives proposed and exhibited form the fundamental essence of entropy that portrays its importance in seemingly unrelated fields.

The conceptual application and subsequent inception of entropy as a factor in non-thermodynamic systems and situations began with a philosophical editorial [13]. In this revolutionary article, entropy is placed amidst economics, human behavior, and sustainability. Posing significant questions and establishing abstract facts on how entropy can be used to compare any two wildly unrelated processes, the article served as a spark for introducing entropy as a natural parameter that can be used in optimization. The article also questions the purpose of the human technological development and its place in the timeline of earth.

*The Principle of Minimum Entropy Production* [14] theoretically interprets the characteristic of entropy relating to temporal equilibrium of process parameters as the minimum rate of entropy production. The included mathematical proof of the same is evident of a natural optimization of processes with entropy as an indicative parameter. Although the proof described is for a simple example of diffusion, the fact that there exists a mathematical framework for entropy in statistical mechanics emphasizes its natural significance. The mathematical convergence of entropy in a state of equilibrium was further developed [15]. The essentiality of the equivalence of minimum entropy production and maximum power production has been studied and verified to be satisfiable only under certain special constraints [16].

The integration of different products and processes under the same roof of entropy to increase the overall efficiency has been devised [17]. The formulation pertains to the integration of components involving combined heat transfer, fluid mechanics, and thermodynamics. The fundamental notion to separately determine individual irreversibility and aggregate the change in entropy concerned with the irreversibility is emphasized as a particularly powerful tool to alter energy optimization.

The impact of entropic analysis on economics and certain exceptional cases has been argued [18]. The argument emphasizes how entropy of matter determines the feasibility for conversion of that matter into its usable form. It also affirms Georgescu-Roegen's notion that entropic efficiency is against anthropological development and how restrictions posed by entropic optimization may be in opposition to the ideals of technological growth. The dependency of technology on entropy, and not the other way around, is strongly accentuated in this argument.

Entropy and its significance in the field of economics have been hugely debated, with a few biased toward positive influence of entropic analysis in modern day economics, while others portray it as an over-constraining parameter to economic resource optimization. Similar to its impact on economics, entropy has also been associated with sustainability and ecological growth. Several perspectives of entropy on economics have been compiled and presented [19].

Entropy has been involved in several ambivalent debates in the field of economics and sustainability. Notable arguments and several important points have been put forth from various eminent personalities of the field. The definition and the concept of entropy have been postulated as a natural law of optimization, albeit one involving purely energy and not resources.

As economics requires a uniform framework for efficient comparison and analyzes, entropy may prove to a suitable parameter to bring about such comparisons and analyzes in the field of economics. Numerous literatures have discussed entropy being an indicator of the natural direction of phenomena and hence a natural optimization parameter in the field of economics [20]. This substantiates its use for an integrated analysis.

Although several studies have already been made to derive a similar economic parameter [21, 22] integrating entropy with economics, there exists scope for application of the same in a broader sense. This article has a two-pronged intention of (1) proposing a holistic approach to design of systems (or processes) and (2) triggering a global level networking analysis for sustainable development.

## 2 Methodology

The performance of a system or a process is not isolated and, in reality, dependent on its associations and connections. This necessitates for integration of the system and its associations to make the analyzes more realistic.

### 2.1 *Need For Holistic Analysis*

Let a comparison be attempted between an automobile propelled by fossil fuel and battery on their influence on environment. Both will need many common and uncommon components. The latter is of significance here, the usage of which can be categorized as components within the vehicle—like fuel, electrical systems, etc., and outside the vehicle—like production and disbursement of propellants (fuel/battery) and propulsion systems, recharging, waste disposal, etc. The battery propelled vehicle is undeniably a zero-emission one, nevertheless only at a local level. However, it could involve pollution not only at the source of power generation but also at places of manufacture and disposal. Thus, a realistic judgment of which is a better system can be arrived at only by considering the other systems it is linked to, which in turn

**Table 1** Specifications for modes of transport

| Mode | Capacity | No. of trips | FE[2](km/l) | Approx. FC (l) for one way | TFC (l) | Avg. speed (kmph) | Time (h) |
|---|---|---|---|---|---|---|---|
| Airplane | 72 | 1 | 0.55 | 900 | 1800 | 500 | 1 |
| Bus | 36 | 3 | 6.5 | 77 | 231 | 60 | 25 |
| Car | 4 | 35 | 20 | 25 | 875 | 60 | 292 |

by themselves are linked in more than one direction, possibly. A small quantitative analysis regarding this is explained in the ensuing paragraph, using a demonstrative example of choosing mode of transport.

Consider two locations A and B which are 500 km apart. Each location has 72 people who are to be transported to the other. The modes of transport considered here are airplane, bus, and a sedan. This problem can be approached from a classical optimization path. The optimization objective dictates the preferred mode of transport. A quick look at Table 1 indicates the bus to have the least TFC, and the least time is exhibited by the airplane.

These values are not actual, but assumed considering typical values for ATR72, Isuzu AC 36, Ford Freestyle for airplane, bus, and car, respectively. Considering TFC, the airplane is not the efficient choice. But, considering time, the bus and the car are not efficient choices. So, the airplane or the bus is choices that are optimum, in a way. Hence, the choice of the transport is based on what the user would prioritize or in other words the objective of the user.

Although the airplane performs the worst when compared to the other two in terms of emissions, it nevertheless provides a saving of 7 h which could be utilized for various activities apart from transportation. Hence, it may not be wise to omit time as a parameter while dealing with optimization of entropy. Here, it is implicit that the user had no other limitations to reach the destination at the earliest.

Stated otherwise, this is the usual unconstrained optimization problem. The aforementioned problem is an optimization problem with discrete options or solutions. The constraints, alike the objective(s), are entirely dependent on the priorities of the user, and generalization might be misleading. For example, if the constraints of the user are temporal, choosing the airplane as the mode of transport seems the rightful choice. The bus is the optimal choice if the constraints are energetic.

From an economic standpoint, the monetary position of the user is important. On similar lines, maximum reduction of emission will be of significance from an energy perspective. It is also possible that these two may agree the third mode of transport, viz. car. Further, localizing the optimization to only the objective and the specified modes of transport offers analytical simplicity. Nevertheless, analytical techniques provide ready solutions to cases close to ideal, which are generally associated with many assumptions. Real systems can hardly be ever isolated in an absolute sense. As indicated previously, these interconnections could happen in a myriad of ways and a few of these are listed in the following sections. Henceforth, from the discussions on choice of propulsion and mode of transport, it is clear that a more generic approach is necessary for arriving at sustainable solutions. An attempt to generalize is discussed in Sect. 2.2.

## 2.2 Holistic Entropy Analysis

The traditional method of analyzing systems based on the II law of thermodynamics can be extended to network of systems too, thus forming an entropy network. A comparison of two products "A" and "B" using such a network is shown as a schematic in Fig. 1.

The flowchart depicted in Fig. 1 describes the different ways through which entropy associated with a process or a product may increase. Determination of a holistic value for entropy increase may be a better way to choose between two or more choices based on the least entropy increase. This calls for an integral effort of energy scholars worldwide, which obviously is easier said than done, yet not as much as adopting the philosophical solution.

Applying the generalized analysis methodology to the problem of choice of transport, though the airplane is temporally efficient, the emissions are observed to be

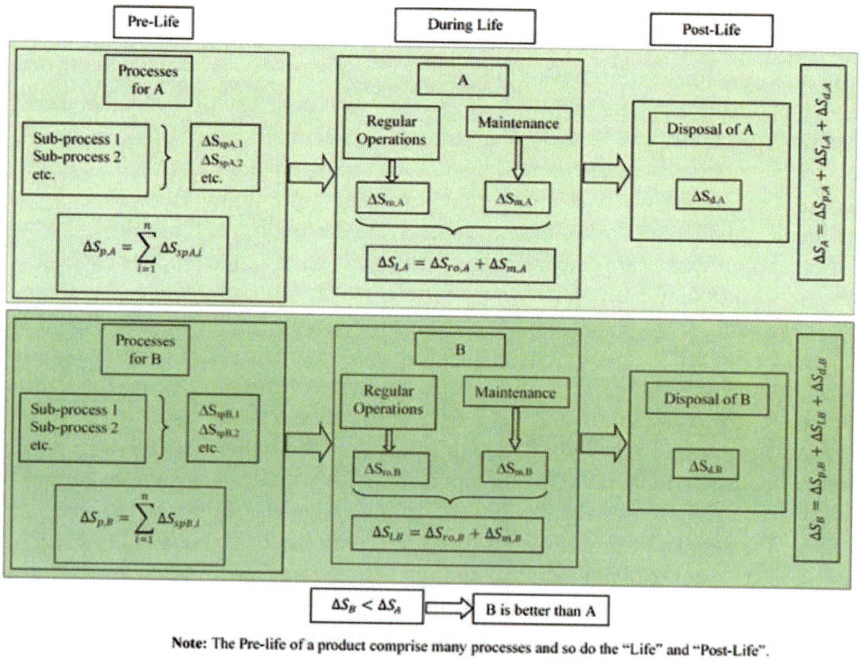


**Fig. 1** Methodology for holistic entropy analysis

higher than the other modes of transport. The pre-life and the post-life entropy increase due to an airplane are also expectedly exceptionally high.

In order to equalize the humongous rise in entropy in pre-life and post-life, the consequences of faster travel must be considered. Contradictorily, though the bus and the car are likely to have minimal pre-life and post-life operations that increase the absolute entropy, they are temporally inefficient compared to the airplane. Hence, time-sensitive events cannot be performed with these modes of transport. Simply put, time and hence, sequentiality are compromised to some extent, unless otherwise intentionally neglected.

## 2.3 *Layered Approach*

In many cases, the entropy increase during the primary operation associated with a process or a product is enough to determine the better choice of the given choices. However, the primary operation may not be the only way through which entropy may increase. A layered approach, illustrated in Fig. 2, is proposed to estimate the entropy increase associated with a process/product. The increase in entropy is categorized into layers based on the level of association with the primary operation of the process/product. The approach is illustrated by a simple example.

Let us consider an example of a flight traveling from a point P to a point Q. The entropy increase associated with this example can be calculated by the following approach.

**Layer 1: The primary operational part of a process/product**

In this example, it is the entropy increase during transport. Typically, this comprises entropy increase due to combustion of the fuel, and its immediate consequences like heat rejection to the environment, drag, vibrations, etc. Other processes like air conditioning of the flight also contribute to increase of entropy. It is implicit that the entropy increase is also enhanced by unaccounted losses. Hence, the total entropy increase associated with layer 1 can be described by Eq. 1.

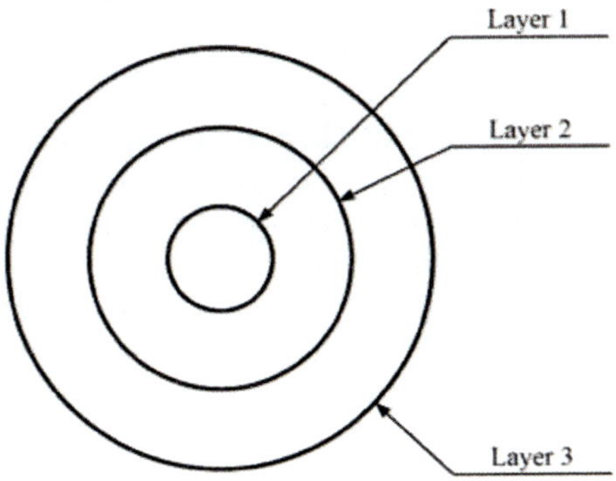


**Fig. 2** An illustration of the layered approach

$$\Delta S = \left[\int_P^Q \frac{dQ}{T}\right]_{\text{e.g.}} + \left[\int_P^Q \frac{dQ}{T}\right]_{\text{air}} + \left[\int_P^Q \frac{dQ}{T}\right]_{\text{plane}} + \Delta S_{\text{unaccounted}} \quad (1)$$

It may be noted that considering an isolated system, i.e., determining the associated materials to the primary operation for layer 1 greatly simplifies determining the total entropy change since the total heat transfer can be determined as in Eq. 2.

$$Q_{\text{in,fuel}} = Q_{\text{out,e.g.}} + Q_{\text{out,air}} + Q_{\text{out,plane}} + W_{PQ} + Q_{\text{unaccounted}} \quad (2)$$

(2).

**Layer 2: Production, disposal, and maintenance**

In the current illustration, this concerns with the increase in entropy associated not only with production and maintenance of the specific mode but also its disposal after its life.

It may be noted that the holistic entropy increase as determined from the flowchart contributes only to the layers 1 and 2 of the total entropy increase associated with a process or a product. In reality, the total entropy increase has infinitely growing layers each a consequence of the previous.

**Layer 3: Spatial and temporal consequences of direct association with the process/product**

**Layer 4: Spatial and temporal consequences associated with Layer 3 and so on**

This approach can be extended to any problem consisting of two or more choices. Selecting the optimal choice which would result in the least entropy increase can be done through the aforementioned approach by breaking down the primary operations and the cascading consequences and associations.

## 2.4 A Simpler Application of the Layered Approach

The approach is illustrated for a fairly pragmatic example. Consider two choices for a water bottle—a 1-L one and a 5-L one. Choosing the optimal choice with the least entropy increase will depend on several parameters. The layered approach is then applied to this problem, and the individual layers can be identified as per the following discussion.

**Layer 1: The primary operational part of a process/product**

In this case, the primary purpose of a water bottle is to store water. The primary operations involved in the purpose of the water bottle is refilling the bottle and lifting the bottle to drink water.

The increase in entropy, as a consequence of work done in moving to the reservoir, is linearly proportional to the distance to the refilling reservoir. Longer the distance, higher is the increase in entropy. A 5-L bottle requires lesser number of refills than a 1-L bottle. The increase in entropy, as a consequence of drinking water from the bottle, is directly proportional to the weight of the bottle. Increase in entropy is inversely proportional to the quantity of water consumed per lift. A lesser quantity consumed per lift would result in a higher number of lifts and hence, higher energy spent.

Hence, the increase in entropy is a function of the distance to the refilling reservoir, weight of bottle and the quantity consumed per lift. If the quantity consumed per lift and the distance to the refilling reservoir are higher, the 5-L bottle may prove optimal.

**Layer 2: Production, disposal, and maintenance**

Applying the holistic entropy analysis to both the bottles, the results obtained from layer 1 may vary. If the entropy increase per unit mass of the 5-L bottle is higher than that of the 1-L bottle, combining it with the average life of the bottle may result in the 1-L bottle leading to a higher entropy rise than the 5-L bottle.

**Layer 3: Spatial and temporal consequences**

Predicting real-life consequences is difficult unless the system in consideration is an isolated system containing very few number components. A suggested alternative is to consider a plausible environment and to determine the consequential entropy increase of products that may be associated to a water bottle.

## 2.5 *Networking and Consequentiality*

The holistic analysis discussed through layers 1 and 2 is upfront fairly comprehensive, as it involves a summation of all entropy changes associated with the product or a process, from the inception through the operation, ending at its disposal. However, realistically, any entropy change is not independent and is almost always linked to multitudinous consequences that may be direct and/or indirect. One of the better ways is to begin with the closest neighborhood and gradually increase the radius of influence and/or dependence. The optimum solution too is expected to change as this radius grows. The possibility of obtaining a truly sustainable solution increases as more and more information is shared among the participants of the network.

Hence, following the primary intended consequences of the operation of a product or the execution of a process yields only a truncated picture. Any operation or execution would likely involve unanticipated secondary consequences, which may not only be unintentional but also be unnoticeable and hence, neglected most often citing the order of magnitude of such consequences. While such consequences are generally diminutive in the immediate time frame, it may lead to unpredictable scenarios through cascading.

Thus, while it may be empirically adequate to estimate the entropy changes of a product or a process locally, it may be largely different theoretically, owing to the impossibility of isolation of a process with the surroundings. Hence, maximizing the efficiency of a process through layer 1 alone may definitely reduce entropy change concerning the process, but it may lead to an increase in entropy in the surroundings, in tune with the Clausius inequality.

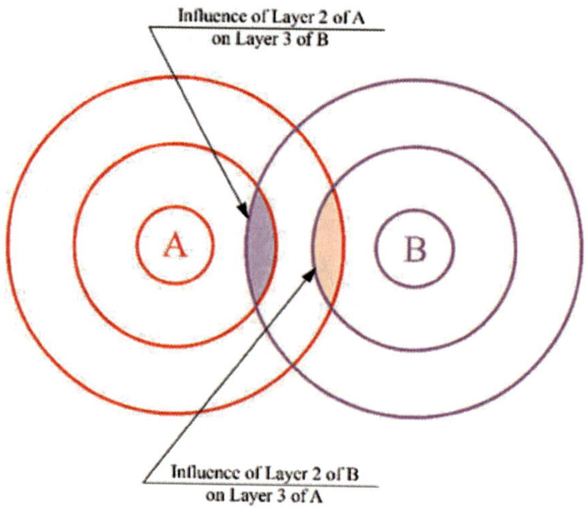


**Fig. 3** Influence of an associated component

The solution can be improved by considering sufficient enough consequences into the picture. Though, in order to expand the order of accuracy and to push the analysis to more realistic terms, a deeper observation of networking of all the consequences, not just spatial but also temporal, should be considered. Thus, though the entropy increase from layers 1 and 2 are easy to estimate owing to their direct association with the process/product, a realistic picture is obtained only with the inclusion of temporal and spatial consequences as more layers.

The increase in entropy of a specific process related with a product may be influenced by the increase in entropy of an associated component as shown in Fig. 3. Also, it is possible that two components or product may share a part of their layers owing to their association. In such cases, estimation of entropy increase of one product may be used to form a part of the entropy increase of its associated component or product. This can be used to vaguely estimate the higher layers without involving much complexity.

It is possible to break down larger layers using associated components and estimating the entropy increase of the initial layers of the individual components. The total entropy can be integrated to estimate a more accurate value of the entropy increase of the product.

In Fig. 4, A is the primary product, and different variations of B are its immediate associations. By estimating the layer 1 of B, an estimate of the layer 2 of A can be obtained.

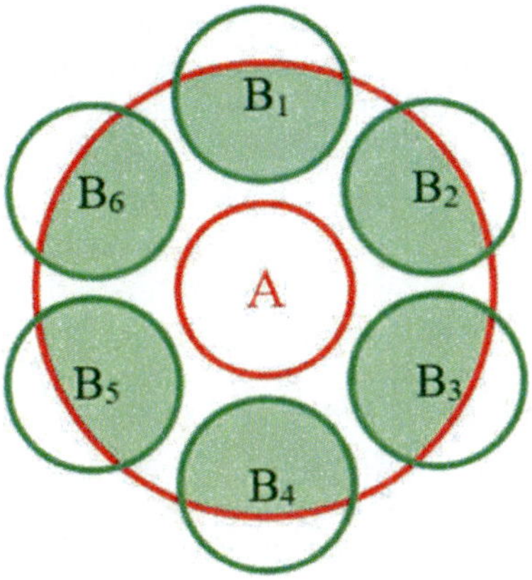


**Fig. 4** Estimating higher layers through associations

## 3 Conclusions

This article presented a methodology for design of systems and processes in a conceptual manner. The following points could be inferred from the preceding discussions:

1. Overtly, ranking a process or an equipment to be energy efficient at the system level may well be a beautiful portrait far from factual.
2. The method of holistic networking and a layered approach to determine the entropy increase spatially and temporally for design of systems and processes could possibly provide a sustainable solution.
3. Optimization of a process must not only involve making the process more efficient but also aim at reducing the global increase in entropy in all of the concerning associations and consequences.

It is now palpable that what earlier appeared to be a chain from energy conversion through utilization is in fact a multi-dimensional energy network forming a complex chain of linked processes. The method of analyzing individual systems based on the II law of thermodynamics can be extended to network of systems too, thus forming an entropy network.

## 4 Scope for Further Work

A little thought will suffice to understand that even though this methodology while being conceptually comprehensive, an actual analysis based on this methodology is too humongous a task to fall in the scope of this article. The methodology presented in the article is primitive and conceptual at best. Hence, a direct application of the methods discussed might be feasible for simple isolated systems where the energy budget is strictly monitored. For instance, if it is proposed to modify a networked process to decrease the entropy at the process level, a simulation needs to be carried out to estimate the change in entropy at all other processes to justify the modifications. As previously discussed, the wider the networking included for simulation, the closer are the predictions to reality. In an open system, the network may exponentially increase in size to encompass a very large domain of associations. A little more thought also helps in realizing that for the execution of this humongous task, a multi-disciplinary and collaborative effort, from energy enthusiasts and economists across the globe, is indispensable.

The methodology draws parallel to exergy analysis of industrial systems and processes with similar use cases in the industry. However, an entropic analysis may provide an absolute standard encompassing all associations of a product or a process with any other product or a process. The study on the associations and its effects on the performance of the system or process is emphasized as future work.

## References


1. Rizzi F, van Eck NJ, Frey M (2014) The production of scientific knowledge on renewable energies: Worldwide trends dynamics and challenges and implications for management. Renew Energy 62:657–671
2. Mourtzis D (2020) Simulation in the design and operation of manufacturing systems: state of the art and new trends. Int J Prod Res 58(7):1927–1949
3. Lumley S, Armstrong P (2004) Some of the nineteenth century origins of the sustainability concept. Environ Dev Sustain 6:367–378
4. Srikanth R (2018) India's sustainable development goals–Glide path for India's power sector. Energy Policy 123:325–336
5. Ulucak R, Khan SUD, Baloch MA, Li N (2020) Mitigation pathways toward sustainable development: is there any trade-off between environmental regulation and carbon emissions reduction? Sustain Dev 28(4):813–822
6. Carson R (1962) Silent spring. Houghton Mifflin, Boston
7. Hardin G (1968) The tragedy of the commons. Science 162:1243–1248
8. United Nations (1972) Report of the United Nations conference on the human environment, 1–6
9. Brundtland G (1987) Our common future: report of the 1987 World Commission on Environment and Development. United Nations, Oslo, vol 1, p 59
10. Ahern JE (1980) Exergy method of energy systems analysis. United States government
11. Marcella TV (1992) Entropy production and the second law of thermodynamics: an introduction to second law analysis. Am J Phys 60(10):888–895
12. Schrodinger E (1992) What is life? Reprinted by Cambridge University Press 1995, 87
13. Georgescu-Roegen N (1993) The entropy law and the economic problem. Valuing Earth: Econ Ecol Ethics:75–88
14. Klein MJ, Meijer PH (1954) Principle of minimum entropy production. Phys Rev 96(2):250–254
15. Bertola V, Cafaro E (2008) A critical analysis of the minimum entropy production theorem and its application to heat and fluid flow. Int J Heat Mass Transfer 51(7–8):1907–1912
16. Salamon P, Hoffmann KH, Schubert S, Berry RS, Andresen B (2001) What conditions make minimum entropy production equivalent to maximum power production?
17. Bejan A (1996) Entropy generation minimization: the new thermodynamics of finite-size devices and finite-time processes. J Appl Phys 79(3):1191–1218

18. Daly HE (1992) Is the entropy law relevant to the economics of natural resource scarcity? Yes, of course it is! J Environ Econ Manag 23(1):91–95
19. McMahon GF, Mrozek JR (1997) Economics, entropy and sustainability. Hydrol Sci J 42(4):501–512
20. Young JT (1991) Is the entropy law relevant to the economics of natural resource scarcity? J Environ Econ Manag 21(2):169–179
21. Kovalev AV (2016) Misuse of thermodynamic entropy in economics. Energy 100:129–136
22. Purvis B, Mao Y, Robinson D (2017) Thermodynamic entropy as an indicator for urban sustainability? Procedia Eng 198:802–812